ADVANCES IN SECOND MESSENGER
AND PHOSPHOPROTEIN RESEARCH

Volume 29

Molecular and Cellular Mechanisms of Neurotransmitter Release

Advances in Second Messenger and Phosphoprotein Research

ADVANCES IN SECOND MESSENGER
AND PHOSPHOPROTEIN RESEARCH
Volume 29

Molecular and Cellular Mechanisms of Neurotransmitter Release

Editors

Lennart Stjärne, M.D.
Professor of Physiology
Department of Physiology
and Pharmacology
Karolinska Institute
Stockholm, Sweden

Paul Greengard, Ph.D.
Vincent Aster Professor
Laboratory of Molecular
and Cellular Neuroscience
The Rockefeller University
New York, New York

Sten E. Grillner, M.D., Ph.D.
Chairman
Department of Neurophysiology
Nobel Institute for Neurophysiology
Karolinska Institute
Stockholm, Sweden

Tomas G. M. Hökfelt, M.D.
Professor
Department of Neuroscience
Karolinska Institute
Stockholm, Sweden

David R. Ottoson, M.D.
Professor
Department of Physiology
Karolinska Institute
Stockholm, Sweden

Raven Press New York

Raven Press, Ltd., 1185 Avenue of the Americas, New York, New York 10036

Made in the United States of America

International Standard Book Number 0-7817-0220-8
ISSN 1040-795-5

The material contained in this volume was submitted as previously unpublished material, except in the instances in which credit has been given to the source from which some of the illustrative material was derived.

Great care has been taken to maintain the accuracy of the information contained in the volume. However, neither Raven Press nor the editors can be held responsible for errors or for any consequences arising from the use of the information contained herein.

Materials appearing in this book prepared by individuals as part of their official duties as U.S. Government employees are not covered by the above-mentioned copyright.

Printed and bound in the United Kingdom
Transfered to Digital Printing, 2011

Contents

Contributing Authors

Cristina Alberini, Ph.D.
Department of Psychiatry
Center for Neurobiology and Behavior
Columbia University College of Physicians and Surgeons
722 West 168th Street
New York, New York 10032

Wolfhard Almers, Ph.D.
Department of Molecular Cell Research
Max-Planck-Institute for Medical Research
29 Jahnstrasse
D-69120 Heidelberg, Germany

Per Andersen, M.D., Ph.D.
Department of Neurophysiology
University of Oslo
Sognsvannsvn 9
0317 Oslo, Norway

Per Åstrand, M.D.
Department of Physiology and Pharmacology
Karolinska Institute
Solnavägen 1
S-17177 Stockholm, Sweden

Harold L. Atwood, Ph.D., D.S.
Department of Physiology
University of Toronto
Toronto, Ontario M5S 1A8, Canada

George J. Augustine, Ph.D.
Department of Neurobiology
Duke University
Durham, North Carolina 27710

Craig H. Bailey, Ph.D.
Department of Psychiatry
Center for Neurobiology and Behavior
Columbia University
College of Physicians and Surgeons
New York State Psychiatric Institute
722 West 168th Street
New York, New York 10032

Sandra M. Bajjalieh, Ph.D.
Department of Molecular and Cellular Physiology
Howard Hughes Medical Institute
Stanford University
Stanford, California 94305

Jian-Xin Bao, M.D.
Department of Physiology and Pharmacology
Karolinska Institute
Solnavägen 1
S-17177 Stockholm, Sweden

Trent A. Basarsky, B.S.
Department of Zoology and Genetics
Iowa State University
Ames, Iowa 50011

Anja Baumeister, Dipl. Biol.
Department of Microbiology
Federal Research Center for Virus Diseases of Animals
Paul-Ehrlich Street 28
D-72076 Tübingen, Germany

John M. Bekkers, Ph.D.
Department of Neuroscience
John Curtin School of Medical Research
G.P.O. Box 334
Canberra ACT 2601, Australia

Fabio Benfenati, M.D.
Department of Experimental Medicine and Biochemical Sciences
University of Rome "Tor Vergata" via Orazio Raimondo 1
I-00173 Rome, Italy

Max R. Bennett, B.Eng., M.Sc., Ph.D., D.Sc., F.A.A.
Department of Physiology
University of Sydney
Sydney, N.S.W. 2006, Australia

Heinrich Betz, M.D.
Department of Neurochemistry
Max-Planck-Institute for Brain Research
Deutschordenstrasse 46
60528 Frankfurt, Germany

Thomas Binz, Dr.
Department of Microbiology
Federal Research Center for Virus Diseases of Animals
Paul-Ehrlich Street 28
D-72076 Tübingen, Germany

Juan Blasi, Ph.D.
Department of Cell Biology and Pathology
University of Barcelona
Casanova 143
08036 Barcelona, Spain

Kurt Bommert, Dr.
Department of Neurochemistry
Max-Planck-Institute for Brain Research
Deutschordenstrasse 46
60528 Frankfurt, Germany

Lennart Brodin, M.D., Ph.D.
Nobel Institute for Neurophysiology
Karolinska Institute
S-17177 Stockholm, Sweden

Janet L. Burton, B.S.
Department of Cell Biology
Howard Hughes Medical Institute
Yale University School of Medicine
295 Congress Avenue
New Haven, Connecticut 06510

Pablo E. Castillo, M.D.
Departments of Pharmacology and Physiology
University of California, San Francisco
San Francisco, California 94143

Edwin R. Chapman, Ph.D.
Howard Hughes Medical Institute
Yale University School of Medicine
295 Congress Avenue
New Haven, Connecticut 06510

Milton P. Charlton, Ph.D.
Department of Physiology
University of Toronto
8 Taddle Creek Road
Toronto, Ontario M5S 1A8, Canada

Stephane Charpier, M.S.
Department of Cellular Neurobiology
INSERM U261
Institut Pasteur
25 Rue du Dr Roux
75724 Paris Cedex 15, France

Bo-Ming Chen, M.D.
Department of Physiology
Jerry Lewis Neuromuscular Research Center
University of California, Los Angeles
Los Angeles, California 90024

Eleanor T. Coffey, Ph.D.
Department of Biochemistry
University of Dundee
Dundee, Scotland

Robin L. Cooper, Ph.D.
Department of Physiology
University of Toronto
Toronto, Ontario M5S 1A8, Canada

Thomas C. Cunnane, Ph.D.
University Department of Pharmacology
Mansfield Road
Oxford 0X1 3QT, England

William M. DeBello, B.S.
Department of Neurobiology
Duke University
Durham, North Carolina 27710

Pietro V. De Camilli, M.D.
Department of Cell Biology
Howard Hughes Medical Institute
Yale University School of Medicine
295 Congress Avenue
New Haven, Connecticut 06510

Sally Durgerian, B.S.
Department of Zoology and Genetics
Iowa State University
Ames, Iowa 50011

Lambert Edelmann, M.S.
Howard Hughes Medical Institute
Yale University School of Medicine
295 Congress Avenue
New Haven, Connecticut 06510

Donald S. Faber, Ph.D.
Department of Anatomy and Neurobiology
The Medical College of Pennsylvania
3300 Henry Avenue
Philadelphia, Pennsylvania 19129

Yu Fang, M.D.
Department of Zoology and Genetics
Iowa State University
Ames, Iowa 50011

Anne C. Field, Ph.D.
Division of Neuroscience
John Curtin School of Medical Research
Australian National University
Canberra ACT 0200, Australia

Anders Franco-Cereceda, M.D., Ph.D.
Department of Physiology and Pharmacology
Karolinska Institute
S-17177 Stockholm, Sweden

Mirella Ghirardi, M.D.
Departments of Psychiatry
Center for Neurobiology and Behavior
Columbia University College of Physicians and Surgeons
722 West 168th Street
New York, New York 10032

François Gonon, Ph.D.
CNRS URA 1195
University of Lyon 1
8 Avenue Rockefeller
69373 Lyon (Cedex 08), France

Paul Greengard, Ph.D.
Laboratory of Molecular and Cellular Neuroscience
The Rockefeller University
1230 York Avenue
New York, New York 10021

Sten E. Grillner, M.D., Ph.D.
Department of Neurophysiology
Nobel Institute for Neurophysiology
Karolinska Institute
S-17177 Stockholm, Sweden

Alan D. Grinnell, Ph.D.
Department of Physiology
Jerry Lewis Neuromuscular Research Center
Univeristy of California
Los Angeles, California 90024

Michael Hans, Ph.D.
The Salk Institute Biotechnology/Industrial Associates
505 Coast Boulevard South
La Jolla, California 92037

Philip G. Haydon, Ph.D.
Department of Zoology and Genetics
Iowa State University
Ames, Iowa 50011

Julian J. B. Jack, Ph.D., B.M., B.Ch.
University Laboratory of Physiology
Oxford University
Parks Road
Oxford 0X1 3PT, England

Reinhard Jahn, Ph.D.
Departments of Pharmacology and Cell Biology
Howard Hughes Medical Institute
Yale University School of Medicine
295 Congress Avenue
New Haven, Connecticut 06510

Vidar Jensen, M.S.
Department of Neurophysiology
University of Oslo
Sognsvannsvn 9
0317 Oslo, Norway

Peter Jonas, Prof. Dr.
Department of Cell Physiology
Max-Planck-Institute for Medical Research
29 Jahnstrasse
D 69120 Heidelberg, Germany

Eric R. Kandel, M.D.
Departments of Psychiatry, Physiology and Cellular Biophysics and Biochemistry and Moleculer Biophysics
Center for Neurobiology and Behavior
Columbia University College of Physicians and Surgeons
722 West 168th Street
New York, New York 10032

Eberhard von Kitzing, Dr.
Department of Cell Physiology
Max-Planck-Institute for Medical Research
29 Jahnstrasse
D-69120 Heidelberg, Germany

Henri Korn, M.D., Ph.D.
Department of Cellular Neurobiology
INSERM U261
Institut Pasteur
25 Rue du Dr Roux
75724 Paris Cedex 15, France

Alan U. Larkman, Ph.D.
University Laboratory of Physiology
Oxford University
Parks Road
Oxford 0X1 3PT, England

Andy K. Lee, B.S.
Department of Molecular Cell Research
Max-Planck-Institute for Medical Research
29 Jahnstrasse
D-69120 Heidelberg, Germany

Pascal Legendre, Ph.D.
Department of Cellular Neurobiology
INSERM U261
Institut Pasteur
25 Rue du Dr Roux
75724 Paris Cedex 15, France

Manfred Lindau, Dipl. Phys., Dr. rer. nat.
Department of Molecular Cell Research
Max-Planck-Institute for Medical Research
29 Jahnstrasse
D-69120 Heidelberg, Germany

Egenhard Link, Ph.D.
Howard Hughes Medical Institute
Yale University School of Medicine
295 Congress Avenue
New Haven, Connecticut 06510

Rodolfo R. Llinás, M.D., Ph.D.
Departments of Physiology and Neuroscience
New York University Medical Center
New York, New York 10016

Ya-ping Lou, Ph.D.
Department of Physiology and Pharmacology
Karolinska Institute
S-17177 Stockholm, Sweden

Jan M. Lundberg, Ph.D., M.B.
Department of Physiology and Pharmacology
Karolinska Institute
S-17177 Stockholm, Sweden

Daniel V. Madison, Ph.D.
Department of Molecular and Cellular Physiology
Beckman Center for Molecular and Genetic Medicine
Stanford Medical School
Stanford, California 94306

Guy Major, Ph.D.
The University Laboratory of Physiology
Oxford University
Parks Road
Oxford OX1 3PT, England

Agnes Modin, Ph.D.
Department of Physiology and Pharmacology
Karolinska Institute
S-17177 Stockholm, Sweden

Mussie Msghina, M.D.
Department of Physiology and Pharmacology
Karolinska Institute
Solnavägen 1
S-17177 Stockholm, Sweden

Erwin Neher, Ph.D.
Department of Membrane Biophysics
Max-Planck-Institute for Biophysics Chemistry
Am Faßberg
D-37077 Göttingen, Germany

David G. Nicholls, Ph.D.
Department of Biochemistry
University of Dundee
Dundee, Scotland

Roger A. Nicoll, M.D.
Departments of Pharmacology and Physiology
University of California, San Francisco
San Francisco, California 94143

Heiner Niemann, Prof. Dr.
Department of Microbiology
Federal Research Center of Virus Diseases of Animals
Paul-Ehrlich Street 28
D-72076 Tübingen, Germany

Jean Nordmann, Dr.
Centre de Neurochimie
5 rue Blaise Pascal
F-67084 Strasbourg, France

John Pernow, M.D., Ph.D.
Department of Physiology and Pharmacology
Karolinska Institute
S-17177 Stockholm, Sweden

Mu-ming Poo, Ph.D.
Department of Biological Sciences
Columbia University
913 Fairchild Center
New York, New York 10027

Andrew Randall, Ph.D.
Department of Molecular and Cellular Physiology
Stanford University
Stanford, California 94305

Stephen J. Redman, M.E., Ph.D, D.S.
Division of Neuroscience
John Curtin School of Medical Research
Australian National University
Canberra ACT 0200, Australia

Hendrik Rosenboom, Ph.D.
Department of Neurobiology
Freie Universität Berlin
Königin-Luise-Strasse 28-30
D-14195 Berlin, Germany

James E. Rothman, Ph.D.
Department of Cellular Biophysics and Biochemistry
Memorial Sloan-Kettering Cancer Center
1275 York Avenue
New York, New York 10021

Ludolf von Rüden, Ph.D.
Department of Molecular and Cellular Physiology
Howard Hughes Medical Institute
Stanford University
Stanford, California 94305

Bert Sakmann, Prof. Dr.
Department of Cell Physiology
Max-Planck-Institute for Medical Research
29 Jahnstrasse
D-69120 Heidelberg, Germany

William A. Sather, Ph.D.
Department of Molecular and Cellular Physiology
Stanford University
Stanford, California 94305

Richard H. Scheller, Ph.D.
Department of Molecular and Cellular Biology
Howard Hughes Medical Institute
Stanford University
Stanford, California 94305

Erin M. Schuman, Ph.D.
Division of Biology
California Institute of Technology
Pasadena, California 91125

Tim J. Searl, Ph.D.
University Department of Pharmacology
Mansfield Road
Oxford OX1 3QT, England

Yoko Shoji-Kasai, Dr. Sci.
Laboratory of Neurochemistry
Mitsubishi Kasei Institute of Life Sciences
11 Minamiooya
Machida, Tokyo 194, Japan

Oleg Shupliakov, Ph.D.
Nobel Institute for Neurophysiology
Karolinska Institute
S-17177 Stockholm, Sweden

Robert B. Silver, M.D.
Department of Physiology and Biophysics
New York University Medical Center
550 First Avenue
New York, New York 10016

Charles F. Stevens, M.D., Ph.D.
The Salk Institute, MNL/S
Laboratory for Molecular Neurobiology
10010 North Torrey Pines Road
La Jolla, California 92037

Eivor Stjärne
Department of Physiology and Pharmacology
Karolinska Institute
Solnavägen 1
S-17177 Stockholm, Sweden

Lennart Stjärne, M.D.
Department of Physiology and Pharmacology
Karolinska Institute
Solnavägen 1
S-17177 Stockholm, Sweden

Ken J. Stratford, Ph.D.
University Laboratory of Physiology
Oxford University
Parks Road
Oxford OX1 3PT, England

Christian Stricker, M.D.
Division of Neuroscience
John Curtin School of Medical Research
Australian National University
Canberra ACT 0200, Australia

Mutsuyuki Sugimori, M.D., Ph.D.
Departments of Physiology and Neuroscience
New York University Medical Center
550 First Avenue
New York, New York 10016

Cyrille Sur, M.S.
Department of Cellular Neurobiology
INSERM U261
Institut Pasteur
25 Rue du Dr Roux
75724 Paris Cedex 15, France

Dieter Swandulla, M.D., Ph.D.
Department of Pharmacology
Institute for Experimental and Clinical Pharmacology
University of Erlangen-Nürnberg
D 91054 Erlangen, Germany

Masami Takahashi, Dr. Sci.
Laboratory of Neurochemistry
Mitsubishi Kasei Institute of Life Sciences
11 Minamiooya
Machida, Tokyo 194, Japan

Paul Thomas, Ph.D.
Department of Human Physiology
University of California School of Medicine
Davis, California 95616

Mari Trommald, M.D.
Department of Neurophysiology
University of Oslo
Sognsvannsvn 9
0317 Oslo, Norway

Frederick W. Tse, Ph.D.
Department of Pharmacology
University of Alberta
Edmonton, Alberta T6G 2H7, Canada

Richard W. Tsien, Ph.D.
Department of Molecular and Cellular Physiology
Stanford University
Stanford, California 94305

Flavia Valtorta, M.D.
Department of Pharmacology
University of Milan School of Medicine
Dibit via Olgettina 58
20132 Milan, Italy

Marc G. Weisskopf, Ph.D.
Departments of Pharmacology and Physiology
University of California, San Francisco
San Francisco, California 94143

David B. Wheeler, B.A., M.S.
Department of Molecular and Cellular Physiology
Stanford University
Stanford, California 94305

J. Martin Wojtowicz, Ph.D.
Department of Physiology
University of Toronto
Toronto, Ontario M5S 1A8, Canada

Shinji Yamasaki, Ph.D.
Department of Microbiology
Federal Research Center for Virus Diseases of Animals
Paul-Ehrlich Street 28
D 72076 Tübingen, Germany

Hinc ſequitur, quòd ab illa levi motione ſpirituum, qua actus imperii voluntatis in cerebro exercentur, poſſint fibræ, ſeu ductus ſpongioſi ſucco ſpirituoſo turgidi aliquorum nervorum concuti, aut vellicari; & proindè convulſiva irritatione, concutiendo totam nervi longitudinem, poſſunt ab eorum extremis orificiis exprimi & eructari guttulæ aliquæ ſpirituoſæ intra correſpondentem muſculum, unde ebullitio & diſploſio, qua muſculus contrahitur & tenditur, ſubſequatur.

(Giovanni Alfonso Borelli, 1685)

"Consequently, this slight motion of the spirits provoked by the will in the brain can shake or excite the fibres or spongy ducts of some nerves turgid with spirituous juice. As a result of this convulsive irritation which shakes all the length of the nerves spirituous droplets can be expressed and disgorged through orifices in their endings into the corresponding muscles. This results in the boiling and bursting by which muscle is contracted." (Modified from a translation by Paul Maquet, 1989.)

Preface

The quotation from Borelli, 1685 puts *Molecular and Cellular Mechanisms of Neurotransmitter Release* in its historical context. To our knowledge, it is the first reference in the literature that explicitly anticipates the modern view that nerve impulses evoke muscle contraction by exocytotic release of transmitter quanta (*guttulae aliquae spirituosae*) from the nerve terminals. In Chapter 1 which describes the origin of the concept of synaptic transmission and its growth during the twentieth century, Bennett shows that the tremendous and steadily accelerating development in this field from Borelli to the present day has followed three main lines. All three eventually focus very much on the *guttulae*, either their contents or the "skin" in which they are wrapped before they are disgorged. The first line, which emerged at the beginning of this century, concerns the identity of the "animal spirits" (i.e., the transmitters and the receptors and transduction mechanisms by which they control the effector). The second line, which started in the early 1950s, questions whether the *guttulae* are uniform in size and composition (i.e., if different transmitters are stored together and released in fixed amounts and proportions and, furthermore, if and why the effector responds to them in a quantal fashion). The third line, initiated in the mid 1970s, is preoccupied with the internal "skeleton" of nerve terminals and the composition and properties of "skins" (i.e., the membrane which limits the vesicles that store transmitters, and which forms the wall of nerve terminals).

The current breathtaking pace of the development of research along these separate but complementary lines adds to the excitement but makes it increasingly difficult for the individual neuroscientist to keep up with the news and appreciate its implications for neuroeffector transmission. We were, therefore, extremely happy that leading authorities within the field accepted our invitation to come to Stockholm to compare notes at a Wenner-Gren International Symposium on "Molecular and Cellular Mechanisms of Neurotransmitter Release" held September 1–4 in 1993. The main emphasis of the meeting was on the second and third research lines described above. This book reflects the tremendous recent progress in these areas discussed at the Symposium. It can be seen that research along the second line, quantal release, has accelerated greatly as the result of new technical approaches including patch clamp analysis of single channel conductances as well as of exocytosis, calcium imaging with high temporal and spatial resolution, and direct recording of the release of single quanta from visualized boutons. It is now evident that calcium signaling serves multiple functions and that transmitter exocytos follows more than one pathway. It has been shown beyond reasonable doubt, at least for "fast" transmitters, that active zones in different neurons function as binary units (i.e., permit each nerve impulse to release the contents of only one of the docked vesicles) but also differ by orders of magnitude in probability of monoquantal release. The recent finding that transmitter exocytosis in nerves shares a number of features with the constitutive exocytosis that has been demonstrated in many kinds of cells has greatly accelerated the development of research along the third line. Comparison between model systems such as yeast cells and isolated components of the secretory machinery in animal neuronal and nonneuronal cells has turned out to be extremely fruitful. This approach has helped reveal how soluble and membrane-bound factors enable the nerve impulse to cause the two "skins" to transiently fuse and create a channel through which the *guttula* is disgorged. In more modern terminology, this research is now beginning to clarify both the molecular and cellular mechanisms that make a single vesicle at the active zone "exocytosis-competent," such that the nerve impulse can release its contents as a "transmitter quantum," and cause the vast majority of the vesicles, even those apparently docked at the active zone, to ignore the nerve impulse. The multiple analytical breakthrough lines described in this volume will obviously have far-reaching implications for our views of many forms of plasticity of synaptic transmission (e.g., long-term potentiation in the hippocampus) and ultimately for our understanding of higher functions such as learning and memory.

This book will be of broad interest to graduate students and seasoned neuroscientists, and we hope that the excitement felt by the participants in the Symposium will be shared by the readers.

Lennart Stjärne
Paul Greengard
Sten Grillner
Tomas Hökfelt
David Ottoson

REFERENCES

Borelli, Joh. Alphonsi, Neapolitani Matheseos Professoris. *De Motu Animalium*, Pars Secunda, Lugduni in Batavis, Apud Cornelium Boutesteyn, Petrum vander Aa, Johannem de Vivie & Danielem à Gaesbeeck. Anno M DC LXXXV.

Borelli, Giovanni Alfonso. *On the movements of animals*. Translated by Paul Maquet. New York, Berlin: Springer Verlag, 1989.

Acknowledgments

We thank the Wenner-Gren Foundation, The Swedish Medical Research Council, Astra Arcus, Kabi-Pharmacia AB, and The Medical Nobel Committee for financial support of the Symposium that made this volume possible. We also wish to express our gratitude to the authors for their thoughtprovoking contributions, to Jasna Markovac and Mark Placito, Science Editors, and Erika Conner, Developmental Assistant at Raven Press, for their advice and assistance during the development of this volume. We thank Dr. Lennart Brodin for valuable editorial assistance.

ADVANCES IN SECOND MESSENGER
AND PHOSPHOPROTEIN RESEARCH

Volume 29

Molecular and Cellular Mechanisms of Neurotransmitter Release

Molecular and Cellular Mechanisms of Neurotransmitter Release, edited by Lennart Stjärne, Paul Greengard, Sten Grillner, Tomas Hökfelt, and David Ottoson, Raven Press, Ltd., New York © 1994.

1

The Concept of Neurotransmitter Release

Max R. Bennett

Department of Physiology, University of Sydney, Sydney, N.S.W. 2006, Australia

RESEARCH ON THE SYNAPSE IN THE LABORATORIES OF SHERRINGTON AND LANGLEY BEFORE THE GREAT WAR

Ninety years ago Charles Sherrington gave his Silliman Lectures at Yale University which were later published as "The Integrative Action of the Nervous System" (1). In that great work Sherrington laid the conceptual foundations for much that was to dominate research on the central nervous system for the rest of the century. Sherrington had begun his studies on the central nervous system at Cambridge in the physiological laboratory of John Langley and later at Liverpool. In his Silliman Lectures Sherrington pointed out that:

> *In view, therefore, of the probable importance physiologically of this mode of nexus between neurone and neurone it is convenient to have a term for it. The term introduced has been "synapse."*

Sherrington had already defined the synapsis in Foster's *Textbook of Physiology* some ten years earlier (2). He went on to say in the Silliman Lectures that:

> *The neurone itself is visibly a continuum from end to end, but continuity, as said above, fails to be demonstrable where neurone meets neurone—at the synapse. There a different kind of transmission may occur. The delay in the gray matter may be referable, therefore, to the transmission at the synapse.*

Regarding how synapses operate, he said:

> *It would be a mechanism where nervous conduction, especially if predominantly physical in nature, might have grafted upon its characters just such as those differentiating reflex-arc conduction from nerve-trunk conduction.*

Sherrington had developed these ideas as a consequence of his studies on the reflex contractions of muscles following stimulation of muscle and skin afferents. His summary diagram of the place of excitation and inhibition in reflex pathways for

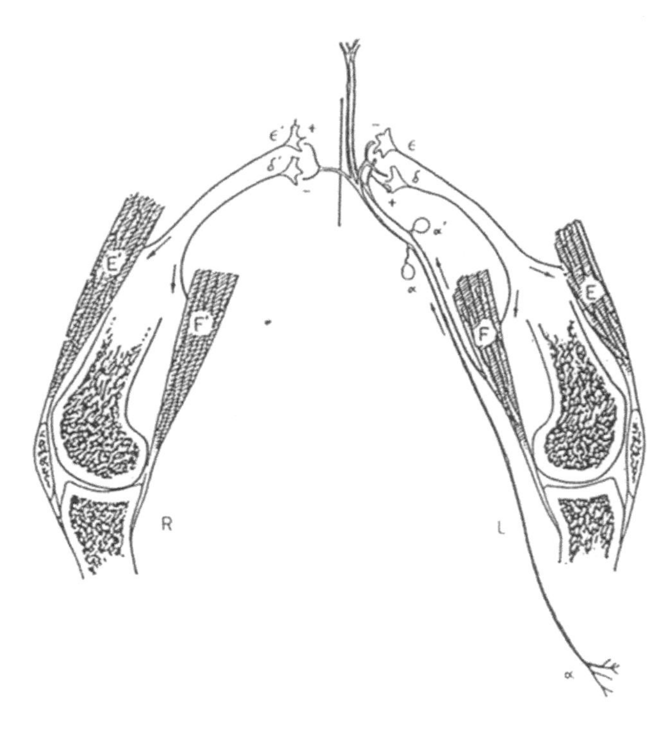

FIG. 1. Sherrington's 1906 diagram "indicating connections and actions of two afferent spinal root-cells (dorsal root ganglia), α and α'" in regard to their reflex influence on the extensor and flexor muscles of the two knees. Flexor and extensor muscles of the knee joint on the right (R) and left (L) sides are shown together with the inputs to the spinal cord by a cutaneous afferent (α) and a muscle spindle afferent (α'). The reflex pathways postulated show flexor (F) excitation (+) and extensor (E) inhibition (−) ipsilaterally and flexor (F) inhibition (−) and extensor (E) excitation (+) contralaterally. In Sherrington's 1906 words, "the sign + indicates that at the synapse which it marks the afferent fibre α (and α') excites the motor neurone to discharge activity, whereas the sign − indicates that at the synapse which it marks the afferent fibre α (and α') inhibits the discharging activity of the motor neurones. The effect of strychnine and of tetanus toxin is to convert the minus sign into a plus sign."

flexor activation and extensor inhibition, shown in Fig. 1, is a masterpiece of fruitful speculation. This diagram not only indicates the concept of excitatory and inhibitory synapses, developed clearly by 1908, but draws the experimentalist into Sherrington's line of inquiry as to what other nervous pathways may be delineated by this approach: in particular, how is the information transferred at the nerve terminal across the synaptic gap in the drawing, and what is the mechanism of inhibition.

A research program entirely different from Sherrington's was directed by his mentor J. N. Langley at Cambridge. In 1903 Langley, who had introduced Sherrington to neurophysiology (3), was at that time laying the foundations for our understanding of the chemical nature of transmission at synapses. In 1901 Langley published a remarkable paper (4) showing that stimulation of the sympathetic component of the autonomic nervous system, which Gaskell and he had already defined, resulted in changes in the effectors that in many cases could be mimicked by application of suprarenal extract (i.e., adrenaline). In Langley's words:

> *I have formerly divided the autonomic nervous system into sympathetic, cranial, sacral and enteric. It is a noteworthy fact that the effect of supra-renal extract in no case corresponds to that which is produced by stimulation in normal conditions of a cranial autonomic or of a sacral autonomic nerve. . . . It is equally noteworthy that the effects produced by supra-renal extract are almost all such as are produced by stimulation of some one or other sympathetic nerve. In many cases the effects produced by the extract and by electrical stimulation of the sympathetic nerve correspond exactly (see Fig. 2A). . . . It is hardly possible to avoid the conclusion that in these cases the extract acts directly on the unstriated muscle, and if this is so, it is probable that in all cases the action is direct. The theory of direct action cannot, however, be regarded as more than provisional until it is shown experimentally that the inhibitory action of supra-renal extract on certain unstriated muscle, and its stimulating action on salivary gland cells take place in the absence of nerve-endings. These points I propose to consider in a later paper (4).*

These experiments were carried out by Langley's student T. R. Elliott who concluded in a note to *The Journal of Physiology* (owned and edited by Langley) in 1904 that:

> *Adrenalin might then be the chemical stimulant liberated on each occasion when the impulse arrives at the periphery (5).*

The idea of chemical transmission at the synapse, and indeed of receptors on the effector organ for receiving the chemical substance released by the nerves, was already a central part of the research program in Cambridge physiology under Langley (6). This research was furthered in 1906 by W. E. Dixon. Working in the Cambridge physiology laboratory on the effect of suprarenal extracts on the lung (7) Dixon decided to perform an experiment in which he took an extract of a dog's heart that had undergone vagal stimulation and applied it to the exposed heart of a frog, obtaining an interruption of the heart beat (Fig. 2B). This work was similar in design to Otto Loewi's famous 1921 experiment (see Fig. 3A) some 14 years later, establishing the idea of chemical transmission in the heart unequivocally (8,9). Henry Dale had observed these experiments of Dixon's. Dale came across acetylcholine accidentally in 1914, as a constituent of a particular sample of ergot. Here he describes his finding:

> *I was led to make a detailed study of its action. This, I think, gave the first hint that acetylcholine might have an interest for physiology. Then I was struck by the remarkable fidelity with which it reproduced the various effects of parasympathetic nerves, inhibitor on some organs and augmentor on others—a fidelity which I compared to that with which adrenaline reproduces the effects of the other, true sympathetic, division of the autonomic system (10).*

At the beginning of the century Sherrington had already placed both excitatory and inhibitory synapses at center stage in the integrative behavior of the spinal cord. Furthermore, Langley's school had shown that synapses at the autonomic neuroeffector junctions were likely to operate by the secretion of a chemical substance,

A

Table 1

Rise of blood-pressure.

Inhibition of the sphincter of the stomach and of the intestine (rabbit).
Inhibition of the bladder.
Dilation of the pupil (cat).
Withdrawal of nictitating membrane (cat) } slightly less readily
Separation of the eyelids (cat) } than the foregoing.

Contraction of uterus, vas deferens, seminal vesicles, etc. (rabbit).
Salivary and lachrymal secretion.
Inhibition of the stomach.
Inhibition of the gall-bladder and increased bile secretion.
Dilation of pupil (rabbit).
Inhibition of internal anal sphincter (rabbit).
Contraction of internal anal sphincter (cat) } effects relatively
Contraction of internal generative organs (cat) } slight.

Contraction of the muscles of the hairs.

Contraction of tunica dartos of scrotum } no certain effect.
Secretion of sweat }

B

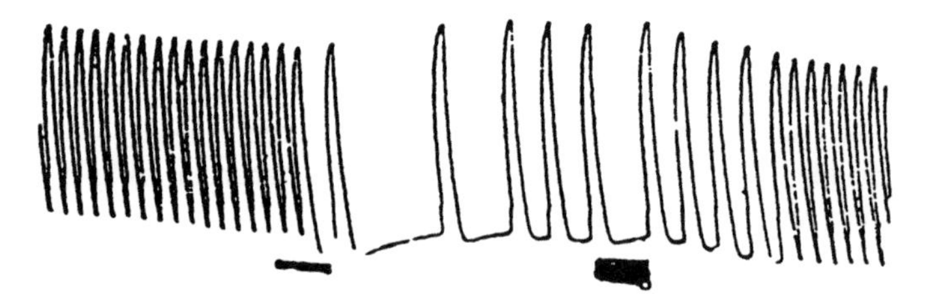

which in the case of the sympathetic was related to adrenaline. Research over the next 25 years attempted to unravel the principles of operation of synapses within a conceptual framework that originated in the great research schools formed by Langley and Sherrington.

SHERRINGTON'S CONCEPT OF THE INHIBITORY AND EXCITATORY STATES OF CENTRAL SYNAPSES

John Eccles was born in 1903 and he entered Melbourne University Medical School at the very young age of 15, in the year that saw the end of the Great War, and later won a Rhodes Scholarship to Oxford in 1925 to work with Sherrington. Eccles entered an intellectual environment on synaptic transmission that was now dominated by Loewi's recent experiments (8) indicating that chemical transmission occurred between the vagus and the heart, and Dale's work indicating a role for acetylcholine in synaptic transmission (10,11,12) (Fig. 3A). Eccles entered Oxford the year J. N. Langley died at the age 73 after completing a 6-hour experiment in Cambridge. He arrived at a time when Sherrington was engaged in research with Liddell on the characteristics of the myotatic reflex (13,14) and with Creed on the flexion reflex (15). Sherrington had just produced a masterly summary of work on inhibition, in which he concluded that:

> *In relation to inhibition at the synapse that it might be mediated by an agent, moreover, one whose existence lies outside the intrinsic properties of pure nerve-fibre and with a, so to say, more chemical mode of origin and function than the nerve impulse per se (16).*

This comment was made only 4 years after the experiments of Loewi. Eccles joined Creed in his first experimental work, which was on the subject destined to dominate his research life for over 40 years: the mechanism of inhibitory synaptic transmission (17). Then in 1929 he joined Sherrington in a technical improvement of the torsion myograph (18) in preparation for a collaboration (that lasted but a few years,

FIG. 2. The evolution of the idea of chemical transmission at synapses. **A,** Langley's 1901 table showing the effect of suprarenal extract (adrenaline) in the cat and rabbit arranged roughly in the order of the amount of extract required per body weight to produce an obvious effect. Langley notes that: *I have formerly divided the autonomic nervous system into sympathetic, cranial, sacral and enteric. It is a noteworthy fact that the effect of supra-renal extract in no case corresponds to that which is produced by stimulation in normal conditions of a cranial autonomic or of a sacral autonomic nerve. It is equally noteworthy that the effects produced by supra-renal extract are almost all such as are produced by stimulation of some one or other sympathetic nerve. It is hardly possible to avoid the conclusion that in these cases the extract acts directly on the unstriated muscle, and if this is so, it is probable that in all cases the action is direct.* **B,** unpublished record from a 1906 experiment by W. E. Dixon showing the beat of the exposed heart of a frog. At the first mark, extract from a dog's heart that had been inhibited by vagal stimulation was applied; at the second mark, atropine was applied (from Dale, 1934). This is the first known record of an attempt to determine if a nerve secretes a substance that, when placed on another organ, will mimic the effects of nerve stimulation to that organ.

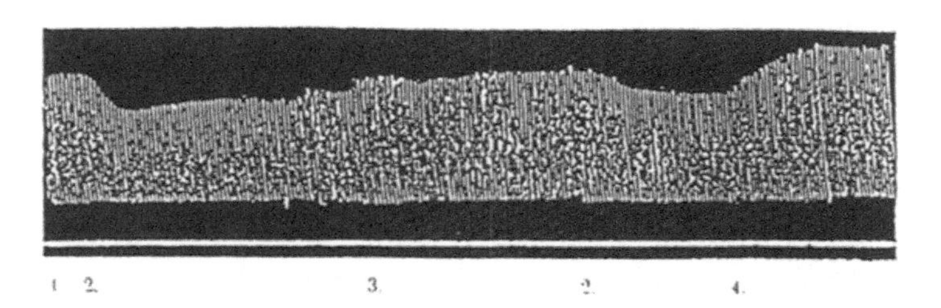

B

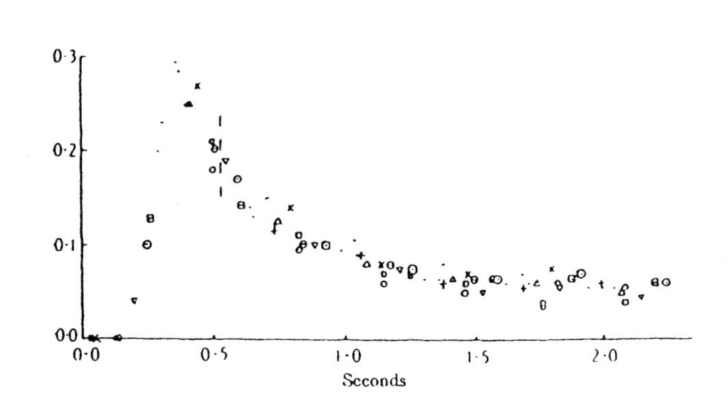

C

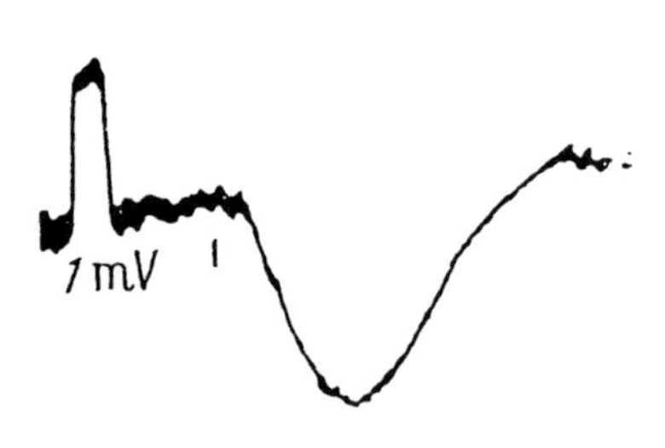

1929–1931) concerned with research on the flexion reflex and inhibition (19,20). These experiments were to see the last flowering of Sherrington's scientific genius at the age of 75. The work on the ipsilateral spinal flexion reflex introduced Eccles to the technique of stimulating first with a just threshold conditioning volley, then at later intervals with a subsequent test volley in order to tease out the time course of the central excitatory state (18) (Fig. 4A). This approach, when applied to the mechanism of monosynaptic transmission in the spinal cord, gave a very precise measure of the time course of the central excitatory state or, as we now know, the excitatory postsynaptic potential (Fig. 4). With hindsight it might be expected that chemical transmission would seem to be the most likely mechanism for determining the central excitatory state (c.e.s.) and the central inhibitory state (c.i.s.), based on the experiments of Langley and his school, along with those of Loewi and Dale. This was certainly not the case, as is shown in the next section.

LUCAS, ADRIAN, AND THE ELECTRICAL CONCEPT OF THE INHIBITORY STATE OF CENTRAL SYNAPSES

Towards the end of Langley's career, the Cambridge School of Physiology came to be dominated by those such as Keith Lucas and E. D. Adrian who were introducing electrophysiological techniques into the study of how impulses conduct in excitable tissue. Lucas had published a Physiology Monograph entitled "The Conduction of the Nervous Impulse" in which he argued that central inhibition might be brought about by the interference of high-frequency discharges in the nerves as they approach their synaptic connections on neurones (21). In this way the refractory state of the axon following an impulse could operate to produce inhibition. This idea was followed up in detail in 1924 by Adrian, (Fig. 5A) who was skeptical about the recent research of Loewi and Dale, commenting:

FIG. 3. The first apparently unequivocal demonstration of chemical transmission at a synapse and characterization of the accompanying synaptic potential. This was performed for the vagal inhibition of the heart beat. **A,** Otto Loewi's 1921 original record in which at 1 the heart beat is shown in normal ringer; at 2, the decline in the heartbeat is due to the addition of a ringer that had been in contact with another heart whose vagus had been stimulated for 15 minutes; at 3, the heartbeat is normal in the presence of a ringer from another heart in which the vagus had not been stimulated; finally, in 4, atropine was added in a normal ringer, increasing the heartbeat. **B,** curve showing the extent of inhibition of the heartbeat (on the vertical axis) due to a single stimulus in an experiment performed by Brown and Eccles (1934). A single stimulus is applied to the vagus nerve and the lengthening of each cardiac cycle (i.e., the amount by which it exceeds a normal cycle) is expressed as a fraction of the normal cycle (of about 305 msec) on the ordinate; the abscissa gives the interval between the vagal stimulus and the end of that particular cycle. There is a latent period of rather more than 100 msec before a volley in the postganglionic fibers produces an inhibitory effect on the pacemaker. **C,** intracellular records of the hyperpolarization in the arrested frog's heart due to a single vagal volley by del Castillo and Katz (1957). Note that the latency between the vagal volley and the hyperpolarization is several hundred msec. Calibration is (small vertical bar) 1mV and indicates the moment of stimulation, which occurs 400 msec before the hyperpolarization commences.

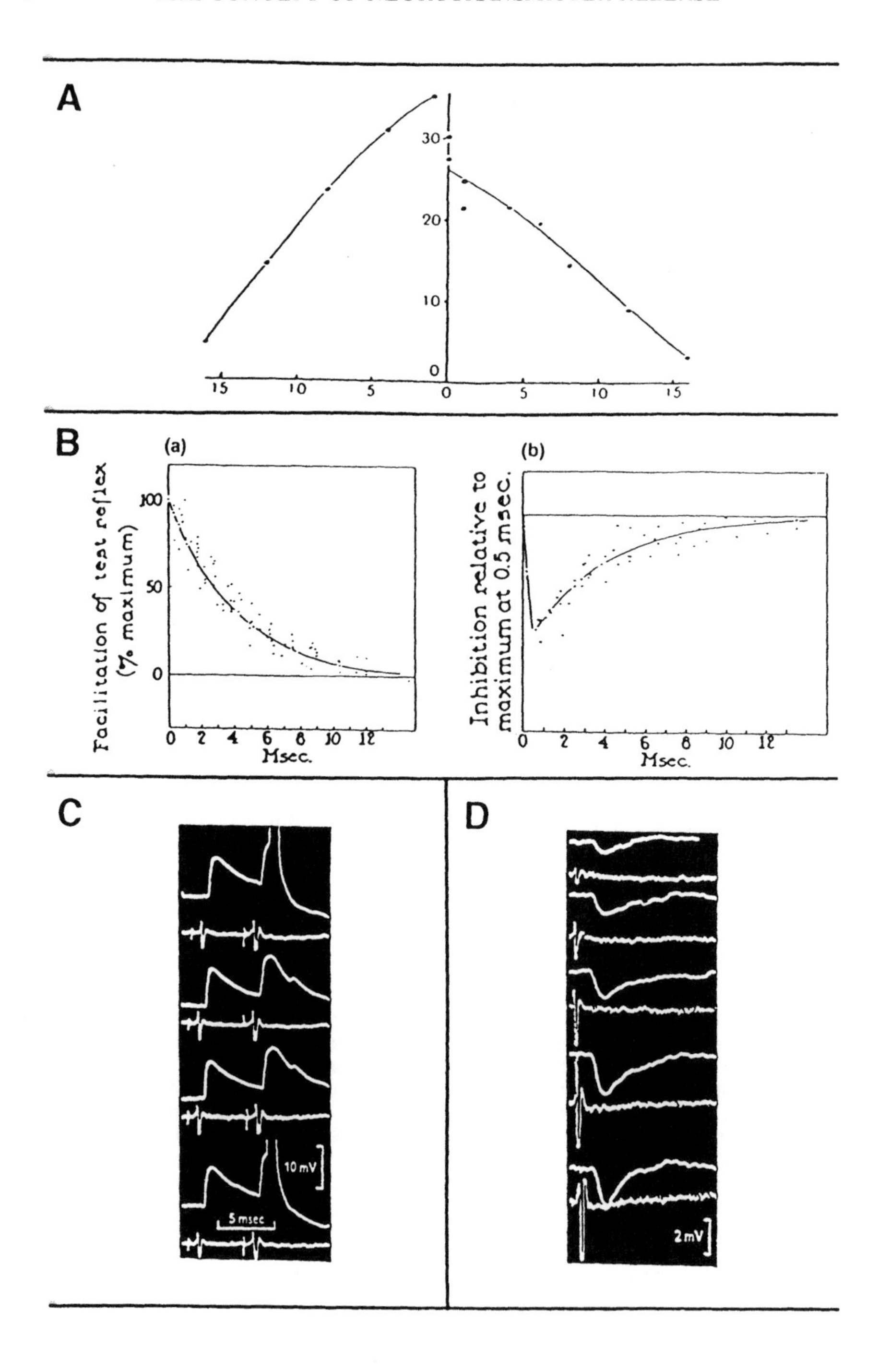
A
30
20
10
0
15
10
5
0
5
10
15
B
(a)
(b)
Facilitation of test reflex (% maximum)
100
50
0
0 2 4 6 8 10 12
Msec.
Inhibition relative to maximum at 0.5 msec.
0 2 4 6 8 10 12
Msec.
C
D
10 mV
5 msec
2 mV

It appears, then that the fluid coming from the stimulated organ reproduces the characteristic effects of the vagus or sympathetic on different tissues, though whether every detail of the nervous effect is copied by the fluid remains an open question. The nature of the "vagus substance" is uncertain. If these results can come to be generally accepted we shall have a new and extremely interesting picture of the action of the autonomic system. The difficulty is that the effects seem to be capricious and are not easily reproduced. Some observers have failed to satisfy themselves that they occur at all outside the margins of the experimental error. In view of this uncertainty we can only wait until there is a more general agreement as to the experimental basis of the humoral theory (22).

In expanding on the Lucas theory of the electrical basis of spinal cord inhibition, Adrian went on to say that:

If it (the humoral theory) is correct, the explanation of peripheral inhibition resolves itself into that of (a) the secretory mechanism which produces the inhibiting substance whenever impulses pass along certain fibres to the muscle, and (b) the way in which an inhibiting substance, adrenalin for instance, exerts its effect on the muscle fibre . . . If we compare the present theory (the electrical theory of Figure 5A), or some modifica-

FIG. 4. The time courses of the central excitatory states (c.e.s.) and central inhibitory states (c.i.s.) compared with those of the excitatory and inhibitory postsynaptic potentials. **A,** time course of the central excitatory state (c.e.s.) determined using muscle reflexes. It shows the reflex response of the tibialis anticus muscle to two stimuli (each one of which alone is just subthreshold) to medial gastrocnemius nerve and lateral gastrocnemius nerve at various intervals, as measured by Eccles and Sherrington (1930). The abscissa gives the time, approximately in msec, and the ordinate the tension in grams. Right of zero shows the interval by which the stimulus to the lateral gastrocnemius nerve is leading; left of zero shows the interval by which stimulus to the medial gastrocnemius nerve is leading to. Each point plotted shows the tension developed at the indicated interval between stimuli. The excitatory state, set up by the conditioning volley, lasts for about 15 msec and is called the "central excitatory state" (c.e.s.). **B,** time course of the central excitatory state (c.e.s.) and central inhibitory state (c.i.s.) determined by measuring the compound action potential in a muscle nerve. **a,** the extent to which the response to a test reflex is enhanced by a prior conditioning reflex (facilitation) at the different intervals given in the abscissa, as measured by Lloyd (1946). The facilitation of the biceps reflex by afferent volleys in the semitendinosus nerve and of one head of the gastrocnemius by afferent volleys in nerves to the other head are given. Conditioning volleys of near reflex threshold strength were used. The relative facilitation, expressed in percent maximum, is plotted as a function of time and gives the c.e.s.; this is similar to that given by the method used by Eccles and Sherrington in A. **b,** the extent to which the response to a test reflex is inhibited by a prior conditioning reflex at the different intervals given in the abscissa, as measured by Lloyd (1946). The inhibition of the reflex of tibialis anterior by weak volleys to the gastrocnemius afferent nerve are given. The ordinate gives the degree of inhibition, in percent of maximum, to the time interval between volleys on the abscissa. This relative inhibition, expressed in percent maximum, is plotted as a function of time and gives the c.i.s. **C,** intracellular potentials recorded by Brock, Coombs, and Eccles (1952) from a biceps semitendinosus motoneuron due to two afferent volleys in the biceps semitendinosus nerve; note that the dorsal root spikes accompanying each volley are shown beneath the intracellular recordings. These excitatory postsynaptic potentials give the electrical signs of the central excitatory state measured by Eccles and Sherrington in 1930 and shown in A; note the similar time course. **D,** intracellular potentials recorded by Brock, Coombs, and Eccles (1952) from a biceps semitendinosus motoneuron due to a single volley in a quadriceps nerve, of increasing size downward; note the increasing size of the dorsal root spikes shown beneath the successive intracellular recordings. These inhibitory postsynaptic potentials give the electrical signs of the central inhibitory state shown in B,b; note the similar time course.

tion of it, with the view which supposes that inhibition is due to the production of a special substance which blocks the excitatory paths to the motor neurone, it will be seen that there is actually not very much difference between them. If an inhibiting substance is produced, its production must be almost instantaneous and it must disappear very rapidly; what the present theory assumes is that the "substance" is to be

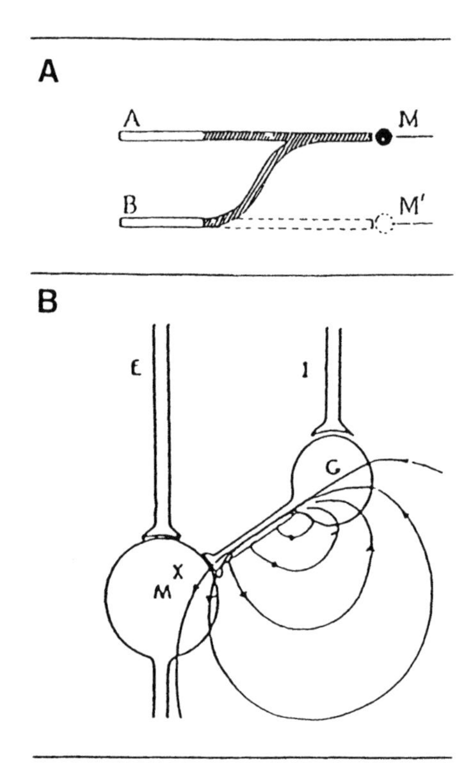

FIG. 5. Electrical theories of central inhibition. **A,** scheme due to Adrian (1924) based on an idea of Keith Lucas (1917; see his Fig. 22) in which inhibition occurs as a consequence of the refractory state left in a nerve following an impulse. In this diagram, A is the excitatory pathway and B the inhibitory pathway converging on the motoneuron M; the shaded areas conduct with a decrement. By making the decremental path from B to M longer than that from A, we can account for the fact that an impulse from B can never succeed in reaching M, whereas impulses from A can do so provided the path is given time for complete recovery between each impulse. With such an arrangement, an impulse from B, though not itself exciting M, would leave the common pathway in a refractory state, absolute or relative, which would hinder the passage of impulses from A for a time depending on the rate of recovery of the path and the extent of the decrement in it. A rapid succession of impulses from B would produce continuous inhibition by never giving time for the complete recovery of the common pathway. If the impulses from B also pass by a more direct route, dotted in the figure, to the antagonistic motoneuron M′, the periods of inhibition of M would synchronize with the discharge of motor impulses from M′. **B,** scheme due to Brooks and Eccles (1947) of how electrical inhibition could occur in the spinal cord. The diagram indicates current flow at a schematic synapse of a Golgi cell G on a motoneuron M according to this electrical theory of inhibition. E shows the excitatory line to M, and I the inhibitory line that subliminally excites G and so generates the current flow producing an electrotonic focus on M. According to this theory, an intracellular electrode would record a brief positively going electrical field at X.

identified with the refractory state and that sustained inhibition is due to a series of refractory states and not to a steady production of an inhibiting substance (22).

Adrian arrived at this conclusion as a consequence of experiments performed with Bronk on the frequency of discharges in motoneurones accompanying reflex and voluntary contractions. It was not until 1929 that Adrian felt able to abandon the idea that central inhibition could be explained along the lines suggested by Lucas (i.e., by the depressant effects produced by high-frequency impulse discharges) (23). Adrian carried great authority on matters concerned with electrical activity in nerves at the time Eccles arrived at Oxford in 1925. He had just successfully recorded for the first time the trains of nerve impulses traveling in single sensory and motor nerve fibers which, according to A. L. Hodgkin, "marks a turning point in the history of physiology " (24).

LOEWI, DALE AND ECCLES EXAMINE THE INHIBITORY STATE AT AUTONOMIC NEUROMUSCULAR JUNCTIONS

The chemical or electrical nature of synaptic transmission in either the peripheral or the central nervous system was an entirely unsettled issue when Eccles arrived at Oxford. It seemed likely, despite Adrian's skepticism, that vagal inhibition of the heartbeat was chemical and that the "vagus substance" or "Vagusstoff" was acetylcholine. However, it must be remembered that Otto Loewi and Henry Dale did not win the Nobel Prize for their investigations until 1936 and Loewi still felt constrained to defend this idea as late as 1935. Eccles, after completing his last work with Sherrington in 1931 on spinal cord inhibition, sought to delineate the characteristics of this chemical inhibition; he used the only appropriate preparation available at the time, namely the vagus to the heart. Together with G. L. Brown he determined that the time course of the inhibitory state (analogous to the c.i.s.) following a single stimulus to the vagus nerve was of the order of a second or more and did not arise for over 100 msec after a stimulus (25) (Fig. 3B). This determination was supported 20 years later with the introduction of intracellular recording of the inhibitory junction potential in the heart by del Castillo and Katz (26) (Fig. 3C). Brown and Eccles commented that:

If the vagal volley is set up late in a cardiac cycle, that cardiac cycle is not inhibited, the latent period of the inhibition being usually 100 to 160 ms. Of this amount the conduction time to the region of the pacemaker probably only accounts for about 10 ms, ie. the greater part of the latent period appears to occur after the arrival of the inhibitory impulses at the nerve fibres of the pacemaker. It is probable that most of this time is occupied in the liberation of the acetylcholine substance and its diffusion to the point of its action (25).

This slow time course of the only reasonably well established chemical synapse seemed to set the temporal characteristics of chemical transmission. This was particularly so for other synapses at which nerve terminals were shown to secrete acetylcholine.

ECCLES DEVELOPS THE ELECTRICAL CONCEPT OF THE EXCITATORY STATE AT AUTONOMIC SYNAPSES

Eccles then turned his attention to the only readily accessible synapse on neurons for which acetylcholine was known to be released on nerve stimulation, namely that in sympathetic ganglia (27,28) (Fig. 6A). It was only natural that he should first approach the problem of defining the excitatory state in the ganglion by the same methods developed to study the time course and other characteristics of the c.e.s. of motoneurons in the spinal cord (Fig. 4A). Examination was made of the interaction of submaximal volleys to each of two preganglionic inputs to a ganglion delivered at different intervals apart in test-conditioning pairs, and the ganglionic action potential measured (Fig. 6B–a). In this case the time course of compound action potentials was being determined rather than the time of reflex contractions of muscles used to determine the c.e.s. of motoneurons (Fig. 4A). Compound action potential waveforms had to be subtracted in the manner indicated in the legend to Fig. 6B–a before the c.e.s. could be estimated; this figure shows that the test compound action potential seems to be elevated compared with the compound action potential in the absence of a conditioning impulse. This elevation occurs from the earliest times for the test-conditioning interval, then declines to zero at an interval of 4.5 msec, increases again, and reaches a peak at 17.6 msec from which it slowly declines over the next 40 msec or so to zero. The time course of these events is indicated in Figure 6B–b which shows the early fast phase of the potentiation of the test compound action potential followed by the later developing slow component. The fast phase, termed the "detonator response" (28) was not affected by anticholinesterases as is vagal inhibition of the heart; furthermore, it is much faster than the action of acetylcholine on the heart (Fig. 3B). The detonator response was not then attributed to the action of acetylcholine but rather to the action currents in the preganglionic nerve terminals that produced a potential response in the ganglion cells; the later phase was identified as the excitatory state. This analysis led to the proposition that:

> *On present evidence however, it seems that the action-current hypothesis offers a more probable explanation for direct synaptic transmission, the acetylcholine liberated in sympathetic ganglia possibly having a secondary excitatory action as already suggested (29).*

The analysis of synaptic transmission from the postganglionic nerves to the smooth muscle of the nictitating membrane also revealed an excitatory state that had an initial fast component followed by a slower late component (30,31,32).In this case the anti-adrenaline drug 933F (piperidonethyl–3–benzodioxane) blocked the slow electrical response and the associated contraction but not the fast response, suggesting that the latter might also indicate the signs of electrical transmission (29,31). Similar difficulties were arising with the parasympathetic innervation of the bladder. In this case only the slow phase of contraction could be blocked by atropine and mimicked by applied acetylcholine; this left the fast phase to be accounted for in terms of an electrical component of transmission (33). In order to escape this awkward fact, Dale and Gaddum (34) suggested that the concentration of adrenaline secreted at the sympathetic postganglionic nerve terminal was very high, giving the fast response, and was therefore insensitive to 933F; the slow reaction was attributed to the escape of adrenaline and its secondary diffusion from the synaptic cleft to act on other cells that were sensitive to 933F. As late as 1937 Dale was reiterating that:

> *This antagonist (atropine) cannot similarly intervene, when acetylcholine is liberated from nerve endings in immediate contact with, or even inside the cell membrane (35).*

Dale thought that the fast phase of contraction of smooth muscle on nerve stimulation was due to the high transmitter concentration reached at the nerve endings, with the slow contraction attributed to its later diffusion to other muscle cells (35). He had often used different renditions of this argument to escape the unpalatable fact that neither adrenaline nor acetylcholine seemed suitable candidates for transmission to some smooth muscles. For example, in 1934 Dale said that:

> *There are some parasympathetic effects, such as the action of the vagus on the intestine, and the vaso-dilator action of parasympathetic nerves in general, which are resistant to atropine, though the otherwise similar actions of injecting or applying acetylcholine are readily abolished by it. . . . Gaddum and I suggested that in such cases the nerve impulses liberate acetylcholine so close to the reactive structures that atropine cannot intervene, whereas it can prevent its access to them when it is artificially applied from without (9).*

Following the discovery of inhibitory junction potentials in smooth muscle in 1963 (36) that are mediated by nonadrenergic noncholinergic (NANC) synapses (37) we now know of many synapses that do not conform to Dale's paradigm.

Eccles developed his ideas on electrical transmission in Oxford between 1934 and 1937. He then returned to Australia in 1937 to become Director of the Kanematsu Memorial Institute of Pathology at Sydney Hospital. His first studies there were concerned with the possibility of electrical transmission at the somatic neuromuscular junction. In 1936 G. L. Brown had shown that close interarterial injection of acetylcholine into the cat's gastrocnemius gave repetitive impulse firing and contraction of the muscle (Fig. 7B). Protection of the metabolism of acetylcholine with eserine allowed it to appear in a perfusate after stimulation of motor nerves (38,39). By 1938 both Eccles and O'Connor (40) as well as Göpfert and Schaefer (41) had recorded the endplate potential with extracellular electrodes in curarized muscles (Fig. 7C). This was probably the first time a synaptic potential had been observed without distortion due to electrotonic conduction. The possibility that the endplate potential was due to the release of acetylcholine was not grasped. In 1939 Eccles and O'Connor used the method of applying conditioning-test volleys to mammalian motor nerves and recording the impulses in a muscle in order to determine the time course of the excitatory state. They concluded that a nerve impulse exerts two excitatory actions at the neuromuscular junction:

> *(1) Newborn muscle impulses are set up by a brief excitatory action probably no more than 1 msec. in duration and analogous to the detonator action described for synaptic transmission. (2) The much more prolonged end-plate potential is set up independently of the newborn impulses, but, if the growth of these impulses is sufficiently delayed, it*

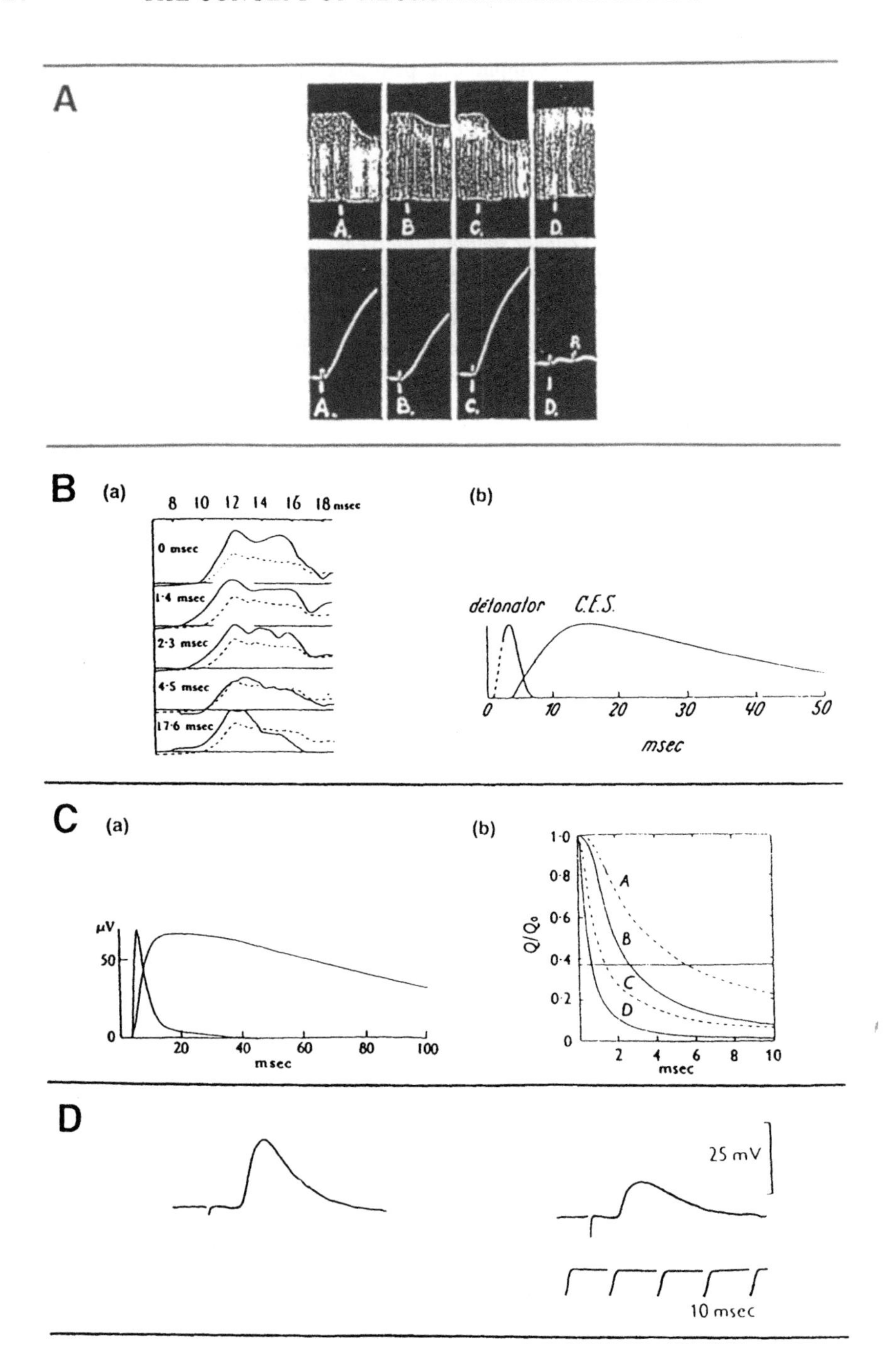
A
A.
B
C.
D.
A.
B.
C.
R
D.
B
(a)
8 10 12 14 16 18 msec
0 msec
1·4 msec
2·3 msec
4·5 msec
17·6 msec
(b)
détonator
C.E.S.
0
10
20
30
40
50
msec
C
(a)
μV
50
0
20
40
60
80
100
msec
(b)
1·0
0·8
0·6
0·4
0·2
0
Q/Q₀
A
B
C
D
2
4
6
8
10
msec
D
25 mV
10 msec

appears to aid in their growth to the fully propagated size. It is analogous to the N wave and the associated central excitatory state of synaptic transmission, and analogous responses have also been described at the neuro-muscular junction of smooth muscle (42).

Thus the test-conditioning volley approach used so successfully to determine the time course of the c.e.s. for motoneurons now led to the erroneous conclusion that a very fast response, much faster than the endplate potential recorded at the neuromuscular junction in curarized preparations, was the primary means of transmission.

KATZ, KUFFLER AND ECCLES ESTABLISH THE MOTOR ENDPLATE AS THE PARADIGM SYNAPSE FOR ELECTROPHYSIOLOGY

J. N. Langley and Keith Lucas initiated the studies of A. V. Hill on the biophysics of nerve and muscle at Cambridge in 1909, and Hill in turn supervised the

FIG. 6. Claims for electrical and chemical transmission at synapses in sympathetic ganglia. **A,** demonstration of how stimulation of the cervical sympathetic causes liberation from the ganglion of a substance pharmacologically identified by Feldberg and Gaddum (1934) as acetylcholine. The top panels show the response of the beating frog's heart and the bottom panels that of contraction of leech muscle as a consequence of adding Ringer's fluid collected from a ganglion during preganglionic stimulation (A); adding Ringer's with different concentrations of acetylcholine (B and C), and adding Ringer's fluid from an unstimulated ganglion (D). Note that the fluid from the stimulated ganglion has identical effects on slowing the beating heart and contracting the leech muscle, as does acetylcholine. **B,** electrophysiological evidence involving the study of action potentials interpreted by Eccles (1937) as showing an electrical component to synaptic transmission. In (a), single submaximal stimuli were applied to two different preganglionic nerve branches at various intervals apart; the continuous lines show the ganglionic action potentials produced by a second stimulus at the indicated intervals and the broken lines show the action potential set up by the second stimulus alone. It will be noted that when the two volleys are simultaneous, spatial facilitation is maximal (i.e., it is effective in producing a discharge from the largest number of ganglion cells); this number is still large at the interval of 2.3 msec. At 4.5 msec, effective summation occurs in very few, if any, ganglion cells. At intervals longer than 4.5 msec, spatial and temporal facilitation again develop and the time course of decay of this second facilitation wave is due to the excitatory state of the ganglion cells, which would now be called the excitatory postsynaptic potential. In (b) is shown the time course of the "detonator response" and the excitatory state derived from experiments such as those in (a); the former is attributed to electrical transmission at the synapse and the latter to chemical transmission using acetylcholine to give the synaptic potential (from Eccles, 1936). **C,** electrophysiological evidence involving the direct study of synaptic potentials in the absence of action potentials interpreted by Eccles (1943) as showing that only chemical transmission occurs in ganglia. In (a) is shown a single synaptic potential, recorded with extracellular electrodes, which lasts for about 100 msec. The faster curve is a theoretical estimate of the time course of transmitter action that gives rise to the synaptic potential, based on the "local potential" theory of A. V. Hill. Ordinates for transmitter actions are in arbitrary units. In (b) is shown the theoretical times for the decline in the amount of transmitter remaining within a sphere of either radius 2 μm (B) or 1 μm (D), as well as a cylinder of radius 2 μm (A) or 1 μm (C) following the transmitter's instantaneous deposition (from Ogston, 1955). The curves show that the concentration declines with a similar time course to that of the time course of transmitter action given in (a), suggesting that free diffusion of acetylcholine out of a synaptic region with the dimensions of 1 μm or 2 μm could account for the observed results. **D,** first synaptic potentials recorded with an intracellular electrode from a mammalian sympathetic ganglion (from R. Eccles, 1955).

A

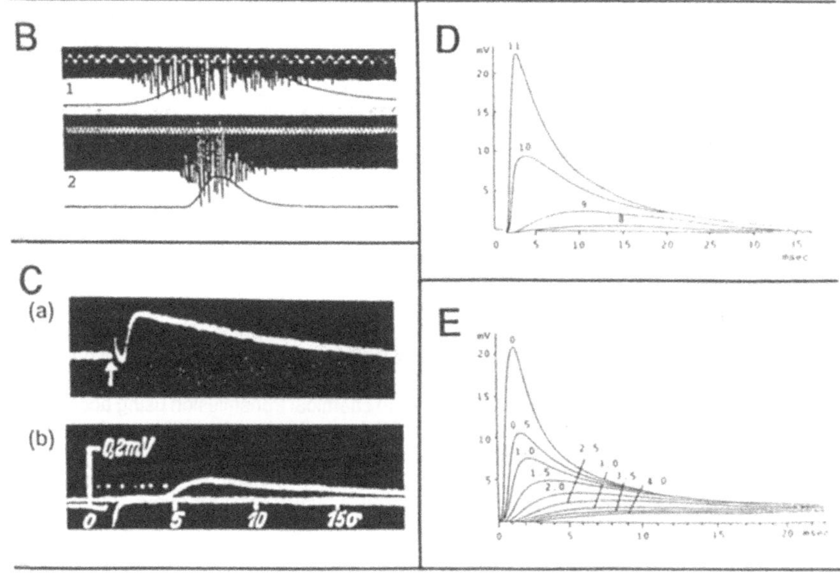
B
1
2
C
(a)
(b)
0,2mV
0
5
10
15σ
D
mV
11
10
9
8
msec
E
mV
0
0.5
1.0
1.5
2.0
2.5
3.0
3.5
4.0
msec

first research of Bernard Katz on electrical excitation and conduction of the nerve impulse at University College London in 1935 (43,44). The frog isolated gastrocnemius-sciatic nerve preparation was used by Katz in his investigations of Hill's theory of excitation in order to provide an index for the duration of maintained nerve excitation (45). However it was the Cambridge zoologist Carl Pantin, who had acted as a guide to A. L. Hodgkin's studies, that was responsible for the first explicit research by Katz on neuromuscular transmission (46). In this work it was shown that magnesium ions could block neuromuscular transmission in crabs. By 1939 Katz was using the frog isolated sartorius-nerve preparation after curarization (47) to confirm the work of Göpfert and Schaefer that showed:

> *A small non-conducted potential change is to be found in the myoneural region, which reaches a maximum 4 msec after arrival of the nerve impulse, and then falls at a slow rate, similar to the electrotonic potential (41).*

It was natural that Katz should have wanted to continue research on this "small non-conducted potential change" or endplate potential, using the frog sartorius-nerve preparation, when he joined Eccles and Stephen Kuffler at the Sydney Kanematsu Institute in 1939 (Fig. 7A). Rather than working on the innervation zones of the cat's soleus muscle with Eccles, Katz did the following:

> *I ganged up with Stephen Kuffler, and I was very pleased when we succeeded in getting hold of some nice Australian tree frogs whose sartorius muscles proved to be very suitable for the experiments we wanted to do, and this kept me busy and moderately happy for two years (48).*

The work of Katz, Kuffler, and Eccles in Sydney (78,81) marks the beginnings of a new era in synaptic physiology after the one begun 50 years earlier by Langley and Sherrington. It is characterized by the use of progressively more sophisticated electrical techniques to probe the mechanism of synaptic transmission. The first experiments of Eccles, Katz, and Kuffler did not involve test-conditioning volleys to estimate the time course of the excitatory state but rather concentrated on the properties of the extracellular signs of the endplate potential made subthreshold by a

FIG. 7. The electrical and chemical analysis of synaptic transmission at the neuromuscular junction. **A,** S. W. Kuffler, J. C. Eccles, and B. Katz in 1941 at the time of their experiments on elucidating the nature of the endplate potential. They are shown in Martin Place, Sydney, walking from the Kanematsu Memorial Institute of Pathology at Sydney Hospital to catch a tram to the University of Sydney to give a lecture in the Physiology Department. **B,** first recordings of the electrical and mechanical responses of the cat's gastrocnemius muscle to close intra-arterial injections of two different concentrations of acetylcholine, taken by G. L. Brown and communicated to J. C. Eccles (1936). **C,** first endplate potentials to be recorded with extracellular electrodes in curarized (a) cat soleus muscle (Eccles and O'Connor, 1938) and (b) frog sartorius muscle (Göpfert and Schaefer, 1938). **D,** endplate potential recorded with an extracellular electrode in a curarized frog sartorius muscle by Eccles, Katz, and Kuffler (Eccles et al., 1941); the numbers on the records refer to the distance in mm from the pelvic end of the muscle. **E,** tracings of endplate potentials recorded with an intracellular electrode from the frog sartorius by Fatt and Katz (1951); the numbers refer to different distances in mm of the recording electrode from the endplate (compare with D above).

suitable dose of curare (Figure 7D). They showed, using an analysis provided by A. V. Hill (49), that the time course of the underlying transmitter action lasted for only a few msec. As Eccles, Katz, and Kuffler stated:

> *Thus it seems that most of the declining phase of the e.p.p. is a passive decay of a negative membrane charge after the depolarizing agent has ceased to act. The earlier suggestion, therefore, that the decline of the endplate potential follows the time course of a passively decaying electrotonic potential is confirmed* (50).

They stated further:

> *By making plausible assumptions it is shown that the observed curare and eserine actions are reconcilable with the hypothesis that acetylcholine is responsible for all the local potential changes set up by nerve impulses* (51).

Eccles then abandoned the hypothesis of an early "detonator" electrical response followed by a slower endplate potential due to the secretion of acetylcholine. In the following year, Eccles used an analysis similar to that applied to the neuromuscular junction when he evaluated extracellular recordings of the synaptic potential in curarized sympathetic ganglia. This led to the conclusion that (Fig. 6C–a):

> *The results conform well with the postulate of a single depolarizing agent. . . . it is concluded that most and possibly all of the evidence for the detonator action may now be attributed to the brief transmitter action* (52).

The time course of transmitter action following a single impulse here could be shown to conform to free diffusion of acetylcholine from the synaptic cleft (53) rather than to the hydrolysis of acetylcholine by cholinesterase (Fig. 6C–b).

Katz returned to A. V. Hill's laboratory in 1946. Using the frog sartorius muscle-nerve preparation once more, and as a result of the introduction of the microelectrode in 1949 by Ling and Gerard (80), he was able to confirm with Fatt that the endplate potential alone initiated the muscle action potential (54); (compare Fig. 7D with Fig. 7E). This idea was soon generalized for the nervous system when Rosemary Eccles (Fig. 6D) showed that the synaptic potential in sympathetic ganglia alone initiated the action potential (55).

ECCLES ELUCIDATES THE ELECTRICAL SIGNS OF THE INHIBITORY AND EXCITATORY STATES OF CENTRAL SYNAPSES

The introduction of the microelectrode also allowed for the first time an investigation of whether the c.e.s. and c.i.s. of a motoneuron could be described in terms of synaptic potentials, and also whether these synaptic potentials were likely to be due to a "detonator" electrical effect or the secretion of a transmitter. Lloyd (56) had already utilized the test-conditioning volley method described 16 years earlier by Eccles and Sherrington (18) to determine this c.e.s. He used the stretch reflex and the more accurate method of electrical recording from the muscle nerves rather than muscle contraction. Lloyd's results for the time course of the c.e.s. of motoneurons were similar to those of Eccles and Sherrington (compare Fig. 4B–a with Fig. 4A);

he also gave the time course of the c.i.s. using this method (Fig. 4B–b). Eccles still thought it possible that electrical transmission might account for the c.i.s. of motoneurons; he provided a Golgi cell model of this as late as 1947 (57) (Fig. 5B), even though he had abandoned the idea of electrical transmission in the peripheral nervous system. It must be remembered that at this stage the chemical transmitters acting on motoneurons were unknown, research having shown that neither acetylcholine nor adrenaline were likely to be secreted at inhibitory or excitatory synapses. The first intracellular recordings of synaptic potentials in motoneurons were awaited with considerable interest.

Eccles had recently left Dunedin and taken up the foundation Chair of Physiology at the John Curtin School of Medical Research in Canberra. While there, he, Brock, and Coombs published their classic paper "The recording of potentials from motoneurones with an intracellular electrode" in 1952. As in the peripheral nervous system, the c.e.s. and the c.i.s. were identified with monophasic synaptic potentials, one in the depolarizing direction (Fig. 4C) and the other in the hyperpolarizing direction (Fig. 4D). Furthermore, these potentials had time courses similar to those predicted for the c.e.s. and the c.i.s. in the original work of Eccles and Sherrington in 1930 (compare Fig. 4C with Fig. 4A) and with that of Lloyd in 1946 (compare Fig. 4D with Fig. 4B–b). On observing the inhibitory postsynaptic potential, Eccles concluded that:

> *the potential change observed is directly opposite to that predicted by the Golgi-cell hypothesis, which is thereby falsified. . . . It may therefore be concluded that inhibitory synaptic action is mediated by a specific transmitter substance that is liberated from the inhibitory synaptic knobs and causes an increase in polarization of the subjacent membrane of the motoneurone (58).*

So concluded the long saga that established the supremacy of chemical transmission at peripheral and central synapses. It enabled Eccles to finally show that Sherrington's c.e.s. and the c.i.s. of motoneurons were due to chemical transmission (Fig. 10A).

In hindsight, Keith Lucas and E. D. Adrian, the founding fathers of the biophysical approach to the study of impulse conduction, had inappropriately tried to transfer knowledge gained on conduction to the theoretical analysis of the mechanisms of transmission. Langley, Loewi, and Dale had exerted a counteracting influence that was founded on a tremendous amount of ingenious experimentation on transmission. However, it was the analysis of the endplate potential offered by the biophysicists A. V. Hill and Bernard Katz that finally clarified the whole matter.

KATZ'S CONCEPT OF QUANTAL TRANSMITTER RELEASE AT THE MOTOR ENDPLATE AND THE VESICLE HYPOTHESIS

The discovery of spontaneous miniature endplate potentials in the frog sartorius muscle by Fatt and Katz revolutionized our understanding of neurotransmitter release (59) (Fig. 8A). Although the endplate potential in response to a nerve impulse had been identified in 1938, it was not possible to observe spontaneous endplate

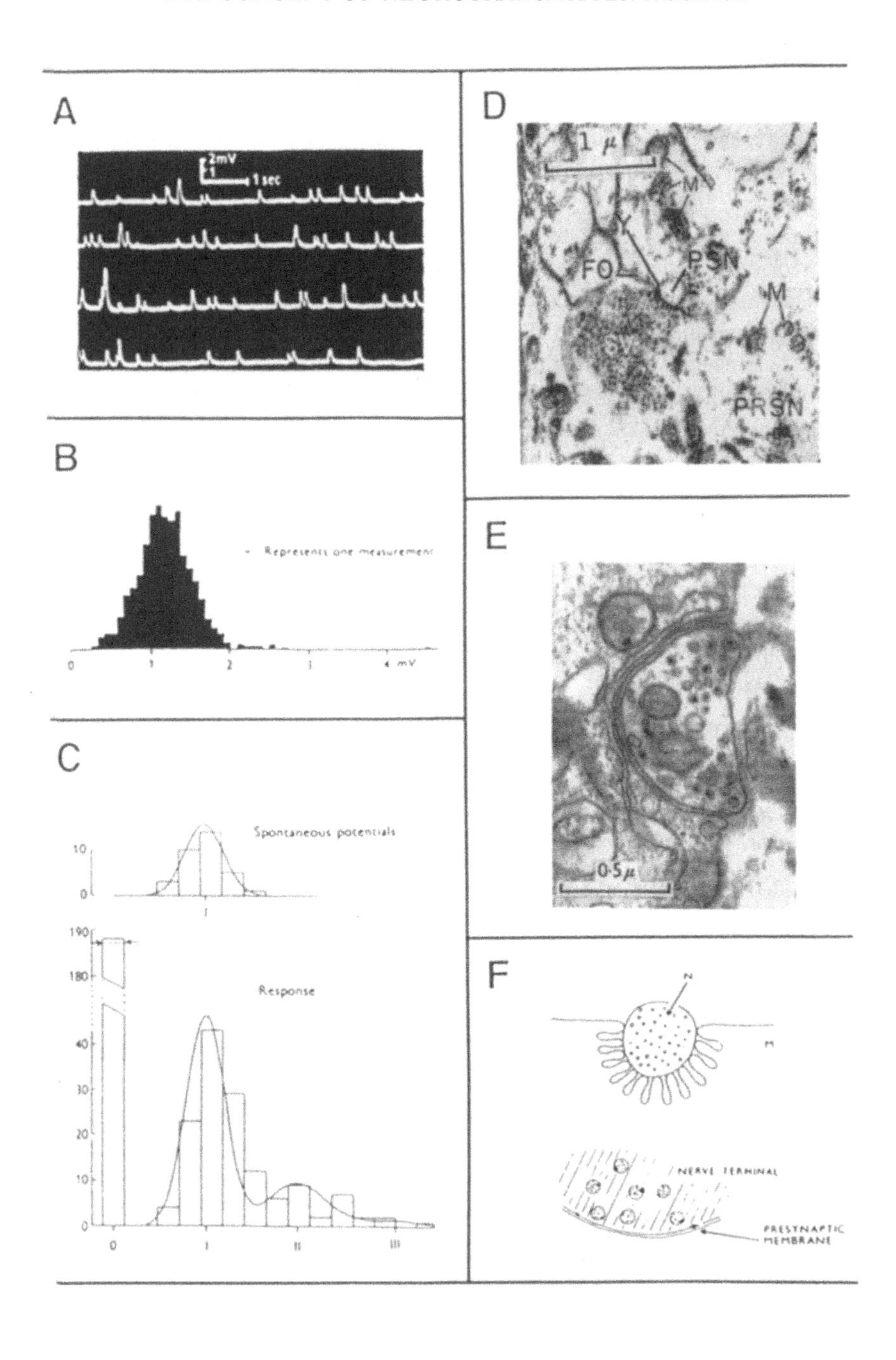
A
2mV
1 sec
B
0
1
2
3
4 mV
C
Spontaneous potentials
Response
10
0
190
180
40
30
20
10
0
D
1 μ
M
FO
PSN
M
PRSN
E
0·5 μ
F
NERVE TERMINAL
PRESYNAPTIC MEMBRANE

potentials without intracellular microelectrodes. The amplitude-frequency distribution of the spontaneous potentials was approximately Gaussian (Fig. 8B), although Fatt and Katz did note that:

> *there is indication of several discharges of about twice the mean amplitude, and of one isolated discharge of three or four times the mean size (59).*

They attributed this to the coincidence of two (or three) unitary discharges that could not be resolved given that detection was only possible down to 5 msec; the calculated chances of units occurring at such small intervals apart supported their conclusion. This question concerning the composition of the unitary discharges is still a matter of great interest.

It was natural to consider if the endplate potential was composed of these spontaneous unitary discharges. Del Castillo and Katz showed that this was likely to be the case. They determined that the amplitude-frequency distribution of the endplate potential under conditions of low transmitter release could be built of units whose mean size and amplitude distribution were identical to those of the spontaneous unitary discharges (Fig. 8C). They concluded that:

> *statistical analysis indicates that the end-plate potential is built up of small all-or-none quanta which are identical in size and shape with the spontaneous occurring miniature potentials (60).*

FIG. 8. The discovery of transmitter quanta and synaptic vesicles. **A,** spontaneous miniature endplate potentials in frog sartorius muscle treated with 10^{-6} prostigmine bromide (Fatt and Katz, 1952). **B,** distribution of amplitudes of spontaneous miniature endplate potentials from an endplate in the frog sartorius muscle treated with prostigmine; there are 800 miniature potentials in the sample (Fatt and Katz, 1952). **C,** histogram showing distribution of amplitudes of spontaneous miniature potentials and endplate responses at a low calcium endplate in the frog sartorius muscle. The lower part shows a continuous curve that has been calculated according to the hypothesis that the responses are built up statistically of units whose mean size and amplitude distribution are identical to those of the spontaneous potentials. The expected number of failures are shown by the arrows. Abscissae: scale units = mean amplitude of spontaneous potentials (0.875 mV) (del Castillo and Katz, 1954). **D,** electron micrograph of a section of an area of complex axonal entanglement from the neuropile of the earthworm. Numerous profiles of axonal membranes can be distinguished, varying in density from place to place. Mitochondria (M) and endoplasmic reticulum are distinguishable in several places. Y denotes an area of specialized axonal contact identified as synaptic in nature. Numerous synaptic vesicles (SV) are seen in the presynaptic neuron (PRSN), whose profile is of irregular outline. The profile of the postsynaptic member (PSN) is identified as a section through a finger-like axonal projection indenting the presynaptic axon, producing puckerings or folds (FO) in the presynaptic axonal membrane. Enlarged 25,000×. (de Robertis and Bennett, 1955). **E,** electron micrograph of catecholamine-containing granules in synaptic vesicles of sympathetic nerve terminals in the rat vas deferens. The distance between nerve and muscle membranes is about 25 nm (Richardson, 1962; 79). **F,** vesicle hypothesis as first enunciated (del Castillo and Katz, 1956). A diagram is shown of a nerve-muscle junction, with several features described after Robertson's (1956) electron micrographs of the junction. N, nerve terminal; M, muscle fiber. In the lower part, an enlarged part of the nerve terminal is shown, containing "ACh-carrier corpuscles," as then described by del Castillo and Katz, or synaptic vesicles. Release of ACh is supposed to occur as a result of critical collisions between these synaptic vesicles and the membrane. This is indicated formally by labeling certain "critical spots" on the surface of both.

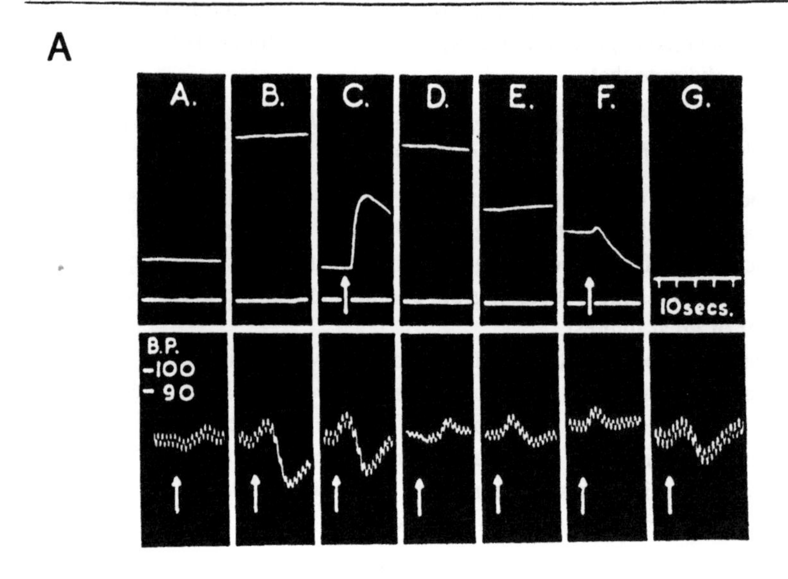
A
A.
B.
C.
D.
E.
F.
G.
10secs.
B.P.
-100
- 90

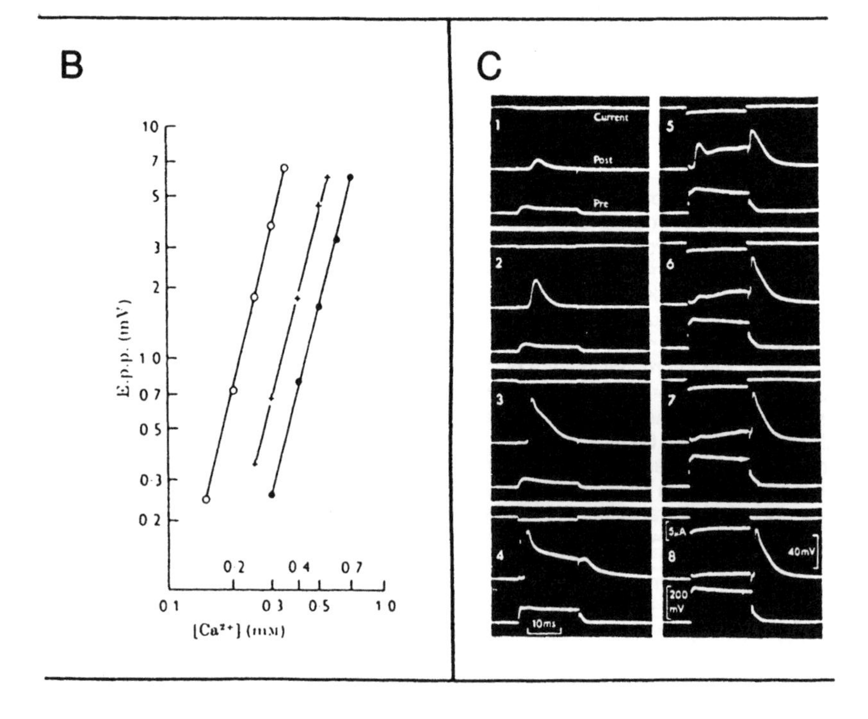
B
E.p.p. (mV)
[Ca²⁺] (mM)
C
Current
Post
Pre
10ms
40mV
200 mV

Del Castillo and Katz noted that the statistical analysis failed under conditions of reasonably high-evoked transmitter release, which they thought may occur because some synaptic units respond more readily than others. With the application of more refined electrophysiological techniques it is now known that there is indeed non-uniformity in the probability of secretion of a quantal unit at different release sites within a nerve terminal. Attempts to both measure this nonuniformity and see if the Katz statistical paradigm for quantal secretion holds for different peripheral and central synapses constitute a major research effort at this time.

The use of the microelectrode to study neurotransmitter release was complemented by the development of refined biochemical techniques for determining the constituents of nerve terminals, as well as by introduction of the electronmicroscope. The major biochemical contributions came from the Karolinska Institute in Stockholm, where U. S. von Euler, who had trained with H. H. Dale in London in 1934, first showed definitively in 1946 that the catecholamine noradrenaline was a transmitter at sympathetic nerve terminals (61). In a letter to Dale in 1945, von Euler commented that:

> *perhaps it will interest you to hear about the sympathetomimetic substance in the spleen which I have been working with lately. It appeared that ordinary alcoholic extracts of cattle spleen contain the somewhat surprising amount of some 10 mg adrenaline pressor equivalent per kg. After perfusion the active substance was found to differ somewhat from adrenaline, and, on the basis of your admirable analysis with Berger in 1910 of the action of sympathomimetic amines, it emerged that it resembled definitely more an amino-base like nor-adrenaline than adrenaline or methylated compounds (62).*

FIG. 9. The calcium dependence of neurotransmitter release. **A,** evidence that calcium is required for the release of acetylcholine (ACh) at sympathetic preganglionic nerve terminals. Both contraction of the cat's nictitating membrane (above) and the cat's blood pressure (below) were used to assay acetylcholine in the venous effluent collected during perfusion of the superior cervical ganglion during the following corresponding periods. **A–C,** perfusion with normal Locke's solution containing eserine: **A,** no stimulation; **B,** maximal preganglionic stimulation, 10 per sec; **C,** at the arrow, injection of 2 mg KCl. **D–F,** perfusion with Ca-free Locke's solution containing eserine: **D,** no stimulation; **E,** 10 min later, maximal preganglionic stimulation, 10 per sec, producing no further contraction of the nictitating membrane; **F,** at the arrow, injection of 2 mg KCl. **G,** time signal, and effect of 0.005 μg of Ach (Harvey and MacIntosh, 1940). **B,** relationship between calcium concentration and amplitude of endplate potential in the frog sartorius muscle. Each symbol gives the results for a different magnesium concentration (open circles, 0.5 mM; crosses, 2.0 mM; filled circles, 4.0 mM). The coordinates are logarithmic, giving straight lines with a slope of approximately 4 (Dodge and Rahamimoff, 1967). **C,** suppression of transmitter release during a large "positive voltage step" of the presynaptic membrane potential of the giant synapse in the stellate ganglion of the squid, treated with tetrodotoxin to block nerve impulses. The presynaptic terminal is loaded with tetraethylammonium ions. Blocks 1 to 8 show increasing pulse intensity. In each block of records, the bottom trace shows the presynaptic voltage step, the middle trace shows postsynaptic response, and the top trace monitors the current pulse. There is a progressive suppression of "on"-response and replacement by "off"-response, as presynaptic voltage is increased from 100 to 200 mV (records 4 to 8), indicating that, if the movement of calcium ions into the terminal is blocked by depolarization to 200 mV, then transmitter release fails to occur until after the depolarization is removed (Katz and Miledi, 1967).

In 1953, Hillarp, Lagerstedt, and Nilsson in Lund, Sweden (63), as well as Blaschko and Welch in Oxford, showed that catecholamines such as noradrenaline were stored in granules within the adrenal medulla. It was not long before von Euler together with Hillarp (65) showed that noradrenaline was also stored in the particulate fraction of sympathetic nerves. Noradrenaline was therefore likely to be stored in granules within sympathetic nerve terminals.

This concept of the packaging of neurotransmitters in particles or granules within nerve terminals was greatly enhanced by the first electronmicroscope images of the terminals by Palade (66) as well as de Robertis and Bennett in 1954 (67) (Fig. 8D). De Robertis and Bennett commented in 1955 that:

> *A granular or vesicular component, here designated the synaptic vesicles, is encountered on the presynaptic side of the synapse and consists of numerous oval or spherical bodies 20 to 50 nm in diameter, with dense circumferences and lighter centers. Synaptic vesicles are encountered in close relationship to the synaptic membranes (68).*

At sympathetic nerve terminals these synaptic vesicles were seen to actually contain the granules of catecholamines (Fig. 8E), as predicted by von Euler and Hillarp.

By 1956 the quantal unit of transmitter release had been discovered at the motor endplate, along with the storage of catecholamines in granules within sympathetic nerves, and synaptic vesicles within both central and peripheral nerve terminals. That year del Castillo and Katz enunciated the vesicle hypothesis, attributing quantization of transmitter release to its association with synaptic vesicles (Fig. 8F). In their own words:

> *Recent electron microscope studies (69) have shown that the motor nerve terminals contain a fairly dense population of microsomes, granules or vesicles, of less than 0.1 μ diameter, which may well be the intracellular corpuscles to which ACh is attached. It has been known since Loewi's investigations that most of the ACh which is present in "homogenised" nerve tissue can only be extracted into an aqueous solution after chemical destruction of the cell proteins. It appears then that the discharge of ACh from a nerve terminal requires the disruption of more than one diffusion barrier: first the release from its intracellular attachment, and secondly a passage through a nerve membrane. One might suppose that when a "critical" collision occurs between an intra-cellular ACh-carrier and the membrane of the nerve terminal, the two barriers are opened simultaneously and the ACh-contents of the carrier particle are suddenly discharged. This picture, though purely speculative, is nevertheless in accord with recent experimental findings; it takes account of the evidence discussed below that the release of ACh from nerve terminals occurs in multi-molecular units or "quanta" and of the evidence, already cited, for the bound state of intracellular ACh content (70).*

The mechanism by which the "contents of the carrier particle are suddenly discharged" is perhaps the major focus of research on neurotransmitter release at the present time.

Katz next tackled the problem of determining the necessary and sufficient conditions for the ACh-contents of the carrier particle to be suddenly discharged. It had been known since the work of Locke reported in 1894 that calcium was necessary for transmission at the neuromuscular junction (71).The reason for this, as Harvey and MacIntosh showed in 1940, was that the release of acetylcholine at nerve terminals required calcium (72).This transmitter was only released in the perfused superior cervical ganglion of the cat upon stimulating the preganglionic nerves in the presence of calcium (Fig. 9A). Kuffler and Eccles had shown by 1942 that the amplitude of the endplate potential was affected by the calcium concentration in accord with the observations of Harvey and MacIntosh on the calcium-dependence of acetylcholine release. However, it was not until 1967 that Katz's laboratory produced two most important observations on how calcium might govern transmitter release. The first was due to Dodge and Rahamimoff, who showed that the endplate potential in low concentrations of calcium ions increased as the fourth power of the calcium concentration (73) (Fig. 9B). This observation gave rise to the idea that the cooperative action of about four calcium ions is necessary for release of each quantal packet of transmitter by the nerve impulse. The second observation was due to Katz and Miledi, who determined that the site of this cooperative action of calcium ions was on the inside of the nerve terminal rather than on the outside (74) (Fig. 9C). The basis of this cooperative action of calcium ions on the inside of the nerve terminal membrane to release the contents of synaptic vesicles is a theme of much current research.

CONCLUSION: THE ESTABLISHMENT OF SHERRINGTON'S CONCEPT OF THE SYNAPSE IN THE CENTRAL NERVOUS SYSTEM AND CENTRAL SYNAPTIC TRANSMISSION

While Katz was researching the mechanism of neurotransmitter release, Eccles and his colleagues in Canberra explored synaptic mechanisms at successively higher levels of the central nervous system. Studies were carried out on inhibition of Purkinje cells in the cerebellum (75) (Fig. 10D), on thalamic-cortical relay cells (76) (Fig. 10B) and on the CA3 pyramidal cells in the hippocampus (77) (Fig. 10C). This work was revolutionary in as much as it supplied a functional microanatomy of the synaptic connections to be found in these different nerve centers. In addition, it provided insights into the ionic basis of inhibition at different synapses and the role this inhibition plays in controlling the excitability of the major neuron types. In this way the research program initiated by Sherrington's concepts of inhibition and excitation in 1906 was brought to fruition.

The research program on transmitter release at peripheral synapses was initiated in Langley's laboratory in 1901 and established by Dale and Loewi. Our present concept of the mechanisms of transmitter release is still largely due to the elegant experiments of Bernard Katz. There are now two great challenges. One is to elucidate the molecular machinery responsible for the release of transmitter quanta. The other is to determine the extent to which the paradigm for transmitter release established for peripheral synapses holds for central synapses that subserve the integrative action of the nervous system.

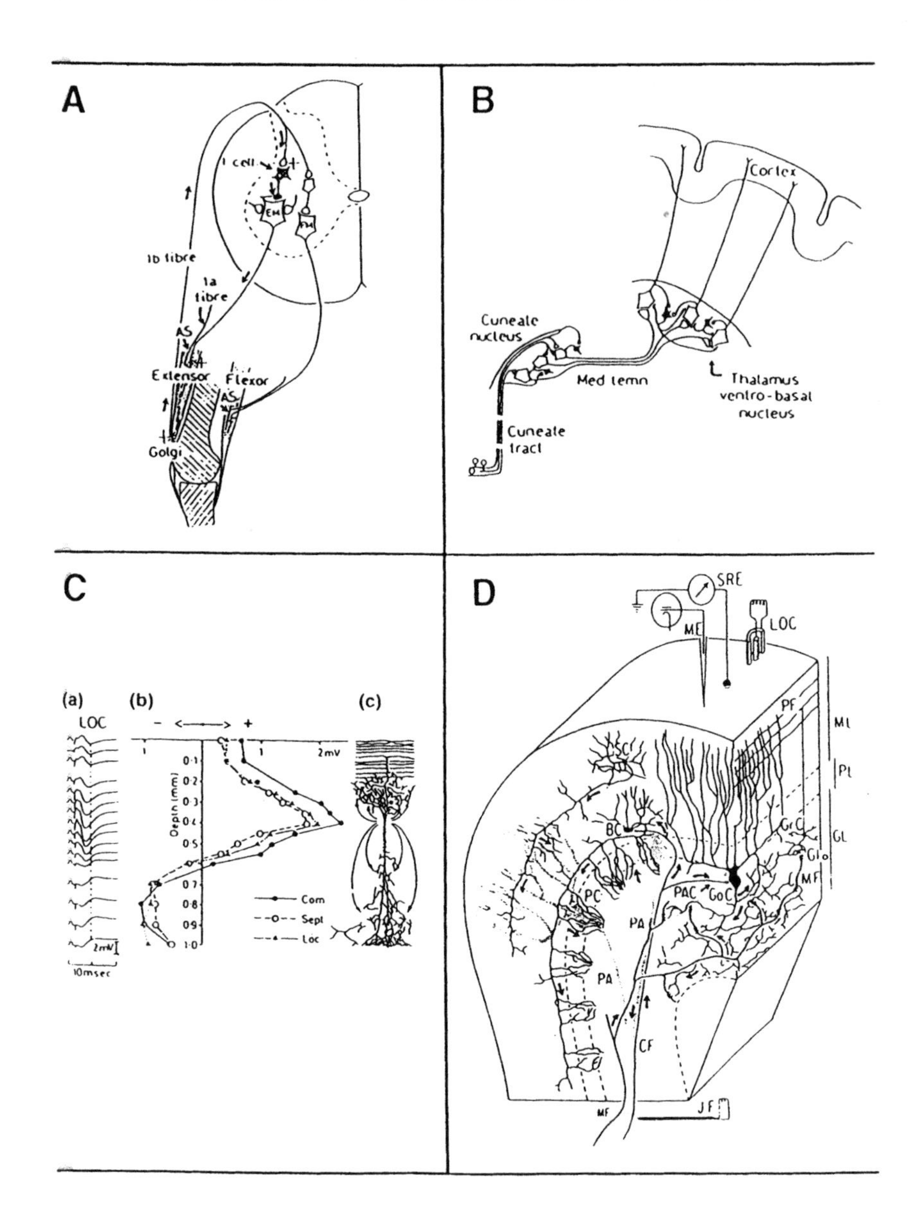

REFERENCES

1. Sherrington CS. *Integrative action of the nervous system*. New Haven: Yale University Press; 1906.
2. Sherrington CS. *The central nervous system* vol. 3. In: Foster M, ed. *A Text-book of Physiology*, 7th ed. London: Macmillan; 1897.
3. Langley JN, Sherrington CS. Secondary degeneration of nerve tracts following removal of the cortex of the cerebrum in the dog. *J Physiol* 1884;5:49–65.
4. Langley JN. Observations on the physiological action of extracts of the supra-renal bodies. *J Physiol* 1901;27:237–256.

5. Elliott TR. On the action of adrenalin. *J Physiol* 1904;31:Pxx–xxi.
6. Langley JN. Croonian lecture of the royal society: On nerve endings and on special excitable substances in cells. *Proc Roy Soc London*. 1906;B.LXXVIII:170–194.
7. Brodie TG, Dixon WE. Contributions to the physiology of the lungs. Part II. On the innervation of the pulmonary blood vessels; and some observations on the action of suprarenal extract. *J Physiol* 1904;30:476–502.
8. Loewi O. Uber humorale Ubertragbarkeit der Herznervenwirkung. *Pflugers Arch ges Physiol* 1921; 189:239–242.
9. Dale HH. Chemical transmission of the effects of nerve impulses. *Br Med J* 1934;1:835–841.
10. Dale HH. The occurrence in ergot and action of acetylcholine. *J Physiol* 1914;48:Piii–iv.
11. Dale HH. The active principles of ergot. *Proc Internat Physiol Congress*. Vienna, 1910;27–30.
12. Loewi O. The Ferrier lecture. Problems connected with the principle of humoral transmission of nervous impulses. *Proc Roy Soc* London. 1935;B.118:299–316.
13. Liddell EGT, Sherrington CS. Reflexes in response to stretch (myotatic reflexes). *Proc Roy Soc London* 1924;B.96:212–242.
14. Liddell EGT, Sherrington CS. Further observations on myotatic reflexes. *Proc Roy Soc London*. 1925;B.97:267–283.
15. Creed RS, Sherrington CS. Observations on concurrent contraction of flexor muscles in the flexor reflex. *Proc Roy Soc* London. 1926;B.100:258–267.
16. Sherrington CS. Remarks on some aspects of reflex inhibition. *Proc Roy Soc London*. 1925;B.97: 519–545.
17. Creed RS, Eccles JC. The incidence of central inhibition on restricted fields of motor units. *J Physiol* 1928;66:109–120.
18. Eccles JC, Sherrington CS. Improved bearing for the torsion myograph. *J Physiol* 1930;69:Pi.
19. Eccles JC, Sherrington CS. Reflex summation in the ipsilateral spinal flexion reflex. *J Physiol* 1930;69:1–28.
20. Eccles JC, Sherrington CS. Studies on the flexor reflex VI. Inhibition. *Proc Roy Soc London*. 1931;B.109:91–113.

FIG. 10. Inhibitory pathways in the central nervous system elucidated by Eccles (1969) and his colleagues. **A,** synaptic connection involved in the inhibitory action produced in extensor motoneurons by afferent volleys in the Group 1b afferent fibers from the Golgi tendon organs in the extensor muscles (Eccles, 1969). **B,** pathway to the sensory-motor cortex for cutaneous fibers from the forelimb; inhibitory neurons are shown in black in both the cuneate nucleus and the ventrobasal nucleus of the thalamus. Note that the inhibitory pathway is of the feed-forward type in the cuneate nucleus and feed-back type in the thalamus (Eccles, 1969). **C,** results of recording from a CA3 hippocampal pyramidal cell in response to commissural (Com), septal (Sept), and local (Loc) stimulation. **a,** responses recorded by a microelectrode penetrating CA3 following local stimulation. **b,** graph in which the size of the positive waves of the responses to commissural, septal, and local stimulation is plotted against depth, with the positivities measured at a time indicated by the stippled line in (a). **c,** a CA3 pyramidal cell, semi-diagrammatically drawn to scale to facilitate comparison with (b). The arrows indicate the extracellular flow of current generated by the inhibitory postsynaptic potential (from Andersen, Eccles, and Loyning, 1963). **D,** perspective drawing of a cerebellar folium to show the synaptic connections of the inhibitory interneurons. The cerebellar cortex is seen to be divided into three layers: molecular layer (ML), Purkinje cell layer (PL), and granular layer (GL). The input to the cortex is by two types of fiber: mossy fiber (MF) and climbing fiber (CF). Single examples are shown of four types of interneurons: granule cells (GrC), Golgi cells (GoC), basket cells (BC), and outer stellate cells (SC). Also shown are two Purkinje cells, one (PC) with its dendritic ramifications, and both axons (PA), one with two collaterals (PAC) ending on the Golgi cell and the basket cell. The mossy fiber is shown with numerous branches and thickenings at the sites of its synapses on granule cell dendrites, so forming the glomeruli (Glo). Collaterals of the climbing fiber (CF) are shown making synapses on the Golgi cell and basket cell. The axons of the granule cells bifurcate to give rise to the parallel fibers (PF) in the molecular layer. Arrows show directions of normal propagation in: the mossy fiber, climbing fiber, and its collaterals; the Purkinje axons and collaterals; and the axons of the interneurons BC, SC, and GoC. (Eccles, Llinas, and Sasaki, 1966).

21. Lucas K. *The conduction of the nervous impulse. Monographs on Physiology*. London: Longmans; 1917.
22. Adrian ED. Some recent work on inhibition. *Brain* 1924;47:399–416.
23. Adrian ED, Bronk DW. The discharge of impulses in motor nerve fibres. Part II. The frequency of discharge in reflex and voluntary contractions. *J Physiol* 1929;67:119–151.
24. Hodgkin AL. ED Adrian, Baron Adrian of Cambridge. Biographical Memoirs of Fellows of the Royal Society, 1978;25:1–73.
25. Brown GL, Eccles JC. The action of a single vagal volley on the rhythm of the heart beat. *J Physiol* 1934;82:211–241.
26. del Castillo J, Katz B. Modifications de la membrane produites par des influx nerveux dans la region du pace-maker du coeur. In Microphysiologie comparee des elements excitable. *Coll Int Centre National Recherche Sci* 1957;67:271–279.
27. Feldberg W, Gaddum JH. The chemical transmitter at synapses in a sympathetic ganglion. *J Physiol* 1934;81:305–309.
28. Eccles JC. The discharge of impulses from ganglion cells. *J Physiol* 1937;91:1–22.
29. Eccles JC. Synaptic and neuro-muscular transmission. *Ergebnise der Physiologie* 1936;38:339–444.
30. Monnier AM, Bacq ZM. Recherches sur la physiologie et la pharmacologie du systeme nerveux autonome. XVI Dualite du mecanisme de la transmission neuromusculaire de l'excitation chez le muscle lisse. *Arch Int Physiol* 1935;40:485–510.
31. Eccles JC, Magladery JW. Pharmacological investigation on smooth muscle. *J Physiol* 1936;87: 87P.
32. Eccles JC, Magladery JW. The excitation and response of smooth muscle. *J Physiol* 1937;90:31–67.
33. Henderson VE, Roepke MH. The role of acetylcholine in bladder contractile mechanisms and in parasympathetic ganglia. *J Pharmacol* 1934;51:97–110.
34. Dale HH, Gaddum JH. Reactions of denervated voluntary muscle, and their bearing on the mode of action of parasympathetic and related nerves. *J Physiol* 1930;70:109–144.
35. Dale HH. Transmission of nervous effects by acetylcholine. *Harvey Lectures* 1937;32:229–245.
36. Bennett MR, Burnstock G, Holman ME. The effect of potassium and chloride ions on the inhibitory potential recorded in the guinea-pig taenia coli. *J Physiol* 1963;169:33–4P.
37. Bennett MR, Burnstock G, Holman ME. Transmission from intramural inhibitory nerves to the smooth muscle of the guinea-pig taenia coli. *J Physiol* 1966;182:541–558.
38. Dale HH, Feldberg W, Vogt M. Release of acetylcholine at voluntary motor nerve endings. *J Physiol* 1936;86:353–380.
39. Brown GL, Dale HH, Feldberg W. Reaction of the normal mammalian muscle to acetylcholine and to eserine. *J Physiol* 1936;87:394–424.
40. Eccles JC, O'Connor WJ. Action potentials evoked by indirect stimulation of curarized muscle. *J Physiol* 1938;9P.
41. Göpfert H, Schaefer H. Uber den direkt und indirekt erregten Aktionsstrom und die Funktion der motorischen Endplatte. *Pflugers Arch ges Physiol* 1938;239:597–619.
42. Eccles JC, O'Connor WJ. Responses which nerve impulses evoke in mammalian striated muscle. *J Physiol* 1939;97:44–102.
43. Hill AV, Katz B, Solandt DY. Nerve excitation by alternating current. *Proc Roy Soc London*. 1936;B.121:74–133.
44. Katz B. Archibald Vivian Hill. Biographical Memoirs of Fellows of the Royal Society of London. 1978;24:71–149.
45. Katz B. Multiple response to constant current in frog's medullated nerve. *J Physiol* 1936;88:239–255.
46. Katz B. Neuromuscular transmission in crabs. *J Physiol* 1936;87:199–221.
47. Katz B. The 'anti-curare' action of a subthreshold catelectrotonus. *J Physiol* 1939;95:286–304.
48. Katz B. Bayliss-Starling Memorial Lecture (1985). Reminiscenes of a Physiologist, 50 years after. *J Physiol* 1986;370:1–12.
49. Hill AV. Excitation and accommodation in nerve. *Proc Roy Soc London*. 1933;B.119:305–355.
50. Eccles JC, Katz B, Kuffler SW. Nature of the 'endplate potential' in curarized muscle. *J Neurophysiol* 1941;4:362–387.
51. Kuffler SW. Electrical potential changes at an isolated nerve-muscle junction. *J Neurophysiol* 1942; 5:18–26.
52. Eccles JC. Synaptic potentials and transmission in sympathetic ganglion. *J Physiol* 1943;101:465–483.

53. Ogston AG. Removal of acetylcholine from a limited volume by diffusion. *J Physiol* 1955;128:222–223.
54. Fatt P, Katz B. An analysis of the end-plate potential recorded with an intra-cellular electrode. *J Physiol* 1951;115:320–370.
55. Eccles RM. Intracellular potentials recorded from a mammalian sympathetic ganglion. *J Physiol* 1955;130:572–584.
56. Lloyd DPC. Facilitation and inhibition of spinal motoneurones. *J Neurophysiol* 1946;9:421–438.
57. Brooks CMcC, Eccles JC. An electrical hypothesis of central inhibition. *Nature* 1947;159:760–764.
58. Brock LG, Coombs JS, Eccles JC. The recordings of potentials from motoneurones with an intracellular electrode. *J Physiol* 1952;117:431–460.
59. Fatt P, Katz B. Spontaneous subthreshold activity at motor nerve endings. *J Physiol* 1952;117:109–128.
60. del Castillo J, Katz B. Quantal components of the end-plate potential. *J Physiol* 1954;124:560–573.
61. von Euler US. A specific sympathomimetic ergone in adrenergic nerve fibers (Sympathin) and its relations to adrenaline and nor adrenaline. *Acta Physiol Scand* 1946;12:73–97.
62. Blaschko HKF. U.S. von Euler. Biographical Memoirs of Fellows of the Royal Society 1985; 31:143–170.
63. Hillarp NA, Lagerstedt S, Nilsson B. The isolation of a granular fraction from the suprarenal medulla, containing the sympathomimetic catechol amines. *Acta Physiol Scand* 1953;29:251.
64. Blaschko H, Welch AD. Localization of adrenaline in cytoplasmic particles of the bovine adrenal medulla. *Naunyn-Schmiedebergs Arch exp Path Pharmakol* 1953;219:17–22.
65. von Euler US, Hillarp NA. Evidence for the presence of noradrenaline in submicroscopic structures of adrenergic axons. *Nature* 1956;177:44–45.
66. Palade GE. Electron microscope observations of interneuronal and neuromuscular synapses. *Anat Rec* 1954;118:335–336.
67. de Robertis EDP, Bennett HS. Submicroscopic vesicular component in the synapse. *Fed Proc* 1954;13:35.
68. de Robertis EDP, Bennett HS. Some features of the submicroscopic morphology of synapses in frog and earthworm. *J Biophys Biochem Cytol* 1955;1:47–58.
69. Robertson JD. The ultrastructure of a reptilian myoneural junction. *J Biophysic & Biochem Cytol* 1956;2:381–394.
70. del Castillo J, Katz B. Biophysical aspects of neuromuscular transmission. *Prog Biophys Biophys Chem* 1956;6:121–170.
71. Locke FS. Notizuber den Einfluss physiologischer Kochsalzlosemg auf die elektrische Erregbarkeit von uskelund Nerv. *Centralblatt Physiol* 1894;VIII:166–167.
72. Harvey AM, MacIntosh FC. Calcium and synaptic transmission in a sympathetic ganglion. *J Physiol* 1940;97:408–416.
73. Dodge FA, Rahamimoff R. Co-operative action of calcium ions in transmitter release at the neuromuscular junction. *J Physiol* 1967;193:419–432.
74. Katz B, Miledi R. A study of synaptic transmission in the absence of nerve impulses. *J Physiol* 1967;192:407–436.
75. Eccles JC, Llinas R, Sasaki K. The inhibitory interneurones within the cerebellar cortex. *Exp Brain Res* 1966;1:1–16.
76. Eccles JC. *The inhibitory pathways of the central nervous system*. The Sherrington Lectures IX, Liverpool University Press; 1969.
77. Andersen P, Eccles JC, Løyning Y. Recurrent inhibition in the hippocampus with identification of the inhibitory cell and its synapses. *Nature* 1963;198:541–542.
78. Eccles JC, Kuffler SW. Initiation of muscle impulses at neuro-muscular junction. *J Neurophysiol* 1941;4:402–417.
79. Richardson KC. The fine structure of autonomic nerve endings in smooth muscle of the rat vas deferens. *J Anat* 1962;96:427–442.
80. Ling G, Gerard RW. The normal membrane potential of frog sartorius fibres. *J Cell Comp Physiol* 1949;34:383–396.
81. Eccles JC, Katz B, Kuffler SW. Effect of eserine on neuromuscular transmission. *J Neurophysiol* 1942;5:211–230.

Molecular and Cellular Mechanisms of Neurotransmitter Release, edited by Lennart Stjärne, Paul Greengard, Sten Grillner, Tomas Hökfelt, and David Ottoson, Raven Press, Ltd., New York © 1994.

2

Synapsin I, an Actin-Binding Protein Regulating Synaptic Vesicle Traffic in the Nerve Terminal

Paul Greengard, *Fabio Benfenati, and †Flavia Valtorta

*Laboratory of Molecular and Cellular Neuroscience, The Rockefeller University, New York, New York 10021; *Department of Physiology, Experimental Medicine and Biochemical Sciences, University of Rome "Tor Vergata," F-00173 Rome, Italy; and Department of Pharmacology, University of Milan School of Medicine, 20132 Milan, Italy*

Small synaptic vesicles are neuronal organelles that specialize in the storage and release of classical neurotransmitters. Synaptic vesicles reach the axon terminal via fast axonal transport and spend most of their life cycle in this compartment. They are loaded with locally synthesized neurotransmitter, fuse with the presynaptic membrane at the active zone upon depolarization-induced Ca^{2+} influx, and are retrieved to undergo a new cycle of exo-endocytosis (1–3).

Recent results obtained with complementary approaches ranging from molecular biology to biochemistry, genetics, and microbiology support the idea that neuronal exocytosis is a particular aspect of the more general processes of intracellular vesicular fusion and constitutive secretion shared by all eukaryotic cells (4).The same or closely related proteins appear to be involved in neurotransmitter release in neurons and to play a role in the constitutive exo-endocytotic pathway, as well as in the intracellular transport of vesicles from the endoplasmic reticulum to the plasma membrane through the Golgi apparatus (5). These proteins appear to be highly conserved during evolution from yeast to mammalian neurons (6). However, the unique features of neurotransmitter release from neuronal cells imply that this process, besides sharing the basic elements of the machinery with other secretory processes, must require some specific additional components.

Neurotransmitter release is the most efficient and plastic example of regulated secretion. Among its remarkable characteristics are the extreme rapidity (it occurs in a fraction of a millisecond after the invasion of the axon terminal by the action potential and the subsequent influx of Ca^{2+}), a high resistance to exhaustion (fatigue) during prolonged periods of intense secretory activity, and the ability to vary its efficiency on the basis of the recent history of the nerve terminal and of its physicochemical environment (7).These characteristics suggest that the nerve terminal has both preassembled complexes between ready-to-fuse synaptic vesicles and the presynaptic membrane (which can account for the rapidity of exocytosis) and an adequate reservoir of vesicles to be recruited upon activity (which can account for the resistance to fatigue).

Electrophysiological studies have suggested the existence of functional pools of synaptic vesicles within the nerve terminal, namely a reserve pool of vesicles that are presumably embedded in the cytoskeletal meshwork and a releasable pool of vesicles that are not restrained by cytoskeletal structures and may be already docked at the presynaptic membrane. Generally the releasable pool accounts for only a small percentage of the total vesicle content of the terminal (0.5% to 10%, depending on the type of synapse analyzed), but its size is functionally very important to define the magnitude of the secretory response to the action potential. The partial depletion of the releasable pool upon electrical activity of the neuron is overcome by recruitment of vesicles from the reserve pool. Moreover, the ability to vary the respective sizes of the reserve and releasable pools of vesicles represents a mechanism by which neurons can modulate the efficiency of neurotransmitter release. (7,8).

The transition of synaptic vesicles between these two functional pools is thought to be controlled by local changes in the intracellular concentrations of second messengers and by the activation/inhibition of protein phosphorylation/dephosphorylation processes. Several phosphoproteins and protein kinases found to be associated with synaptic vesicles are key candidates for transducing the second messenger signals into a modulation of neurotransmitter release (9).This paper will focus on the synapsins, a group of abundant synaptic vesicle-specific, actin-binding phosphoproteins that have been proposed to mediate the interactions between synaptic vesicles and the actin-based cytoskeleton of the nerve terminal and to regulate the availability of the vesicles for exocytosis (9–11).

THE SYNAPSINS: A FAMILY OF SYNAPTIC VESICLE-ASSOCIATED PHOSPHOPROTEINS

The synapsins are a family of phosphoproteins specific for the nervous system, where they are present in virtually all neurons, irrespective of the transmitter released. While some of the other synaptic vesicle proteins have been reported to be associated to some extent with neuropeptide-containing large dense-core vesicles both in neurons and in endocrine cells, or with synaptic-like microvesicles of neuroendocrine cells, the synapsins are specific for small synaptic vesicles that are responsible for the storage and release of classical neurotransmitters. (10,12).

The synapsins were originally discovered as major substrates for cAMP-dependent and Ca^{2+}/calmodulin-dependent phosphorylation processes in brain tissue (13–15).They comprise four homologous proteins: synapsins Ia and Ib (collectively referred to as synapsin I) and synapsins IIa and IIb (collectively referred to as synapsin II). The primary transcripts of the genes encoding for synapsins I and II are differentially spliced to generate the a and b isoforms of each protein. In the rat, the four mRNAs code for proteins of 704, 668, 586, and 479 amino acids, respectively. A high degree of similarity, accounting for more than half of each protein starting from the NH_2-terminal, exists between the amino acid sequences of synapsins I and II, with a high degree of conservation among various mammalian species. (16) The synapsins are excellent substrates for both cAMP-dependent protein kinase and for Ca^{2+}/calmodulin-dependent protein kinase I, which phosphorylate a serine residue in the NH_2-terminal region (head). In addition, synapsin I, but not synapsin II, is the best known substrate for Ca^{2+}/calmodulin-dependent protein kinase II (CaM kinase II), which phosphorylates two serines in the COOH-terminal region (tail) of synapsin I (17–21).

The binding of purified synapsins to synapsin-depleted synaptic vesicles has been extensively characterized for synapsin I. In mature neurons, synapsin I is concentrated in the nerve terminal in close association with small synaptic vesicles (22,23). However, it is not an integral membrane protein and can be dissociated from the membrane of purified synaptic vesicles by relatively mild treatments (24). Purified exogenous synapsin I reversibly binds to synapsin I-depleted synaptic vesicles with high affinity and saturability (approximately 10 molecules per vesicle). The binding affinity is selectively decreased after phosphorylation of synapsin I by CaM kinase II (25). Due to an extremely high surface activity and to the very large limiting surface area of the synapsin I molecule, it has been suggested that synapsin I covers a large percentage of the vesicle surface (26).

Structure-function analysis has demonstrated that distinct sites of synapsin I bind to vesicle phospholipids and vesicle proteins. The hydrophobic head region is responsible for the interaction with the acidic phospholipids of the cytoplasmic leaflet of the vesicle membrane and for partial penetration of synapsin I into the core of the membrane (27,28). Recent results suggest that these interactions of synapsin I with vesicle phospholipids may play a role in the formation of clusters within the nerve terminal and in the stabilization of the bilayer structure of the vesicle membrane (29). The hydrophilic tail region binds to a protein component of the vesicles and its binding is greatly reduced by phosphorylation of the tail sites. (27) A protein to which synapsin I binds is a vesicle-associated form of the α-subunit of CaM kinase II. The binding involves the autoregulatory domain of the kinase, indicating that this enzyme may serve dual functions, both structurally as a binding protein for synapsin I and catalytically to phosphorylate synapsin I and bring about its dissociation from the vesicles (30). The existence of a preformed complex of synapsin I and CaM kinase II on vesicles could account for a fast rate of CaM kinase II-dependent phosphorylation of synapsin I upon increases in the intraterminal Ca^{2+} concentration.

The substantial homology with the head region of synapsin I suggests that synapsin II might also be able to interact with vesicle phospholipids. As a matter of fact, synapsin II also binds to synaptic vesicles, and its association appears to be tighter than that of synapsin I, possibly because of the lack of the highly charged tail region present in synapsin I (31).

THE ACTIN-BASED CYTOSKELETON IN NEURONS: DYNAMICS, ASSEMBLY, AND REGULATION OF NEUROEXOCYTOSIS

The neuronal cytomatrix is composed of three kinds of polymers: microfilaments, microtubules, and neurofilaments, whose subunits are actin, tubulin, and neurofilament proteins, respectively. While microfilaments and microtubules are metastable polymers (i.e., they are in equilibrium with monomers and undergo regulated assembly and disassembly processes), neurofilaments are inherently stable. Cytoskeletal proteins have important structural and dynamic roles in all cells and these properties are emphasized in highly polarized and compartmentalized cells such as neurons. Cytoskeletal proteins represent a backbone for the cell that defines its shape and physical resistance. They play a fundamental role in the transport, guidance, and accumulation of organelles within functional compartments. In addition, they regulate the trafficking of organelles both by directly binding to the organelles and by changing the viscosity of the cytosol through variations in their polymerization state (32).

All three kinds of polymers are present in dendrites and axons. However, microfilaments are the main polymers of the subplasmalemmal cytomatrix as well as of the cytoplasm of specialized neuronal compartments such as dendritic spines, growth cones, and nerve terminals (33). Actin and actin-binding proteins are therefore critical for the accumulation of synaptic vesicles within the nerve terminal and for regulating their traffic, actions that ultimately affect the processes of exocytosis and endocytosis.

The ability of actin monomers to polymerize into filaments is essential for many biological functions in both muscle and nonmuscle cells (34). Nonmuscle actin filaments are highly dynamic structures and undergo self-assembly and disassembly processes that are finely regulated by an array of specific actin-binding proteins. The polymerization of actin involves an energetically unfavorable reaction of nucleation of three to four monomers, followed by a much more favorable process of elongation of the nuclei by endwise addition of monomers. At steady-state, polymeric actin is in dynamic equilibrium with monomeric actin (i.e., the processes of monomer addition to and dissociation from the filaments are balanced). Actin filaments are polarized structures and the two ends are not equivalent in their monomer binding constants; monomer addition is favored at the so-called "barbed" (fast growing) end, whereas dissociation is more likely to occur at the so-called "pointed" (slowly growing) end of the filament. The concentration of actin monomers required to maintain the steady state is known as the "critical concentration" and differs for the two ends of the filament (34–37).

A variety of actin-binding proteins can affect either the assembly or the extent of polymerization of actin by various mechanisms. The actin-binding activity of some of these proteins (e.g., gelsolin, caldesmon, profilin, synapsin I, etc.) is modulated either directly or indirectly by the intracellular levels of second messengers (35).

Actin-binding proteins can be divided into those interacting with actin monomers and those interacting with actin filaments. The first group includes proteins such as profilin that prevent polymerization by binding and sequestering actin monomers, as well as proteins that stimulate nucleation of actin monomers either by catalyzing the nucleation reaction (e.g., the *Acanthamoeba* capping protein (38)) or by binding actin monomers and leading to the formation of an actively elongating "pseudo-nucleus" (e.g., brevin (39)). It has been calculated that a large percentage (up to 50%) of the total actin in cells is monomeric, partly sequestered by G-actin-binding proteins such as profilin (40). The second group of actin-binding proteins includes: 1) proteins that bind to either the barbed or the pointed end of the filament and prevent addition or dissociation of monomers (capping proteins), and 2) proteins that bind to the sides of actin filaments and may either organize them in bundles or sever them (an action generally accompanied by a barbed end capping) (35).

The meshwork of actin filaments present in the nerve terminal can represent both a viscous physical barrier limiting synaptic vesicle motion and targeting vesicles to the presynaptic membrane and a synaptic vesicle-capturing apparatus (through specific and reversible interactions with vesicle-associated actin-binding proteins). Both properties would be expected to limit the random diffusion of the vesicles away from the release sites and the chances of spontaneous interactions with the presynaptic membrane (41).

In neuroendocrine cells, the stimulation of exocytosis is accompanied by a disaggregation of the sub-plasmalemmal F-actin meshwork, attributable to the activation of an actin-severing activity (42). This event, which facilitates the contact and subsequent fusion of the secretory granules with the plasma membrane, is prevented by treatment of the cells with tetanus or botulinum toxins that also block secretion (43). Candidates for this effect are actin-binding proteins such as gelsolin and scinderin, whose filament-severing activity appears to be regulated by the intracellular levels of Ca^{2+} (42).

In nerve terminals isolated from brain, actin assembly also appears to be dynamically regulated, exhibiting large fluctuations during the exo-endocytotic cycle. This indicates that actin depolymerization accompanies normal neurotransmitter release and that subsequent repolymerization limits excessive release during sustained depolarization (44).

Actin polymerization-depolymerization also plays a fundamental role during development, governing the growth of neuronal processes and formation of synaptic contacts. Actin filaments and bundles of actin filaments constitute the highly mobile backbone of growth cone lamellipodia and filopodia, whose extension and retraction may involve controlled addition or removal of actin monomers at the leading edges of the processes.

In the sections that follow, we will describe the interactions of synapsin I with monomeric and polymeric actin, and how these activities might play a functional role in the regulation of neurotransmitter release in mature neurons as well as in the formation of functional synaptic contacts during development.

SYNAPSIN I INTERACTS WITH ACTIN IN A COMPLEX MANNER

Interaction of Synapsin I with Actin Filaments

The possibility of an interaction between synapsin I and actin was first tested in a reconstituted system by monitoring the effects induced by the various purified phosphoforms of synapsin I on the organization of a solution of actin filaments. The results indicated that dephosphosynapsin I can cause extensive bundling of actin filaments. This effect is modulated by phosphorylation, being slightly decreased after synapsin I phosphorylation by cAMP-dependent protein kinase and abolished after synapsin I phosphorylation by CaM kinase II (45,46).

The binding constants for the interaction of synapsin I with actin filaments have been determined using a centrifugation assay. Dephosphosynapsin I binds to actin filaments with an affinity similar to that reported for other actin-binding proteins (1–2 μM) (45,46). The binding affinity is not modulated by phosphorylation of synapsin I. However, phosphorylation by CaM kinase II decreases the B_{max} from 0.14 to 0.08 mol synapsin I/mol actin (45).

Synapsin I has been shown to decrease the critical concentration for actin polymerization measured at steady state (47). This effect, which is dependent on the concentration of synapsin I, is compatible with the idea that synapsin I, by binding to the sides of actin filaments, inhibits the dissociation of monomers from the filament ends and thereby stabilizes the filaments (Chieregatti et al., manuscript in preparation).

The kinetics of the interaction between synapsin I and actin filaments has been analyzed using the technique of fluorescence resonance energy transfer between purified components labeled with different fluorochromes. At 37°C, the interaction was too rapid to be measured with the available instrumentation. At 7°C, the half-time for the binding was of the order of 2 seconds. Synapsin I phosphorylation appeared to influence the extent of energy transfer but to have little effect on the time constant of the binding (48). The rapidity of the binding appears to be compatible with a role of synapsin I/actin interactions in synaptic physiology.

In principle, an actin-binding protein might achieve bundle formation by either of two mechanisms: 1) the protein might possess two or more domains that bind F-actin, or 2) the protein might bear a single actin-binding site and achieve bundling through self-association (49). Analysis of purified synapsin I fragments indicates that one major actin-binding site is located in the central, hydrophobic portion of the synapsin I molecule. The COOH-terminal fragment of the molecule does not bind actin significantly, but appears necessary to achieve bundle formation, thus raising the possibility that it contains a second binding site that becomes functional only in the holomolecule (50,51). However, self-association of two or more synapsin I molecules has also been reported (50,52,53).

Interaction of Synapsin I with Actin Monomers

The first suggestion of a possible effect of synapsin I on the process of actin monomer assembly came from experiments in which the number of cytochalasin B binding sites in an actin solution was measured as an indication of the number of filament ends present. The results of these experiments indicated that synapsin I increases the number of actin filaments and that this effect is not due to severing of preexisting filaments (47). Subsequent studies on the kinetics of actin polymerization in the presence of synapsin I clarified the molecular basis of this effect.

When actin polymerization is triggered by the addition of nucleating salts to a solution of actin monomers, synapsin I abolishes the lag phase that is usually required to achieve monomer condensation and increases the initial rate of actin polymerization in a dose-dependent manner. The final level of fluorescence reached is higher in the presence of synapsin I, and the polymerization curve is biphasic (49,54). The development of analytical equations derived from previously reported biochemical models of actin polymerization has allowed a mathematical analysis of the polymerization curves. Such analysis indicates that synapsin I can bind to actin monomers and change their conformation, giving rise to the formation of pseudonuclei that actively elongate into filaments. The process of synapsin I-induced nucleation predominates at initial time points in polymerization, whereas at later time points this process is overshadowed by spontaneous (salt-induced and synapsin I-independent) nucleation (55).

This interpretation is supported by the results of experiments in which actin monomers were incubated with various concentrations of synapsin I in the absence of nucleating salts. Under these conditions, synapsin I can still induce actin polymerization. However, the final extent of polymerization is lower than in the presence of salts and essentially depends on the synapsin I concentration. The bi-exponential time-course of actin polymerization observed under these conditions suggests that the pseudonuclei formed by synapsin I are able to undergo elongation, although the elongation rate is slower and the critical concentration appears to be higher than in the presence of salts (54,55).

The fraction of actin polymerized by synapsin I in the absence of nucleating salts is directly proportional to the synapsin I concentration and inversely related to the actin concentration. The proportion between the fluorescent levels reached before and after the addition of nucleating salts, measured at various synapsin I/actin ratios, indicates a binding stoichiometry of 1:3–4 (synapsin I:actin monomer) (49, 54). This value is quite different from those reported for most other "actin-nucleating" proteins, which are generally active at stoichiometries of 1–3 molecules per actin filament (each filament usually being formed by several hundred monomers) (56). This difference may be explained by a model in which synapsin I complexes with actin monomers to form pseudonuclei and also binds to sides of the filaments generated by the elongation of these pseudonuclei, possibly stabilizing the filaments. Lateral binding of synapsin I to the filaments formed in the absence of nucleating salts is also suggested by the observation that the filaments formed under these conditions are organized in large bundles (54).

Analysis of the kinetics of the interaction of synapsin I with actin monomers by fluorescence resonance energy transfer indicates that at physiological ionic strength the interaction is rapid (half-time of 9 sec at 7°C). Decreasing the ionic strength increases the extent of energy transfer, but makes the interaction slower (48).

The ability of synapsin I to increase the number of actin filaments and to nucleate actin polymerization is regulated by phosphorylation, being strongly decreased and virtually abolished after phosphorylation by cAMP-dependent protein kinase and CaM kinase II, respectively. In contrast, the concentration of Ca^{2+} in the medium has no direct influence on either effect (47,54,55). It appears therefore that intraterminal Ca^{2+} modulates the activity of synapsin I indirectly by activating CaM kinase II.

A Ternary Complex of Synaptic Vesicle-Synapsin I-Actin

In nerve terminals, most synapsin I is associated with the surface of synaptic vesicles (22–24). It is therefore important to know whether synapsin I is able to bind to actin filaments and influence the dynamics of actin polymerization even when it is in the vesicle-bound form. Circumstantial evidence suggests that this is indeed the case:

1. The number of actin filament ends is significantly increased when actin is polymerized in the presence of purified synaptic vesicles. This effect is abolished when synapsin I has been previously removed from the vesicle surface and is restored upon incubation of synapsin I-depleted synaptic vesicles with purified exogenous synapsin I (47).
2. In the fluorescence resonance energy transfer assay, the presence in the medium of unlabeled synaptic vesicles deprived of endogenous synapsin I does not modify the transfer of energy between labeled exogenous synapsin I and actin, even when the vesicles have been preincubated with synapsin I prior to adding actin (48).
3. The presence of synaptic vesicles in the medium does not inhibit the ability of synapsin I to nucleate actin filaments (Chieregatti et al., manuscript in preparation).

In these experiments, however, the possibility that these effects are due to synapsin I released from synaptic vesicles cannot be ruled out. Recently, direct evidence of the formation of a ternary complex consisting of synaptic vesicle-synapsin I-actin has come from analysis by video-enhanced microscopy of a reconstituted system consisting of purified labeled components. We have thus obtained evidence involv-

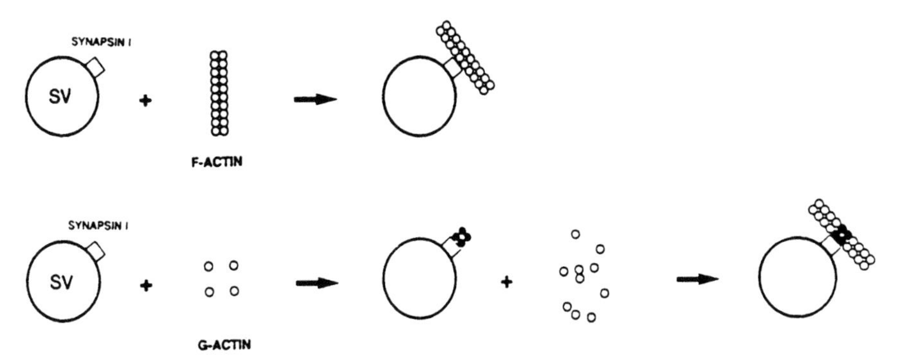

FIG. 1. Schematic model depicting possible mechanisms of interaction between synaptic vesicles and actin. Top: synaptic vesicle (SV)-bound synapsin I binds to preformed actin filaments. Bottom: SV-bound synapsin I binds actin monomers, thus forming a pseudonucleus that then further elongates to form actin filaments.

ing direct visualization that synapsin I bound to synaptic vesicles is able to promote actin polymerization and crosslink synaptic vesicles to actin filaments (Fig. 1). These interactions are virtually abolished upon phosphorylation of synapsin I by CaM kinase II (57).

SYNAPSINS AND NEUROTRANSMITTER RELEASE

Several lines of evidence suggest that synapsin I is involved in the regulation of neurotransmitter release from nerve terminals. Virtually all physiological and pharmacological manipulations known to induce or facilitate Ca^{2+}-dependent neurotransmitter release increase the state of phosphorylation of synapsin I. These manipulations include electrical stimulation of nerve cells, veratridine- or K^+-induced depolarization, application of α-latrotoxin in the presence of extracellular Ca^{2+}, and activation of certain classes of presynaptic receptors (58).

Direct evidence of involvement of synapsin I in neurotransmitter release comes from injection experiments. Microinjection of dephosphorylated synapsin I into the preterminal digit of the squid giant axon or into goldfish Mauthner neurons inhibits both spontaneous and evoked neurotransmitter release. (In contrast, synapsin I phosphorylated by CaM kinase II is completely ineffective) (59–62). These effects occur with no detectable change in the presynaptic inward calcium current, indicating that the number of quanta of neurotransmitter released in response to a given Ca^{2+} influx is decreased. In agreement with these data, microinjection of CaM kinase II induces a three- to sevenfold increase in the number of neurotransmitter quanta released in response to presynaptic depolarization (59,60). Similar results were observed in experiments in which synapsin I or CaM kinase II was introduced into rat brain synaptosomes by a freeze-thaw procedure and depolarization-induced glutamate or norepinephrine release was monitored (63,64).

Given the ability of synapsin I to crosslink synaptic vesicles to actin in a phosphorylation-dependent manner, it is tempting to speculate that its inhibitory role on synaptic vesicle exocytosis is causally related to this property. Thus, we have developed a model according to which dephosphorylated synapsin I bound to synaptic vesicles induces the growth of actin filaments, embedding the vesicles in a cytoskeletal meshwork and making them unavailable for exocytosis. Upon influx of Ca^{2+}, CaM kinase II becomes activated and the state of phosphorylation of synapsin I is increased. Phosphorylated synapsin I tends to dissociate from synaptic vesicles and from F-actin, thus reducing the number of crossbridges between synaptic vesicles and actin, and increasing the number of synaptic vesicles that are not tethered to the cytoskeleton and are therefore available for release. According to this model, synapsin I regulates the efficiency of neurotransmitter release by shifting the proportion of vesicles in the available and reserve pools (i.e., in the pool of "free" vesicles and the pool of vesicles crosslinked to the cytoskeleton, respectively) (9,11,65).

One attempt to test the validity of this model comes from the results of a computer simulation procedure that analyzes the impact of synapsin I phosphorylation on the proportion of free vesicles and vesicles linked to actin filaments. The procedure is based on the experimentally determined binding constants of synapsin I for synaptic vesicles and actin filaments. It calculates the mean number of crossbridges per vesicle and uses binomial distribution statistics to extrapolate, from this number, the frequency of vesicles bearing no crossbridge (free vesicles). The results indicate that the proportion of free vesicles is strongly influenced by the actin concentration assumed to be present in the nerve terminal (an unknown parameter). At all actin concentrations tested, however, phosphorylation of synapsin I significantly increased the proportion of vesicles that are not bound to actin (Fig. 2). This effect is small in absolute terms, but quite relevant with respect to the limited number of vesicles forming the pool of free vesicles. Indeed, at actin concentrations ranging between 10 and 30 μM (a reasonable concentration, based on measurements carried out in other cell types), stoichiometric phosphorylation of synapsin I would induce a two- to sixfold increase in the number of vesicles available for release (66).

Experimental results in support of a partial dissociation of the ternary complex upon stimulation of neurotransmitter release have been obtained by analyzing the distribution of synapsin I in resting and stimulated nerve terminals. These results show that K^+-induced depolarization of rat brain synaptosomes increases the state of phosphorylation of synapsin I and the fraction of the protein found in the cytosol (67). In addition, immunoelectron microscopy of frog neuromuscular junctions fixed under various conditions of enhanced neurotransmitter release has revealed that, at high frequencies of stimulation, synapsin I dissociates from the synaptic vesicle membrane during exocytosis and reassociates with it during the retrieval process (68).

SYNAPSINS AND SYNAPTOGENESIS

The appearance of synapsins in neurons and their subcellular distribution correlate with the onset of synaptogenesis and formation of synaptic contacts (69). A

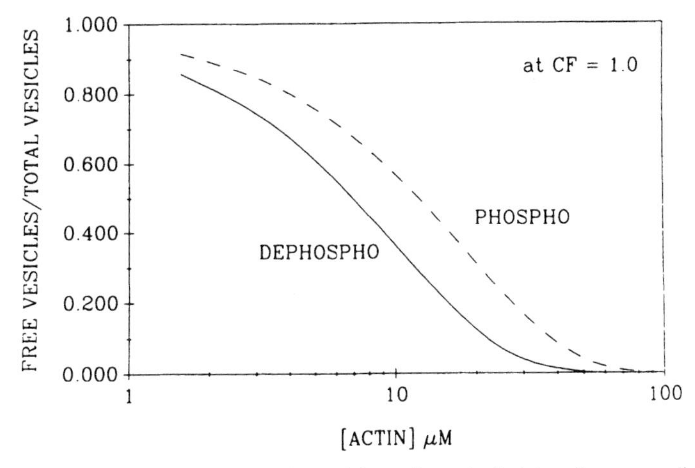

FIG. 2. Increase in synaptic vesicles released from the cytoskeleton (i.e., synaptic vesicles bearing no cross-bridge with actin filaments) induced by stoichiometric phosphorylation of synapsin I. The ratio of free vesicles to total vesicles was calculated as a function of the actin concentration in the nerve terminal, assuming no cooperativity (cooperativity factor = 1.0) in the binding of synapsin I to synaptic vesicles and actin filaments. The frequency of vesicles devoid of crossbridges was calculated from the average number of crossbridges by applying binomial distribution statistics, assuming an average of 10 synapsin I binding sites per vesicle. Calculations were made for dephosphosynapsin I and for synapsin I stoichiometrically phosphorylated by CaM kinase II. (Benfenati et al., 1991)

redistribution of synapsin I immunoreactivity with accumulation of the protein in nerve terminals has been observed in hippocampal cells in culture shortly after the formation of synaptic contacts (70).

Overexpression of synapsin IIb in neuroblastoma x glioma hybrid cell lines induces, upon differentiation, increases in the number of varicosities per cell and of synaptic vesicles per varicosity, in the level of synaptic vesicle proteins (71) and in the number of functional synapses formed upon co-culture with myoblasts (72). Injection of either synapsin I or synapsin II (73,74) into *Xenopus* early embryos accelerates the maturation of developing neuromuscular synapses in culture, as indicated by an increase in the frequency, and an earlier appearance of bell-shaped distribution of spontaneous synaptic currents, as well as an increase in the amplitude of evoked synaptic currents. This maturation of quantal secretion mechanisms induced by the synapsins is accompanied by morphological changes in the developing synapses (e.g., an increase in the number of synaptic vesicles that cluster in proximity to precociously developed presynaptic specializations of the plasma membrane) (75). Finally, transfection of embryonic hippocampal cells in culture with synapsin II antisense mRNA prevents their normal morphological development, involving axonal outgrowth and nerve terminal development. These cellular changes are associated with a drastic reorganization of the actin-cytoskeleton (76).

The molecular mechanisms by which synapsins promote functional and structural maturation of developing synapses is unknown. It is possible that the effect of synapsins on maturation occurs through a mechanism related to that used by synapsin I to regulate release from adult nerve terminals. Thus, the synapsins, by interactions with the actin-based cytoskeleton, might direct the assembly of a macromolecular network that triggers the formation and maturation of axons and nerve terminals. In addition, the recently reported ability of synapsin I to aggregate vesicles (29) might facilitate their clustering close to the plasma membrane.

ACKNOWLEDGMENTS

This work has been supported by grants USPHS MH39327 and USEPA CR819506 (P. G.), Telethon and CNR (F. B. and F. V.).

REFERENCES

1. Valtorta F, Fesce R, Grohovaz F, Haimann C, Hurlbut WP, Iezzi N, Torri Tarelli F, Villa A, Ceccarelli B. Neurotransmitter release and synaptic recycling. *Neuroscience* 1990;35:477–489.
2. Kelly RB. Storage and release of neurotransmitters. *Cell (supplement)* 1993;72:43–53.
3. Valtorta F, Benfenati F. Protein phosphorylation and the control of exocytosis in neurons. *Ann NY Acad Sci*, in press.
4. Barinaga M. Secrets of secretion revealed. *Science* 1993;260:487–489.
5. Söllner T, Whiteheart SW, Brunner M, Erdjument-Bromage H, Geromanos S, Tempst P, Rothman JE. SNAP receptors implicated in vesicle targeting and fusion. *Nature* 1993;362:318–324.
6. Bennett MK, Scheller RH. The molecular machinery for secretion is conserved from yeast to neurons. *Proc Natl Acad Sci USA* 1993;90:2559–2563.
7. Fesce R. Stochastic approaches to the study of synaptic function. *Progr Neurobiol* 1990;35:85–133.
8. Elmqvist D, Quastel DMJ. A quantitative study of end-plate potentials in isolated human muscle. *J Physiol* 1965;178:505–529.
9. Greengard P, Valtorta F, Czernik AJ, Benfenati F. Synaptic vesicle phosphoproteins and regulation of synaptic function. *Science* 1993;259:780–785.
10. De Camilli P, Benfenati F, Valtorta F, Greengard P. The synapsins. *Annu Rev Cell Biol* 1990; 6:433–460.
11. Valtorta F, Benfenati F, Greengard P. Structure and function of the synapsins. *J Biol Chem* 1992; 267:7195–7198.
12. De Camilli P, Jahn R. Pathways to regulated exocytosis in neurons. *Annu Rev Physiol* 1990;52:625–645.
13. Johnson EM, Ueda T, Maeno H, Greengard P. Adenosine 3′:5′-monophosphate-dependent phosphorylation of a specific protein in synaptic membrane fractions from rat cerebrum. *J Biol Chem* 1972;247:5650–5652.
14. Ueda T, Maeno H, Greengard P. Regulation of endogenous phosphorylation of specific proteins in synaptic membrane fractions from rat brain by adenosine 3′:5′-monophosphate. *J Biol Chem* 1973; 248:8295–8305.
15. Schulman H, Greengard P. Stimulation of brain membrane protein phosphorylation by calcium and an endogenous heat-stable protein. *Nature* 1978;271:478–479.
16. Südhof TC, Czernik AJ, Kao H, Takei K, Johnston PA, Horiuchi A, Wagner M, Kanazir SD, Perin MS, De Camilli P, Greengard P. The synapsins: mosaics of shared and unique domains in a family of synaptic vesicle phosphoproteins. *Science* 1989;245:1474–1480.
17. Huttner WB, Greengard P. Multiple phosphorylation sites in Protein I and their differential regulation by cyclic AMP and calcium. *Proc Natl Acad Sci USA* 1979;76:5402–5406.
18. Huttner WB, DeGennaro LJ, Greengard P. Differential phosphorylation of multiple sites in purified Protein I by cyclic AMP-dependent and calcium-dependent protein kinases. *J Biol Chem* 1981; 256:1482–1488.

19. Kennedy MB, McGuinness T, Greengard P. A calcium/calmodulin-dependent protein kinase activity from mammalian brain that phosphorylates synapsin I: partial purification and characterization. *J Neurosci* 1983;3:818–831.
20. Nairn AC, Greengard, P. Purification and characterization of Ca^{2+}/calmodulin-dependent protein kinase I from bovine brain. *J Biol Chem* 1987;262:7273–7281.
21. Czernik AJ, Pang DT, Greengard, P. Amino acid sequences surrounding the cAMP-dependent and calcium/calmodulin-dependent phosphorylation sites in rat and bovine synapsin I. *Proc Natl Acad Sci USA* 1987;84:7518–7522.
22. De Camilli P, Cameron R, Greengard P. Synapsin I (Protein I), a nerve terminal-specific phosphoprotein. I. Its general distribution in synapses of the central and peripheral nervous system demonstrated by immunofluorescence in frozen and plastic sections. *J Cell Biol* 1983;96:1337–1354.
23. De Camilli P, Harris SM, Huttner WB, Greengard P. Synapsin I (Protein I), a nerve terminal-specific phosphoprotein. II. Its specific association with synaptic vesicles demonstrated by immunocytochemistry in agarose-embedded synaptosomes. *J Cell Biol* 1983;96:1355–1373.
24. Huttner WB, Schiebler W, Greengard P, De Camilli P. Synapsin I (Protein I), a nerve terminal-specific phosphoprotein. III. Its association with synaptic vesicles studied in a highly purified synaptic vesicle preparation. *J Cell Biol* 1983;96:1374–1388.
25. Schiebler W, Jahn R, Doucet JP, Rothlein J, Greengard P. Characterization of synapsin I binding to small synaptic vesicles. *J Biol Chem* 1986;261:8383–8390.
26. Ho M, Bähler M, Czernik AJ, Schiebler W, Kezdy FJ, Kaiser ET, Greengard P. Synapsin I is a highly surface active molecule. *J Biol Chem* 1991;266:5600–5607.
27. Benfenati F, Bähler M, Jahn R, Greengard P. Interactions of synapsin I with small synaptic vesicles: Distinct sites in synapsin I bind to vesicle phospholipids and vesicle proteins. *J Cell Biol* 1989; 108:1863–1872.
28. Benfenati F, Greengard P, Brunner J, Bähler M. Electrostatic and hydrophobic interactions of synapsin I and synapsin I fragments with phospholipid bilayers. *J Cell Biol* 1989;108:1851–1862.
29. Benfenati F, Valtorta F, Rossi MC, Onofri F, Sihra T, Greengard P. Interactions of synapsin I with phospholipids: possible role in synaptic vesicle clustering and in the maintenance of bilayer structures. *J Cell Biol*, 123:1845–1855.
30. Benfenati F, Valtorta F, Rubenstein JR, Gorelick FS, Greengard P, Czernik AJ. Synaptic vesicle-associated Ca^{2+}/calmodulin-dependent protein kinase II is a binding protein for synapsin I. *Nature* 1992;359:417–420.
31. Siow YL, Chilcote TJ, Benfenati F, Greengard P, Thiel G. Synapsin IIa: expression in insect cells, purification, and characterization. *Biochemistry* 1992;31:4268–4275.
32. Burgoyne RD, ed. *The neuronal cytoskeleton*. New York: Wiley-Liss; 1991.
33. Fifkova E. Actin in the nervous system. *Brain Res* 1985;356:187–215.
34. Korn ED. Actin polymerization and its regulation by proteins from nonmuscle cells. *Physiol Rev* 1982;62:672–737.
35. Pollard TD, Cooper JA. Actin and actin-binding proteins. A critical evaluation of mechanisms and functions. *Ann Rev Biochem* 1986;55:987–1035.
36. Gaertner A, Ruhnau K, Schröer E, Selve N, Wanger M, and Wegner A. Probing nucleation, cutting and capping of actin filaments. *J Muscle Res Cell Motil* 1989;10:1–9.
37. Carlier MF. Actin: protein structure and filament dynamics. *J Biol Chem* 1991;266:1–4.
38. Cooper JA, Pollard TD. Effect of capping protein on the kinetics of actin polymerization. *Biochemistry* 1985;24:793–799.
39. Doi Y, Frieden C. Actin polymerization: the effect of brevin on filament size and rate of polymerization. *J Biol Chem* 1984;259:11868–11875.
40. Stossel TP. From signal to pseudopod. *J Biol Chem* 1989;264:18261–18264.
41. Linstedt AD, Kelly RB. Overcoming barriers to exocytosis. *Trends Neurosci* 1987;10:446–448.
42. Trifaro JM, Vitale ML, del Castillo AR. Cytoskeleton and molecular mechanisms in neurotransmitter release by neurosecretory cells. *Eur J Pharmacol* 1992;225:83–104.
43. Marxen P, Bigalke H. Tetanus and botulinum A toxins inhibit stimulated F-actin rearrangements in chromaffin cells. *Neuroreport* 1991;2:33–36.
44. Bernstein BW, Bamburg IR. Cycling of actin assembly in synaptosomes and neurotransmitter release. *Neuron* 1989;3:257–265.
45. Bähler M, Greengard P. Synapsin I bundles F-actin in a phosphorylation-dependent manner. *Nature* 1987;326:704–707.
46. Petrucci TC, Morrow J. Synapsin I: an actin-bundling protein under phosphorylation control. *J Cell Biol* 1987;105:1355–1363.

47. Benfenati F, Valtorta F, Chieregatti E, Greengard P. Interaction of free and synaptic vesicle-bound synapsin I with F-actin. *Neuron* 1992;8:377–386.
48. Ceccaldi PE, Benfenati F, Chieregatti E, Greengard P, Valtorta F. Rapid binding of synapsin I to F- and G-actin. A study using fluorescence resonance energy transfer. *FEBS Letters* 1993;329:301–305.
49. Valtorta F, Ceccaldi PE, Grohovaz F, Chieregatti E, Fesce R, Benfenati F. Fluorescence approaches to the study of the actin-nucleating and bundling activities of synapsin I. *J Physiol (Paris)* 1993; 87:117–122.
50. Bähler M, Benfenati F, Valtorta F, Czernik AJ, Greengard P. Characterization of synapsin I fragments produced by cysteine-specific cleavage: A study of their interactions with F-actin. *J Cell Biol* 1989;108:1841–1849.
51. Petrucci TC, Morrow IS. Actin and tubulin binding domains of synapsins Ia and Ib. *Biochemistry* 1991;30:413–421.
52. Ueda T, Greengard P. Adenosine 3′:5′-monophosphate-regulated phosphoprotein system of neuronal membranes. I. Solubilization, purification, and some properties of an endogenous phosphoprotein. *J Biol Chem* 1977;252:5155–5163.
53. Font B, Aubert-Foucher E. Detection by chemical cross-linking of bovine brain synapsin I self-association. *Biochem J* 1989;264:893–899.
54. Valtorta F, Greengard P, Fesce R, Benfenati F. Effects of the neuronal phosphoprotein synapsin I on actin polymerization. I. Evidence for a phosphorylation-dependent nucleating effect. *J Biol Chem* 1992;267:11281–11288.
55. Fesce R, Benfenati F, Greengard P, Valtorta F. Effects of the neuronal phosphoprotein synapsin I on actin polymerization. II. Analytical interpretation of kinetic curves. *J Biol Chem* 1992;267:11289–11299.
56. Glenney JR, Kaulfus P, Weber K. F-actin assembly modulated by villin: Ca^{2+}-dependent nucleation and capping of the barbed end. *Cell* 1981;24:471–480.
57. Ceccaldi PE, Grohovaz F, Benfenati F, Chieregatti E, Greengard P, Valtorta F. Phosphorylation-dependence of the synapsin I-mediated actin-synaptic vesicle interactions: direct evidence by video imaging. Abstract presented at the 5th International Congress on Cell Biology, Madrid (Spain), 1992.
58. Nestler EJ, Greengard P. *Protein Phosphorylation in the Nervous System*, New York: John Wiley and Sons; 1984.
59. Llinas R, McGuinness T, Leonard CS, Sugimori M, Greengard P. Intraterminal injection of synapsin I or calcium/calmodulin-dependent protein kinase II alters neurotransmitter release at the squid giant synapse. *Proc Natl Acad Sci USA* 1985;82:3035–3039.
60. Llinas R, Gruner JA, Sugimori M, McGuinness TL, Greengard P. Regulation by synapsin I and Ca^{2+}/calmodulin-dependent protein kinase II of transmitter release in squid giant synapse. *J Physiol* 1991;436:257–282.
61. Lin JW, Sugimori M, Llinas R, McGuinness TL, Greengard P. Effects of synapsin I and calcium/calmodulin-dependent protein kinase II on spontaneous neurotransmitter release in the squid giant synapse. *Proc Natl Acad Sci USA* 1990;87:8257–8261.
62. Hackett JT, Cochran SL, Greenfield LJ, Brosius DC, Ueda T. Synapsin I injected presynaptically into goldfish Mauthner axons reduces quantal synaptic transmission. *J Neurophysiol* 1990;63:701–706.
63. Nichols RA, Sihra TS, Czernik AJ, Nairn AC, Greengard P. Calcium/calmodulin-dependent protein kinase II increases glutamate and noradrenaline release from synaptosomes. *Nature* 1990;343:647–651.
64. Nichols RA, Chilcote TJ, Czernik AJ, Greengard P. Synapsin I regulates glutamate release from rat brain synaptosomes. *J Neurochem* 1992;58:783–787.
65. Bähler M, Benfenati F, Valtorta F, Greengard P. The synapsins and the regulation of synaptic function. *BioEssays* 1990;12:259–263.
66. Benfenati F, Valtorta F, Greengard P. Computer modeling of synapsin I binding to synaptic vesicles and F-actin: implications for regulation of neurotransmitter release. *Proc Natl Acad Sci USA* 1991; 88:575–579.
67. Sihra TS, Wang JK, Gorelick FS, Greengard P. Translocation of synapsin I in response to depolarization of isolated nerve terminals. *Proc Natl Acad Sci USA* 1989;86:8108–8112.
68. Torri Tarelli F, Bossi M, Fesce R, Greengard P, Valtorta F. Synapsin I partially dissociates from synaptic vesicles during exocytosis induced by electrical stimulation. *Neuron* 1992;9:1143–1153.

69. Lohmann SM, Ueda T, Greengard P. Ontogeny of synaptic phosphoproteins in brain. *Proc Natl Acad Sci USA* 1978;75:4037–4041.
70. Fletcher TL, Cameron P, De Camilli P, Banker G. The distribution of synapsin I and synaptophysin in hippocampal neurons developing in culture. *J Neurosci* 1991;11:1617–1626.
71. Han HQ, Nichols RA, Rubin MR, Bähler M, Greengard P. Induction of formation of presynaptic terminals in neuroblastoma cells by synapsin IIb. *Nature* 1991;349:697–700.
72. Higashida H, Noda M, Zhong DG, Kimura Y, Han HQ, Rubin MR, Li L, Czernik AJ, Chen W, Horiuchi A, Greengard P. The level of synapsin II in NG108-15 cells regulates the rate of formation of functional synapses with muscle cells. (submitted).
73. Lu B, Greengard P, Poo M-m. Synapsin I promotes functional maturation of developing neuromuscular synapses. *Neuron* 1992;9:759–768.
74. Schaeffer E, Alder J, Greengard P, Poo M-m. Synapsin IIa accelerates functional development of neuromuscular synapses. *Proc Natl Acad Sci USA* 1994;91:3882–3886.
75. Valtorta F, Iezzi N, Benfenati F, Lu B, Poo M-m, Greengard P. Accelerated structural maturation induced by synapsin I at developing neuromuscular synapses. 1994 (submitted).
76. Ferreira A, Kosik KS, Greengard P, Han HQ. Aberrant neurites and synaptic vesicle protein deficiency in synapsin II-depleted neurons. *Science* 1994;264:977–979.

Molecular and Cellular Mechanisms of Neurotransmitter Release, edited by Lennart Stjärne, Paul Greengard, Sten Grillner, Tomas Hökfelt, and David Ottoson, Raven Press, Ltd., New York © 1994.

3

Tetanus and Botulinal Neurotoxins

Tools to Understand Exocytosis in Neurons

*Egenhard Link, †Juan Blasi, *Edwin R. Chapman, *Lambert Edelmann, ‡Anja Baumeister, ‡Thomas Binz, ‡Shinji Yamasaki, ‡Heiner Niemann, and *§Reinhard Jahn

*§Department of Pharmacology and Cell Biology, *Howard Hughes Medical Institute, Yale University School of Medicine, New Haven, Connecticut 06510; †Department of Cell Biology and Pathology, University of Barcelona, 08036 Barcelona, Spain; and ‡Departments of Microbiology, Federal Research Center for Virus Diseases of Animals, D-72076 Tübingen, Germany*

Neurons release their neurotransmitters by means of exocytosis of synaptic vesicles. To accomplish this goal, neurons possess a highly specialized pathway of membrane recycling in their nerve terminals. Following exocytosis, this pathway enables neurons to regenerate exocytosis-competent synaptic vesicles from membranes that undergo endocytosis within the nerve terminal at a rate faster than any other secretory system (1,2).

The molecular mechanisms underlying synaptic vesicle exocytosis are still enigmatic. However, rapid progress in recent months provides a first glimpse into the biochemistry of this process. It appears that eucaryotic cells possess a basic fusion machine for the fusion of exocytotic carrier vesicles with the plasma membrane that is modulated, but not principally different, in all differentiated cells. The mechanism of action of this fusion machine is not yet understood. It appears that the fusion machine is composed of proteins that are members of distinct protein families with a general structure preserved from yeast to mammals (3,4).Progress in understanding the mechanism was made possible by a convergence of four independent lines of research: 1) study of yeast mutants deficient in secretion (sec mutants) (5) 2) establishment of assays for fusion of intracellular organelles in cell-free systems (6) 3) characterization of synaptic vesicle and synaptic membrane proteins that allowed a direct biochemical and molecular characterization of the proteins participating in membrane fusion (7,2) and 4) study of clostridial neurotoxins that led to identification of proteins involved in fusion (see next section).

In the section that follows, we will briefly review recent work on the mechanism of action of clostridial neurotoxins that has contributed to the progress just mentioned, centered around work in our own laboratories.

CLOSTRIDIAL NEUROTOXINS: POWERFUL INHIBITORS OF NEURONAL EXOCYTOSIS

It has been known for many years that the anaerobic bacteria *Clostridium tetani* and *Clostridium botulinum* produce a group of powerful neurotoxins that are responsible for the clinical manifestations of tetanus and botulism (8). Whereas *C. tetani* produces a single toxin species, strains of *C. botulinum* synthesize at least seven serologically distinct neurotoxins, designated BoNT/A, B, C1, D, E, F, and G. In contrast to tetanus toxin (TeTx), the botulinal toxin (BoNT) proteins are complexed with nontoxic proteins that include proteins with hemagglutinating properties. These complexes are resistant to degradation by the catabolic processes in the gastrointestinal tract, allowing them to be resorbed in intact form. This explains why botulism can develop upon ingestion of contaminated food whereas tetanus toxin requires direct release into the bloodstream, mostly via colonization of wounds by *C. tetani* (8,9).

Molecular characterization of tetanus and botulinal toxins has revealed that these proteins possess a common structure and share a significant degree of sequence homology (Table 1). All toxins are synthesized as precursor polypeptides of 150 kDa that are nontoxic. For activation, the toxin proteins are proteolyzed at a single site yielding a heavy (H) chain of approximately 100 kDa and a light (L) chain of approximately 50 kDa. The H and L chains remain linked by a disulfide bond. For most toxins, proteolytic activation is carried out by proteases endogenous to the bacterium, but occasionally host proteases are required (e.g., BoNT/E) (9).

Despite some controversial details, it is generally accepted that all clostridial

TABLE 1. *Sequence comparison of tetanus and botulinum toxin light chains (% identity)*

	TeTx	BoNT/A	BoNT/B	BoNT/C1	BoNT/D	BoNT/E	BoNT/F	BoNT/G
TeTx	100							
BoNT/A	32.1	100						
BoNT/B	51.6	32.4	100					
BoNT/C1	34.8	34.5	33.6	100				
BoNT/D	34.5	34.1	34.5	46.5	100			
BoNT/E	43.9	34.3	36.5	34.1	36.4	100		
BoNT/F	45.1	35.0	39.2	35.3	36.4	57.6	100	
BoNT/G	49.0	35.1	61.1	35.3	36.6	38.5	40.5	100

The data are compiled from the following references: TeTx, (Eisel et al., 1986); BoNT/A, (Binz et al., 1990); BoNT/B, (Kurazono et al., 1992); BoNT/C1, (Hauser et al., 1990); BoNT/D, (Binz et al., 1992); BoNT/E, (Poulet et al., 1992); BoNT/F, (East et al., 1992); BoNT/G, (Campbell et al., 1993).

neurotoxins share a principal mechanism of action that ultimately results in a block of neurotransmitter release from presynaptic nerve endings. According to the current models, active toxin proteins bind with high selectivity and affinity to receptors exposed on the outer surface of presynaptic nerve terminals. This binding is mediated by the H chains of the toxin proteins. The neuromuscular endplates of skeletal muscles are the primary targets. The high selectivity and affinity of the toxin-receptor interaction is largely responsible for the high toxicity of these toxins, which are among the most toxic compounds known (e.g., LD50 for BoNT/A in mice is in the range of 5×10^{-12}g or 3.3×10^{-17} mol; (10).

After surface binding, clostridial neurotoxins are internalized, most probably by endocytosis, and ultimately reach the cytoplasm of the synapse. The mechanism of toxin translocation across the membrane is not clear. Recent evidence indicates that the H chains may form proteinaceous channels that allow the L chains to escape into the cytoplasm. This translocation appears to be less efficient for tetanus toxin than for botulinal toxins. Tetanus toxin is transported retrogradely by vesicular transport into the CNS and does not appear to act at the neuromuscular endplate. After reaching the cell bodies/dendrites of the motor neurons, tetanus toxin is released, probably by exocytosis (transcytosis). Tetanus toxin then enters presynaptic neurons, most prominently the terminals of inhibitory glycinergic interneurons, and ultimately reaches higher centers of the CNS. This difference between the peripheral action of botulinal toxins and the central action of tetanus toxin is primarily responsible for the differential clinical manifestations of tetanus and botulism. Tetanus poisoning results in muscle cramps due to loss of inhibitory control, whereas botulinum poisoning results in flaccid muscular paralysis (8,9).

After reaching the cytoplasm, the L chains are released from the H chains due to cleavage of the disulfide bond in the reductive intracellular environment. The free L chains then cause an essentially irreversible block of exocytosis in virtually all neurons examined. This block is slow in onset and lasts for long periods of time. Careful analysis of poisoned synapses indicates that the block affects membrane fusion itself, since all other aspects of nerve terminal function such as membrane potential, voltage-gated Ca^{2+}-currents, and the morphology of intracellular structures remained unchanged.

Although the L chains of all clostridial neurotoxins inhibit exocytosis, a careful analysis of transmitter release at the neuromuscular endplate revealed interesting differences beween BoNT/A and BoNT/B or TeTx, respectively. Although BoNT/A results in a potent depression of both spontaneous and evoked transmitter release, its action can be counteracted by 4-aminopyridine to a significantly greater extent than that of BoNT/B or TeTx. The K-channel blocker 4-aminopyridine is known to cause depolarization of the presynaptic plasma membrane, resulting in increased transmitter release upon arrival of an action potential (11). Furthermore, black widow spider venom, a neurotoxin that elicits massive exocytosis at the neuromuscular endplate, causes exocytosis of synaptic vesicles from BoNT/A but not from BoNT/B or TeTx poisoned synapses (12). In Aplysia neurons, inhibition of neurotransmitter release due to BoNT/A microinjection can be transiently reversed by microinjection of a toxin-neutralizing antibody (13). This effect is observed only in the early phases of poisoning and indicates that, in contrast to some of the other toxins, the action of BoNT/A may be transiently reversible.

Until recently, the molecular mechanism(s) of L-chain action was not understood. The potency of the toxins was underlined by the observation that only few (perhaps even a single) L-chain molecules are sufficient to block transmission at a synapse. On the other hand, efforts to identify binding sites for any of the tested L chains proved futile. These observations led to the suggestion that the toxin L chains exert their action by a catalytic rather than a stoichiometric process. However, the mechanism of action remained enigmatic until Jongeneel and colleagues (14) made the seminal observation that the L chain of tetanus toxin possesses a sequence motif, HExxH, that is common to Zn^{2+}-dependent proteases. Sequence comparison then revealed that all clostridial toxin L chains share this motif. Tetanus toxin has been demonstrated to contain bound Zn^{2+} (15). In the Zn-protease thermolysin, the histidyl-residues of the HExxH motif coordinate a Zn^{2+}-ion in the catalytic center, whereby the glutamyl residue holds a water molecule by a hydrogen bond whose free electron pairs occupy the fourth position in the tetrahedral complex (16). Thus, it appeared that the L chains exerted their actions as Zn^{2+}-dependent proteases within the nerve terminal, an assumption that since then has been confirmed by the identification of the target proteins, characterization of the cleavage reaction, and mutagenesis of amino acids of the HExxH motif (17–23).

SYNAPTOSOMES AS TOOLS FOR STUDYING THE MOLECULAR MECHANISMS OF BOTULINAL AND TETANUS TOXIN ACTION

To identify the targets of clostridial neurotoxins, we utilized synaptosomes as a model system. Synaptosomes are isolated nerve terminals that are sheared off their axons and resealed upon homogenization of brain tissue. For a few hours, synaptosomes maintain an intact membrane potential and the ability to both synthesize ATP and respond to depolarization with exocytotic (i.e., Ca^{2+}-dependent) release of neurotransmitter (24). Glutamate release by synaptosomes can be conveniently monitored using an online photometric assay (25) that is compatible with subsequent analysis of synaptic proteins. Thus, all analyzed preparations have a recorded physiological history that can be qualitatively and quantitatively correlated with the effect of the toxins on the respective target proteins. This approach was chosen to avoid artefacts associated with in vitro experiments involving isolated membrane fragments and purified toxins that led to erroneous conclusions in the past.

Previous work has shown that neurotransmitter release from synaptosomes is efficiently blocked upon preincubation with TeTx, BoNT/B, BoNT/C1, and BoNT/D, although in some cases (e.g., with BoNT/A) only partial inhibition could be obtained (26). Isolated L chains were ineffective, in agreement with the notion that the H chains are required for cell entry.

Careful electrophoretic analysis of synaptic proteins upon toxin poisoning resulted in the following picture: TeTx as well as BoNT/D caused degradation of the synaptic vesicle protein synaptobrevin (VAMP) (19,26a). This degradation was highly selective since no other change was observed by one- and two-dimensional electrophoresis or by immunoblotting using a multitude of antibodies directed against other synaptic proteins. The degree of synaptobrevin breakdown correlated with the degree of inhibition (19,26a). Closer analysis of TeTx action revealed that inhibition of release and synaptobrevin breakdown was only seen when holotoxin (i.e., disulfide bonded complex of H chain and L chain) was used. Neither isolated H nor isolated L chains were effective, in agreement with earlier physiological observations (19,26a). In addition, synaptosomes were subfractionated after TeTx-incubation to yield a fraction containing plasma membranes and tightly bound synaptic vesicles, and a synaptic vesicle-enriched fraction devoid of plasma membranes. No difference in the degree of breakdown was observed, suggesting that the toxin does not have a preference for free or docked synaptic vesicles (unpublished observations). Despite the use of a panel of antibodies, no breakdown product of synaptobrevin was observed, indicating that the protein is completely degraded under these conditions (19,26a).

In contrast to the toxins just mentioned, neither BoNT/A nor BoNT/C1 caused significant breakdown of synaptobrevin in these experiments, although a slight effect ("cross talk") cannot be excluded (see sentences that follow). A search for potential targets of these toxins revealed that two other recently characterized synaptic membrane proteins were cleaved; BoNT/A poisoning resulted in the selective breakdown of the protein SNAP-25, whereas BoNT/C1 poisoning caused the cleavage of the protein syntaxin (21,22). Again, cleavage appears to be highly specific since no other protein was found to be affected. In both cases, breakdown products were observed but the amount was low indicating that, similar to synaptobrevin cleavage, the initial cleavage by the toxin is followed by complete degradation due to endogenous proteases. Once again, a good correlation between BoNT/C1-induced inhibition of transmitter release and syntaxin breakdown was observed.

SNAP-25 does not contain a transmembrane domain but rather appears to be anchored to the membrane by means of palmitoyl side chains (27). In contrast, syntaxin (originally discovered in the retina, named HPC-1 (28)) is an integral membrane protein that is predominantly localized on the axonal and synaptic plasma membrane (29,30). A smaller pool, however, exists in recycling organelles of nerve endings, including synaptic vesicles. Interestingly, BoNT/C1 appears to predominantly cleave the vesicular pool, whereas no breakdown was detectable in the plasmalemmal pool (Walch-Solimena et al., unpublished observations). The significance of this finding is not clear at present, but it is possible that syntaxin participates in protein-protein interactions in the plasma membrane that protect it against toxin attack.

The experiments on synaptosomes clearly show that: 1) inhibition of neurotransmitter release by clostridial neurotoxin leads to the selective degradation of synaptobrevin, SNAP-25, and syntaxin, respectively, while individual toxins are specific for only one protein; 2) cleavage appears to correlate well with inhibition of transmitter release; and 3) both inhibition of transmitter release and proteolysis are observed only when holotoxin is used, whereas isolated H or L chains are ineffective. The substrate difference between BoNT/A and TeTx, respectively, may form the molecular basis for the difference observed in the physiological action of these toxins (see previous section).

INTERACTION OF TOXIN L CHAINS WITH THE TARGET MOLECULES

Whereas the synaptosome experiments were crucial in establishing the functional significance of toxin-induced proteolysis, the mechanism of the cleavage reaction can be more conveniently studied with purified components. Two approaches have been used: 1) incubation of isolated synaptic vesicles or synaptic membranes with purified toxin L chains or reduced holotoxin, and 2) incubation of recombinant substrate proteins or protein fragments with purified or recombinant toxin L chains.

Synaptobrevin is an integral membrane protein of synaptic vesicles with an M_r of approximately 13,000 (larger in invertebrates) that contains a single transmembrane domain at the C-terminal end of the molecule (31–33). Synaptobrevin is highly conserved throughout evolution, containing a central conserved domain with 80% invariant amino acid residues between mammals and Drosophila (32). In neurons, two isoforms have been described. Synaptobrevin 1 (VAMP-1) is predominantly expressed in the spinal cord whereas synaptobrevin 2 (VAMP-2) is the predominant form in the frontal brain (33). In addition, a third isoform, termed cellubrevin, recently has been described that appears to be ubiquitously expressed in all cells. This isoform is enriched on organelles involved in membrane traffic between the plasma membrane and endosomal compartments. Apparently, the synaptobrevin protein family comprises trafficking proteins that are involved in exocytosis in all cells regardless of whether it is regulated by Ca^{2+} or not (34).

Cleavage of synaptobrevin by TeTx and BoNT/B L chains in isolated synaptic vesicles was the first to be discovered (18,19) and has been characterized extensively. The cleavage sites of BoNT/D and BoNT/F were also determined, generating the following findings (20, Yamasaki et al., submitted).

1. Each toxin cleaves synaptobrevin at a single site. Surprisingly, individual toxins cleave synaptobrevin at different sites that are clustered in a small segment of the central conserved domain. TeTx and BoNT/B cleave at Gln^{76}-Phe^{77}, BoNT/D at Lys^{59}-Leu^{60}, and BoNT/F at Gln^{58}-Lys^{59}, respectively (cleavage sites refer to isoform 2). Interestingly, the two resulting fragments are observed only when purified synaptic vesicles or recombinant synaptobrevin are used. Apparently, the initial attack of the toxins is followed by further degradation due to endogenous proteases in intact nerve terminals (see previous section and Fig. 1).
2. Despite the high degree of homology between the synaptobrevin isoforms, TeTx, BoNT/B, and BoNT/D are less active towards synaptobrevin-1 than towards synaptobrevin-2 and cellubrevin (20,26a). In contrast, BoNT/F cleaves all

isoforms with comparable potency (20,26a). Replacement of individual amino acid residues in synaptobrevin-2 with those found in synaptobrevin-1 revealed that Met^{46} is required for optimal proteolysis by BoNT/D and Gln^{76} for optimal proteolysis by TeTx.

3. Analysis of recombinant synaptobrevin mutants revealed that N-terminally located sequences that are distal from the cleavage sites are required for cleavage, whereas C-terminally located domains are not important, including the transmembrane domain (26a).
4. Recombinant L chain of TeTx is as active as native L chain. Substitutions of amino acid residues contained in the HExxH motif as well as $glutamate^{270}$ and $glutamate^{271}$ resulted in the loss of activity, confirming the requirement of these residues for catalytic activity (Fig. 2) (26a).

In contrast to synaptobrevin, SNAP-25 does not contain a putative transmembrane domain but is palmitoylated at one or more of four cysteine residues clustered in the middle of the molecule. Using recombinant soluble SNAP-25, the cleavage site for BoNT/A was determined to be Gln^{197}-Arg^{198} (i.e., only nine amino acids from the C-terminus of the molecule). Apparently, the C-terminal end of the molecule is crucial for function. In addition, recently we have found that BoNT/E also cleaves SNAP-25 with high specificity. Interestingly, the cleavage site (Arg^{180}-Ile^{181}) is different from that of BoNT/A (44). Again, mutagenesis of SNAP-25

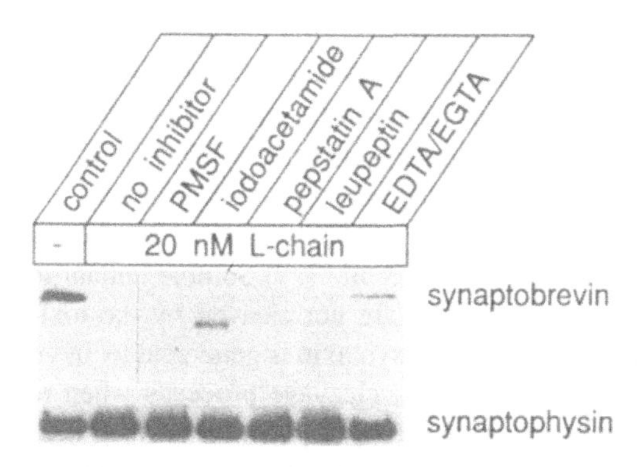

FIG. 1. Protease inhibitors do not block TeTx L-chain-induced cleavage of synaptobrevin in crude membranes but may prevent further degradation. Isolated synaptic vesicles (0.4 mg/ml) were incubated in the absence (control) or in the presence of TeTx L chain (20 nM). Incubations were carried out for 1 h at 37°C (see Link et al., 1992, for details). The following inhibitor concentrations were used: PMSF (1 mM), iodoacetamide (0.1 mM), pepstatin A (1 μM), leupeptin (0.1 mM). As control, we used EGTA/EDTA (0.1 mM/1 mM) that caused some slowing of the breakdown reaction due to its partial complexation of the cofactor Zn^{2+}. All samples were analyzed by SDS-PAGE and immunoblotting (5 μg protein/lane) using previously published antibodies as detectors (Link et al., 1992). The synaptic vesicle protein synaptophysin was monitored, as a control that is not cleaved by TeTx. (Note that no synaptobrevin fragments are observed unless iodoacetamide is present, indicating that, following the initial attack by TeTx L chain, synaptobrevin is further degraded by endogenous, -SH-sensitive proteases.)

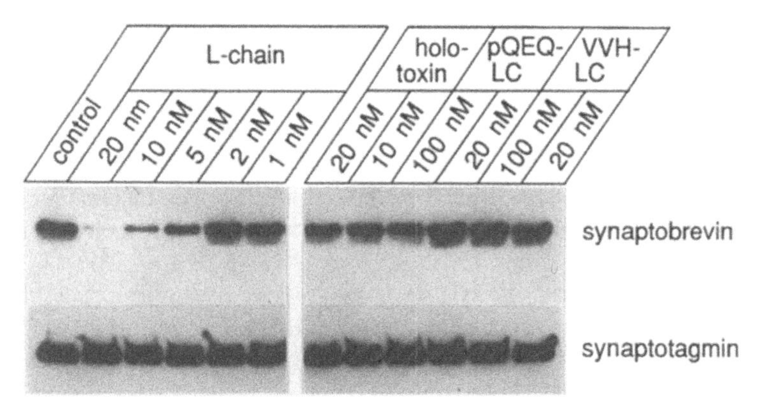

FIG. 2. L-chain of TeTx cleaves synaptobrevin in a dose-dependent manner whereas TeTx holotoxin and two L-chain mutants bearing substitutions in the Zn^{2+}-binding motif are inactive. Isolated synaptic vesicles (0.4 mg/ml) were incubated in the absence or presence of varying amounts of TeTx L-chain (left panel). In addition, incubations were performed with unreduced toxin or with two recombinant L-chain mutants in which amino acids present in the $H^{233}E^{234}xxH^{237}xxH^{240}$ motif were changed (VVH-LC: H^{233} and H^{237} to Val^{233} and Val^{237}; pQEQ-LC: E^{234} to Q^{234}) (Yamasaki et al., 1994; McMahon et al., 1993) (right panel). All samples were analyzed by SDS-PAGE and immunoblotting (5 μg protein/lane, see Fig. 1). No change was observed in the vesicle protein synaptotagmin, a protease-sensitive membrane protein that served as a control (see Link et al., 1992 for details).

revealed that domains that are located distally at the N-terminal side of the cleavage site are required for cleavage (Binz et al., submitted).

Whereas cleavage of both synaptobrevin and SNAP-25 progresses to completion, proceeds readily with recombinant protein fragments, and does not require membrane association, BoNT/C1-induced cleavage of syntaxin requires an intact membrane anchor (Fig. 3) (22). Similar to synaptobrevin, syntaxin occurs in two highly homologous isoforms in the brain (a and b) that have a single transmembrane domain at the C-terminal end of the molecule (29). Soluble mutant forms of syntaxin that lack the transmembrane domain are not cleaved by the toxin. Similarly, no cleavage is observed when full-length syntaxin is generated by in vitro translation in the absence of membranes. However, cleavage proceeds when recombinant syntaxin is incorporated into membranes (co- or posttranslationally) whereby both isoforms are affected equally. The cleavage site has not yet been determined but appears to be close to the C-terminus (Fig. 4), although somewhat distant from the transmembrane domain (22).

Although the existence of additional substrate proteins cannot be excluded at present, all available data suggest that clostridial neurotoxins exert their action by the selective cleavage of any of the proteins synaptobrevin, SNAP-25, and syntaxin. The extraordinary substrate specificity is in sharp contrast to the low specificity of the cleavage site. It appears that all toxins recognize the substrate based on domains at the N-terminal of the respective cleavage site. The specifity of the catalytic site may be quite low, determined more by the positioning of the polypeptide

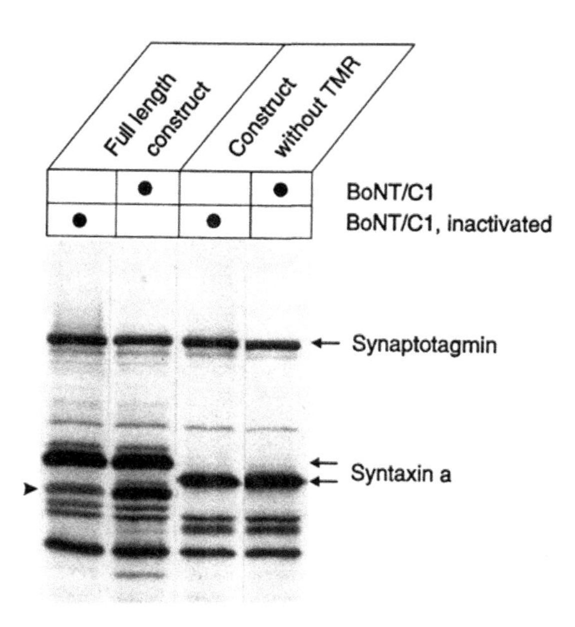

FIG. 3. BoNT/C1 cleaves syntaxin translated in vitro; cleavage requires the transmembrane domain. Syntaxin or a construct of syntaxin lacking the putative transmembrane region was translated in vitro in the presence of microsomes. The synaptic vesicle protein synaptotagmin I was co-translated in the same assay, as a control. At the end of the translation reactions, samples were treated with active or heat-inactivated BoNT/C1. All samples were then analyzed by Tricine SDS-PAGE and [^{35}S]methionine labeled proteins were visualized by fluorography (see (Blasi et al., 1993, for details). The major breakdown product of syntaxin is indicated by an arrowhead. (Note that translation resulted in additional labeled bands below syntaxin, some of which may represent additional syntaxin fragments.)

chain in the active center than by the nature of the amino acid residues surrounding the cleavage sites. However, further experiments are required to resolve these issues.

SYNAPTOBREVIN, SYNTAXIN, AND SNAP-25: CONSTITUENTS OF A NEURONAL FUSION MACHINE?

The identification of synaptobrevin, syntaxin, and SNAP-25 as the proteins responsible for the block of synaptic vesicle exocytosis by clostridial neurotoxins coincides in an astounding manner with a completely independent line of research that led to the identification of the same set of proteins essential for membrane fusion.

During the last decade, Rothman and collaborators utilized in vitro fusion assays of intracellular organelles (mostly traffic within the Golgi apparatus) to identify soluble proteins required for docking and/or fusion (6).This work led to the discovery of a highly conserved protein complex, consisting of an N-ethylmaleimide-sensitive fusion protein (NSF) and two soluble NSF attachment proteins (αSNAP

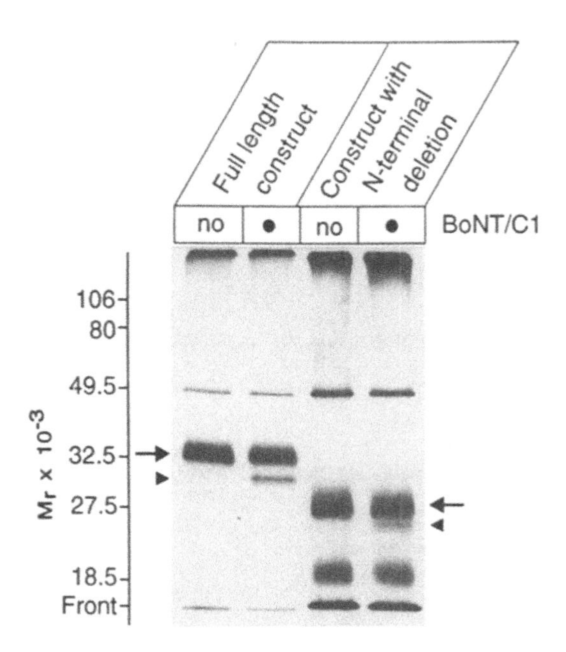

FIG. 4. An N-terminally truncated fragment of syntaxin is cleaved by BoNT/C1, demonstrating that the cleavage site is located close to the C-terminus. Wild-type syntaxin a and a mutant form that lacked the 80 N-terminal amino acids were translated in vitro in the absence of membranes and subsequently incorporated into liposomes. Analysis was performed by SDS-PAGE and fluorography as described in Fig. 3. The arrows point to the primary translation products. Arrowheads point to the position of the toxin-generated fragments. The two right lanes were exposed longer due to reduced incorporation of radioactivity. (Note that both the wild-type and the mutant form are cleaved by the toxin, although cleavage appears to be less efficient with the mutant form.)

and γSNAP). Interestingly, the yeast homologs of NSF and αSNAP (Sec18p and Sec17p, respectively) were genetically identified as gene products required for intracellular membrane traffic and growth. A detailed study of these protein factors demonstrated that they exert their action by specific interaction with membrane proteins in the fusing membranes and that ATP-cleavage by NSF is required (6). Using an elegant affinity-ligand approach, Söllner et al. demonstrated that in brain extracts, the membrane receptors of the NSF-SNAP complex are identical with the proteins synaptobrevin, syntaxin a and b and SNAP-25 (35) (see chapter five). Further work in our laboratory revealed that the three proteins exist in a complex also in the absence of NSF and SNAP, indicating that interaction with these soluble factors occurs after the complex between the membrane has been formed (4, Walch-Solimena et al., submitted).

While these findings are essentially based on protein-protein interactions in detergent extracts, the "knockout" of each of the proteins in the complex by clostridial neurotoxins directly proves their requirement in membrane fusion. Together, all these findings strongly suggest that synaptobrevin, syntaxin, and SNAP-25 form the core of an exocytotic fusion complex functioning in neurotransmitter release, allowing access to one of the most basic problems of cell biology. It should now be possible to elucidate the step-by-step mechanics of the fusion machine in order to obtain a description of membrane fusion at the molecular level.

REFERENCES

1. Augustine GJ, Charlton MP, Smith SJ. Calcium action in synaptic transmitter release. *Annu Rev Neurosci* 1987;10:633–693.
2. Jahn R, Südhof TC. Synaptic vesicles and exocytosis. *Annu Rev Neurosci* 1994;17:219–246.
3. Bennett MK, Scheller RH. The molecular machinery for secretion is conserved from yeast to neurons. *Proc Natl Acad Sci USA* 1993;90:2559–2563.
4. Südhof TC, De Camilli P, Niemann H, Jahn R. Membrane fusion machinery: Insights from synaptic proteins. *Cell* 1993;75:1–4.
5. Pryer NK, Wuestehube LJ, Schekman R. Vesicle-mediated protein sorting. *Annu Rev Biochem* 1992;61:471–516.
6. Rothman JE, Orci L. Molecular dissection of the secretory pathway. *Nature* 1992;355:409–415.
7. Trimble WS, Llinial M, Scheller RH. Cellular and molecular biology of the presynaptic nerve terminal. *Annu Rev Neurosci* 1991;14:93–122.
8. Simpson LL, ed. *Botulinum neurotoxin and tetanus neurotoxin*. New York: Academic Press; 1989.
9. Niemann H. Molecular biology of clostridial neurotoxins. In: Alouf JE, Freer JH, eds. *Sourcebook of bacterial toxins*. New York: Academic Press; 1991;303–348.
10. Maisey EA, Wadsworth JDF, Poulain B, et al. Involvement of the constituent chains of botulinum neurotoxins A and B in the blockade of neurotransmitter release. *Eur J Biochem* 1988;177:683–691.
11. Gansel M, Penner R, Dreyer F. Distinct sites of action of clostridial neurotoxins revealed by double-poisoning of mouse motor nerve terminals. *Pflügers Arch* 1987;409:533–539.
12. Dreyer F, Rosenberg F, Becker C, Bigalke H, Penner R. Differential effects of various secretagogues on quantal transmitter release from mouse motor nerve terminals treated with botulinum A and tetanus toxin. *Naunyn-Schmiedebergs Arch Pharmacol* 1987;335:1–7.
13. Di Bello IC, Poulain B, Shone CC, Tauc L, Dolly JO. Antagonism of the intracellular action of botulinum neurotoxin type A with monoclonal antibodies that map to L chain epitopes. *Eur J Biochem* 1993;219:161–169.
14. Jongeneel CV, Bouvier J, Bairoch A. A unique signature identifies a family of zinc-dependent metallopeptidases. *FEBS Lett* 1989;242:211–214.
15. Wright JF, Pernollet M, Reboul A, Aude C, Colomb MG. Identification and partial characterization of a low affinity metal-binding site in the light chain of tetanus toxin. *J Biol Chem* 1992;267;9053–9058.
16. Colman PM, Jansonius JN, Matthews BW. The structure of thermolysin: an electron density map at 2-3 A. *J Mol Biol* 1972;70:701–724.
17. Schiavo G, Poulain B, Rossetto O, Benfenati F, Tauc L, Montecucco C. Tetanus toxin is a zinc protein and its inhibition of neurotransmitter release and protease activity depend on zinc. *EMBO J* 1992;11:3577–3583.
18. Schiavo G, Benfenati F, Poulain, B, et al. Tetanus and botulinum-B neurotoxins block neurotransmitter release by proteolytic cleavage of synaptobrevin. *Nature* 1992;359:832–835.
19. Link E, Edelmann L, Chou JH, et al. Tetanus toxin action: Inhibition of neurotransmitter release linked to synaptobrevin proteolysis. *Biochem Biophys Res Commun* 1992;189:1017–1023.
20. Schiavo G, Shone CC, Rossetto O, Alexander FC, Montecucco C. Botulinum neurotoxin serotype F is a zinc endopeptidase specific for VAMP/synaptobrevin. *J Biol Chem* 1993;268:11516–11519.
21. Blasi J, Chapman ER, Link E, et al. Botulinum neurotoxin A selectively cleaves the synaptic protein SNAP-25. *Nature* 1993;365:160–163.
22. Blasi J, Chapman ER, Yamasaki S, Binz T, Niemann H, Jahn R. Botulinum neurotoxin C1 blocks neurotransmitter release by means of cleaving HPC-1/syntaxin. *EMBO J* 1993;12:4821–4828.
23. Yamasaki S, Hu Y, Binz T, et al. Synaptobrevin (VAMP) of *Aplysia californica:* Structure and proteolysis by tetanus toxin and botulinal neurotoxins type D and F. *Proc Natl Acad Sci USA* 1994.

24. Nicholls DG. Release of glutamate, aspartate, and gamma-aminobutyric acid from isolated nerve terminals. *J Neurochem* 1989;52:331–341.
25. Nicholls DG, Sihra TS. Synaptosomes possess an exocytotic pool of glutamate. *Nature* 1986;321: 772–773.
26. McMahon HT, Foran P, Dolly JO, Verhage M, Wiegant VM, Nicholls DG. Tetanus toxin and botulinum toxins type A and B inhibit glutamate, γ-aminobutyric acid, aspartate, and Met-enkephalin release from synaptosomes. *J Biol Chem* 1992;267:21336–21343.
26a. Yamasaki S, Baumeisten A, Binz T, et al. Cleavage of members of the synaptobrevin family by types D and F botulinal neurotoxins and tetanus toxin. *J Biol Chem* 1994;269:12764–12772.
27. Oyler GA, Higgins GA, Hart RA, et al. The identification of a novel synaptosomal-associated protein, SNAP-25, differentially expressed by neuronal subpopulations. *J Cell Biol* 1989;109:3039–3052.
28. Barnstable CJ, Hofstein R, Akagawa K. A marker of early amacrine cell development in rat retina. *Dev Brain Res* 1985;20:286–290.
29. Bennett MK, Calakos N, Scheller RH. Syntaxin: a synaptic protein implicated in docking of synaptic vesicles at presynaptic active zones. *Science* 1992;257:255–259.
30. Inoue A, Obata K, Akagawa K. Cloning and sequence analysis of cDNA for a neuronal cell membrane antigen, HPC-1. *J Biol Chem* 1992;267:10613–10619.
31. Trimble WS, Cowan DM, Scheller RH. VAMP-1: A synaptic vesicle-associated integral membrane protein. *Proc Natl Acad Sci USA* 1988;85:4538–4542.
32. Südhof TC, Baumert M, Perin MS, Jahn R. A synaptic vesicle membrane protein is conserved from mammals to Drosophila. *Neuron* 1989;2:1475–1481.
33. Elferink LA, Trimble WS, Scheller RH. Two vesicle-associated membrane protein genes are differentially expressed in the rat central nervous system. *J Biol Chem* 1989;264:11061–11064.
34. McMahon HT, Ushkaryov YA, Edelmann L, Link E, Binz T, Niemann H, Jahn R, Südhof TC. Cellubrevin is a ubiquitous tetanus-toxin substrate homologous to a putative synaptic vesicle fusion protein. *Nature* 1993;364:346–349.
35. Söllner T, Whiteheart SW, Brunner M, et al. SNAP receptors implicated in vesicle targeting and fusion. *Nature* 1992;362;318–324.
36. Eisel U, Jarausch W, Goretzky K, et al. Tetanus toxin: primary structure, expression in *E coli*, and homology with botulinum toxins. *EMBO J* 1986;5:2495–2502.
37. Binz T, Kurazono H, Wille M, Frevert J, Wernars K, Niemann H. The complete sequence of botuminum neurotoxin type A and comparison with other clostridial neurotoxins. *J Biol Chem* 1990; 265:9153–9158.
38. Kurazono H, Mochida S, Binz T, et al. Minimal essential domains specifying toxicity of the light chains of tetanus toxin and botulinum neurotoxin type A. *J Biol Chem* 1992;267:14721–14729.
39. Hauser D, Eklund MW, Kurazono H, et al. Nucleotide sequence of clostridium botulinum C1 neurotoxin. *Nucl Acids Res* 1990;18:4924.
40. Binz T, Kuraxono H. Popoff MR, et al. Nucleotide sequence of the gene encoding *Clostridium botulinum* neurotoxin type D. *Nucl Acids Res* 1990;18:5556.
41. Poulet S, Hauser D, Quanz M, Niemann H, Popoff MR. Sequences of the botulinal neurotoxin E derived from *Clostridium botulinum* type E (strain Beluga) and *Clostridium butyricum* (strains ATCC 43181 and ATCC 43755). *Biochem Biophys Res Commun* 1992;183:107–113.
42. East AK, Richardson PT, Allaway D, Collins MD, Roberts TA, Thompson DE. Sequence of the gene encoding type F neurotoxin of *Clostridium botulinum*. *FEMS Microbiol Lett* 1992;96:225–230.
43. Campbell K, Collins MD, East AK. Nucleotide sequence of the gene coding for *Clostridium botulinum* type G neurotoxin: genealogical comparison with other clostridial neurotoxins. *Biochim Biophys Acta* 1993;1216:487–491.
44. Binz T, Blasi J, Yamasaki S, et al. Proteolysis of SNAP-25 by types E and A botulinal neurotoxins. *J Biol Chem* 1944;269:1617–1620.

Molecular and Cellular Mechanisms of Neurotransmitter Release, edited by Lennart Stjärne, Paul Greengard, Sten Grillner, Tomas Hökfelt, and David Ottoson, Raven Press, Ltd., New York © 1994.

4

Synaptic Vesicle Proteins and Exocytosis

*Sandra M. Bajjalieh and †Richard H. Scheller

*Departments of Molecular and *Cellular Physiology, and †Cellular Biology, Howard Hughes Medical Institute, Stanford University, Stanford, California 94305*

Electron micrographic studies of neurons have revealed that synaptic endings are filled with small, clear vesicles apparently "docked" at the presynaptic membrane (1). The quantal release of neurotransmitter is achieved by calcium-dependent fusion of these vesicles with the synaptic plasma membrane.

Characterization of these vesicles, referred to as synaptic vesicles, is a first step toward understanding the molecular events of neurosecretion. The high concentration of secretory vesicles in brain has allowed the purification of quantities large enough for biochemical characterization. Proteins specific to synaptic vesicles have been identified in highly purified vesicle preparations by comparative gel electrophoresis and by using vesicles as an antigen source for the generation of antibodies (2,3). Several synaptic vesicle proteins identified in this manner have been characterized at a molecular level. These include the small guanosine 5′-triphosphate (GTP)-binding protein rab3, and the integral membrane proteins VAMP (synaptobrevin), synaptotagmin (p65), synaptophysin, synaptic vesicle protein 2 (SV2), and a large, multi-subunit proton adenosinetriphosphatase (ATPase) (Fig. 1). With the apparent exception of the proton ATPase, all of the synaptic vesicle proteins are members of small gene families that are differentially expressed in brain.

Synaptic vesicles undergo multiple rounds of exocytosis at the synapse and therefore must contain proteins mediating several functions, including vesicle formation, filling with neurotransmitter, association with cytoskeletal elements, docking at the presynaptic membrane, and membrane fusion. By combining knowledge of protein structure with mutational and biochemical analyses, we are beginning to describe the molecular events underlying these phases of vesicle cell functioning. This discussion will attempt to summarize our current understanding of how synaptic vesicle proteins may act at each of these stages.

CYTOPLASM

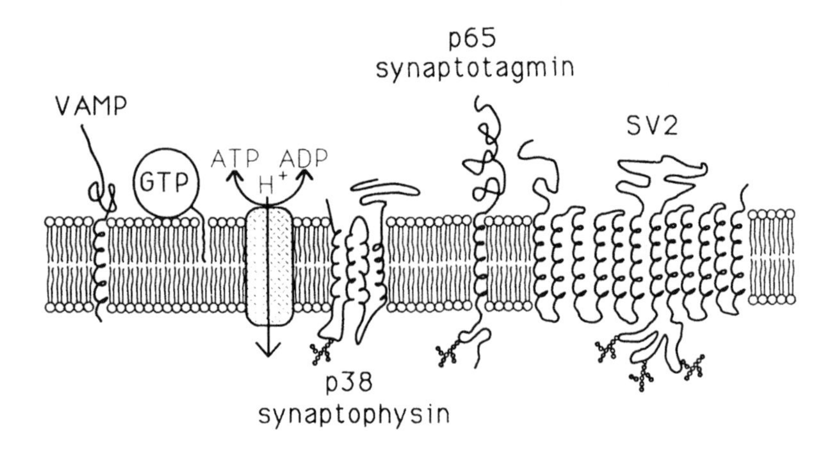

LUMEN

FIG. 1. Proteins of the synaptic vesicle membrane. Shows the predicted membrane topologies of synaptic vesicle membrane proteins. The proteins spanning the membrane once or not at all (VAMP, rab3 and synaptotagmin) appear to be involved in vesicle targeting and/or docking. Proteins with multiple membrane domains (SV2 and synaptophysin) most likely function as transporters, channels, or pores. Not depicted are the neurotransmitter transporters, which have a similar predicted membrane topology to that of SV2. See text for discussions of each.

SV, synaptic vesicle
VAMP, synaptobrevin

SYNAPTIC VESICLE BIOGENESIS

Morphological analysis of secretory vesicles from both brain and endocrine cells has revealed two classes of vesicles, distinguishable on the basis of size, contents, and their appearance in electron micrographs. Large (>100 nm diameter) neuropeptide- and amine-containing vesicles appear to have dense cores, whereas small (50 nm) transmitter-containing vesicles appear clear. While both types of vesicles have several components in common, only small, clear vesicles are released at the active zone of synapses.

There are two tasks to creating a synaptic vesicle: 1) the correct proteins must be sorted together and 2) a vesicle of very small diameter must be generated. It has been hypothesized that the large membrane curvature of 50 nm vesicles is thermodynamically unfavorable, suggesting that a special process is required to create and maintain it. Constitutive and neuropeptide-containing (dense core) secretory vesicles are formed in the trans-Golgi network, suggesting that this structure is a sorting center for different vesicle classes. Two observations suggest that synaptic vesicles are also synthesized in the Golgi and move to the synapse via fast axonal transport. First, mutation of a kinesin-like molecule in *C elegans* results in the accumulation of synaptic-like vesicles in the cell body and a dearth of vesicles in synaptic regions, consistent with the interpretation that axonal transport is required for the delivery of synaptic vesicles to the synapse (4). Second, rat synaptic vesicles injected into the axon of the squid giant synapse are transported to synaptic regions, indicating that vesicles contain the information needed to be targeted to these regions (5). While these observations may indicate that vesicles are formed in the cell body, they are also consistent with the interpretation that individual vesicle proteins are transported to the synapse in vesicle-like structures and sorted into synaptic vesicles at the synapse. Studies of synaptophysin localization in neuroendocrine cells corroborate the latter interpretation. Newly synthesized synaptophysin is associated first with constitutive secretory vesicles and then with early endosomes before emerging in synaptic vesicle-like structures (6). This suggests that the full complement of vesicle proteins is assembled during membrane recycling. In neuroendocrine cells, but not in transfected fibroblasts, synaptophysin appears to be segregated to endosomal compartments distinct from those involved in receptor-mediated endocytosis. This suggests that neuroendocrine cells contain a specialized endocytic compartment that exists for the synthesis of synaptic vesicles (7,8). Indeed, endosomal vesicles in the axons of cultured hippocampal neurons appear to be insensitive to disruption by Brefeldin A, whereas endosomes in the dendrites and cell body collapse into tubule-like structures (9).

Comparisons of dense core and synaptic vesicles have revealed that synaptophysin is absent from dense core vesicles (10). This raises the possibility that it plays a role in synaptic vesicle biosynthesis. When synaptophysin is expressed in non-neuronal cells, it is segregated to small clear vesicles that contain few other proteins (11). However, when synaptophysin is co-expressed with synaptotagmin and SV2 in non-neuroendocrine cells, all three proteins are segregated to different membrane compartments (123) indicating that synaptophysin alone is not sufficient for the assembly of synaptic vesicles.

ASSOCIATION WITH CYTOSKELETAL ELEMENTS AT THE SYNAPSE

Synaptic vesicles are stored at the synapse in one of two pools: 1) a releasable pool that is free to interact with plasma membrane docking sites or 2) a reserve pool that associates with the actin cytoskeleton. Association of vesicles with the cytoskeleton is mediated by the synapsins, a family of four phosphoproteins that are peripherally associated with the cytoplasmic surface of vesicles (see review (12)). The association of synapsin I with synaptic vesicles is dependent on its phoshorylation state; unphosphorylated synapsin associates with vesicles, whereas phosphorylated synapsin does not. A number of observations, reviewed elsewhere in this issue, indicate that the calcium-dependent phosphorylation of synapsin can alter the amount of exocytosis in vivo. Thus, the actin network, via its interaction with synapsins, creates a reserve that can be called upon to produce changes in the number of vesicles available for release.

SYNAPTIC VESICLE FILLING: TRANSPORTER PROTEINS

Vesicle accumulation of neurotransmitter is driven by an electrochemical gradient generated by a H^+-ATPase. The electric potential created by this H^+-ATPase is dispelled to some extent by chloride uptake, permitting the acidification of the vesicle interior. Therefore, synaptic vesicles contain at least three transport activities, for protons, chloride, and transmitter. Regulation of any of these could alter the amount of transmitter contained per vesicle and therefore the efficacy of neurotransmission. In addition to the known transporters, vesicles also contain proteins whose predicted structure suggest that they may function as transporters, perhaps indicating novel vesicle contents.

The Proton ATPase (H^+-ATPase)

The H^+-ATPase of synaptic vesicles is most likely identical to the one isolated from brain clathrin-coated vesicles (see review) (13,14). It has a molecular mass of approximately 700 kd and comprises 20% of the protein in synaptic vesicles (15,16). The H^+-ATPase has been extensively characterized biochemically; it is constructed of nine polypeptides segregated into two functional domains: 1) a membrane domain that mediates the transport of protons and 2) a peripheral domain that hydrolyzes ATP. The subunit structure of the coated vesicle H^+-ATPase is similar to the H^+-ATPase of chromaffin granules and yeast vacuoles and to bacterial F-type H^+-ATPase. Functionally these proteins are similar; in fact, the F-type H^+-ATPase from *E. coli* can drive the uptake of neurotransmitter when reconstituted with synaptic vesicle proteins (17).

The cDNAs encoding four of the nine proteins that make up the H^+-ATPase have been cloned. The 73 kd and 58 kd (18–20) subunits of the peripheral domain demonstrate a high degree of homology to corresponding subunits from archaebacteria, suggesting that these molecules are quite ancient (20). cDNAs encoding the membrane-associated 17 kd subunit have been cloned from chromaffin granules (21) and from cerebellum (22). This subunit is predicted to be very hydrophobic and the presence of six copies per H^+-ATPase suggests that it may form a pore through which protons pass. Finally, a 116 kd subunit not common to all species of H^+-ATPase is present in the pump of brain clathrin-coated vesicles. Its predicted structure includes six transmembrane domains, suggesting that it may also mediate movement of protons through the membrane (23).

The H^+-ATPase has been found to be inactive in vesicles both entering and leaving the endosome, suggesting that pump activity is regulated (24). A recent report that the ATP-binding, 58 kd subunit is phosphorylated by a protein of the clathrin adaptor complex suggests a possible mechanism for this regulation (25).

Neurotransmitter Transporters

Transporters mediating the uptake of catecholamines, acetylcholine, γ-aminobutyric acid (GABA), glycine, and glutamate have been biochemically characterized in synaptic vesicle preparations. All require the electrochemical gradient generated by the vesicular H^+-ATPase. However, transport activity appears not to require the presence of the H^+-ATPase itself, since uptake can be observed in reconstituted systems with artificial gradients generated either chemically (26) or by bacteriorhodopsin, which produces a proton gradient in response to light (27). Various pharmacological agents have been known for some time to differentially affect the transport of different neurotransmitters. These observations suggested that a different transporter exists for each class of transmitter, a conclusion supported by the identification of cDNAs encoding amine and acetylcholine vesicular transporters.

Two vesicular catecholamine transporters have been identified: 1) one specific to brain (28,29) and 2) one localized to endocrine tissues (29). These transporters are predicted to have 12 transmembrane domains and therefore are structurally similar to the neurotransmitter transporters of the plasma membrane (30). However, the vesicular transporters demonstrate no significant homology to the plasma membrane transporters. Instead, they weakly resemble several bacterial antibiotic resistance proteins believed to act by transporting drugs across the cell membrane (29).

A putative acetylcholine transporter from *C. elegans* has also been identified (31). While functional evidence has not yet been reported, this protein is 37% to 39% identical to the rat vesicular amine transporters and is localized to cholinergic neurons.

Less is known about the transporters of amino acid neurotransmitters (see review (32)). Competition experiments suggest that the inhibitory amino acid transmitters, GABA and glycine, may be transported by a single carrier (33). Characterization of both native and reconstituted transport indicates that the glutamate and GABA/glycine transporters may utilize different components of the electrochemical gradient to effect uptake. Under conditions that maintain a potential gradient while dispelling the proton gradient, glutamate transport is observed. However, no transport is observed when the proton gradient is maintained in the absence of a potential gradient (34,35,36). This suggests that glutamate transport utilizes primarily the potential gradient created by the proton pump. Transport of GABA, on the other hand, utilizes both potential and proton gradients (26,36). These observations suggest that the relative contribution of proton and potential gradients to the electrochemical gradient may differ in excitatory and inhibitory amino acid-containing vesicles.

Chloride Transporter

The balance of proton and chloride transport determines the composition of the electrochemical gradient across vesicle membranes. The observation that the transport of excitatory and inhibitory amino acids utilizes different components of the electrochemical gradient suggests that chloride transport might be regulated to accommodate the requirements of the resident neurotransmitter transporter. Both chloride-transporter (37) and chloride-channel (38) activities have been reported for clathrin-coated vesicles and synaptic vesicles, respectively. Neither of these activities has been characterized in detail, though a chloride transporter activity has been purified away from the proton ATPase, indicating that it is a separate entity.

Synaptic Vesicle Protein 2 (SV2)

SV2 was identified with a monoclonal antibody generated against purified cholinergic vesicles from the electric organ of the electric fish *Discopyge ommata* (39). The cDNAs encoding SV2 (40–42) predict a protein homologous to a large family of transporter proteins typified by bacterial sugar transporters. Conserved motifs of this family are also evident in bacterial and fungal citrate transporters, antibiotic resistance proteins, and a family of mammalian facilitative glucose transporters (43). This homology suggests that SV2 may be a vesicle-specific transporter.

SV2 is predicted to have 12 transmembrane domains with the six-loop-six structure of the sugar transporters (Fig. 1). Sequence homology to the sugar transporters is greatest in the first half of the protein (40,41) although some characteristic motifs are present in the second half. SV2 is distinguished from other members of the sugar transporter family by a large loop between the seventh and eighth membrane domains that is predicted to be in the lumen of the vesicle and that contains three N-glycosylation consensus sites. Immunoprecipitation of SV2 from different tissues indicates that these sites may be differentially glycosylated (44).

There are two known isoforms of SV2. They are 65% identical and ~80% similar to one another, with most of the differences localized to the amino terminus. Northern blot analyses indicate that both are expressed throughout the brain (45) suggesting that if the SV2 proteins are specific neurotransmitter transporters, they could only mediate the transport of the amino acid transmitters, which are widely distributed. However, both immunocytochemical and in situ hybridization analyses indicate that neither isoform is limited in its expression to either excitatory or inhibitory neurons (46). One or both forms of SV2 is expressed in all brain regions, indicating that it performs a function not limited to neurons of any transmitter type. Immunoprecipitation of synaptic vesicles with isoform-specific antibodies suggests that both SV2 isoforms may be present on a single vesicle, perhaps indicating that SV2 forms oligomers that vary in their isoform composition.

Since the expression data indicate that SV2 is most likely not a specific neurotransmitter transporter, what else might it be? While there is no known active uptake of ATP by vesicles, the high concentration of this nucleotide in their lumen may be due to a transporter. Alternatively, SV2 may move ions, most notably chloride or calcium, as either a transporter or channel. A precedent for this hypothesis is the cystic fibrosis gene product, CFTR, which resembles a multidrug transporter

(MDR) and yet displays chloride channel activity (47). However, the assignment of biological activity to proteins with transporter-like structure may be complex. For example, MDR appears to conduct chloride in addition to transporting drugs (48) calling into question its primary role in cells. Finally, since the family of proteins with homology to SV2 includes transporters that move molecules both into and out of cells, the direction of any putative transport activity by SV2 cannot be assumed. Therefore, it is equally possible that SV2 moves molecules out of vesicles, perhaps as part of an exocytotic fusion pore (40).

Synaptophysin (p38)

Synaptophysin, like SV2, was also identified with a monoclonal antibody generated against purified synaptic vesicles (49). Unlike other synaptic vesicle proteins, synaptophysin appears unique to small, clear vesicles (10). In addition, it has not been detected in invertebrates, suggesting perhaps that its function is specific to synaptic secretion in higher species. The deduced amino acid sequence of synaptophysin predicts four transmembrane domains and a charged carboxy terminus oriented toward the cytoplasm (50,51) (Fig. 1). There are two isoforms of synaptophysin, both of which have hydropathy profiles similar to gap junction proteins (52). When reconstituted into liposomes, synaptophysin conducts current, suggesting that it may act as a pore or channel in vivo (53). While this may indicate that synaptophysin functions as a transporter, its association with a putative plasma membrane protein, physophillin (54), and the finding that antisynaptophysin antibodies block neurotransmitter release (55) suggest instead that it may be involved in exocytosis (see next section). If so, this would imply that small, clear vesicles have a unique fusion mechanism since synaptophysin is not a component of dense core secretory vesicles.

DOCKING AT THE PRESYNAPTIC MEMBRANE

The synaptic vesicle proteins synaptotagmin, rab3, and VAMP are implicated in vesicle targeting and docking. Amino acid sequence analysis has revealed that rab3 and VAMP have homologues in other cellular compartments as well as in species as diverse as yeast. Recent biochemical analyses have identified plasma membrane proteins and soluble factors that appear to interact with vesicle proteins to form sequential protein complexes that mediate vesicle attachment and fusion. These observations indicate that a basic targeting and docking mechanism is common to all membrane trafficking and that the regulated fusion of secretory vesicles in neurons is an embellished version of a system common to all eukaryotes. Yeast homologues of the synaptic vesicle proteins rab3 and VAMP, of the synaptic plasma membrane proteins syntaxin and (SNAP)-25, and of two soluble factors required for inter-Golgi transport, N-ethylmaleimide-sensitive fusion factor (NSF) and α-SNAP, implicate them in a basic system of membrane transport. The absence of known homologues of synaptotagmin suggest that it may play a modulatory role required for regulation of constitutive processes.

Synaptotagmin (p65)

Synaptotagmin (p65) is present on both dense core and synaptic vesicles (56,57) suggesting that it performs a function essential to regulated secretion. The most striking characteristic of the protein is two repeated domains that have significant homology to the second conserved (C2) domain of protein kinase C (58–60). This region has been implicated in the binding of protein kinase C to membranes and is found in other membrane-associating proteins, leading to the hypothesis that it represents a membrane-binding motif (61). A role for synaptotagmin in membrane binding is supported by the observation that it binds acidic phospholipids (58) in a calcium-dependent fashion (62) suggesting that synaptotagmin forms a protein-calcium-phospholipid complex similar to the one reported to occur between protein kinase C, phosphatidylserine, and calcium (63). Synaptotagmin has been reported to interact with proteins in vitro, including RACKS (receptors of activated C kinase) (64). Phosphorylation of synaptotagmin by casein kinase II (65,66) and calcium calmodulin protein kinase II (67,68) has been reported. Whether the phosphorylation of synaptotagmin alters its lipid- or protein-binding properties has not been determined.

If synaptotagmin participates in the association of vesicles with the plasma membrane, its depletion from cells would be expected to alter secretion. Similarly, microinjection of antibodies or synaptotagmin peptide fragments would be expected to disrupt the interaction between synaptotagmin and other proteins or lipids. Depletion experiments have produced conflicting results. Clonal cell lines selected for the absence of synaptotagmin were not deficient in regulated secretion, suggesting that exocytosis does not require synaptotagmin (69). A slight increase in stimulated exocytosis in these cells may, however, suggest that synaptotagmin plays a negative modulatory role in secretion. If it does, depletion effects would be most apparent in basal levels of release, which have not been reported. Unlike the results from cultured cells, the production of synaptotagmin null mutants in drosophila (70) and *C. elegans* (71) produced profoundly affected animals. Drosophila lacking synaptotagmin did not develop past the larval stage, and *C. elegans* null mutants lacked basic motor capacity. The observation that these animals are capable of movement may indicate that some synaptic transmission occurs in the absence of synaptotagmin, perhaps due to a still unidentified redundancy in synaptotagmin function. However, the severe phenotype apparent in both null mutants indicates that synaptotagmin is required for normal synaptic transmission.

The results of microinjection experiments have been more consistent. Two groups have observed that microinjection of antisynaptotagmin antibodies (72) or of peptides corresponding to the C2 domains of synaptotagmin (72,73) produced a decrease in stimulated exocytosis. These observations suggest a role for synaptotagmin, and more specifically for the C2 domain, in regulated exocytosis. Care in interpreting these experiments may be warranted since, at high concentrations, the C2 domain of synaptotagmin might interact with receptors for the C2 domains of other proteins. While the C2 peptides did not alter the activity of protein kinase C in one of these preparations (73) these peptides may have affected the function of yet other C2 domain-containing proteins, most notably rabphilin. Rabphilin was identified on the basis of its ability to interact with the synaptic vesicle protein rab3 (see discussion that follows), and has C2 domains similar to those of synaptotagmin (74). Alternatively, the C2 peptides may have disrupted interactions between synaptic vesicle proteins (72). However, since injection of antisynaptotagmin Fab fragments produced an inhibition similar to that obtained with the injection of C2 domain peptides, it seems likely that these observations reflect a role for synaptotagmin in exocytosis.

Synaptotagmin has been reported to co-purify with the receptor for α-latrotoxin, a spider venom that induces exocytosis. Using partial amino acid sequence of the purified receptor, several cDNAs have been isolated. The predicted proteins constitute a family, termed neurexins, that have a short intracellular domain, a single transmembrane region, and large extracellular domain homologous to extracellular matrix proteins (75). While the neurexins have not been reported to bind latrotoxin, they do bind synaptotagmin, suggesting that an interaction between synaptotagmin and the neurexins may play a role in synapse formation.

rab3

The rab, or smg, proteins are a family of GTP-binding proteins similar to the oncogene product ras. Like the ras proteins, rabs contain four highly conserved domains that constitute a GTP-binding motif. They are hydrophilic yet membrane-associated, due in part to the addition of isoprenyl moieties, a modification that appears to be required for biological activity (76,77). At least 20 rab proteins have been identified in animal cells. Some are tissue specific; rab3A and rab 15, for example, appear to be expressed largely in brain (78,79). Most appear to be localized to a single compartment of the secretory pathway (80).

In brewer's yeast, two proteins homologous to the rab proteins, sec4p and ypt1p, are required for secretory function. Like the rab proteins, sec4p and ypt1p are localized to distinct cellular compartments (sec4p to secretory vesicles and the plasma membrane, ypt1p to the Golgi). Mutations of these proteins disrupt membrane traffic; sec4 mutants accumulate post-Golgi secretory vesicles (81) whereas ypt1 mutants appear to accumulate Golgi transport vesicles (82). Analyses of disrupted cell preparations revealed that ypt1 mutants generate endoplasmic reticulum (ER) transport vesicles, but that these vesicles do not target to and/or fuse with Golgi membranes (83,84). Mutations that disrupt the ability of sec4p to bind or hydrolyze GTP result in the accumulation of secretory vesicles, suggesting that cycling between guanosine 5′-diphosphate (GDP)- and GTP-bound states is critical to sec4p function (85). These observations, coupled with the compartmental localization of ypt1p and sec4p, led to the proposal that rab-like proteins mediate the vectorial movement of vesicles through the secretory pathway by specifying compartmental fidelity and insuring directional transport (86). However, this conclusion is not supported by the observation that sec4p, engineered with a small internal portion of ypt1p, can rescue ypt1 mutants, even though the protein does not demonstrate significant Golgi localization (87,88). This observation argues against the interpretation that small GTP-binding proteins specify compartmental identity, but is still consistent with a role in insuring vectorial transport.

A rab protein localized to synaptic vesicles is rab 3 (78) which, like other synaptic vesicle proteins, is present in differentially expressed isoforms. Depletion of rab3B produces decreased calcium-stimulated exocytosis in pituitary cells (89). This suggests a role for rab proteins in regulated secretion similar to that of sec4p in yeast constitutive secretion. To determine whether rab3, like sec4p, requires cycling between GDP- and GTP-bound states, mouse endocrine (AtT-20) cells were transfected with rab3 mutants deficient in GTP binding or hydrolysis. These cells demonstrated no alteration in basal or regulated secretion of hormone (90). However, cells expressing mutant rab3 protein did incorrectly sort hormone-containing vesicles that are usually sequestered in the tips of cell processes. This observation is compatible with a role for rab3 in vectorial transport and suggests that some rab proteins may mediate interactions with cytoskeletal elements. Interestingly, the endogenous synaptic vesicle protein VAMP was also incorrectly sorted, even though the endogenous rab 3 and the synaptic vesicle proteins SV2 and synaptotagmin were still localized to cell processes. This abbarent sorting of VAMP is interesting since several suppressors of rab3 mutants in yeast are VAMP-like proteins, indicating that these proteins could interact directly.

A protein purified on the basis of its ability to bind rab3 in a GTP-dependent fashion has been identified (74). This protein, termed rabphilin, contains two C2 domains homologous to those of synaptotagmin, suggesting that rabphilin may mediate the association of rab3 with proteins or lipids. The similarity of rabphilin to synaptotagmin may indicate a redundancy in synaptic vesicle docking mechanisms and explain persistence of some synaptic transmission in synaptotagmin null mutants.

VAMP

Vesicle associated membrane protein (VAMP, also synaptobrevin) is one of the more abundant proteins of synaptic vesicle membranes. As with the other synaptic vesicle proteins, there are two highly homologous isoforms of VAMP, VAMP-1 and VAMP-2 (91–93). These isoforms are differentially expressed, with VAMP-2 being the more prevalent isoform in mammalian brain (94). The VAMP proteins are largely cytoplasmic and anchored in the membrane by a single transmembrane domain near the carboxy terminus (Fig. 1). The hypothesized role for VAMP in exocytosis is supported by the discovery that two potent inhibitors of neural transmission, the clostridial toxin tetanus and the B serotype of botulinum toxin, are proteases that hydrolyze VAMP-2 (95–97). The cleavage site of both toxins has been localized to a region preceding the transmembrane domain, indicating that toxin activity removes the cytoplasmic portion of VAMP.

Several VAMP-like proteins have been localized to two stages of the secretory pathway in yeast. Most similar to VAMP is a protein that can suppress deletions of CAP (cyclase-associated protein), a protein required for ras activation of adenylate cyclase in yeast (98). This VAMP-like protein, termed SNC1 (supressor of null allele of CAP), is approximately 35% identical and 50% similar to mammalian VAMP. While SNC1 null mutants display no discernible phenotype (98), deletion of both SNC1 and a homologous gene SNC2 produces an accumulation of secretory vesicles reminiscent of sec4 mutants (99). This suggests that SNC1p acts in the late stages of secretion. Two other proteins, SLY2 (Sec22) and SLY12 (bet1), display a somewhat weaker homology to VAMP and are also linked to both ras-like proteins and secretion. These proteins were identified both as multicopy supressors of ypt1 deletions (100) and as genes required for ER to Golgi transport (101). Amino acid sequence homology between these VAMP-like proteins is localized to a 65 amino acid region preceding the membrane domain, which is predicted to form an alpha helix (100). Since overexpression of these VAMP-like proteins can overcome mutations in rab-like proteins, it suggests that the action of VAMP is downstream of rab action. It will be interesting to determine whether overexpression of SNC1p and SNC2p will suppress a sec4 mutation.

Anti-VAMP immunoreactivity has been detected in the synaptic-like vesicles of adipocytes that deliver the glucose transporter GLUT4 to the plasma membrane in response to insulin (102) indicating that VAMP is not specific to the nervous system. In addition, two other non-neural VAMP proteins have also been identified: 1) a VAMP homologue that is expressed exclusively in the gut of drosophila (103) and 2) a VAMP-like protein, cellubrevin, that is expressed in all cells and tissues tested and appears to localize to the Golgi (104). Therefore, like the small GTP-binding proteins, VAMP-like proteins appear to be a ubiquitous feature of eukaryotic membrane trafficking.

Syntaxin

Syntaxin was identified as a protein that co-immunoprecipitates with synaptotagmin from detergent extracts of crude synaptic vesicle preparations (105). The cDNAs encoding two highly homologous syntaxin proteins were subsequently identified. The predicted syntaxin proteins have a charged cytoplasmic domain and a hydrophobic region at the carboxy terminus that is predicted to be membrane spanning. Immunolabeling of primary neuronal cultures with antisyntaxin antibodies revealed that syntaxin is localized to the plasma membrane of synapses. Given its plasma membrane localization and association with synaptotagmin, syntaxin was hypothesized to serve as a vesicle-docking protein. This hypothesis is supported by the observation that immunoprecipitation of syntaxins from detergent-solubilized synaptosomes co-precipitates the N-type calcium channel (105). Also, one of several monoclonal antibodies generated against a protein complex containing an N-type calcium channel recognizes syntaxin (106) suggesting an interaction between these proteins. The localization of syntaxin to the plasma membrane, along with its association with both synaptotagmin and calcium channels, suggests that syntaxins dock synaptic vesicles near calcium channels where they will be exposed to high concentrations of calcium upon membrane depolarization.

Northern blot analyses revealed that syntaxins are expressed only in brain and spinal cord, suggesting that their function is neural specific. However, shortly after syntaxins were described, three yeast proteins with homology to syntaxin were identified; sed5p, pep12p, and sso1p. Most homologous to syntaxin is the sso1 gene product, a protein identified by its ability to suppress mutations of sec1, a late-acting secretory mutant (107). Sed5 mutants are deficient in ER to Golgi transport, (108) whereas pep12 mutants incorrectly sort vesicles targeted to the vacuole. Therefore, analogous to rab3 and VAMP, there appears to be a syntaxin-like molecule specific to each stage of the secretory pathway. The syntaxin-like proteins are most similar in a 70 amino acid region that precedes the predicted transmembrane domain. As in the shared domains of VAMP, this region is predicted to form an amphipathic helix.

The presence of syntaxin-like proteins at different stages of the secretory pathway suggests that a family of syntaxins is part of a generic vesicle transport machine. This interpretation was confirmed with the identification of several non-neuronal syntaxin proteins in rat (109). The brain-specific isoforms are denoted syntaxin 1A and 1B and the others syntaxin 2 through 5. Examination of the cellular localization of non-neuronal syntaxins revealed that syntaxins 2 and 4 are sorted to the plasma membrane when transfected into fibroblast cells, suggesting that some syntaxin isoforms may function in constitutive secretion. Unlike syntaxins 1–4, however, syntaxin 5 co-localizes with β-COP, a coatamer protein of the cis-Golgi, suggesting that syntaxin 5 functions earlier in the secretory pathway. The compartmental specificity of syntaxins along with the presence of several plasma membrane isoforms suggest that syntaxins may serve as vesicle receptor proteins. If so, disruption of syntaxin activity would be expected to prevent vesicle docking. Indeed, microinjection of anti-syntaxin 1A Fab fragments or a peptide corresponding to the conserved domain of the syntaxin proteins greatly diminished the stimulation-induced appearance of dopamine β-hydroxylase on the surface of neuroendocrine cells, suggesting that exocytosis was disrupted (109). It will therefore be interesting to determine whether syntaxin 5 peptides disrupt another stage of vesicular transport.

The presence of non-neuronal and yeast homologues of rab, VAMP, and syntaxin suggested that these proteins perform a generic function in the secretory pathway of animal cells. This conclusion was confirmed for syntaxin and VAMP with the discovery that they interact with soluble factors known to be required for membrane trafficking through the Golgi (110).

Transport of proteins from the cis- to medial Golgi requires at least three soluble proteins: an NEM-sensitive ATPase and two proteins required for the attachment of this ATPase to membranes. Termed NSF, and α- and γ-SNAP, these proteins can

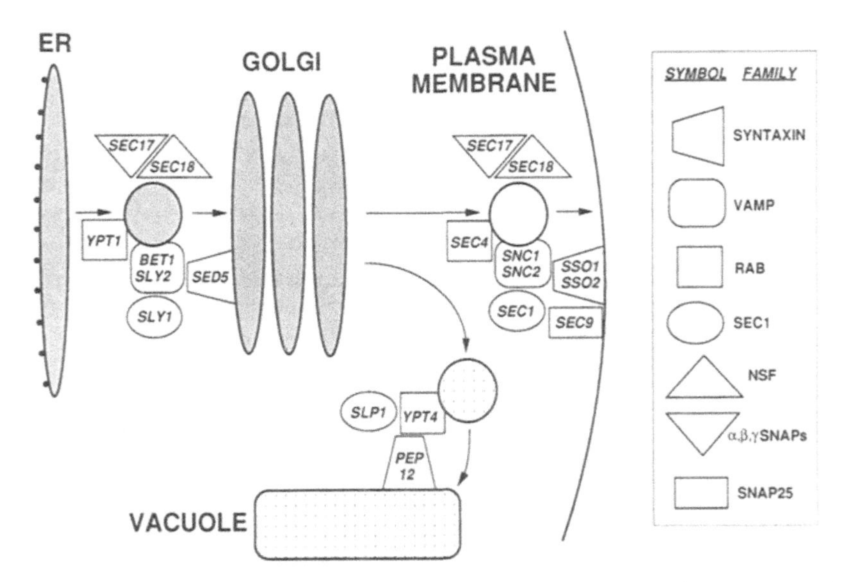

FIG. 2. The proteins mediating membrane traffic are conserved throughout the secretory pathway and through evolution. Proteins involved at each stage of the yeast secretory pathway are identified in terms of homologous families. Corresponding mammalian homologues are illustrated on the right. With the exception of the soluble proteins sec17p (SNAP) and sec18p (NSF), each stage of membrane transport has a specific version of each family member.

NSF, N-ethylmaleimide-sensitive fusion protein
SNAP, soluble NSF attachment protein

reconstitute Golgi transport in vitro. The observation that SNAP proteins bind membranes in a saturable manner suggested the presence of a SNAP receptor (111). To purify this receptor, epitope-tagged recombinant NSF and SNAPs were mixed with detergent-solubilized brain membranes and resolved on an antiepitope affinity column. This procedure isolated a large, 20S protein particle (110) that contained NSF, SNAPs, syntaxin, VAMP, and a protein previously found to be a component of synaptic terminals, SNAP 25 (synapse-associated protein 25) (112). Interestingly, there are yeast homologues of all members of the 20S particle and all appear to function in the yeast secretory pathway. The NSF and SNAP homologues are the sec18 and sec17 gene products, respectively. A yeast SNAP25 homologue has also been identified and is, as might be expected, a sec protein, sec9p (113).

Components of the yeast secretory pathway are displayed schematically in Fig. 2. It should be noted that the yeast proteins most homologous to synaptic vesicle proteins mediate transport from the late Golgi to the plasma membrane (Table 1), reinforcing the interpretation that the components of both constitutive and regulated secretion rely on the same basic mechanism. One of the proteins depicted in this schematic, sec1, has yet to be localized to nerve terminals, though mutations in a sec1 homologue in *C. elegans* produce animals apparently deficient in acetylcholine release (114,115). Other sec gene products have yet to be fit into this picture. Most

TABLE 1. *Yeast homologues of nerve terminal proteins*

Nerve terminal	Yeast
VAMP	SNC1, SNC2
syntaxin	SED5, PEP12, SSO1, SSO2
Rab	YPT1, SEC4
NSF	SEC18
αSNAP	SEC17
SNAP25	SEC9
Rop, Unc18	SEC1

NSF, N-ethylmaleimide-sensitive fusion protein
SNAP, soluble NSF attachment protein
VAMP, synaptobrevin

notable is sec14p, which has been identified as a phosphotidylinositol transfer protein (116) and which can substitute for a protein required to reconstitute secretion in semi-intact endocrine cells (117 and J. Hay, personal communication).

There are no known yeast homologues of synaptotagmin, which may indicate that it performs a function specific to regulated secretion. Most interesting is the possibility that synaptotagmin is both a targeting signal and inhibitor of constitutive membrane fusion. The absence of synaptotagmin from the 20S particle suggests that the syntaxin-synaptotagmin association must be disrupted for the 20S particle to form. One can imagine, then, two stages of vesicle association with the plasma membrane. In the first, a syntaxin-synaptotagmin complex exists as the vesicle docks with the membrane. A second stage, perhaps triggered by a signaling event, removes synaptotagmin and allows the assembly of the constitutive vesicle transport system (20S particle). This sequence of events is depicted in Fig. 3.

VESICLE FUSION—EXOCYTOSIS

The emerging similarities between the components of constitutive and regulated secretion suggest that regulation of exocytosis occurs via inhibition of constitutive processes. Several lines of evidence suggest that a series of readying steps precede regulated exocytosis, perhaps reflecting the sequential removal of inhibitors. A series of calcium requirements has been reported for regulated secretion in chromaffin cells. When cells are injected with caged calcium, exocytosis is observed to occur at a range of rates. Interestingly, secretion is faster in cells that experience a transient exposure to calcium prior to a calcium burst (118). A protein purified on the basis of its ability to restore calcium-stimulated secretion in endocrine cells may have a role in this process (119).

Examination of cell capacitance changes during exocytosis in mast cells has revealed that the earliest event is a small conductance similar in magnitude to that of an ion channel, and indicative of a pore-like structure between the lumen of the

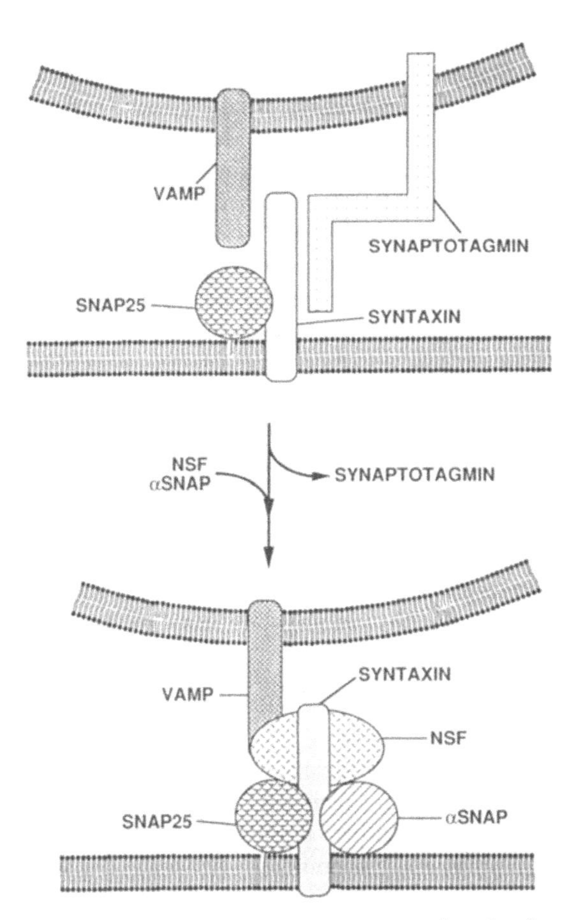

FIG. 3. Assembly of the fusion machinery at the synapse requires the dissociation of synaptotagmin and syntaxin. Synaptic vesicles are hypothesized to progress through two stages of membrane association; the first is characterized by a synaptotagmin-syntaxin interaction, and the second by a protein complex that includes soluble factors required for membrane fusion.

secretory vesicle and the extracellular domain (120). The conductance of this pore gradually increases though it can, up to a point, reverse. These observations suggest that exocytosis is initiated by a pore-like structure that precedes membrane continuity. Indeed, calculations of the percentage of vesicle contents that can diffuse through a pore during a reversible opening suggest that synaptic vesicles need not undergo complete exocytosis to effect secretion (117).

What is the nature of the fusion pore? If pore-mediated fusion is unique to regulated exocytosis, then any candidate pore protein would be expected to be a constituent unique to secretory vesicles. There are two candidates for such a fusion pore among the synaptic vesicle proteins: SV2 and synaptophysin. SV2 is homologous to a family of transporters that move substrates both inward (uptake proteins) and outward (antibiotic resistance proteins). It may function in conjunction with a cohort in the plasma membrane to create a pore. Synaptophysin has been found to form current-conducting structures when reconstituted into lipid bilayers (53). The conductance observed is similar in character to a channel activity measured in secreting primary pituitary cultures (121). Both secretion and channel activity in these cells demonstrated an identical biphasic calcium dependence, and both secretion and channel activity were inhibited by equal concentrations of TEA analogues and alcohols. Most significantly, channel activity could be inhibited by antisynaptophysin antibodies (122). Synaptophysin has been reported to interact with a plasma membrane protein that could serve as the other half of a fusion pore (54). However, nothing is known about the structure of this protein.

Alternatively, a fusion pore may be a common component of membrane trafficking. In this case, one might imagine that the protein components mediating targeting and docking would induce a lipidic pore by membrane disruption. Such disruption could be produced by a conformational change in the members of the 20S particle, which contains proteins spanning both vesicle and target membranes. The hydrolysis of ATP by NSF, which is required for vesicle transport, is a candidate for such an event. This hypothesis predicts that all NSF-mediated fusion is fast and that the rate-limiting step in regulated secretion is the removal of an inhibitory influence. In the case of neural secretion, this inhibitor may be synaptotagmin.

CONCLUSION

The identification of synaptic vesicle constituents has laid the foundation for functional analyses of secretion. Continued examination of the interactions and activities of the proteins discussed here promises a deeper understanding of the molecular events underlying neurotransmission. The speed of recent advances is testament to the progress that can be made when approaches to a problem converge. As a better picture of the molecular basis of transmitter release becomes available, new insights into its modulation are sure to arise. This modulation is likely to form the cellular basis of learning and memory.

ACKNOWLEDGMENTS

We thank Drs. Beverly Wendland and Tony Ting for reviewing the manuscript and Dr. Mark Bennett for helpful discussions.

REFERENCES

1. Heuser JE, Reese TS. Structure of the Synapse. In: Kandel ER, ed. *Handbook of physiology; the nervous system*. Bethesda, MD: American Physiological Society; 1977;261–294.
2. Trimble WS, Linial M, Scheller RH. Cellular and molecular biology of the presynaptic nerve terminal. *Annu Rev Neurosci* 1991;14:93–122.
3. Sudhof TC, Jahn R. Proteins of synaptic vesicles involved in exocytosis and membrane recycling. *Neuron* 1991;6:665–677.

4. Hall DH, Hedgecock EM. Kinesin-related gene unc-104 is required for axonal transport of synaptic vesicles in C elegans. *Cell* 1991;65:837–847.
5. Llinas R, Sugimori M, Lin JW, Leopold PL, Brady ST. ATP-dependent directional movement of rat synaptic vesicles injected into the presynaptic terminal of squid giant synapse. *Proc Natl Acad Sci USA* 1989;86:5656–5660.
6. Regnier-Vigouroux A, Tooze SA, Huttner WB. Newly synthesized synaptophysin is transported to synaptic-like microvesicles via constitutive secretory vesicles and the plasma membrane. *Embo J* 1991;10:3589–3601.
7. Cameron PL, Sudhof TC, Jahn R, De Camilli P. Colocalization of synaptophysin with transferrin receptors: implications for synaptic vesicle biogenesis. *J Cell Biol* 1991;115:151–164.
8. Linstedt AD, Kelly RB. Synaptophysin is sorted from endocytotic markers in neuroendocrine PC12 cells but not transfected fibroblasts. *Neuron* 1991;7:309–317.
9. Mundigl O, Matteoli M, Daniels L, et al. Synaptic vesicle proteins and early endosomes in cultured hippocampal neurons: differential effects of brefeldin A in axon and dendrites. *J Cell Biol* 1993; 122:1207–1221.
10. Cutler DF, Cramer LP. Sorting during transport to the surface of PC12 cells: divergence of synaptic vesicle and secretory granule proteins. *J Cell Biol* 1990;110:721–730.
11. Leube RE, Wiedenmann B, Franke WW. Topogenesis and sorting of synaptophysin: synthesis of a synaptic vesicle protein from a gene transfected into nonneuroendocrine cells. *Cell* 1989;59:433–446.
12. Greengard P, Valtorta F, Czernik AJ, Benfenati F. Synaptic vesicle phosphoproteins and regulation of synaptic function. *Science* 1993;259:780–785.
13. Forgac M. Structure and properties of the coated vesicle proton pump. *Ann N Y Acad Sci* 1992; 671:273–283.
14. Forgac M. Structure, function and regulation of the coated vesicle V-ATPase. *J Exp Biol* 1992; 172:155–169.
15. Floor E, Leventhal PS, Schaeffer SF. Partial purification and characterization of the vacuolar H(+)-ATPase of mammalian synaptic vesicles. *J Neurochem* 1990;55:1663–1670.
16. Moriyama Y, Maeda M, Futai M. The role of V-ATPase in neuronal and endocrine systems. *J Exp Biol* 1992;172:171–178.
17. Moriyama Y, Iwamoto A, Hanada H, Maeda M, Futai M. One-step purification of Escherichia coli H(+)-ATPase (F0F1) and its reconstitution into liposomes with neurotransmitter transporters. *J Biol Chem* 1991;266:22141–22146.
18. Puopolo K, Kumamoto C, Adachi I, Forgac M. A single gene encodes the catalytic "A" subunit of the bovine vacuolar H(+)-ATPase. *J Biol Chem* 1991;266:24564–24572.
19. Puopolo K, Kumamoto C, Adachi I, Magner R, Forgac M. Differential expression of the "B" subunit of the vacuolar H(+)-ATPase in bovine tissues. *J Biol Chem* 1992;267:3696–3706.
20. Südhof TC, Fried VA, Stone DK, Johnson PA, Xie X-S. Human endomembrane pump strongly resembles the ATP-synthetase of archaebacteria. *Proc Natl Acad Sci USA* 1989;86:6067–6071.
21. Mandel M, Moriyama Y, Hulmes JD, Pan Y-C, Nelson N. cDNA sequence encoding the 16-kDa proteolipid of chromaffin granules implies gene duplication in the evolution of H+-Atpases. *Proc Natl Acad Sci USA* 1988;85:5521–5524.
22. Hanada H, Moriyama Y, Maeda M, Futai M. Kinetic studies of chromaffin granule H+-ATPase and effects of bafilomycin A1. *Biochem Biophys Res Commun* 1990;170:873–878.
23. Perin MS, Fried VA, Stone DK, Xie X-S, Südhof TC. Structure of the 116-kDa polypeptide of the clathrin-coated vesicle/synaptic vesicle proton pump. *J Biol Chem* 1991;266:3877–3881.
24. Mellman I. The importance of being acid: the role of acidification in intracellular membrane traffic. *J Exp Biol* 1992;172:39–45.
25. Myers M, Forgac M. The coated vesicle vacuolar (H+)-ATPase associates with and is phosphorylated by the 50-kDa polypeptide of the clathrin assembly protein AP-2. *J Biol Chem* 1993;268: 9184–9186.
26. Hell JW, Edelmann L, Hartinger J, Jahn R. Functional reconstitution of the gamma-aminobutyric acid transporter from synaptic vesicles using artificial ion gradients. *Biochemistry* 1991;30:11795–11800.
27. Maycox PR, Deckwerth T, Jahn R. Bacteriorhodopsin drives the glutamate transporter of synaptic vesicles after co-reconstitution. *Embo J* 1990;9:1465–1469.
28. Erickson JD, Eiden LE, Hoffman BJ. Expression cloning of a reserpine-sensitive vesicular monoamine transporter. *Proc Natl Acad Sci USA* 1992;89:10993–10997.

29. Liu Y, Peter D, Roghani A, et al. A cDNA that suppresses MPP+ toxicity encodes a vesicular amine transporter. *Cell* 1992;539–51.
30. Amara SG, Kuhar MJ. Neurotransmitter transporters: recent progress. *Annu Rev Neurosci* 1993; 16:73–93.
31. Alfonso A, Grundahl K, Duerr JS, Han H-P, Rand JB. The caenorhabditis elegans unc-17 gene: A putative vesicular acetylcholine transporter. *Science* 1993;261:617–619.
32. Maycox PR, Hell JW, Jahn R. Amino acid neurotransmission: spotlight on synaptic vesicles. *Trends in Biochem* 1990;13:83–87.
33. Burger PM, Hell J, Mehl E, Krasel C, Lottspeich F, Jahn R. GABA and glycine in synaptic vesicles: storage and transport characteristics. *Neuron* 1991;7:287–293.
34. Shioi J, Ueda T. Artificially imposed electrical potentials drive L-glutamate uptake into synaptic vesicles of bovine cerebral cortex. *Biochem J* 1990;267:63–68.
35. Tabb JS, Kish PE, Van Dyke R, Ueda T. Glutamate transport into synaptic vesicles. Roles of membrane potential, pH gradient, and intravesicular pH. *J Biol Chem* 1992;267:15412–15418.
36. Hell JW, Maycox PR, Jahn R. Energy dependence and functional reconstitution of the gamma-aminobutyric acid carrier from synaptic vesicles. *J Biol Chem* 1990;265:2111–2117.
37. Xie X-S, Crider BP, Stone DK. Isolation and reconstitution of the chloride transporter of clathrin-coated vesicles. *J Biol Chem* 1989;264:18870–18873.
38. Rahamimoff R, DeRiemer SA, Sakman B, Stadler H, Yakir N. Ion channels in synaptic vesicles from Torpedo electric organ. *Proc Natl Acad Sci USA* 1988;85:5310–5314.
39. Buckley K, Kelly RB. Identification of a transmembrane glycoprotein specific for secretory vesicles of neural and endocrine cells. *J Cell Biol* 1985;100:1284–1294.
40. Bajjalieh SM, Peterson K, Shinghal R, Scheller RH. SV2, a brain synaptic vesicle protein homologous to bacterial transporters. *Science* 1992;257:1271–1273.
41. Feany MB, Lee S, Edwards RH, Buckley KM. The synaptic vesicle protein SV2 is a novel type of transmembrane transporter. *Cell* 1992;70:861–867.
42. Gingrich JA, Anderson PH, Tiberi M, et al. Identification, characterization, and molecular cloning of a novel transporter-like protein localized to the central nervous system. *FEBS* 1992;312:115–122.
43. Maiden MCJ, Davis EO, Baldwin SA, Moore DCM, Henderson PJF. Mammalian and bacterial sugar transport proteins are homologous. *Nature* 1987;325:641–643.
44. Scranton TW, Iwata M, Carlson SS. The SV2 protein of synaptic vesicles is a keratan sulfate proteoglycan. *J Neurochem* 1993;61:29–44.
45. Bajjalieh SM, Peterson K, Linial M, Scheller RH. Brain contains two forms of synaptic vesicle protein 2. *Proc Natl Acad Sci USA* 1993;90:2150–2154.
46. Bajjalieh SM, Franz G, Weimann JM, McConnell SK, Scheller RH. Differential expression of synaptic vesicle protein 2 (SV2) isoforms. *J Neurosci* 1994; in press.
47. Collins FS. Cystic fibrosis: molecular biology and therapeutic implications. *Science* 1992;256: 774–779.
48. Valverde MA, Diaz M, Sepulveda FV, Gill DR, Hyde SC, Higgins CF. Volume-regulated chloride channels associated with the human multidrug-resistance P-glycoprotein. *Nature* 1992;355:830–833.
49. Wiedenmann B, Franke WW. Identification and localization of synaptophysin, and integral membrane glycoprotein of Mr38,000 characteristic of presynaptic vesicles. *Cell* 1985;41:1017–1028.
50. Buckley KM, Floor E, Kelly RB. Cloning and sequence analysis of cDNA encoding p38, a major synaptic vesicle protein. *J Cell Biol* 1987;105:2447–2456.
51. Südhof TC, Lottspeich F, Greengard P, Mehl E, Jahn R. A synaptic vesicle protein with a novel cytoplasmic domain and four transmembrane regions. *Science* 1987;238:1142–1144.
52. Knaus P, Marqueze-Pouey B, Scherer H, Betz H. Synaptoporin, a novel putative channel protein of synaptic vesicles. *Neuron* 1990;5:453–462.
53. Thomas L, Hartung K, Langosssch D, et al. Identification of synaptophysin as a hexameric channel protein of the synaptic vesicle membrane. *Science* 1988;242:1050–1053.
54. Thomas L, Betz H. Synaptophysin binds to physophilin, a putative synaptic plasma membrane protein. *J Cell Biol* 1990;111:2041–2052.
55. Alder J, Xie ZP, Valtorta F, Greengard P, Poo M. Antibodies to synaptophysin interfere with transmitter secretion at neuromuscular synapses. *Neuron* 1992;9:759–768.
56. Lowe AW, Madeddu L, Kelly RB. Endocrine secretory granules and neuronal synaptic vesicles have three integral membrane proteins in common. *J Cell Biol* 1988;106:51–59.

57. Floor E, Feist BE. Most synaptic vesicles isolated from rat brain carry three membrane proteins, SV2, synaptophysin, and p65. *J Neurochem* 1989;52:1433–1437.
58. Perin MS, Fried VA, Mignery GA, Jahn R, Sudhof TC. Phospholipid binding by a synaptic vesicle protein homologous to the regulatory region of protein kinase C. *Nature* 1990;345:260–263.
59. Wendland B, Miller KG, Schilling J, Scheller RH. Differential expression of the p65 gene family. *Neuron* 1991;6:993–1007.
60. Geppert M, Archer B, Sudhof TC. Synaptotagmin II. A novel differentially distributed form of synaptotagmin. *J Biol Chem* 1991;266:13548–13552.
61. Clark JD, Lin LL, Kriz RW, et al. A novel aracidonic acid-selective cytosolic PLA2 contains a Ca2+-dependent translocation domain with homology to PKC and GAP. *Cell* 1991;65:1043–1051.
62. Brose N, Petrenko AG, Sudhof TC, Jahn R. Synaptotagmin: a calcium sensor on the synaptic vesicle surface. *Science* 1992;256:1021–1025.
63. Hannun YA, Bell RM. Phorbol ester binding and activation of protein kinase C on triton X-100 mixed micelles containing phosphatidylserine. *J Biol Chem* 1986;261:9341–9347.
64. Mochly-Rosen D, Miller KG, Scheller RH, Khaner H, Lopez J, Smith BL. p65 fragments, homologous to the C2 region of protein kinase C, bind to the intracellular receptors for protein kinase C. *Biochemistry* 1992;31:8120–8124.
65. Davletov B, Sontag JM, Hata Y, et al. Phosphorylation of synaptotagmin I by casein kinase II. *J Biol Chem* 1993;268:6816–6822.
66. Bennett MK, Miller KG, Scheller RH. Casein kinase II phosphorylates the synaptic vesicle protein p65. *J Neurosci* 1993;13:1701–1707.
67. Popoli M. Synaptotagmin is endogenously phosphorylated by Ca2+/calmodulin protein kinase II in synaptic vesicles. *FEBS* 1993;317:85–88.
68. Takahashi M, Arimatsu Y, Fujita S, et al. Protein kinase C and Ca2+/calmodulin-dependent protein kinase II phosphorylate a novel 58-kDa protein in synaptic vesicles. *Brain Res* 1991;551: 279–292.
69. Shoji-Kasai Y, Yoshida A, Sato K, et al. Neurotransmitter release from Synaptotagmin-deficient clonal variants of PC12 cells. *Science* 1992;256:1820–1823.
70. DiAntonio A, Parfit K, Schwarz TL. Synaptic transmission persists in synaptotagmin mutants of drosophila. *Cell* 1993;73:1281–1290.
71. Nonet ML, Grundahl K, Meyer BJ, Rand JB. Synaptic function is impaired but not eliminated in C *elegans* mutants lacking synaptotagmin. *Cell* 1993;73:1291–1305.
72. Elferink LA, Peterson MR, Scheller RH. A role for synaptotagmin (p65) in regulated exocytosis. *Cell* 1993;72:153–159.
73. Bommert K, Charlton MP, DeBello WM, Chin GJ, Betz H, Augustine GJ. Inhibition of neurotransmitter release by C2-domain peptides implicates synaptotagmin in exocytosis. *Nature* 1993; 363:163–165.
74. Shirataki H, Kaibuchi K, Sakoda T, et al. Rabphilin-3A, a putative target protein for smg p25A/rab3A p25 small GTP-binding protein related to synaptophysin. *Mol Cell Biol* 1993;13:2061–2068.
75. Ushkaryov YA, Petrenko AG, Geppert M, Sudhof TC. Neurexins: synaptic cell surface proteins related to the alpha-latrotoxin receptor and laminin. *Science* 1992;257:50–56.
76. Pfeffer SR. GTP-binding proteins in intracellular transport. *Trends Cell Biol* 1992;2:41–46.
77. Bourne HR, Sanders DA, McCormick F. The GTPase superfamily: conserved structure and molecular mechanism. *Nature* 1991;349:117–127.
78. Fischer von Mollard G, Mignery GA, Baumert M, et al. rab3 is a small GTP-binding protein exclusively localized to synaptic vesicles. *Proc Natl Acad Sci USA* 1990;87:1988–1992.
79. Elferink LA, Anzai K, Scheller RH. rab15, a novel low molecular weight GTP-binding protein specifically expressed in rat brain. *J Biol Chem* 1992;267:5768–5775.
80. Chavrier P, Parton RG, Hauri HP, Simons K, Zerial M. Localization of low molecular weight GTP binding proteins to exocytic and endocytic compartments. *Cell* 1990;62:317–329.
81. Salminen A, Novick PJ. A ras-like protein is required for a post-Golgi event in yeast secretion. *Cell* 1987;49:527–538.
82. Segev N, Mulholland J, Botstein D. The yeast GTP-binding YPT1 protein and a mammalian counterpart are associated with the secretion machinery. *Cell* 1988;52:915–924.
83. Bacon RA, Salminen A, Ruohola H, Novick P, Ferro-Novick S. The GTP-binding protein Ypt1 is required for transport in vitro: the Golgi apparatus is defective in ypt1 mutants. *J Cell Biol* 1989; 109:1015–1022.

84. Segev N. Mediation of the attachment or fusion step in vesicular transport by the GTP-binding Ypt1 protein. *Science* 1991;252:1553–1556.
85. Walworth NC, Goud B, Kabcenell AK, Novick PJ. Mutational analysis of SEC4 suggests a cyclical mechanism for the regulation of vesicular traffic. *Embo J* 1989;8:1685–1693.
86. Bourne HR. Do GTPases direct membrane traffic in secretion? *Cell* 1988;53:669–671.
87. Brennwald P, Novick P. Interactions of three domains distinguishing the Ras-related GTP-binding proteins Ypt1 and Sec4. *Nature* 1993;362:560–563.
88. Dunn B, Sterns T, Botstein D. Specificity domains distinguish the RAS-related GTPases Ypt1 and Sec4. *Nature* 1993;362:563–565.
89. Lledo P-M, Vernier P, Vincent J-D, Mason W, Zorec R. Inhibition of Rab3B expression attenuates Ca2+-dependent exocytosis in rat anterior pituitary cells. *Nature* 1993;364:540–544.
90. Ngsee JK, Fleming AM, Scheller RH. A rab protein regulates the localization of secretory granules in AtT-20 cells. *Mol Biol Cell* 1993;4:747–756.
91. Trimble WS, Cowan DM, Scheller RH. VAMP-1: a synaptic vesicle-associated integral membrane protein. *Proc Natl Acad Sci USA* 1988;85:4538–4542.
92. Elferink LA, Trimble WS, Scheller RH. Two vesicle-associated membrane protein genes are differentially expressed in the rat central nervous system. *J Biol Chem* 1989;264:11061–11064.
93. Baumert M, Maycox PR, Navone F, De Camilli P, Jahn R. Synaptobrevin: an integral membrane protein of 18,000 daltons present in small synaptic vesicles of rat brain. *Embo J* 1989;8:379–384.
94. Trimble WS, Gray TS, Elferink LA, Wilson MC, Scheller RH. Distinct patterns of expression of two VAMP genes within the rat brain. *J Neurosci* 1990;10:1380–1387.
95. Schiavo G, Benfenati F, Poulain B, et al. Tetanus and botulinum-B neurotoxins block neurotransmitter release by proteolytic cleavage of synaptobrevin [see comments]. *Nature* 1992;359: 832–835.
96. Schiavo G, Poulain B, Rossetto O, Benfenati F, Tauc L, Montecucco C. Tetanus toxin is a zinc protein and its inhibition of neurotransmitter release and protease activity depend on zinc. *Embo J* 1992;11:3577–3583.
97. Link E, Edelmann L, Chou JH, et al. Tetanus toxin action: inhibition of neurotransmitter release linked to synaptobrevin proteolysis. *Biochem Biophys Res Commun* 1992;189:1017–1023.
98. Gerst JE, Rodgers L, Riggs M, Wigler M. SNC1, a yeast homolog of the synaptic vesicle-associated membrane protein/synaptobrevin gene family: genetic interactions with the RAS and CAP genes [published erratum appears in *Proc Natl Acad Sci USA* 1992 Aug 1;89(15):7287]. *Proc Natl Acad Sci USA* 1992;89:4338–4342.
99. Protopopov V, Govindan B, Novick P, Gerst JE. Homologs of the synaptobrevin/VAMP family of synaptic vesicle proteins function on the late secretory pathway in S. cerevisiae. *Cell* 1993;74:855–861.
100. Dascher C, Ossig R, Gallwitz D, Schmitt HD. Identification and structure of four yeast genes (SLY) that are able to suppress the functional loss of YPT1, a member of the RAS superfamily. *Mol Cell Biol* 1991;11:872–885.
101. Newman AP, Shim J, Ferro-Novick S. BET1, BOS1, and SEC22 are members of a group of interacting yeast genes required for transport from the endoplasmic reticulum to the Golgi complex. *Mol Cell Biol* 1990;10:3405–3414.
102. Cain CC, Trimble WS, Lienhard GE. Members of the VAMP family of synaptic vesicle proteins are components of glucose transporter-containing vesicles from rat adipocytes. *J Biol Chem* 1992; 267:11681–11684.
103. Chin AC, Burgess RW, Wong BR, Schwarz TL, Scheller RH. Differential expression of transcripts from syb, a Drosophila melanogaster gene encoding VAMP (synaptobrivin) that is abundant in non-neuronal cells. *Gene* 1993 Sept 15; 131(2):175–181.
104. McMahon HT, Ushkaryov YA, Edelmann L, et al. Cellubrevin is a ubiquitous tetanus-toxin substrate homologous to a putative synaptic vesicle fusion protein [see comments]. *Nature* 1993; 364:346–349.
105. Bennett MK, Calakos N, Scheller RH. Syntaxin: a synaptic protein implicated in docking of synaptic vesicles at presynaptic active zones. *Science* 1992;257:255–259.
106. Yoshida A, Oho C, Omori A, Kuwahara R, Ito T, Takahashi M. HPC-1 is associated with synaptotagmin and omega-conotoxin receptor. *J Biol Chem* 1992;267:24925–24928.
107. Bennett MK, Scheller RH. The molecular machinery for secretion is conserved from yeast to neurons. *Proc Natl Acad Sci USA* 1993;90:2559–2563.
108. Hardwick KG, Pelham HR. SED5 encodes a 39-kD integral membrane protein required for vesicular transport between the ER and the Golgi complex. *J Cell Biol* 1992;119:513–521.

109. Bennett MK, Garcia-Arraras JE, Elferink LA, et al. The syntaxin family of vesicular transport receptors. *Cell* 1993;74(5):863–873.
110. Sollner T, Whiteheart SW, Brunner M, et al. SNAP receptors implicated in vesicle targeting and fusion [see comments]. *Nature* 1993;362:318–324.
111. Whiteheart SW, Brunner M, Wilson DW, Wiedmann M, Rothman JE. Soluble N-ethylmaleimide-sensitive fusion attachment proteins (SNAPs) bind to a multi-SNAP receptor complex in Golgi membranes. *J Biol Chem* 1992;267:12239–12243.
112. Oyler GA, Higgins GA, Hart RA, et al. The identification of a novel synaptosomal-associated protein, SNAP-25, differentially expressed by neuronal subpopulations. *J Cell Biol* 1989;109: 3039–3052.
113. Cleves AE, Bankaitis VA. Secretory pathway function in *Saccharomyces cerevisiae*. *Adv Micro Physiol* 1992;33:73–144.
114. Hosono R, Hekimi S, Kamiya Y, et al. The unc-18 gene encodes a novel protein affecting the kinetics of acetylcholine metabolism in the nematode caenorhabditis elegans. *J Neurochem* 1992; 58:1517–1525.
115. Pelham HRB. Neurotransmission and Secretion. *Nature* 1993,364:582.
116. Bankaitis VA, Aitken JR, Cleves AE, Dowhan W. An essential role for a phospholipid transfer protein in yeast Golgi function [see comments]. *Nature* 1990;347:561–562.
117. Alvarez de Toledo G, Fernandez-Chacon R, Fernandez JM. Release of secretory products during transient vesicle fusion [see comments]. *Nature* 1993;363:554–558.
118. Neher E, Zucker RS. Multiple calcium-dependent processes related to secretion in bovine chromaffin cells. *Neuron* 1993;10:21–30.
119. Walent JH, Porter BW, Martin TFJ. A novel 145 kd brain cytosolic protein reconstitutes Ca^{2+}–regulated secretion in permeable neuroendocrine cells. *Cell* 1992;70:765–775.
120. Almers W, Tse FW. Transmitter release from synapses: does a preassembled fusion pore initiate exocytosis? *Neuron* 1990;4:813–818.
121. Lee CJ, Dayanithi G, Nordmann JJ, Lemos JR. Possible role during exocytosis of a Ca(2+)-activated channel in neurohypophysial granules. *Neuron* 1992;8:335–342.
122. Lemos JR, Lee CJ, Dayanithi G, Nordmann JJ. Possible role during exocytosis of synaptophysin-like ca^{2+}-activated channel in neurosecretory granules of neurohypophysis. *Soc Neurosci Abs* 1992;18:248.5.
123. Feany MB; Yee AG; Delvy ML; Buckley KM. The synaptic vesicle proteins SV2, synaptotagmin and synaptophysin are sorted to separate cellular compartments in CHO fibroblasts. *J Cell Biol* 1993;123:575–584.

Molecular and Cellular Mechanisms of Neurotransmitter Release, edited by Lennart Stjärne, Paul Greengard, Sten Grillner, Tomas Hökfelt, and David Ottoson, Raven Press, Ltd., New York © 1994.

5

Intracellular Membrane Fusion

James E. Rothman

Department of Cellular Biophysics and Biochemistry, Memorial Sloan-Kettering Cancer Center, New York, New York 10021

The secretory pathway of eukaryotic cells underlies the targeting of newly synthesized proteins, generation and maintenance of subcellular compartments and organelles, and proper cell growth and division. Insight into this complex maze of vesicular transport pathways has come in several waves, beginning with the pioneering work of the 1960s (1) in which the fundamental relationship of the endoplasmic reticulum, Golgi, and secretory storage vesicles in regulated secretion was deduced, and the need for vesicular carriers to ferry cargo between topologically separate compartments was first recognized. These now central paradigms of cell biology were extended to the formation of plasma membranes, lysosomes, and other organelles in the 1970s, and continue to be refined as ever more sophisticated techniques of immunocytochemistry flesh out and better define subcompartments. In the 1980s, elucidation of the molecular machinery of vesicular transport began with the reconstitution of intracellular transport in cell-free systems from animal cells (2) and the isolation of secretory mutants in yeast (3).

The fruits of genetics and biochemistry are teaching us that many of the components of the secretory pathway are universal; the same machinery operates in both yeast and animals. The same enzyme system that fuses a vesicle with the Golgi also fuses endocytic vesicles. The coats that pinch off endocytic and Golgi vesicles have underlying similarities not evident from their morphology. Here I will focus on the transport pathways that operate within the endoplasmic reticulum (ER) and Golgi, and only touch on the related subjects of secretory granule formation, regulated exocytosis, lysosome biogenesis, and sorting to apical and basolateral domains in polarized cells.

MACHINERY FOR VESICULAR TRANSPORT IN THE GOLGI

The central mechanistic questions posed by the cell biology just discussed concern how transport vesicles bud and fuse. These questions are the same whether the direction of movement is anterograde or retrograde, and independent of the exact number of transfers needed to transit the Golgi. What force would drive a vesicle to bud off from a membrane? On what basis does each vesicle choose its target for fusion to achieve the specificity inherent in a strictly ordered series of unidirectional transfers? What accounts for the rapid fusion of the transport vesicle with its target membrane, whereas spontaneous fusion between lipid bilayers is extraordinarily slow under physiological conditions? The resolution of these questions in the language of protein-protein interactions constitutes a fundamental understanding of the secretory pathway. The identification and study of the proteins responsible for budding and fusion, based on biochemical approaches using cell-free reconstituted transport systems from animal cells, and/or genetic approaches with the yeast *S. cerevisiae*, are rapidly leading to concordant answers.

The molecular nature of the transport vesicles and information concerning the mechanisms of budding and fusion has been most clearly delineated for the process of intercisternal transport in the Golgi stack, and so will be reviewed first, although studies are proceeding rapidly on many other fronts (4,5). When isolated Golgi stacks are incubated with cytosol (100,000 g supernatant) and adenosine 5′-triphosphate (ATP), numerous vesicles approximately 75 nm in diameter form from every cisterna (6). These are likely to be transport vesicles, as they contain a transported protein (the VSV-encoded G protein) present when Golgi fractions are prepared from VSV-infected cells (7). The concentration of G protein in the 75 nm vesicles is indistinguishable from that of the parental cisternae, and vesicle budding is unlinked to the presence of cargo molecules (8) both of these situations are hallmarks expected of carriers of the anterograde bulk flow. Completion of a round of transport of VSV G protein between cisternae can be monitored by transport-coupled glycosylation (8) (i.e., the maturation of the glycoprotein's saccharide chains following transfer to the next compartment, which routinely employs N-Acetylglucosamine as the sugar tag, added in the *medial* (9) compartment of the Golgi stack). Transport-coupled glycosylation, like the production of 75 nm VSV G protein-containing vesicles, requires both cytosol and ATP. A considerable body of evidence implies that transport in the in vitro system involves vesicle movements between Golgi stacks that themselves remain intact, and maintains the compartmental specificity that exists in vivo (10–12).

The 75 nm vesicles come in two varieties: 1) those coated on their outer (cytoplasmic) surface with an electron-dense material (not containing clathrin), and 2) those uncoated (7). The coated vesicles are the precursors of the uncoated vesicles. Transport-coupled glycosylation is blocked by nonhydrolyzable analogues of guanosine 5′-triphosphate (GTP), like GTPγS (13), and also by pretreatment of the Golgi membranes with the sulphydryl reagent N-ethylmaleimide (NEM) under mild conditions (14). The GTPγS blockade leads to the accumulation of coated buds and coated vesicles (13), whereas NEM treatment blocks differently, leading to the accumulation of the uncoated variety of 75 nm vesicles (15). It follows that coated vesicles are precursors to uncoated vesicles, because biochemical experiments showed that the GTPγS block precedes the NEM-block to transport, and combining both treatments leads to the exclusive accumulation of coated vesicles (16).

Altogether, these findings suggest a pathway in which coat proteins assemble on a Golgi cisterna to form coated vesicles that then transfer to the next cisternal compartment, lose their coats (in a reaction that likely requires or is linked to GTP hydrolysis), and fuse (in a reaction that requires an NEM-sensitive protein). The predicted NEM-sensitive fusion protein (NSF) was purified (17) from cytoplasmic extracts by using its ability to restore transport (e.g., fusion) to NEM-blocked Golgi membranes as a guide, yielding a homotetramer of 76 kd subunits that requires ATP for stability. Both budding (18) and (19) fusion require ATP, and each also requires fatty acyl-coenzyme A, for purposes that are still unclear.

The Golgi-derived coated vesicles have been purified from salt extracts of Golgi fractions incubated in vitro to accumulate the coated vesicle with GTPγS (20,21). They contain a number of major and stoichiometric components that are externally disposed and together comprise the coat. These coat proteins (COPS) include α-COP (170 kd), β-COP (110 kd), γ-COP (98 kd), δ-COP (61 kd), and a small GTP-binding protein called (22) adenosine 5′-diphosphate (ADP)-ribosylation Factor (ARF, 21 kd) because of its discovery as a cofactor in the cholera toxin-catalyzed ADP-ribosylation of stimulatory G protein alpha subunits (23). The ARF was first suggested to be involved in secretion because of secretory phenotypes in yeast and because of its localization (when bound to membranes) to the Golgi stack in animal cells (24). Immunocytochemistry has localized β-COP and ARF to the coated but not the uncoated 75 nm Golgi-derived vesicles (21,22,25). The coat is virtually absent from Golgi membranes before in vitro transport begins, and little if any β-COP or ARF is present on membranes, the bulk being in cytosol (7,22,26). Thus, assembly of coated vesicles must occur from a dispersed pool of cytosolic subunits at the Golgi membrane surface.

FUSION OF TRANSPORT VESICLES WITH GOLGI CISTERNAE AND FUSION IN GENERAL

What is the basis for the specificity of intracellular membrane compartments? As for transport between Golgi cisternae, transactions among the membrane-bound compartments in the cytoplasm of eukaryotic cells are generally executed by transport vesicles that bud from one membrane and then fuse selectively with another (1). The mechanism by which each vesicle chooses its target is currently unknown, but must embody the essence of compartmental specificity. Choice of target is implicit in every cellular fusion process, from the ubiquitous steps in the constitutive secretory and endocytic pathways (such as the fusion of an ER-derived transport vesicle with the Golgi) to specialized and tightly regulated forms of exocytosis (such as the triggered fusion of a neurotransmitter-containing synaptic vesicle at an axonal synapse).

We now describe (27) a fundamental relationship among these various fusion processes, suggesting that a general machinery is employed to execute the act of fusion, with specificity being established by a complementary pairing of compartment-specific partner proteins of the transport vesicle with counterparts at the target membrane. In particular, we have utilized three purified components of the general fusion machinery (28) (NSF, α-soluble NSF attachment protein (SNAP), and γ-SNAP) to isolate from brain membranes those proteins (SNAP receptors) required to form the 20S particles previously implicated in fusion in the secretory pathway (29,30). We obtain (27) in highly purified form a complex containing integral membrane proteins originating from both neurotransmitter-containing synaptic vesicles (synaptobrevin (VAMP) (31–33) and from the target presynaptic membranes (syntaxin (34) as well as synaptosome-associated protein-25 (35) a palmitylated protein (36) also localized to the synapse). Syntaxin has been previously implicated in vesicle docking, as it co-immunoprecipitates with synaptotagmin, a synaptic vesicle membrane component (34). Synaptobrevin has been implicated in vesicle fusion, as it represents a site of action of certain proteolytic neurotoxins (37). These SNAP receptors (abbreviated SNAREs, also connoting "a device for capturing small game" (38)) are members of multigene families (39–41) that we now propose are specialized into two subtypes: 1) v-SNAREs attached to transport vesicles and 2) t-SNAREs that define target membranes.

The N-ethylmaleimide sensitive fusion protein (NSF), a soluble tetramer of 76 kd subunits, was discovered (42) and purified (43) based on its ability to restore intercisternal Golgi transport in a cell-free system (44,45). The NSF is required for fusion, because in its absence transport vesicles accumulate at the acceptor membrane (46,47). In yeast, NSF is encoded by the sec18 gene (48) originally defined according to its need in ER-to-Golgi transport in living cells (49) and subsequently shown to cause the accumulation of transport vesicles when defective (50). The sec18 protein can functionally replace NSF in animal cell-free Golgi transport (48) and recently sec18 function has been found to be required at every discernable step of the secretory pathway (51) in yeast in vivo, extending earlier findings that NSF is required for ER-to-Golgi transport and in endocytic vesicle fusion in animal cell-free systems (52,53). Thus, compelling evidence exists that NSF is generally required for intracellular fusion processes.

The NSF requires distinct cytoplasmic factors to attach to Golgi membranes (54). Three species of monomeric SNAP (α-SNAP (35 kd), β-SNAP (36 kd), and γ-SNAP (39 kd)), were purified from brain (55,56) and SNAP activity is required together with NSF for fusion in vitro (56). As α-SNAP (but not β- or γ-SNAPs) can restore animal cell Golgi transport activity to cytosol prepared from sec17 mutant yeast, we proposed (56) that the sec17 gene encodes α-SNAP in yeast. This conclusion was strengthened by the recent finding that sec17 protein is functionally equivalent to α-SNAP (57) and is now directly established (58). Together with the fact that sec17 function is needed for ER-to-Golgi transport in yeast, accumulating transport vesicles when defective (50) this establishes that SNAPs, like NSF, are components of a general intracellular membrane fusion machinery that is employed in living cells and conserved in evolution.

The SNAP proteins bind to distinct sites in membranes that up until now have only been operationally defined in Golgi membranes. The NSF does not bind to SNAPs in solution, but rather requires prior attachment of SNAPs to SNAP receptors in membranes to interact (30). The α-SNAP and β-SNAP compete for binding to the same SNAP receptor site with low nm affinity; γ-SNAP binds to a noncompetitive site in the same complex (30). While not essential for NSF binding to membranes, γ-SNAP increases the affinity of SNAP-SNAP receptor complexes for NSF (29). Crosslinking studies suggested that the SNAP receptor contains an α-SNAP binding subunit of 30–40 kd (30). When the membrane-bound NSF-SNAP-SNAP receptor complexes are solubilized with detergent, they sediment as a distinct multisubunit particle at 20S (29) that has been proposed to function as the core of a generalized "fusion machine" capable of catalyzing bilayer fusion at its point of assembly (46,56). The 20S fusion particles can also assemble following solubilization of the SNAP receptor from Golgi membranes when the detergent extract is mixed with SNAPs and NSF. When NSF and SNAPs are added in excess, all of the "assembly factor" activity (i.e., SNAP receptor content) of the membrane extract is incorporated into 20S particles, from which it can be released to promote another round of particle assembly (29). The stoichiometry and total number of subunits in 20S particles is unknown.

The NSF is an adenosinetriphosphatase (ATPase) containing two ATP binding sites residing in separate domains (59) and the binding and hydrolysis of ATP are critical in determining the stability of NSF (43) and its attachment to membranes (29,42). The 20S fusion particles will form and are stable when prepared and maintained in the presence of either MgATPγS (a nonhydrolyzable analogue of ATP) or in the presence of ATP itself but in the absence of magnesium. However, particles rapidly dissociate (even at 0°C) in the presence of MgATP in a process that requires hydrolysis of ATP, liberating NSF from SNAP-SNAP receptor complexes (29). It has been proposed that the fusion particle disassembles in the course of an ATP-dependent fusion reaction as an intrinsic step in the fusion mechanism (29).

PURIFICATION OF SNAP RECEPTORS

We took advantage (27) of the two layers of biochemical specificity inherent in the successive processes of assembly and ATP hydrolysis-dependent disassembly of 20S fusion particles to enable the affinity purification of SNAP receptors according to their function. Figure 1 illustrates the method used. The essential idea is to attach the fusion particles to a solid matrix to which SNAP receptor will become bound as a result of particle assembly, and then to release SNAP receptor (in purified form) as particles disassemble when ATP hydrolysis is later permitted.

We employed (27) a recombinant form of NSF, epitope-tagged with an myc peptide (60) (EQKLISEEDL) at its carboxy terminus. This NSF-myc can be expressed in *E. coli* and is functionally active (61). The 20S particles are formed in solution by mixing (at 0°C) a Triton X-100 extract of membranes with NSF-myc and with pure, recombinant *E. coli*-expressed α- and γ-SNAPs (58) in the presence of ATPγS and EDTA (to chelate any magnesium) to ensure that particles can form, but to prevent ATP hydrolysis. Then, protein G beads harboring a covalently at-

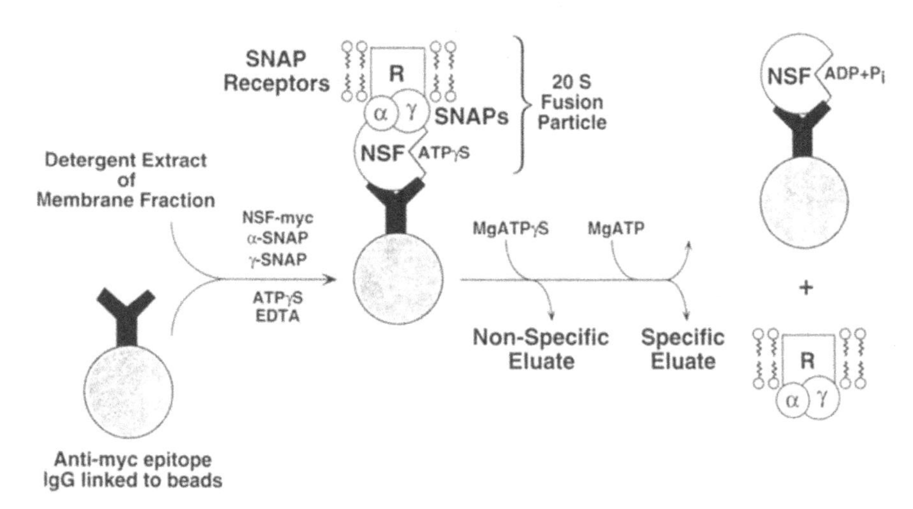

FIG. 1. Outline of procedure used to purify SNAP receptors (SNAREs). Recombinant NSF, α-SNAP, and γ-SNAP are assembled into 20S particles by SNAREs present in a crude detergent extract of membranes. The SNAREs are incorporated stoichiometrically into the particles, which are then bound to beads via NSF. For this purpose, the NSF is epitope-tagged with myc, and an anti-myc monoclonal IgG is linked to the beads. The beads are washed, and eluted first with MgATPγS (nonspecific eluate) and then with MgATP (specific eluate). The bound 20S particles disassemble in the presence of MgATP (but not Mg ATPYγS), releasing stoichiometric amounts of SNAPs and SNAREs. The NSF remains bound to the beads.

NSF, N-ethylmaleimide-sensitive fusion protein
SNAP, soluble NSF attachment protein
SNARE, SNAP receptor

tached monoclonal anti-myc IgG (9E10) (62) are added to adsorb the 20S particles via their NSF-myc subunits. The anti-myc IgG does not inhibit the function of NSF-myc in cell-free transport assays (unpublished results).

The beads are then formed into a column, which is washed extensively before a first (nonspecific) elution with MgATPγS. This serves to elute any remaining proteins that are attached in an Mg-sensitive fashion, or by MgATPγS per se. Finally, a second (specific) elution is performed with MgATP (replacing MgATPγS). Only proteins released as the immediate consequence of an ATP-hydrolysis reaction occurring on the column will be recovered in this specific eluate fraction. This scheme is based on procedures in an earlier report (29) but includes a number of key modifications detailed in the Figure legends.

Because NSF remains on the beads, the specific eluate should consist of a fraction of the added SNAPs, together with additional polypeptides that would represent SNAP receptors. Ideally, stoichiometric amounts of receptors and SNAPs would be released.

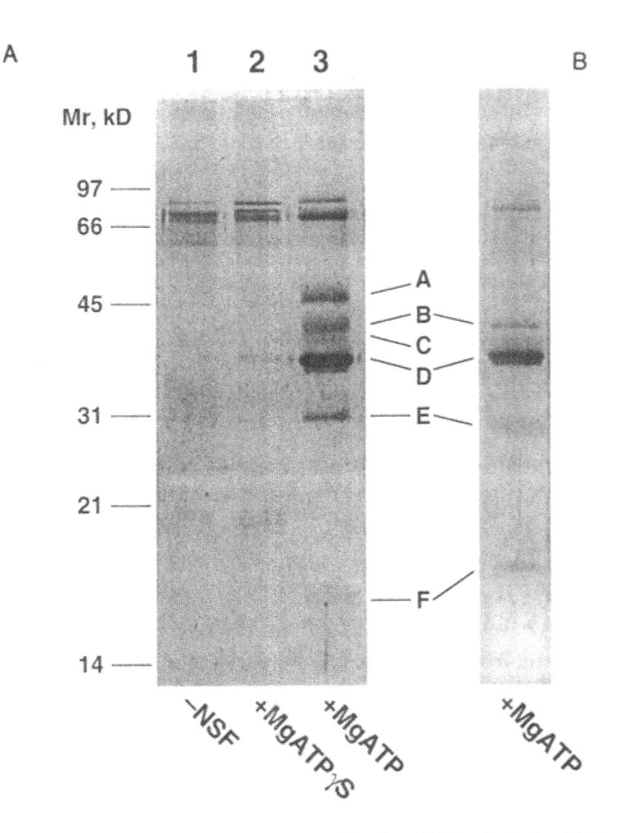

FIG. 2. Identification of proteins released from NSF after ATP hydrolysis. A) Coomassie-stained gel. Lane 1: control. MgATP eluate of control-binding reaction performed in absence of NSF. Lane 2: nonspecific eluate from complete binding reaction performed with NSF-myc and MgATPγS. Lane 3: specific eluate of the same column (as for Lane 2) following the exchange of ATP for ATPγS (in the presence of EDTA) and then addition of Mg^{2+} to allow ATP hydrolysis; see Fig. 1. B) Silver-stained Laemmli (7953) gel of the specific (MgATP) eluate.

ATP, adenosine 5′-triphosphate
EDTA, ethylenediaminetetraaceticacid
NSF, N-ethylmaleimide-sensitive fusion protein

A detergent extract was prepared from a crude, salt-washed total particulate fraction from the gray matter of bovine brain, and utilized as the source of potential SNAP receptors. Membranes were washed with 1 M KCl before use to remove most of their endogenous SNAP supply (55). As a great excess of recombinant SNAPs is added in the binding reaction (Fig. 1), these proteins should compete out remaining endogenous SNAPs in forming particles on the beads.

Figure 2A shows a Coomassie blue stained SDS-urea-high tris-polyacrylamide gel of the specific eluate (lane 3) and the nonspecific eluate fraction (lane 2), along with the specific eluate from a control experiment in which NSF-myc was omitted (lane 1). A number of bands (labeled A–F) are recovered only in the specific (MgATP) eluate and depend upon the presence of NSF. A set of bands at approximately 70 kd are present in all three lanes and, apart from these, the specific bands (A–F) are present in substantially pure form in the specific eluate fraction. Band F runs as a band in a standard Laemmli gel (Fig. 2B; specific eluate, stained with silver) this is sharper than the band in the high tris-urea gel (Fig. 2A). Bands A and C virtually coelectrophorese with the abundant band D in a Laemmli gel (Fig. 2B; see bottom).

IDENTIFICATION OF SNAP RECEPTORS BY AMINO ACID SEQUENCING

Amino acid microsequencing (Fig. 3) was used (27) to confirm the identity of the SNAP bands and determine whether additional receptor bands might correspond to any known proteins. The specific bands were excised from blots and digested with trypsin, and the resulting peptides were separated by high-pressure liquid chromatography and sequenced. Bands B and D coelectrophoresed with recombinant γ-SNAP and α-SNAP, respectively, which are both His_6-tagged (58) and are thus slightly larger than their endogenous counterparts (not shown in Fig. 3). These identifications were confirmed by microsequencing peptides from bands B and D.

The identity of syntaxins, SNAP-25, and synaptobrevin as SNAP receptors seems clear (27) based on: 1) the dependence of their affinity purification on NSF; 2) the fact that their elution from beads requires the act of ATP hydrolysis and not the mere presence of ATP; 3) the approximately stoichiometric abundance of SNAP receptors in toto relative to SNAP itself; and 4) the substantial purity of the SNAP receptor preparation obtained. The fact that each of the molecules purified originates from the synapse offers additional reassurance concerning the specificity and biological relevance of our procedures.

CONCLUSIONS

Cells can be expected to house a multiplicity of SNAP receptors because they employ a common pool of NSF and SNAP proteins (58) to help fuse vesicles to many distinct compartments. The SNAP receptors have been purified from a crude brain membrane fraction according to their function of assembling 20S particles from SNAPs and NSF, and on the basis of their release from NSF upon hydrolysis of ATP. Thus, each of these polypeptides must have either the capacity to assemble a 20S particle or a very high affinity for one that has been assembled. Most likely, we have isolated only a sampling of the receptor types that may be present, as the method is biased to isolating the highest abundance and highest affinity receptors. We think it is the subtlety and the biological specificity of this ATP hydrolysis-dependent elution, exploiting the same enzyme cycle used in the fusion process, that accounts for the clarity of the results of the fractionation (27).

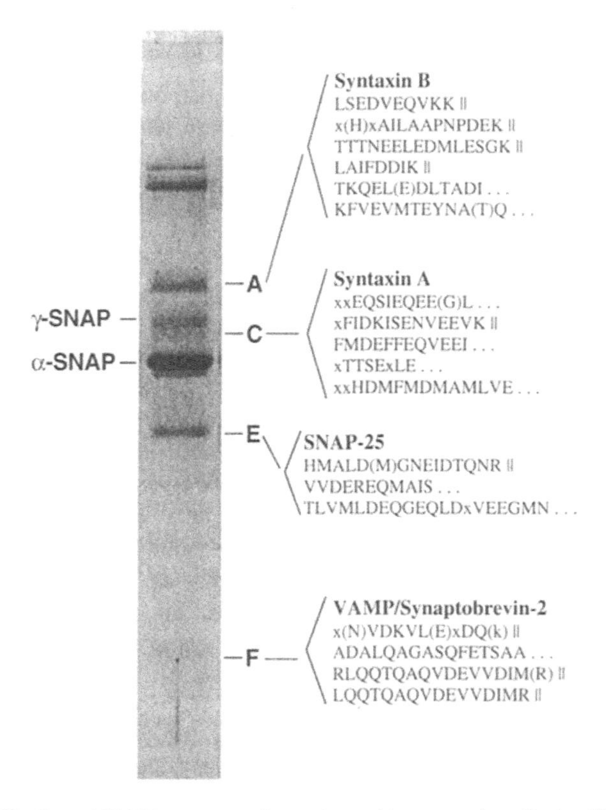

FIG. 3. Identification of SNAP receptors by amino acid sequencing. Each of bands A–F was excised from an electroblot (onto a nitrocellulose membrane) after staining with Ponceau S, digested with trypsin, followed by separation of fragments by Reverse Phase HPLC.

SNAP, soluble N-ethylmaleimide-sensitive fusion protein (NSF) attachment protein

It is a remarkable fact and a testimony to the specificity of the purification procedure that from the crudest possible extract of brain our method has culled out only a handful of proteins, all of which originate from synapses. This striking selection for synaptic components probably reflects the degree of specialization of the brain towards this aspect of membrane fusion. It is also notable that each SNAP receptor purified corresponds to a previously cloned and sequenced gene, no doubt providing a measure of the exhaustive structural characterization of synaptic vesicle and other synaptic membrane proteins by molecular biologists who have focused on the synapse because of its central importance in neuronal development and function (63–65). Despite this comprehensive effort, very little is actually known concerning the precise roles that these proteins perform in fusion because of the lack of functional assays.

What is known about the SNAP receptors we have identified from the synapse?

The SNAP-25 is found only in the presynaptic terminals of neurons (35) but has not been precisely localized. Although the sequence of SNAP-25 predicts a 25 kd hydrophilic protein, the polypeptide behaves as an integral membrane protein (35 and data not shown); and has been reported to be palmitylated at Cysteines (36). The SNAP-25 has only four Cysteine residues, clustered between positions 85 and 92 (murine sequence). We cite our earlier report that fatty acyl-CoA is required for NSF-dependent fusion (66). Possibly, SNAP-25 or equivalent proteins are the relevant acyl acceptors. The use of multiple acylations to create a hydrophobic surface useful for triggering fusion seems an attractive possibility.

Synaptobrevin (VAMP) (31–33) is inserted into synaptic (neurotransmitter-containing) vesicles by a single hydrophobic spanning segment at its C terminus, and the remainder of the protein, including its N terminus, is in the cytoplasm. The fact that both tetanus and botulinum B toxins, potent inhibitors of neurotransmitter release, are proteases that cleave only VAMP/synaptobrevin-2 (and not VAMP/synaptobrevin-1) in synaptic vesicles provides important evidence that this molecule is needed for fusion in vivo (37) as well as vividly demonstrates the physiological relevance of SNAP receptors for neurosecretion. Syntaxin (34) has the same topography as VAMP, but is instead localized to the presynaptic membrane, the target membrane for synaptic vesicle fusion following the influx of calcium ions as occurs when an action potential is received. In particular, syntaxin is concentrated in "active zones" of the presynaptic membrane (34) at which a subpopulation of synaptic vesicles is docked (67) to enable exceedingly rapid fusion, within 200 μsec of the influx of calcium (68). Thus, the components of the fusion machinery would apparently have to be largely preassembled at these sites to enable such a rapid response.

In light of the evidence just mentioned, the fact that 20S fusion particles contain VAMP and syntaxin is of major interest, because these complexes would seem ideal candidates for a preassembled fusion machine that could dock the vesicle to its target in the active zone. It is of course not yet possible to say whether both syntaxin and VAMP exist together in a single 20S particle (Fig. 4A) or whether separately assembled particles might join together to constitute an attachment site (Fig. 4B), perhaps with the aid of additional proteins.

From a different perspective, our discovery that molecules previously implicated in regulated exocytosis at the synapse are SNAP receptors that assemble with SNAPs and NSF clearly implies that the same NSF and SNAP-dependent fusion machinery needed for many constitutively-operating fusion events also underlies triggered release of neurotransmitters at synapses. By extension, we would expect that NSF and SNAP are similarly employed in the many varieties of triggered (regulated) exocytosis (69) that are of broad physiological importance.

How can a constitutively operating fusion machinery also be used in triggered exocytosis? It would seem that one or more inhibitory components would be necessary to clamp the 20S fusion particle to prevent them from either engaging or completely assembling in the first place. Such a fusion clamp could either cover an active site in NSF, SNAP, or the SNAP receptors, or prevent any of the additional

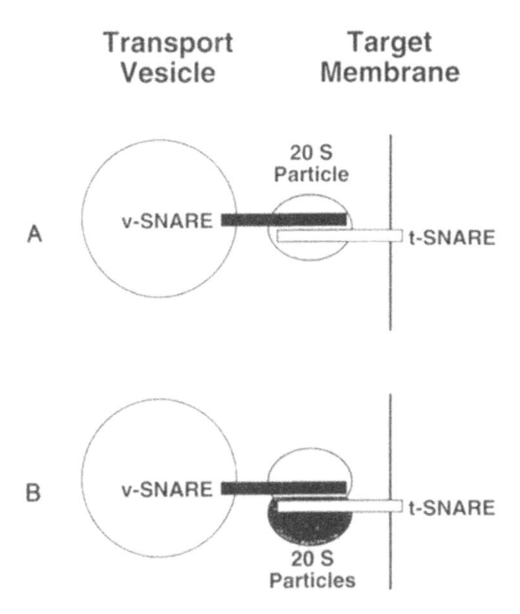

FIG. 4. Models to explain vesicle targeting based on the finding that SNAREs isolated in 20S fusion particles can originate from either the transport vesicle (v-SNAREs) or from the target membrane (t-SNAREs). A 20S particle (containing NSF and SNAPs) that simultaneously binds a v-SNARE and a t-SNARE (A) would attach a vesicle to its target. Alternatively (B), 20S particles, each capable of binding only one SNARE at a time, could interact to attach vesicle to target, perhaps requiring other proteins to assemble together.

NSF, N-ethylmaleimide-sensitive fusion protein
SNAP, soluble NSF attachment protein
SNARE, SNAP receptor

cytoplasmic subunits of the fusion machinery (70) from binding. In the case of the synapse, a calcium trigger would remove the clamp. As has been suggested, a strong candidate for a calcium clamp is the synaptic vesicle membrane protein synaptotagmin (71,72). Synaptotagmin coimmunoprecipitates with the presynaptic plasma membrane protein syntaxin (54) shown here to be a SNAP receptor, and the recombinant proteins can bind each other (34). Moreover, synaptotagmin itself undergoes a calcium-dependent conformational change (73).

It is notable that synaptotagmin does not appear to be among the major components present in 20S particles that contain syntaxin. This is not suprising, however, since the interaction between syntaxin and synaptotagmin was reported to be disrupted when membranes were washed with 0.5 M salt (34) we employed a 1 M salt wash in these studies.

Rab3A, a ras-related GTP-binding protein, is another candidate for a clamp, since this protein dissociates from synaptic vesicles upon stimulation of fusion (74). Fusion clamps could potentially respond to a variety of second messengers (i.e., adenosine 3′,5′-cyclic phosphate (cAMP), activated G proteins, protein phosphorylation, etc.), in addition to calcium, depending on cell type and physiological context.

THE SNARE HYPOTHESIS

Both syntaxin and VAMP have structurally related homologues in yeast as originally pointed out (39,40). The sed5 gene in yeast encodes a molecule with significant homology to syntaxin (39). The sed5 function is required for ER-to-Golgi transport, accumulating transport vesicles in its absence, and the sed5 protein most likely resides in the Golgi (39). Another homologue, pep12p, is found on the vacuole membrane, and is required for transport from the Golgi to this organelle (75). The yeast sec22, bet1, and bos1 gene products are structurally related to VAMP, are required for fusion with the Golgi in vivo, and of these at least bos1p is a component of ER-derived transport vesicles (40,50,76,77,78). Additional recent examples are discussed by Bennett and Scheller (41). The SNAP-25 is distantly related to sed5p, pep12p, and syntaxins. For example, SNAP-25 is identified in a data base search in comparison with sed5.

The SNAP receptors (SNAREs) we have identified from the synapse are thus members of a family of proteins that to date appear to be distributed in compartmentally-specific fashion, with one set attached to the transport vesicles (v-SNAREs) and another set attached to target membranes (t-SNAREs). Our finding that 20S fusion particles can assemble on/interact with both a v-SNARE and a t-SNARE of the synapse therefore raises the general possibility that complementary donor and acceptor compartment-specific members of the v- and t-SNARE families pair with each other in 20S particles to provide specificity to an otherwise general fusion machinery (Fig. 4). There may of course be additional layers of specificity involved in vesicle targeting.

It is therefore possible (27) that each transport vesicle is endowed with one or more members of the v-SNARE superfamily, which it would obtain when it buds from a corresponding donor compartment. Likewise, every target compartment in a cell could be endowed with one or more members of the t-SNARE superfamily. Specificity in membrane transactions would be assured because of the unique and nonoverlapping distribution of v-SNAREs and t-SNAREs among subcellular membrane-bound compartments. In the simplest view (i.e., if there were no other source of specificity), a productive fusion event would be initiated only when complementary v-SNARE and t-SNARE pairs engage. Thus, a v-SNARE from the ER (putatively sec22p) would have to be designed to engage a t-SNARE of the Golgi (perhaps sed5p) in 20S particles, but not one of the lysosomes (putatively pep12p).

According to this model, the 20S fusion particle assembly reaction can be employed as a cell-free "readout" system to test whether candidate proteins are SNAREs (in the manner demonstrated here for SNAP-25) and to establish which SNAREs, if any, form specific cognate pairs. Localization of the SNARE proteins in situ would then establish which are v-SNARES and which are t-SNAREs, as well as which compartments are involved in each pairing. For the proposed mechanism to work (again, assuming no other source of specificity), mismatched pairs of SNAREs would have to be excluded from the same 20S particle (Fig. 4A) or multi-particle assemblies (Fig. 4B), even though any one SNARE might be able to form a 20S particle by itself while waiting for a partner to arrive.

A related mechanistic question is whether the SNAPs (and NSF) are associated primarily with a v-SNARE or with a t-SNARE, or whether an essentially symmetrical structure is formed at the junction of vesicle and target. That NSF and SNAPs can associate with Golgi membranes to form 20S particles in the absence of transport vesicles (i.e., at ice temperature in the absence of coat proteins). Wilson et al. suggest that 20S Particles can be initiated with only one SNARE partner (29). Since NSF is a tetramer and since each monomer in NSF has two ATPase domains, it is easy to imagine several ways in which a 20S particle containing one NSF could form a symmetrical vesicle-target junction by simultaneously binding a v-SNARE and a t-SNARE (Fig. 4A).

Although many important details need to be established, the broad meaning of our findings seems clear, implying a general role for NSF and SNAPs in regulated as well as constitutive intracellular fusion processes, and in synaptic transmission in particular. That the SNAP receptors we have identified appear to be members of the compartmentally-specific membrane protein multigene families just mentioned raises the possibility that SNAREs encode targeting information. The fact that SNAREs can reside in both vesicle and target membranes implies that, together with NSF and SNAPs, SNAREs can form attachment sites between a vesicle and its target membrane.

SUMMARY

The NSF, SNAP, and SNAP receptors are key elements of the intracellular membrane fusion machinery. We use an affinity purification scheme, based on the function of SNAP receptor in assembling 20S fusion particles from NSF and SNAP proteins, to purify SNAP receptors from brain. Remarkably, each of the four SNAP receptors (or, SNAREs) thus delineated resides in synapses, with one receptor originating in the synaptic vesicle and another in the presynaptic plasma membrane that is targeted for fusion. This suggests a simple mechanism in which the general NSF/SNAP fusion machinery can assemble to bridge partner membranes in a complex containing elements of both vesicle and target membranes, and implies that similar fusion machines drive both constitutive fusion (ER→Golgi→surface and endocytosis) and regulated exocytosis. The vesicle (v-SNARE) and the target-associated t-SNAREs from the synapse are each members of compartmentally-specific families of membrane proteins found in yeast, animal cells, and neurons, thus raising the possibility that v-SNAREs and t-SNAREs encode specificity in membrane fusion processes that utilize a common mechanism.

REFERENCES

1. Palade G. Intracellular aspects of the process of protein synthesis. *Science* 1975;189:347–358.
2. Fries E, Rothman JE. Transport of vesicular stomatitis virus glycoprotein in a cell-free extract. *Proc Natl Acad Sci USA* 1980;77:3870–3814.
3. Novick P, Ferro S, Schekman R. Order of events in the yeast secretory pathway. *Cell* 1981;25:461–469.
4. Balch WE. Biochemistry of interorganelle transport. *J Biol Chem* 1989;264(29):16965–16968.
5. Goda Y, Pfeffer SR. Cell-free systems to study vesicular transport along the secretory and endocytic pathways. *Fed Am Soc Exp Biol* 1989;3:2488–2495.
6. Balch WE, Glick BS, Rothman JE. Sequential intermediates in the pathway of intercompartmental transport in a cell-free system. *Cell* 1984;39:525–536.
7. Orci L, Glick BS, Rothman JE. A new type of coated vesicular carrier that appears not to contain clathrin: its possible role in protein transport within the Golgi stack. *Cell* 1986;46:171–184.
8. Balch WE, Dunphy WG, Braell WA, Rothman JE. Reconstitution of the transport of protein between successive compartments of the Golgi measured by the coupled incorporation of N-acetylglucosamine. *Cell* 1984;39(2 Pt 1):405–416.
9. Dunphy WG, Brands R, Rothman JE. Attachment of terminal N-acetylglucosamine to asparagine-linked oligosaccharides occurs in central cisternae of the Golgi stack. *Cell* 1985;40(2):463–472.
10. Dunphy WG, Fries E, Urbani LJ, Rothman JE. Early and late functions associated with the Golgi apparatus reside in distinct compartments. *Proc Natl Acad Sci USA* 1981;78(12):7453–7457.
11. Rothman JE, Urbani LJ, Brands R. Transport of protein between cytoplasmic membranes of fused cells: correspondence to processes reconstituted in a cell-free system. *J Cell Biol* 1984;99(1 Pt 1):248–259.
12. Braell WA, Balch WE, Dobbertin DC, Rothman JE. The glycoprotein that is transported between successive compartments of the Golgi in a cell-free system resides in stacks of cisternae. *Cell* 1984.39(3 Pt 2):511–24.
13. Melançon P, Glick BS, Malhotra V, Weidman PJ, Serafini T, Gleason ML, Orci L, Rothman JE. Involvement of GTP-binding "G" proteins in transport through the Golgi stack. *Cell* 1987;51(6): 1053–1062.
14. Glick BS, Rothman JE. Possible role for fatty acyl-coenzyme A in intracellular protein transport. *Nature* 1987;326:309–312.
15. Malhotra V, Orci L, Glick BS, Block MR, Rothman JE. Role of an N-ethylmaleimide-sensitive transport component in promoting fusion of transport vesicles with cisternae of the Golgi stack. *Cell* 1988;54(2):221–227.
16. Orci L, Malhotra V, Amherdt M, Serafini T, Rothman JE. Dissection of a single round of vesicular transport: sequential intermediates for intercisternal movement in the Golgi stack. *Cell* 1989;56: 357–368.
17. Block MR, et al. Purification of an N-ethylmaleimide-sensitive protein catalyzing vesicular transport. *Proc Natl Acad Sci USA* 1988;85:7852–7856.
18. Pfanner N, et al. Fatty acyl-coenzyme A is required for budding of transport vesicles from Golgi cisternae. *Cell* 1989;59:95–102.
19. Pfanner N, et al. Fatty acylation promotes fusion of transport vesicles with Golgi cisternae. *J Cell Biol* 1990;110:955–961.
20. Malhotra V, et al. Purification of a novel class of coated vesicles mediating biosynthetic protein transport through the Golgi stack. *Cell* 1989;58:329–336.
21. Serafini T, Stenbeck G, Brecht A, Lottspeich F, Orci L, Rothman JE, Wieland FT. A coat subunit of Golgi-derived non-clathrin-coated vesicles with homology to the clathrin-coated vesicle coat protein beta-adaptin. *Nature* 1991;349(6306):215–220.
22. Serafini T, Orci L, Amherdt M, Brunner M, Kahn RA, Rothman JE. ADP-ribosylation factor is a subunit of the coat of Golgi-derived COP-coated vesicles: a novel role for a GTP-binding protein. *Cell* 1991;67(2):239–253.
23. Kahn RA, Gilman AG. The protein cofactor necessary for ADP-ribosylation of Gs by cholera toxin is itself a GTP binding protein. *J Biol Chem* 1986;261(17):7906–7911.
24. Stearns T, Willingham MC, Botstein D, Kahn RA. ADP-ribosylation factor is functionally and physically associated with the Golgi complex. *Proc Natl Acad Sci USA* 1990;87:1238–1242.
25. Duden R, et al. Beta-COP, a 110 kd protein associated with non-clathrin-coated vesicles and the Golgi complex, shows homology to Beta-Adaptin. *Cell* 1991;64:649–665.

26. Orci L, Tagaya M, Amherdt M, Perrelet A, Donaldson JG, Lippincott-Schwartz J, Klausner RD, Rothman JE. Brefeldin A, a drug that blocks secretion, prevents the assembly of non-clathrin-coated buds on Golgi cisternae. *Cell* 1991;64(6):1183–1195.
27. Söllner T, Whiteheart SW, Brunner M, Erdjument-Bromage H, Geromanos S, Tempst P, Rothman JE. SNAP receptors implicated in vesicle targeting and fusion. *Nature* 1993;362(6418):318–324.
28. Wilson DW, Whiteheart SW, Orci L, Rothman JE. Intracellular membrane fusion. *Trends Biochem Sci* 1992;16(9):334–337.
29. Wilson DW, Whiteheart SW, Wiedmann M, Brunner M, Rothman JE. A multisubunit particle implicated in membrane fusion. *J Cell Biol* 1992;117(3):531–538.
30. Whiteheart SW, Brunner M, Wilson DW, Wiedmann M, Rothman JE. Soluble N-ethylmaleimide-sensitive fusion attachment proteins (SNAPs) bind to a multi-SNAP receptor complex in Golgi membranes. *J Biol Chem* 1992;267(17):12239–12243.
31. Trimble WS, Cowan DM, Scheller RH. VAMP-1: a synaptic vesicle-associated integral membrane protein. *Proc Natl Acad Sci USA* 1988;85(12):4538–4542.
32. Baumert M, Maycox RP, Navone F, DeCamilli P, Jahn R. Synaptobrevin: an integral membrane protein of 18,000 daltons present in small synaptic vesicles of rat brain. *EMBO J* 1989;8(2):379–384.
33. Südhof TC, Baumert M, Perin MS. Jahn R. A synaptic vesicle membrane protein is conserved from mammals to Drosophila. *Neuron* 1989;2(5):1475–1481.
34. Bennett MK, Calakos N, Scheller RH. Syntaxin: a synaptic protein implicated in docking of synaptic vesicles at presynaptic active zones. *Science* 1992;257(5067):255–259.
35. Oyler GA, et al. The identification of a novel synaptosomal-associated protein, SNAP-25, differentially expressed by neuronal subpopulations. *J Cell Biol* 1989;109:3039–3052.
36. Hess DT, Slater TM, Wilson MC, Skene JH. The 25 kDa synaptosomal-associated protein SNAP-25 is the major methionine-rich polypeptide in rapid axonal transport and a major substrate for palmitoylation in adult CNS. *J Neurosci* 1992;12(12):4634–4641.
37. Schiavo G, et al. Tetanus and botulinum-B neurotoxins block neurotransmitter release by proteolytic cleavage of synaptobrevin. *Nature* 1992;359:832–835.
38. Flexner SB (ed.) *Random House dictionary of the English language*. 2nd ed. Unabridged. New York: Random House; 1987:1807.
39. Hardwick KG, Pelham HR. SED5 encodes a 39-kD integral membrane protein required for vesicular transport between the ER and the Golgi complex. *J Cell Biol* 1991;119(3):513–521.
40. Dascher C, Ossig R, Gallwitz D, Schmitt HD. Identification and structure of four yeast genes (SLY) that are able to suppress the functional loss of YPT1, a member of the RAS superfamily. *Mol Cell Biol* 1991;11(2):872–885.
41. Bennett MK, Scheller RH. The molecular machinery for secretion is conserved from yeast to neurons. *Proc Natl Acad Sci USA* 1993;90:2559–2563.
42. Glick BS, Rothman JE. *Nature* (Lond.) 1987;326;309–312.
43. Block MR, Glick BS, Wilcox CA, Wieland FT, Rothman JE. *Proc Natl Acad Sci USA* 1988;85: 7852–7856.
44. Fries E, Rothman JE. *Proc Natl Acad Sci USA* 1980;77:3870–3874.
45. Balch WE, Dunphy DW, Braell WA, Rothman JE. *Cell* 1984;39:405–416.
46. Malhotra V, Orci L, Glick BS, Block MR, Rothman JE. *Cell* 1988;54:221–227.
47. Orci L, Malhotra V, Amherdt M, Serafini, T, Rothman JE. *Cell* 1989;56:357–368.
48. Wilson DW, Wilcox CA, Flynn GC, Chen E, Kuang WJ, Henzel WJ, Block MR, Ulrich A, Rothman JE. A fusion protein required for vesicle-mediated transport in both mammalian cells and yeast. *Nature* 1989;339:355–359.
49. Novick P, Ferro S, Schekman R. Order of events in the yeast secretory pathway. *Cell* 1981;25:461–469.
50. Kaiser CA, Schekman R. Distinct Sets of SEC genes govern transport vesicle formation and fusion early in the secretory pathway. *Cell* 1990;61:723–733.
51. Graham TR, Emr SD. Compartmental organization of Golgi-specific protein modification and vacuolar protein sorting events defined in a yeast sec18 (NSF) mutant. *J Cell Biol* 1991;114(2):207–218.
52. Beckers CJ, Block MR, Glick BS, Rothman JE, Balch WE. Vesicular transport between the endoplasmic reticulum and the Golgi stack requires the NEM-sensitive fusion protein. *Nature* 1989; 339:397–398.
53. Diaz R, Mayorga LS, Weidman PJ, Rothman JE, Stahl PD. Vesicle fusion following receptor-mediated endocytosis requires a protein active in Golgi transport. *Nature* 1989;339:398–400.

54. Weidman PJ, Melançon P, Block MR, Rothman JE. Binding of an N-Ethylmaleimide-sensitive fusion protein to Golgi membranes requires both a soluble protein(s) and an integral membrane receptor. *J Cell Biol* 1989;108:1589–1596.
55. Clary DO, Rothman JE. Purification of three related peripheral membrane proteins needed for vesicular transport. *J Biol Chem* 1990;265:10109–10117.
56. Clary DO, Griff IC, Rothman JE. SNAPs, a family of NSF attachment proteins involved in intracellular membrane fusion in animals and yeast. *Cell* 1990;61:709–721.
57. Griff IC, Schekman R, Rothman JE, Kaiser CA. The yeast SEC17 gene product is functionally equivalent to mammalian alpha-SNAP protein. *J Biol Chem* 1992;267(17):12106–12115.
58. Whiteheart SW, Griff IC, Brunner M, Clary DO, Mayer T, Buhrow SA, Rothman JE. SNAP family of NSF attachment proteins includes a brain-specific isoform. *Nature* 1993;362(6418):353–355.
59. Tagaya M, Wilson DW, Brunner M, Arango N, Rothman JE. Domain structure of an N-ethylmaleimide-sensitive fusion protein involved in vesicular transport. *J Biol Chem* 1993;268(4):2662–2666.
60. Munro S, Pelham HRB. A C-terminal signal prevents secretion of luminal ER proteins. *Cell* 1987;48:899–907.
61. Wilson DW. Rothman JE. Expression and purification of recombinant N-ethylmaleimide-sensitive fusion protein from Escherichia coli. *Methods Enzymol* 1992;219:309–318.
62. Evan GI, Lewis GK, Ramsay G, Bishop JM. Isolation of monoclonal antibodies specific for human c-myc proto-oncogene product. *Mol Cell Biol* 1985;5(12):3610–3616.
63. Südhof TC, Jahn R. Proteins of synaptic vesicles involved in exocytosis and membrane recycling. *Neuron* 1991;6:665–677.
64. Greengard P, Valtorta F, Czernik AJ, Benfenati F. Synaptic vesicle phosphoproteins and regulation of snyaptic function. *Science* 1993;259(5096):780–785.
65. Trimble WS, Linial M, Scheller RH. Cellular and molecular biology of the presynaptic nerve terminal. *Annu Rev Neurosci* 1991;14:93–122.
66. Pfanner N, et al. Fatty acylation promotes fusion of transport vesicles with Golgi cisternae. *J Cell Biol* 1990;110:955–961.
67. Landis DM, Hall AK, Weinstein LA, Reese TS. The organization of cytoplasm at the presynaptic active zone of a central nervous system synapse. *Neuron* 1988;1(3):201–209.
68. Llinas R, Steinberg IZ, Walton K. Relationship between presynaptic calcium current and postsynaptic potential in squid giant synapse. *Biophys J* 1981;33(3):323–351.
69. Kelly RB. Pathways of protein secretion in eukaryotes. *Science* 1985;230:25–32.
70. Waters MG, Clary DO, Rothman JE. A novel 115-kD peripheral membrane protein is required for intercisternal transport in the Golgi stack. *J Cell Biol* 1992;118(5):1015–1026.
71. Südhof TC, et al. Phospholipid binding by a synaptic vesicle protein homologous to the regulatory region of protein kinase C. *Nature* 1990;345:260–263.
72. Wendland B, Miller KG, Schilling J, Scheller RH. Differential expression of the p65 gene family. *Neuron* 1991;6(6):993–1007.
73. Brose N, Petrenko AG, Südhof TC, Jahn R. Synaptotagmin: a calcium sensor on the synaptic vesicle surface. *Science* 1992;256(5059):1021–1025.
74. Fischer von Mollard G, Südhof TC, Jahn R. A small GTP-binding protein dissociates from synaptic vesicles during exocytosis. *Nature* 1991;349(6304):79–81.
75. Preston RA, Manolson MF, Becherer K, Weidenhammer E, Kirkpatrick D, Wright R, Jones EW. Isolation and characterization of PEP3, a gene required for vacuolar biogenesis in Saccharomyces cerevisiae. *Mol Cell Biol* 1991;11(12):5801–5812.
76. Shim J, Newman AP, Ferro-Novick S. The BOS1 gene encodes an essential 27-kD putative membrane protein that is required for vesicular transport from the ER to the Golgi complex in yeast. *J Cell Biol* 1991;113(1):55–64.
77. Newman AP, Groesch ME, Ferro-Novick S. Bos1p, a membrane protein required for ER to Golgi transport in yeast, co-purifies with the carrier vesicles and with Bet1p and the ER membrane. *EMBO J* 1992;11(10):3609–3617.
78. Newman AP, Graf J, Mancini P, Rossi G, Lian JP, Ferro-Novick S. SEC22 and SLY2 are identical. *Mol Cell Biol* 1992;12(8):3663–3664.

Molecular and Cellular Mechanisms of Neurotransmitter Release, edited by Lennart Stjärne, Paul Greengard, Sten Grillner, Tomas Hökfelt, and David Ottoson, Raven Press, Ltd., New York © 1994.

6

Final Steps in Ca^{2+}-Triggered Exocytosis in Neuroendocrine Cells

Wolfhard Almers, Andy K. Lee, *Yoko Shoji-Kasai, *Masami Takahashi, †Paul Thomas, and ‡Frederick W. Tse

*Department of Molecular Cell Research, Max-Planck-Institute for Medical Research, 69120 Heidelberg, Germany; *Laboratory of Neurochemistry, Mitsubishi Kasei Institute of Life Sciences, Machida, Tokyo 194, Japan; †Department of Human Physiology, University of California School of Medicine, Davis, California 95616; and ‡Department of Pharmacology, University of Alberta, Edmonton, Alberta T6G 2H7, Canada*

Exocytosis is the last in a sequence of membrane fusion events that accompany the transport of membranes and soluble cargo from the endoplasmic reticulum to the cell surface and the extracellular space. Each fusion event in turn is a cascade of biochemical reactions that may be separated conceptually into several steps: 1) the specific docking of the vesicle with the correct target membrane; 2) the maturation of the docking complex into a fully competent fusion machine; 3) fusion of the vesicle with the target membrane; and, in regulated exocytosis, 4) a mechanism to prevent fusion with the plasma membrane until the correct messenger (Ca^{2+} in neurons and neuroendocrine cells) appears in the cytosol. Genetic (1) and molecular biologic studies (2,3,4) have identified many, and perhaps most, of the molecules required. Except for minor differences, the same molecular machinery may well be used in many or all fusion events in intracellular membrane traffic, including exocytosis. Much progress is currently being made in cataloging and characterizing the participating molecules and their biochemical interactions. However, what these proteins do and the sequence in which they do it remain unknown.

STEPS IN THE EXOCYTIC CASCADE

The most successful attempts to distinguish individual steps in the exocytic cascade have been made in neurons and neuroendocrine cells, where exocytosis is arrested at a late step, and where the arrest can be relieved by cytosolic Ca^{2+}. Earlier studies were done in cells whose cytosol was made accessible by permeabilizing the plasma membrane, either by mechanical disruption, by a high voltage electric discharge, or by treatment with digitonin. In chromaffin cells so treated, it was found that extensive secretion requires magnesium adenosine 5′-triphosphate (MgATP) (5) but also that Ca^{2+} triggers exocytosis of some vesicles even in the absence of MgATP (6) (see Hay and Martin (7) for similar results on PC-12 cells). The small amount of MgATP-independent exocytosis occurs more quickly than the exocytosis that continues in the presence of MgATP (6). Evidently, exocytic membrane fusion itself needs no MgATP; instead, MgATP is required to "prime" vesicles so they may later undergo exocytosis in response to an increase in Ca_i. Further kinetic studies of catecholamine release from chromaffin cells led Bittner and Holz (8) to suggest that a steeply temperature-sensitive step must follow the MgATP-dependent reaction to generate a primed vesicle. Their results are summarized in the reaction scheme of Fig. 1A. Ca^{2+}-triggered exocytosis of a primed vesicle needs a 145 kd cytosolic protein as a homodimer (9).

EXOCYTOSIS OF A PRIMED VESICLE IS A MULTISTEP REACTION

Given that Ca^{2+} triggers transmitter release at synapses in fractions of a millisecond, it seems likely that the final steps in the exocytic cascade are fast also in neuroendocrine cells, probably faster than can be resolved by biochemical assays of catecholamine release. We therefore have tracked the surface area of single cells by measuring the plasma membrane capacitance (10). This method provides an assay for exo- and endocytosis with millisecond resolution. We view the exocytic cascade as a sequence of stations that each vesicle must visit in turn before undergoing exocytosis (Fig. 1). Suppose that S_1 in Fig. 1B is the last station in the exocytic cascade that is stable in the absence of Ca^{2+}. Then, after a step increase in Ca_i, the vesicles at station S_1 will undergo exocytosis first. Vesicles at station S_2 will take longer, namely by the time required to perform the reaction S_2 to S_1. Vesicles at station S_3 will take longer still, and so on. We predict that a step rise in Ca_i will trigger exocytosis in phases occurring at different speeds, the fastest representing vesicles that have advanced the farthest along the exocytic pathway. Indeed, evidence for multiple kinetic components was already provided by earlier results (melanotrophs (11), chromaffin cells (12), gonadotrophs (13)).

Most of our work was done at 21–26°C on melanotrophs, a neuroendocrine cell of the rat pituitary (14,15). To achieve rapid increases in Ca_i, cells were loaded through a glass micropipette with a photolabile Ca chelator, DM-nitrophen, and Ca^{2+} was released from the chelator by flash photolysis. Ca-DM-nitrophen is no larger than ATP and is expected to distribute uniformly in the cytoplasm. Hence, a flash-induced increase in Ca_i is expected to be uniform throughout the cell. Ca_i was measured with either of two fluorescent Ca-indicator dyes: fura-2 to measure Ca_i before the flash, and furaptra, an indicator of lower affinity, to measure the much higher Ca_i after the flash. Since DM-nitrophen has significant affinity not only for Ca^{2+} but also for Mg^{2+}, we dialyzed Mg^{2+} out of the cytosol before raising Ca^{2+}. As a consequence, cytosolic $[Mg^{2+}]$ in our experiments was in the 100 nM range,

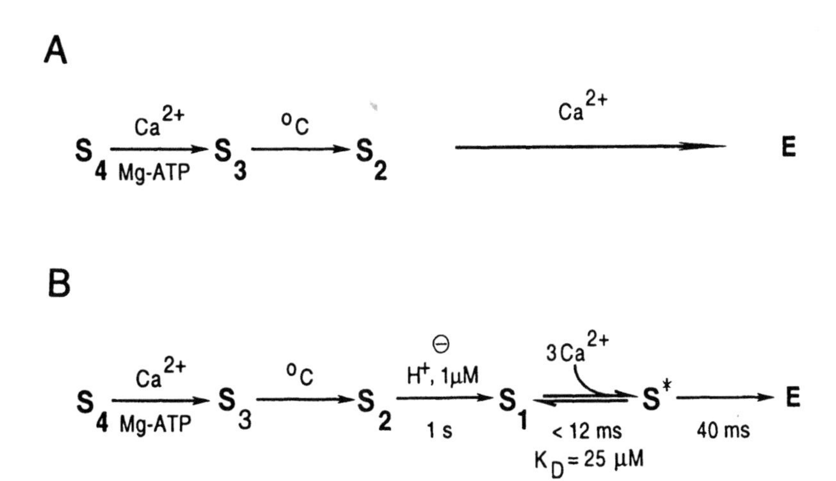

FIG. 1. Final stations on the Ca-triggered exocytic pathway in neuroendocrine cells. **A,** from permeabilized chromaffin cells according to Baker and Knight, 1981. **B,** from kinetic analysis of the exocytic response to a step rise in Ca_i in melanotrophs. For details, see text.

too low to support MgATP-dependent steps in the exocytic cascade. With this method, exocytosis after a step increase in Ca_i is seen to occur in three phases, schematically drawn in Fig. 2 (also see Neher and Zucker (16)).

The fastest is an exocytic burst that starts with a 6–10 ms delay, runs its course in about 100 milliseconds and adds about 200 fF to the plasma membrane capacitance (corresponding to 200 vesicles of 1 fF capacitance). During the exocytic burst, melanotrophs secrete at rates of 5000 fF/s (or 5000 vesicles/s, see Fig. 3A), but this high rate declines rapidly even though Ca_i remains elevated. Evidently, the exocytic burst represents a small pool of rapidly releasable vesicles that is quickly exhausted. The rate constant of exhaustion depends strongly on Ca_i, suggesting that three or more Ca^{2+} must combine with a regulatory molecule to trigger exocytosis. At high Ca_i, it reaches a limit of 25/s at 21–26°C, suggesting that exocytosis takes 40 milliseconds after all Ca^{2+} have bound to the vesicle's regulatory site(s). The rate of exhaustion is half-maximal at around 27 μM Ca_i, suggesting a relatively low-affinity Ca-receptor. The exocytic burst represents the fastest exocytic response so far observed in a neuroendocrine cell; hence we attribute it to S_1 in Fig. 1B, the last station of the secretory pathway that is stable in the absence of Ca^{2+}. We envisage that the reversible binding of Ca^{2+} to S_1 generates an activated state, S*, that results in exocytosis within 40 milliseconds.

After the burst, exocytosis continues in a slower phase lasting seconds. Mild cytosolic acidification to pH 6.2 inhibits the slower phase but does not affect the exocytic burst. The pH-sensitive component adds around 1,000 fF to the membrane capacitance, corresponding to about 1,000 vesicles. The number of vesicles released in this slower phase of exocytosis is underestimated because this phase occurs simultaneously with a rapid phase of membrane retrieval. We attribute the

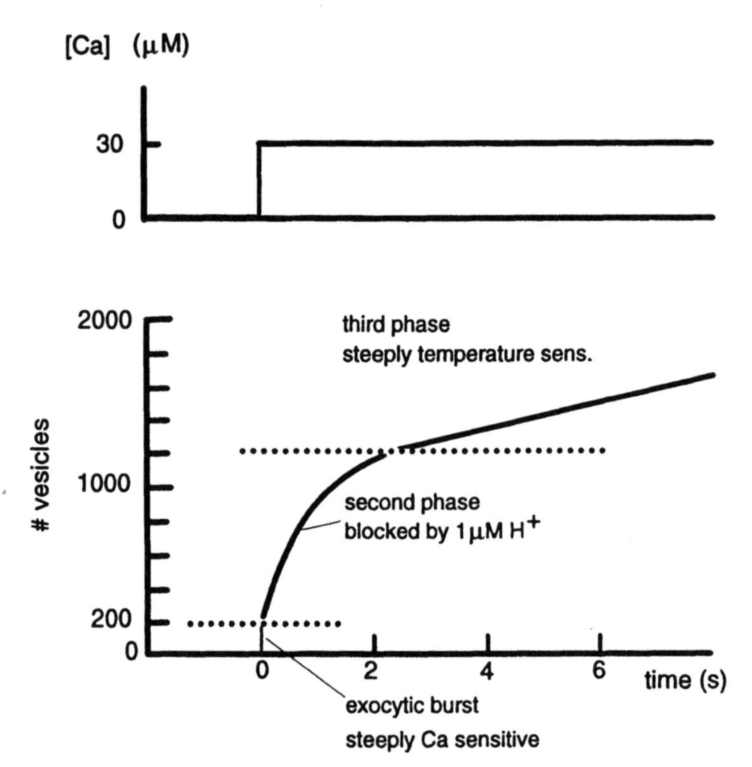

FIG. 2. Schematic drawing of the exocytic response to a step change in Ca_i. Upper trace, Ca_i after flash photolysis; lower trace, C_m. C_m was converted into the number of vesicles undergoing exocytosis by assuming each contributes 1 fF, as expected for a vesicle diameter of 180 nm and a specific membrane capacitance of 10 fF/μm^2.

reaction S_2–S_1 in Fig. lb to the pH-sensitive component. Figure 3A shows examples of the exocytic burst and the subsequent slower phase of exocytosis.

After the pH-sensitive phase, exocytosis continues for tens of seconds at a slow rate that is heavily dependent on temperature. The membrane capacitance added by this component is well over 1000 fF, corresponding to well over 1,000 vesicles. Because of its slow time course and temperature dependence, it is tempting to suggest that this phase represents the similarly temperature-sensitive, final step in "priming" a chromaffin granule for release (8) as represented by the reaction S_3–S_2 in Fig. 1A,B.

We suggest that the exocytic burst and the pH-sensitive phase represent sequential steps in the exocytosis of a primed vesicle. Consistent with this idea, both proceed in the absence of significant amounts of Mg^{2+}, and also when the cytosol is dialyzed with solutions lacking ATP. Figure lB builds on the reaction scheme of Bittner and Holz (8), and summarizes our current view of the exocytic cascade. Note that there are two Ca^{2+}-dependent reactions. One controls the exocytic burst, requires three or more Ca^{2+}, and is due to a Ca-receptor of relatively low affinity.

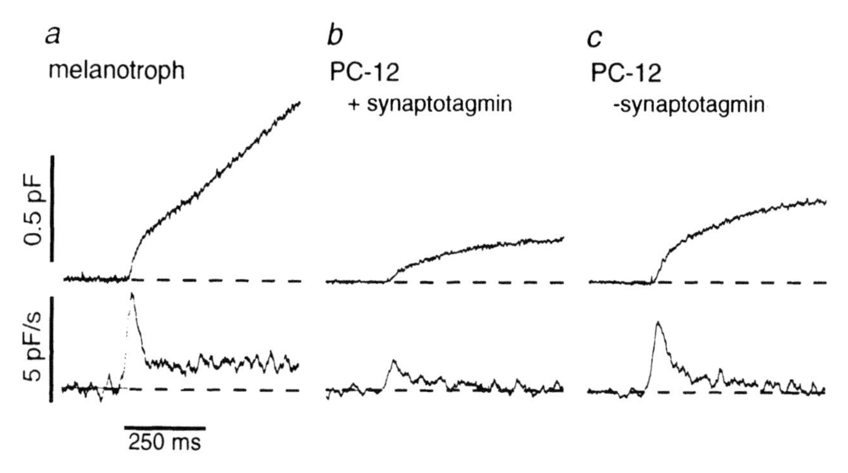

FIG. 3. Examples of experimental recordings in three different cells. Upper traces, C_m; lower traces, dC_m/dt calculated point by point from the upper trace as the slope of a regression line extending 20 ms to the left and 20 ms to the right. Rat melanotrophs were harvested and maintained as described by Thomas et al., 1990; PC-12 cells were subclones described by Shoji-Kasai et al., 1992; one of them expressed synaptotagmin I and II, the other did not. Methods used are the same as described by Thomas et al., 1993a, 1993b. External solution: 125 mM NaCl, 20 mM TEACl, 10 mM NaHEPES, 5.5 mM glucose, 3 mM KCl, 2 mM $CaCl_2$ and 1 μM tetrodotoxin, pH 7.4. Internal solution: 100 mM Cs-glutamate, 16 mM tetraethylammonium (TEA)-Cl, 10 mM Na_4DM-nitrophen, 8.5 mM $CaCl_2$, 2 mM NaATP, 0.1 mM fura-2 or furaptra and 8 mM Cs-HEPES (pH 7.2). Temperature 21–26°C.

The other is attributed to a Ca-receptor of micromolar affinity (6). Evidence for an early, high-affinity Ca-dependent step was also obtained by von Rüden et al. (17) who called it augmentation.

HOW MANY VESICLES ARE DOCKED?

In all likelihood, the vesicles participating in the exocytic burst are molecularly docked beneath the plasma membrane. These 200–250 vesicles would decorate the 500 μm^2 of plasma membrane at an average density of 0.5 vesicles/μm^2. In chromaffin cells, a similarly small number is seen morphologically subjacent to the plasma membrane (18).

In melanotrophs the amplitude of the pH-sensitive component would suggest that "primed" vesicles number in the thousands and, on kinetic grounds, this may also apply to chromaffin cells (16). Does "priming" (i.e., the last ATP-dependent reaction) precede docking? Studies of membrane traffic through the Golgi stack suggest otherwise, because ATP-dependent reactions follow the morphologic docking of transport vesicles to the acceptor compartment (e.g., as shown by Clary et al. (19)). In fact, N-ethylmaleimide-sensitive fusion protein (NSF)-mediated fusion requires MgATP. Clearly, the role of ATP deserves careful reinvestigation. In a melanotroph, thousands of vesicles are located within a vesicle radius of the plasma membrane (see discussion by Thomas et al. (14)), hence it is possible that all "primed" vesicles are docked. Exocytosis of a primed vesicle being a multistep reaction would mean that docking in itself does not immediately confer fusion competence. Instead, the fusion machinery of a docked vesicle must undergo further reaction steps. This idea would be consistent with findings at the neuromuscular junction, where only about 200 vesicles are released during an action potential while the total number of vesicles directly subjacent to the presynaptic plasma membrane is some fiftyfold higher. Yet, statistical arguments suggest that exocytosis after a single action potential significantly exhausts a small store of release-ready vesicles (20,21,22).

FLASH PHOTOLYSIS REVEALS MULTIPLE COMPONENTS OF EXOCYTOSIS ALSO IN OTHER CELLS

Flash photolysis of Ca-DM-nitrophen causes massive exocytosis also in chromaffin cells (16) and PC-12 cells, a cell line derived from an adrenal medullary tumor. Figure 3B,C shows the membrane capacitance (upper) and its time derivative (lower traces) in two PC-12 subclones. Their responses are smaller than those of melanotrophs (Fig. 3A). Nonetheless, Fig. 3B,C shows an exocytic burst that is kinetically similar to that in melanotrophs; the difference in amplitudes between Fig. 3B and Fig. 3C may be due to variability between different cells. Figure 3B shows the wild-type response, while Fig. 3C shows the response from a subclone (23) selected to lack synaptotagmin I and II, isoforms of a protein found in the membranes of both small synaptic vesicles and dense-core secretory vesicles. Synaptotagmin undergoes a Ca-dependent conformational change enabling it to bind phospholipids, and has been hypothesized to be the Ca sensor for exocytosis (24). Nonetheless, PC-12 cells lacking isoforms I and II are capable of Ca-triggered catecholamine release that appears normal on the time scale of minutes (23) like in melanotrophs and wild-type cells, they respond to step rise in Ca^{2+} with an exocytic burst. Evidently the triggering of the exocytic burst by Ca^{2+} does not require synaptotagmin I or II. Recently (24a) a new isoform prevalent in endocrine and PC-12 cells was cloned and termed synaptotagmin III. If a synaptotagmin serves as the final trigger in exocytosis in PC-12 cells, the isoform may be synaptotagmin III. However, the role of synaptotagmins in exocytosis is not yet known with certainty (25,26).

Mast cells contain large, 0.8 μm diameter secretory vesicles that are controled by a still incompletely understood signaling cascade that involves G-proteins and is not triggered directly by an increase in Ca_i. Because the vesicles are so large, individual exocytic fusion events are readily recognized as capacitance steps (27) (Fig. 4A). In addition, mast cells can increase their membrane capacitance purely in response to cytosolic Ca^{2+}, but this capacitance increase is gradual (28) (Fig. 4B). Evidently, there are two independently regulated exocytic pathways in mast cells: 1) one that releases large granules and is triggered by G-proteins and 2) one that releases

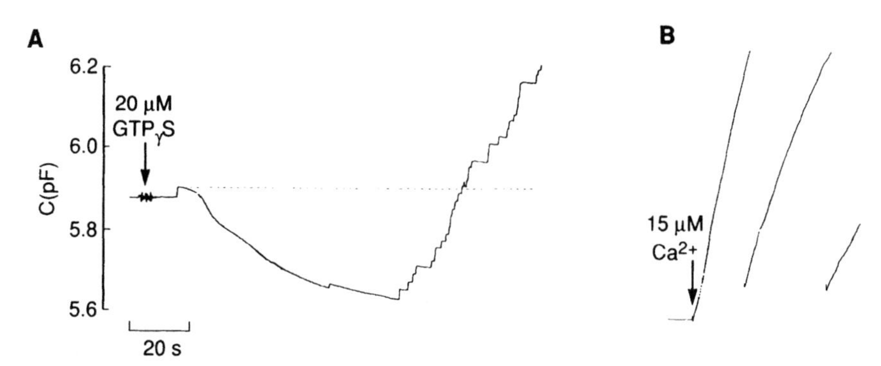

FIG. 4. Graded and stepwise capacitance changes in rat peritoneal mast cells. **A,** GTPγS-induced exocytosis of large dense-core vesicles. The pipette contained 140 mM NaCl, 10 mM Na-HEPES, 0.1 mM EGTA, 50 μM $CaCl_2$, 20 μM GTPγS. Cl was chosen as the major anion to slow exocytosis so the steps become more clearly visible on this time scale. Similar steps are also seen with more physiological internal solutions (Fernandez et al., 1984). **B,** Ca-triggered exocytosis of vesicles too small to make visibly stepwise contributions to the capacitance increase. The trace is divided into three segments, each of which continues the previous one. The pipette contained 155 K-glutamate, 10 mM Na-HEPES (pH 7.2), 10 mM HEDTA, 8 mM $CaCl_2$, and no GTPγS. The Ca-HEDTA mixture was adjusted to establish $Ca_i = 15$ μM. Throughout, the external solution contained 140 mM NaCl, 2.5 mM KCl, 2 mM $MgCl_2$, 2 mM $CaCl_2$, 10 mM Na-HEPES (pH 7.2), 5 mM glucose. Arrows indicate when the pipette was connected to the cytosol. Calibration bars refer to both panels. Reprinted with permission from Kirillova et al., 1993).

smaller granules (<0.3 μm diameter) and is regulated directly by Ca_i. We have found that flash photolysis of Ca-DM-nitrophen also elicits exocytosis in mast cells (29). Interestingly, the exocytic response is kinetically almost indistinguishable from that in melanotrophs (29) showing an exocytic burst followed by a slower phase. Probably the exocytic cascade of neuroendocrine cells is also present in mast cells, where it releases an unknown mediator and operates in parallel with the better known G-protein regulated pathway.

A TRIGGERED MEMBRANE RETRIEVAL MECHANISM OPERATING ON A TIME SCALE OF SECONDS

Morphologic studies and studies with extracellular markers have suggested that cells take minutes to hours to retrieve membrane added to the plasma membrane by exocytosis (30). Yet, exocytosis takes only milliseconds at synapses and tens of milliseconds in neuroendocrine cells. It seems possible that faster mechanisms of membrane retrieval might be discovered if one uses assays capable of subsecond time resolution, such as capacitance measurements.

When the pH-sensitive component of exocytosis was suppressed at pH 6.2, a transient rise in Ca_i was found to trigger the exocytic burst in isolation. Under these conditions, exocytosis added 150–250 vesicles to the cell surface within 100 milliseconds and then stopped. Interestingly, the C_m increase during the exocytic burst was transient (30a). The membrane inserted was completely retrieved (time constant, 3.8 seconds) even through repeated, flash-triggered cycles of exo- and endocytosis (not shown). Evidently the exocytosis continuing after the exocytic burst in Fig. 3a masks an episode of rapid endocytosis.

Surprisingly, more membrane was retrieved after the first flash than the exocytic burst had inserted (excess retrieval; see Thomas et al. (30a)). After subsequent flashes, however, exo- and endocytosis were found to be closely matched (unpublished). The finding suggests that excess retrieval after the first flash internalizes membrane that was left stranded on the cell surface after a previous episode of exocytosis. Such exocytosis occurs when Ca-DM-nitrophen first enters the cell from a glass micropipette and encounters the Mg^{2+} within. Mg^{2+} then displaces some of the Ca^{2+}, causing Ca_i to rise transiently to a few μM ("loading transient") and to trigger exocytosis (16). Indeed, the amount of excess retrieval was similar to, and statistically correlated with, the membrane inserted during the loading transient. Evidently the cell surface can accumulate membrane marked for rapid retrieval. Because retrieval is temporally correlated with the flash and not with the insertion of the membrane, we call it triggered. The identity of the trigger is unknown.

Further studies showed excess retrieval to occur in a kinetically distinct phase that was ten times faster (time constant, 350 milliseconds) than the retrieval observed after second and subsequent flashes where exo- and endocytosis were matched (30a). Probably the membrane internalized during excess retrieval had already been prepared extensively. We suggest that exocytosis at low Ca_i (<10 μM) initiates the first steps of the endocytic reaction sequence, but that endocytosis is arrested at a late step and does not go to completion. Instead, the membrane thus prepared must wait for a subsequent trigger, either a large rise in Ca_i or a substance released as a consequence thereof. Once the trigger substance is present, the final steps in membrane retrieval can proceed, which takes about 0.35 seconds.

THE MECHANISM OF EXOCYTIC FUSION

Among the various steps in the exocytic cascade, membrane fusion is the one we know least about. In all likelihood, fusion is initiated by a structure that contains only a few molecules, that arises rarely (once per fusing vesicle) and survives for less than 1 millisecond. These constraints create difficulties for most biochemical assays.

Some of us have studied single fusion events with electrophysiologic methods suitable for observing millisecond events. These studies have focused on mast cells from mutant "beige" mice with their unusually large secretory vesicles (31,32). They have shown that the first aqueous connection between the inside of the vesicle and the outside of the cell (the "fusion pore") has an electrical conductance of about 300 pS, similar to that of a large ion channel (33). Within milliseconds, the fusion pore dilates (34) by incorporation of lipids into its circumference (34,35).

Unfortunately, the molecular mechanism of fusion remains a mystery. The most recent hypothesis (36) envisages "hemifusion," a structure where the cytosolic leaflets of the fusing bilayers have merged long before an action potential triggers synaptic transmission. Transmitter release is prevented because the vesicle lumen is still separated from the extracellular space by a lipid bilayer fashioned from the external leaflet of the plasma membrane and the luminal leaflet of the vesicle membrane. When an action potential triggers a rise in Ca_i, this bilayer somehow ruptures, an aqueous connection forms, and transmitter escapes into the extracellular space.

As with alternatives, the hypothesis of hemifusion in exocytosis has been hard to test experimentally. To avoid the difficulty arising from the presumably short life of early fusion intermediates, we have investigated fusion mediated by the envelope protein, HA, of the influenza virus. Fibroblasts transfected with HA bind erythrocytes, and HA can be triggered by mild acidification to fuse the erythrocyte with the fibroblast (37). Single fusion events can then be studied by measuring the capacitance of the fibroblast and watching its increase during fusion with the erythrocyte (38). As in exocytosis, fusion creates an aqueous channel between the fusing cells that is narrow at first and later dilates. Unlike in mast cells, however, fusion pores of 500 pS sometimes persist for minutes before dilating.

If hemifusion occurs in this system, the continuity between the extracellular leaflets of the two plasma membranes may be expected to allow their lipids to mix long before a fusion pore opens. To test this point, the extracellular leaflet of the erythrocyte plasma membrane was labeled with a fluorescent lipid, di-I, and lipid flux from erythrocyte to fibroblast was monitored by quantitative image analysis. Contrary to our expectation, fusion pores of 500 pS conductance opened minutes before lipid flux was observed. Lipid flux was observed only once the conductance had increased significantly and the pore had dilated (39). Evidently, lipid flux occurs only in dilated pores; only in a dilated pore is there a continuous path of mobile lipids that allows lipid flux between the fusing membranes. These results do not fit with the hemifusion models, at least not in their simplest form (36). In small fusion pores, the outer bilayer leaflets are either not continuous (e.g., as discussed by Almers and Tse (22)), or their lipid molecules are locally immobilized by a ring of protein around the fusion junction. Pores dilate by breaking this ring, and by recruiting lipid molecules into their circumference.

One must recognize, however, that HA-mediated fusion may be molecularly different from the "endoplasmic" fusion of which exocytosis is an example. HA-mediated fusion is probably more akin to "ectoplasmic fusion" (i.e., the fusion of a sperm to an egg (40)) or, perhaps, to the fission of intracellular membrane compartments. In all of the events just mentioned, the membrane leaflets facing each other before fusion are the non-cytosolic, "ectoplasmic" leaflets.

Clearly, new experimental systems are needed. We expect that biophysical approaches, because of their sensitivity and time resolution, will be well suited to elucidate the mechanism of fusion during exocytosis.

REFERENCES

1. Schekman R. Genetic and biochemical analysis of vesicular traffic in yeast. *Curr Opin Cell Biol* 1992;4:587–592.
2. Rothman JE et al., this volume.
3. Sudhof TC, Jahn R. Proteins of synaptic vesicles involved in exocytosis and membrane recycling. *Neuron* 1991;6:665–677.
4. Bennet MK, Scheller RH. The molecular machinery for secretion is conserved from yeast to neurons. *Proc Natl Acad Sci USA* 1993;90:2559–2563.
5. Baker PF, Knight DE. Calcium control of exocytosis and endocytosis in bovine adrenal medullary cells. *Phil Trans R Soc Lond* 1981;B296:83–103.
6. Bittner MA, Holz RW. Kinetic analysis of secretion from permeabilized adrenal chromaffin cells reveals distinct components. *J Biol Chem* 1992a;267:16219–16225.
7. Hay KC, Martin TFJ. Resolution of regulated secretion into sequential MgATP-dependent and calcium-dependent stages mediated by distinct cytosolic proteins. *J Cell Biol* 1992;119:139–151.
8. Bittner MA, Holz RW. A temperature-sensitive step in exocytosis. *J Biol Chem* 1992b;267:16226–16229.
9. Walent JH, Porter BW, Martin TFJ. A novel 145 kd brain cytosolic protein reconstitutes Ca^{2+}-regulated secretion in permeable neuroendocrine cells. *Cell* 1992;70:765–775.
10. Neher E, Marty A. Discrete changes of cell membrane capacitance observed under conditions of enhanced secretion in bovine adrenal chromaffin cells. *Proc Natl Acad Sci USA* 1982;79:6712–6716.
11. Thomas P, Surprenant A, Almers W. Cytosolic Ca^{2+}, exocytotsis, and endocytosis in single melanotrophs of the rat pituitary. *Neuron* 1990;5:723–733.
12. Augustine GJ, Neher E. Calcium requirements for secretion in bovine chromaffin cells. *J Physiol* 1992;450:247–271.
13. Tse A, Tse FW, Almers W, Hille B. Rhythmic exocytosis stimulated by GnRH-induced calcium oscillations in rat gonadotrops. *Science* 1993;260:82–84.
14. Thomas P, Wong JF, Almers W. Millisecond studies of secretion in single rat pituitary cells stimulated by flash photolysis of caged Ca^{2+}. *EMBO J* 1993a;12:303–306.
15. Thomas P, Wong JF, Almers W. A low affinity Ca^{2+} receptor controls the final steps in peptide secretion from pituitary melanotrophs. *Neuron* 1993b;11:93–104.
16. Neher E, Zucker RS. Ca^{2+}-dependent steps in chromaffin cell secretion. *Neuron* 1993;10:21–30.
17. Neher, E, this volume.
18. Burgoyne RD. Control of exocytosis in adrenal chromaffin cells. *Biochim Biophys Acta* 1991;1071: 174–202.
19. Clary DO, Griff IC, Rothman JE. SNAPs, a family of NSF attachment proteins involved in intracellular membrane fusion in animals and yeast. *Cell* 1990;61:709–721.
20. Miyamoto MD. Binomial analysis of quantal transmitter release at glycerol treated frog neuromuscular junctions. *J Physiol* 1975;250:121–142.
21. Steinbach JH, Stevens CF. Neuromuscular transmission. In: Llinas R, Precht W, eds. *Frog Neurobiology*. New York: Springer Verlag; 1976:33–92.
22. Almers W, Tse FW. Transmitter release from synapses: does a preassembled fusion pore initiate exocytosis? *Neuron* 1990;4:813–818.
23. Shoji-Kasai Y, Yoshida A, Kazuki S, Hoshino T, Ogura A, Kondo S, Fujimoto Y, Kuwahara R, Kato R, Takahashi M. Neurotransmitter release from synaptotagmin deficient clonal variants of PC12 cells. *Science* 1992;256:1820–1823.
24. Brose N, Petrenko AG, Südhof TC, Jahn R. Synaptotagmin: a calcium sensor on the synaptic vesicle surface. *Science* 1992;256:1021–1025.

24a. Mizuta M, Inagaki N, Nemoto Y, Matsukara S, Takahashi M, Seino S. Synaptotagmin III is a novel isoform of rat synaptotagmin expressed in endocrine and neuronal cells. *J Biol Chem* 1994;269: 11675–11678.

25. DiAntonio A, Parfitt KD, Schwarz TL. Synaptic transmission persists in *synaptotagmin* mutants of drosophila. *Cell* 1993;73:1281–1290.
26. Nonet M, Grundahi K, Meyer BJ, Rand JB. Synaptic function is impaired but not eliminated in *C elegans* mutants lacking synaptotagmin. *Cell* 1993;73:1291–1305.

27. Fernandez JM, Neher E, Gomperts BD. Capacitance measurements reveal stepwise fusion events in degranulating mast cells. *Nature* 1984;312:453–455.
28. Almers W, Neher E. Gradual and stepwise changes in the membrane capacitance of rat peritoneal mast cells. *J Physiol* 1987;386:205–217.
29. Kirillova J, Thomas P, Almers W. Two independently regulated secretory pathways in mast cells. *J Physiol (Paris)* 1993;87:203–208.
30. Watts C, Marsh M. Endocytosis: what goes in and how? *J Cell Sci* 1992;103:1–8.
30a. Thomas P, Lee AK, Wong JG, Almers W. A triggered mechanism retrieves membrane in seconds after Ca-stimulated exocytosis in single pituitary cells. *J Cell Biol* 1994;124:667–675.
31. Breckenridge LJ, Almers W. Final steps in exocytosis observed in a cell with giant secretory granules. *Proc Natl Acad Sci USA* 1987a;84:1945–1949.
32. Zimmerberg J, Curran M, Cohen FS, Brodwick M. Simultaneous electrical and optical measurements show that membrane fusion precedes secretory granule swelling during exocytosis of beige mouse mast cells. *Proc Natl Acad Sci USA* 1987;84:1585–1589.
33. Breckenridge LJ, Almers W. Currents through the fusion pore that forms during exocytosis of a secretory vesicle. *Nature* 1987b;328:814–817.
34. Spruce AE, Breckenridge LJ, Lee AK, Almers W. Properties of the fusion pore that forms during exocytosis of a mast cell secretory vesicle. *Neuron* 1990,4:643–654.
35. Monck JR, Alvarez de Toledo G, Fernandez JM. Tension in secretory granule membranes causes extensive membrane transfer through the exocytic fusion pore. *Proc Natl Acad Sci USA* 1990; 87:7804–7808.
36. Monck JR, Fernandez JM. The exocytotic fusion pore. *J Cell Biol* 1992;119:1395–1404.
37. Doxsey SJ, Sambrook J, Helenius A, White J. An efficient method for introducing macromolecules into living cells. *J Cell Biol* 1985;101:19–27.
38. Spruce AE, Iwata A, White JM, Almers W. Patch clamp studies of single cell-fusion events mediated by a viral fusion protein. *Nature* 1989;342:555–558.
39. Tse FW, Iwata A, Almers W. Membrane flux through the pore formed by a fusogenic viral envelope protein during cell fusion. *J Cell Biol* 1993;121:543–552.
40. White JM. Membrane fusion. *Science* 1992;258:917–924.

Molecular and Cellular Mechanisms of Neurotransmitter Release, edited by Lennart Stjärne, Paul Greengard, Sten Grillner, Tomas Hökfelt, and David Ottoson, Raven Press, Ltd., New York © 1994.

7

A Novel Mammalian Guanine Nucleotide Exchange Factor (GEF) Specific for Rab Proteins

Janet Burton and Pietro De Camilli

Department of Cell Biology, Howard Hughes Medical Institute, Yale University School of Medicine, New Haven, Connecticut 06510

Intracellular transport involves the budding, targeting, and fusion of carrier vesicles from one intracellular compartment to another (1). During the transport process, subcellular compartments maintain their unique protein compositions, indicating that protein sorting is occurring and that membrane fusion reactions proceed in an orderly and regulated fashion. The biogenesis and exocytosis of synaptic vesicles may be seen as a highly specialized form of vesicular transport. Increasing evidence suggests that mechanisms involved in synaptic vesicle fusion are closely related to mechanisms underlying membrane fusion in all cells, and at all stations of the secretory and endocytic pathways.

It has recently been proposed that fusion events between intracellular membranes are specified by the interaction of integral vesicle membrane proteins (v-SNAREs) with integral target membrane proteins (t-SNAREs), (reviewed in other chapters of this book) (2,3). However, work in yeast has suggested that other protein factors participate in fusion and contribute to the fidelity of v-SNARE and t-SNARE interactions (4).

Several cytosolic factors, unlike the general fusion machinery of NSF (n-ethylmolemide-sensitive fusion protein) and SNAP (soluble NSF attachment protein), have been shown to mediate fusion at a specific vesicular transport step in yeast. For example, the yeast sec8 and sec15 proteins are part of a large cytosolic complex that is proposed to participate in the fusion of post-Golgi secretory vesicles with the plasma membrane (5,6). Another protein, sec 1, which does not appear to be part of this large cytosolic complex, has also been shown to be important in the exocytosis of these vesicles (5,6). Mutations in any of these three proteins disrupts vesicular transport between the Golgi complex and the plasma membrane but does not affect fusion reactions in other vesicular transport steps (5,7). Sec 1 has been shown to genetically interact with sso1 and sso2, yeast homologues of the mammalian plasmalemma t-SNARE, syntaxin (8).

Two yeast proteins related to sec1—Sly1 and Slp1—have been implicated in the fusion of transport vesicles between the endoplasmic reticulum (ER) and the Golgi complex and the Golgi complex and the vacuole, respectively (9–11). The vacuole is the yeast equivalent of the mammalian lysosome. The sly1 and slp1 proteins genetically interact with sed5 and pep12, yeast syntaxin-like proteins located respectively on membranes of the Golgi complex and the vacuole (12). Furthermore, sec1, sly1, and slp1 proteins, although related, are not functionally interchangeable, suggesting that they function at distinct fusion points in the secretory pathway (8). Recently, a sec1 homologue from mammalian brain that is enriched in tissues of neuroendocrine origin and specifically interacts with syntaxin has been identified (13–15). It is likely that mammalian homologues for sly1 and slp1 also exist.

The rab proteins—which refers to *ra*t *b*rain proteins—represent another class of proteins that are proposed to participate in the docking and fusion of transport vesicles to the appropriate target membrane within the cell.

THE RAB FAMILY OF SMALL GTPASES

The rab proteins are a subfamily of the ras guanosine 5′-triphosphatease (GTPase) superfamily. There are approximately 30 rab GTPases that have been identified to date (16,17). The essential role of rab GTPases in vesicular transport was first uncovered in genetic studies in yeast. Two small GTPases, sec4 and ypt1, were identified and found to be essential for transport from the Golgi apparatus to the plasma membrane and from the ER to the Golgi apparatus, respectively (18–20). Ypt1 and sec4 have been found to genetically interact with SNAREs and cytosolic factors that participate in the corresponding stations of the secretory pathway (4,8,9,18).

Molecules related to sec4 and ypt1 proteins were isolated using the polymerase chain reaction with oligonucleotides corresponding to the GTP-binding domains conserved between sec4, ypt1, and other ras-like GTPases (21,22). The acronym rabs is now commonly used to define all members of this protein subfamily regardless of their origin.

Rab proteins have additional regions of sequence homology with sec4 and ypt1 that are not present in ras and ras-related GTPases. Therefore, rab proteins were hypothesized to behave analogously to sec4 and ypt1 proteins in higher eukaryotes to mediate vesicular transport (23–25). The multiplicity of rab proteins and their diverse subcellular localizations agreed with the idea that each of these proteins monitor a distinct subcellular sorting event (17,26).

Several lines of experimental evidence have now supported a role for rab proteins in the docking and fusion of transport vesicles. First, GTP-binding proteins had been implicated in these processes because the nonhydrolyzable analogue of GTP,

GTPγS, was shown to affect membrane fusion reactions. Second, small peptides corresponding to the rab effector domain were also shown to affect vesicular transport (27). The effector domain is a short stretch of amino acids within the GTPase that is very homologous between the different rab protein members, and has been shown to be the site of interaction between the GTPase and its GTPase-activating protein (GAP, see section that follows). For example, the rab3a effector peptide has been shown both to inhibit early transport steps between the ER and the Golgi apparatus and to stimulate exocytosis in mast cells and several other cell types (27,28). Third, results of cell transfection studies in which wild-type rab proteins are overexpressed, or a dominant negative mutant of the rab protein is expressed, supported the role of these proteins at specific steps of the secretory and endocytic pathway (29–31). Additionally, studies with antisense oligonucleotides or antibodies to specific rab proteins interfered with distinct transport events (32–35).

The combination of all these studies, in addition to immunocytochemical findings, has provided strong evidence for the involvement of a distinct rab protein at each station of the secretory and endocytic pathway. Rab1 and rab2 proteins are involved in transport between the ER and the Golgi apparatus (17). Rab3 proteins function in the exocytosis of vesicles with the plasma membrane (35–38). Rab4, rab5, rab7, and rab9 regulate endocytic trafficking steps (30,31,39–41). Rab6 has been proposed to regulate intra-Golgi transport (42). Rab8 protein appears to mediate transport from the trans-Golgi network to the basolateral domain in polarized MDCK cells and to the somatodendritic domain in polarized hippocampal neurons in culture (33,34). Rab17 is specifically expressed in epithelial cells and may play a role in transcytosis from the apical to basolateral domains in these cells (43). Other rab proteins and isoforms of rab proteins just mentioned have been identified, but their precise role in subcellular sorting has not been elucidated.

THE RAB PROTEINS CYCLE BETWEEN A GDP- AND A GTP-BOUND STATE

The exact mechanism by which rab proteins operate in the vesicular transport process has remained elusive. However, it has been postulated that rab proteins, like ras proteins, undergo a conformational change depending upon whether guanosine 5′-diphosphate (GDP) or GTP is bound, and are therefore thought to behave as molecular switches to regulate intracellular transport events (23,44). When the GDP-GTP cycle is disrupted (e.g., by GTPγS), vesicular transport is affected, strongly suggesting that GTP hydrolysis is essential and that rab proteins must cycle between the two guanine nucleotide states to carry out their function (17,23). Results from experiments using GTPγS are consistent with the possibility that the rab-GTPγS protein is capable of one round of docking and fusion, but cannot be reutilized for subsequent rounds of docking and fusion.

The intrinsic rates of GDP-release and GTP-hydrolysis for the small GTPases are quite low, suggesting that accessory proteins that modulate these properties are required. Several accessory proteins that influence the GDP-GTP cycle of rab proteins have been identified through genetic and biochemical approaches. Such factors include the GAPs (GTPase activating proteins) that stimulate the intrinsic GTP hydrolysis rate of the GTPases, GEFs (guanine-nucleotide-exchange factors) that stimulate GDP-release and subsequent binding of GTP to the rab protein, and GDI (guanine-nucleotide dissociation inhibitor). The GDI binds to the membrane associated GDP-form of rab proteins and promotes their dissociation from the membrane. The cytosolic pool of rab proteins exists as a complex with GDI (4,26).

THE IDENTIFICATION OF MSS4, A GEF FOR RAB PROTEINS

The high degree of conservation of the GTPases and their role in vesicular traffic (see previous section) prompted us to undertake a genetic screen using the conditional yeast secretory mutant, sec4-8, to isolate mammalian components that are important in the exocytic process (5). The sec4-8 cells contain a point mutation in the sec4 protein that results in an arrest of cell growth and an accumulation of post-Golgi secretory vesicles at the restrictive temperature of 37°C (18). A rat brain cDNA library was constructed in a yeast expression vector. The cDNAs were then expressed in the sec4-8 cells and assessed for the property of restoring growth to these cells at the restrictive temperature (Fig. 1, (45)). The rationale behind this approach was to identify molecular components of the exocytic targeting and fusion machinery that might be functionally conserved between yeast and the nerve terminal, an idea that has now been further supported by more recent studies (see above; Bennett and Scheller, 1993 (12) other chapters in this book). By this method, we isolated a novel mammalian protein, mss4, which has been shown biochemically to stimulate GDP-release from members of the rab GTPase subfamily and is therefore most likely a GEF (45).

The mss4 protein has a yeast homologue, dss4, which is 27% identical and 51% similar at the amino acid level (45,46). Both proteins are hydrophilic and of similar size, display similar biochemical properties, and are able to genetically interact with the sec4-8 mutant protein. Results from genetic studies suggested that dss4 facilitates but does not replace sec4 function in vivo (46). Using an in vitro GDP-release assay, recombinant mss4 and dss4 proteins were shown to stimulate GDP-release from recombinant sec4 (45,46). Recombinant mss4 protein (s100) and mss4 fusion protein (mss4 maltose-binding protein (MBP)) were found to be equally effective in promoting GDP-release from sec4 protein (Fig. 2, (45)). Mss4 was also shown to stimulate GDP-release from rab3a and ypt1, but not from the yeast ras2, suggesting that mss4 acts on more than one rab but does not interact with other subfamilies of the ras GTPase superfamily. Accordingly, preliminary evidence using an overlay assay technique demonstrates that mss4 binds selectively to recombinant rab but not to mammalian ras or ral proteins that are immobilized on nitrocellulose (Burton et al., in preparation). These findings suggest that mss4 is promiscuous in its interaction with several of the rab proteins, and may therefore influence several intracellu-

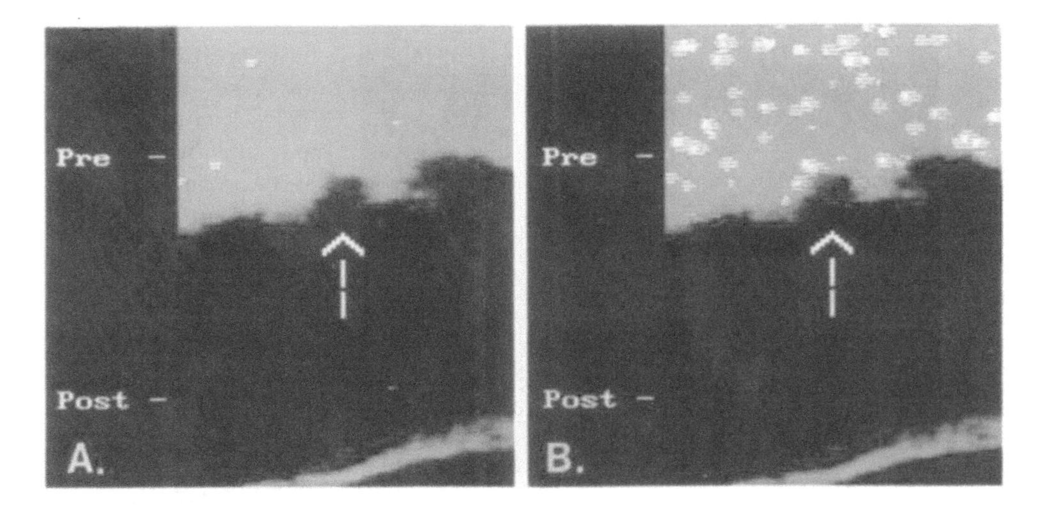

FIG. 1. Calcium QEDs and calcium entry into the synaptic pre-terminal of the squid giant axon occur within microdomains before (*Panel A*) and during (*Panel B*) an action potential. **A,** calcium entry to a portion of a synaptic pre-terminal labeled with *n*-aequorin-*J* during a quiescent period [0.33 sec (10 NTSC video frames)] prior to stimulation. **B,** calcium QEDs occurring in the same region of that pre-terminal upon a single action potential. Activity is clustered within several microdomains across the pre-terminal. The pre-terminal (Pre-) was labeld with *n*-aequorin-*J* as described earlier (Llinás et al., 1992) and is pseudocolored blue in this Fig. The post-terminal (Post-) was injected with procion black and is pseudocolored black. Calcium QEDs are pseudocolored qualitatively in white and shades of pink or red. The striped appearance of these calcium QEDs indicates that the lifetime of these single events was detected within a single NTSC video field (i.e., 16.5 milliseconds or less). An arrow points to a single dendritic spine, which is shown in detail in Fig. 2.

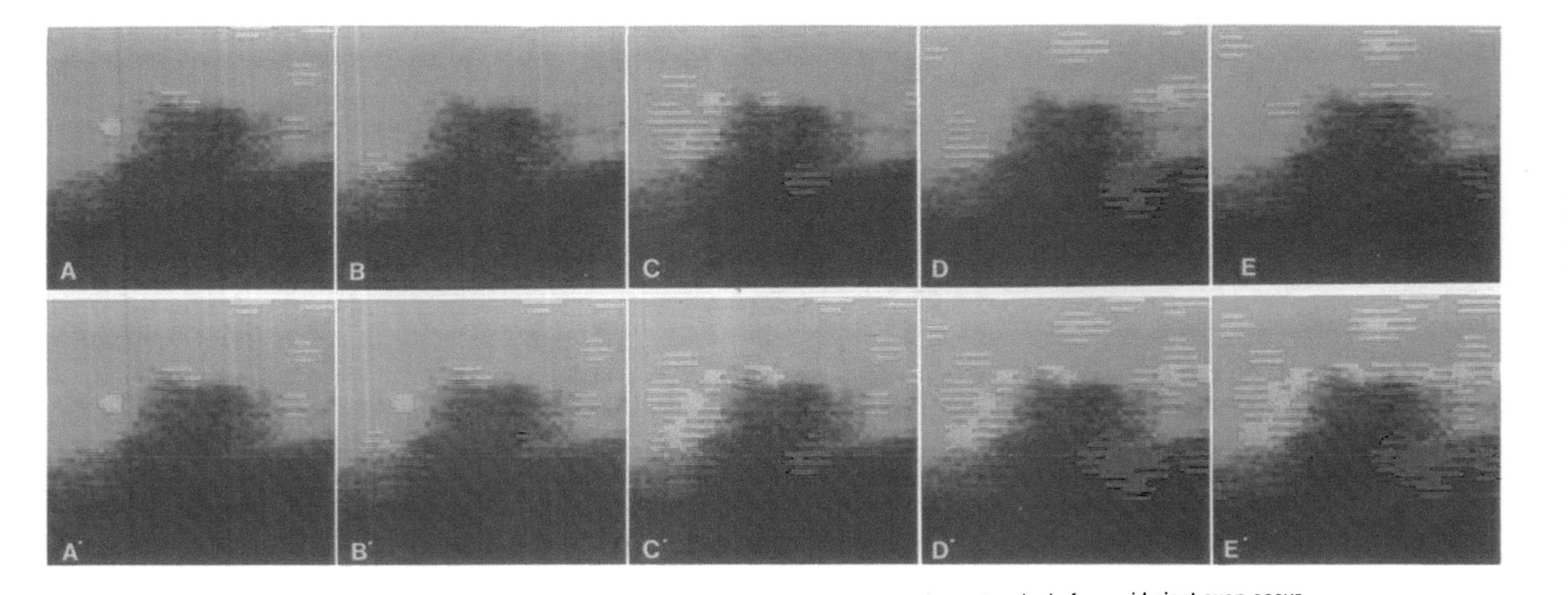

FIG. 2. Calcium QEDs and calcium entry at a single dendritic spine at a synaptic pre-terminal of a squid giant axon occur within spatially limited microdomains during five separate, randomly applied action potentials. Panels **A** through **E** show the recordings of calcium QEDs for each of five separate action potentials applied successively. The QEDs are clustered at the interface between the pre-terminal and the post-terminal. **A** through **E** show progressive integrations of individual recordings of calcium QEDs for the five separate action potentials seen in **A** through **E**. **A,** integration of the single action potential (a repeat of the image seen in **A**). **B,** integration of the two separate action potentials seen in **A** and **B**. **C,** integration of the three separate action potentials seen in **A** through **C**. **D,** integration of the four separate action potentials seen in **A** through **D**. **E,** integration of the five separate action potentials seen in **A** through **E**. Note the clustering of QEDs into microdomains, and the chevron appearance of the individual QEDs discussed in the text and the legend of Fig. 1. This is the dendritic spine indicated with an upward pointing arrow in Fig. 1. All other conditions are as described in the legend for Fig. 1 and the text.

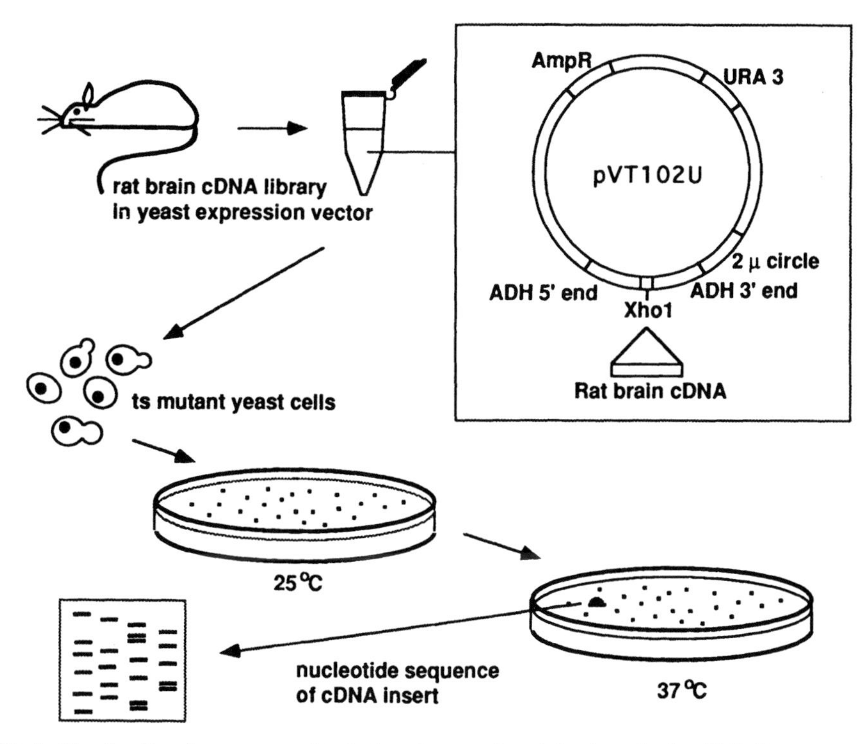

FIG. 1. The functional screen used to isolate the mss4 protein. A rat brain cDNA library was constructed from total brain mRNA that was random primed. The cDNAs were inserted into the yeast expression vector pVT102U in the XhoI restriction site (Burton et al., 1993). The library was then transformed into the sec4-8 yeast strain and selected for the uptake of plasmid by the ability to grow in selective media (lacking uracil) at the permissive temperature of 25°C. Plates were then switched to the restrictive temperature of 37°C. Cells capable of growing at the restrictive temperature appear as large colonies. Large colonies were isolated and the corresponding cDNAs were sequenced. (AmpR, ampicillin resistance gene; ADH, alcohol dehydrogenase gene; URA3, uracil 3 gene; cDNA, complementary DNA; ts, temperature sensitive.)

lar targeting steps. In this respect, mss4 appears to be analogous to the rab GDI, in that it can carry out its biological function on several or all of the rab GTPases, but not on members of other GTPase subfamilies (17,47). The precise mechanism of action of mss4 in the cycle of rab proteins needs to be further investigated. For instance, does mss4 interact with the cytosolic rab-GDI complex to displace GDI and then promote exchange, or is GDI first dissociated from the rab protein by another as yet unidentified protein? Thus far, no other exchange factors for the rab subfamily have been cloned. An exchange activity for the rab3a protein has been identified, but only partially purified, and the corresponding cDNA has not yet been isolated. (48) Whether or not this activity is specific for rab3a has not been determined. It will be of interest to see if this activity corresponds to the mss4 protein or some other as yet unidentified protein.

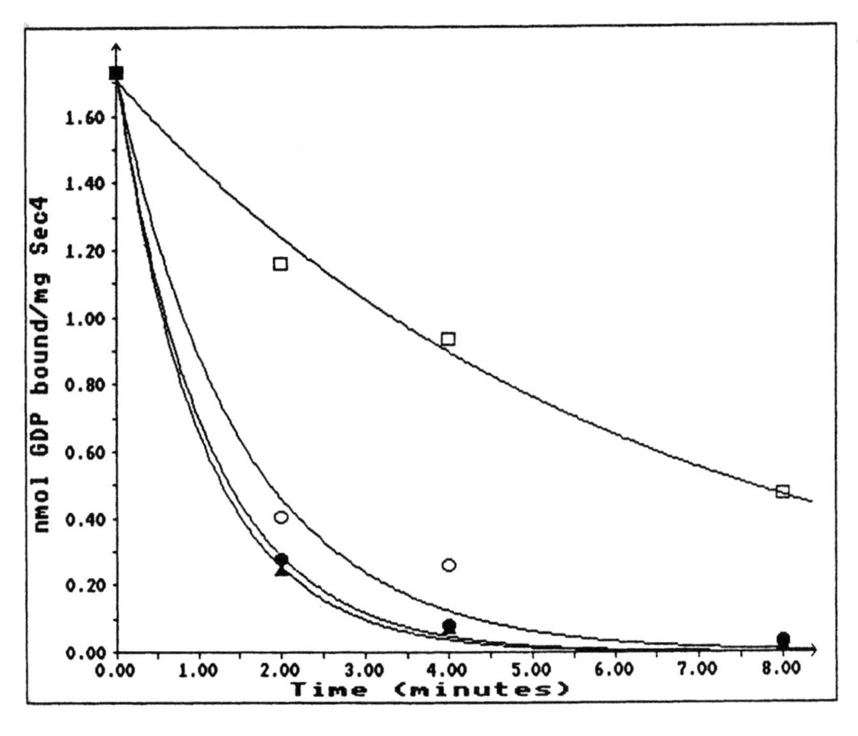

□ Vector only
○ 3X Molar excess Mss4 S100
● 6X Molar excess Mss4 S100
▲ 6X Molar excess Mss4-MBP

FIG. 2. Mss4 S100 and mss4-MBP are equally effective in promoting GDP-release from recombinant Sec4p. Mss4 recombinant protein was produced in bacteria by using either the pET11d vector system (Studier et al., 1990) and purified using an S100 gel filtration column (mss4 S100), or using the pMal-p vector (New England Biolabs) producing a protein in which mss4 is fused to the 3′-end of the MBP (mss4-MBP, (Burton et al., 1993)). GDP-release from the sec4p was measured over time using the filter binding assay (Kabcenell et al., 1990). Both proteins stimulate the release of ^{3}H-GDP from sec4 approximately sixfold relative to the intrinsic release rate when at a concentration of sixfold molar excess relative to the sec4p. (MBP, maltose-binding protein; S100, sephacryl-100)

HOW DOES MSS4 COMPARE WITH OTHER KNOWN GEFS?

The GEF proteins were first discovered and characterized for the ras GTPases. cdc25, a 180 kd protein in the yeast Saccharomyces cerevisiae, was initially shown to play a role in the RAS signaling pathway. In this organism, RAS is known to control adenosine 3′,5′-cyclic phosphate (cAMP) production (49). Mutations in the CDC25 gene resulted in phenotypes analogous to those of RAS inactivation mutations, such as arrest in the G1 phase of the cell cycle. Furthermore, activated alleles of the RAS gene were capable of suppressing cdc25 mutations (49). From these and other genetic data, it was postulated that the cdc25 protein behaves as a guanine-nucleotide exchange factor for ras, a hypothesis that was later confirmed biochemically (49,50). Another yeast protein, bud5, is a putative exchange factor that was isolated based on its ability to suppress strains with a dominant-negative allele of the RAS2 gene. The bud5 protein is believed to be an exchange factor not for ras2, but for a ras-like protein that participates in bud-site selection on the yeast cell membrane (51,52).

Exchange factors for the ras and ras-related GTPases in other organisms have been isolated and many, including the bud5 protein, have been shown to have limited homology (20% to 30% identity) with the carboxy-terminal 380 amino acids of the cdc25 protein. Several studies have confirmed that this domain of homology is important in catalyzing the exchange of GDP for GTP on the ras proteins. Mammalian ras-GRF, a 140 kd brain-specific exchange factor, was initially identified using cdc25 yeast mutants (53). The cDNA that was isolated in this study turned out to be a partial clone of the full-length protein now referred to as the ras-GRF. This truncated cDNA, which contained the portion of mammalian ras-GRF that is homologous to cdc25, was isolated by its ability to restore growth to the cdc25 mutant yeast cells. The corresponding truncated protein (containing the carboxy-terminal domain) was found to promote cAMP production in isolated yeast membranes in an ras2-dependent manner (53). In a separate study, the same domain of ras-GRF was shown biochemically to mediate the exchange of GDP for GTP on both h-ras and n-ras but not on the ras-related GTPases, ralA or cdc24H (54). Another exchange factor, the dbl protein, was originally isolated as an oncogene that resulted from an amino-terminal truncation of about 500 amino acids from the corresponding proto-oncogene. This region was replaced by several amino acids from a different protein, suggesting that the carboxyterminal region of dbl contains a functional domain which, when expressed in a different context, was capable of inducing cell transformation (55). This carboxyterminal region of the dbl protein was later found to be similar to the cdc25 protein, and to mediate guanine-nucleotide exchange on the ras-like protein cdc24H but not on rac1, c-H-ras or rapla GTPases (56). Cdc24H belongs to the rho-subgroup of GTPases, which may influence cell morphology and cytoskeletal organization (56).

A ras exchange factor in Drosophila melanogaster was identified through genetic studies (57). This exchange factor, sos, and the Drosophila ras protein were shown genetically to interact in a tyrosine kinase receptor signaling pathway in the developing photoreceptors of the eye (57). It was found that a region in the middle of the sos molecule was similar to the carboxyterminal domain of the cdc25 protein. It was therefore postulated that sos mediates guanine-nucleotide exchange to activate the ras protein in this pathway (57). Mammalian homologues of the sos protein, msos1 and msos2, have since been isolated; in contrast to the brain-specific ras-GRF, they are ubiquitously expressed in all cells (58). None of the cdc25-related GEFs just mentioned were reported to act on rab proteins.

A GEF with little if any significant homology to the cdc25 protein referred to as smgGDS, was identified by Kaibuchi and collegues (59,60). This protein acts on

Ki-ras and other selected ras-like proteins (59,60). SmgGDS was also not reported to exert its activity on rab proteins. In contrast to the cdc25-related GEFs, smgGDS binds stoichiometrically to its substrates and dissociates them from membranes (60–62).

Other recently identified exchange factors for small GTPases are sec12 and rcc1. These two proteins are unrelated to each other and to other GEFs. The sec12 is a yeast protein that genetically interacts with the sar1 protein. Both the sec12 and sar1 genes are essential for budding of secretory vesicles from the ER (17). Sar1 encodes a GTPase and the sec12 protein behaves as an exchange factor for sar1 (17,63). The sar1 protein is soluble and quite distinct in amino acid sequence from the rab GTPases. Sec12 is a transmembrane protein of 70 kd located in ER membranes that is unable to stimulate GDP release from ras2, ypt1 or arf proteins (63). Rcc 1 is a nuclear protein of 47 kd that stimulates GDP-release from the mammalian ran GTPase. The latter protein is believed to be involved in inhibiting the onset of mitosis until DNA replication is complete (64).

The mss4 and dss4 exchange proteins do not contain significant homology to cdc25 or any of the exchange factors just mentioned, suggesting that mss4 and dss4 may mediate GDP-release from the rab proteins in a different manner (45,46).

In conclusion, there are many families of guanine-nucleotide exchange factors and, in most cases, each one appears to exert its catalytic activity on only one subgroup of the ras GTPase superfamily. The low level of homology between the different types of GEFs suggests that they may mediate exchange via different molecular mechanisms. A more precise understanding of the molecular mechanisms behind mss4 action may further our understanding of how the rab proteins influence membrane interactions, including synaptic vesicle exocytosis.

REFERENCES

1. Palade G. Intracellular aspects of the process of protein secretion. *Science* 1975;189:347–358.
2. Sollner T, Whiteheart SW, Brunner M, et al. SNAP receptors implicated in vesicle targeting and fusion. *Nature* 1993;362:318–324.
3. Sollner T, Bennett MK, Whiteheart SW, Scheller R, Rothman JE. A protein assembly-disassembly pathway in vitro that may correspond to sequential steps of synaptic vesicle docking, activation, and fusion. *Cell* 1993;75:409–418.
4. Novick P, Brennwald P. Friends and family: the role of the rab GTPases in vesicular traffic. *Cell* 1993;75:597–601.
5. Novick P, Field C, Schekman R. Identification of 23 complementation groups required for post-translational events in the yeast secretory pathway. *Cell* 1980;21:205–215.
6. Bowser R, Muller H, Govindan B, Novick P. Sec8p and Sec15p are components of a plasma membrane-associated 19.5S particle that may function downstream of Sec4p to control exocytosis. *J Cell Biol* 1992;118:1041–1056.
7. Salminen A, Novick PJ. The Sec15 protein responds to the function of the GTP binding protein, Sec4, to control vesicular traffic in yeast. *J Cell Biol* 1989;109:1023–1036.
8. Aalto MK, Ronne H, Keranen S. Yeast syntaxins Sso1p and Sso2p belong to a family of related membrane proteins that function in vesicular transport. *EMBO J* 1993;12(11):4095–4104.
9. Ossig R, Dascher C, Trepte HH, Schmitt HD, Gallwitz D. The yeast SLY gene products, suppressors of defects in the essential GTP-binding Ypt1 protein, may act in endoplasmic reticulum-to-Golgi transport. *Mol Cell Bio* 1991;11:2980–2993.

10. Dascher C, Ossig R, Gallwitz D, Schmitt HD. Identification and structure of four yeast genes (SLY) that are able to suppress the functional loss of YPT1, a member of the RAS superfamily. *Mol Cell Bio* 1991;11:872–885
11. Wada Y, Katsuhiko K, Kanbe T, Tanaka K, Anraku Y. The SLP1 gene of *Saccharomyces cerevisiae* is essential for vacuolar morphogenesis and function. *Mol Cell Bio* 1990;10(5):2214–2223.
12. Bennett MK, Scheller RH. The molecular machinery for secretion is conserved from yeast to neurons. *Proc Natl Acad Sci USA* 1993;90:2559–2563.
13. Hata Y, Slaughter CA, Sudhof TC. Synaptic vesicle fusion complex contains unc-18 homologue bound to syntaxin. *Nature* 1993;366:347–351.
14. Garcia EP, Gatti E, Butler M, Burton J, De Camilli P. A rat brain Sec1 homologue related to Rop and UNC18 interacts with syntaxin. *Proc Natl Acad Sci USA* 1994;91:2003–2007.
15. Pevesner J, Hsu SC, Scheller R. N-Sec1: A neurospecific syntaxin binding protein. *Proc Natl Acad Sci USA* 1994 (in press).
16. Pfeffer SR. GTP-binding proteins in intracellular transport. *Trends Cell Biol* 1992;2:41–46.
17. Ferro-Novick S, Novick P. The Role of GTP-binding proteins in transport along the exocytic pathway. *Annu Rev Cell Biol* 1993;9:575–599.
18. Salminen A, Novick P. A ras-like protein is required for a post-Golgi event in yeast secretion. *Cell* 1987;49:527–538.
19. Gallwitz D, Donath C, Sander C. A yeast gene encoding a protein homologous to the human *c-has/bas* proto-oncogene product. *Nature* 1983;306:704–707.
20. Segev N, Mulholland J, Botstein D. The yeast GTP-binding YPT1 protein and a mammalian counterpart are associated with the secretion machinery. *Cell* 1988;52:915–924.
21. Touchot N, Chardin P, Tavitan A. Four additional members of the *ras* gene superfamily isolated by an oligonucleotide strategy: molecular cloning of YPT-related cDNAs from a rat brain library. *Proc Natl Acad Sci USA* 1987;84:8210–8214.
22. Chavrier P, Simons K, Zerial M. The complexity of the Rab and Rho GTP-binding protein subfamilies revealed by a PCR cloning approach. *Gene* 1992;112:261–264.
23. Bourne HR, Sanders DA, McCormick F. The GTPase superfamily: a conserved switch for diverse cell functions. *Nature* 1990;348:125–132.
24. Hall A. The cellular functions of small GTP-binding proteins. *Science* 1990;249:635–640.
25. Balch WE. Small GTP-binding proteins in vesicular transport. *TIBS* 1990;15:473–477.
26. Simons K, Zerial M. Rab proteins and the road maps for intracellular transport. *Neuron* 1993; 11:789–799.
27. Plutner H, Schwaninger R, Pind S, Balch WE. Synthetic peptides of the Rab effector domain inhibit vesicular transport through the secretory pathway. *EMBO J* 1990;9:2375–2383.
28. Oberhauser AF, Monck JR, Balch WE, Fernandez JM. Exocytotic fusion is activated by Rab3a peptides. *Nature* 1992;360:270–273.
29. Tisdale EJ, Bourne JR, Khosravi-Far R, Der CJ, Balch WE. GTP-binding mutants of rab1 and rab2 are potent inhibitors of vesicular transport from the endoplasmic reticulum to the Golgi complex. *J Cell Biol* 1992;119:749–761.
30. Bucci C, Parton RG, Mather IH, et al. The small GTPase rab5 functions as a regulatory factor in the early endocytic pathway. *Cell* 1992;70:715–728.
31. van der Sluijs P, Hull M, Webster P, Male P, Goud B, Mellman I. The small GTP-binding protein rab4 controls an early sorting event in the endocytic pathway. *Cell* 1992;70:729–740.
32. Plutner H, Cox AD, Pind S, et al. Rab1b regulates vesicular transport between the endoplasmic reticulum and successive Golgi compartments. *J Cell Biol* 1991;115:31–43.
33. Huber LA, Pimplikar S, Parton RG, Virta H, Zerial M, Simons K. Rab8, a small GTPase involved in vesicular traffic between the TGN and the basolateral plasma membrane. *J Cell Biol* 1993; 123(1):35–45.
34. Huber LA, de Hoop MJ, Dupree P, Zerial M, Simons K, Dotti C. Protein transport to the dendritic plasma membrane of cultured neurons is regulated by rab8p. *J Cell Biol* 1993;123:47–55.
35. Lledo PM, Vernier P, Vincent JD, Mason WT, Zorec R. Inhibition of rab3B expression attenuates Ca^{2+}-dependent exocytosis in rat anterior pituitary cells. *Nature* 1993;364:540–544.
36. Fischer von Mollard G, Sudhof TC, Jahn R. A small GTP-binding protein dissociates from synaptic vesicles during exocytosis. *Nature* 1991;349:79–81.
37. Fischer von Mollard G, Mignery GA, Baumert M, et al. Rab3 is a small GTP-binding protein exclusively localized to synaptic vesicles. *Proc Natl Acad Sci USA* 1990;87:1988–1992.
38. Matteoli M, Takei K, Cameron R, et al. Association of rab3A with synaptic vesicles at late stages of the secretory pathway. *J Cell Biol* 1991;115(3):625–633.

39. Gorvel JP, Chavrier P, Zerial M, Gruenberg J. Rab5 controls early endosome fusion in vitro. *Cell* 1991;64:915–925.
40. Lombardi D, Soldati T, Riederer MA, Goda Y, Zerial M, Pfeffer SR. Rab9 functions in transport between late endosomes and the trans Golgi network. *EMBO J* 1993;12:677–682.
41. Shapiro AD, Riederer MA, Pfeffer SR. Biochemical analysis of rab9, a ras-like GTPase involved in protein transport from late endosomes to the trans Golgi network. *J Biol Chem* 1993;268:6925–6931.
42. Goud B, Zahraoui A, Tavitian A, Saraste J. Small GTP-binding protein associated with Golgi cisternae. *Nature* 1990;345:553–556.
43. Lutcke A, Jansson S, Parton RG, et al. Rab17, a novel small GTPase, is specific for epithelial cells and is induced during cell polarization. *J Cell Biol* 1993;121:553–564.
44. Bourne HR. Do GTPases direct membrane traffic in secretion? *Cell* 1988;53:669–671.
45. Burton J, Roberts D, Montaldi M, Novick P, De Camilli P. A mammalian guanine-nucleotide-releasing protein enhances function of yeast secretory protein Sec4. *Nature* 1993;361:464–467.
46. Moya M, Roberts D, Novick P. DSS4-1 is a dominant suppressor of sec4-8 that encodes a nucleotide exchange protein that aids sec4p function. *Nature* 1993;361:460–463.
47. Ullrich O, Stenmark H, Alexandrov K, et al. Rab GDP dissociation inhibitor as a general regulator for the membrane association of rab proteins. *J Biol Chem* 1993;268:18143–18150.
48. Burstein ES, Macara IG. Characterization of a guanine nucleotide-releasing factor and a GTPase-activating protein that are specific for the ras-related protein p25rab3A. *Proc Natl Acad Sci USA* 1992;89:1154–1158.
49. Broek D, Toda T, Michaeli T, et al. The S. cerevisiae *CDC25* gene product regulates the *RAS*/adenylate cyclase pathway. *Cell* 1987;48:789–799.
50. Jones S, Vignais ML, Broach JR. The *CDC25* protein of *Saccharomyces cerevisiae* promotes exchange of guanine nucleotides bound to ras. *Mol and Cell Biol* 1991;11(5):2641–2646.
51. Powers S, Gonzales E, Christensen T, Cubert J, Broek D. Functional cloning of BUD5, a CDC25-related gene from *S. cerevisiae* that can suppress a dominant-negative ras2 mutant. *Cell* 1991;65: 1225–1231.
52. Chant J, Corrado K, Pringle JR, Herskowitz I. Yeast bud5, encoding a putative GDP-GTP exchange factor, is necessary for bud site selection and interacts with bud formation gene bem1. *Cell* 1991:65: 1213–1224.
53. Martegani E, Vanoni M, Zippel R, et al. Cloning by functional complementation of a mouse cDNA encoding a homologue of *CDC25*, a *Saccharomyces cerevisiae* ras activator. *EMBO J* 1992;11(6): 2151–2157.
54. Shou C, Farnsworth CL, Neel BG, Feig LA. Molecular cloning of cDNAs encoding a guanine-nucleotide-releasing factor for ras p21. *Nature* 1992;358:351–354.
55. Eva A, Vecchio G, Rao CD, Tronick SR, Aaronson SA. The predicted *DBL* oncogene product defines a distinct class of transforming proteins. *Proc Natl Acad Sci USA* 1988;85:2061–2065.
56. Hart MJ, Eva A, Evans T, Aaronson SA, Cerione RA. Catalysis of guanine nucleotide exchange on the CDC42Hs protein by the *dbl* oncogene product. *Nature* 1991;354:311–314.
57. Simon MA, Bowtell D, Dodson GS, Laverty TR, Rubin GM. Ras1 and a putative guanine nucleotide exchange factor perform crucial steps in signaling by the sevenless protein tyrosine kinase. *Cell* 1991;67:701–716.
58. Bowtell D, Fu P, Simon M, Senior P. Identification of murine homologues of the *Drosophila* son of sevenless gene: potential activators of *ras*. *Proc Natl Acad Sci USA* 1992;89:6511–6515.
59. Kaibuchi K, Mizuno T, Fujioka H, et al. Molecular cloning of the cDNA for stimulatory GDP/GTP exchange protein for *smg* p21s (*ras* p21-like small GTP-binding proteins) and characterization of stimulatory GDP/GTP exchange protein. *Mol and Cell Biol* 1991;11(5):2873–2880.
60. Kikuchi A, Kaibuchi K, Hori Y, et al. Molecular cloning of the human cDNA for a stimulatory GDP/GTP exchange protein for c-Ki-ras p21 and smg21. *Oncogene* 1992;7(2):289–293.
61. Yamamoto T, Kaibuchi K, Mizuno T, Hiroyoshi M, Shirataki H, Takai Y. Purification and characterization from bovine brain cytosol of proteins that regulate the GDP/GTP exchange reaction of *smg* p21s, *ras* p21-like GTP-binding proteins. *J Biol Chem* 1990;265(27):16626–16634.
62. Kawamura S, Kaibuchi K, Hiroyoshi M, Hata Y, Takai Y. Stoichiometric interaction of smgp21 with its GDP/GTP exchange protein and its novel action to regulate the translocation of smgp21 between membrane and cytoplasm. *Biochem Biophys Res Commun* 1991;174(3):1095–1102.
63. Barlowe C, Schekman R. SEC12 encodes a guanine-nucleotide-exchange factor essential for transport vesicle budding from the ER. *Nature* 1993;365:347–349.

64. Bischoff FR, Ponstingl H. Catalysis of guanine nucleotide exchange on ran by the mitotic regulator rcc1. *Nature* 1991;354:80–82.
65. Studier WF, Rosenberg AH, Dunn JJ, Dubendorff JW. Use of T7 rna polymerase to direct expression of cloned genes. *Meth in Enzymol* 1990;185:61–89.
66. Kabcenell AK, Goud B, Northup JK, Novick PJ. Binding and hydrolysis of guanine nucleotides by sec4p, a yeast n protein involved in the regulation of vesicular traffic. *J Biol Chem* 1990;265:9366–9372.

Molecular and Cellular Mechanisms of Neurotransmitter Release, edited by Lennart Stjärne, Paul Greengard, Sten Grillner, Tomas Hökfelt, and David Ottoson, Raven Press, Ltd., New York © 1994.

8

GTP-Binding Proteins: Necessary Components of the Presynaptic Terminal for Synaptic Transmission and Its Modulation

Yu Fang, Sally Durgerian, Trent A. Basarsky, and Philip G. Haydon

Department of Zoology and Genetics, Iowa State University, Ames, Iowa 50011

Since the work of Sir Bernard Katz and the articulation of the vesicle hypothesis for synaptic transmission, many studies have been directed toward identifying the mechanism of synaptic transmission. Due to the extensive work in several laboratories, many synaptic proteins have been identified and characterized (1–17). To elucidate the role of these proteins in synaptic transmission, preparations must be available that permit direct access to the presynaptic terminal for local microperturbation. The giant synapse of the squid (18,19) the presynaptic calyx of the chick ciliary ganglion (20,21) and the neuromuscular preparation of the crayfish (22,23) have yielded information critical for our understanding of calcium action in synaptic transmission. We have chosen a different approach: cultured synapses that form directly between somata of identified neurons (24). Such reconstructed synapses consist of spherical nerve terminals that are ideal for voltage-clamp and for microinjection of antibodies, proteins, and peptides. We are using this experimentally tractable system of synaptic terminals to study presynaptic mechanisms of synaptic transmission (25–32).

In this review we summarize studies using the neuromodulator, FMRFamide, a molluscan neuropeptide, and provide evidence for the involvement of heterotrimeric GTP-binding proteins and the arachidonic acid signaling cascade in regulating presynaptic inhibition. Then we discuss the function of the vesicle associated GTP-binding protein, rab3A, in the synapse.

To gain direct access to the presynaptic terminal, we constructed a chemical synapse between pairs of spherical neuronal somata. Neurons were plated together in a nonadhesive culture environment where soma-soma adhesion was maintained in the absence of neurite extension or adhesion to the culture dish. In this configuration, somata formed chemical synapses with similar properties to the synapse that forms between neurites of the same cell types. Synaptic transmission in this preparation is both quantal and calcium-dependent (30).

MODULATION OF SYNAPTIC TRANSMISSION

The neuropeptide FMRFamide is present throughout the nervous system of *Helisoma*, where it has multiple modulatory actions. In addition to effects on neuronal excitability, FMRFamide causes a presynaptic inhibition of synaptic transmission from presynaptic neuron B5 (24,26,27,30–33). FMRFamide reversibly reduces the magnitude of synaptic transmission through presynaptic effects. Our studies have demonstrated that FMRFamide causes these actions through the regulation of ion channels as well as by reducing the secretory response to internal calcium.

Using whole cell voltage clamp, Bahls et al. (32) demonstrated that FMRFamide enhances the magnitude of a potassium current. Studies using inhibitors of arachidonic acid generation and subsequent metabolism have identified the lipoxygenase pathway of arachidonic acid metabolism as mediating the activation of this potassium current. Coincident with the activation of the potassium current, FMRFamide reduces the magnitude of a high-voltage-activated (HVA) calcium current (26,34). We have tested whether arachidonic acid or its metabolites mediate the inhibitory actions of FMRFamide on calcium current. Addition of arachidonic acid does not change the magnitude of the HVA calcium current and inhibitors of arachidonic acid metabolism do not attenuate the inhibitory effect of FMRFamide on calcium current magnitude (34). Therefore, two distinct signal transduction cascades mediate the modulatory actions of FMRFamide on calcium and potassium currents. Thus, FMRFamide has inhibitory and stimulatory effects on calcium and potassium currents respectively, and acts on the potassium current by activating PLA_2. This coordinate modulation of calcium and potassium currents reduces the magnitude of the calcium signal for exocytosis (35).

REGULATION OF SECRETORY RESPONSE TO CALCIUM

A second route to reducing neurotransmitter release is to reduce the secretory response to internal calcium. Several investigators have documented a correlated change in the frequency of miniature synaptic currents that accompanies a presynaptic modulation of transmitter release. This change in miniature frequency is often used as evidence to demonstrate that the secretory apparatus is a site of modulation underlying the change in evoked transmission. We have directly determined whether FMRFamide acts at this locus, downstream of the regulation of the calcium signal, by calcium clamping synaptic terminals (26). Soma-soma synapses were voltage- and calcium-clamped. The presynaptic neuron was dialyzed with the photosensitive calcium cage, Nitr-5, and the secretory response, which was detected in the postsynaptic cell by the frequency of miniature inhibitory postsynaptic currents (MIPSCs), was assessed during the application of FMRFamide. Nitr-5 was dialyzed presynaptically and ultraviolet (UV) light was used to photolyze the calcium cage to cause a step change in calcium level that was sustained for tens of seconds. Photolysis of nitr-5 caused an acceleration of the frequency of MIPSCs. After recording the pre-test MIPSC frequency, FMRFamide was applied to the synapse. FMRFamide drastically reduced the frequency of MIPSCs. This reduction in MIPSC frequency was not accompanied by changes in internal calcium and was detected in the absence of external calcium. Thus, FMRFamide reduces the secretory response to calcium, by actions on the secretory apparatus (30,36).

Since performing these initial experiments, we have examined secretory apparatus using two additional technical approaches. First we used the calcium cage DM-Nitrophen which, due to its unique properties, is able to cause a rapid transient rise in calcium to levels reaching or exceeding 100μM (37). Flash-photolysis of calcium-loaded DM-Nitrophen causes the synchronized release of neurotransmitter (30,38). Under this calcium clamp condition, FMRFamide reduced flash-photolysis-induced release of transmitter (30). In nonclamped conditions, we have also demonstrated that the frequency of MIPSCs is reduced following FMRFamide application, with no detectable change in resting calcium level as monitored by Fura-2 (30).

Our studies show that a decrease in miniature frequency is related to a reduction in secretory response to calcium that we have demonstrated with calcium clamp methods. However, an observed change in miniature frequency alone cannot be taken as proof of a change in secretory response to calcium. For example, the injection of specific peptides can change the frequency of miniature synaptic currents without affecting the release of transmitter due to action potentials (see next section). Thus, miniature frequency can change without consequences for evoked synaptic transmission. Even when a change in miniature frequency accompanies a simultaneous modulation of evoked transmission, one cannot conclude that the change in evoked release is due to a change in the secretory response to calcium. Nonetheless, there is growing evidence supporting the possibility that the secretory apparatus is a site for regulation during synaptic modulation. For example, in *Aplysia* FMRFamide and serotonin both act on ion channels (39–41) and secretory apparatus (42). Adenosine regulates the frequency of MIPSCs in hippocampal neurons (43,44). Following long-term potentiation (LTP) the frequency of MEPSCs is increased (45) and 5-hydroxytryptamine (5-HT), which enhances transmitter release at crayfish neuromuscular junctions, causes an increase in MEPSP frequency (23).

PTX-SENSITIVE G PROTEINS ARE REQUIRED FOR THE MODULATION OF SYNAPTIC TRANSMISSION

In recent years considerable attention has been paid to the role of GTP-binding proteins in regulating secretion and vectorial membrane transport. We have asked whether pertussis toxin (PTX)-sensitive G proteins are required for the modulation of the synapse. Using a variety of experimental perturbations that include the use of guanine nucleotide analogues and PTX, our studies have demonstrated that PTX-sensitive G proteins are required for the modulatory effect of FMRFamide on calcium and potassium currents and on secretory apparatus. For example, microinjection of the A protomer of PTX into neuron B5 blocks the ability of FMRFamide to activate the potassium current (F. Bahls, unpublished observation). Similarly, PTX injection blocks the FMRFamide induced modulation of the calcium current (34) and secretory apparatus (30). Furthermore, PTX blocked the FMRFamide modulation of action-potential-evoked synaptic transmission (30). Thus, PTX-sensitive G proteins are involved in mediating all of the modulatory effects of FMRFamide in evoked synaptic transmission. The PTX injection into the synapse not only demonstrated that PTX-sensitive G proteins are required for the modulation of the synapse but also showed that the activation of these PTX-sensitive G proteins is not required for functional synaptic transmission.

In addition to heterotrimeric GTP-binding proteins, synapses contain low-molecular-weight, small GTP-binding (smg) proteins. Rab3A is a brain-smg protein that is associated with synaptic vesicles. By analogy with the documented role of smg proteins in membrane trafficking between cellular compartments, rab3A has been implicated in playing a pivotal role in vesicle transport.

GUANINE NUCLEOTIDE ANALOGUES BLOCK SYNAPTIC TRANSMISSION

Since PTX does not ADP-ribosylate rab3A or other smg proteins, we have asked whether a GTP-binding protein is necessary for synaptic transmission by injecting GDPβS. When a sufficient concentration of GDPβS is injected it competes with GTP, and as a result GTP-binding proteins (both heterotrimeric and smg proteins) retain a GDP-bound-state. Presynaptic microinjection of GDPβS, to yield an intracellular concentration of $\geq$ 100μM blocked action-potential-evoked transmitter release, while control injections of GTP did not (Haydon and Fang, unpublished observations). In the squid, GDPβS reduces synaptic transmission without reducing calcium influx, indicating that a GTP-binding protein is a required component of the secretory apparatus (46). Additionally, GDPβS reduces calcium-stimulated capacitance increases in melanotrophs (47). In support of a site of action at secretory apparatus we have demonstrated that GDPβS reduces the frequency of MIPSCs in synapses without reducing the resting level of presynaptic free calcium. Taken together, these data demonstrate that one or more GTP-binding protein is necessary for synaptic transmission.

We have performed experiments complementary to those with GDPβS, where we have activated the cellular complement of GTP-binding proteins with GTPγS. Presynaptic injection of GTPγS in *Helisoma* caused complex actions on synaptic transmission. GTPγS inhibited action-potential-evoked transmitter release but stimulated an increase in the frequency of MIPSCs (48). It is difficult to assign the stimulatory effect of GTPγS on miniature frequency to a specific class of G protein in the *Helisoma* synapse because of the presence of heterotrimeric and smg proteins.

These data are consistent with experiments performed in the squid giant synapse where GTPγS reduced action-potential-evoked synaptic transmission without a reduction in calcium influx. Since the effects of GTPγS in squid were not mimicked by A1Fn-, the inhibitory action of GTPγS is likely due to the stimulation of a smg protein. Due to difficulties in measuring miniature synaptic potential frequency in the squid synapse, the effect of GTPγs on this parameter is unknown. Nonetheless, GTPγS has a stimulatory effect on secretion in neurons that was detected as an increased frequency of MIPSCs (*Helisoma*) similar to its stimulatory action in non-neuronal cells, even though GTPγS blocks action-potential-evoked transmitter release (*Helisoma* and squid). Thus, in many cell types—including mast cells, melanotrophs, chromaffin cells, and neurons—GTP-binding protein activation can stimulate exocytosis.

RAB3A: A SYNAPTIC SMALL GTP-BINDING PROTEIN

Small GTP-binding proteins play critical roles in vectorial membrane transport between cellular compartments (49,50) raising the possibility that rab3A regulates vesicle transport to docking sites at the plasma membrane in the nerve terminal (9,49–54). Rab3A, which is found exclusively in secretory cells, is associated with synaptic vesicles (6,55). Studies using synaptosomes have demonstrated that the association of rab3A with the vesicle is regulated during synaptic transmission. The K^+-induced depolarization of synaptosomes causes a calcium-dependent release of glutamate that ultimately results in a dissociation of rab3A from the vesicle membrane (7). The dissociation of rab3A from the vesicular membrane is regulated by a GDP-dissociation inhibitor protein (GDI). In the GDP-bound-state, rab3A and GDI form a complex that results in dissociation of rab3A from the vesicular membrane (51). Since the association of rab3A with the vesicle membrane is regulated by stimuli that promote transmitter release, it is likely that this protein plays a role in exocytosis. In support of this possibility, treatment of rat anterior pituitary cells with antisense oligonucleotides specific for rab3B—the only rab3 sub-type of protein in these cells—attenuated secretion, (55a) demonstrating a critical role for this protein in exocytosis.

RAB3 IN HELISOMA

In order to elucidate the role of rab3 and other smg proteins in the synapse, a series of experimental manipulations is needed in a fast-transmitting synaptic system. Toward this goal we have asked whether a rab3A homologue is present in the nervous system of *Helisoma* so that it would then be possible to perform microperturbations of this protein in the soma-soma synapse. Using an antibody directed against rat rab3A, we have demonstrated rab3 immunofluorescence in neurons of *Helisoma*, including in presynaptic neuron B5. We have also recently cloned and sequenced rab3 from a *Helisoma* central ganglia cDNA library (Fang, Durgerian and Haydon, unpublished observations). Rab3 from *Helisoma* is 85% identical to *Drosophila* and 77% identical to rat rab3A. Thus, there is a homologue of rab3A in *Helisoma*. Since we have only identified one form of rab3 in *Helisoma*, we call the *Helisoma* homologue rab3 and do not denote it as either A or B. Having identified rab3 in the *Helisoma* nervous system, we can now use the experimentally tractable preparation of giant spherical synapses to elucidate the role of this protein in synaptic transmission.

RAB3A EFFECTOR DOMAIN PEPTIDES

To begin testing the role of smg proteins such as rab3A in synaptic transmission, it is necessary to selectively manipulate specific GTP-binding protein systems. Peptides have been shown to be powerful mimetics of protein action. For example, peptides with the sequence of portions of transducin stimulate phosphodiesterase activity (56). We have used peptide fragments with the sequence of rat rab3A to perturb the holoprotein system in the synapse (57). The ras protein has a defined region, the effector domain, which, when the parent ras protein binds GTP, is exposed and interacts with effector proteins. Rab3A and other homologues of ras have a similar effector domain. Several groups have determined that rab effector domain sequence peptides can stimulate secretion in non-neuronal systems (48, 52,53,58). For example, Oberhauser et al. (53) have demonstrated that rab effector peptides stimulate mast cell degranulation in a manner similar to that of GTPγS. Injection of rab3A effector domain peptides into the soma-soma synapse stimulated an increase in secretion by increasing the frequency of MIPSCs, without affecting action-potential-evoked transmitter release (57) (Fig. 1). Control peptides with the effector domain sequence of a non-synaptic smg protein, ral, or with rab3 sequence in another region of the protein (Fang and Haydon, unpublished observations), did not increase MIPSC frequency. This sequence-specific stimulatory action on MIPSC frequency is not accompanied by a change in resting calcium level, demonstrating that it is exerted at the level of secretory apparatus.

What do guanine nucleotide and peptide injections inform us about the role of GTP-binding proteins in the synapse? Fast synaptic transmission is controlled by vesicles that cycle locally in the nerve terminal. Vesicles fill with neurotransmitters and dock with the presynaptic plasma membrane in preparation for release. Unlike constitutive secretion, vesicles accumulate at this docked stage until elevated calcium stimulates fusion and exocytosis. Following this release of neurotransmitter, membrane is retrieved and, after an endosomal intermediate, the vesicle cycle commences again. The ability of GDPβS to block both MIPSCs and evoked synaptic transmission demonstrates that a GTP-binding protein is necessary for the synapse to function, but a question remains: at which step in the vesicle cycle is the GTP-binding protein essential for synaptic transmission?

The smg proteins have been shown to regulate membrane trafficking in various cellular compartments, a finding that has led to the suggestion that the synapse

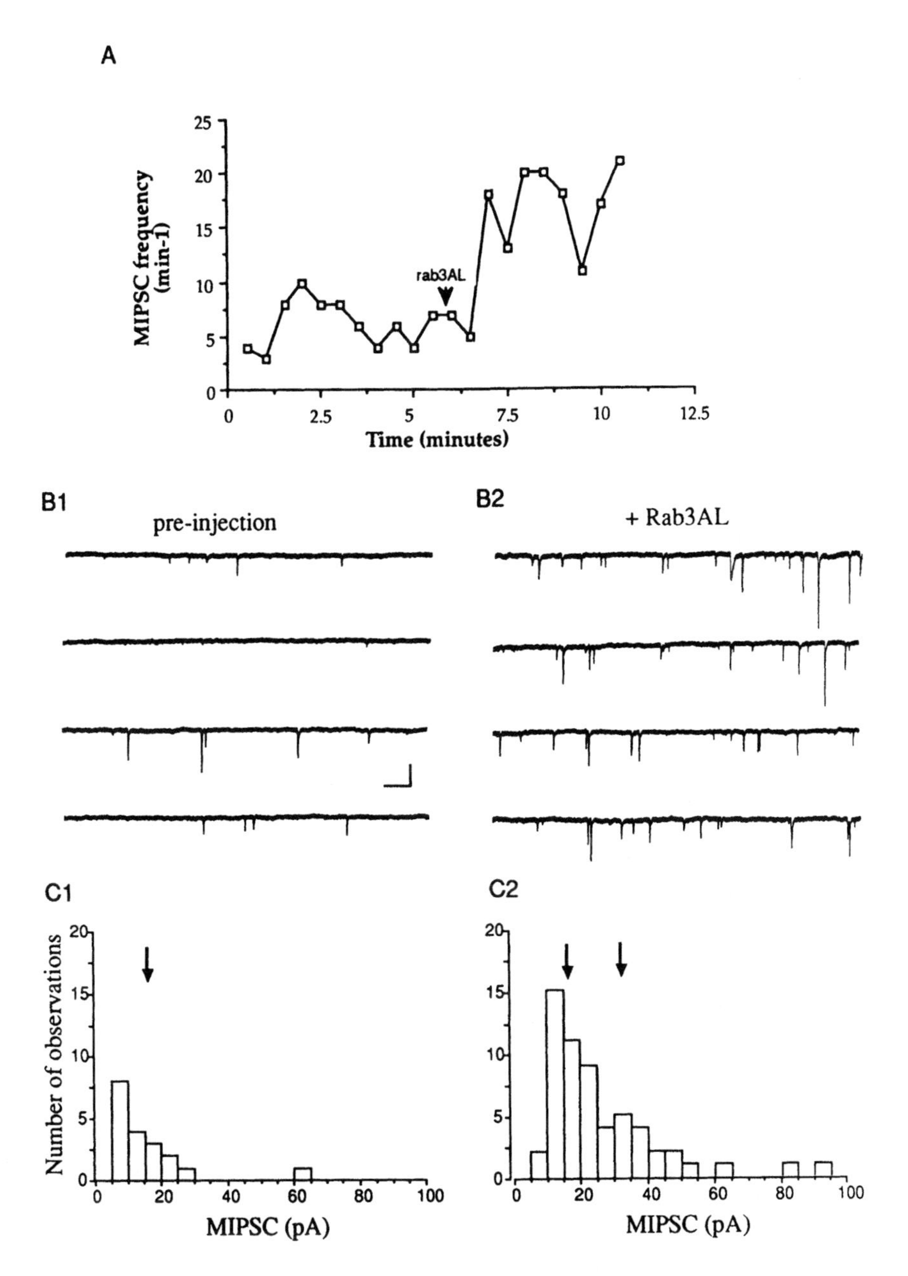

FIG. 1. Microinjection of rab3AL(33–48) accelerates the frequency of MIPSCs. **A.** The frequency of MIPSCs in a single preparation. At the arrow, rab3AL(33–48) injection caused a rapid increase in MIPSC frequency that was sustained for the recording period. **B.** Examples of the MIPSCs recorded before (B1) and after rab3AL(33–48) injection (B2). **C.** Amplitude distribution histograms for the MIPSCs before (C1) and following (C2) rab3AL(33–48) injection. Calibration in B 20pA and 1s. From Richmond and Haydon 1993, FEBS Letters.

specific smg protein, rab3A, plays a similar role in vesicle transport to the plasma membrane. Both rab1 and rab2 are crucial for vesicular transport between the endoplasmic reticulum and the Golgi complex. Rab effector domain peptides and GTPγS block vesicle transport between the ER and Golgi and within the Golgi stack (59). However, in secretory systems, rab3 effector domain peptides and GTPγS stimulate basal secretion (48,52,53,58) indicating that they may act in different manners in the ER and Golgi compared to the plasma membrane. This suggests a fundamental difference between rab protein action in ER to Golgi transport and in secretion.

The rab3 effector domain peptides stimulate an enhanced miniature frequency without changing evoked transmitter release in the synapse. This is contrary to expectations if rab3 controls the initiation of vesicle docking, where one would expect simultaneous actions on miniature frequency and on action-potential-evoked transmitter release. Dialysis of rab3 effector domain peptides and GTPγS into mast cells led Oberhauser et al. (53) to conclude that a rab3 protein is a critical component of the fusion scaffold intimately involved in the normal triggering of secretion. In neurons, rab3A peptides also increase basal release, suggesting that they similarly act on, or with, a target protein in controlling exocytosis.

The possibility that rab3A is intimately associated with the fusion scaffold after docking is supported by immunoelectron microscopy and immunofluorescence studies reveal that rab3A is associated with synaptic vesicles in synaptic terminals. Following stimulation of frog neuromuscular terminals with α-latrotoxin, vesicles incorporate into the plasmalemma. Immunocytochemistry reveals a pattern of rab3A immunoreactivity consistent with it remaining associated with the synaptic vesicle immediately after exocytosis (60). Taken together, these data suggest that rab3A (or rab-like GTP-binding protein) is a necessary component of the fusion scaffold of the nerve terminal that acts at a late step in transmitter release. While we cannot yet eliminate a potential role in docking per se, these data raise two other possible roles for this GTP-binding protein:

1. Perhaps rab3, or other GTP-binding proteins, maintain vesicles in a docked state awaiting a calcium signal to activate calcium-binding proteins such as the rab3A binding protein rabphilin or, alternatively, synaptotagmin.
2. It has been hypothesized that synaptotagmin and rabphilin, a rab3A-binding protein that has high homology with the calcium-binding protein, synaptotagmin, (61) inhibit vesicle exocytosis and that calcium stimulates synaptic transmission by relieving this inhibition. Since rab3A-GTP, but not rab3A-GDP, binds rabphilin, it is likely that rab3A is tightly coupled to proteins that are the calcium trigger for synaptic transmission. Since GDPβS injections reveal a requirement for GTP-bound proteins to permit synaptic transmission, it is possible that rab3A is an essential molecular component of the fusion apparatus of the vesicle.

Studies of the synapse are at an exciting stage where the interface between molecular and cell biology is permitting new insights into the roles of proteins in synaptic transmission. We currently have glimpses at the potential role of a protein such as rab3A in synaptic transmission. However, multiple manipulations are needed where this protein is selectively perturbed to gain an understanding of its precise role in synaptic transmission.

CONCLUSION

The GTP-binding proteins play crucial roles in the synapse. Heterotrimeric G proteins modulate the release of transmitter by regulating the calcium stimulus for exocytosis, as well as the secretory response to calcium. Since PTX blocks synaptic modulation but not transmission, PTX-sensitive heterotrimeric GTP-binding proteins are unlikely to be necessary for the process of synaptic transmission per se. Small GTP-binding proteins such as rab3 are intimately associated with vesicles awaiting release at the synapse. This class of GTP-binding protein likely plays an essential role in the synapse by acting in the macromolecular docking and fusion machinery. The exact role of rab3 remains to be determined. However, we hypothesize that it is required for exocytosis, and cooperates with other proteins to maintain pre-docked vesicles in preparation for exocytosis.

SUMMARY

Using synapses that form between somata of *Helisoma* neurons in cell culture, we have studied the presynaptic regulation of synaptic transmission. Guanosine 5′-triphosphate (GTP)-binding proteins play critical roles in regulating synaptic transmission. Injection of guanine nucleotide analogues has demonstrated that one or more GTP-binding protein is necessary for transmitter release. Heterotrimeric G proteins continuously regulate the amount of transmitter released at the synapse by modulating potassium and calcium channels, and by controlling the secretory response to calcium. Perturbations of the synapse using guanosine 5′-diphosphate (GDP)βS, GTPγS, and rab effector domain peptides suggest that small GTP-binding proteins also play critical roles in the synapse. We discuss the possibility that rab3, or related proteins, are required for exocytosis, and by cooperating with other proteins maintain vesicles in a docked state in the synapse.

ACKNOWLEDGMENTS

This work was supported by an NIH grant (NS26650) and by the MRC of Canada (TAB) and by the NSF (SJD). The continuing support of the Iowa State University Biotechnology Council is gratefully acknowledged.

REFERENCES

1. Valtorta F, Jahn R, Fesce R, Greengard P, Ceccarelli B. Synaptophysin (p38) at the frog neuromuscular junction: its incorporation into the axolemma and recycling after intense quantal secretion. *J Cell Biol* 1988;107:2717–2727.
2. Linial M, Miller K, Scheller RH. VAT-1: an abundant membrane protein from torpedo cholinergic synaptic vesicles. *Neuron* 1989;2:1265–1273.
3. Südhof TC, Baumert M, Perin MS, Jahn RA. A synaptic vesicle membrane protein is conserved from mammals to Drosophila. *Neuron* 1989;2:1475–1481.
4. Ngsee, JK, Miller K, Wendland B, Scheller RH. Multiple GTP-binding proteins from cholinergic synaptic vesicles. *J Neurosci* 1990;10:317–322.
5. Perin MS, Fried VA, Mignery GA, Jahn R, Südhof TC. Phospholipid binding by a synaptic vesicle protein homologous to the regulatory region of protein kinase C. *Nature* 1990;345:260–263.
6. Fischer von Mollard G, Mignery GA, Baumert M, et al. rab3 is a small GTP-binding protein exclusively localized to synaptic vesicles. *Proc Natl Acad Sci USA* 1990;87:1988–1992.
7. Fischer von Mollard G, Südhof TC, Jahn R. A small GTP-binding protein dissociates from synaptic vesicles during exocytosis. *Nature* 1991;349:79–81.
8. Petrenko AG, Perin MS, Davletov BA, Ushkaryov YA, Geppert M, Südhof TC. Binding of synaptotagmin to the alpha-latrotoxin receptor implicates both in synaptic vesicle exocytosis. *Nature* 1991;353:65–68.
9. Südhof TC, Jahn R. Proteins of synaptic vesicles involved in exocytosis and membrane recycling. *Neuron* 1991;6:665–677.
10. Perin MS, Brose N, Jahn R, Südhof TC. Domain structure of synaptotagmin (p65) *J Biol Chem* 1991;266:623–629.
11. Südhof TC, Czernik AJ, Kao HT, et al. Synapsins: mosaics of shared and individual domains in a family of synaptic vesicle phosphoproteins. *Science* 1989;245:1474–1480.
12. Baumert M, Maycox PR, Navone F, De Camilli P, Jahn R. Synaptobrevin: an integral membrane protein of 18,000 daltons present in small synaptic vesicles of rat brain. *EMBO J* 1989;8:379–384.
13. Jahn R, Hell J, Maycox PR. Synaptic vesicles: key organelles involved in neurotransmission. *J Physiol (Paris)* 1990;84:128–133.
14. De Camilli P, Jahn R. Pathways to regulated exocytosis in neurons. *Annu Rev Physiol* 1990;52: 625–645.
15. Elferink LA, Anzai K, Scheller RH. Rabl5, a novel low molecular weight GTP-binding protein specifically expressed in rat brain. *J Biol Chem* 1992;267:5768–5775.
16. Brose N, Petrenko AG, Südhof TC, Jahn R. Synaptotagmin: a calcium sensor on the synaptic vesicle surface. *Science* 1992;256:1021–1025.
17. Bennett MK, Calakos N, Kreiner T, Scheller RH. Synaptic vesicle membrane proteins interact to form a multimeric complex. *J Cell Biol* 1992;116:761–775.
18. Llinas R, Steinberg IZ, Walton K. Presynaptic calcium currents in squid giant synapse. *J Gen Physiol* 1981;33:289–322.
19. Llinas R, Steinberg IZ, Walton K. Relationship between presynaptic calcium current and postsynaptic potential in squid giant synapse. *J Gen Physiol* 1981;33:323–352.
20. Stanley EF. Calcium currents in a vertebrate presynaptic nerve terminal: the chick ciliary ganglion calyx. *Brain Res* 1989;505:341–345.
21. Stanley EF, Atrakchi AH. Calcium currents recorded from a vertebrate presynaptic nerve terminal are resistant to the dihydropyridine nifedipine. *Proc Natl Acad Sci USA* 1990;87:9683–9687.
22. Wojtowicz JM, Atwood HL. Presynaptic long-term facilitation at the crayfish neuromuscular junction: voltage-dependent and ion-dependent phases. *J Neurosci* 1988;8:4667–4674.
23. Zucker RS, Delaney KR, Mulkey R, Tank DW. Presynaptic calcium in transmitter release and posttetanic potentiation. *Ann NY Acad Sci* 1991;635:191–207.
24. Haydon PG. The formation of chemical synapses between cell-cultured neuronal somata. *J Neurosci*. 1988;8:1032–1038.
25. Zoran MJ, Haydon PG, Matthews PJ. Aminergic and peptidergic modulation of motor function at an identified neuromuscular junction in Helisoma. *J Exp Biol* 1989;142:225–243.
26. Man-Son-Hing HJ, Zoran MJ, Lukowiak K, Haydon PG. A neuromodulator of synaptic transmission acts on secretory apparatus as well as ion channels. *Nature* 1989;341:237–239.
27. Man-Son-Hing HJ, Codina J, Abramowitz J, Haydon PG. Microinjection of the a-subunit of the G protein Go2, but not Gol, reduces a voltage-sensitive calcium current. *Cellular Signalling* 1992; 4:429–441.

28. Richmond JE, Funte LR, Smith LR, Price DA, Haydon PG. Activation of a peptidergic synapse locally modulates postsynaptic calcium influx. *J Exp Biol* 1991;161;257–271.
29. Bahls FH, Haydon PG. FMRFamide activates a potassium conductance in an identified neurone of Helisoma. *J Physiol (Lond)* 1991; in press.
30. Haydon PG, Man-Son-Hing H, Doyle RT, Zoran M. FMRFamide modulation of secretory machinery underlying presynaptic inhibition of synaptic transmission requires a pertussis toxin-sensitive G-protein. *J Neurosci* 1991;11:3851–3860.
31. Man-Son-Hing H, Haydon PG. Modulation of growth cone calcium current is mediated by a PTX-sensitive G protein. *Neurosci Lett* 1992;137:133–136.
32. Bahls FH, Richmond JE, Smith WL, Haydon PG. A lipoxygenase pathway of arachidonic acid metabolism mediates FMRFamide activation of a potassium current in an identified neuron of *Helisoma*. *Neurosci Lett* 1992;138:165–168.
33. Haydon PG, Zoran MJ. Chemical synapses in cell culture. In: Chad J, Wheal H, eds. *Cellular neurobiology: a practical approach*. Oxford University Press,1991,
34. Man-Son-Hing HJ, Codina J, Abramowitz J, Haydon PG. Microinjection of the a-subunit of the G protein G_{o2}, but not G_{o1}, reduces a voltage-sensitive calcium current. *Cell Signal* 1992;4:429–441.
35. Durgerian S, Bahls FH, Richmond JE, Doyle RT, Larson DD, Haydon PG. Roles for arachidonic acid and GTP-binding proteins in synaptic transmission *J Physiol (Paris)* 1993;87:123–137.
36. Man-Son-Hing H, Zoran MJ, Lukowiak K, Haydon PG. A neuromodulator of synaptic transmission acts on the secretory apparatus as well as on ion channels. *Nature* 1989;341:237–239.
37. Zucker RS. The calcium concentration clamp: spikes and reversible pulses using the photolabile chelator DM-nitrophen. *Cell Calcium* 1993;14:87–100.
38. Delaney KR, Zucker RS. Calcium released by photolysis of DM-nitrophen stimulates transmitter release at squid giant synapse. *J Physiol (Lond)* 1990;426:473–498.
39. Siegelbaum SA, Camardo JS, Kandel ER. Serotonin and cAMP close single K channels in *Aplysia* sensory neurones. *Nature* 1982;299:413–417.
40. Belardetti F, Kandel ER, Siegelbaum SA. Neuronal inhibition by the peptide FMRFamide involves opening of S K+ channels. *Nature* 1987;325:153–156.
41. Siegelbaum SA, Belardetti F, Camardo JS, Shuster MJ. Modulation of the serotonin-sensitive potassium channel in *Aplysia* sensory neurone cell body and growth cone. *J Exp Biol* 1986;124:287–306.
42. Dale N, Kandel ER. Facilitatory and inhibitory transmitters modulate spontaneous transmitter release at cultured *Aplysia* sensorimotor synapses. *J Physiol (Lond)* 1990;421:203–222.
43. Scholz KP, Miller RJ. Inhibition of quantal transmitter release in the absence of calcium influx by a G protein-linked adenosine receptor at hippocampal synapses. *Neuron* 1992;8:1139–1150.
44. Scanziani M, Capogna M, Gähwiler BH, Thompson SM. Presynaptic inhibition of miniature excitatory synaptic currents by baclofen and adenosine in the hippocampus. *Neuron* 1992;9:919–927.
45. Malgaroli A, Tsien RW. Glutamate-induced long-term potentiation of the frequency of miniature synaptic currents in cultured hippocampal neurons. *Nature* 1992;357:134–139.
46. Hess SD, Doroshenko PA, Augustine GJ. A functional role for GTP-binding proteins in synaptic vesicle cycling. *Science* 1993;259:1169–1172.
47. Okano K, Monck JR, Fernandez JM. GTPgammaS stimulates exocytosis in patch-clamped rat melanotrophs. *Neuron* 1993;11:165–172.
48. Richmond JE, Haydon PG. Rab effector domain peptides stimulate the release of neurotransmitter from cell cultured synapses. *Febs Lett* 1993; in press.
49. Balch WE. Small GTP-binding proteins in vesicular transport. *TIBS* 1990;15:473–477.
50. Pfeffer SR. GTP-binding proteins in intracellular transport. *TICB* 1992;2:41–45.
51. Takai Y, Kaibuchi K, Kikuchi A, Kawata M. Small GTP-binding proteins. *Int Rev Cytol* 1992; 133:187–222.
52. Senyshyn J, Balch WE, Holz RW. Synthetic peptides of the effector-binding domain of rab enhance secretion from digitonin-permeabilized chromaffin cells. *FEBS Lett* 1992;309:41–46.
53. Oberhauser AF, Monck JR, Balch WE, Fernandez JM. Exocytotic fusion is activated by Rab3a peptides. *Nature* 1992;360:270–273.
54. Hess SD, Doroshenko PA, Augustine GJ. A functional role for GTP-binding proteins in synaptic cycling. *Science* 1993;259:1169–1172.
55. Mizoguchi A, Arakawa M, Masutani M, et al. Localization of *smg* p25A/*rab*3A p25, a small GTP-binding protein, at the active zone of the rat neuromuscular junction. *Biochem Biophys Res Commun* 1992;186:1345–1352.
55a. Lledo PM, Vernier P, Vincent J-D, Mason WT, Zorec R. Inhibition of rab3B expression attenuates Ca2+-dependent exocytosis in rat anterior pituitary cells. *Nature* 1993;364:540–544.

56. Rarick HM, Artemyev NO, Hamm HE. A site on rod G protein a subunit that mediates effector activation. *Science* 1992;256:1031–1033.
57. Richmond J, Haydon PG. Rab effector domain peptides stimulate the release of neurotransmitter from cell cultured synapses. *FEBS Lett* 1993;326:124–130.
58. Padfield PJ, Balch WE, Jamieson JD. A synthetic peptide of the rab3a effector domain stimulates amylase release from permeabilized pancreatic acini. *Proc Natl Acad Sci USA* 1992;89:1656–1660.
59. Plutner H, Schwaninger R, Pind S, Balch WE. Synthetic peptides of the rab effector domain inhibit vescular transport through the secretory pathway. *EMBO* 1990;9:2375–2383.
60. Matteoli M, Takei K, Cameron R, et al. Association of rab3A with synaptic vesicles at late stages of the secretory pathway. *J Cell Biol* 1991;115:625–633.
61. Shirataki H, Kaibuchi K, Sakoda T, et al. Rabphilin-3A, a putative target protein for *smg* p25A/*rab*3A p25 small GTP-binding protein related to synaptotagmin. *Mol Cell Biol* 1993;13:2061–2068.

Molecular and Cellular Mechanisms of Neurotransmitter Release, edited by Lennart Stjärne, Paul Greengard, Sten Grillner, Tomas Hökfelt, and David Ottoson, Raven Press, Ltd., New York © 1994.

9

Localization of Calcium Concentration Microdomains at the Active Zone in the Squid Giant Synapse

Rodolfo R. Llinás, Mutsuyuki Sugimori, and Robert B. Silver

Departments of Physiology and Neuroscience, New York University Medical Center, New York, New York 10016

The question of transmitter release in chemical synapses is an area of biological research that continues to be full of surprises. A decade ago the question of synaptic transmitter release would have been addressed simply as a process triggered by calcium entry into the presynaptic terminal, producing transmitter efflux from synaptic vesicles by exocytosis. Today such a statement would be at best a caricature, as much detail was not only unknown but also unfathomed ten years ago. For example, as we have seen in this conference, the very docking of vesicles onto the release site at the presynaptic plasmalemma and the release event itself seem to involve a large number of molecular steps (1,2). What is more, since much of the information is yet to be discovered, new surprises are probably in store for us.

Here, we address the question of calcium entry and how it may be related to transmitter release. Over the last decade, this field has changed by the realization that the term "calcium entry" falls short of describing the event on many levels, among other things. The type of model considered today speaks of compartments and of three-dimensional molecular constraints that trigger, by well-placed and rapidly occurring intracellular calcium-concentration profiles, fast low-calcium-affinity events with the resulting rapid binding and unbinding of kinetics (3).

In agreement with such a view, recent histological evidence confirmed the morphological prerequisite of synaptic vesicles being positioned in direct proximity to the calcium channels (1). Voltage-clamp data calculated that the maximum $[Ca^{2+}]_i$ against the membrane was about 10^{-4} M (4). The data described in the previous paragraphs led the experimental focus to shift from that of cytosolic residual calcium to the study of calcium microdomains (5,6,7,8). The phrase "calcium microdomains" refers to a definite distribution of spatially limited sites for $[Ca^{2+}]_i$

change. These microdomains, which may be described as small domes of increased $[Ca^{2+}]_i$, were predicted to occur at the cytoplasmic surface of the presynaptic terminal membrane within an active zone. In fact, each active calcium channel was expected to produce a rapid (microseconds) increase in $[Ca^{2+}]_i$, lasting throughout the average open time of the channel (7) and then rapidly returning to the pre-opening value when the Ca^{2+} channel closes (7). This calcium influx is thought to generate a $[Ca^{2+}]_i$ profile as high as 200–300 μM in the proximity of the calcium channels. In this case, transmitter release would be triggered by high transient $[Ca^{2+}]_i$ changes localized within the immediate area of the presynaptic release sites where the vesicles are lodged.

A special type of signaling methodology was introduced to test this hypothesis and enable direct study of the existence of such $[Ca^{2+}]_i$ microdomains. A hybrid synthetic *n*-aequorin-J (9) having a sensitivity to $[Ca^{2+}]_i$ in the order of 10^{-4} M, was developed by Drs. Shimomura, Inoye, and Kishi, and was provided through Dr. Shimomura. Aequorin is a photoprotein that emits light upon binding-free Ca^{2+}. Of the total presynaptic injection of this aequorin in the squid giant synapse, 5% was a fluorescent apo-aequorin whose fluorescence emission facilitated intracellular localization (10). Use of the "low-sensitivity" aequorin preparation permitted selective detection of high-calcium-concentration microdomains (10). Distribution of the injected aequorin after impalement and injection of the presynaptic terminal of the giant synapse of the squid *Loligo pealii* was visualized with a fluorescence microscope using a 40X water-immersion lens. Aequorin luminescence was then measured with a VIM camera operated in the photon-counting mode; the images were stored on videotape and characterized by digital image-processing and analysis methods (10) The synapses for such study were bathed in artificial seawater (10 mM Ca^{2+}).

Upon tetanic stimulation, small points of light localized in time and space were detected over the preterminal region in the area of the "active zone " (10). The average diameter of these points was approximately 0.5 μm and they were distributed over roughly 5–10% of the total area of the presynaptic membrane (with an average of 8.4 μm^2 per 100 μm^2). Results from all synapses were quite similar; they demonstrated light emission during presynaptic activation from that portion of the presynaptic terminal that forms the active zone, thus suggesting that calcium concentration is elevated in the range of 10^{-4} M during such an active period. These transient light points were named "quantum emission domains" or QEDs.

These Ca^{2+} microdomains were found to be distributed in such a way that the sites could coincide with active zones where increased $[Ca^{2+}]_i$ triggers neurotransmitter release. By binding to a low-affinity Ca^{2+} site at the presynaptic vesicles, they could initiate the release process (10).

MICRODOMAINS AT THE ACTIVE ZONE

The findings described in the previous section, following the presynaptic injection of the photoprotein *n*-aequorin-*J*, while indicating the presence of light-emitting sites in the presynaptic membrane having the distribution and size consistent with the number and distribution of active zones at that junction (10) were nevertheless not a direct demonstration of coincidence with active zones. However, a direct demonstration of the correspondence of QED with the actual active zones had not yet been attained. In order to generate the necessary data for this assertion to be made unambiguously, we proceeded to localize the contact site between pre- and postsynaptic membranes in the squid giant synapse and to determine whether such a place was also the site of origin of the QEDs.

The techniques utilized consisted of marking the "dendritic" branches that arise from the giant postsynaptic axon and contact the presynaptic plasmalemma (11) using either fluorescent dye or a dark stain. Two substances were found to be useful: 1) dextran-bound fluorescein, which could be injected postsynaptically, resulted in clearly demarcated postdendritic terminals contacting the prefiber, and 2) injection of procion black postsynaptically, which gave a clear image and direct determination of contact points of the postsynaptic dendrites on the pre-axon.

Following postsynaptic staining, the prefiber was injected with fluorescent *n*-aequorin-*J* and the site and amount of protein injected was directly determined, using Texas Red as a flurophor. Under these conditions, as shown in Fig. 1, a clear image could be obtained of the presynaptic terminal (blue) and postsynaptic fibers (black) and the actual site of contact could be visualized at different magnifications. Following this procedure, the presynaptic fiber was activated and synaptic transmission was observed. In order to increase the likelihood of light emission by the *n*-aequorin-*J* triggered due to the influx of calcium into the prefiber, the duration of the prespike was increased to 10 milliseconds by injecting TEA with the *n*-aequorin-*J* presynaptically. The results presented in the next section were obtained with experimental paradigm.

DISTRIBUTION OF CALCIUM-CONCENTRATION MICRODOMAINS AT THE ACTIVE ZONE

The first set of experiments using action potentials was shown to generate distinct light-emission profiles similar to those previously described following injection of *n*-aequorin-*J* (10). In order to visualize in more detail the correlation between calcium entry and the site of synaptic contact, a 60X objective lens with high numerical aperture was used, allowing visualization of the junctional sites.

Experiments were performed in such a manner that the light emission by single action potentials could be compared to determine the constancy and/or variability of the QED sites. The results shown in Fig. 2 indicate that superposition of five different action potentials produced well-localized calcium microdomains at the actual contact point between pre- and postsynaptic structures. In addition, particular QED sites were repeatedly observed from one action potential to the next. On the other hand, it also became apparent that, while the distributions of QEDs were quite similar, the actual sites activated could vary from one spike to the next. These findings indicated that, while action potentials tend to activate calcium entry at the same active site, the precise distribution of microdomains fluctuates from one release event to the next such that the same groups of channels do not seem to be always activated.

DURATION OF LIGHT EMISSION

From analysis of microdomains described in the previous section, a second important measurement could be obtained relating to QED duration during transmitter release. Given the amount of time required for the imaging system to generate a light spot in the video system, these spots were almost exclusively imaged as chevrons. This indicated that duration of light emission was 16 milliseconds or less, which is the amount of time required for half the scan to occur in this video system. From this, and considering that *n*-aequorin-*J* requires 5–10 milliseconds to be activated, the overall duration for the calcium concentration microdomain was on the order of a few milliseconds at the most. Since in no instance did the one QED last more than 16 milliseconds, we may conclude that, as expected from theoretical calculations (5) calcium microdomains appear and disappear quickly. More significantly, however, the probability of the second QED being activated by the same set of calcium channels was found to be very low, even during prolonged action potential. This suggests that calcium microdomains have some self-regulating attribute, as *n*-aequorin-*J* availability in the intracellular space is not the limiting factor in this slow reactivation.

The possibility of self-regulation has already been considered (10) as calcium-dependent channel inactivation seen at the squid synapse (12). From this we can propose that some of the properties of synaptic transmission relating to the probability of synaptic release are governed by the availability of transmission and biochemical events regulating fusion—as well as docking—but that those mechanisms that regulate the ability of the active zone to generate calcium profiles in the microdomain range must also be considered as a significant factor. This is important when considering that the hypothesis of "one active zone–one vesicle" as proposed by Korn and Faber (13,14) seems the most likely explanation for much of the electrophysiological phenomena seen in many central synapses (e.g., a single vesicle contains sufficient transmitter to saturate the very small postsynaptic target areas, especially in the central nervous system (see Chapter 19).

ACKNOWLEDGMENT

Research support to RL and MS NIH-NS13742, F49620-92-J-0363DEF; and for RBS from NSF DCB-9005343 and DCB-9308024.

REFERENCES

1. Jahn R, Südhof TC. Synaptic vesicle traffic: rush hour in the nerve terminal. *J Neurochem* 1993;61: 12–21.
2. Bennet MK, Scheller RH. The molecular machinery for secretion is conserved from yeast to neurons. *Proc Natl Acad Sci USA* 1993;90:2559–2563.
3. Llinás, R. Depolarization release coupling: an overview. *Annals New York Acad of Sci* 1991;635:3–17.
4. Llinás R, Steinberg IZ, Walton K. Presynaptic calcium currents in squid giant synapse. *Biophys J* 1981;33:289–322.
5. Chad JE, Ekert R. Calcium domains associated with individual channels can account for anomalous voltage relations of Ca-dependent responses. *Biophys J* 1984;45:993–999.
6. Simon SM, Sugimori M, Llinás R. Modelling of submembranous calcium-concentration changes and their relation to rate of presynaptic transmitter release in the squid giant synapse. *Biophys J* 1984;45:264a.
7. Simon SM, Llinás R. Compartmentalization of the submembrane calcium activity during calcium influx and its significance in transmitter release. *Biophys J* 1985;48:485–498.
8. Fogelson AL, Zucker RS. Presynaptic calcium diffusion from various arrays of single channels. *Biophys J* 1985;48:1003–1007.
9. Shimomura O, Musicki B, Kishi V. Semi-synthetic aequorins with improved sensitivity to Ca^{2+} ions. *Biochem J* 1989;261:913–920.
10. Llinás R, Sugimori M, Silver RB. Microdomains of high calcium concentration in a presynaptic terminal. *Science* 1992;256:677–679.
11. Young JZ. Structure of nerve fibers and synapses in some invertebrates. *Cold Spring Harbor Symposia Quantitative Biology* 1936;6:1–6.
12. Augustine GJ, Charlton MP, Smith S. Calcium action in synaptic transmitter release. *Ann Rev Neurosci* 1987;10:633–693.
13. Korn J, Mallet A, Triller A, Faber DS. Transmission at a central inhibitory synapse. II. Quantal description of release, with a physical correlate for Binomial *n*. *J Neurophysiol* 1982;48:679–707.
14. Triller A, Korn H. Transmission at a central inhibitory synapse. III. Ultrastructure of physiologically identified and stained terminals. *J Neurophysiol* 1982;48:708–736.

Molecular and Cellular Mechanisms of Neurotransmitter Release, edited by Lennart Stjärne, Paul Greengard, Sten Grillner, Tomas Hökfelt, and David Ottoson, Raven Press, Ltd., New York © 1994.

10

Molecular Pathways for Presynaptic Calcium Signaling

George J. Augustine, *Heinrich Betz, *Kurt Bommert, †Milton P. Charlton, William M. DeBello, ‡Michael Hans, and §Dieter Swandulla

*Department of Neurobiology, Duke University, Durham, North Carolina 27710; *Department of Neurochemistry, Max-Planck-Institute for Brain Research, 60528 Frankfurt, Germany; †Department of Physiology, University of Toronto, Toronto, Ontario M5S 1A8, Canada; ‡The Salk Institute Biotechnology/Industrial Associates, La Jolla, California 92037; and §Department of Pharmacology, Institute for Experimental and Clinical Pharmacology, University of Erlangen-Nürnberg, D-91054 Erlangen, Germany*

One of the most important actions of calcium (Ca) in neuronal signaling is the triggering of secretion of neurotransmitter agents (1). It is now clear that this secretion is due to an elaborate sequence of events in the nerve terminal of the presynaptic (secreting) neuron. This sequence begins with an action potential propagating into the presynaptic terminal, which causes the opening of voltage-gated calcium channels and entry of calcium ions into the terminal (2). These events culminate in rapid accumulation of calcium ions in the presynaptic cytoplasm; the resultant rise in presynaptic calcium concentration, $[Ca]_i$, acts as a signal for the fusion of docked synaptic vesicles with the presynaptic plasma membrane (3,4). Because these vesicles are storage organelles filled with neurotransmitter, their fusion results in the exocytotic secretion, or release, of neurotransmitter on to receptors in the membrane of the postsynaptic neuron.

More recently it has been appreciated that a rise in $[Ca]_i$ serves as a signal for many presynaptic processes in addition to transmitter release (Fig. 1). A short list of Ca-sensitive processes in presynaptic terminals includes activation of Ca-dependent potassium channels (5) inactivation of Ca channels (6,7), regulation of endocytotic retrieval of synaptic vesicle membrane (8), enhancement of the synthesis of certain neurotransmitters (9) regulation of pools of synaptic vesicles available for secretion (10) and the triggering of several forms of synaptic plasticity (11) including synaptic facilitation and post-tetanic potentiation (PTP). Such a large number of Ca-

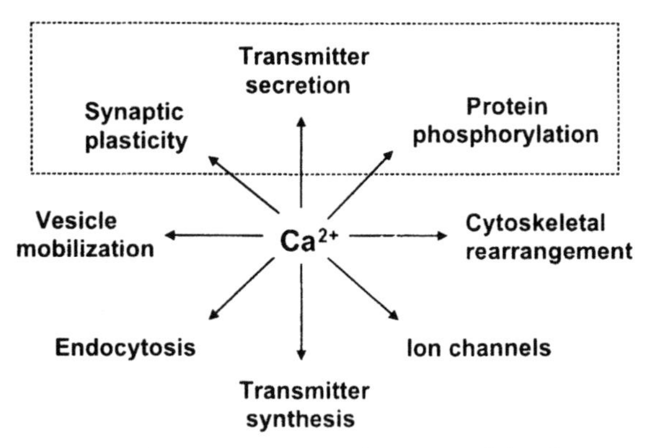

FIG. 1. A rise in intracellular Ca concentration activates a variety of processes inside nerve terminals. The dashed rectangle indicates processes considered in this article.

dependent processes leads to the question of how one signal can differentially regulate so diverse a set of processes. In this chapter we explore this question by evaluating the molecular pathways that mediate two different Ca-dependent processes of presynaptic terminals: 1) the triggering of neurotransmitter secretion, and 2) the production of PTP, a separate process that enhances the ability of the terminal to secrete neurotransmitter. We focus on recent experiments employing the unique "giant" presynaptic terminal of squid to elucidate the molecular pathways for these two processes.

SPATIOTEMPORAL COMPARTMENTALIZATION OF Ca SIGNALING

Although both neurotransmitter secretion and PTP are triggered by rises in $[Ca]_i$, these two processes differ markedly in their kinetic properties. Neurotransmitter release is triggered by single action potentials, occurs within a millisecond of action potential arrival, and is terminated within a few milliseconds after the action potential ends. On the other hand, PTP is only evident after trains of action potentials, builds up over seconds during these trains, and also decays over seconds after the trains end (12). These kinetic differences imply differences in the regulation of these two Ca-dependent processes.

One important clue as to how Ca ions differentially trigger transmitter secretion and PTP comes from the actions of the calcium chelator, EGTA, on these two processes. Microinjection of EGTA into presynaptic terminals can completely eliminate PTP (13,14) an example of the effects of EGTA microinjection into the squid giant nerve terminal is shown in Fig. 2A. In control conditions, a 50 Hz, 10 second train of action potentials causes an enhancement of synaptic transmission that rises slowly and decays over several seconds. The kinetics of this enhancement identifies it as the augmentation phase of PTP. (12) Injection of EGTA completely eliminates this augmentation without affecting basal synaptic transmission. The block of PTP

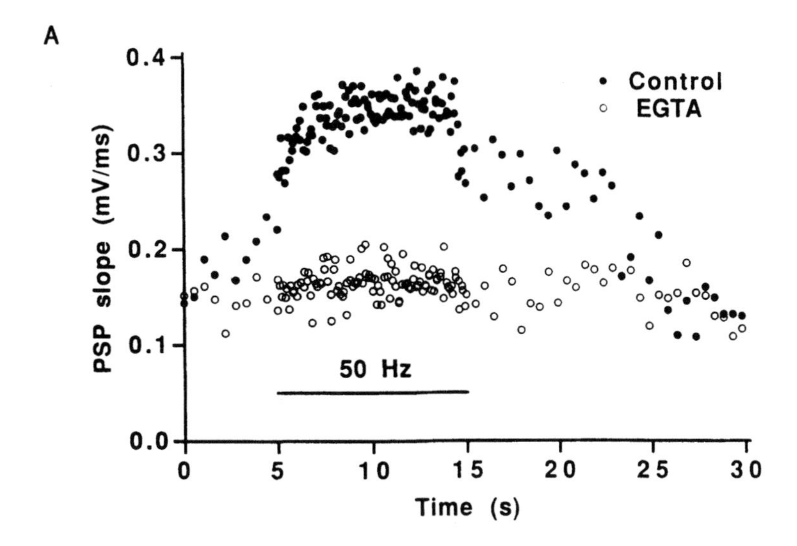

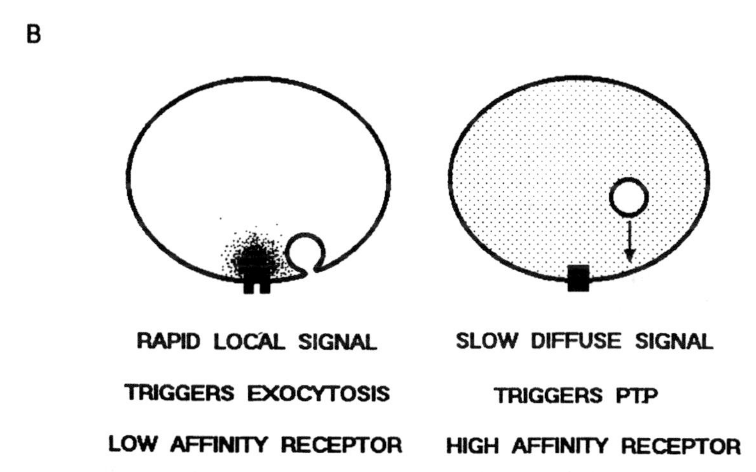

FIG. 2. Ca regulation of post-tetanic potentiation (PTP). **A,** abolition of PTP by intracellular injection of the calcium chelator EGTA. In control conditions, a train of action potentials (50 Hz, during the time indicated by bar) produces a transient rise in transmitter release, as measured by an increase in the slope of the resultant postsynaptic potential (PSP) changes. This increase in transmitter release, and its relaxation following the train, is the build-up and decay of PTP. Microinjection of EGTA eliminated PTP but has no effect on responses to single action potentials evoked before the train. (From Swandulla et al., 1991.) **B,** diagram of Ca signaling during transmitter secretion and PTP. While Ca channels are open, local rises in Ca concentration trigger secretion (left). After the channels close, diffusion of Ca leads to a more widespread rise in Ca that causes PTP (right).

presumably is due to the EGTA binding to Ca ions that would otherwise be available to trigger PTP. The complete lack of effect of EGTA on basal synaptic transmission shows that EGTA does not affect transmitter release evoked by single action potentials (14–16). The ability of EGTA to block PTP, but not release, is strong evidence that these two processes are mechanistically distinct.

Why is it that neurotransmitter secretion, which is known to be activated by a rise in $[Ca]_i$, is not blocked by EGTA? The Ca chelator BAPTA, which has a Ca-binding affinity similar to that of EGTA but binds Ca more rapidly than EGTA, *is* capable of blocking release (16). Thus, it appears that the inability of EGTA to block release is related to the slow kinetics of Ca binding to this chelator. At the concentrations injected into the presynaptic terminals, BAPTA will bind entering Ca ions within a fraction of a μsecond and while EGTA takes tens of μseconds (17). This leads to three interesting conclusions about the Ca signals responsible for neurotransmitter secretion and PTP: 1) it means that Ca signaling of transmitter release must occur within tens of μseconds after the Ca ions enter the terminal, while Ca signaling of PTP takes place over a slower time course; 2) because Ca ions can diffuse no more than a few tens of nanometers in a few tens of μseconds, this also means that the Ca signal for transmitter release must be restricted to a very small spatial compartment while the Ca signal for PTP may be more spatially diffuse; and 3) because of the differences in temporal and spatial dimensions of the Ca signals for these two responses, diffusion will cause $[Ca]_i$ to be much lower when PTP is triggered than when secretion is triggered. Thus, the rise in $[Ca]_i$ that triggers secretion appears to be much higher than the rise in $[Ca]_i$ that triggers PTP. In summary, it appears that Ca ions can differentially activate secretion and PTP because of differences in the spatiotemporal distribution of these ions during the two responses. These differences are shown schematically in Fig. 2B.

In order for the scheme shown in Fig. 2B to work, there must be separate molecular pathways to transduce the two kinds of rises in $[Ca]_i$ into signals for secretion and PTP. In particular, the Ca signal for secretion should be sensed by a receptor that has a low affinity for Ca (i.e., one that responds only to the high, local rises in $[Ca]_i$ that cause secretion), while the signal for PTP should be transduced by a receptor that has a relatively high affinity for Ca ions. Physiological experiments suggest that these two receptors have a number of other important differences in their functional properties (Tables 1 and 2). The remainder of this chapter summarizes experiments that substantiate this notion of separate pathways for secretion and PTP by identifying the distinct molecular mechanisms that produce these two processes.

SYNAPTOTAGMIN AS A MEDIATOR OF NEUROTRANSMITTER SECRETION

Synaptic vesicles are known to contain a number of proteins that are not found on other organelles. (18) Given the critical role that these vesicles play in neuro-

TABLE 1. *Comparison of the properties of the Ca receptor for transmitter secretion and synaptotagmin*

Ca receptor for release	Synaptotagmin
Needs hundreds of μM Ca	Low affinity for Ca (tens–hundreds of μM)
Cooperative Ca binding (n = 4)	Cooperative Ca binding (n = 4)
Rapid Ca binding (tens of μsec)	Binding kinetics unknown
Positioned close to Ca channels	Linked to Ca channels
Divalent selectivity Ca>Sr>Ba	Binds divalents Ca>Sr,Ba
Activated/bypassed by α-latrotoxin	Binds to α-latrotoxin receptor

transmitter secretion, it is reasonable to hypothesize that some of these unique proteins are involved in secretion. One protein that looks particularly promising as the Ca receptor for secretion is synaptotagmin. Synaptotagmin is an integral protein of the synaptic vesicle membrane (18). Molecular cloning of the gene for synaptotagmin indicates that this protein contains two repeats of a motif, called the C2 domain, that is found in a family of Ca- and lipid-binding proteins (19,20). The C2 domain presumably imparts to synaptotagmin an ability to bind Ca ions; recent analysis of the Ca-binding properties of synaptotagmin reveal that not only does this protein bind Ca (21) but that its Ca-binding properties closely resemble those expected of the Ca receptor that triggers transmitter release (Table 1 (22)). The close parallels between the properties of synaptotagmin and the Ca receptor for transmitter secretion have led to the hypothesis that these two entities may be the same (22).

One clear prediction of the hypothesis that synaptotagmin is the Ca receptor is that perturbation of synaptotagmin should disrupt normal synaptic transmission. Indeed, mutation of the synaptotagmin gene in *Drosophila* and *Caenorhabditis* appears to alter synaptic function (23,24,25) although the mechanism of action of these mutations is not well understood. As an alternative approach to test this hypothesis, we have used the squid giant synapse system to define the importance of synaptotagmin for neurotransmitter secretion.

The first issue to be addressed when considering the role of synaptotagmin in secretion from squid nerve terminals is whether or not these nerve terminals possess synaptotagmin. PCR-based molecular cloning reveals that the squid nervous system contains a gene with high homology to synaptotagmin (26). The deduced amino acid sequence of the protein produced by this gene is shown in Fig. 3. The sequence of this protein is more than 50% identical to that of mammalian synaptotagmin. Antibodies generated against a recombinant version of the protein bind to synaptic vesicles purified from squid optic lobe, demonstrating the vesicular location of squid synaptotagmin. Like mammalian synaptotagmin, this protein has a single stretch of hydrophobic amino acids that presumably act as a domain that spans the vesicle membrane (Fig. 3). Most importantly, the squid synaptotagmin protein also possesses two copies of the C2 domain, which is highly conserved in all species studied thus far (26). This fact, in combination with the known Ca- and lipid-binding properties of mammalian synaptotagmin, strongly implies that the C2 domains are functionally significant.

```
Drosophila   MPPNAKSETDAKPEAEPAPASEPAAELESVDQKLEETHHSKFREVDRQEQEVLAEKAAEA   60
Human                                                              MVSESH    6

Squid                EADEITTAVDDTAGSAMETLTKAAPKNGIDKIMDEALHELEKLPIWAIILICA   53
Drosophila   ASQRIAQVESTTRSATTEAQESTTTAVPVIKKIEHVGEVVTEVIAERTG..TWGVVA.II  120
Human        HEALAAPPVTTVATVLPSNATEPASPGEGKEDAFSKLKEKFMNELHKIP..PWALIA.AI   66

Squid        GVLLFLVCGTYCCCKR-ICRRRGKKDGK-KGLKGAVDLRGVQLLGNSIKEK--PDLEELP  109
Drosophila   LVF.VVFGIIFFCVRR-FLKKRRT.D..-G--.KGVDMKS.QL..SAY.EK--P.MEELT  174
Human        VAV.LVLTCCFCICKKCLFKKKNK.K..EKGG.NAINMKD.KD..KTM.DQALK.-DDAE  125

                                                                 Pep11
Squid        MNMEDNEDAESTKSEVKLGKLQYSMDYDFQKGELTVNVIQAADLPGMDMSGTSDPYVKVY  169
Drosophila   ENAEEGDE-.DKQS.Q...R.NFKLE...NSNS.A.TV...EE..AL..G.........Y  233
Human        TGLTDGEEK.EPKE.E...K.QYSLD...QNNQ.L.GI...AE..AL..G.........F  185

                  Pep15
Squid        LMPDKKKKFETKVHRKTLNPVFNESFTFKNVPYADITGKTLVFAIYDFDRFSKHDQIGQV  229
Drosophila   .L................S.....T....SL..ADAMN....FAIF.........Q..EV  293
Human        .L................N.....Q....-V..SELGG....MAVY.........I..EF  244

Squid        QVAMNSIDLGSVMEEWRDLTSPDDDAEKENKLGDICFSLRYVPTAGKLTVVILEAKNLKK  289
Drosophila   K.PLCTI.LAQTI......V.VEGEGGQE-..............................  352
Human        K.PMNTV.FGHVT......Q.AEKE--EQE..............................  302

                                          Pep20
Squid        MDVGGLSDPYVKISLMLNGKRIKKKKTTVKKCTLNPYYNESFAFEVPFEQIQKVSLYVTV  349
Drosophila   .............AI.Q....L.....SV..C..........S.......M..ICLV...  412
Human        .............HL.Q....L.....TI..N..........S.......I..VQVV...  362

Squid        VDYDRHWTSEPIGRTFLGCNSTGTGLRHWSDMLANPRRPIAQWHTLQEVPEKN         402
Drosophila   V...RIGTSEP..RCIL.CMG..TE.........S...........KDPE.TDEILKNMK  472
Human        L...KIGKNDA..KVFV.YNS..AE.........N...........QVEE.VDAMLAVKK  422
```

FIG. 3. Alignment of synaptotagmin amino acid sequence from squid, *Drosophila*, and human. The location of the three C2 domain peptides are indicated (short bars), as are both extravesicular C2 domains (long bars). Boxed region indicates hydrophobic transmembrane anchor. (Modified from Bommert et al., 1993.)

To explore the role of synaptotagmin in squid nerve terminals, peptides that mimic portions of the C2 domains of synaptotagmin were synthesized and microinjected into living nerve terminals. The rationale for this approach was that synaptotagmin is known to bind to a number of presynaptic proteins (see review (27)); thus, injecting peptides might disrupt such protein-protein interactions if the peptides represented regions involved in these interactions (28). These peptides were designed to mimic portions of the C2 domains that are most highly conserved and that are predicted to have a high probability of surface exposure. The sequences selected are shown by the three stretches of amino acids indicated in Fig. 3.

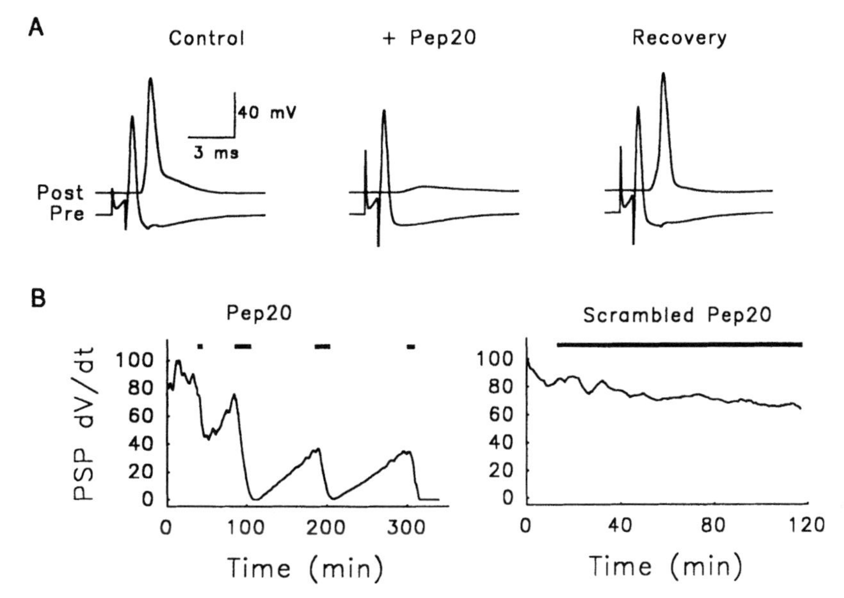

FIG. 4. Synaptotagmin Pep20 inhibits transmitter release. **A,** recordings of presynaptic (Pre) and postsynaptic (Post) electrical signals before (control), during (Pep20), and at long times after (recovery) Pep20 injection. **B,** time course of changes in transmitter release caused by injection of Pep20 (left) or scrambled Pep20 (right). Transmitter release was determined by measuring the postsynaptic potential (PSP) slope (dV/dt), which is normalized to peak preinjection value. Pep20 inhibits transmitter release while the scrambled peptide has little effect.

Microinjection of the longest of these peptides, a 20-mer termed Pep20, produced a dramatic inhibition of synaptic transmission evoked by single action potentials (Fig. 4A). Because the peptide was injected only into the presynaptic neuron, the decrease in synaptic transmission can be attributed to a decline in the amount of neurotransmitter secreted from the presynaptic terminal. Several lines of evidence suggest that the inhibitory effect of this peptide is due to a specific disruption of synaptotagmin function. First, the effect of this peptide often could be reversed; acute injection of peptide inhibited transmitter secretion, but this inhibitory effect gradually subsided over a time course of approximately 1 hour (Fig. 4B). This reversal, which presumably is due to diffusion of the peptide out of the terminal and into the adjacent axonal region of the presynaptic neuron, indicates that the injection was not simply damaging the terminal. Second, injection of another "scrambled" peptide that was identical to Pep20 in both length and amino acid composition, but with an altered sequence of amino acid residues, did not inhibit secretion (Fig. 4B). This indicates that the effect of Pep20 is not simply a consequence of excess free peptides in the nerve terminal. Finally, injection of Pep15 (another synaptotagmin-specific peptide (Fig. 3)) was also capable of inhibiting secretion while injection of Pep11, a peptide representing a sequence common to all C2 domain proteins, had no inhibitory effect (26). Thus, the effect of Pep20 appears to be due to impairment of a protein-protein interaction that involves synaptotagmin binding to an unidentified acceptor protein within the presynaptic terminal.

Three kinds of experiments were performed to identify the mechanism of action of Pep20 and thereby define the role of synaptotagmin in neurotransmitter secretion:

1. The time courses of both Pep20 diffusion and the resultant inhibition of transmitter release were compared by using a confocal laser scanning microscope to visualize the spread of fluorescently-labelled Pep20. It was found that Pep20 slowly spread through the presynaptic terminal over a period of tens of minutes (Fig. 5A) and that the degree of inhibition of secretion was closely correlated with the spread of the peptide (Fig. 5B). This indicates that the rate-limiting step in inhibition is diffusion of the peptide, so that block of secretion is faster than the speed at which the peptide moves through the terminal. This implies that the peptide acts, within the local vesicle cycling pathway, at a step near exocytosis because agents that block upstream steps are slower in inhibiting secretion (29).
2. Video imaging of the intracellular fluorescent Ca indicator, fura-2 (30) was performed to determine whether Pep20 inhibits secretion by affecting the rise in $[Ca]_i$ that triggers release. Injection of Pep20 into fura-filled terminals had no effect on the changes in $[Ca]_i$ evoked by trains of presynaptic action potentials (Fig. 6), even though transmitter release was virtually abolished from these terminals. This indicates that Pep20 does not interfere with presynaptic Ca channels or Ca buffers, but instead works downstream from the steps that lead to elevation of $[Ca]_i$.
3. Electron microscopy was performed on Pep20-injected terminals to determine whether the peptide inhibited synaptic vesicle cycling. Pep20 appeared to have little effect on the gross morphology of active zones, the presynaptic sites of transmitter secretion (Fig. 7A, B). Morphometric methods (29) were used to quantify the actions of Pep20 upon the spatial distribution of synaptic vesicles within the presynaptic terminal. The Pep20-injected terminals had the usual complement of synaptic vesicles, except for a striking, selective accumulation of vesicles in very close proximity to the plasma membrane (Fig. 7C). This subpopulation of vesicles presumably represents vesicles that are docked and ready to undergo exocytosis. Thus, it appears that Pep20 does not inhibit vesicle docking but, instead, blocks secretion at a step that follows docking.

In summary, these experiments indicate that Pep20 inhibits transmitter secretion by preventing a protein-protein interaction that: (1) is downstream from the events that produce entry and accumulation of Ca within the presynaptic terminal, and (2) follows the docking of synaptic vesicles. These results suggest that synaptotagmin is intimately involved in the Ca-regulated fusion of synaptic vesicles with the presynaptic plasma membrane, an event presumably triggered by the Ca-regulated binding of synaptotagmin to an acceptor protein. Although the identity of this acceptor is not yet known, it may be one of the presynaptic plasma membrane proteins known to bind synaptotagmin (e.g., the syntaxins (31), the α-latrotoxin receptor (32), or voltage-gated Ca channels (33)). An additional and provocative candidate

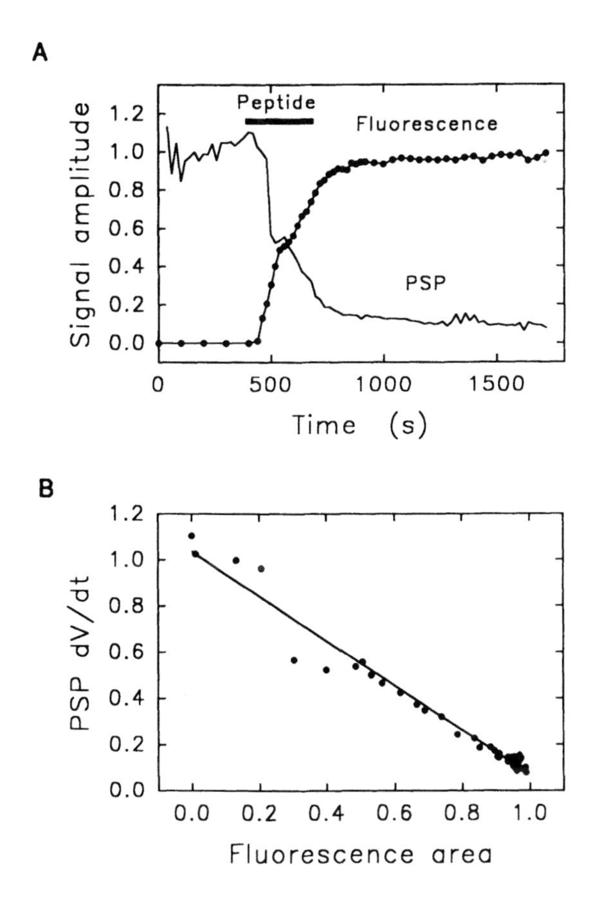

FIG. 5. Pep20 effect is diffusion-limited. **A,** time courses of Pep20 inhibition and Pep20 diffusion. Confocal laser scanning microscopy was used to image the diffusion of Texas-red labeled Pep20 within the presynaptic terminal. Fluorescence was normalized to its peak value, which occurred when the entire presynaptic terminal was filled with labeled Pep20. The postsynaptic potential (PSP) slope was normalized to its mean preinjection value. **B,** high correlation between diffusional spread of peptide and inhibition of release, measured for each of the time points shown in **A.**

for the synaptotagmin acceptor is provided by the recent demonstration that synaptotagmin interacts with proteins of the SNAP receptor (SNARE)/soluble NSF attachment protein (SNAP)/N-ethylmalemide-sensitive fusion protein (NSF) complex, a putative "fusion machine" for constitutive cellular secretion (34). This suggests an attractive mechanism for transmitter release in which the Ca-regulated binding of synaptotagmin to its acceptor protein(s) causes secretion either by activating the SNARE/SNAP/NSF fusion complex or by removing a synaptotagmin-imposed barrier to fusion. However, many more critical experimental tests will be needed to distinguish among these many possibilities.

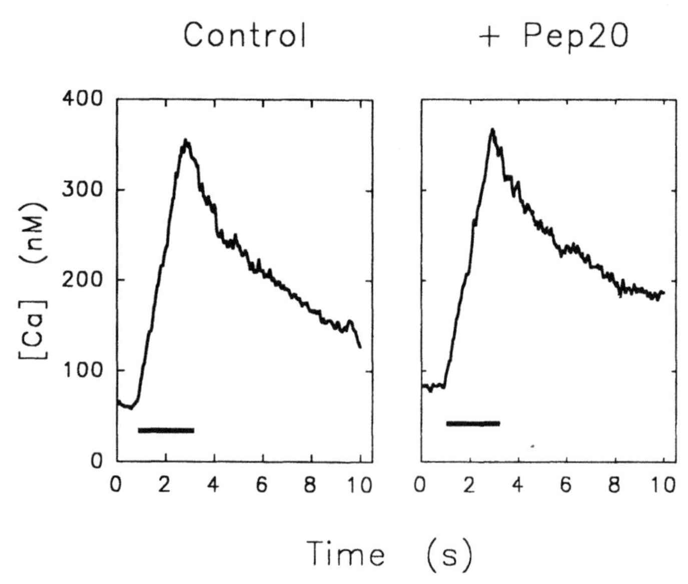

FIG. 6. Presynaptic Ca signals are not changed by Pep20 injection. Terminals were filled with fura-2 and brief trains of action potentials (2 seconds, 50 Hz) were used to evoke changes in presynaptic Ca concentration ([Ca]) both before (control) and after (Pep20) Pep20 injection.

CALCIUM/CALMODULIN-DEPENDENT KINASE AS A MEDIATOR OF PTP

Another protein found on synaptic vesicles, and elsewhere within presynaptic terminals, is the Ca/calmodulin-dependent protein kinase, type II (CaMKII; 35). The functional properties of this kinase parallel those of the Ca receptor responsible for PTP (Table 2). Further, microinjection of CaMKII into nerve terminals is known to potentiate neurotransmitter secretion (36). Thus, it is possible that this kinase could mediate PTP.

To evaluate the possible involvement of CaMKII in PTP, we microinjected a peptide inhibitor of this kinase (37) into the squid giant nerve terminal (38). This inhibitor selectively eliminated PTP without affecting basal synaptic transmission

TABLE 2. *Comparison of the properties of the Ca receptor for PTP and Calcium/Calmodulin-Dependent Protein Kinase, type II (CaMKII)*

Ca receptor for PTP	CaMKII
High affinity for Ca (μM sufficient)	High affinity for Ca (via calmodulin)
Possible cooperative binding	Cooperative Ca binding by calmodulin
Slow Ca binding kinetics (μsec–sec)	Activates in msec–sec time range
Not clustered at released sites	Found on vesicles and in cytoplasm

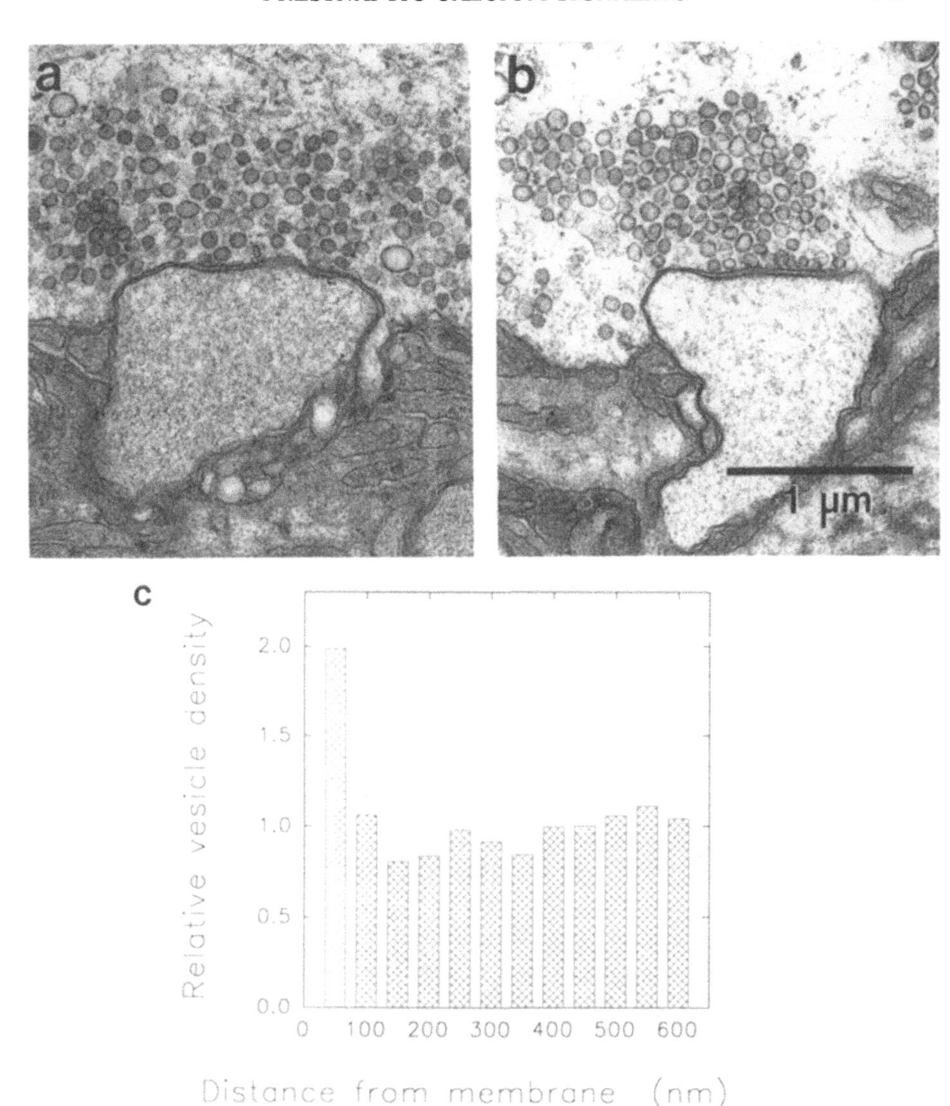

FIG. 7. Electron microscopy reveals that Pep20 injection selectively increases docked synaptic vesicles. **A,** single active zone from a terminal injected with a control, noninhibitory peptide (Pep11) exhibits typical presynaptic morphology. **B,** single active zone from a Pep20 injected terminal exhibits similar morphology. **C,** spatial distribution of synaptic vesicles around each active zone determined by measuring the shortest distance from each vesicle to the plasma membrane. A histogram, obtained by dividing the total number of vesicles per active zone per spatial bin for Pep20 terminals by values for control terminals, indicates that Pep20 increased the number of vesicles in close proximity to the presynaptic plasma membrane.

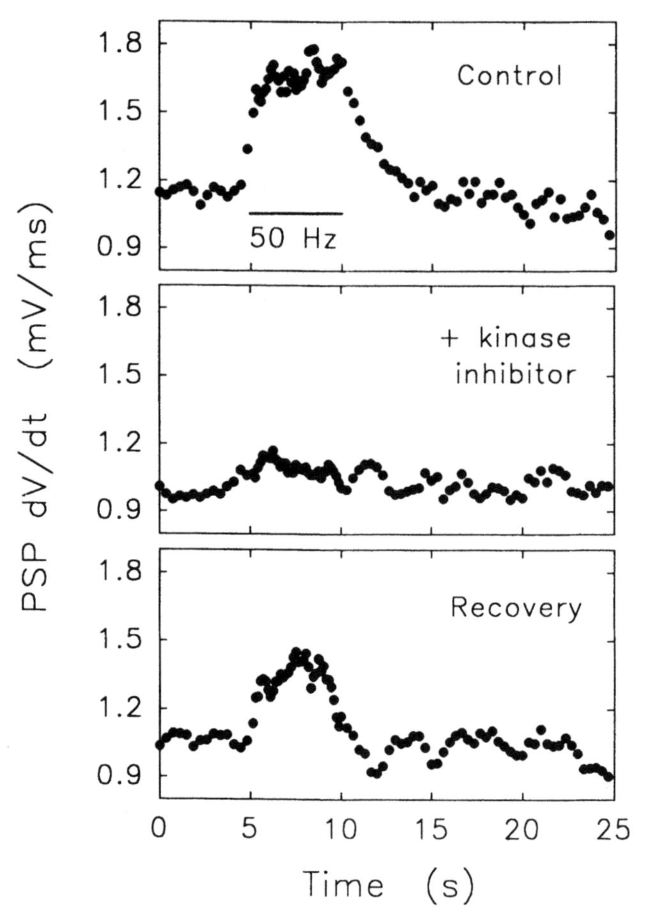

FIG. 8. Calcium/calmodulin-dependent protein kinase (CaMKII) inhibitory peptide selectively and reversibly abolishes postsynaptic potentiation (PTP). Trains of action potentials were delivered to the synapse before (control), during (+ kinase inhibitor), and at long times after (recovery) injection of CaMKII autoinhibitory domain peptide. (From Hans et al., 1994.)

(Fig. 8). Like the inhibition of secretion by Pep20, the inhibitory effect of this peptide upon PTP could be progressively reversed over a time span of 1 hour. Simultaneous fura-2 measurements indicated that the rise in $[Ca]_i$ produced by the trains of action potentials that elicited PTP were not affected by peptide injection (38). Thus, the inhibitory effect of the peptide was not due to alterations in Ca entry or accumulation.

Although the CaMKII inhibitor peptide represents a sequence of amino acids found in this kinase, it has been reported that this peptide also has some ability to inhibit other kinases (39). Because activators of protein kinase C (PKC) potentiate neurotransmitter release (40,41), it is possible that the inhibitor peptide blocks PTP by inhibiting PKC rather than by inhibiting CaMKII. To distinguish between these

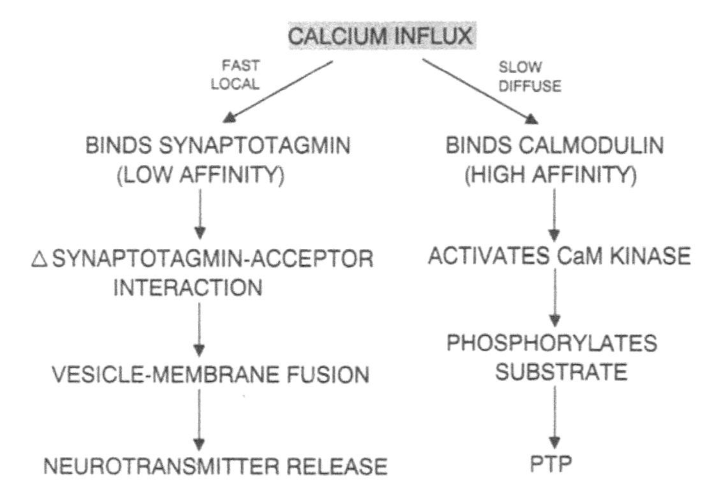

FIG. 9. A model for differential regulation of transmitter secretion and postsynaptic potentiation (PTP) by Ca ions. A fast, local Ca signal triggers transmitter release by binding to synaptotagmin while the slower, more diffuse consequent of this signal triggers PTP by binding to calmodulin and subsequently activating calcium/calmodulin-dependent protein kinase (CaMKinase).

two possibilities, we examined the sensitivity of PTP to staurosporine, a compound that is more potent in inhibiting PKC than in inhibiting CaMKII (42). This compound had no effect on PTP, or on basal synaptic transmission, at a concentration (5 μM) that should be capable of inhibiting PKC. Thus, we conclude that the ability of the CaMKII inhibitor peptide to prevent PTP is due to an interaction with CaMKII rather than with PKC.

These data are consistent with the idea that PTP is mediated by CaMKII. Presumably this kinase is activated by a rise in $[Ca]_i$, which allows calmodulin to activate the kinase and phosphorylate one or more of the many substrate proteins found within presynaptic terminals (43). One particularly interesting substrate for the CaMKII is synapsin, a phosphorylation-sensitive link between synaptic vesicles and the presynaptic cytoskeleton (44,45). A number of lines of evidence suggest that phosphorylation of synapsin by CaMKII enhances neurotransmitter release (36,46). However, further work will be needed to conclusively identify the CaMKII substrate involved in PTP.

SUMMARY

The results presented in this article describe two distinct, Ca-regulated molecular pathways in presynaptic terminals and implicate these two pathways in differentially mediating neurotransmitter secretion and PTP. Our current view of the Ca-dependent triggering of secretion and PTP is shown in Fig. 9. According to this scheme, differential activation of these two pathways is achieved by a combination of diffusion-based dilution of Ca that enters the terminal through voltage-gated Ca channels and by coupling these pathways to Ca receptors with different affinities for Ca ions. A simple way to achieve these conditions is to position these two receptors at different distances from the Ca channels, as shown in Fig. 2. Given that Ca ions are involved in activating many different presynaptic processes (Fig. 1), we propose that closer scrutiny of the molecular physiology of nerve terminals will reveal a wide variety of Ca-activated pathways responsible for producing these diverse processes.

ACKNOWLEDGMENTS

We thank Marisa Rose and Michelle Balow for help in preparing this article and Marie E. Burns for helpful comments. Supported by NIH grant NS-21624 to GJA, DFG funds to HB and DS, and MRC funds to MPC.

REFERENCES

1. Katz B. *The release of neurotransmitter substances*. Liverpool, England: Liverpool University Press; 1969.
2. Augustine GJ, Charlton MP, Smith SJ. Calcium action in synaptic transmitter release. *Ann Rev Neurosci* 1987;10:633–693.
3. Heuser JE, Reese TS, Dennis MJ, Jan Y, Jan L, Evans L. Synaptic vesicle exocytosis captured by quick freezing and correlated with quantal transmitter release. *J Cell Biol* 1979;51:275–300.
4. Torri-Tarelli F, Grohovaz F, Fesce R, Ceccarelli B. Temporal coincidence between synaptic vesicle fusion and quantal secretion of acetylcholine. *J Cell Biol* 1985;101:1386–1399.
5. Robitaille R, Garcia ML, Kaczorowski GJ, Charlton MP. Functional colocalization of calcium and calcium-gated potassium channels in control of transmitter release. *Neuron* 1993;11:645–655.
6. Augustine GJ, Eckert R. Calcium-dependent inactivation of presynaptic calcium channels. *Soc Neurosci Abs* 1984;10:194.
7. Yawo H, Momiyama A. Re-evaluation of calcium currents in pre- and postsynaptic neurones of the chick ciliary ganglion. *J Physiol* 1993;460:153–172.
8. Ceccarelli B, Hurlbut WP. Ca-dependent recycling of synaptic vesicles at the frog neuromuscular junction. *J Cell Biol* 1980;87:297–303.
9. Zigmond RE, Schwarzschild MA, Rittenhouse AR. Acute regulation of tyrosine hydroxylase by nerve activity and by neurotransmitters via phosphorylation. *Ann Rev Neurosci* 1989;12:415–462.
10. Greengard P, Valtorta F, Czernik AJ, Benfenati F. Synaptic vesicle phosphoproteins and regulation of synaptic function. *Science* 1993;259:780–785.
11. Zucker RS. Short term synaptic plasticity. *Ann Rev Neurosci* 1989;12:13–31.
12. Magleby KL, Zengel JE. A quantitative description of tetanic and post-tetanic potentiation of transmitter release at the frog neuromuscular junction. *J Physiol* 1975;245:183–208.
13. Delaney K, Tank DW, Zucker RS. Presynaptic calcium and serotonin-mediated enhancement of transmitter release at crayfish neuromuscular junction. *J Neurosci* 1991;11:2631–2643.
14. Swandulla D, Hans M, Zipser K, Augustine GJ. Role of residual calcium in synaptic depression and posttetanic potentiation: fast and slow calcium signaling in nerve terminals. *Neuron* 1991;7:915–926.
15. Adams DJ, Takeda K, Umbach JA. Inhibitors of calcium buffering depress evoked transmitter release at the squid giant synapse. *J Physiol* 1985;369:145–159.
16. Adler EM, Augustine GJ, Duffy SN, Charlton MP. Alien intracellular calcium chelators attenuate neurotransmitter release at the squid giant synapse. *J Neurosci* 1991;11:1496–1507.
17. Augustine GJ, Adler EM, Charlton MP. The calcium signal for transmitter secretion from presynaptic nerve terminals. *Ann NY Acad Sci* 1991;635:365–381.

18. Sudhof TC, Jahn R. Proteins of synaptic vesicles involved in exocytosis and membrane recycling. *Neuron* 1991;6:657.
19. Perin MS, Fried VA, Mignery GA, Jahn R, Sudhof TC. Phospholipid binding by a synaptic vesicle protein homologous to the regulatory region of protein kinase C. *Nature* 1990;345:260–263.
20. Clark CD, Lin LL, Kriz RW, Ramesha CS, Sultzman LA, Lin AY, Milona N, Knopf JL. A novel arachidonic acid-selective cystolic PLA_2 contains a Ca^{2+}-dependent translocation domain with homology to PKC and GAP. *Cell* 1991;65:1043–1051.
21. Brose N, Petrenko AG, Sudhof TC, Reinhard J. Synaptotagmin: a calcium sensor on the synaptic vesicle surface. *Science* 1992;256:1021–1025.
22. DeBello WM, Betz H, Augustine GJ. Synaptotagmin and neurotransmitter release. *Cell* 1993;74: 947–950.
23. DiAntonio A, Parfitt KD, Schwarz TL. Synaptic transmission persists in synaptotagmin mutants of drosophila. *Cell* 1993;73:1281–1290.
24. Nonet ML, Grundahl KM, Meyer BJ, Rand JB. Synaptic function is impaired but not eliminated in C. elegans mutants lacking synaptotagmin. *Cell* 1993;73;1291–1305.
25. Littleton JT, Stern M, Schulze K, Perin M, Bellen HJ. Mutational analysis of *Drosophila* synaptotagmin demonstrates its essential role in Ca-activated neurotransmitter release. *Cell* 1993;74: 1125–1134.
26. Bommert K, Charlton MP, DeBello WM, Chin GJ, Betz H, Augustine GJ. Inhibition of neurotransmitter release by C2-domain peptides implicates synaptotagmin in exocytosis. *Nature* 1993; 363:163–165.
27. Popov SV, Poo M. Synaptotagmin: a calcium-sensitive inhibitor of exocytosis? *Cell* 1993;73:1247–1249.
28. Plutner H, Schwaninger R, Pind S, Balch WE. Synthetic peptides of the Rab effector domain inhibit vesicular transport through the secretory pathway. *EMBO J* 1990;9:2375–2383.
29. Hess SD, Doroshenko PA, Augustine GJ. A functional role for GTP-binding proteins in synaptic vesicle cycling. *Science* 1993;259:1169–1172.
30. Grynkiewicz G, Poenie M, Tsien RY. A new generation of Ca indicators with greatly improved fluorescence properties. *J Biol Chem* 1985;260:3440–3450.
31. Bennett MK, Calakos N, Scheller RH. Syntaxin: a synaptic protein implicated in docking of synaptic vesicles at presynaptic active zones. *Science* 1992;257:255–259.
32. Petrenko AG, Perin MS, Davletov BA, Ushkaryov YA, Geppert M, Sudhof TC. Binding of synaptotagmin to the alpha-latrotoxin receptor implicates both in exocytosis. *Nature* 1991;353:65–68.
33. Leveque C, Hoshing T, David P, Shoji-Kasai Y, Leys K, Omori A, Lang B, El Far O, Sato K, Martin-Moutot N, Newsom-Davis J, Takahashi M, Seagar MJ. The synaptic vesicle protein synaptotagmin associates with calcium channels and is a putative Lambert-Eaton myasthenic syndrome antigen. *Proc Natl Acad Sci USA* 1992;89:3625–3629.
34. Söllner T, Bennett MK, Whiteheart SW, Rothman JE. A protein assembly-disassembly pathway in vitro that may correspond to sequential steps of synaptic vesicle docking, activation, and fusion. *Cell* 1993;75:409–418.
35. Befenati F, Valtorta F, Rubenstein JL, Gorelick FS, Greengard P, Czernik AJ. Synaptic vesicle-associated Ca/calmodulin-dependent protein kinase II is a binding protein for synapsin I. *Nature* 1992;359:417–420.
36. Llinas R, Gruner JA, Sugimori M, McGuinness TL, Greengard P. Regulation by synapsin I and Ca-calmodulin-dependent protein kinase II of transmitter release in squid giant synapse. *J Physiol* 1991; 436:257–282.
37. Malinow R, Schulman H, Tsien R. Inhibition of postsynaptic PKC or CaMKII blocks induction but not expression of LTP. *Science* 1989;245:862–866.
38. Hans M, Swandulla D, Augustine GJ. Presynaptic calcium/calmodulin-dependent protein kinase mediates posttetanic potentiation at the squid giant synapse. *J Physiol* 1994.
39. Smith KM, Roger JC, Soderling R. Specificities of autoinhibitory domain peptides for four protein kinases. *J Biol Chem* 1990;265:1837–1840.
40. Osses LR, Barry SR, Augustine GJ. Protein kinase C activators enhance transmission at the squid giant synapse. *Biol Bull* 1989;177:146–153.
41. Shipira R, Silverberg SD. Ginsburg S, Rahamimoff R. Activation of protein kinase C augments evoked transmitter release. *Nature* 1987;325:58–60.
42. Hidaka H, Kobayashi R. Pharmacology of protein kinase inhibitors. *Ann Rev Pharmacol Toxicol* 1992;32:377–397.

43. Nestler EJ, Greengard P. *Protein phosphorylation in the nervous system*. New York: John Wiley and Sons; 1984.
44. Bahler M, Greengard P. Synapsin I bundles F-actin in a phosphorylation dependent manner. *Nature* 1987;326:704.
45. Baines AJ, Bennett V. Synapsin I is a spectrin-binding protein immunologically related to erythrocyte protein 4.1. *Nature* 1985;31:410.
46. Nichols RA, Chilcote TJ, Czernik AJ, Greengard P. Synapsin I regulates glutamate release from rat brain synaptosomes. *J Neurochem* 1992;58:783–785.

Molecular and Cellular Mechanisms of Neurotransmitter Release, edited by Lennart Stjärne, Paul Greengard, Sten Grillner, Tomas Hökfelt, and David Ottoson, Raven Press, Ltd., New York © 1994.

11

Distinctive Properties of a Neuronal Calcium Channel and Its Contribution to Excitatory Synaptic Transmission in the Central Nervous System

David B. Wheeler, William A. Sather, Andrew Randall, and Richard W. Tsien

Department of Molecular and Cellular Physiology, Stanford University, Stanford, California 94305

INTRODUCTION AND OVERVIEW OF THE DIVERSITY OF NEURONAL Ca^{2+} CHANNELS

Calcium channels open in response to membrane depolarization and support the Ca^{2+} entry that initiates transmitter release at presynaptic terminals (1–3). This renowned function of voltage-gated Ca^{2+} channels is a prime example of how voltage-gated Ca^{2+} channels link electrical signaling to intracellular biochemistry (4). Ca^{2+} channels also serve as an interface between electrical activity at the surface membrane and other key neuronal functions such as the control of excitability, metabolism, and gene expression. Despite their importance, many questions about Ca^{2+} channels in the brain remain open. How diverse are these channels? How is this diversity generated? How can the contributions of individual channel types be distinguished? Which of the Ca^{2+} channels support synaptic transmission in various brain regions? Are individual Ca^{2+} channel types selectively modulated during the course of short-term or long-term synaptic plasticity?

We are now in a position to begin to answer some of these questions, thanks to considerable advances in electrophysiology and molecular biology. These approaches have yielded much information about Ca^{2+} channel types and the diversity commensurate with the wide range of their neuronal functions. Functional studies from our laboratory and other groups have uncovered several types of Ca^{2+} channels in vertebrate neurons, distinguished by their electrophysiology, pharmacology, and tissue localization. These channels are commonly called L-, T-, N- and P-type (5–9). More recently, molecular cloning has revealed an even greater diversity, arising from multiple genes for Ca^{2+} channel subunits as well as alternative splicing (10,11). Voltage-gated Ca^{2+} channels are composed of a large transmembrane protein (200–260 kd) called the α_1 subunit, and two ancillary subunits called α_2-δ and β. The α_1 subunit contains the voltage sensor, gating machinery, and channel pore, all of which may be modified by the other subunits (12–14). Much of the diversity seen among Ca^{2+} channels may arise from the existence of multiple forms of α_1 subunit (15–18). Genes encoding five different calcium channel α_1 subunits have been identified in mammalian brain (19,20), and they have been labeled classes A, B, C, D, and E in the most widely used nomenclature (9,19). These α_1 subunit genes have been organized into two major subfamilies on the basis of sequence homology (Fig. 1). The first subfamily includes α_1 cDNAs encoding channels classified as L-type because they are responsive to 1,4-dihydropyridines (DHPs). These include the class C and class D clones (21,22) as well as the L-type channel of skeletal muscle encoded by the class S cDNA. The second α_1 subfamily consists thus far of cDNAs derived from nervous tissue, including mammalian brain (classes A, B, and E) and marine ray electric lobe (doe-1 and doe-4) (23). Individual genes within this subfamily show upwards of 60% identity with each other but only approximately 45% with members of the L-type subfamily. Furthermore, when expressed, they lack the characteristic DHP response of L-type channels. The most extensively characterized member of this subfamily is known as α_{1B}. When expressed in HEK293 cells (17) or dysgenic muscle cells (24) this subunit generates an N-type Ca^{2+} channel with appropriately high sensitivity to the conesnail toxin ω-CTx-GVIA.

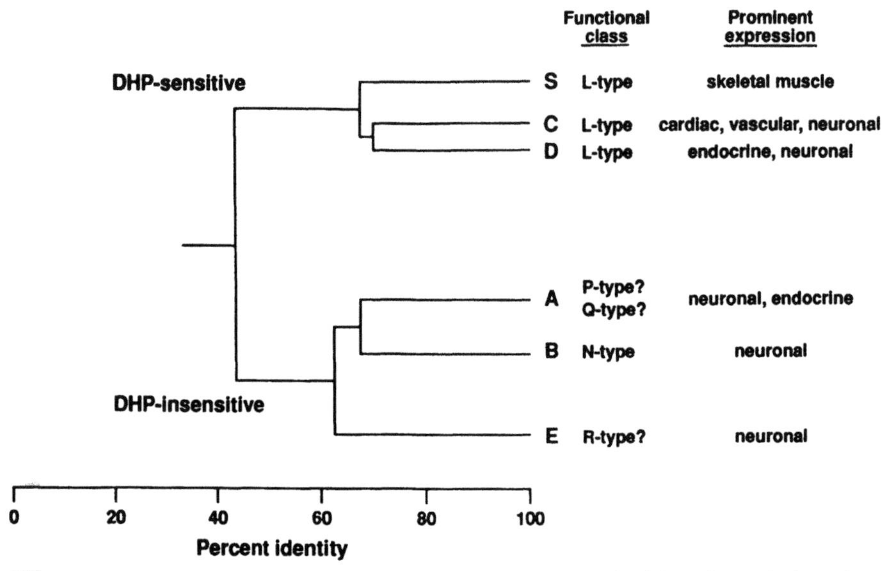

FIG. 1. Structural relationships among α_1 subunits of voltage-gated calcium channels, based on cDNA sequence identity. The correspondence between α_1 subunit classes and functional calcium channel types is shown on the right side of the figure along with an indication of those tissues expressing high levels of particular classes of α_1 subunits. The nomenclature for the cloned calcium channel α_1 subunits follows that of Snutch et al., 1990. Adapted from Zhang et al., 1993.

DISTINCTIVE FEATURES OF Ca^{2+} CHANNEL CURRENTS GENERATED BY α_{1A}

Information about the functional properties of the class A and class E α_1 subunits is more limited. It is clear, however, that neither encodes classical T-, N-, or L-type channels (15,16,25,26). Initial observations about the class A α_1 subunit suggested that it corresponded to the P-type channel (9,15). This view was supported by three findings: 1) class A transcripts are particularly abundant in the cerebellum, which expresses a high level of P-type current; 2) Ca^{2+} channel currents generated by α_{1A} subunits expressed in *Xenopus* oocytes are inhibited by the crude venom of the funnel web spider (15), which also blocks P-type channels; and 3) the level of cerebellar α_{1A} transcripts was found to be greatly reduced in mice with a hereditary deficiency of Purkinje cells (15) neurons that normally express P-type Ca^{2+} current almost exclusively (27,28).

Closer comparison of the properties of α_{1A} subunits expressed in *Xenopus* oocytes (29) and those of P-type channels in cerebellar Purkinje cells (9,30,31) revealed clear differences. P-type channels support a sustained, noninactivating current during depolarizing pulses longer than 1 second, whereas α_{1A} subunits expressed in *Xenopus* oocytes exhibit profound inactivation of macroscopic currents within a few hundred milliseconds (Fig. 2A). This discrepancy held true regardless of which β subunit was co-expressed with the α_{1A} subunit (29) although variations in the speed of inactivation were seen. Inactivation was also examined at the level of unitary currents, to be certain that the apparent inactivation of the global current in oocytes did not arise from inadequate voltage clamp, or from contamination by currents other than those generated by α_{1A}. Figure 3A shows a series of cell-attached patch records of single channel activity. Class A channel openings appear near the onset of the step depolarization and often cease before the end of the voltage pulse, indicating that the channel undergoes a transition into an inactivated state. The time course of inactivation was determined by averaging single channel records from cell-attached patches, particularly ones showing multiple channel activity (Fig. 3B). A pronounced and relatively rapid inactivation is evident, very similar to that seen in whole-cell currents. Thus, inactivation appears to be a consistent property of the class A gene product in this expression system, in sharp contrast with P-type current that displays little or no inactivation over a period of seconds.

Another distinction between α_{1A} channels in oocytes and P-type channels in Purkinje cells is pharmacological and involves the peptide toxin ω-Aga-IVA, an active component of the funnel web spider venom (32). P-type current in cerebellar Purkinje neurons is almost completely blocked by 20 mM ω-Aga-IVA (Fig. 4B). In fact, the IC_{50} is probably ~1 mM (27,28). In contrast, class A channels expressed in oocytes are much less sensitive, ~200 mM ω-Aga-IVA being required for 50% block (Fig. 4A) (29).

Table 1 provides a systematic comparison of the available information on class A channels in oocytes and P-type currents in cerebellar Purkinje neurons. These components are alike in many ways, including similar sensitivity to the recently described peptide toxin, ω-CTx-MVIIC (33–35). On the other hand, they differ in

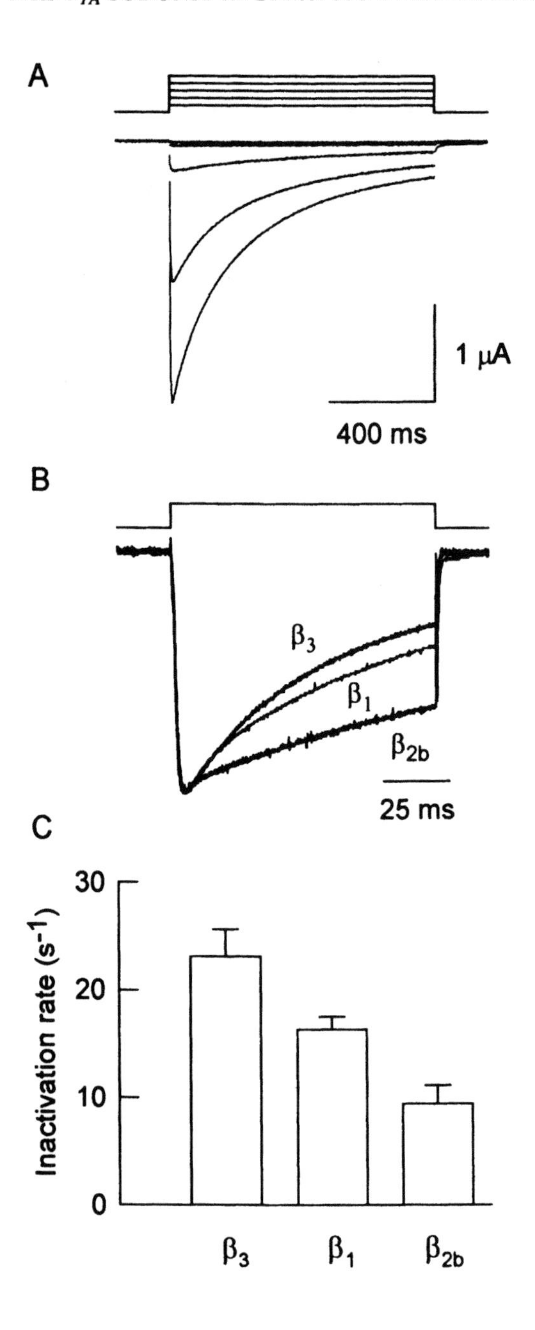
A
1 μA
400 ms
B
β3
β1
β2b
25 ms
C
Inactivation rate (s-1)
30
20
10
0
β3
β1
β2b

their single channel current-voltage properties (29,30) in addition to the contrasting properties of inactivation and ω-Aga-IVA sensitivity already mentioned.

Q-TYPE Ca^{2+} CHANNELS IN CEREBELLAR GRANULE NEURONS

Do these biophysical and pharmacological differences arise as artifacts of the oocyte expression system, or do they in fact reflect genuine differences in functional properties relevant to CNS neurons? If class A channels are truly distinguishable from P-type, then a component of Ca^{2+} current with the phenotype of α_{1A} subunit expressed in oocytes would be expected to be present in the brain, particularly in the cerebellum, since it contains class A transcripts in such abundance (15,16). Indeed, pharmacological dissection of currents in cerebellar granule neurons has revealed such a current, designated Q-type to distinguish it from P-type channels (34). Like P-type current, Q-type current is high voltage-activated, insensitive to dihydropyridines, insensitive to ω-CTx-GVIA, and potently blocked by ω-CTx-MVIIC. Unlike P-type current, the Q-type current exhibits an inactivating waveform even with Ba^{2+} as a charge carrier, and requires >>1 nM ω-Aga-IVA for half-block. In each respect, the Q-type current resembles α_{1A} channels heterologously expressed in oocytes (34). It is worth noting that the cerebellar granule neurons also display a classical P-type current that is noninactivating and half-blocked by 2–5 nM ω-Aga-IVA.

Q-TYPE Ca^{2+} CHANNELS SUPPORT SYNAPTIC TRANSMISSION

How diverse Ca^{2+} channels contribute to transmission in the CNS is not entirely clear (7). We have used conventional extracellular recordings from hippocampal slices to investigate the synapse between hippocampal CA3 and CA1 neurons, a focus of considerable interest in the examination of glutamatergic transmission and synaptic plasticity (36). At this synapse, inhibition of L-type channels by nifedipine has little effect (37–39) and much of the transmission remains after blockade of

FIG. 2. Inactivation of macroscopic currents carried by class A calcium channels. **A,** family of voltage-clamp records obtained from *Xenopus* oocytes co-injected with synthetic RNA encoding α_{1A}, α_2/δ and β_1 subunits. The holding potential was −80 mV, and test potentials ranged from −50 mV to −10 mV in 10 mV increments. **B,** the effect of various β subunits on inactivation kinetics of the class A channel. Oocytes were co-injected with synthetic RNAs encoding α_{1A}, α_2/δ subunits, together with β_1 or β_{2b} or β_3 subunits. The respective currents were evoked by steps from −80 mV to −10 mV and normalized to their peak amplitudes in order to facilitate comparison of their waveforms. **C,** mean inactivation rates (± standard error of measurement (s.e.m.)) for combinations of α_{1A}, α_2/δ and various β subunits. Inactivation rates (τ^{-1}) were determined from single exponential fits to the decay phases of the currents, and nonzero steady-state currents were allowed. In **A, B,** and **C,** data were obtained using a two-electrode voltage clamp and oocytes bathed in 2 mM Ba^{2+}, 0 Cl^- solutions. Records have been leak-subtracted and remaining capacitance transients have been blanked. Reproduced with permission from Sather et al., 1993.

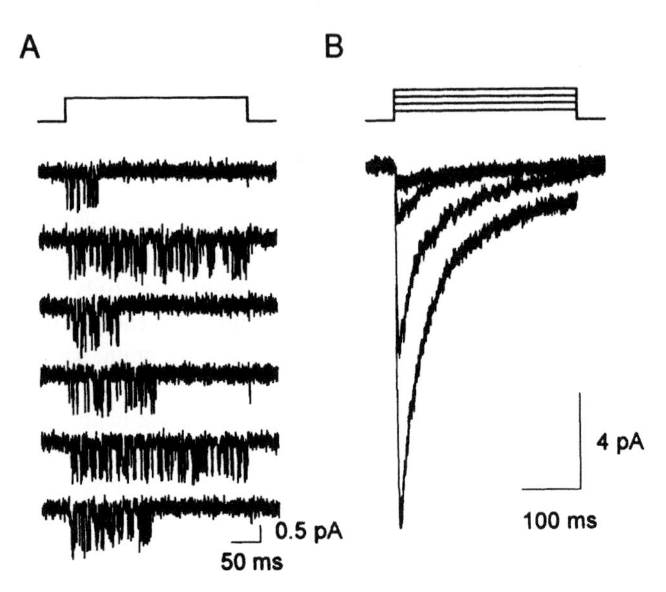

FIG. 3. Inactivation of class A channels at the single-channel level. **A,** single-channel records obtained from a cell-attached patch on an oocyte co-injected with RNA encoding α_{1A}, α_2/δ and β_1 subunits. Channel activity was elicited by voltage steps from a holding potential of −80 mV to a test potential of 0 mV. **B,** ensemble-averaged currents from a cell-attached patch showing multiple-channel activity. The oocyte was co-injected with RNA encoding α_{1A}, α_2/δ and β_1 subunits. The holding potential was −80 mV and test potentials ranged from −10 mV to +20 mV in 10 mV increments, with Ba^{2+} current increasing with larger depolarizations. Ensemble averages were constructed from 23 to 43 records at each potential. In **A** and **B**, currents were carried by 110 mM Ba^{2+}. Records were filtered at 2 kHz, sampled at 10 kHz, and leak subtracted using a P/4 protocol. Reproduced with permission from Sather et al., 1993.

N-type channels by ω-CTx-GVIA (37–40). We found that the Ca^{2+} channels that mediated the remaining transmission were pharmacologically distinct from classical L- and P-type Ca^{2+} channels (41). Instead, their pharmacological profile resembled that of α_{1A} in oocytes (29) and Q-type current in cerebellar granule neurons (34).

The effects of selective blockade of different Ca^{2+} channel types on field excitatory postsynaptic potentials (EPSPs) recorded in area CAl of the rat hippocampal slice are summarized in Fig. 5. Synaptic transmission was unaffected by nimodipine, a specific blocker of L-type channels (n = 6). Likewise, application of 30 nM ω-Aga-IVA, which potently blocks P-type Ca^{2+} channels (27,42) had no effect on transmission even after exposures of up to 20 minutes (n = 6). On the other hand, block of N-type Ca^{2+} channels with 1 μM ω-CTx-GVIA (43, 44) caused a rapid but incomplete depression of synaptic transmission (n = 30). The ω-CTx-GVIA- and ω-Aga-IVA-resistant transmission was completely and reversibly eliminated by removal of external Ca^{2+} ions. These experiments demonstrate that substantial excitatory synaptic transmission can be supported by a Ca^{2+} channel that is not N-, L- or P-type (41).

Q-type Ca^{2+} channels in cerebellar granule neurons are resistant to blockade by

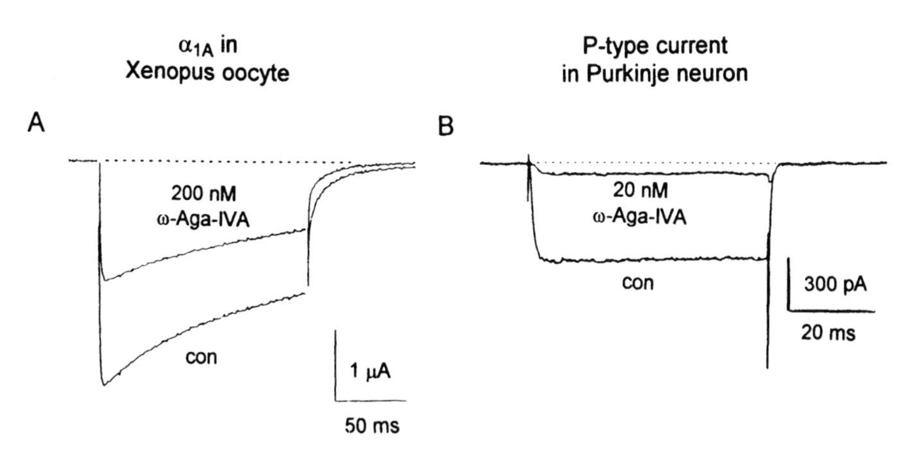

FIG. 4. Comparison of ω-Aga-IVA sensitivity and inactivation kinetics for the class A channel expressed in *Xenopus* oocytes and the P-type channel of cerebellar Purkinje neurons. **A,** example of the block of class A channels by 200 mM ω-Aga-IVA (Sather et al., 1993). The oocyte was co-injected with RNA encoding α_{1A}, α_2/δ and β_1 subunits, and was perfused with a 2 mM Ba^{2+}, 0 Cl^- solution supplemented with 0.1 mg/ml cytochrome C to saturate nonspecific peptide binding sites. The ω-Aga-IVA was applied until steady-state block was attained. The holding potential was −90 mV and the test potential was 0 mV. **B,** example of ω-Aga-IVA block of P-type current in an acutely dissociated cerebellar Purkinje neuron (Mintz, Venema, et al., 1992). Whole-cell patch clamp recording was carried out in 5 mmol Ba^{2+} solution containing 1 mg/ml cytochrome C, and the currents were evoked by steps from −90 mV to −20 mV. Parts **A** and **B** are adapted from Mintz, Venema, et al., 1992; Sather et al., 1993; and Zhang et al., 1993.

ω-CTx-GVIA, nimodipine, and ω-Aga-IVA at concentrations sufficient to eliminate N-, L-, and P-type channels respectively (34,45), and appear to be generated by the α_{1A} subunit (23,29,45,46). Q-type channels are completely blocked by 1.5 μM ω-CTx-MVIIC and are largely suppressed by ω-Aga-IVA at 1 μM (29,34) a concentration 100–1,000 times higher than that needed to block P-type channels (27,42). These characteristics enabled us to delineate the contribution of Q-type channels to hippocampal synaptic transmission. Application of ω-CTx-MVIIC (5 μM) completely abolished the transmission that was not mediated by N-, L- or P-type Ca^{2+} channels (n = 14) (Fig. 6A). The blockade by ω-CTx-MVIIC increased in speed and completeness as its concentration was raised from 150 nM to 5 μM (n = 2 to 10) (Fig. 6B). The dose-dependence resembled that found for ω-CTx-MVIIC-blockade of Q-type current in cultured cerebellar granule cells (34) and α_{1A} currents expressed in *Xenopus* oocytes (29).

At doses up to ~1 μM, ω-Aga-IVA blocks Q-type (and P-type) channels but spares N-type channels (42). Thus, ω-Aga-IVA provides a means of isolating transmission supported by N-type channels alone. The ω-Aga-IVA (1 μM) inhibited synaptic transmission by 85% ± 4% (n = 5) (Fig. 5). Lower doses of the toxin (200–300 nM) caused partial blockade of transmission that did not attain steady-state, even after a 90-minute application (n = 4) (41). The requirement for relatively high toxin concentrations and the slow time course of block stand in contrast to the effects of ω-Aga-IVA in systems where P-type channels have been implicated in

TABLE 1. *Comparison of α_{1A} in oocytes and native P-type channel*

Property	α_{1A} (in oocytes)	P-type (native) (9,28,30)
Activation		
Voltage range	HVA	HVA
Time to peak	8 ms	~4–6 ms
Inactivation		
Rate of onset	Faster	Slower
Single channel properties		
Conductance[a]	16 pS	9, 14, 19 pS
Unitary current (0 mV)	0.8 pA	0.8, 1.1, 1.4 pA
DHP sensitivity		
Nifedipine	Insensitive	Insensitive
Nimodipine	~Insensitive	Insensitive
ω-Aga-IIIA	~40% block (at 50 nM)	Max. block: 40% (at 200 nM)
ω-Aga-IVA (IC_{50})[b]	200 nM	0.7 nM
ω-CTx-MVIIC (IC_{50})[c]	<150 nM	~10 nM (59)
ω-CTx-GVIA	Insensitive	Insensitive

HVA, high-voltage activated
[a]Value of dominant conductance state for α_{1A}
[b]2 mM Ba^{2+} for α_{1A}, 5 mM Ba^{2+} for P-type
[c]2 mM Ba^{2+}

transmitter release. For example, ω-Aga-IVA reduces synaptic transmission at mouse endplates by 50% at ~20 nM (46). Similarly, ω-Aga-IVA reduces inhibitory transmission onto deep cerebellar neurons with an IC_{50} value of 9.4 nM (47), and decreases dopamine release from striatal synaptosomes with an IC_{50} of ~30 nM (48). Thus, the kinetics of ω-Aga-IVA inhibition found here are consistent with properties of Q-type channels (29,34,41).

PHYSIOLOGY OF TRANSMISSION SUPPORTED BY Q-TYPE Ca^{2+} CHANNELS

Synaptic transmission supported by Q-type Ca^{2+} channels alone was sensitive to modulation by neurotransmitter receptor stimulation or by activation of protein kinase C (PKC). Thus, application of agonists of mGlu, γ-aminobutyric acid type B ($GABA_B$), adenosine, or ACh receptors greatly depressed synaptic transmission remaining after ω-CTx-GVIA (Fig. 7A). In contrast, the PKC-activating compound phorbol-12, 13-dibutyrate (PDBu) first produced a small, transient depression (14% ± 2%), followed by a large, sustained potentiation of the ω-CTx-GVIA-insensitive synaptic response (222% ± 31%, n = 4) (Fig. 7B). Neither of these changes were seen with non-PKC-activating congener, 4α-PDBu. Prior application of PDBu reduced or abolished modulation of transmission by stimulation of neurotransmitter receptors (Fig. 7C). These effects were not observed when 4α-PDBu was applied

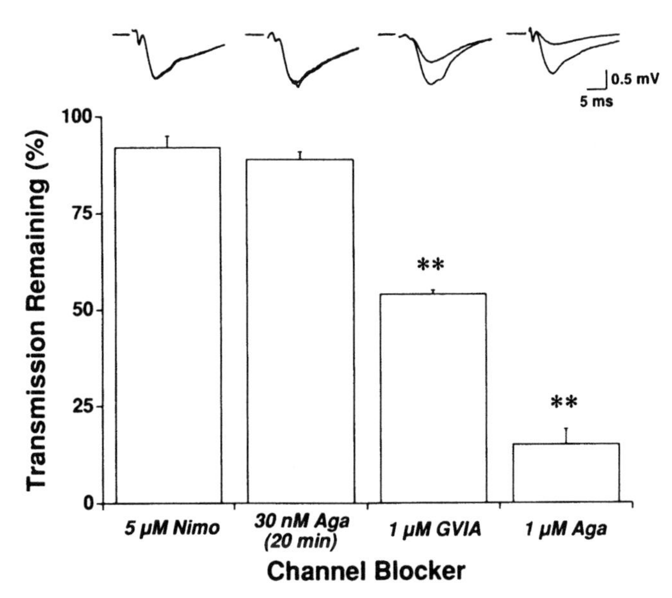

FIG. 5. Effects of selective blockade of L-, P-, N-, or Q-type Ca^{2+} channels on the strength of synaptic transmission recorded from area CA1 of a rat hippocampal slice (Wheeler et al., 1994). Percentage of control EPSP slope remaining (mean ± s.e.m.) in the presence of various Ca^{2+} channel inhibitors is shown. Blockade of L-type channels with 5 μM nimodipine, P-type channels with 30 nM ω-Aga-IVA, N-type channels with 1 μM ω-CTx-GVIA, and P- and Q-type channels with 1 μM ω-Aga-IVA. Representative traces illustrating each procedure are shown above.

**$p < 0.01$, paired Student's *t*-test.

(n = 2). An interaction between PKC and neurotransmitter effects has been found for transmission at other synapses (49,50) and has been attributed to modulatory effects on Ca^{2+} channels, primarily N-type (49–51). A similar convergence of signaling pathways may also hold for Q-type channels, although changes in the Ca^{2+} responsiveness of the transmitter release machinery are not excluded (52).

Hippocampal synapses display several forms of plasticity, including paired-pulse facilitation (PPF) and longterm potentiation (LTP). We assessed the ability of synaptic transmission mediated by individual Ca^{2+} channel types to undergo such changes in synaptic strength. Synaptic responses to a pair of stimuli were recorded before and after isolation of transmission mediated by Q-type or N-type channels (Fig. 8A). The PPF was taken as the percent increase in the slope of the second response relative to the first. PPF was increased from 40% ± 4% to 61% ± 5% by application of 1 μM ω-CTx-GVIA (n = 14), and from 39% ± 7% to 77% ± 8% by exposure to 1 μM ω-Aga-IVA (n = 5). The increase in PPF after N-type blockade appears not to be a nonspecific effect of diminished EPSP amplitude (41). Stimulus strength was decreased, reducing the slope of the EPSP to about the same extent as

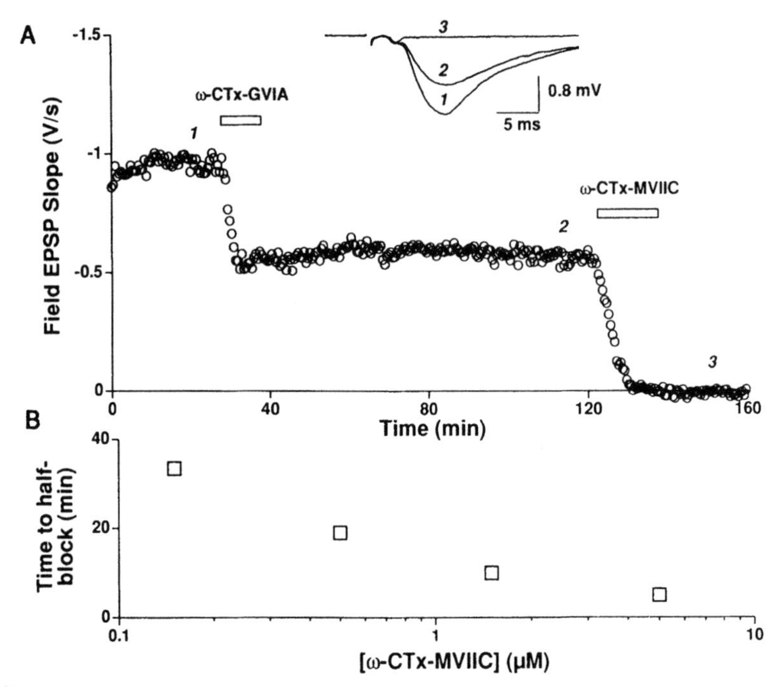

FIG. 6. Dose-dependent inhibition of ω-CTx-GVIA-insensitive synaptic trannsmission by ω-CTx-MVIIC (Wheeler et al., 1994). **A,** responses to 1 μM ω-CTx-GVIA and 5 μM ω-CTx-MVIIC in a representative recording. Traces show EPSPs recorded at time points as indicated. **B,** time to half-block of ω-CTx-GVIA-insensitive synaptic transmission by various concentrations of ω-CTx-MVIIC. Each symbol represents mean determinations from two to ten experiments.

would be expected with application of ω-CTx-GVIA, and PPF was 43% ± 2%. Subsequent addition of a ω-CTx-GVIA after returning the stimulus strength to its original level significantly increased PPF to 66% ± 5% (ω-CTx-GVIA versus size matched control, $p<0.05$ paired Student's t-test, n = 6). It seems probable, therefore, that the effect of the toxins on PPF is a direct consequence of the reduction in Ca^{2+} influx during the action potential. This was supported by the observed effect of lowering extracellular Ca^{2+}, which also significantly enhanced PPF (35% ± 3% in 2.0 mM Ca^{2+} versus 75% ± 4% in 1.3 mM Ca^{2+}, $p<0.0001$, paired Student's t-test, n = 6). Additionally, blockade of N-type Ca^{2+} channels caused no obvious change in the dependence of PPF on the interstimulus interval. Thus, PPF is more pronounced for synaptic transmission mediated by either Q- or N-type channels alone than in the case of transmission supported by both channel types together.

After application of ω-CTx-GVIA to isolate synaptic transmission mediated solely by Q-type Ca^{2+} channels, tetanic stimulation in one pathway induced a consistent and sustained synaptic enhancement (38% ± 1% at 30 minutes post-tetanus, n = 5), whereas transmission remained stable in an untetanized, independent pathway (Fig. 8B). The synaptic transmission in both control and potentiated pathways

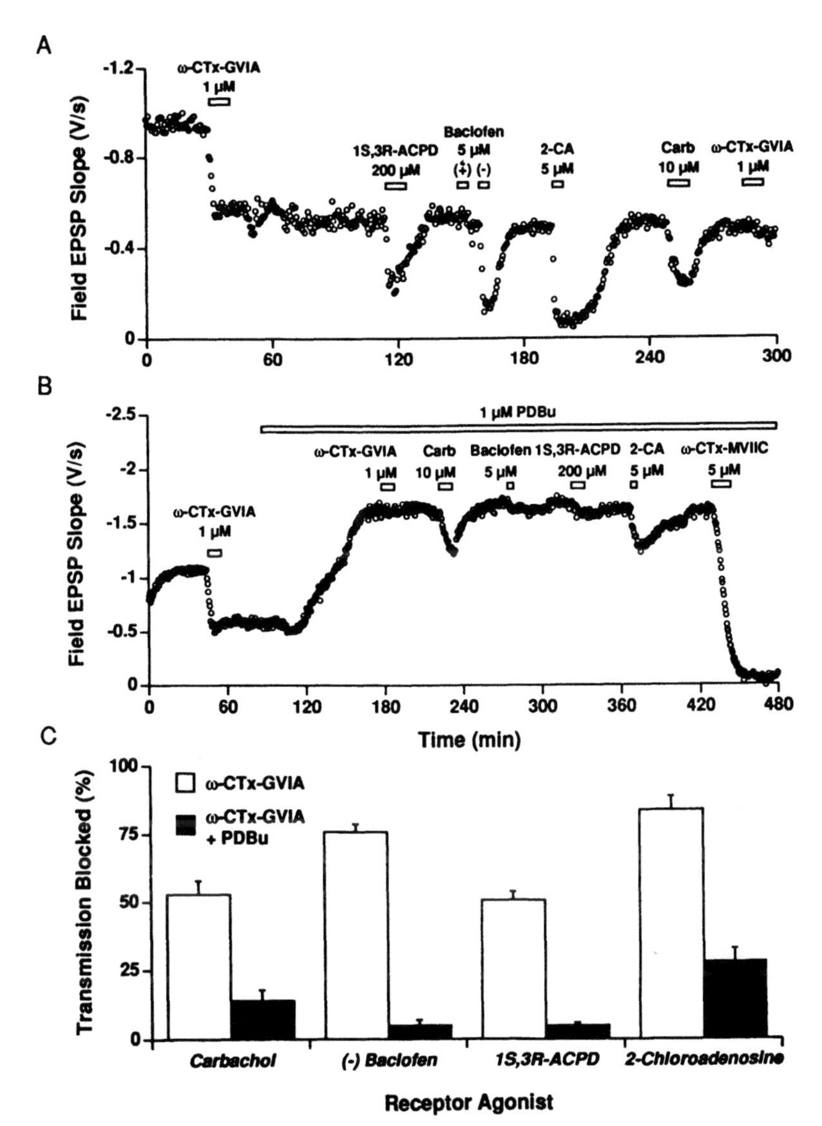

FIG. 7. Modulation of synaptic transmission supported by Q-type Ca^{2+} channels by stimulation of neurotransmitter receptors of PKC (Wheeler et al., 1994). **A,** Modulation by agonists of common neurotransmitter receptors of synaptic transmission recorded after blockade of N-type Ca^{2+} channels with ω-CTx-GVIA. Effects of stimulating mGlu receptors (1S, 3R-ACPD), $GABA_B$ receptors ((−)-baclofen but not (+)-baclofen), adenosine receptors (2-chloroadenosine), and ACh receptors (carbachol). **B,** A similar experiment, except that receptor agonists were applied following PKC-stimulation with PDBu. **C,** PKC stimulation significantly reduces the effect of receptor stimulation on synaptic strength. Mean ± s.e.m. of responses to: 1) 10-minute application of carbachol in the absence (n = 6) or presence (n = 4) of PDBu; 2) 5-minute application of (−)-baclofen in the absence (n = 12) or presence (n = 4) of PDBu; 3) 10-minute application of 1S,3R-ACPD in the absence (n = 10) or presence (n = 4) of PDBu; and 4) 5-minute application of 2-chloroadenosine in the absence (n = 6) or presence (n = 4) of PDBu.

$p < 0.01$ for all ω-CTx-GVIA vs. ω-CTx-GVIA + PDBu in C, paired Student's *t*-tests.

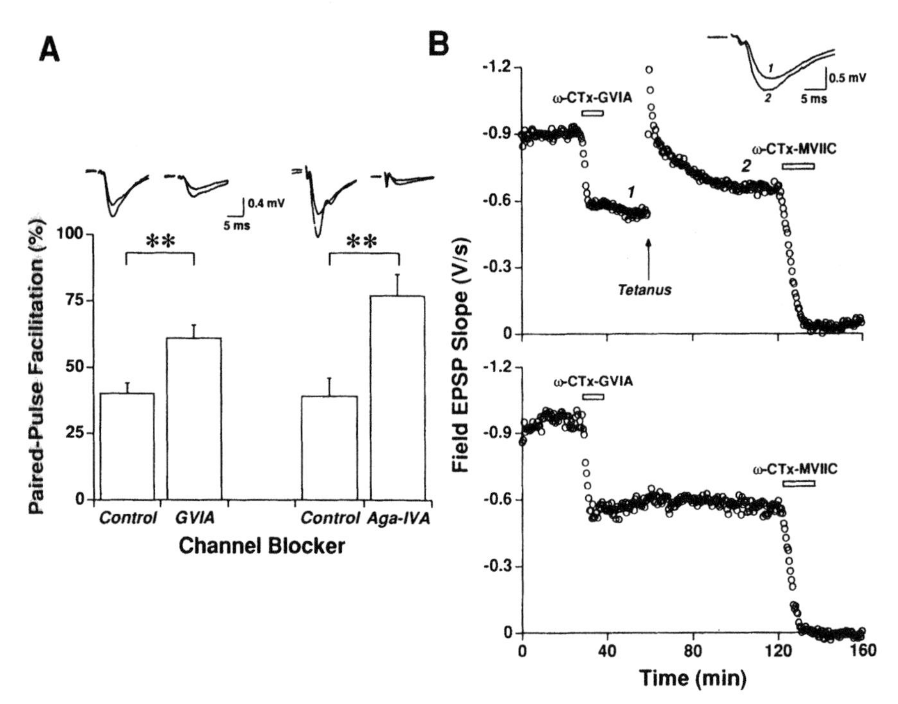

FIG. 8. Synaptic plasticity of transmission supported by N- or Q-type Ca^{2+} channels alone (Wheeler et al., 1994). **A,** Effect of ω-CTx-GVIA or ω-Aga-IVA on paired-pulse facilitation (PPF). Traces show superimposed responses to two stimuli, applied 40 milliseconds apart. The response to the first stimulus is invariably smaller than the response to the second stimulus. Exemplar traces are averages of five consecutive responses while bars represent the mean ± s.e.m. of several experiments. **Left,** PPF in the absence of toxin (Control) or after blockade of N-type channels with 1 μM ω-CTx-GVIA (GVIA); **right,** PPF in the absence of toxin (Control) or after inhibition of Q-type channels with 1 μM ω-Aga-IVA (Aga-IVA). **B,** induction and expression of LTP after application of 1 μM ω-CTx-GVIA. **Upper panel,** responses in the pathway that was tetanized at the indicated time (two trains of 100 Hz stimuli for 1 second, separated by 30 seconds); **lower panel,** responses to stimulation of a separate pathway in the same slice that did not undergo tetanic stimulation.

**Control vs. GVIA, $p<0.01$ paired Student's *t*-test, n = 14; control vs. Aga-IVA, $p<0.0001$ paired Student's *t*-test, n = 5.

was completely eliminated by ω-CTx-MVIIC (n = 3). This suggests that the glutamate release triggered following Q-type channel activation is sufficient to allow both the induction and expression of LTP (41).

DISCUSSION

Recent findings from molecular cloning of Ca^{2+} channels have greatly increased our understanding of Ca^{2+} channel diversity. This has allowed: 1) a new perspective on the familial relationships between various channel types; 2) an expansion of the classification of Ca^{2+} channels beyond those types uncovered by earlier biophysical and pharmacological analysis; and 3) a more precise description of the pharmacological properties of individual channels (albeit under somewhat different conditions than in their native environments). Our studies of the α_{1A} subunit are an interesting case in point. When expressed in oocytes, this subunit generates a Ca^{2+} channel phenotype that differs from classical P-type channels in its inactivation kinetics and pharmacology.

The properties of α_{1A} in oocytes are not simply a peculiarity of this expression system. A current component with similar properties is very prominent in cerebellar granule neurons, and is designated Q-type to distinguish it from P-type currents in the same cells (34). A detailed study of the effects of ω-Aga-IVA demonstrates distinct kinetics of blockade for these two components (Randall and Tsien, in preparation). The P-type current in granule neurons strongly resembles the classical P-type current in cerebellar Purkinje cells.

We do not yet know the basis of the P-type current itself. One possible explanation is that this component is generated by a novel α_1 subunit that has not yet been cloned. An alternative and more likely hypothesis is that class A products underly *both* Q- and P-type currents. The differences in their functional properties might arise from alternative RNA splicing, divergent posttranslational processing, or preferential association with different, as yet uncharacterized, auxiliary subunits.

The characterization of Ca^{2+} channels in oocytes and cerebellar granule neurons formed a basis for understanding the pharmacological profile of synaptic transmission at hippocampal CA3-CAl synapses. Here our evidence strongly suggests a physiological role for Q-type channels, presumably encoded by the α_{1A} subunit. While there was no indication of any involvement of L- or P-type channels, blockade of Q-type Ca^{2+} channel caused a drastic reduction of synaptic transmission. This Ca^{2+} channel appears dominant under our experimental conditions, but N-type Ca^{2+} channels also participate in the transmitter release. Together, Q- and N-type channels account entirely for the synaptic transmission.

Our experiments address the question of whether the N- and Q-type channels act in concert at the same transmitter release sites or at two separate populations of sites. Two observations stand out: 1) selective blockade of Q- or N-type channels produces reductions in synaptic strength whose sum exceeds 100%, and 2) paired-pulse facilitation of synaptic transmission mediated by either channel type in isolation is greater than PPF of control transmission. These findings appear inconsistent with the two classes of channels acting independently at nonoverlapping populations of release sites (e.g., on different axons or axon branches). If this were the case, the effects of selective and complete blockade of N- and Q-type channels would be additive. Additionally, if PPF at one kind of synaptic input were greater than average, PPF in the other pathway should be below average. Neither of these predictions was supported experimentally. On the other hand, our results can be understood if Q- and N-type Ca^{2+} channels are localized in close proximity and cooperate in the delivery of Ca^{2+} to individual release sites. Since transmitter release depends upon the local Ca^{2+} concentration raised to a power $n>2$ (53,54)

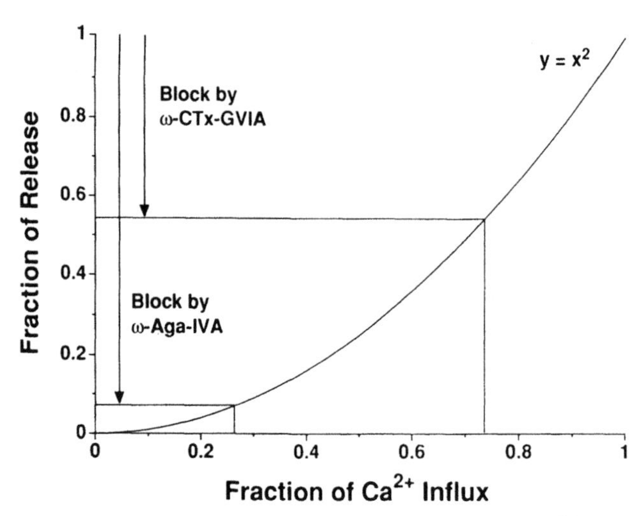

FIG. 9. Power-law relationship between neurotransmitter release and Ca^{2+} influx (Wheeler et al., 1994). The fractional release in the presence of toxin relative to control (R) should be proportional to the nth power of Ca^{2+} entry near release sites, where $n \geq 2$ (Dodge et al., 1967; Wu and Saggau, 1994). If Q- and N-type channels contribute fractions q and 1-q respectively of the total Ca^{2+} entry in the absence of toxin, their contributions in the presence of toxin will be $f_Q q + f_N (1-q)$, where f_Q and f_N are the fractions of Q- and N-type Ca^{2+} channels remaining unblocked. Then $R = [f_Q q + f_N(1-q)]^n$. Taking $n = 2$ as a conservative estimate, and bearing in mind that 1 μM ω-CTx-GVIA completely blocks N-type but spares Q-type Ca^{2+} channels, $R = [q]^2 = 0.54$, and $q = 0.735$ (q would be as high as 0.814 if $n = 3$). The proportion of transmission that would be left after complete and selective blockade of Q-type channels would then be estimated as $R = [1-q]^2 = (0.265)^2 = 0.070$. The expected effect of block of Q-type channels by 1 μM ω-Aga-IVA can be estimated by noting that ω-Aga-IVA blocks Q-type current in oocytes with half-maximal block at ~200 nM (Sather et al., 1993). Thus, $f_Q = 0.167$, and $R = [0.167q + (1-q)]^2 = 0.15$, in good agreement with the fraction of transmission remaining after ω-Aga-IVA (see Fig. 5).

elimination of Ca^{2+} entry through either type of channel would produce a disproportionately large reduction in synaptic strength (Fig. 9). Furthermore, when transmission is diminished by isolation of a single class of channels, PPF would be enhanced by the reduction in Ca^{2+} entry, much as it is with lowered extracellular Ca^{2+} $[Ca^{2+}]_o$ (55,56).

Our results set limits on possible mechanisms of LTP. The induction, maintenance, and expression of LTP were qualitatively unaffected by the presence of ω-CTx-GVIA (41). This rules out any essential role in LTP of N-type channels in either presynaptic terminals or postsynaptic dendrites. Additional experiments are needed to test for more subtle influences such as a change in the threshold for LTP induction. Another finding is that blockade of N-type channels by ω-CTx-GVIA reduces potentiated synaptic transmission to the same degree as basal transmission (41). This finding excludes any sizable change in the relative contributions of Q- or N-type channels, although it leaves open the possibility of an overall enhancement of both channel types. Our results are complementary to fura-2 measurements in Schaffer collateral axons, which showed no change with LTP in the intracellular Ca^{2+} concentration transient (54). Together, these findings argue strongly against any appreciable change in presynaptic Ca^{2+} delivery during LTP, while leaving open the possibility of a change in Ca^{2+} sensitivity or the extent of the Ca^{2+}-dependent response.

Synaptic transmission solely mediated by Q-type channels is highly susceptible to various forms of modulation. Activators of mGlu, $GABA_B$, adenosine, or ACh receptors strongly inhibit the transmission, while PKC enhances it and also opposes the receptor-mediated inhibition. All of these characteristics have previously been attributed to N-type channels. Our results indicate that this cannot be the sole explanation, and suggest that modulation of Q-type Ca^{2+} channels may also be important for regulation of synaptic strength (however, see Scholz and Miller (52)). These results are consistent with evidence for modulation of non-L-, N- or P-type Ca^{2+} channel currents in cerebellar granule cells (57) hippocampal pyramidal cells, and spinal interneurons (27,58). The finding that neurotransmitter release depends heavily on the cooperative actions of more than one class of Ca^{2+} channel, including a hitherto unknown class, increases our knowledge about possible mechanisms for physiological regulation of synaptic transmission by neuroactive substances. The power-law relationship between intracellular Ca^{2+} and synaptic output would give any of the classes of cooperating channels considerable leverage for modulation of neurotransmitter release (47,48). It will be interesting to see if various transmitters act concurrently or independently on individual classes of Ca^{2+} channels. New agents with greater specificity for individual channel types will facilitate the delineation of their importance in neuronal communication and the development of therapeutic strategies for manipulating specific synaptic pathways.

ACKNOWLEDGMENT

David B. Wheeler is a Howard Hughes Medical Institute Predoctoral Fellow, William A. Sather is a postdoctoral Fellow of the American Heart Association (California Affiliate), and Andrew Randall is a Lucille P. Markey Visiting Scholar. We acknowledge G. Miljanich of Neurex Corporation for ω-CTx-MVIIC (SNX-230). This work was supported by NIH grant NS24067, and an NIMH-Silvio Conte Center for Neuroscience research grant.

REFERENCES

1. Katz B. *The release of neurotransmitter substances*. Springfield, IL: C.C. Thomas; 1969.
2. Augustine G, Charlton MP, Smith SJ. Calcium action in synaptic transmitter release. *Annu Rev Neurosci* 1987;10:633–693.
3. Zucker RS. Calcium and transmitter release. *J Physiol (Paris)* 1993;87:25–36.
4. Hille B. *Ionic channels of excitable membranes*. Sunderland MA: Sinauer, 1992.
5. Tsien RW, Lipscombe D, Madison DV, Bley KR, Fox AP. Multiple types of neuronal calcium channels and their selective modulation. *Trends Neurosci* 1988;11:431–438.
6. Bean BP. Classes of calcium channels in vertebrate cells. *Annu Rev Physiol* 1989;51:367–384.

7. Hess P. Calcium channels in vertebrate cells. *Annu Rev Neurosci* 1990;13:1337–1356.
8. Miller RJ, Fox AP. Voltage sensitive calcium channels. In: Bronner F, ed. *Intracellular Calcium Regulation*. New York: Wiley-Liss; 1990:97–138.
9. Llinás R, Sugimori M, Hillman DE, Cherksey B. Distribution and functional significance of the P-type, voltage-dependent Ca^{2+} channels in the mammalian central nervous system. *Trends Neurosci* 1992;15:351–355.
10. Snutch TP, Reiner PB. Ca^{2+} channels: diversity of form and function. *Curr Op Neurobiol* 1992;2:247–253.
11. Tsien RW, Ellinor PT, Horne WA. Molecular diversity of voltage-dependent Ca^{2+} channels. *Trends Pharmacol Sci* 1991;12:349–354.
12. Catterall WA. Structure and function of voltage-sensitive ion channels. *Science* 1988;242:50–61.
13. Campbell KP, Leung AT, Sharp AH. The biochemistry and molecular biology of the dihydropyridine-sensitive calcium channel. *Trends Neurosci* 1988;11:425–430.
14. Catterall WA, Striessnig J. Receptor sites for Ca^{2+} channel antagonists. *Trends Pharmacol Sci* 1992;13:256–262.
15. Mori Y, Friedrich T, Kim M-S, et al. Primary structure and functional expression from complementary DNA of a brain calcium channel. *Nature* 1991;350:398–402.
16. Starr TVB, Prystay W, Snutch TP. Primary structure of a calcium channel that is highly expressed in the rat cerebellum. *Proc Natl Acad Sci USA* 1991;88:5621–5625.
17. Williams ME, Brust PF, Feldman DH, et al. Structure and functional expression of an ω-conotoxin-sensitive human N-type calcium channel. *Science* 1992;257:389–395.
18. Dubel SJ, Starr TVB, Hell J, Ahlijanian MK, Enyeart JJ, Catterall WA, Snutch TP. Molecular cloning of the α-1 subunit of an ω-conotoxin-sensitive calcium channel. *Proc Natl Acad Sci USA* 1992;89:5058–5062.
19. Snutch TP, Leonard JP, Gilbert MM, Lester HA, Davidson N. Rat brain expresses a heterogeneous family of calcium channels. *Proc Natl Acad Sci USA* 1990;87:3391–3395.
20. Perez-Reyes E, Wei X, Castellano A, Birnbaumer L. Cloning and expression of a cardiac/brain β subunit of the L-type calcium channel. *J Biol Chem* 1990;265:20430–20436.
21. Mikami A, Imoto K, Tanabe T, et al. Primary structure and functional expression of the cardiac dihydropyridine-sensitive calcium channel. *Nature* 1989;340:230–233.
22. Williams ME, Feldman DH, McCue AF, Brenner R, Velicelebi G., Ellis SB, Harpold MM. Structure and functional expression of α_1, α_2, and β subunits of a novel human neuronal calcium channel subtype. *Neuron* 1992;8:71–84.
23. Ellinor PT, Zhang J-F, Randall AD, Zhou M, Schwarz TL, Tsien RW, Horne WA. Functional expression of a rapidly inactivating neuronal calcium channel. *Nature* 1993;363:455–458.
24. Fujita Y, Mynlieff M, Dirksen RT, et al. Primary structure and functional expression of the ω-conotoxin-sensitive N-type channel from rabbit brain. *Neuron* 1993;10:585–598.
25. Niidome T, Kim M-S, Friedrich T, Mori Y. Molecular cloning and characterization of a novel calcium channel from rabbit brain. *FEBS Lett* 1992;308:7–13.
26. Soong TW, Stea A, Hodson CD, Dubel SJ, Vincent SR, Snutch TP. Structure and functional expression of a member of the low voltage-activated calcium channel family. *Science* 1993;260: 1133–1136.
27. Mintz IM, Adams ME, Bean BP. P-type calcium channels in rat central and peripheral neurons. *Neuron* 1992;9:85–95.
28. Mintz IM, Venema VJ, Swiderek K, Lee T, Bean BP, Adams ME. P-type calcium channels blocked by the spider toxin ω-Aga-IVA. *Nature* 1992;355:827–829.
29. Sather WA, Tanabe T, Mori Y, Adams ME, Tsien RW. Distinctive biophysical and pharmacological properties of class A (BI) calcium channel α_1 subunits. *Neuron* 1993;11:291–303.
30. Usowicz M, Sugimori M, Cherksey B, Llinás R. P-type calcium channels in the somata and dendrites of adult cerebellar Purkinje cells. *Neuron* 1992;9:1185–1199.
31. Regan LJ. Voltage-dependent calcium currents in Purkinje cells from rat cerebellar vermis. *J Neurosci* 1991;11:2259–2269.
32. Adams ME, Bindokas VP, Hasegawa L, Venema VJ. Omega-agatoxins: novel calcium channel antagonists of two subtypes from funnel web spider (*Agelenopsis aperta*) venom. *J Biol Chem* 1990;265:861–867.
33. Hillyard DR, Monje VD, Mintz IM, et al. A new Conus peptide ligand for mammalian presynaptic Ca^{2+} channels. *Neuron* 1992;9:69–77.
34. Randall AD, Wendland B, Schweizer F, Miljanich G, Adams ME, Tsien RW. Five pharma-

cologically distinct high voltage-activated Ca^{2+} channels in cerebellar granule cells. *Soc Neurosci Abstr* 1993;19:1478.
35. Swartz KB, Mintz IM, Boland LM, Bean BP. Block of calcium channels in central and peripheral rat neurons by ω-conotoxin-MVIIC. *Soc Neurosci Abstr* 1993;19:1478.
36. Bliss TVP, Collingridge GL. A synaptic model of memory: long-term potentiation in the hippocampus. *Nature* 1993;361:31–39.
37. Kamiya H, Sawada S, Yamamoto C. Synthetic ω-conotoxin blocks synaptic transmission in the hippocampus in vitro. *Neurosci Lett* 1988;91:84–88.
38. Dutar P, Rascol O, Lamour Y. ω-Conotoxin GVIA blocks synaptic transmission in the CA1 field of the hippocampus. *Eur J Pharmacol* 1989;174:261–266.
39. Horne AL, Kemp JA. The effect of ω-conotoxin GVIA on synaptic transmission within the nucleus accumbens and hippocampus of the rat in vitro. *Br J Pharmacol* 1991;103:1733–1739.
40. Parfitt KD, Madison DV. Phorbol esters enhance synaptic transmission by a presynaptic, calcium-dependent mechanism in rat hippocampus. *J Physiol (Lond)* 1993;471:245–268.
41. Wheeler DB, Randall A, Tsien RW. Roles of N-type and Q-type Ca^{2+} channels in supporting hippocampal synaptic transmission. *Science* 1994;264:107–111.
42. Mintz IM, Bean BP. Block of calcium channels in rat neurons by synthetic ω-Aga-IVA. *Neuropharmacology* 1993;32:1161–1169.
43. Kasai H, Aosaki T, Fukuda J. Presynaptic Ca-antagonist omega-conotoxin irreversibly blocks N-type Ca-channels in chick sensory neurons. *Neurosci Res* 1987;4:228–235.
44. McCleskey EW, Fox AP, Feldman DH, Cruz LJ, Olivera BM, Tsien RW, Yoshikami D. ω-Conotoxin: direct and persistent blockade of specific types of calcium channels in neurons but not muscle. *Proc Natl Acad Sci USA* 1987;84:4327–4331.
45. Zhang J-F, Randall AD, Ellinor PT, et al. Distinctive pharmacology and kinetics of cloned neuronal Ca^{2+} channels and their possible counterparts in mammalian CNS neurons. *Neuropharmacology* 1993;32:1075–1088.
46. Kim YI, Longacher JM, Viglione MP. Evidence for P-type calcium channels at the mammalian neuromuscular junction: a study with omega-agatoxin IVA, omega-conotoxin GVIA, nimodipine and autoantibodies. *Biophys J* 1994;66:A16.
47. Takahashi T, Momiyama A. Different types of calcium channels mediate central synaptic transmission. *Nature* 1993;366:156–158.
48. Turner TJ, Adams ME, Dunlap K. Multiple Ca^{2+} channel types coexist to regulate synaptosomal neurotransmitter release. *Proc Natl Acad Sci USA* 1993;90:9518–9522.
49. Swartz KJ, Merritt A, Bean BP, Lovinger DM. Protein kinase C modulates glutamate receptor inhibition of Ca^{2+} channels and synaptic transmission. *Nature* 1993;361:165–168.
50. Swartz KJ. Modulation of Ca^{2+} channels by protein kinase C in rat central and peripheral neurons: disruption of G protein-mediated inhibition. *Neuron* 1993;11:305–320.
51. Yang J, Tsien RW. Enhancement of N- and L-type calcium channel currents by protein kinase C in frog sympathetic neurons. *Neuron* 1993;10:127–136.
52. Scholz KP, Miller RJ. Inhibition of quantal transmitter release in the absence of calcium influx by a G protein-linked adenosine receptor at hippocampal synapses. *Neuron* 1992;8:1139–1150.
53. Dodge Jr FA, Rahamimoff R. Co-operative action of calcium ions in transmitter release at the neuromuscular junction. *J Physiol (Lond)* 1967;193:419–432.
54. Wu LG, Saggau P. Presynaptic calcium is increased during normal synaptic transmission and paired-pulse facilitation, but not in long-term potentiation in area CAl of hippocampus. *J Neurosci* 1994:14:645–654.
55. Creager R, Dunwiddie T, Lynch G. Paired-pulse and frequency facilitation in the CA1 region of the in vitro rat hippocampus. *J Physiol (Lond)* 1980;299:409–424.
56. Manabe T, Wyllie DJA, Perkel DJ, Nicoll RA. Modulation of synaptic transmission and long-term potentiation: effects on paired pulse facilitation and EPSC variance in the CAl region of the hippocampus. *J Neurophysiol* 1993;70:1451–1459.
57. Randall AD, Wheeler DB, Tsien RW. Modulation of Q-type Ca^{2+} channels and Q-type Ca^{2+} channel-mediated synaptic transmission by metabotropic and other G protein-linked receptors. *Functional Neurol* 1993;8(suppl.):44.
58. Regan LJ, Sah DWY, Bean BP. Ca^{2+} channels in rat central and peripheral neurons: high-threshold current resistant to dihydropyridine blockers and ω-conotoxin. *Neuron* 1991;6:269–280.
59. Swartz KJ, Mintz IM, Bean BP. personal communication.

Molecular and Cellular Mechanisms of Neurotransmitter Release, edited by Lennart Stjärne, Paul Greengard, Sten Grillner, Tomas Hökfelt, and David Ottoson, Raven Press, Ltd., New York © 1994.

12

Exocytosis and Endocytosis in Single Peptidergic Nerve Terminals

Manfred Lindau, *Hendrik Rosenboom, and †Jean Nordmann

*Department of Molecular Cell Research, Max-Planck-Institute for Medical Research, D-69120 Heidelberg, Germany; *Department of Neurobiology, Freie Universität Berlin, D-14195 Berlin, Germany; and †Centre de Neurochimie, F-67084 Strasbourg, France*

Release of neurotransmitters and neuropeptides occurs in specialized organelles, the nerve terminals. The nonapeptides vasopressin and oxytocin are released from nerve terminals located in the posterior pituitary. These nerve terminals belong to so-called magnocellular neurons, which have their cell bodies in the hypothalamus. In the cell body, precursors of the hormones are synthesized and packed into secretory vesicles together with the appropriate processing enzymes. While the processing enzymes cleave the precursors, the vesicles are transported along the axon to the nerve terminals where they accumulate in large numbers and where they may be released in response to action potential bursts (1).The electrical stimulation leads to opening of different calcium channel types present in the plasma membrane of the nerve terminals (2,3) and the Ca^{2+} entry induces exocytotic peptide release followed by reuptake of the membrane by endocytic processes (2–4).

Oxytocin regulates ejection of milk by stimulating contraction of myoepithelial cells (5) whereas vasopressin affects blood pressure by effects on vascular smooth muscle and regulates water reabsorption in the kidney (6). However, the two peptides are also found in nerve fibers throughout the CNS (7) and there is certainly a role for vasopressin as a neurotransmitter or neuromodulator. The observation that vasopressin administration could prevent extinction of avoidance reactions in conditioned rats has suggested that vasopressin may influence memory (8). Exocytosis of neuropeptides by pituitary nerve terminals may thus be considered as a general model for nerve terminals in the brain.

ISOLATED NERVE TERMINALS FOR PATCH CLAMP EXPERIMENTS

The posterior pituitary of the rat can be readily isolated. When it is gently homogenized in Ca^{2+}-free sucrose buffer, the nerve terminals break off the axon and reseal. This preparation allows the study of stimulus-secretion coupling in isolated nerve terminals (9). The individual nerve terminals have approximately spherical shape. Most of them have a diameter of 2–3 μm but each preparation contains about 50 very large terminals with a diameter of 8–15 μm. These large nerve terminals secrete to the same extent as the small ones (10) and have been extensively used for patch clamp experiments to characterize their ionic channels and currents (9,11–17).

In the patch clamp "whole terminal" configuration, the exocytotic activity of single nerve terminals has been investigated (4,18,19) using the lock-in amplifier technique (20) that was introduced by Erwin Neher and Alain Marty to demonstrate exocytosis and endocytosis in single chromaffin cells (21). Exocytosis and endocytosis lead to small changes of the plasma membrane area. Since biological membranes have a rather constant specific electrical capacitance of ~0.7–1 $\mu F/cm^2$, an increase in plasma membrane area, as occurs when secretory vesicles fuse with the plasma membrane, generates a proportional increase of the plasma membrane capacitance. As shown in the pioneering work of Neher and Marty, it is possible to separate changes in membrane capacitance from changes in conductance by proper phase adjustment of a lock-in amplifier (21).

DEPOLARIZATION INDUCES CAPACITANCE INCREASE

Figure 1 shows a nerve terminal with a patch pipette. Assuming spherical shape and a smooth surface, the surface area can be estimated. The relationship between

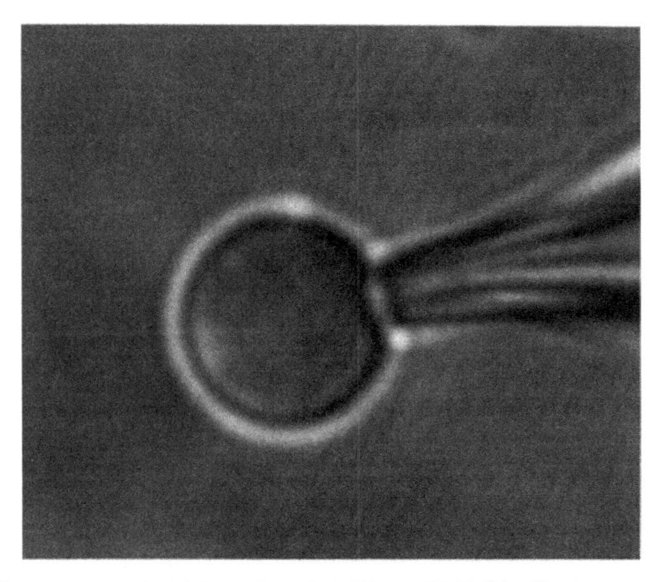

FIG. 1. A pituitary nerve terminal (10 μm diameter) with a patch pipette. The nerve terminals are spherical and appear smooth under nomarski optics.

surface area and capacitance of resting nerve terminals is 0.76 $\mu F/cm^2$ or 7.6 $fF/\mu m^2$ (19). Assuming that the specific capacitance of the vesicle membrane has a similar value, we estimate that a single peptidergic secretory vesicle with a mean diameter of ~180 nm has a capacitance of ~0.8 fF. The recordings shown in Fig. 2 (4) are from a nerve terminal with an initial capacitance of 2.6 pF corresponding to a diameter of ~10 μm. When the nerve terminal was depolarized, large conductance and capacitance changes were induced. Figure 2A shows that after repolarization the conductance was unchanged but the capacitance had increased. This particular nerve terminal generated comparatively large capacitance changes. During the 7–10-second depolarizations, the capacitance increased by 200–300 fF corresponding to exocytosis of ~300 vesicles. The total increase in this nerve terminal was 2 pF corresponding to ~5% of all the vesicle membrane present in a nerve terminal of this size. In such experiments with prolonged or repeated depolarizations, we were never able to release more than 5% percent of the vesicles. However, this amount of vesicles approximately doubles the plasma membrane area and corresponds to fusion of about 10 vesicles with 1 μm^2 of plasma membrane.

EXOCYTOSIS CAN BE MEASURED DURING DEPOLARIZATION

In Fig. 2A the lock-in signal during the depolarizations is omitted. However, the capacitance change can also be measured *during* depolarization as demonstrated in Fig. 2B, which shows the measurement during the third depolarization on an expanded time scale. Since this recording was done at high gain, the amplifier was saturated during period 2 due to activation of large conductances. However, during the last 3 seconds the conductance had decreased, the amplifier was no longer saturated; the lock-in outputs provide the traces shown during period 3. At the end of the depolarizing pulse the conductance decreased almost instantaneously without a concomitant change in the capacitance trace. This shows that the activation of this conductance does not distort the capacitance measurements. The capacitance measurement is insensitive to membrane potential and to the activated conductance. We are thus able to follow the exocytotic activity also during the depolarization period.

A FAST AND SLOW PHASE OF EXOCYTOSIS

Figure 2C shows a subsequent but much longer depolarization from the same nerve terminal. The amplifier gain was reduced such that capacitance and conductance could be recorded throughout the 30-second depolarization. The trace reveals an initial jump followed by a slower increase with a decreasing slope. The existence of an initial jump due to very high exocytotic activity was confirmed using short (80 milliseconds) depolarizations (4,18). The rate of exocytosis during the fast increase is higher than 1,000 vesicles/s, whereas the initial rate of the slow phase is about 30–40 vesicles/s. Following step changes in $[Ca^{2+}]_i$, a rapid burst of exocytosis followed by a long-lasting slow phase was also observed in rat pituitary melano-

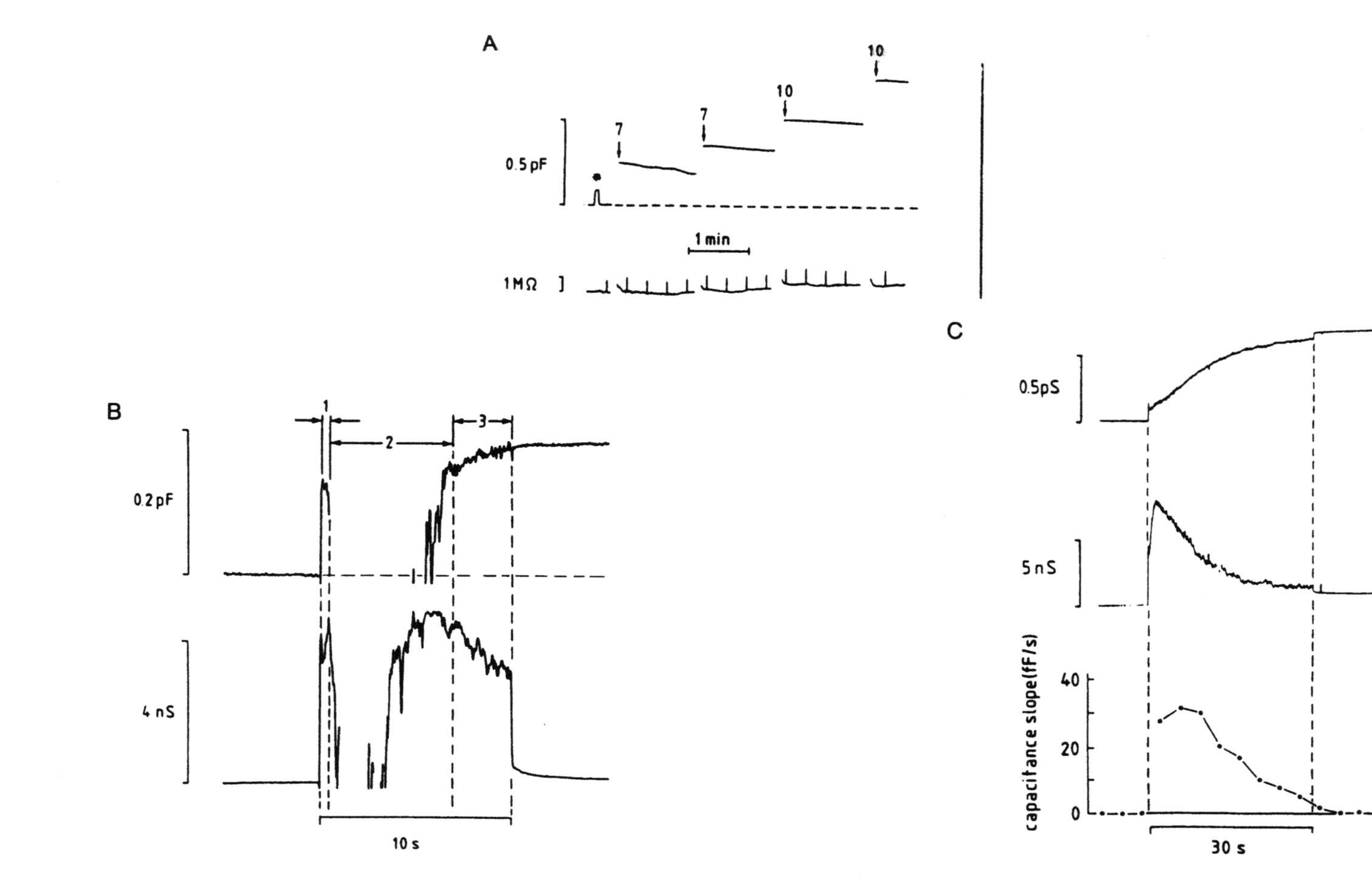
A
0.5 pF
7
7
10
10
1 min
1 MΩ
B
1
2
3
0.2 pF
4 nS
10 s
C
0.5pS
5 nS
capacitance slope(fF/s)
40
20
0
30 s

trophs and bovine adrenal chromaffin cells (22–25). This pattern of exocytosis is believed to represent the sequential release of at least two vesicle populations, one that is docked at the plasma membrane and immediately available for secretion, and other vesicle populations that must pass additional slow steps before they become available for secretion (24,26).

CORRELATION OF EXOCYTOSIS WITH Ca^{2+}-DEPENDENT K^+ CURRENT

The bottom trace of Fig 2C shows the slope of the slow increase. After the initial jump the slope is about 32 fF/s corresponding to exocytosis of about 35 vesicles/s. The conductance in the middle trace is presumably mainly due to Ca^{2+}-dependent K channels (13,14). It was recently reported that the decay of Ca^{2+}-dependent K^+ current during prolonged depolarization depends on $[Ca^{2+}]_i$ (13). The decrease here may thus reflect either a decrease of $[Ca^{2+}]_i$ or may be due to inactivation or desensitization of the channels. If it was a decrease of $[Ca^{2+}]_i$, then this decrease would directly explain the parallel decrease of the capacitance slope.

CORRELATION OF EXOCYTOSIS WITH $[Ca^{2+}]_i$

To estimate $[Ca^{2+}]_i$ we have used the fura 2 method in conjunction with the capacitance measurements. In the experiment shown in Fig. 3, the pipette contained 100 μM fura 2 and no additional Ca^{2+} buffer. The nerve terminal was depolarized for 4 seconds (4). The capacitance (top trace) shows again the initial jump followed by a slower increase of 30 fF/s. The conductance (middle trace) increased and then started to decrease. $[Ca^{2+}]_i$ (bottom trace), however, continued to increase throughout the depolarization period. Either the conductance decrease is due to inactivation of the conductance or the Ca^{2+} concentration at the inner membrane surface has a time course that is very different from the average concentration within the nerve terminal's cytoplasm. Although Ca^{2+} equilibrates rapidly, large gradients may be

FIG. 2. Depolarization-induced capacitance changes in a pituitary nerve terminal. Initial capacitance was 2.6 pF. Holding potential −80 mV, depolarization to −10 mV for variable periods of time. A ±20 mV sine wave was superimposed on the potential throughout the recording. **A.** Capacitance (upper trace) and conductance (lower trace). At the asterisk the capacitance compensation was transiently reduced by 0.1 pF. At the arrows the nerve terminal was depolarized for the indicated number of seconds. The upward deflections in the conductance trace reflect artificial series resistance changes used for phase adjustment. The sections during depolarization were deleted in this panel. **B.** Recording of capacitance (upper trace) and conductance (lower trace) during the third depolarization. During period 2 the amplifier was saturated due to activation of large conductance. During period 3 the conductance had decreased and the amplifier was not saturated. Note that the large conductance decrease upon repolarization does not induce a change in the capacitance trace. **C.** A 30-second depolarization at reduced gain allowing capacitance (top trace) and conductance (middle trace) to be recorded throughout the depolarization. The bottom trace shows the time course of the capacitance slope.

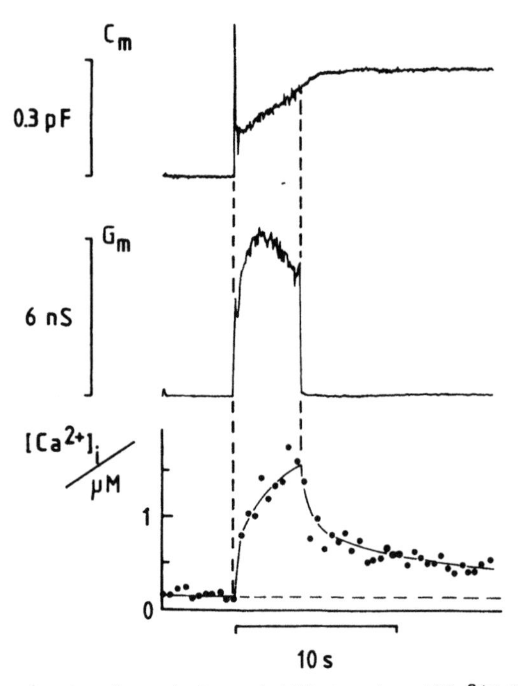

FIG. 3. Capacitance (top trace), conductance (middle trace), and $[Ca^{2+}]_i$ (bottom trace). Same conditions as in Fig. 2 except that the pipette contained 100 μM fura 2 and no Ca chelator (EGTA). Duration of the depolarization was 4 seconds. $[Ca^{2+}]_i$ increases throughout the depolarization period, but the conductance already decreases.

maintained during depolarization while Ca^{2+} channels are open. During Ca^{2+} entry through voltage-dependent Ca^{2+} channels, the Ca^{2+} concentration at the mouth of the channel and at the inner membrane surface may be much higher than the average concentration in the cytoplasm.

EXOCYTOSIS STIMULATED WITH PIPETTE APPLICATION OF Ca^{2+}

To avoid this complication of Ca^{2+} gradients, we have introduced Ca^{2+} via the patch pipette. With this method we observed a capacitance increase with an initial slope very similar to the slow phase induced by depolarization (Fig. 4A). Following patch disruption and beginning of internal perfusion of the nerve terminal with high Ca^{2+}, the capacitance increases with an initial slope of 33 fF/s. To get such a slope, however, it was not sufficient to elevate $[Ca^{2+}]_i$ to a few μM. In this experiment the pipette contained 50 μM free Ca^{2+}, and Fig. 4B shows the concentration dependence of the maximal slope. The concentrations required to obtain exocytotic activity comparable to that induced by depolarization are apparently much higher than those which we measured inside nerve terminals during depolarization. It should be

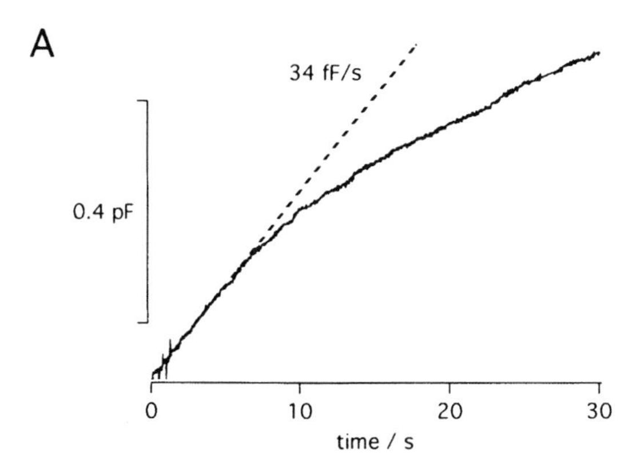

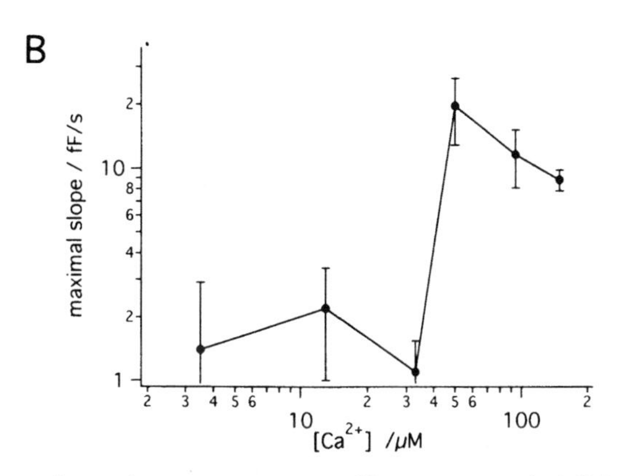

FIG. 4. A. Initial part of capacitance trace measured in a nerve terminal dialyzed with a pipette solution containing approximately 50 μM free Ca^{2+}. The recording starts 2–5 seconds after patch disruption and the beginning of internal perfusion with the pipette solution. The initial slope is indicated by the dashed line. **B.** Dependence of the maximal capacitance slope on pipette $[Ca^{2+}]$.

noted that in the experiments with pipette application of Ca^{2+}, $[Ca^{2+}]$ was not very strongly buffered in the pipette solution and the actual $[Ca^{2+}]_i$ may have been somewhat lower than the nominal values given in Fig. 4B. However, the actual concentration was certainly not smaller than half of these values; this means that more than 15 μM free Ca^{2+} are required for an increase of about 30 fF/s, which is about ten times more than what may be achieved during depolarization.

ENDOCYTOSIS OF LARGE VACUOLES AT HIGH $[Ca^{2+}]_i$

When the capacitance is followed for longer times as shown in Fig. 5A, sudden drops are seen between phases of capacitance increase (19). Such downward capacitance steps have previously been observed in strongly stimulated chromaffin cells by Neher and Marty (21) and are believed to represent a decrease in plasma mem-

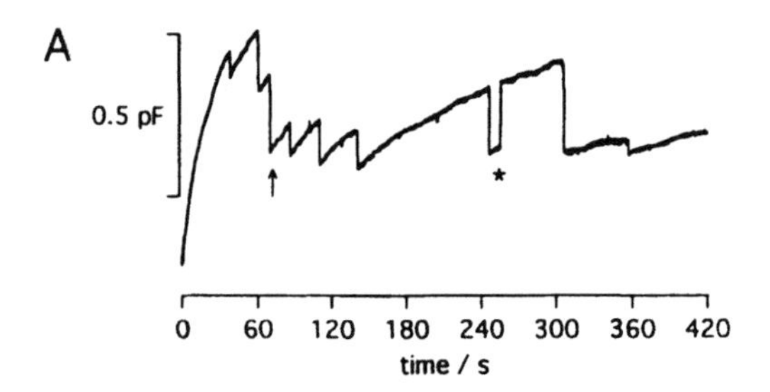

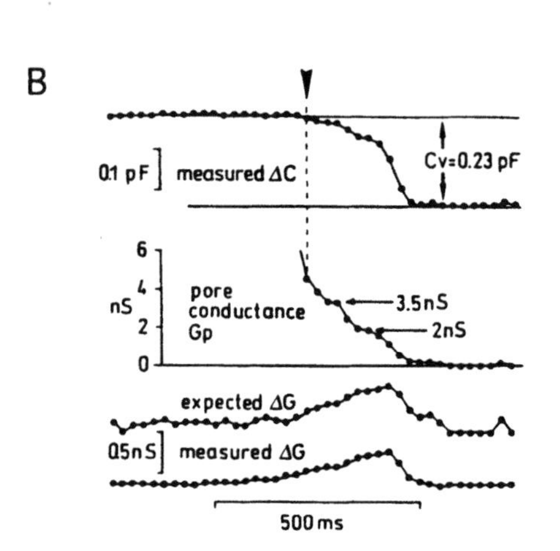

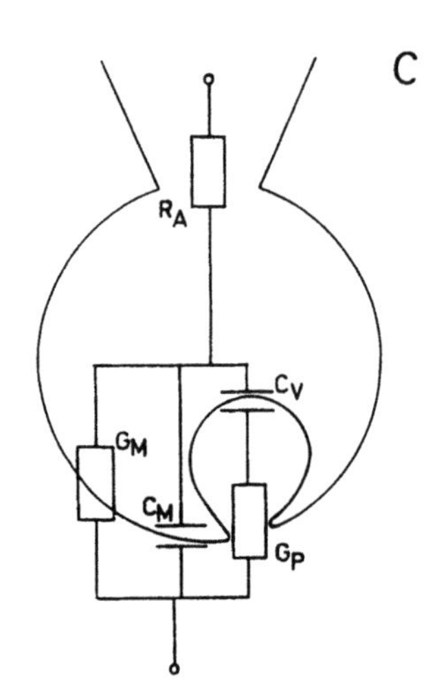

FIG. 5. A. Time course of capacitance measured in the same nerve terminal on a longer time scale. At the asterisk the capacitance compensation was transiently reduced by 0.2 pF. Superimposed on the capacitance increase are large downward steps indicating endocytosis of large vacuoles. **B.** Top trace: time course of capacitance trace during the third downward step on an expanded time scale. The apparent capacitance decrease is not instantaneous. The total decrease C_V corresponds to the capacitance of the endocytosed vacuole. Second trace: the time course of fission pore conductance G_p calculated from the ΔC trace using the equivalent circuit shown in B with $C_V = 0.23$ pF. Third trace: expected change in the in-phase signal assuming constant C_V and the above time course of G_P. Bottom trace: measured in-phase signal of the lock-in amplifier. **C.** Equivalent circuit of a nerve terminal during endocytosis of a single vacuole. R_A, access resistance of the pipette tip; G_M, plasma membrane conductance; C_M, plasma membrane capacitance; C_V, vacuole capacitance; G_P, fission pore conductance.

brane area due to endocytosis of large vacuoles. In the nerve terminals, such endocytotic events were rare but allowed us to analyze the dynamics of fission in more detail.

THE LAST 500 MILLISECONDS OF THE FISSION PORE

In the top trace of Fig. 5B, the third capacitance drop is shown on an expanded time scale. The measured capacitance decrease is apparently not abrupt but requires at least 300 millisecond. This could be due to either uptake of many small vesicles in rapid succession or due to endocytosis of a large vacuole pinching off from the plasma membrane. During the final stages preceding the fission event, the endocytic vacuole may be connected to the extracellular space by a narrow pore with low electrical conductance (Fig. 5C). We call this the fission pore in analogy to the fusion pore, which opens during the reverse process of exocytosis (27,28). The nerve terminal should thus be described by the equivalent circuit shown in Fig. 5C. During the actual fission event, only the pore conductance (G_P) should change while the other elements of the equivalent circuit are constant. The time course of the pore conductance can then be calculated from the time course of the measured capacitance change as previously described for the opening of an exocytotic fusion pore (27).

Knowing the vacuole capacitance (C_V) of the endocytosed vesicle from the size of the capacitance decrease, the deviation of the capacitance trace from an abrupt down step allows us to calculate the pore conductance G_P (27,29). When the capacitance shows a significant decrease below the initial value (arrowhead and dashed line in Fig. 5B), the pore conductance can be calculated; the result is shown in the second trace. In this example the decrease of pore conductance occurs in phases. Between these phases the conductance halts at 3.5 and 2 nS for about 25 milliseconds (arrows). The decrease down from 5 nS occurs with an overall slope of about -16 pS/ms.

To confirm that the gradual capacitance decrease is indeed due to a gradual decrease of pore conductance and not due to endocytosis of many small vesicles in rapid succession, the in-phase output of the lock-in amplifier was examined. If a single vacuole is endocytosed with decreasing fission pore conductance, then the in-phase output of the lock-in amplifier can be predicted (29) to have the shape shown in the third trace of Fig. 5B. The measured in-phase signal shown as the bottom trace agrees very well with the prediction, confirming that our analysis is correct. A similar agreement between the predicted and measured in-phase component was observed for all capacitance off steps. If during a capacitance drop many small vesicles would be rapidly endocytosed in sequence, the conductance trace would be flat. The analysis showed that the events of rapid capacitance decrease always represented endocytosis of single large vacuoles.

THE FISSION EVENT SHOWS A REPRODUCIBLE TIME COURSE

To compare the individual events with each other, they are all superimposed in Fig. 6A by shifting them in time such that they coincide at a conductance of 500 pS. Although there is some variability, the overall shape of all traces is quite similar. Below 400 pS the noise makes it very difficult to determine the time course of fission pore conductance (see inset). To determine the mean time course at improved signal-to-noise ratio, the traces were averaged as shown in Fig. 6B as the solid line. The dashed line shows the event of Fig. 5B, which resembles the average behavior quite well, at least up to $+85$ milliseconds where the pore conductance is as low as 140 pS (inset). This event is thus representative for the average behavior.

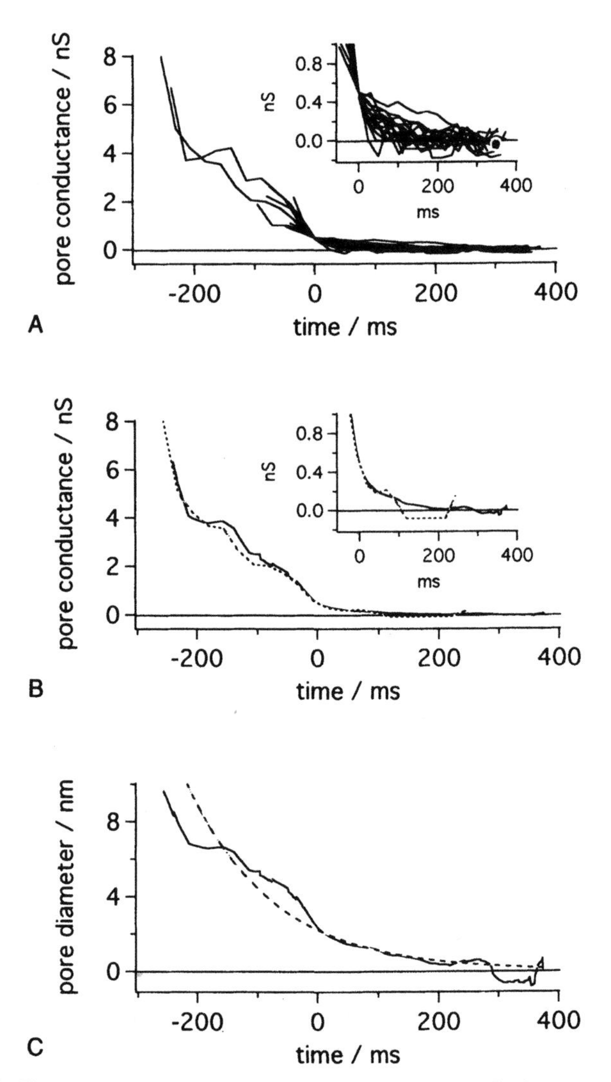

FIG. 6. A. Time course of pore conductance measured in 17 fission events. The traces were superimposed such that zero time was chosen as the value where the pore conductance had a value of 500 pS. The lines start at different times since large conductance values can only be determined for large endocytic vacuoles. The inset shows the part below 1 nS on an expanded scale. **B.** Average time course of fission pore closure obtained by averaging the traces shown in **A** (solid line). The time course of the single fission event of Fig. 5B is shown for comparison (dashed line). The inset shows the final part expanded as in **A. C.** Estimated time course of fission pore diameter d_P assuming a pore length corresponding to the thickness of two membranes ($L = 15$ nm) and a conductance of the bath solution of $\rho = 60$ Ωcm, using the relationship $d_P^2 = 4 \cdot \rho \cdot L \cdot G_P / \pi$. The time course was fitted with a single exponential (dashed line). For the fit the individual points in the average curve were weighted with the number of values, which were averaged at any particular time point. As shown in **A,** there are more data at low than at high conductance values.

THE PORE CLOSES WITH A TIME CONSTANT OF ~200 MILLISECONDS

To estimate the size of the fission pore certain assumptions have to be made regarding the length and the shape of the pore. The simplest model is a cylindrical pore with a length L of 15 nm equivalent to the thickness of two membrane bilayers. For a given pore conductance this assumption provides the time course of the pore diameter as shown in Fig. 6C. The diameter decreases from 8 nm to less than 0.6 nm within 500 milliseconds. The size of a lipid head group is about 0.8 nm; we are thus indeed observing the properties of the very final stages of membrane fission. If the closing is fitted with a single straight line, an average closing rate of 20 nm/s is obtained. However, the slope appears to change, suggesting an approximately exponential decrease of the diameter. A simple model would be to assume that the rate at which lipids are removed from the circumference depends on the number of lipids present and should thus be proportional to the length of the circumference as well as to the pore diameter. According to this model the pore diameter should decrease exponentially. A single exponential fit shown as the dashed line in Fig. 6C fits the time course quite well, giving a time constant of 150–200 milliseconds that corresponds to removal of 10% of the lipids from the circumference in 15–20 milliseconds.

HIGH $[Ca^{2+}]_i$ STIMULATES FUSION AMONG VESICLES

When the nerve terminals were dialyzed internally with high Ca^{2+} concentrations, we observed the formation of vacuoles inside as shown in Fig. 7A. Initially we thought that these vacuoles are the morphological manifestation of the large endocytic events. However, such structures were also observed in nerve terminals that did not show large capacitance drops as shown by the capacitance trace of this nerve terminal (Fig. 7B), and were thus formed inside the nerve terminal. We believe that these were formed by granule-granule fusion, which apparently occurs at similar Ca^{2+} concentrations as the granule-plasma membrane fusion. Although in the cytoplasm 50 or 100 μM free Ca^{2+} may never be reached under physiological conditions, this may well be the case beneath the plasma membrane. If granule-plasma membrane fusion and granule-granule fusion occur at similar Ca^{2+} concentrations, then this would question the absolute requirement of specific plasma membrane components for exocytotic fusion, supporting the idea that all components required for fusion are contained in the granule membranes (30).

FUSION OF MULTIVESICULAR COMPOUNDS REVEALS OPENING OF THE FUSION PORE

Figure 8 shows that after intense stimulation, large vacuoles can also fuse with the plasma membrane (31).In this recording we see a fusion event developing over

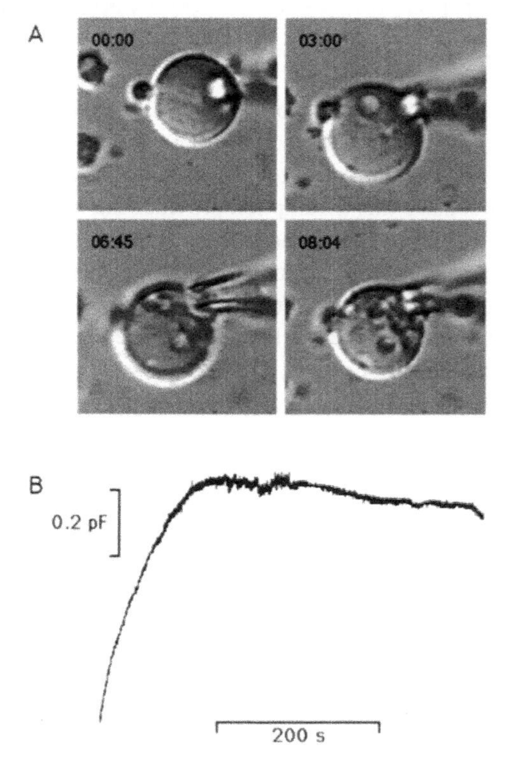

FIG. 7. A. Video images of a nerve terminal internally dialyzed with 150 μM Ca^{2+}. The images are taken from a continuous video recording at the indicated times (in min:sec). The displayed area has a size of 17 × 17 μm^2. The diameter of the nerve terminal is ~8 μm. **B.** The capacitance trace of this nerve terminal shows exocytosis but no endocytosis of large vacuoles.

several seconds. First a capacitance flicker is seen (top trace) where the vacuole fuses transiently with the plasma membrane. The fusion pore conductance (second trace) has a value between 0.8 and 1 nS for 1.7 seconds and closes again completely before the vacuole eventually fuses irreversibly. The agreement between the predicted (third trace) and measured (bottom trace) in-phase signal of the lock-in amplifier again confirms that we observe fusion of a single very large vesicle. It should be noted that the final irreversible fusion starts with the same conductance as the transient pore. The agreement is indicated in the conductance trace by the dashed line. The pore then expands gradually, fluctuates between 2 and 4 nS, transiently returns to the initial size, and eventually opens completely. The fusion properties of such compound vesicles allow us to investigate fusion pore formation between the vesicle membrane and plasma membrane of nerve terminals.

It is still an open question if neurosecretion occurs by complete vesicle-plasma membrane fusion or by release through transient fusion pore openings (32,33). In pituitary nerve terminals there is substantial evidence that the vesicles are rapidly

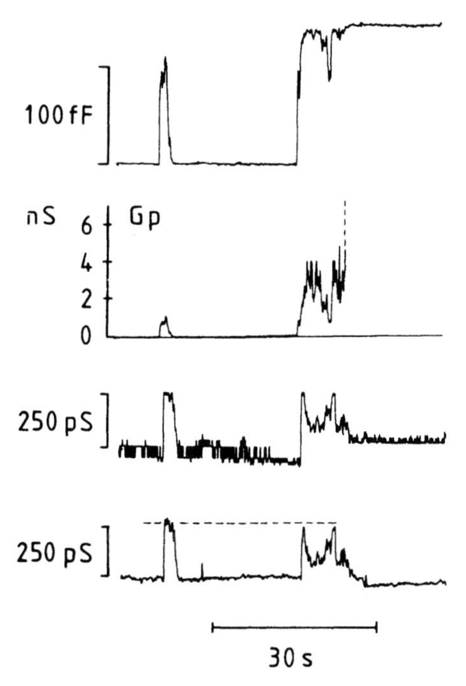

FIG. 8. Transient and complete fusion of a multigranular compound with the plasma membrane. From top to bottom traces are capacitance, fusion pore conductance, expected in-phase signal, measured in-phase signal. For details of the analysis, compare the legend of Fig. 5. The dashed line in the bottom trace (measured in-phase signal) was drawn by eye parallel to the baseline.

internalized after transient fusion (2,3,34). The properties of the associated transient fusion pore may be characterized by studying fusion of multigranular compounds with the plasma membrane of the nerve terminal.

CONCLUSION

Pituitary nerve terminals can be easily isolated and are large enough for patch clamp experiments allowing the investigation of relationships between depolarization, Ca^{2+} entry, ionic channels, and exocytosis. The experiment revealed distinct phases of exocytosis suggesting the presence of a small population of secretory vesicles *ready for release* and another population that must be slowly recruited. The local Ca^{2+} concentration at the site of fusion must be much higher than what is usually observed in the cytoplasm of a stimulated nerve terminal. The present data strongly suggest that an about tenfold higher Ca^{2+} concentration exists at the site where granule-plasma membrane fusion occurs. The observation that vesicle-vesicle fusion occurs strongly suggests that the fusion site present on the vesicle membrane is not exclusively designed to find its partner in the plasma membrane. However, under physiological conditions, the Ca^{2+} concentration will be high enough to allow fusion to occur only at the plasma membrane. Furthermore, exocytosis of multivesicular compounds and endocytosis of large vacuoles make it possible to characterize the details of fusion and fission and to observe the formation, size, and disassembly of a transient fusion pore.

ACKNOWLEDGMENT

This work has been supported by the DFG (Sfb 312/B6) and by the DAAD (311-pro-ca).

REFERENCES

1. Cazalis M, Dayanithi G, Nordmann JJ. The role of patterned burst and interval on the excitation-coupling mechanism in the isolated rat neural lobe. *J Physiol* 1985;369:45–60.
2. Nordmann JJ, Dreifuss JJ, Baker PF, Ravazzola M, Malaisse-Lagae F, Orci L. Secretion dependent uptake of extracellular fluid by the rat neurohypophysis. *Nature (Lond)* 1974;250:155–157.
3. Knoll G, Plattner H, Nordmann JJ. Exo-endocytosis in isolated peptidergic nerve terminals occurs in the sub-second range. *Biosci Rep* 1992;12:495–501.
4. Lindau M, Stuenkel EL, Nordmann JJ. Depolarization, intracellular calcium and exocytosis in single vertebrate nerve endings. *Biophys J* 1992;61:19–30.
5. Leake RD, Fisher DA. Oxytocin secretion and milk ejection in the human. In: Amico JA, Robinson AG, eds. *Oxytocin: Clinical and Laboratory Studies*. New York: Elsevier Science Publishers; 1985; 200–206.
6. Handler JS, Orloff J. Antidiuretic hormone. *Ann Rev Physiol* 1981;43:611.
7. Biegon A, Terlous M, Boorhuis B, De Kloet E. Arginine-vasopressin binding sites in rat brain: a quantitative autoradiographic study. *Neurosci Lett* 1984;44:229–234.
8. de Wied D. Peptides and behaviour. *Life Sci* 1977;19:685–692.
9. Nordmann JJ, Dayanithi G, Lemos JR. Isolated neurosecretory nerve terminals as a tool for studying the mechanism of stimulus-secretion coupling. *Biosci Rep* 1987;7:411–426.
10. Nordmann JJ, Dayanithi G. Release of neuropeptides does not only occur at nerve terminals. *Biosci Rep* 1988;8:471–483.
11. Wang X, Treistman SN, Lemos JR. Two types of high-threshold calcium currents inhibited by ω-conotoxin in nerve terminals of the rat neurohypophysis. *J Physiol* 1992;445:181–199.
12. Lemos JR, Nowycky MC. Two types of calcium channels coexist in peptide-releasing vertebrate nerve terminals. *Neuron* 1989;2:1419–1426.
13. Wang G, Thorn P, Lemos JR. A novel large-conductance Ca^{2+}-activated potassium channel and current in nerve terminals of the rat neurohypophysis. *J Physiol* 1992;457:47–74.
14. Bielefeldt K, Rotter JL, Jackson MB. Three potassium channels in rat posterior pituitary nerve terminals. *J Physiol* 1992;458:41–67.
15. Thorn PJ, Wang X, Lemos JR. A fast, transient K^+ current in neurohypophysial nerve terminals of the rat. *J Physiol* 1991;432:313–326.
16. Lemos JR, Nordmann JJ. Ionic channels and hormone release from peptidergic nerve terminals. *J Exp Biol* 1986;124:53–72.
17. Wang G, Lemos JR. Tetrandrine blocks a slow, large-conductance, Ca^{2+}-activated potassium channel besides inhibiting a non-inactivating Ca^{2+} current in isolated nerve terminals of the rat neurohypophysis. *Pflügers Arch Eur J Physiol* 1992;421:558–565.
18. Fidler Lim N, Nowycky MC, Bookman J. Direct measurement of exocytosis and calcium currents in single vertebrate nerve terminals. *Nature (Lond)* 1990;344:449–451.
19. Rosenboom H, Lindau M. Exo-endocytosis and closing of the fission pore during endocytosis in single pituitary nerve terminals internally perfused with high calcium concentrations. *Proc Natl Acad Sci USA* 1994;91:in press.

20. Lindau M, Neher E. Patch-clamp techniques for time-resolved capacitance measurements in single cells. *Pflügers Arch Eur J Physiol* 1988;411:137–146.
21. Neher E, Marty A. Discrete changes of cell membrane capacitance observed under conditions of enhanced secretion in bovine adrenal chromaffin cells. *Proc Natl Acad Sci USA* 1982;79:6712–6716.
22. Augustine GJ, Neher E. Calcium requirements for secretion in bovine chromaffin cells. *J Physiol* 1992;450:247–271.
23. Thomas P, Surprenant A, Almers W. Cytosolic Ca^{2+}, exocytosis and endocytosis in single melanotrophs of the rat pituitary. *Neuron* 1990;5:723–733.
24. Thomas P, Wong JG, Almers W. Millisecond studies of secretion in single rat pituitary cells stimulated by flash photolysis of caged Ca^{2+}. *EMBO J* 1993;12:303–306.
25. Neher E, Zucker RS. Multiple calcium-dependent processes related to secretion in bovine chromaffin cells. *Neuron* 1993;10:21–30.
26. Heinemann C, von Ruden L, Chow RH, Neher E. A two-step model of secretion control in neuroendocrine cells. *Pflügers Arch Eur J Physiol* 1993;424:105–112.
27. Breckenridge LJ, Almers W. Currents through the fusion pore that forms during exocytosis of a secretory vesicle. *Nature (Lond)* 1987;328:814–817.
28. Spruce AE, Breckenridge LJ, Lee AK, Almers W. Properties of the fusion pore that forms during exocytosis of a mast cell secretory vesicle. *Neuron* 1990;4:643–654.
29. Lindau M. Time-resolved capacitance measurements: monitoring exocytosis in single cells. *Quart Rev Biophys* 1991;24:75–101.
30. Vogel SS, Zimmerberg J. Proteins on exocytotic vesicles mediate calcium-triggered fusion. *J Cell Biol* 1992;89:4749–4753.
31. Lindau M, Nordmann JJ. Capacitance and calcium changes during exo- and endocytosis in single vertebrate nerve terminals. *Biophys J* 1991;59:276a.
32. Alvarez de Toledo G, Fernández-Chacòn R, Fernandez JM. Release of secretory vesicle products during transient vesicle fusion. *Nature (Lond)* 1993;363:554–558.
33. Neher E. Secretion without full fusion. *Nature (Lond)* 1993;363:497–498.
34. Nordmann JJ, Artault J-C. Membrane retrieval following exocytosis in isolated neurosecretory nerve endings. *Neurosci* 1992;49:201–207.

Molecular and Cellular Mechanisms of Neurotransmitter Release, edited by Lennart Stjärne, Paul Greengard, Sten Grillner, Tomas Hökfelt, and David Ottoson, Raven Press, Ltd., New York © 1994.

13

Glutamate Exocytosis from Isolated Nerve Terminals

David G. Nicholls and Eleanor T. Coffey

Department of Biochemistry, University of Dundee, Dundee DD1 4HN, Scotland

The isolated nerve terminal, or synaptosome, is the simplest system that retains all the machinery for the uptake, storage, and exocytosis of neurotransmitters (reviewed by Nicholls (1)). The cerebral cortical preparation contains a high proportion of glutamatergic terminals and can be exploited to investigate presynaptic aspects of this dominant excitatory transmitter. Two complementary approaches are currently being undertaken: 1) the functional, where the action of the intact synaptosome is dissected, and 2) the structural, where the proteins that are uniquely expressed in the presynaptic terminal are identified and characterized. The "top-down" approach taken by our laboratory and the "bottom-up" approaches of other groups (reviewed by Südhof and Jahn (2)) are beginning to intermesh to provide an integrated view of presynaptic mechanisms.

It should be emphasized that the in-situ nerve terminal functions largely autonomously from the relatively distant cell body, requiring only the electrical signal from the axonal action potential to trigger release and the replenishment of materials via axonal transport mechanisms for longterm survival. The synaptosome preparation, which remains viable for several hours, therefore allows release to be investigated, subject only to a means of replacing the physiological stimulus of the action potential from the absent axon.

The classic technique of elevated KCl depolarizes synaptosomes and triggers the exocytosis of transmitter, but the resultant clamped depolarization scarcely mimics the transient depolarization of the physiological action potential. A closer approach to the in situ stimulus is provided by the ability of the K^+ channel inhibitor α-dendrotoxin to inhibit a subclass of presynaptic K^+ channels that appear to be responsible for stabilizing the presynaptic membrane potential, thus removing a mechanism that normally functions to prevent fluctuations in membrane potential from initiating spontaneous action potentials in the terminals.

Thus addition of dendrotoxin, or judicious concentrations of 4-aminopyridine,

(3) results in a small time- and population-average depolarization monitored by cyanine dye, an elevation in the mean cytoplasmic free Ca^{2+} concentration monitored by fura-2, and an extensive release of glutamate monitored by the reduction of exogenous nicotinamide adenine dinucleotide phosphate ($NADP^+$) in the presence of glutamate dehydrogenase (Fig.1). Each parameter is inhibited by the Na^+-channel inhibitor tetrodotoxin, supporting the suggestion that the terminals are firing action potentials under these conditions. The classic depolarization stimulus of elevated KCl also causes Ca^{2+} elevation and glutamate release, but in this case tetrodotoxin is without effect, consistent with the negligible contribution of the fast inactivating Na^+ channels to this prolonged clamped depolarization (Fig. 1). As we shall discuss in the next paragraph, the availability of these two complementary means of stimulating synaptosomes has proven invaluable in establishing the distinction between agents that modulate presynaptic action potentials and those that act subsequently at the presynaptic Ca^{2+} channel or aspects of Ca^{2+}-secretion coupling.

Prolonged KCl depolarization of cerebral cortical synaptosomes releases glutamate by two parallel pathways: a Ca^{2+}-dependent exocytosis superimposed upon a Ca^{2+}-independent release of glutamate from cytoplasmic stores. The exocytotic

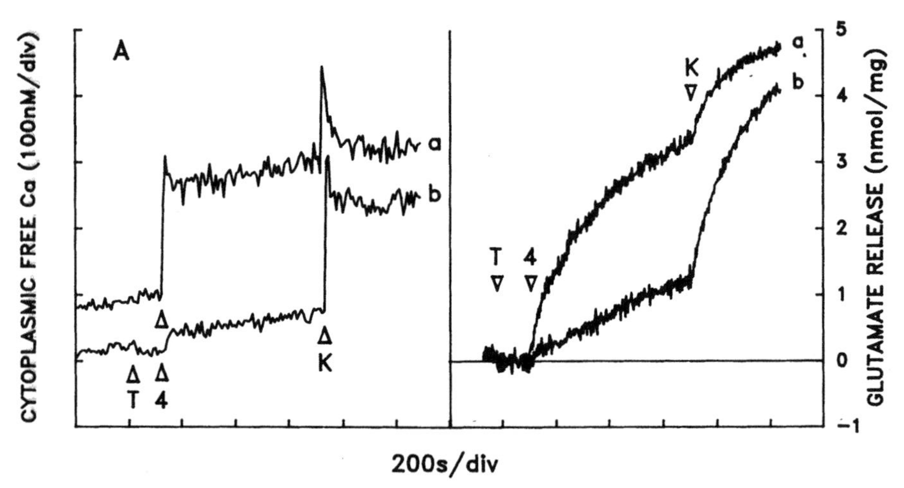

FIG. 1. Tetrodotoxin inhibits the increase in cytoplasmic free Ca^{2+} ($[Ca^{2+}]_c$) and exocytosis of glutamate evoked by 4-aminopyridine but not by elevated KCl. Guinea-pig cerebral cortical synaptosomes were either preloaded with fura-2 for determination of $[Ca^{2+}]_c$ or incubated in the presence of glutamate dehydrogenase and $NADP^+$ for assay of glutamate release. Synaptosomes were depolarized with 1 mM 4-aminopyridine (from Sanchez-Prieto et al., ref. 4) in the absence (upper traces) or presence (lower traces) of 1 μM tetrodotoxin (TTx). Note that the 4-aminopyridine stimulated increase in $[Ca^{2+}]_c$ and Ca^{2+}-dependent release of glutamate are both strongly inhibited by tetrodotoxin, indicating that Na^+ channels are firing repetitively during 4-aminopyridine depolarization. In contrast, the subsequent addition of 30 mM KCl (K) effectively increases $[Ca^{2+}]_c$ and releases glutamate in the presence of tetrodotoxin, showing that Na^+ channel activation is not involved in the KCl-evoked clamped depolarization and exocytosis. (Adapted from Tibbs et al., ref. 3.)

pathway can be distinguished by being adenosine 5'-triphosphate (ATP) dependent, (4) and inhibitable by tetanus toxin and botulinum neurotoxin types A and B (5) as well as by the novel Ca^{2+}-channel antagonist agatoxin-GI (Aga-GI) (6). Ca^{2+}-independent release is a thermodynamic consequence of a reversal of the plasma membrane acidic amino acid transporter and results in the efflux of both glutamate and aspartate from the cytoplasm (7). The latter process is likely only of in vivo significance under pathological conditions such as ischemia.

The Ca^{2+} entry into the synaptosomes in response to KCl depolarization can be monitored by ^{45}Ca while the population average increase in cytoplasmic free Ca^{2+} concentration ($[Ca^{2+}]_c$) can be monitored by fura-2. The isotope reveals a strongly biphasic uptake, with an extensive initial entry (sufficient to increase the total Ca^{2+} concentration in the terminals by 1 mM) followed by a much slower, noninactivating uptake (8). The initial transient uptake is reflected in a modest "spike" in the fura-2 trace that declines to a plateau of about 350 nM $[Ca^{2+}]_c$ (see Fig. 1). Two independent observations argue that it is the slow noninactivating phase of Ca^{2+} entry that is coupled to the release of glutamate, rather than the initial rapid entry: 1) predepolarization of synaptosomes in the absence of Ca^{2+} followed by Ca^{2+} readdition (which inactivates the initial rapid phase of Ca^{2+} entry) does not affect the kinetics of release (9) and 2) depolarization by 4AP results in comparable release, although 4AP does not evoke the first rapid phase of ^{45}Ca entry. (10) The functional significance of the initial massive ^{45}Ca entry for glutamate release is unclear and may possibly involve damaged, or nonglutamatergic, terminals.

Glutamate exocytosis thus appears to be coupled to Ca^{2+} entry through a noninactivating channel. Studies by Pocock on the nature of the glutamate-coupled Ca^{2+} channel have exploited a novel Ca^{2+} channel neurotoxin, Aga-GI, from the venom of the funnel web spider *Agelenopsis aperta* (6). Aga-GI can be distinguished from Aga-IVA, which has been reported to cause a 50% inhibition of glutamate exocytosis from partially depolarized synaptosomes (11) by its ability totally to block glutamate release. Consistent with the noninactivating nature of the presynaptic Ca^{2+} channel in these synaptosomes, Aga-GI lowers the plateau phase of elevated $[Ca^{2+}]_c$ generated by elevated KCI (6).

A detailed analysis of the kinetics of glutamate exocytosis in response to prolonged KCl depolarization reveals a strongly biphasic release (9). Of a total glutamate content of 30 nmol/mg, 1 nmol is released within 1 second and an additional 4 nmol is released much more slowly, with a $t½$ of 60 seconds. Since the release-coupled Ca^{2+} channel is noninactivating, an explanation other than biphasic Ca^{2+} entry is required for the biphasic glutamate release. One attractive possibility is that the relatively rapid phase represents the exocytosis of vesicles already docked at the presynaptic membrane, whereas the slow phase corresponds to the release of those vesicles that require prior transport through the cytoplasm to the active zone.

Cytoskeletal interactions have been implicated in the mobilization and transport of synaptic vesicles to the active zone (for review, see Greengard et al. (12)). The present model therefore provides a means to investigate factors that might modify this transport. The *Clostridial* neurotoxins botulinum neurotoxin serotype B and tetanus toxin have recently been shown to act as Zn-proteases, and to cleave a C-terminal fragment from the integral synaptic vesicle protein synaptobrevin-2 (13). When preincubated with intact synaptosomes, botulinum neurotoxins serotype A and B, and tetanus toxin each partially inhibit glutamate exocytosis (5). Interestingly, it is the second, slow, phase of glutamate release that is selectively inhibited (Fig. 2), consistent with a role of synaptobrevin in vesicle mobilization or targeting. Recently it has been convincingly demonstrated that synaptobrevins are receptors for the NSF/SNAP complex implicated in synaptic vesicle targeting and fusion (14) and it is therefore of interest in the present preparation that targeting or transport, rather than exocytosis itself, appears to be most sensitive to the action of the toxins.

α-Latrotoxin is a 130kd polypeptide neurotoxin that targets the presynaptic terminal (15) interacting with a neurexin (16) that in turn co-extracts with synaptotagmin (17) an integral component of the exocytotic mechanism. At the neuromuscular junction, the toxin causes a massive loss of synaptic vesicles and release of acetyl-

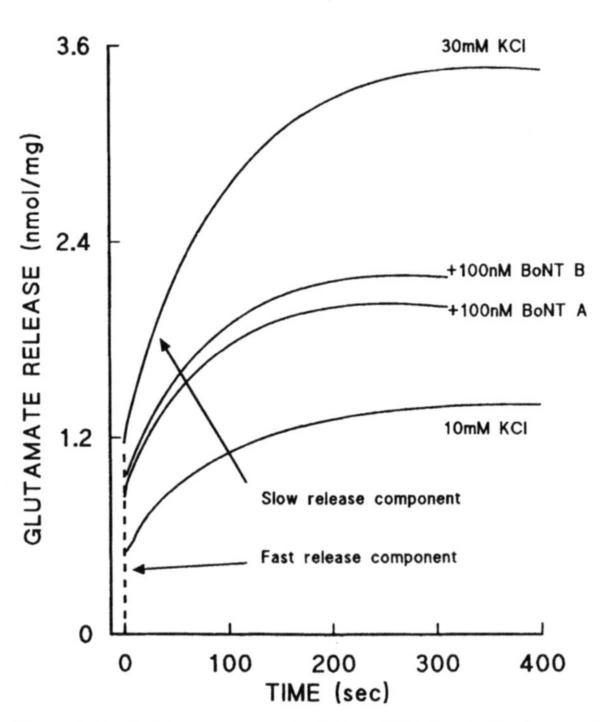

FIG. 2. Kinetic analysis of glutamate exocytosis from KCl-depolarized synaptosomes: botulinum neurotoxins preferentially decrease the slow phase of release. Net Ca^{2+}-dependent glutamate release was corrected for the lag in the enzyme assay, revealing "fast" (<2 seconds) and "slow" components of release. While suboptimal depolarization with 10 mM KCl decreased both components equally, preincubation with botulinum neurotoxins type A and B preferentially decrease the slow component that may be limited by vesicle transport to active zones. (Adapted from McMahon et al., ref. 5.)

choline that is independent of external Ca^{2+} (18). Our investigations of the toxin action upon synaptosomes yielded some surprising results: while a massive Ca^{2+}-independent release of glutamate was observed in the presence of low nanomolar concentrations of the toxin (Fig. 3), this phenomenon was (by a variety of criteria) almost entirely cytoplasmic in origin, being accompanied by aspartate and other cytoplasmic markers (7). The difference between the behavior of the toxin at the neuromuscular junction and at CNS glutamatergic terminals suggests that there may be some subtle differences in the structural organization of the release machinery at these two locations. Certainly the effect of α-latrotoxin is not a nonspecific lysis of the terminal (the concentration dependency correlates well with the high affinity binding of the toxin to the terminals) but may, for example, reflect the premature formation of a plasma membrane fusion pore in the absence of correct vesicular docking, leading to a release of cytoplasmic components.

When glutamate exocytosis is plotted as a function of the population average $[Ca^{2+}]_c$, a very steep relationship is obtained: maximal glutamate release requiring only a doubling of the fura-2 signal from 200 nM to 400 nM. This is either an accurate reflection of the Ca^{2+} concentration at the release site, which would demand a high degree of cooperativity, or alternatively it is possible that there is a high, localized, Ca^{2+} concentration at the release site that is diluted by the global fura-2 signal. Localized Ca^{2+} is directly observable at large presynaptic terminals such as the squid giant synapse, where there is clear evidence that a multiple array

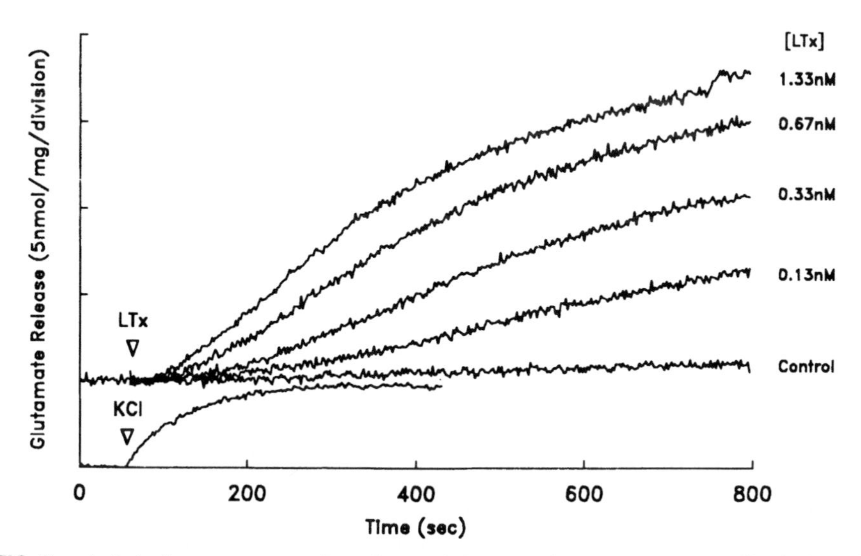

FIG. 3. α-Latrotoxin causes a massive release of glutamate from synaptosomes. The indicated concentrations of α-latrotoxin were added where indicated (Ltx) and glutamate release monitored. The maximal release of glutamate amounts to about 80% of the total content, far greater than evoked by 30 mM KCl, is essentially independent of external Ca^{2+}, and is accompanied by a collapse of adenosine 5′-triphosphate (ATP) levels and a massive rise in $[Ca^{2+}]_c$. (Adapted from McMahon et al., ref. 7.)

of Ca^{2+} channels are synchronously activated to generate a steep gradient of Ca^{2+} concentration away from the membrane (19–21). A similar mechanism appears to function at the mammalian neuromuscular junction (22) where an action potential results in the synchronous release of several hundred synaptic vesicles. Extrapolation to CNS terminals, however, presents some conceptual and practical difficulties: 1) the typical CNS terminal is very small (<1 μm) in relation to the NMJ or squid giant synapse, limiting the volume available for a bulk gradient of Ca^{2+} concentration, and is thus incapable of showing significant resolution by imaging techniques; and 2) the terminal has a low quantal release—an invading action potential may release a single synaptic vesicle, or none at all—rather than the massive multiquantal release seen at giant terminals.

We have employed an indirect approach to investigate whether localized Ca^{2+} gradients were likely to exist within synaptosomes during exocytosis (9,23). The approach consisted of generating the same global average fura-2 signal by two different means: 1) by depolarization and 2) by addition of the Ca^{2+} ionophore ionomycin. The rationale was that in the latter case there would be a uniform entry of Ca^{2+} across the plasma membrane, with no possibility for the formation of local "hot-spots" at the sites of exocytosis. When release of glutamate (and aminobutyric acid (GABA)) was compared by these two protocols, a dramatic difference was found: depolarization-evoked glutamate release was maximal but ionomycin-evoked release was negligible, although the fura-2 signals were identical (Fig. 4). As an essential control, since it had been proposed that depolarization played an inherent role in the release process, the experiment with ionomycin was repeated with depolarized synaptosomes, the Ca^{2+} channels being inhibited by very high Mg^{2+} concentrations. Still no glutamate release occurred, and it was concluded that Ca^{2+} entry through voltage-activated Ca^{2+} channels was essential for release at an indicated $[Ca^{2+}]_c$ of 0.4 μM. Glutamate release could be evoked only by ionomycin when sufficient ionophore was added to increase the average Ca^{2+} concentration to levels that were more than sufficient to saturate the fura-2 signal (>5 μM). The conclusion was therefore drawn that a high localized Ca^{2+} concentration was created in the immediate vicinity of the voltage-activated Ca^{2+} channel, but was masked in the overall fura-2 signal by the much smaller increase in $[Ca^{2+}]_c$ generated in the bulk cytoplasm after the Ca^{2+} had diffused away from the release site (9,23).

The failure of ionomycin to release glutamate could be due to some nonspecific damage to the bioenergetic integrity of the synaptosome in the presence of the ionophore. That this was not the case was indicated by a parallel series of experiments in which the release of a representative neuropeptide, cholecystokinin, was studied in the same preparation (23). The neuropeptide was found to be releasable by low ionomycin concentrations; indeed, for a given fura-2 signal, the neuropeptide was released *more* efficiently by ionomycin than by KCl-depolarization.

Neuropeptides are known to be released ectopically (away from active zones), and furthermore to be released in a frequency-dependent manner in physiological preparations; low stimulation frequencies release classical transmitters, while fre-

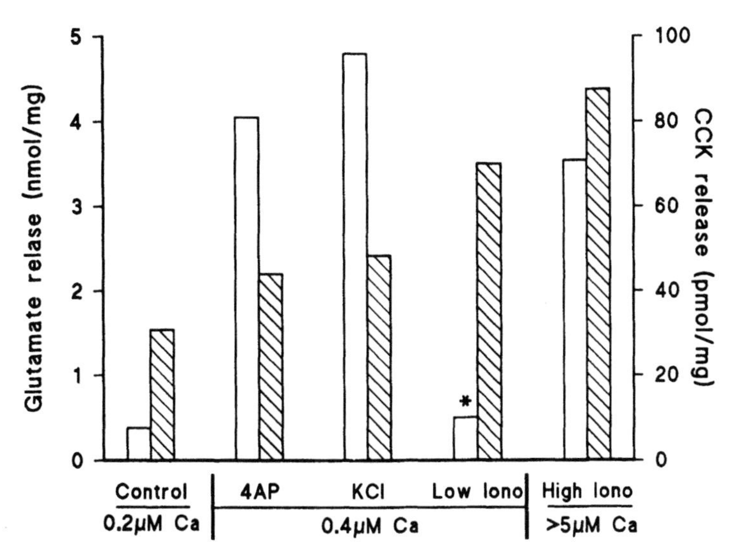

FIG. 4. Ca^{2+} entry via voltage-gated Ca^{2+} channels is much more effective than ionomycin in releasing glutamate (but not cholecystokinin) from synaptosomes. Concentrations of 4AP, KCl, and ionomycin were selected to give the same elevation in fura-2 signal to 0.4 μM. 4AP and KCl were at least eight times as effective as ionomycin (starred) in evoking glutamate release (open bars). Cholecystokinin, however (hatched bars), was released more effectively by ionomycin than by KCl or 4AP. The results are interpreted as indicating the need for a high localized $[Ca^{2+}]$ in the vicinity of the Ca^{2+} channel for glutamate exocytosis, but not for the ectopic release of CCK. (Adapted from Verhage et al., ref. 23.)

quencies of 5–10 Hz release increasing proportions of neuropeptide. Our experiments provided some rationale for these observations. At low stimulation frequencies, glutamate was released by a direct localized interaction between Ca^{2+} entering through the Ca^{2+} channel; however, the Ca^{2+}-homeostatic mechanisms in the terminal were sufficient to prevent the bulk Ca^{2+} from rising sufficiently to release the neuropeptide. Glutamate release would thus be triggered by a relatively low affinity interaction between the high concentration of Ca^{2+} at the mouth of the Ca^{2+} channel, while the neuropeptide would be released by a high affinity interaction of the bulk Ca^{2+} concentration. As the frequency of stimulation increased, the bulk Ca^{2+} would begin to rise and cholecystokinin would begin to be released. The postulated low-affinity Ca^{2+} receptor for which there is now increasing evidence (20) confers great kinetic advantages for the recovery of the terminal following exocytosis; the localized region of high $[Ca^{2+}]$ at the mouth of a Ca^{2+} channel will collapse immediately when the channel closes (without the need for Ca^{2+} transport mechanisms to pump Ca^{2+} out of the cytoplasm), while additionally the low affinity of the postulated receptor allows for a high offrate for the dissociation of Ca^{2+}.

There are two ways to consider a localized Ca^{2+} gradient within a nerve terminal: 1) as a wave of elevated Ca^{2+} generated by the synchronous firing of many Ca^{2+} channels, or 2) by an even more localized interaction between individual Ca^{2+} channels and individual release complexes. Since the release of glutamate from terminals appears to be capable of very precise regulation by the Ca^{2+} entry across the membrane, it is not readily apparent how a wave of Ca^{2+} would create the precision required for a controlled, graded glutamate release, being more suited to the massive synchronous release of many synaptic vesicles characteristic of the neuromuscular junction. At the other extreme, a discrete one-to-one relationship between a single Ca^{2+} channel and a release site would allow a precise gradation of release directly proportional to the number of channel openings.

A rough calculation is relevant here. During the plateau phase of the fura-2 response following KCl, the rate of Ca^{2+} entry is less than 2 nM/min/mg. One mg of synaptosomal protein corresponds to some 10^{10} synaptosomes, and thus the Ca^{2+} entry into an individual depolarized synaptosome will average only about 1,000 ions/s, or far less than the open channel conductance of a single channel. Channel opening thus appears to be only an occasional event for a depolarized synaptosome—more consistent with a series of single interactions than with a synchronous opening of multiple Ca^{2+} channels.

The synaptosome possesses a number of proteins that can be phosphorylated in situ by Ca^{2+}-calmodulin-dependent protein kinase II (CaMKII). The most intensively studied of these proteins is synapsin I, which undergoes a rapid cycle of increased phosphorylation-dephosphorylation following KCI depolarization (12). There is persuasive evidence linking the phosphorylation state of synapsin I with synaptic vesicle attachment to the cytoskeleton, and it has been proposed that elevated presynaptic Ca^{2+}, by initiating synapsin phosphorylation, plays a role in the mobilization of a reserve population of synaptic vesicles from the cytoskeleton, allowing their transport to the active zone for ultimate exocytosis (reviewed by Greengard et al. (12)).

Since the massive glutamate exocytosis evoked by prolonged KCI depolarization is likely to involve mobilization of this reserve pool, as discussed in the previous paragraph, this would appear to be a suitable preparation to investigate the functional relationship between synapsin I phosphorylation and glutamate release. While inhibitors of calmodulin-dependent protein kinases exist that are highly specific when tested against a range of purified kinases, there is always the possibility that the inhibitors may be less specific in the more complex milieu of the intact terminal. Instead of employing inhibitors we have therefore investigated the consequences of substituting Ba^{2+}, which does not activate calmodulin, for Ca^{2+}. Ba^{2+} addition is complicated by its ability to inhibit presynaptic K^+ channels and initiate spontaneous "action potentials" similar to those in the presence of 4AP; therefore the protocol chosen was to inhibit this firing with tetrodotoxin and to depolarize with KCl. Under these conditions the KCl-evoked release of glutamate was virtually identical in the presence of Ca^{2+} or Ba^{2+} (Fig. 5).

It has been suggested that the ability of Ba^{2+} to evoke transmitter release might be contingent upon a Ba^{2+}-evoked release of Ca^{2+} from internal stores (24). There is no evidence that this occurs in synaptosomes—indeed, the terminals are notable for the lack of nonmitochondrial Ca^{2+} stores comparable to those seen in the cell soma. Additionally, the release of mitochondrial Ca^{2+} stores into the synaptosomal

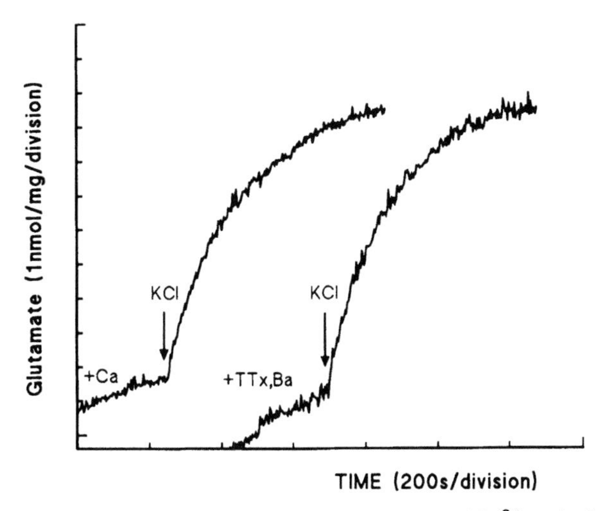

FIG. 5. KCl evokes the same glutamate release in the presence of Ba^{2+} as in the presence of Ca^{2+}. Tetrodotoxin (TTx) is present in the Ba^{2+} trace to prevent the cation from initiating spontaneous action potentials similar to those in the presence of 4AP. (Adapted from McMahon and Nicholls, ref. 38.)

cytoplasm does not trigger glutamate release (McMahon and Nicholls, unpublished observations), as would be predicted since this mode of Ca^{2+} mobilization would not satisfy the need for a high, localized $[Ca^{2+}]_c$ at the active zone. Finally, if Ba^{2+} caused an indirect mobilization of internal Ca^{2+} one would expect to see a cycle of synapsin I phosphorylation comparable to that seen in the presence of external Ca^{2+}. Figure 6 shows however that, in the Ba^{2+} medium, little or no depolarization-evoked increase in phosphorylation of synapsin I can be detected when monitored by *Staphylococcus aureus* protease treatment of the synapsin-containing band. Our results demonstrate therefore that an extensive Ba^{2+}-evoked exocytosis of glutamate from cortical synaptosomes can occur in the absence of changes in phosphorylation of synapsin I.

The facility with which Ba^{2+} triggers glutamate exocytosis places some constraints upon the as yet unidentified intrasynaptosomal Ca^{2+}-binding site responsible for the triggering of exocytosis, since this entity must additionally be capable of interacting with Ba^{2+}. For example, synaptotagmin, which can bind Ca^{2+} and has been suggested as a candidate for such a site, has been reported not to interact with Ba^{2+} at concentrations up to 1 mM (25).

Several isoforms of protein kinase C (PKC) have been detected in presynaptic nerve terminals (26) and PKC activation by phorbol esters has been generally observed to increase transmitter release from synaptosomes, brain slices, and neuronal cultures (reviewed by Nicholls (1)). Based on first principles, there are two classes of mechanism that could be responsible for enhanced transmitter release: 1) modification of the properties of the presynaptic ion channels, resulting directly or indi-

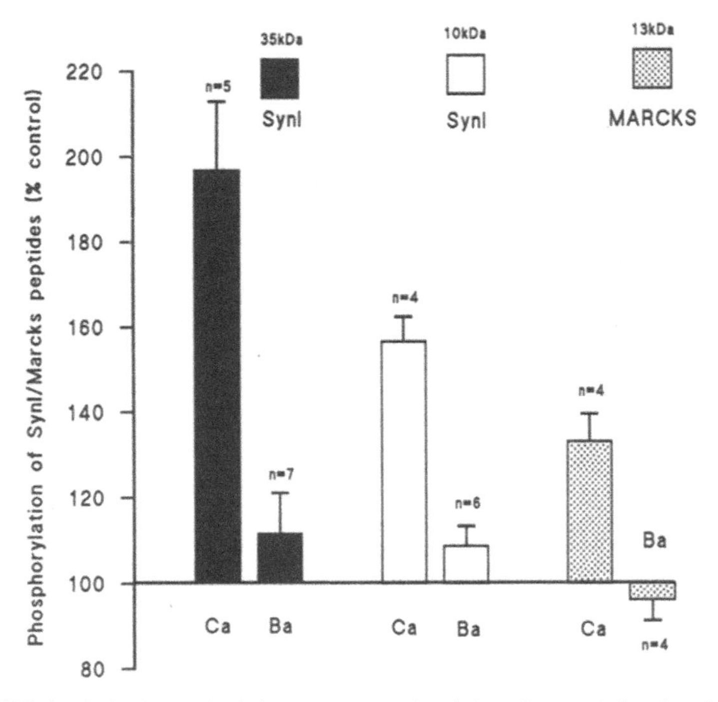

FIG. 6. KCl-depolarization evoked changes in phosphorylation of synapsin-I and myristoylated alanine-rich C-kinase substrate (MARCKS) in Ca^{2+}- or Ba^{2+}-containing media. Synaptosomes prelabeled with ^{32}P were suspended in media containing 1 mM Ca^{2+} or Ba^{2+} and depolarized by 30 mM KCl for 15 seconds before quenching. V8 protease treatment of the synapsin/MARCKS band obtained by PAGE was performed (Wang and Greengard, ref. 39) followed by a second electrophoresis to separate peptides of 35 kD (synapsin-I, CaMKII site), 10 kD (synapsin-I, CaMKI, and PKA site) and 13 kD (MARCKS, PKC site).

rectly in an enhanced entry of Ca^{2+} into the terminal, or 2) facilitation of an intrasynaptosomal locus.

We were able to throw some light upon the mechanism of PKC potentiation when we observed that the sensitivity of synaptosomal glutamate release to phorbol esters was far greater when release was evoked with 4AP rather than with KCl (27). Indeed, we have recently refined these findings such that 4AP-evoked glutamate exocytosis can be modulated from maximal (when PKC is activated by phorbol esters) to essentially zero (when PKC is fully inhibited by the PKC inhibitor Ro 31-8220); (28) see Fig. 7). In total contrast, KCl-evoked glutamate release is entirely insensitive to phorbol ester or Ro 31-8220. That these agents are effective in controlling PKC activity in the intact synaptosome is supported by monitoring the phosphorylation state of the characteristic 13 kD peptide generated by *Staphlococcal* V8 protease treatment of the in situ myristolated alanine-rich C-kinase substrate (MARCKS). The failure of phorbol esters to modulate release in the absence of 4AP is not due to an inactivation of PKC, since MARCKS is still phosphorylated under these conditions.

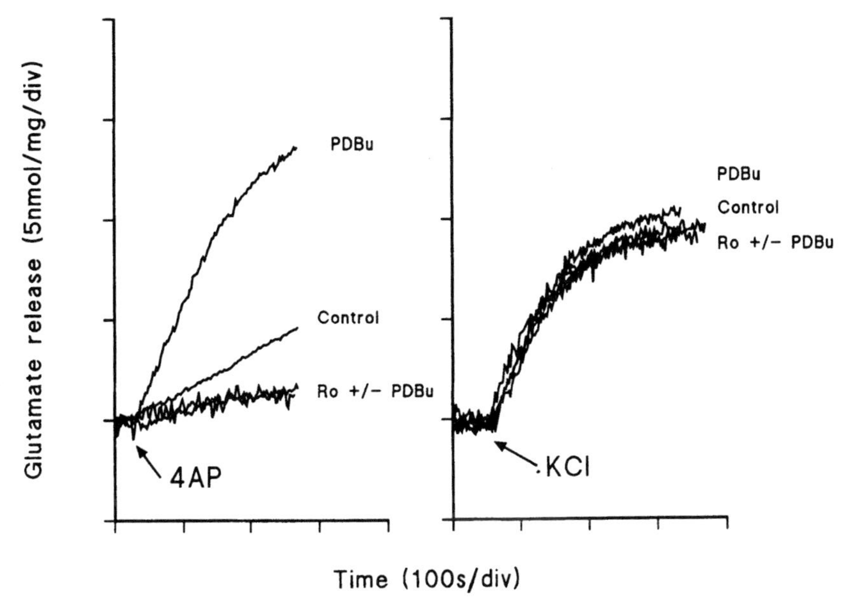

FIG. 7. Protein kinase C (PKC) activators and inhibitors exert complete control over 4AP-evoked glutamate exocytosis but have no effect on that evoked by high KCl. Net Ca^{2+}-dependent release is shown, evoked by either 1 mM 4AP or 30 mM KCl. 1 μM phorbol dibutyrate (PDBu) ± 10 μM Ro-31 8220 (a PKC inhibitor) were added prior to depolarization as indicated. (Adapted from Coffey et al., ref. 28.)

Since KCl causes a clamped depolarization of the plasma membrane and evokes glutamate release that is independent of PKC activity, this would appear to eliminate a locus for the kinase at the release-coupled Ca^{2+} channel (which as just discussed does not inactivate during prolonged KCl depolarization) or downstream of Ca^{2+} entry at an intrasynaptosomal locus. This conclusion differs from that gathered from studies on the exocytosis of large secretory granules from platelets, chromaffin cells, and mast cells where phorbol esters potentiate exocytosis in permeabilized preparations (29,30) and emphasizes that it is essential to avoid unsupported extrapolation from one exocytotic system to another.

If, by elimination, PKC must be acting upstream of the Ca^{2+} channel, this strongly suggests that the kinase might be modulating the ion channels that are responsible for the spontaneous action potentials initiated by 4AP. There are two possibilities: 1) activation of the voltage-activated Na^+ channels that initiate the action potentials or 2) inhibition of the 4AP-resistant K^+ channels that help to terminate the action potentials. While there are PKC phosphorylation sites on Na^+ channels that decrease peak conductance and slow inactivation, (31) a more likely locus is the 4AP-resistant K^+-channel.

The 4AP has limited pharmacological specificity and the delineation of the PKC-sensitive K^+-channel is easier if the high-affinity $K^+{}_A$ channel inhibitor α-dendrotoxin is substituted for 4AP. Dendrotoxin evokes the same release of glutamate as

esters. This indicates that the putative K^+ channel regulated by PKC is dendrotoxin-resistant. The channel is, however, sensitive to 1 mM Ba^{2+} (acting as a K^+-channel inhibitor rather than a secretogogue), since Ba^{2+} mimics the effects of phorbol-12, 13-dibutyrate (PDBu) in stimulating release, while the phorbol esters no longer potentiate in the presence of Ba^{2+}.

The locus for PKC action on glutamate release appears therefore to be a dendrotoxin-insensitive, Ba^{2+}-sensitive K^+ channel. There are electrophysiological precedents for PKC-mediated inhibition of K^+ channels, particularly the delayed rectifier that terminates action potentials (32), while action potential prolongation has been shown to be a highly effective way of enhancing transmission at the squid giant synapse (33).

Although action potential prolongation in the synaptosomal preparation cannot be directly observed, it would be predicted that the average depolarization of the synaptosomal preparation would be enhanced and that this would lead to an increased elevation of the bulk $[Ca^{2+}]_c$ reported by fura-2. Both are observed in practice.

A candidate for the physiological activator of presynaptic PKC has been suggested by Herrero et al. (34). A specific agonist for the metabotropic glutamate receptor, (1S,3R)-ACPD, was able to mimic the effects of phorbol esters on synap-

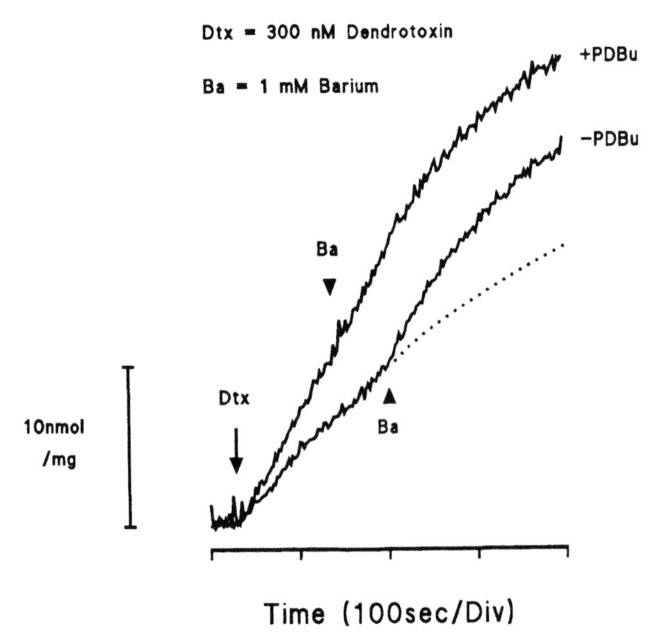

FIG. 8. Phorbol esters and Ba^{2+} each inhibit the dendrotoxin-insensitive channel responsible for the modulation of glutamate release. Dendrotoxin (300 nM), Ba^{2+} (1 mM) and PDBu (1 μM) were present where indicated.

tosomal membrane potential, fura-2 elevation, and glutamate release, but only in the presence of low concentrations of arachidonic acid (AA). Recently (35) it has been possible to confirm that (1S,3R)-ACPD in the presence of AA does activate PKC in intact synaptosomes (monitored by MARCKS phosphorylation). The synergistic requirement for both the glutamate receptor agonist and AA is consistent with findings for purified α- βII- or ϵ-PKC, the activation of each of which by diacylglycerol (DAG) is potentiated by AA (36). If the synaptosomal locus of AA action is at PKC, it would be predicted that DAG production by (1S,3R)-ACPD would be upstream of AA and therefore not require the fatty acid; this has been confirmed (35).

The proposed sequence of events at the presynaptic membrane would thus be: 1) glutamate binds to the presynaptic mGluR, activating a G protein coupled to phospholipase C; 2) the DAG generated will only activate PKC in the presence of AA; 3) activated PKC directly or indirectly phosphorylates a presynaptic K^+ channel, resulting in its inactivation, which prolongs or intensifies physiological action potentials; 4) since the Ca^{2+} channel coupled to glutamate release does not appear to undergo voltage-dependent inactivation, its probability of opening will increase with prolonged depolarization; and 5) the tight linkage between Ca^{2+} entry and glutamate release will result in an increased quantal release of transmitter.

The physiological significance of this positive-feedback autoreceptor can prompt intriguing speculation. First, it should be emphasized that what is being observed must be an extremely widespread phenomenon in the cerebral cortex to produce such a robust control of glutamate release from the isolated synaptosomes; if only a subpopulation of glutamatergic terminals possessed this regulatory pathway, the response would be diluted out by nonresponsive terminals. A number of possible sources for AA in vivo can be proposed; one would be phospholipase A_2 activated postsynaptically in response to NMDA receptor activation—thus providing a mechanistic basis for proposals that AA could function as a retrograde messenger during plastic changes at the synapse. While it is unreasonable that such a mechanism would chronically elevate glutamate release, a shortlasting enhancement during the induction of potentiation could act to ensure that glutamate release remained high until a synapse is securely switched to the potentiated form.

A second major mechanism regulating the release of glutamate from synaptosomes is mediated by the adenosine A1 receptor. In contrast to the facilitatory effect of PKC activation, which is only seen during 4AP-evoked action potentials and is accompanied by an enhanced population depolarization, the adenosine A1 inhibition of release is equally apparent with both KCl and 4AP-evoked depolarization and is without effect on plasma membrane potential with either agent (37). A decreased fura-2 elevation indicates a receptor-mediated inhibition of the presynaptic Ca^{2+} channel.

In conclusion, the synaptosome is proving to be an interface between the neurophysiological approach to synaptic function, with its emphasis on receptors and ion channels, and molecular investigations into the protein chemistry of exocytosis that are the subject of much of the rest of this volume.

ACKNOWLEDGMENTS

Research into synaptosomal exocytosis has been supported by the Medical Re-

butions to this work by Drs. Harvey McMahon, Talvinder Sihra, Jennifer Pocock, Anne Barrie, and Gareth Tibbs.

REFERENCES

1. Nicholls DG. The glutamatergic nerve terminal. *Eur J Biochem* 1993;212:613–631.
2. Sudhof TC, Jahn R. Proteins of synaptic vesicles involved in exocytosis and membrane recycling (Review). *Neuron* 1991;6:665–677.
3. Tibbs GR, Barrie AP, Van-Mieghem F, McMahon HT, Nicholls DG. Repetitive action potentials in isolated nerve terminals in the presence of 4-aminopyridine: effects on cytosolic free Ca^{2+} and glutamate release. *J Neurochem* 1989;53:1693–1699.
4. Sanchez-Prieto J, Sihra TS, Nicholls DG. Characterization of the exocytotic release of glutamate from guinea-pig cerebral cortical synaptosomes. *J Neurochem* 1987;49:58–64.
5. McMahon HT, Foran P, Dolly JO, Verhage M, Wiegant VM, Nicholls DG. Tetanus toxin and botulinum toxins type A and B inhibit glutamate, gamma-aminobutyric acid, aspartate, and met-enkephalin release from synaptosomes. Clues to the locus of action. *J Biol Chem* 1992;267:21338–21343.
6. Pocock JM, Nicholls DG. A toxin (Aga-GI) from the venom of the spider Agelenopsis aperta inhibits the mammalian presynaptic Ca channel coupled to glutamate exocytosis. *Eur J Pharmacol* 1992;226:343–350.
7. McMahon HT, Rosenthal L, Meldolesi J, Nicholls DG. α-Latrotoxin releases both vesicular and cytoplasmic glutamate from isolated nerve terminals. *J Neurochem* 1990;55:2039–2047.
8. Nachshen DA. The early time-course of K-stimulated Ca uptake in presynaptic nerve terminals isolated from rat brain. *J Physiol (Lond)* 1985;361:251–268.
9. McMahon HT, Nicholls DG. Transmitter glutamate release from isolated nerve terminals: evidence for biphasic release and triggering by localized Ca^{2+}. *J Neurochem* 1991;56:86–94.
10. Delaney KR, Zucker RS. Calcium released by photolysis of DM-nitrophen stimulates transmitter release at squid giant synapse. *J Physiol (Lond)* 1990;426:473–498.
11. Turner TJ, Adams ME, Dunlap K. Calcium channels coupled to glutamate release identified by omega-Aga-IVA. *Science* 1992;258:310–313.
12. Greengard P, Valtorta F, Czernik AJ, Benfenati F. Synaptic vesicle phosphoproteins and regulation of synaptic function. *Science* 1993;259:780–785.
13. Schiavo G, Benfenati F, Poulain B, Rossetto O, Polverino-de-Laureto P, Dasgupta BR, Montecucco C. Tetanus and botulinum-B neurotoxins block neurotransmitter release by proteolytic cleavage of synaptobrevin. *Nature* 1992;359:832–835.
14. Iredale PA, Martin KF, Hill SJ, Kendall DA. The control of intracellular calcium and neurotransmitter release in guinea pig-derived cerebral cortical synaptoneurosomes. *Biochem Pharmacol* 1993; 45:407–414.
15. Rosenthal L, Meldolesi J. α-Latrotoxin and related toxins. *Pharmac Ther* 1989;42:115–134.
16. Ushkaryov YA, Petrenko AG, Geppert M, Südhof TC. Neurexins: Synaptic cell surface proteins related to the α-latrotoxin receptor and laminin. *Science* 1992;257:50–56.
17. Petrenko AG, Perin MS, Davletov BA, Ushkaryov YA, Geppert M, Sudhof TC. Binding of synaptotagmin to the alpha-latrotoxin receptor implicates both in synaptic vesicle exocytosis. *Nature* 1991;353:65–68.
18. Ceccarelli B, Hurlbut WP. Vesicle hypothesis of the release of quanta of acetylcholine. *Physiol Rev* 1980;60:396–441.
19. Llinas R, Sugimori M, Silver RB. Microdomains of high calcium concentration in a presynaptic terminal. *Science* 1992;256:677–679.
20. Adler EM, Augustine GJ, Duffy SN, Charlton MP. Alien intracellular calcium chelators attenuate neurotransmitter release at the squid giant synapse. *J Neurosci* 1991;11:1496–1507.

21. Zucker RS, Stockbridge N. Presynaptic calcium diffusion and the time-courses of transmitter release and synaptic facilitation at the squid giant synapse. *J Neurosci* 1983;3:1263–1269.
22. Simon SM, Llinas R. Compartmentalization of the submembrane calcium activity during Ca influx and its significance in transmitter release. *Biophys J* 1985;48:485–498.
23. Verhage M, McMahon HT, Ghijsen WEJM, Boomsma F, Wiegant V, Nicholls DG. Differential release of amino acids, neuropeptides and catecholamines from nerve terminals. *Neuron* 1991;6: 517–524.
24. Przywara DA, Chowdhury PS, Bhave SV, Wakade TD, Wakade AR. Barium-induced exocytosis is due to internal calcium release and block of calcium efflux. *Proc Natl Acad Sci USA* 1993;90:557–561.
25. Brose N, Petrenko AG, Südhof TC, Jahn R. Synaptotagmin: A calcium sensor on the synaptic vesicle surface. *Science* 1992;256:1021–1025.
26. Shearman MS, Shinomura T, Oda T, Nishizuka Y. Synaptosomal protein kinase-C subspecies. A. Dynamic changes in the hippocampus and cerebellar cortex concomitant with synaptogenesis. *J Neurochem* 1991;56:1255–1262.
27. Barrie AP, Nicholls DG, Sanchez-Prieto J, Sihra TS. An ion channel locus for the protein kinase C potentiation of transmitter glutamate release from guinea pig cerebrocortical synaptosomes. *J Neurochem* 1991;57:1398–1404.
28. Coffey ET, Sihra TS, Nicholls DG. Protein kinase C and the regulation of glutamate exocytosis from cerebrocortical synaptosomes. *J Biol Chem* 1993;268:21060–21065.
29. Rink TJ, Sanchez A, Hallam TJ. DAG and phorbol ester stimulate secretion without raising cytoplasmic free Ca in human platelets. *Nature* 1983;305:317–319.
30. Knight DE, Sugden D, Baker PF. Evidence implicating protein kinase C in exocytosis from electropermeabilized bovine chromaffin cells. *J Membrane Biol* 1988;104:21–34.
31. Neumann R, Catterall WA, Scheuer T. Functional modulation of brain sodium channels by protein kinase C phosphorylation. *Science* 1991;254:115–118.
32. Linden DJ, Smeyne M, Sun SC, Connor JA. An electrophysiological correlate of protein kinase C isozyme distribution in cultured cerebellar neurons. *J Neurosci* 1992;12:3601–3608.
33. Augustine GJ. Regulation of transmitter release at the squid giant synapse by presynaptic delayed rectifier potassium current. *J Physiol (Lond)* 1990;431:343–364.
34. Herrero I, Miras-Portugal MT, Sanchez-Prieto J. Positive feedback of glutamate exocytosis by metabotropic presynaptic receptor stimulation. *Nature* 1992;360:163–166.
35. Coffey ET, Herrero I, Sihra TS, Sanchez-Prieto J, Nicholls DG. Synergistic activation of protein kinase C by arachidonic acid and (1S,3R)-ACPD via a presynaptic metabotropic receptor facilitates glutamate release. *J Neurochem* (in press).
36. Asaoka Y, Nakamura S, Yoshida K, Nishizuka Y. Protein kinase C, calcium and phospholipid degradation. *Trends Biochem Sci* 1992;17:414–417.
37. Barrie AP, Nicholls DG. Adenosine A1 receptor inhibition of glutamate exocytosis and protein kinase C-mediated decoupling. *J Neurochem* 1993;60:1081–1086.
38. McMahon HT, Nicholls DG. Barium-evoked glutamate release from guinea-pig cerebrocortical synaptosomes. *J Neurochem* 1993;61:110–115.
39. Wang JKT, Greengard P. Protein phosphorylation in nerve terminals: comparison of Ca/CAM and protein kinase C systems. *J Neurosci* 1988;8:281–288.

Molecular and Cellular Mechanisms of Neurotransmitter Release, edited by Lennart Stjärne, Paul Greengard, Sten Grillner, Tomas Hökfelt, and David Ottoson, Raven Press, Ltd., New York © 1994.

14

Central Glutamatergic Transmission

A View from the Presynaptic Axon

Lennart Brodin, Oleg Shupliakov, and Sten E. Grillner

Department of Neurophysiology, Nobel Institute for Neurophysiology, Karolinska Institute, S-171 77 Stockholm, Sweden

The synaptic transmitter of the last decade is no doubt L-glutamate, as it has proved to be responsible for most excitatory synaptic transmission at all levels of the CNS (1). The properties of the different types of postsynaptic glutamate receptors are well known, and the respective roles of many different receptor subunits have been clarified (2,3). On the presynaptic side, however, the storage and release mechanisms have only now become the center of interest. In this review we will focus on one experimentally amenable central glutamate synapse that has been explored—the giant reticulospinal synapse in the lamprey spinal cord—and compare it to other glutamatergic synapses in the same species that have distinctly different properties. We will first survey the subcellular localization of glutamate and some related amino acids, and discuss the mechanisms involved in the supply of transmitter in individual axons. Thereafter, the physiological mechanisms involved in the modulation of glutamate release in reticulospinal and interneuronal synapses will be considered.

SYNAPTIC VESICLES OF RETICULOSPINAL AXONS STORE GLUTAMATE, BUT NOT ASPARTATE OR HOMOCYSTEATE

The lamprey spinal cord has the general outline of a vertebrate spinal cord, but it is thin and ribbon-like (Fig. 1A), and lacks myelin as well as internal blood vessels (4). The neuronal cell bodies are located in cell columns (Fig. 1A, lateral cell column = LC) that are surrounded by tracts of descending and ascending axons. The ventromedial fiber column contains some very large axons (Fig. 1A, reticulospinal axon = RS) originating from reticulospinal cells in the brainstem. These axons have a large main trunk without collaterals, and form en passant synapses with dendrites of spinal interneurons and motoneurons. The axonal diameter may exceed 50 μm.

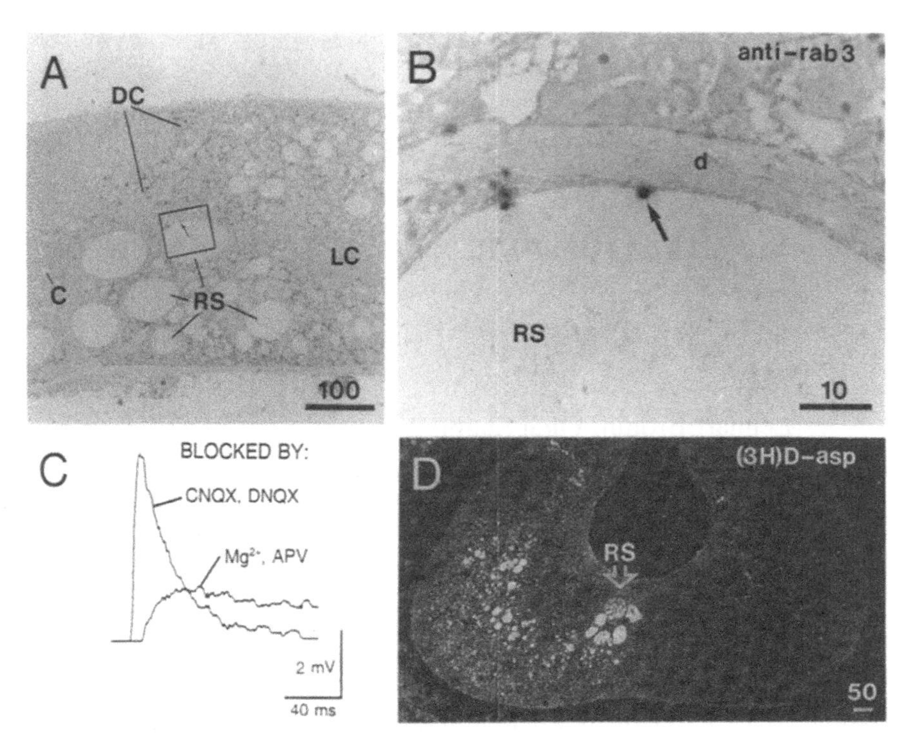

FIG. 1. Location and pharmacological properties of reticulospinal synapses in the lamprey spinal cord. **A, B.** Transverse section of the spinal cord in which the synapses have been visualized by immunohistochemistry for rab3. The section (**A**) includes the central canal area (C), the lateral cell column (LC), the dorsal column with cutaneous afferent axons (DC), and the ventromedial column, which contains the giant reticulospinal axons (RS). The section was incubated with a polyclonal antiserum raised against rab3A (gift from Dr. W. E. Balch, Scripps, La Jolla) and processed with a peroxidase antiperoxidase complex (PAP) protocol. The black spots in the section correspond to labeled synaptic vesicle clusters at active zones inside different axons, as confirmed by electronmicroscopic analysis of adjacent ultrathin sections (Shupliakov and Brodin, unpublished observations). **B.** Framed area in **A** at high magnification. The arrow indicates a labeled synapse established by the axon (RS) on a dendrite (d). **C.** Pharmacological properties of the chemical excitatory postsynaptic potential (EPSP) evoked by a reticulospinal axon. The reticulospinal EPSPs mostly contain an initial fast component that can be depressed by AMPA/kainate receptor blockers (CNQX, DNQX, kynurenic acid), and a slow component, that can be depressed by NMDA receptor blockers (APV, Mg^{2+}). This EPSP was first recorded in Mg^{2+}-free solution; thereafter 1.8 mM Mg^{2+} was applied, followed by application of 2 mM kynurenic acid. The sweeps were obtained by subtraction to show each component in isolation. An initial electrotonic EPSP (Rovainen, ref. 4) has also been subtracted and is not illustrated (for further details see Buchanan et al., ref. 5). **D.** Accumulation of (^{3}H)D-aspartate in giant reticulospinal axons (RS) and other types of descending axons, following a unilateral injection in the rostral spinal cord. Dark-field micrograph of a section from the brainstem at the level of obex (modified from Brodin et al., ref. 6). Scale bars: A—100 μm; B—10 μm; D—50 μm.

The synapses contain active zones with a dense cluster of synaptic vesicles, which can be visualized at the light microscopic level with antisera to synaptic vesicle-associated proteins, such as rab3 (Fig. 1A,B). Functionally, these axons resemble glutamatergic terminals in the brain, as they activate postsynaptic NMDA and AMPA/kainate receptors (5) (Fig. 1C), and take up and transport the glutamate analogue (3H)D-aspartate (6) (Fig. 1D).

The favorable anatomical organization of the reticulospinal synapses has permitted a direct analysis of the amino acid content of the synaptic vesicle clusters. The application of quantitative immunocytochemical techniques (7,8) thus showed that immunoreactivity to fixed glutamate is strongly accumulated over the synaptic vesicle cluster of reticulospinal axons (9) (Fig. 2A). By comparing the labeling density (gold particles/μ^2) over the vesicle cluster (Fig. 2A) with that in coprocessed test conjugates (Fig. 2B), the concentration of fixed glutamate could be estimated to approximately 30 mM (9). The intravesicular glutamate concentration was estimated to exceed 60 mM (cf. Burger et al. (10)), as the extravesicular space represents around half of the area in a section of the cluster and contains a low glutamate concentration. A certain loss of amino acids will occur during the fixation (7). The reticulospinal axons contain only low levels of immunoreactivity to aspartate (Fig. 3) and homocysteate, and there is no clear accumulation of these types of immunoreactivity over the vesicle cluster (9) (cf. Burger et al. (10) and Naito and Ueda (11)).

PLASMA MEMBRANE TRANSPORT OF GLUTAMATE IN AXONS AND GLIA

The glutamate analogue D-aspartate has been used extensively in retrograde tracing experiments with autoradiography (12) to identify neurons that transport glutamate (Fig. 2B). More recently, ultrastructural immunocytochemical methods have been developed that enable a quantitative analysis of the D-aspartate uptake into different synaptic compartments (13). After an incubation of the lamprey spinal cord in D-aspartate-containing solution, followed by immunogold labeling, the giant reticulospinal axons show only a limited D-aspartate uptake, but a prominent uptake occurs into adjacent glial processes (Gundersen, Shupliakov, Brodin, Ottersen, and Storm-Mathisen, unpublished observations). Thus, although an uptake of glutamate occurs into the reticulospinal axons (cf. Fig. 1D), this transport does not appear to play any major role in replenishing the presynaptic glutamate pool in these synapses (cf. other glutamate synapses (13)).

The glial processes encapsulating each synaptic junction provide a system for the removal of glutamate from synaptic areas, which appears to be remarkably efficient. The thin (200 μm), isolated lamprey spinal cord can in fact be bathed in high concentrations of L-glutamate, or D- or L-aspartate (200–300 μM), with little postsynaptic activation occurring, unless the glutamate/aspartate uptake is suppressed (16). The functional significance of this uptake system has been demonstrated under

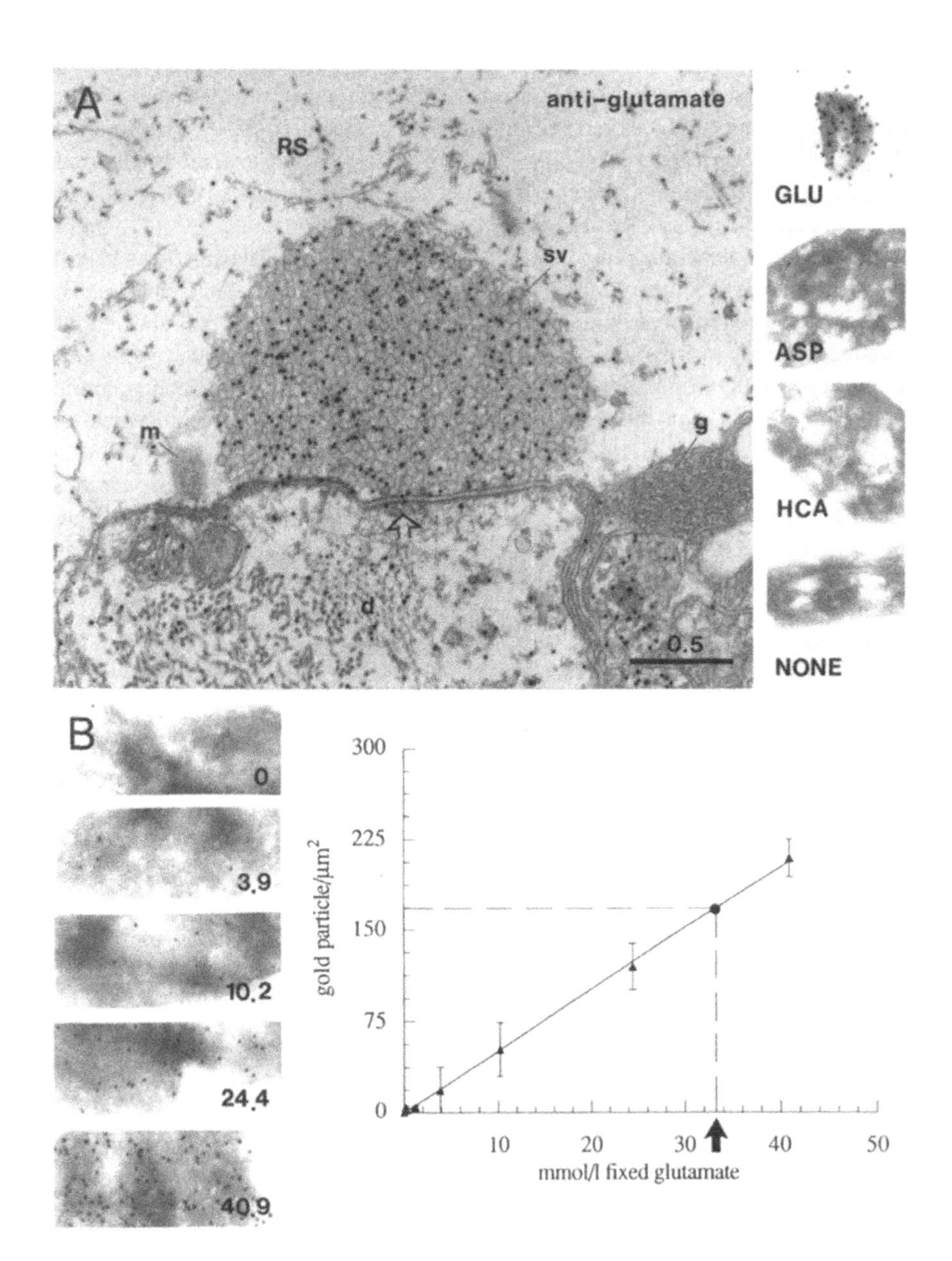
A
anti-glutamate
RS
sv
m
g
d
0.5
GLU
ASP
HCA
NONE
B
0
3.9
10.2
24.4
40.9
300
225
150
75
0
gold particle/µm²
10
20
30
40
50
mmol/l fixed glutamate

conditions of sustained synaptic activity in studies of the spinal interneuronal network generating locomotor activity (see information that follows). When this network, which is driven by glutamatergic synapses (14,15) is activated by sensory stimulation, the application of a glutamate uptake inhibitor (dihydrokainate) causes a marked prolongation of each burst, as well as of the length of the complete motor response (16) (cf. Brodin et al., 1988 (17)). In contrast, single excitatory postsynaptic potentials (EPSPs) are not affected (17) implying that the glutamate uptake does not determine the time course of the unitary synaptic response (cf. Hestrin et al. (18) and Sarantis et al. (19)).

Glutamine is a major precursor of glutamate in central nerve terminals (for review, see Fonnum (20)). Quantitative immunogold studies in reticulospinal (Figs. 2, 4) and other glutamatergic synapses (see information that follows) show that the level of glutamate labeling exceeds that of the glutamine labeling in the presynaptic element (21), which is in agreement with the proposed presynaptic conversion of glutamine into glutamate (20,22,23). In the perisynaptic glial processes (Figs. 2, 4), however, the glutamate/glutamine ratio is low (21) suggesting that after glutamate has been released and taken up into these processes, it is rapidly reconverted to glutamine (17,20,22,23).

LEVELS OF SYNAPTIC MITOCHONDRIA AND GLUTAMATE DIFFER BETWEEN GLUTAMATERGIC AXONS WITH DIFFERENT PATTERNS OF ACTIVITY

Glutamatergic axons in the lamprey spinal cord exhibit distinct and specialized features that appear to relate to their respective activity patterns. The differences are

FIG. 2. Accumulation of immunoreactivity to fixed glutamate over synaptic vesicles. **A.** Electron micrograph of the synaptic junction established by a reticulospinal axon on a dendrite of a spinal neuron. The ultrathin section was incubated with an antiserum to fixed glutamate, followed by secondary antibodies conjugated with 15 nm gold particles. The area containing synaptic vesicles (sv) displays a markedly higher density of gold particles than the axoplasmic matrix of the reticulospinal axon (RS), the postsynaptic dendrite (d) a glial profile (g) and a mitochondria (m). A gap junction is present to the left of the active zone. Insets in **A** show part of the specificity control used in the experiment. The electron micrographs represent test sections containing glutaraldehyde-brain macromolecule conjugates of L-glutamate (GLU), L-aspartate (ASP), and L-homocysteate (HCA) incubated in parallel with the tissue section. The bottom inset ("None") shows a conjugate made by reacting a brain macromolecule extract with glutaraldehyde without addition of amino acids (for details see Ottersen, ref. 7). **B.** Relationship between the concentration of fixed glutamate and the density of gold particles in test conjugates used to estimate the concentration of glutamate in coprocessed tissue sections. The insets to the left show sections of conjugates containing different concentrations of fixed glutamate, which were incubated along with the tissue sections. The concentration of fixed glutamate in mmol/l is indicated in each micrograph. The diagram shows the relation between the gold particle density and the concentration of fixed glutamate in the test conjugates. This relationship was linear within the examined concentration range. The bars represent standard error of (SEM). The arrow indicates the concentration of fixed glutamate, which corresponds to the density of gold particles present over the synaptic vesicle cluster in the synapse shown in **A** (see ref. 9 for further details). Scale bar: **A, B:** 0.5 μm.

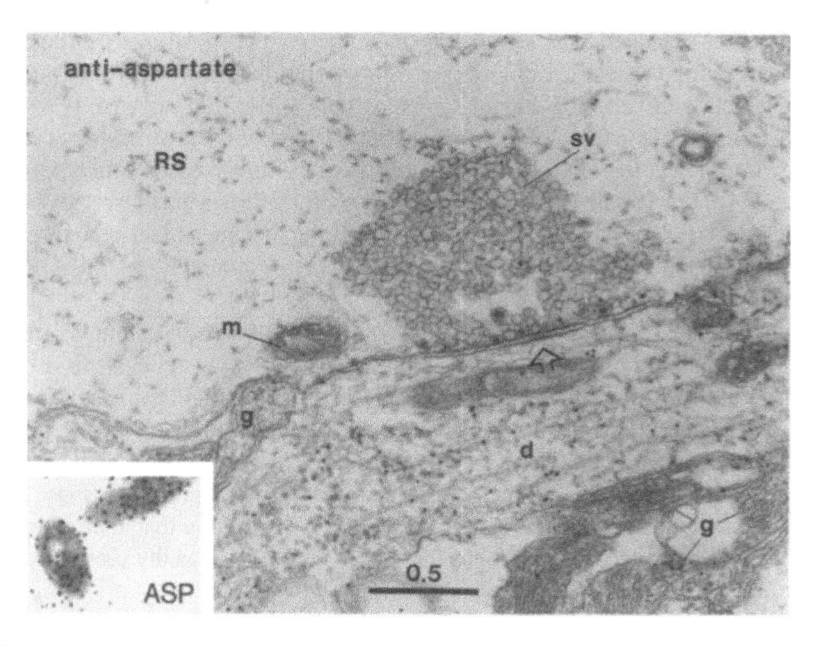

FIG. 3. Immunoreactivity to fixed aspartate in a giant reticulospinal synapse. Electron micrograph of a synapse from an ultrathin section incubated with antiaspartate antiserum. Note the lack of accumulation of gold particles over the synaptic vesicle cluster (sv), while some labeling occurs over the postsynaptic dendrite (d). Other designations are as in Fig. 2. The inset shows the L-aspartate-containing conjugate from the test section incubated together with the tissue section. The conjugates containing other amino acids were unlabeled (for details see Shupliakov et al., ref. 9). Scale bar: 0.5 μm.

most evident if a comparison is made between the giant reticulospinal axons and the cutaneous afferent axons (24) located in the dorsal column (Fig. 1A). The former axons appear to be silent most of the time in the living animal, and they generally fire in brief bursts when recruited (25,26). The dorsal column axons, on the other hand, mainly comprise pressure-sensitive cutaneous afferents, which fire tonically at high frequency during skin stimulation (27). Ultrastructural comparison of these two types of axon shows that the number of mitochondrial profiles in the vicinity of the synaptic vesicle clusters is considerably higher in the dorsal column axon synapses than in the reticulospinal synapses (9) (Fig. 5A,B), indicating that the former have a higher metabolic competence (cf. crustacean motor axons (28)). Moreover, quantitative immunogold analysis of glutamate shows that the level of glutamate labeling in the axoplasmic matrix surrounding the synaptic vesicle clusters is at least four to five times higher in the dorsal column axons than in the reticulospinal axons (Fig. 5A,B), while the synaptic vesicle clusters contain similar levels in both types of axon (9). A high axoplasmic glutamate concentration would seem to be beneficial in the dorsal column axons, as it would facilitate an efficient refilling of the synaptic vesicles during tonic exocytotic release. In the giant reticulospinal axons, on the other hand, a lower level should be sufficient to keep up with the intermittent release occurring in these synapses.

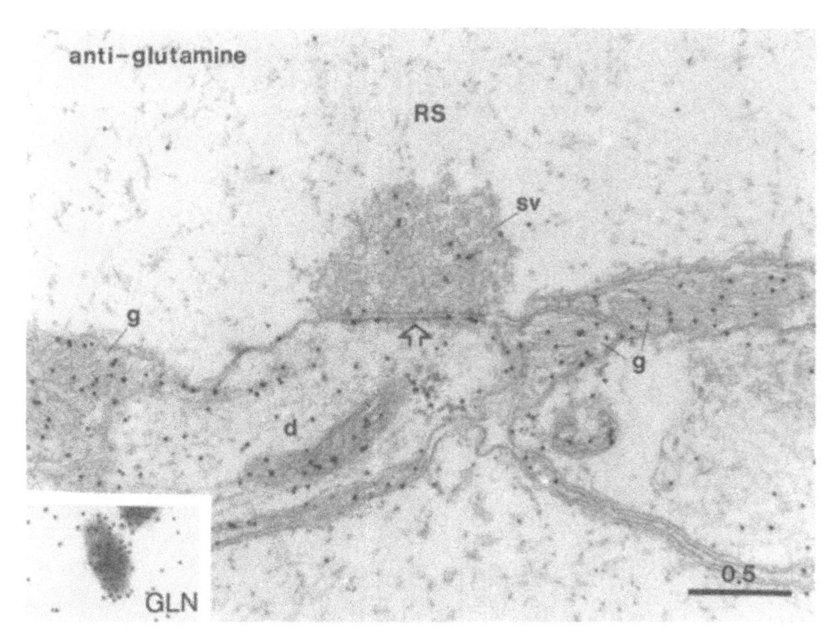

FIG. 4. Glutamine immunoreactivity in a giant reticulospinal synapse. Electron micrograph of a section of the lamprey spinal cord treated with glutamine antiserum. Gold particles representing immunolabeling are mainly accumulated over glial profiles (g) and the postsynaptic dendrite (d). Other designations are as in Fig. 2. The inset shows the glutamine-containing conjugate from the test section incubated together with the tissue section. Scale bar: 0.5 μm.

The axoplasmic glutamate concentration may be controlled either by glutamate reuptake or by synthesis in the axon. Experiments with D-aspartate incubation (see previous paragraph) indicate that the capacity for glutamate reuptake is limited in both reticulospinal and dorsal column axons (Gundersen, Shupliakov, Brodin, Ottersen, and Storm-Mathisen, unpublished observations), suggesting that the difference is primarily due to the level of the presynaptic glutamate synthesis. To obtain a correlate of the rate of conversion from glutamine to glutamate, the ratio between the glutamate and glutamine immunolabeling was compared between the two types of axon. This ratio was several times higher over synaptic mitochondria and axoplasmic matrix in dorsal column axons than in reticulospinal axons (21). This finding provides support for the possibility that the glutamate synthesis is more efficient in the former type. It is interesting to note that phosphate-activated glutaminase, the mitochondrial enzyme that converts glutamine into glutamate, is stimulated by both Ca^{2+} and phosphate (20,22). Hence, tonic synaptic activity, leading to frequent raises in Ca^{2+} in the synaptic region and a large adenosine 5′-triphosphate (ATP) consumption, could in itself cause an increase of the presynaptic glutamate synthesis. Most likely, however, other mechanisms are also involved in the regulation of the glutamate synthesis in nerve terminals (cf. aminobutyric acid (GABA) synthesis (29)).

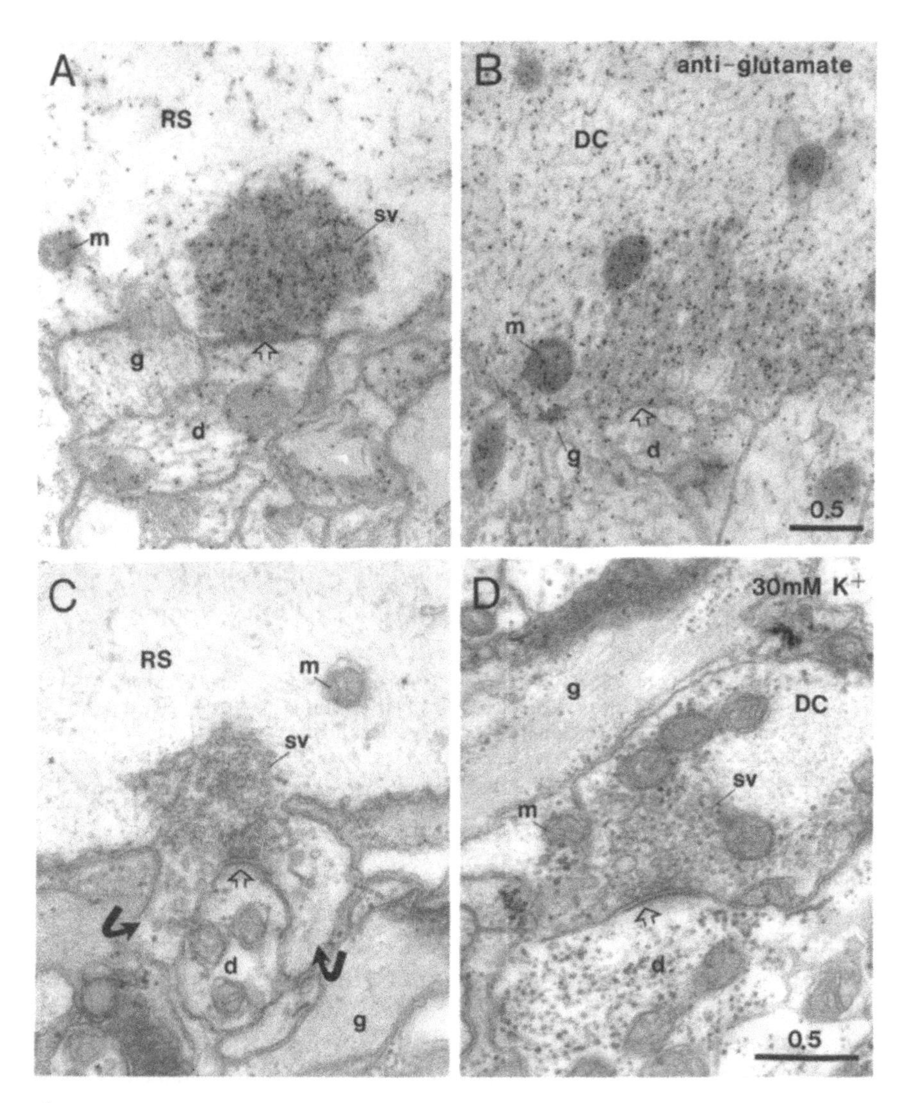

FIG. 5. Characteristics of two types of glutamate axon with different patterns of activity. **A,B.** Sections from a giant reticulospinal axon (**A**) and a dorsal column axon (**B**) from the same ultrathin section incubated with an antiserum to fixed glutamate. Note that the density of gold particles is significantly higher in the axoplasmic matrix and mitochondria (m) in the dorsal column axon (DC) than in the reticulospinal axon (RS). Note also that the number of mitochondrial profiles surrounding the synaptic vesicle cluster is higher in the dorsal column axon (for details see Shupliakov et al., ref. 9). **C,D.** Effect of intense stimulation evoked by 30-minute exposure of the spinal cord to Ringer solution containing 30 mM K^+ on the synaptic ultrastructure in a giant reticulospinal axon (**C**) and a dorsal column axon (**D**). The electron micrographs are from the same ultrathin section. Note that the plasma membrane of the reticulospinal axon shows distinct invaginations that surround the postsynaptic element, indicating that the presynaptic membrane has expanded. In the dorsal column axon, however, no corresponding expansion of the presynaptic membrane is evident. (Shupliakov and Brodin, unpublished observations.) Scale bars: **A, B** and **C, D:** 0.5 μm.

The two types of glutamatergic axon also appear to differ with regard to the capacity for replenishment of synaptic vesicles (30). The giant reticulospinal axon is one of the few central synapses in which it has been shown that a marked reduction in the number of synaptic vesicles can be induced by a prolonged action potential or high K^+ stimulation, in the absence of neurotoxins or treatments that reduce endocytosis (30–32). The reduction of the number of synaptic vesicles is accompanied by an expansion of the presynaptic plasma membrane, which is seen as characteristic membrane protrusions surrounding the postsynaptic element (curved arrows in Fig. 5C). The giant reticulospinal synapse thus appears to have a low capacity for endocytosis and replenishment of synaptic vesicles, which could relate to the physiological activity pattern of this synapse. A tonic firing does not normally occur, and a relatively limited capacity for synaptic vesicle reformation would thus seem to be sufficient. Figure 5C shows a giant reticulospinal synapse, which was fixed after 30-minute exposure to 30 mM K^+. The characteristic expansions of the presynaptic plasma membrane can be seen on each side of the active zone. Figure 5D shows a dorsal column synapse from the same stimulated specimen. In this case the presynaptic plasma membrane has a normal appearance. If we assume that the treatment has caused a similar activation of both types of axon, the latter synapse thus appears to have a higher capacity for replenishment of synaptic vesicles, which would seem to correlate with the demand of synapses with a tonic transmitter release (27,30).

REGULATION OF GLUTAMATE RELEASE— PHASIC MODULATION OF EXCITATORY NETWORK SYNAPSES

Apart from the reticulospinal and cutaneous afferent synapses just discussed, the lamprey spinal cord contains several other types of glutamatergic axons, including excitatory premotor interneurons (15,33). Intracellular recordings from the axons of these neurons during locomotor activity (Fig. 6A) have shown that they are subjected to a phasic modulation, which is mediated both by fast $GABA_A$ receptors and $GABA_B$ receptors (Fig. 6B), the latter acting via a pertussis toxin-sensitive G-protein (34,35). If the presynaptic axon is stimulated while the postsynaptic response is recorded in a target neuron, the glutamatergic EPSP is found to be suppressed by application of GABA, or the $GABA_B$ agonist baclofen (Fig. 6D). In the latter case, the input resistance of the postsynaptic neuron is unchanged, indicating that the effect is presynaptic, while inclusion of GABA activates postsynaptic $GABA_A$ receptors (35). Recording of the presynaptic action potential during drug application shows that GABA causes a reduction of both the spike amplitude and duration (Fig. 6C), while baclofen causes a smaller reduction of the spike duration (35). Ultrastructurally, spinal glutamatergic axons, some of which are likely to belong to premotor interneurons, receive GABA-immunoreactive input synapses (36) originating from segmental bipolar or multipolar GABA neurons (37).

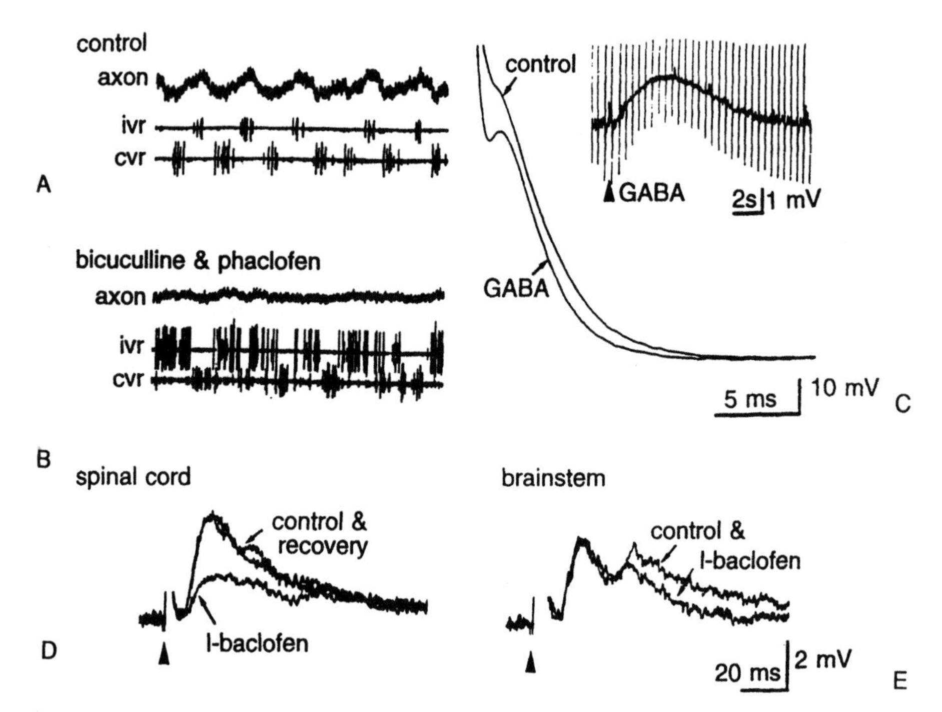

FIG. 6. Phasic GABAergic modulation of interneuronal axons. **A.** Intracellular recording from an interneuronal axon (top trace) during fictive locomotion in an in vitro preparation of the lamprey spinal cord comprising 15 segments. The efferent motor pattern recorded from an ipsilateral (ivr) and a contralateral ventral root (cvr) is shown below. **B.** After blockade of $GABA_A$ as well as $GABA_B$ receptors by simultaneous application of bicuculline (10 μM) and phaclofen (1 mM), the phasic input signals in the axon are markedly depressed. Note also that the efferent motor pattern has become less regular. **C.** Intracellular recording from an interneuronal axon before and during exposure to GABA, applied from an adjacent pressure pipette. Note that the width and the amplitude of the action potential are reduced by GABA. The baseline has been adjusted for the GABA-induced depolarization of the axonal resting membrane potential (see inset). **D.** EPSPs evoked by stimulation of interneuronal axons before and during application of baclofen from a pressure pipette. **E.** A similar application of baclofen has no effect on EPSPs evoked by stimulation of reticulospinal fibers in the brainstem. Both EPSPs could be depressed by CNQX. Strychnine (5 μM) was present in order to suppress inhibitory synaptic potentials (modified from Alford et al., ref. 34; Alford and Grillner, ref. 35).

Taken together, the preceding data show that the synaptic glutamate release from excitatory premotorinterneurons is subjected to a phasic control by a modulation of the presynaptic spike width that presumably controls the presynaptic calcium current. The excitatory premotor interneurons are integral components of the interneuronal network, which generates the rhythmic motor pattern underlying locomotion (Fig. 7) (15). These cells are rhythmically active during locomotion and fire a burst of action potentials during each locomotor cycle. They drive other, gly-

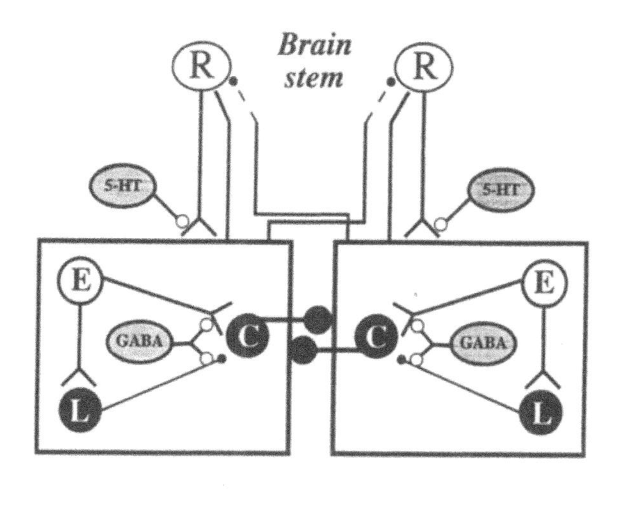

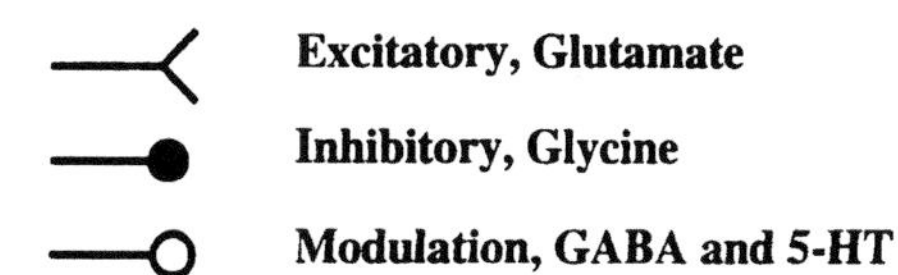

FIG. 7. Schematic diagram of the connectivity between reticulospinal and spinal neurons in lamprey. The two squares contain the basic interneuronal circuitry present at the segmental level, which is controlled by reticulospinal neurons (R). Excitatory premotorinterneurons (E) are glutamatergic. Lateral interneurons (L) and crossing, caudally projecting interneurons (C) are glycinergic. Segmental 5-hydroxytryptamine (5-HT) neurons modulate the transmission in the reticulospinal synapses, while gamma-aminobutyric acid (GABA) interneurons modulate the transmission in excitatory and inhibitory interneuronal synapses (for further details see Grillner et al., refs. 15,38).

cinergic, network interneurons, as well as motoneurons (for detailed descriptions of the spinal locomotor network, see Grillner et al. (15,38)). Increasing the level of GABAergic presynaptic inhibition can make the synaptic transmission from the network interneurons less efficient. Conversely, reduction of the level of GABA interneuronal activity will make the network synaptic transmission more powerful. From a baseline level of activity, the synaptic transmission can thus be adjusted in both a potentiating and a depressing direction by the segmental GABAergic system, which serves as a local control circuit that gates the synaptic transmission in relation to the ongoing locomotor activity. A similar locomotor-related phasic presynaptic modulation has also been observed in glutamatergic sensory afferents in lampreys, as well as in mammals (39,40). This modulation provides a phasic gating of the sensory synaptic transmission, presumably to ascertain that the sensory input can act only when it is appropriate, but be gated away when this is not the case.

SEGMENTAL MODULATION OF RETICULOSPINAL TRANSMISSION—5-HYDROXYTRYPTAMINE (5-HT) AND SHORT-TERM PLASTICITY

The reticulospinal synaptic transmission is not sensitive to GABA (Fig. 6E) (35) but is highly sensitive to 5-HT. Local intraspinal 5-HT neurons are strategically located in the midline of each segment, and these neurons provide a dense 5-HT innervation around the giant reticulospinal axons (41) Application of 5-HT in doses of 1 μM and higher causes a depression of the reticulospinal EPSP (Fig. 8A,B) (42), while the postsynaptic response to exogenously applied glutamate is not reduced even by a tenfold higher 5-HT concentration, implying that the effect is presynaptic. Unlike the GABA effect on interneuron axons, however, 5-HT does not alter the shape of the presynaptic action potential in reticulospinal axons (42). A reduced presynaptic Ca^{2+} current may, however, be difficult to detect, as the Ca^{2+} current in these axons may be small in relation to the total ionic current. To further examine this possibility, experiments were made in which the Na^+ and K^+ channels were blocked with tetrodotoxin, tetraethylammonium, and 4-aminopyridine, and a high concentration of Ba^{2+} (30 mM) was added. Under these conditions, long depolarizing current pulses applied in the axon cause a slow regenerative potential, which most likely represents the current flowing in presynaptic calcium channels (cf. MacVicar and Llinas (43)). Application of high concentrations of 5-HT (20 μmol) was found not to depress this Ba^{2+}-mediated component (Brodin, Shupliakov, unpublished observations), indicating that 5-HT does not cause any major reduction of the presynaptic calcium current. Although we cannot rule out that some reduction may occur under physiological conditions, these data suggest that the 5-HT-mediated modulation of the reticulospinal glutamate release is independent of the Ca^{2+} entry (cf. Man-Soh-Hing et al. (44), and Dale and Kandel (45)). It is interesting to note that this type of presynaptic modulation in reticulospinal axons will not interfere with the conduction of action potentials to more caudal parts of the spinal cord, which would have been the case if the modulation had been accompanied by conductance changes (cf. interneuronal axons just mentioned).

Another factor that contributes to local regulation of the reticulospinal transmission is the level of short-term plasticity at individual synapses (46). If a reticulospinal neuron is stimulated with a brief high-frequency pulse train, the postsynaptic response mostly facilitates, but the level of facilitation varies markedly between individual synapses. This difference is true also for synapses made onto different target cells from one and the same reticulospinal axon, and it applies as well to the synaptic depression seen at intermediate firing rates (46). Conversely, different reticulospinal neurons may evoke synaptic responses with different types of short-term plasticity in the same postsynaptic cell (Fig. 8C,D). As the differentiated short-term plasticity is seen at firing rates that are well below those at which a receptor desensitization may be of importance to determine the synaptic response (47) it can most likely be attributed to presynaptic factors (46). Furthermore, since

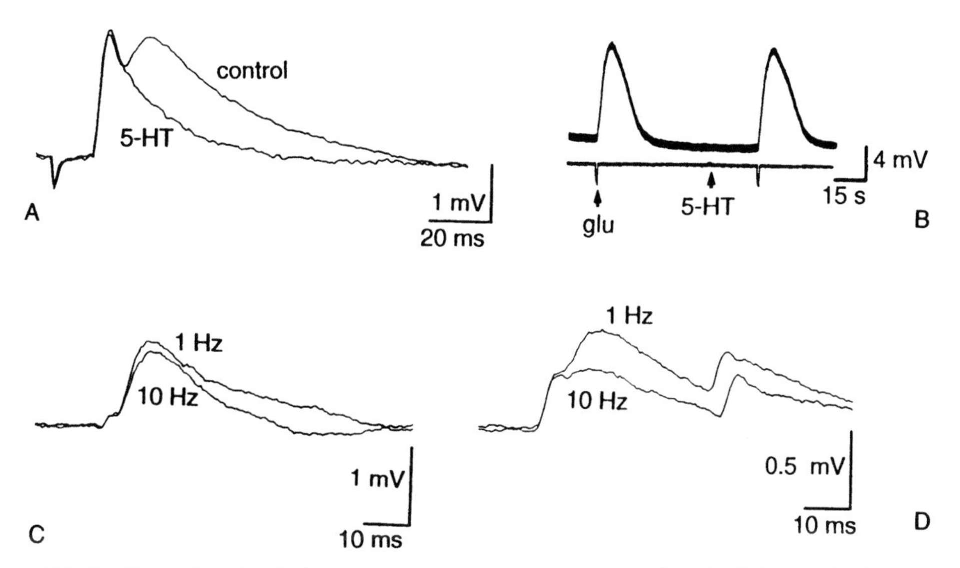

FIG. 8. Modulation of reticulospinal excitatory postsynaptic potentials (EPSPs) by 5-hydroxytryptamine (5-HT) and firing frequency. **A.** Mixed electrotonic and chemical EPSP in a spinal neuron evoked by a giant reticulospinal axon. After pressure ejection of 5-HT (1 mM) from an adjacent pressure pipette, the chemical EPSP was strongly depressed, while the electrotonic component remained unchanged. **B.** Responses to brief applications of glutamate (glu) recorded in a spinal neuron. Application of 5-HT (1 mM) from a second pressure pipette does not alter the response to glutamate (modified from Buchanan and Grillner, ref. 42). **C,D.** Two different EPSPs recorded in a single spinal neuron during subsequent stimulation of two different medium-sized reticulospinal neurons, located in the middle rhombencephalic reticular nucleus (**C**) and posterior rhombencephalic reticular nucleus (**D**), respectively. In both cases the chemical EPSP is preceded by an electrotonic EPSP. Note that there is little change in the amplitude of the monosynaptic chemical EPSP when the firing rate is increased from 1 to 10 Hz in **C,** while a marked depression occurs in **D.** A polysynaptic component is also present in **D.** Averages of 40 sweeps are shown in each case (from Brodin and Shupliakov, unpublished observations).

these axons are unbranched, any uncertainty about the action potential propagation in different branches is not a consideration (cf. e.g., Lüscher and Shiner (48)).

The differentiation of the short-term plasticity will influence the reticulospinal synaptic effects locally, with the strength of the drive signal in individual interneurons and motoneurons being partly controlled by the presynaptic firing rate (cf. Koerber and Mendell (49)). This may be of importance in determining which neurons are to be recruited during a given reticulospinal command. Computer simulation studies (50) also suggest that the short-term plasticity may play a role in regulating the balance between NMDA versus AMPA/kainate receptor activation in the postsynaptic cells. The spinal locomotor network (Fig. 7) generates slow burst activity if an activation of NMDA receptors predominates, while rapid bursting occurs only if an AMPA/kainate receptor activation predominates (15). If synapses with large relative AMPA/kainate components would tend to show more facilitation, and those with large NMDA components would tend to show less facilitation and/or more depression, the same reticulospinal synapses could be used to drive the locomotor network over its entire range of burst frequencies (50).

THE GIANT RETICULOSPINAL AXON AS AN EXPERIMENTAL MODEL OF A CENTRAL SYNAPSE

Invertebrate synapses such as the squid giant synapse, the crustacean neuromuscular junction, and synapses formed between Helisoma neurons in culture are important model systems that have contributed much to our understanding of different presynaptic mechanisms (51–54). These systems are indeed useful in the analysis of basic components of the transmitter release machinery, as many synaptic proteins are highly conserved between species, as well as between different secretory systems (55–57). However, certain synaptic vesicle-associated proteins, including synapsins and synaptophysin/synaptoporin, are probably not well conserved between vertebrates and invertebrates (57,58). Moreover, an understanding of the various regulatory mechanisms present in different types of central synapses will require complementary models from the CNS. The giant reticulospinal axon provides a glutamatergic terminal, with pharmacological properties similar to those of mammalian brain synapses (NMDA, AMPA/kainate receptors) and a set of regulatory mechanisms that controls the synaptic drive of different target cells in the motor system. The size of these axons permits presynaptic microinjections at one or several sites along the axon during simultaneous recording of the postsynaptic response, thus permitting a direct analysis of the function of different presynaptic protein components (59).

REFERENCES

1. Mayer M, Westbrook G. The physiology of excitatory amino acids in the vertebrate central nervous system. *Prog Neurobiol* 1987;28:197–276.
2. Nakanishi S. Molecular diversity of glutamate receptors and implications for brain function. *Science* 1992;258:597–603.
3. Sommer B, Seeburg P. Glutamate receptor channels: novel properties and new clones. *Trends Pharmacol Sci* 1992;13,291–296.
4. Rovainen CM. Neurobiology of lampreys. *Physiol Rev* 1979;59:1007–1077.
5. Buchanan J, Brodin L, Dale N, Grillner S. Reticulospinal neurones activate excitatory amino acid receptors. *Brain Res* 1987;408:321–325.
6. Brodin L, Ohta Y, Hökfelt T, Grillner S. Further evidence for excitatory amino acid transmission in lamprey reticulospinal neurons: retrograde labeling with (3H)D-aspartate. *J Comp Neurol* 1989;281: 225–233.
7. Ottersen OP. Postembedding immunogold labelling of fixed glutamate: an electron microscopic analysis of the relationship between gold particle density and antigen concentration. *J Chem Neuroanat* 1989;2:57–67.
8. Storm-Mathisen J, Ottersen OP, Immunocytochemistry of glutamate at the synaptic level. *J Cytochem Histochem* 1990;38:1733–1743.

9. Shupliakov O, Brodin L, Cullheim S, Ottersen OP, Storm-Mathisen J. Immunogold quantification of glutamate in two types of excitatory synapses with different firing patterns. *J Neurosci* 1992; 12:3789–3803.
10. Burger PM, Mehl E, Cameron PL, Maycox P, Baumert M, Lottspeich F, De Camilli P, Jahn R. Synaptic vesicles immunoisolated from rat cerebral cortex contain high levels of glutamate, *Neuron* 1989;3:715–720.
11. Naito S, Ueda T. Characterization of glutamate uptake into synaptic vesicles. *J Neurochem* 1985; 44:99–109.
12. Cuénod M, Streit P. Neuronal tracing using retrograde migration of labeled transmitter-related compounds. In: Björklund A, Hökfelt T, eds. *Methods in Chemical Neuroanatomy,* Amsterdam: Elsevier, 1983:365–397.
13. Gundersen V, Danbolt N, Storm-Mathisen J, Ottersen OP. Demonstration of glutamate/aspartate uptake activity in nerve endings by use of antibodies recognizing exogenous D-aspartate. *Neuroscience* 1993;57:97–111.
14. Grillner S, McClellan A, Sigvardt K, Wallén P, Wilén M. Activation of NMDA receptors elicits "fictive locomotion" in lamprey spinal cord in vitro. *Acta Physiol Scand* 1981;113:549–551.
15. Grillner S, Wallén P, Brodin L, Lansner A. Neuronal network generating locomotor behavior in lamprey—circuitry, transmitters, membrane properties and simulation. *Annu Rev Neurosci* 1991; 14:169–199.
16. Brodin L, Grillner S. The role of putative excitatory amino acid neurotransmitters in the initiation of locomotion in the lamprey spinal cord. II. The effects of amino acid uptake inhibitors. *Brain Res* 1985;360:149–158.
17. Brodin L, Tossman U, Ungerstedt U, Grillner S. The effect of an uptake inhibitor (dihydrokainate) on endogenous excitatory amino acids in the lamprey spinal cord as revealed by microdialysis, *Brain Res* 1988;458:166–169.
18. Hestrin V, Sah P, Nicoll RA. Mechanisms generating the time course of dual component excitatory synaptic currents recorded in hippocampal slices. *Neuron* 1990;5:247–253.
19. Sarantis M, Ballarini L, Miller B, Silver RA, Edwards M, Atwell D. Glutamate uptake from the synaptic cleft does not shape the decay of the non-NMDA component of the synaptic current. *Neuron* 1993;11:541–549.
20. Fonnum F. Regulation of the synthesis of the transmitter glutamate pool. *Prog Biophys Molec Biol* 1993;60:47–57.
21. Shupliakov O, Brodin L, Storm-Mathisen J, Ottersen OP. Compartmentation of glutamate and glutamine in two types of central synapse. *Eur J Neurosci* 1993;suppl 6:892.
22. Kvamme E, Torgner IA, Roberg B. Evidence indicating that pig renal phosphate-activated glutaminase has a functionally predominant external localization in the inner mitochondrial membrane. *J Biol Chem* 1991;266:13185–13192.
23. Ottersen OP, Zhang N, Walberg F. Metabolic compartmentation of glutamate and glutamine: morphological evidence obtained by quantitative immunocytochemistry in rat cerebellum. *Neuroscience* 1992;46:519–534.
24. Brodin L, Christenson J, Grillner S. Single sensory neurons activate excitatory amino acid receptors in the lamprey spinal cord. *Neurosci Lett* 1987;75:75–79.
25. Kasicki S, Grillner S, Ohta Y, Dubuc R, Brodin L. Phasic modulation of reticulospinal neurones during fictive locomotion and other types of spinal motor activity in lamprey. *Brain Res* 1989;484: 203–216.
26. McClellan A. Brainstem command systems for locomotion in the lamprey: localization of descending pathways in the spinal cord. *Brain Res* 1988;457:338–349.
27. Christenson J, Bohman A, Lagerbäck PÅ, Grillner S. The dorsal cell, one class of primary sensory neurone in the lamprey spinal cord. I. Touch, pressure but no nociception—A physiological study. *Brain Res* 1988;440:1–8.
28. Atwood HL, Wojtowicz JM. Short-term and long-term plasticity and physiological differentiation of crustacean motor synapses. In: Smythies JR, Bradley RJ, eds. *International review of neurobiology, vol. 28,* New York: Academic Press, 1986;275–362.
29. Erlander MG, Tobin AJ. The structural and functional heterogeneity of glutamic acid decarboxylase: a review. *Neurochem Res* 1991;16:215–226.
30. Brodin L, Shupliakov O. Presynaptic regulation of glutamate transmission studied in single large vertebrate synapses. *J Neurochem* 1993;61(suppl.):S269.
31. Wickelgren WO. Leonard JP, Grimes MJ, Clark RD. Ultrastructural correlates of transmitter release in presynaptic areas of lamprey reticulospinal axons. *J Neurosci* 1985;5:1188–1201.

32. Ceccarelli B, Hurlbut WP. Vesicle hypothesis of the release of quanta of acetylcholine. *Physiol Rev* 1980;60:396–441.
33. Buchanan JT, Grillner S. Newly identified "glutamate interneurons" and their role in locomotion in the lamprey spinal cord. *Science* 1987;236:312–314.
34. Alford S, Christenson J, Grillner S. Presynaptic $GABA_A$ and $GABA_B$ receptor-mediated phasic modulation in axons of spinal motor interneurones. *Eur J Neurosci* 1991;3:107–117.
35. Alford S, Grillner S. The involvement of $GABA_B$ receptors and coupled G-proteins in spinal GABAergic presynaptic inhibition. *J Neurosci* 1991;11:3718–3726.
36. Christenson J, Shupliakov O, Cullheim S, Grillner S. Possible morphological substrates for GABA-mediated presynaptic inhibition in the lamprey spinal cord. *J Comp Neurol* 1993;328:463–472.
37. Brodin L, Dale N, Christenson J, Storm-Mathisen J, Hökfelt T, Grillner S. Three types of GABA-immunoreactive cells in the lamprey spinal cord. *Brain Res* 1990;508:172–175.
38. Grillner S, Wallén P, El Manira A. Intrinsic function of neuronal networks—from ion channels to behavior in the lamprey CNS. *J Neurosci.* (in press).
39. Dubuc R, Cabelguen JM, Rossignol S. Rhythmic fluctuations of dorsal root potentials and antidromic discharges of primary afferents during fictive locomotion in the cat. *J Neurophysiol* 1988; 60:2014–2036.
40. Gossard JP, Cabelguen JM, Rossignol S. An intracellular study of muscle primary afferents during fictive locomotion in the cat. *J Neurophysiol* 1991;65:914–926.
41. Van Dongen PAM, Hökfelt T, Grillner S, Rehfeld JF, Verhofstad AAJ, Steinbusch HWM, Cuello AC, Terenius L. Immunohistochemical demonstration of some putative neurotransmitters in the lamprey spinal cord and spinal ganglia: 5-hydroxytryptamine-, tachykinin-, and neuropeptide Y-immunoreactive neurons and fibers. *J Comp Neurol* 1985;234:501–522.
42. Buchanan JT, Grillner S. 5-Hydroxytryptamine depresses reticulospinal excitatory postsynaptic potentials in motoneurons of the lamprey. *Neurosci Lett* 1991;112:71–74.
43. MacVicar BA, Llinas RR. Barium action potentials in regenerating axons of the lamprey spinal cord. *J Neurosci Res* 1985;13:323–335.
44. Man-Son-Hing H, Zoran MJ, Lukowiak K, Haydon PG. A neuromodulator of synaptic transmission acts on the secretory apparatus as well as on ion channels. *Nature* 1989;341:237–239.
45. Dale N, Kandel E. Facilitatory and inhibitory transmitters modulate spontaneous transmitter release at cultured Aplysia sensorimotor synapses. *J Physiol* 1990;421:203–222.
46. Brodin L, Shupliakov O, Hellgren J, Pieribone V, Hill R. The reticulospinal glutamate synapse in lamprey: plasticity and presynaptic variability. *J Neurophysiol.* (in press).
47. Trussell LO, Zhang S, Raman I. Desensitization of AMPA receptors upon multiquantal neurotransmitter release. *Neuron* 1993;10:1185–1196.
48. Lüscher HR, Shiner JS. Simulation of action potential propagation in complex terminal arborizations. *Biophys J* 1990;58:1389–1399.
49. Koerber R, Mendell LM. Modulation of synaptic transmission at Ia-afferent fiber connections on motoneurons during high-frequency stimulation: role of postsynaptic target. *J Neurophysiol* 1991; 65:590–597.
50. Tråvén H, Brodin L, Lansner A, Wallén P, Ekeberg Ö, Grillner S. Computer simulations of NMDA and non-NMDA receptor-mediated synaptic drive: supraspinal and sensory modulation of single neurons and small networks. *J Neurophysiol* 1993;70:695–709.
51. Llinas R, Sugimori M, Silver RB. Microdomains of high calcium concentration in a presynaptic terminal. *Science* 1992;256:677–679.
52. Bommert K, Charlton MP, De Bello WM, Chin GJ, Betz H, Augustine GJ. Inhibition of neurotransmitter release by C2-domain peptides implicates synaptotagmin in exocytosis. *Nature* 1993; 363:163–165.
53. Atwood HL, Cooper RL, Wojtowicz JM. Non-uniformity and plasticity of quantal release at crustacean motor nerve terminals. In: Stjärne L, Greengard P, Grillner S, Hökfelt T, Ottoson D, eds. *Molecular and cellular mechanisms of neurotransmitter release*. New York: Raven Press: 1994.
54. Fang Y, Durgerian S, Basarsky T, Haydon PG. GTP-binding proteins: necessary components of the presynaptic terminal for synaptic transmission and its modulation. In: Stjärne L, Greengard P, Grillner S, Hökfelt T, Ottoson D, eds. *Molecular and cellular mechanisms of neurotransmitter release*. New York: Raven Press: 1994.
55. Bennet MK, Scheller RH. The molecular machinery for secretion is conserved from yeast to neurons. *Proc Natl Acad Sci USA* 1993;90:2559–2563.
56. Söllner T, Whiteheart S, Brunner M, Erdjument-Bromage H, Geromanos S, Tempst P, Rothman J. SNAP receptors implicated in vesicle targeting and fusion. *Nature* 1993;363:318–324.

57. Südhof T, De Camilli P, Niemann H, Jahn R. Membrane fusion machinery: insights from synaptic proteins. *Cell* 1993;75:1–4.
58. Greengard P, Valtorta F, Czernik A, Benfenati F. Synaptic vesicle phosphoproteins and regulation of synaptic function. *Science* 1993;259:780–785.
59. Pieribone VA, Müller TH, Brodin L, Czernik AJ, Shupliakov O, Schaeffer E, Grillner S, Greengard P. Synapsin in Müller axons in the lamprey spinal cord. *Soc Neurosci Abstr* 1993;373:15.

Molecular and Cellular Mechanisms of Neurotransmitter Release, edited by Lennart Stjärne, Paul Greengard, Sten Grillner, Tomas Hökfelt, and David Ottoson, Raven Press, Ltd., New York © 1994.

15

Differential Release of Classical Transmitters and Peptides

Jan M. Lundberg, Anders Franco-Cereceda, Ya-ping Lou, Agnes Modin, and John Pernow

Department of Physiology and Pharmacology, Karolinska Institute, S-171 77 Stockholm, Sweden

The discovery of the coexistence of multiple bioactive peptides and classical low molecular weight transmitters in neurons (1) raises several important issues about mechanisms of storage and release of these agents. In the peripheral nervous system, there is evidence that certain peptides are often found together with a specific classical transmitter, and there is often a functional cooperation between these agents (1).

Thus, vasoactive intestinal polypeptide (VIP) and peptide histidine isoleucine (PHI), two vasodilatory peptides, are present in certain postganglionic cholinergic neurons innervating parenchyma and vessels of exocrine glands (2). In addition, recent evidence suggest that these neurons contain nitric oxide synthase (NOS), indicating that NO may represent another chemical mediator released from these nerves (3,4). The parasympathetic nerves may thus use three types of chemical signals: 1) the classical transmitter acetylcholine (Ach); 2) peptides (e.g., VIP and PHI); and 3) the gas NO. These signals have different characteristics for storage, release, and postjunctional effector mechanisms.

Sympathetic postganglionic nerves contain the vasoconstrictor agent neuropeptide Y (NPY) in addition to noradrenaline (NA) (5,6). Furthermore, a population of sensory nerves sensitive to capsaicin, the pungent agent in hot peppers, synthesizes tachykinins (substance P (SP) and neurokinin A (NKA)) and calcitonin gene-related peptide (CGRP) (7) and most likely also contain the amino acid transmitter glutamate (8). In the present paper we will compare principal mechanisms for classical transmitters, peptides, and NO, giving special attention to the release process and presence of a frequency-dependent chemical coding of neuronal signaling in the peripheral autonomic nervous system.

STORAGE OF MEDIATORS

It seems well established that Ach, NA, and glutamate are stored in small transmitter vesicles (which appear clear in the electron microscope in both sensory and cholinergic motor nerves and dense cored in sympathetic nerves). Peptides (SP, CGRP, VIP, and NPY), on the other hand, are mainly found in large dense cored vesicles (9,10,11,12). These large dense cored vesicles in sympathetic nerves also contain NA.

In contrast, NO is not prestored but only synthesized on demand (i.e., when increased intracellular Ca^{2+} leads to activation of NOS). When formed, NO has been postulated to diffuse out of the nerve endings in a nondirected fashion and into target cells (13).

GENERAL ASPECTS ON RELEASE OF MEDIATORS

The partly separate storage of classical transmitters and peptides has been one explanation suggested for the findings that Ach and NA already are released upon low frequency stimulation, whereas peptide overflow and peptide mediated functional responses are mainly observed upon high frequency stimulation (1). Based on the observations that exocytotic peptide-containing vesicles are not commonly close to active synaptic zones where small vesicles release their low molecular weight transmitters (14) it has been postulated that peptide release is a nondirected exocytotic process. This view may be challenged, however, by the presence of peptide-mediated functional responses upon single pulse stimulation (see information that follows). Furthermore, it has been suggested that neuropeptide release is triggered by uniform elevations in Ca^{2+} concentration in the bulk cytoplasm, while secretion of low molecular weight transmitters (e.g., amino acids) require locally high Ca^{2+} levels as in the vicinity of voltage activated Ca^{2+} channels (15) that may be clustered in active zones (16).

Release of mediators can be detected by a variety of techniques, including rapid electrophysiological or electrochemical recordings. For neuropeptides, electrophysiological techniques of responses in effector cells have mainly been used in sympathetic ganglia where tachykinins released from sensory nerves (17,18) or other peptides (19) evoke slowly, developing long-lasting depolarization. The NA release in the periphery has recently been detected with rapid electrochemical techniques in isolated organs (20). Peptide release, on the other hand, has been studied mainly using biochemical analysis with radioimmunoassay (RIA) of peptide overflow into the superfusate of isolated organs or into the venous effluent of blood perfused tissues, or systemic circulation in vivo. Obvious limitations in studying peptide overflow are problems with local degradation by enzymes, which for tachykinins can be overcome by using neutral endopeptidase inhibitors like phosphoramidon (21). Also, enzymes present in plasma may degrade circulating tachykinins (22). Furthermore, peptides bind in tissue to high-affinity receptors on target cells and possibly other structures (e.g., albumin). The removal of peptides by the lymph may also reduce the amount of peptide entering the venous effluent (23). Finally, diffusion barriers in the capillary endothelium may prevent or delay escape of released peptides into the bloodstream. A typical observation is thus the delay in appearance of neuropeptides in the venous effluent compared to low molecular weight transmitters; (24) the peptide washout from the tissue may be a long-lasting process, although still faster than for even larger vesicle constituents (e.g., dopamine-B-hydroxylase) (25).

With regards to capillary permeability to large molecules (e.g., peptides), the spleen with its specialized vasculature represents a tissue in which diffusion barriers are less pronounced. Also, in the kidney peptide diffusion is relatively well developed, while in other tissues (e.g., skeletal muscle) the capillaries are less permeable. These facts have some important consequences: 1) peptide release into the venous effluent is more easily studied in certain tissues even if the innervation is relatively dense; 2) sampling times for peptide overflow must be extended in relation to sampling (e.g., for NA); and 3) due to "trapping" of the released peptide in the tissues because of diffusion barriers combined with slow degradation, the duration and magnitude of the functional responses to peptides may vary considerably between tissues such as spleen, kidney, and skeletal muscle (26,27).

Even in the absence of biochemically detectable peptide overflow, release can be indirectly estimated using pharmacological bioassays combined with selective peptide receptor antagonists. Recently, potent and selective peptide receptor antagonists have become available, thus increasing our knowledge about neuropeptide mechanisms and finally establishing certain examples that peptides fulfill all transmitter criteria for functions such as: 1) sensory plasma protein extravasation where tachykinins, most likely SP, activate neurokinin 1 (NK1) receptors, as revealed by the nonpeptide antagonist CP96345 and RP67580 (28,29); 2) sensory bronchoconstriction, where tachykinins such as NKA activate NK2 receptors (30,31) as revealed by the nonpeptide antagonist SR48968; 3) antidromic vasodilatation, where the fragment CGRP (8–37) inhibits the CGRP effect but not the response to SP or NKA (32); and 4) capsaicin-evoked tachycardia, which also is inhibited by CGRP (8–37) (33) (Table 1).

CHARACTERISTICS OF SENSORY NEUROPEPTIDE RELEASE

The frequency dependency and ion-channel mechanisms for sensory neuropeptide release evoked by a variety of agents including capsaicin have recently been studied mainly in isolated tissues (e.g., the isolated perfused lung and heart of the guinea pig) (34,35). The main findings can be summarized as follows: low-frequency stimulation (1 Hz) and even single nerve impulses cause bronchoconstriction that is tachykinin mediated, as revealed by inhibitory effects of the NK2 antagonist SR48968 (36). Furthermore, a low frequency maximum is likely to be present for peptide release from sensory nerves, since peptide overflow from the lung is not

TABLE 1. *Sensory neuropeptide transmitters*

	SP	NKA	CGRP
Presence	small neurons (C and AΔ)	small neurons (C and AΔ)	small neurons (C and AΔ)
Synthesis	PPT precursor	PPT precursor	precursor
Storage	LDV	LDV	LDV
Resupply	axonal transport	axonal transport	axonal transport
Release	low frequency	low frequency	low frequency
Preferred receptor	NK1	NK2	CGRP
Second messenger	IP_3, NO	IP_3	cAMP
Response	plasma protein extravasation (high frequency)	bronchoconstriction	vasodilatation tachycardia smooth-muscle relaxation
Degradation	NEP, ACE	NEP	(NEP)
Antagonist	CP96345 RP67580	SR48968	CGRP(8–37)

It has now been established that sensory neuropeptides in C and A delta (Δ) fibers fulfill all established transmitter criteria, including inhibition of nerve-evoked responses by specific neuropeptide receptor antagonists. Since these peptides are coreleased upon activation (e.g., by capsaicin or low pH), receptor specificity and extent of degradation will determine the final functional response. The necessity to stimulate with many impulses to obtain plasma protein extravasation may depend on the fact that the postcapillary venules are not directly innervated by sensory nerves, in contrast to arterioles mediating vasodilatation (76), and the released SP has to diffuse over a considerable distance before reaching NK1 receptors on endothelial cells.

SP, substance P
NKA, neurokinin A
CGRP, calcitonin gene-related peptide
PPT, preprotachykinin
LDV, large dense-cored vesicles
NK, neurokinin
IP_3, inositoltrisphosphate
cAMP, adenosine 3′,5′-cyclic phosphate
NO, nitric oxide
NEP, neutral endopeptidase
ACE, angiotensin converting enzyme

larger at 10 Hz than at 1 Hz, and the maximal bronchoconstrictor response is similar at 1 Hz and 10 Hz and influenced to the same degree by an NK2 antagonist (36) (Table 1).

By using various toxins it has been established that sensory neuropeptide release evoked by nerve stimulation is a classical exocytotic process using tetrodotoxin (TTX)-sensitive action potential propagation and influx of Ca^{2+} through omega-conotoxin (CTX)-sensitive channels (Table 2). The effects of capsaicin depend on the concentration used, possibly depending on the degree of receptor occupancy. Thus a low concentration of capsaicin seems to release peptides via similar mechanisms as antidromic nerve stimulation (Table 2). In high toxic concentrations, the peptide release by capsaicin is mainly triggered via Ca^{2+} entering the cation channel associated with the capsaicin receptor, which is blocked by the dye ruthenium

TABLE 2. *Regulatory mechanisms for peptide release from capsaicin-sensitive sensory nerves*

	Capsaicin 10^{-8} M	Capsaicin 10^{-6} M	Nerve stimulation	Nicotine	Low pH
RR	–	–	0	0	–
Capsazepine	–	–	0	0	–
TTX	–	(–)	–	0	0
CTX	–	(–)	–	(–)	(–)
Nifedipine	0	NT	0	NT	NT
α2-agonists	–	(–)	–	(–)	NT

–: inhibition, 0: no influence, (–): slight inhibition, NT: not tested

Different regulatory mechanisms were revealed by using ruthenium red (RR), an inhibitor of Ca^{2+} influx through nonselective cation channels, the NA^{+} channel blocker tetrodotoxin (TTX), the N-type Ca^{2+} channel blocker omega-conotoxin (CTX), L-type Ca^{2+} channel blocker nifedipine, the capsaicin receptor antagonist capsazepine, and α2-adrenoceptor antagonists. From Lou, ref. 35.

red (RR). The influence of action potential propagation and N-type of Ca^{2+} channels is here negligible, as it is for the response to nicotine (Table 2) (37). Only the CTX-sensitive type of Ca^{2+} influx and peptide release seem to be regulated by prejunctional α2-adrenoceptors (Table 2) (38). Interestingly, low pH triggers sensory neuropeptide release via mechanisms that are similar to the effects of capsaicin regarding sensitivity to RR and the capsaicin receptor antagonist capsazepine, suggesting some common mechanism of action (Table 2) (39,40).

CHARACTERISTICS OF NPY AND NA RELEASE

Detailed studies have been performed regarding mechanisms of NPY release from sympathetic nerves, mainly in the heart and spleen. As seen in Table 3, nerve stimulation-evoked NA and NPY release depends on influx of extracellular Ca^{2+} through CTX-sensitive channels (41,42,43).

When the heart is ischemic, there is a nonexocytotic release of NA that is similar to the effect of tyramine (41) while no outflow of NPY occurs (43). Furthermore, tyramine selectively releases NA but not NPY (41,44) (Table 3). With regards to other sympathoactive drugs, α2-adrenoceptor blocking agents enhance both NA and NPY outflow (45,46,47,48). Guanethidine is a sympatholytic agent that inhibits nerve stimulation-evoked release of both NA and NPY (45,49).

Interesting information has been obtained using reserpine, a classical agent that depletes monoamine stores, including NA. Thus, as seen in Table 4, reserpine also depletes NPY from some, but not all, sympathetic nerves (50). Furthermore, reserpine mainly depletes NPY in terminal regions, while levels in cell bodies and axons if anything increase due to enhanced synthesis (51). Since the depletion of NPY in cardiovascular nerves using a high reserpine dose can be prevented by drugs or procedures that reduce sympathetic nerve impulse traffic that include clonidine (centrally active), chlorisondamine (ganglionic blocker), or surgical transection of

TABLE 3. *NPY differentiates exocytic-nonexocytic NA release*

	X-cell Ca^{2+}	CTX	α2
NA			
nerve stimulation	+	+	+
ischemia	–	–	–
tyramine	–	–	–
NPY			
nerve stimulation	+	+	+
ischemia	no release		
tyramine	no release		

+ : presence of regulation, – : no effect

Comparison between characteristics for noradrenaline (NA) and neuropeptide Y (NPY) release evoked by electrical nerve stimulation, ischemia, or tyramine regarding dependency on extracellular (X-cell) calcium influx, neuronal (N-) type of calcium channels as revealed by inhibition of omegaconotoxin (CTX) and prejunctional inhibitory α2-adrenoceptor feedback regulation. The data suggest that parallel NPY overflow differentiates between exocytic and nonexocytic NA release. From Haass et al., refs. 41, 42; Franco-Cereceda et al., ref. 43; Lundberg et al., ref. 44.

preganglionic nerves, this event can most likely be explained by enhanced release in excess of resupply by axonal transport (6). Thus, it is known that sympathetic nerve impulse traffic is increased after reserpine (52). Furthermore, in the absence of NA the prejunctional autoinhibition of NPY release by α2-adrenoceptors is lost, resulting in enhanced release per nerve impulse (53,54). Taken together, this leads to a progressive depletion of NPY in terminal regions. Interestingly, when sympathetic nerves are stimulated after reserpine treatment in a large animal model (e.g., pig, cat (49) or dog (55)) very little of the usually strong vasoconstrictor responses remain. When reserpine is combined with interrruption of nerve activity by preganglionic decentralization, postganglionic stimulation leads to large, long-lasting vasoconstrictor responses in a variety of tissues (6). Usually these effects are mainly seen upon high-frequency stimulation, but in nasal mucosa (56) and skeletal muscle (27) of the pig vasoconstrictor responses are evoked by single-impulse nerve stimulation even after reserpine. Since these effects are slow and long-lasting, they are likely to depend on a peptide mediator such as NPY rather than rapid short-acting transmitters such as adenosine 5′-triphosphate (ATP) (20,57). Furthermore, ATP-mediated vasoconstriction demonstrated on isolated arteries in vitro is not reduced by reserpine, (58) in contrast to the minor remaining response to nerve stimulation in reserpinized animals in vivo in the absence of interrupted nerve activity (49,55, 56,59). Finally, mATP tachyphylaxis in vivo (44,56) has not yielded similar conclusive results concerning inhibition of sympathetic vasoconstrictor responses as in vitro (20).

Under control conditions, very strong and intense prolonged stimulations can lead to some depletion of tissue NPY and reduced NPY release (24) from sympathetic nerves and presumably NPY-evoked responses (60). However, the NPY outflow can be maintained surprisingly long considering the sole dependency on resup-

TABLE 4. *Reserpine effects*

	NPY	NA
Dosage to deplete	high	low
Onset of depletion	hours	min
Depletion in cardiovascular terminals	50%–90%	95%–99%
Depletion in axon and ganglion cells	no	yes
Depletion in vas deferens, iris, and uterus terminals	no	yes
Prevention of depletion by:		
clonidine	yes	no
chlorisondamine	yes	no
decentralization	yes	no
Upregulation of synthesis	yes	yes

Evidence suggesting that the effect of reserpine on tissue levels of noradrenaline (NA) and neuropeptide Y (NPY) differs regarding mechanisms of action. Certain populations of sympathetic nerves (e.g., to the cardiovascular system) are activated by reserpine, and the increased nerve activity combined with loss of prejunctional α2-adrenoceptor regulation due to NA depletion leads to enhanced NPY release in excess of resupply by axonal transport (Lundberg, Franco-Cereceda et al., 1990, ref. 6). The increased nerve activity leads to upregulation of NPY synthesis as revealed by enhanced NPYmRNA expression. Furthermore, the expression of tyrosine hydroxylase (TH)mRNA is also increased via mechanisms involving nicotinic ganglionic receptor stimulation. From Schalling et al., ref. 51.

ply by axonal transport (61) that contrasts with the situation for sensory nerves (62). A likely explanation for this relative "peptide sparring mechanism" of sympathetic nerves in the presence of NA is probably the powerful prejunctional α2-mediated inhibition of NPY release by NA. After reserpine treatment when NA is depleted and the prejunctional regulation is absent, nerve stimulation-evoked NPY release is markedly enhanced, the vasoconstriction response is prolonged, and repeated stimulation clearly leads to exhaustion of peptide stores available for release (24,61). Stimulation of α2-adrenoceptors in reserpine treated animals reduces the nerve stimulation-evoked NPY overflow to levels similar to those observed in control animals (unpublished data). An important role for the prejunctional regulation of release by the classical transmitter NA may thus be to prevent exhaustion of neuropeptide stores. Similar mechanisms may be present concerning prejunctional muscarinic receptor regulation of VIP release (see next section 63). For sensory nerves on the other hand, there is no clear cut evidence of major prejunctional regulation (e.g., by glutamate) on neuropeptide release in the periphery. The inhibitory influence of α2-adrenoceptors on sensory neuropeptide release (38) may only come into play when sympathetic tone is high.

The involvement of NPY as a mediator of sympathetic vasoconstriction has not definitely been proven, due to lack of specific NPY antagonists, but is supported by several observations: 1) tachyphylaxis to NPY or Y1 receptor agonists (64,65) inhibits nonadrenergic sympathetic nerve responses, and 2) NPY overflow is highly correlated to nonadrenergic vasoconstrictor responses (57), and the NPY levels in plasma (nM) escaping into the splenic venous effluent upon stimulation are clearly within the vasoconstrictor range (44,57,61).

CHARACTERISTICS OF VIP RELEASE AND NO MECHANISMS

The regulation of blood flow in exocrine glands is under major parasympathetic control (63). It has been established that low-frequency stimulation leads to a cholinergic (atropine sensitive) vasodilatation, while the large increase in blood flow upon high-frequency stimulation is atropine resistant and paralleled by a marked vasoactive intestinal polypeptide (VIP) release (63,66). No potent VIP receptor antagonists are available, although the use of immunoneutralization by VIP antiserum has supported VIP's possible involvement as a noncholinergic vasodilatory transmitter (66) (Table 5). Available data with atropine suggest that Ach regulates VIP release in salivary glands via prejunctional muscarinic receptors (66). Recently NOS has been demonstrated in VIP-containing postganglionic parasympathetic neurons (3,4) suggesting that NO could be a noncholinergic vasodilatory mediator; this finding was supported by initial data showing that the NOS inhibitor L-NNA inhibited penile erection evoked by parasympathetic stimulation (67,68) although there were problems in demonstrating reversibility of the L-NNA effect by L-arginine (67) (Table 5). In contrast to VIP and Ach, NO is not likely to be prestored in synaptic vesicles but only synthetized on demand upon neuronal activation and Ca^{2+} influx (13). When the NOS inhibitors L-NNA or L-NAME were given to the submandibular salivary gland, the parasympathetic vasodilatory response was markedly reduced, suggesting major involvement of NO (69,70). The interpretation of these data is further complicated by the view that the Ach-evoked vasodilatation is NO mediated (71) and should be blocked by NOS inhibitors. Furthermore, L-NAME was reported to induce salivation (70) most likely by interfering with muscarinic receptors due to structural similarity with Ach (72) (Table 5). In addition, it has been reported that L-NNA inhibits VIP release (73).

In the salivary gland, the vasodilatory effects of VIP were reduced by the NOS inhibitors L-NNA (69) or L-NAME (70), indicating postjunctional actions of these agents. In fact, some evidence suggests that L-NMMA is not a specific inhibi-

TABLE 5. *NO and VIP as parasympathetic vasodilatory transmitters*

1) NOS inhibitors—endothelial and neuronal
2) L-NNA inhibits VIP release
3) L-NNA and L-NAME inhibit VIP effects
4) L-NAME inhibits Ach effects and muscarinic receptor binding
5) Reversal of NOS inhibitors by L-arginine?
6) Reduction of vasodilatation by VIP antiserum
7) Parallel VIP-release and high frequency-evoked vasodilatation

Different experimental observations illustrating the complex interactions between VIP and putative NO mechanisms in parasympathetic vasodilatory control with special emphasis on the effects of NOS inhibitors. From Lundberg, ref. 63; Lundberg et al., refs. 66, 69; Holmquist et al., ref. 67; Gaw et al., ref. 75.

tor of vascular relaxation (74) and that the endothelium-independent relaxation of blood vessels by VIP is also reduced by L-NMMA (75).

Taken together, the findings just discussed suggest that the role of NO as peripheral parasympathetic nerve transmitter of noncholinergic vasodilatation is not yet firmly established, due to interference at several levels with transmission mechanisms (including VIP) using presently available NOS inhibitors that influence both endothelial and neuronal NOS (77).

ACKNOWLEDGMENTS

The present paper summarizes data supported by the Swedish Medical Research Council (14X-6554). For expert secretarial help we are grateful to Mrs. Ylva Jerhamre.

REFERENCES

1. Lundberg JM, Hökfelt T. Coexistence of peptides and classical neurotransmitters. *Trends Neurosci.* 1983;6:325–333.
2. Lundberg JM, Fahrenkrug J, Hökfelt T, et al. Co-existence of peptide HI (PHI) and VIP in nerves regulating blood flow and bronchial smooth muscle tone in various mammals including man. *Peptides* 1984;5:593–606.
3. Ceccatelli S, Lundberg JM, Fahrenkrug J, Bredt DS, Snyder SH, Hökfelt T. Evidence for involvment of nitric oxide in the regulation of hypothalamic portal blood flow. *Neuroscience* 1992;51, 4:769–772.
4. Kummer W, Fischer A, Mundel P, et al. Nitric oxide synthase in VIP-containing vasodilator nerve fibres in the Guinea-pig. *Neuroreport* 1992;3:653–655.
5. Lundberg JM, Terenius L, Hökfelt T, et al. Neuropeptide Y (NPY)-like immunoreactivity in peripheral noradrenergic neurons and effects of NPY on sympathetic function. *Acta Physiol Scand* 1982; 116:477–480.
6. Lundberg JM, Franco-Cereceda A, Hemsén A, Lacroix JS, Pernow J. Pharmacology of noradrenaline and neuropeptide tyrosine (NPY)-mediated sympathetic cotransmission. *Fundam Clin Pharmacol* 1990;4:373–391.
7. Lundberg JM, Franco-Cereceda A, Hua X-Y, Hökfelt T, Fischer J. Co-existence of substance P and calcitonin gene-related peptide-like immunoreactivities in sensory nerves in relation to cardiovascular and bronchoconstrictor effects of capsaicin. *Eur J Pharmacol* 1985;108:315–319.
8. De Biasi S, Rustioni A. Glutamate and substance P coexist in primary afferent terminals in the superficial laminae of spinal cord. *Proc Natl Acad Sci USA* 1988;85:7820–7824.
9. Lundberg JM, Fried G, Fahrenkrug J, et al. Subcellular fractionation of cat submandibular gland: comparative studies on the distribution of acetylcholine and vasoactive intestinal polypeptide (VIP). *Neuroscience* 1981;6:1001–1010.
10. Johansson O, Lundberg JM. Ultrastructural localization of VIP-like immunoreactivity in large densecored vesicles of cholinergic-type nerve terminals in cat exocrine glands. *Neuroscience* 1981; 5:847–862.
11. Fried G, Lundberg JM, Theodorsson-Norheim E. Subcellular storage and axonal transport of neuropeptide Y (NPY) in relation to catecholamines in the cat. *Acta Physiol Scand* 1985;125:145–154.
12. Gulbenkian S, Merighi A, Wharton J, Varndell JM, Polak JM. Ultrastructural evidence for the coexistence of calcitonin gene related peptide and substance P in the secretory vesicles of peripheral nerves in the guinea-pig. *J Neurocytol* 1986;15:535–542.
13. Bredt DS, Snyder SH. Nitric oxide, a novel neuronal messenger. *Neuron* 1992;8:3–11.
14. Zhu PC, Thureson-Klein Å, Klein RL. Exocytosis from large dense-cored vesicles outside the active synaptic zones of terminals within the trigeminal subnucleus caudalis: a possible mechanism for neuropeptide release. *Neuroscience* 1986;19:43–54.

15. Verhage M, McMahon HT, Ghijsen WEJM, et al. Differential release of amino acids, neuropeptides, and catecholamines from isolated nerve terminals. *Neuron* 1991;6:517–524.
16. Smith SJ, Augustine GJ. Calcium ions, active zones and synaptic transmitter release. *Trends Neurosci* 1988;11:458–464.
17. Tsunoo A, Konishi S, Otsuka M. Substance P as an exitatory transmitter of primary afferent neurons in guinea-pig sympathetic ganglia. *Neuroscience* 1984;7:2025–2037.
18. Saria A, Ma R, Dun N, Theodorsson-Norheim E, Lundberg JM. Neurokinin A in capsaicin-sensitive neurons of the guinea-pig inferior mesenteric ganglion: an additional putative mediator for the noncholinergic excitatory postsynaptic potential. *Neuroscience* 1987;21:951–958.
19. Kuffler SW, Sejnowski TJ. Peptidergic and muscarinic excitation at amphibian sympathetic synapses. *J Physiol (Lond)* 1983;341:257–278.
20. Bao JX. Sympathetic neuromuscular transmission in rat tail artery. *Acta Physiol Scand* 1993;suppl 610:1–58.
21. Kröll F, Karlsson JA, Lundberg JM, Persson CGA. Capsaicin-induced bronchoconstriction and neuropeptide release in guinea pig perfused lungs. *J Appl Physiol* 1990;68:4:1679–1687.
22. Martling C-R, Theodorsson-Norheim E, Lundberg JM. Bronchoconstrictor and hypotensive effects in relation to pharmacokinetics of tachykinins in the guinea-pig. Evidence for extraneuronal cleavage of neuropeptide K to neurokinin A. *Naunyn Schmiedebergs Arch Pharmacol* 1987;336:183–189.
23. Bloom SR, Edwards AV. Effects of autonomic stimulation of the release of vasoactive intestinal peptide from the gastro intestinal tract in the calf. *J Physiol (Lond)* 1980;299:437–452.
24. Lundberg JM, Rudehill A, Sollevi A, Fried G, Wallin G. Co-release of neuropeptide Y and noradrenaline from pig spleen in vivo: importance of subcellular storage, nerve impulse frequency and pattern, feedback regulation and resupply by axonal transport. *Neuroscience* 1989;28:475–486.
25. Muscholl E, Rache K, Ritzel H. Facilitation by low sodium urea medium of the washout of dopamine-B-hydroxylase released by potassium ions from the perfused rabbit heart. *Neuroscience* 1980; 5:453–457.
26. Modin A, Pernow J, Lundberg JM. Evidence for two neuropeptide Y receptors mediating vasoconstriction. *Eur J Pharmacol* 1991;203:165–171.
27. Modin A, Pernow J, Lundberg JM. Sympathetic regulation of skeletal muscle blood flow in the pig: a non-adrenergic component likely to be mediated by neuropeptide Y. *Acta Physiol Scand* 1993;148: 1–11.
28. Delay-Goyet P, Lundberg JM. Cigarette smoke-induced airway oedema is blocked by the NK_1 antagonist, CP-96,345. *Eur J Pharmacol* 1991;203:157–158.
29. Delay-Goyet P, Franco-Cereceda A, Golsalves SF, Clingan CA, Lowe III JA, Lundberg JM. CP-96345 antagonism of NK_1 receptors and smoke-induced protein extravasation is unrelated to its cardiovascular effects. *Eur J Pharmacol* 1992;222:213–218.
30. Satoh H, Lou Y-P, Lee L-Y, Lundberg JM. Inhibitory effects of capsazepine and the NK2 antagonist SR48968 on bronchoconstriction evoked by sensory nerve stimulation in guinea-pigs. *Acta Physiol Scand* 1992;146:535–536.
31. Lou Y-P, Lee L-Y, Satoh H, Lundberg JM. Postjunctional inhibitory effect of the NK2 receptor antagonist, SR-48968, on sensory NANC bronchoconstriction in the guinea-pig. *Br J Pharmacol* 1993;109:765–773.
32. Delay-Goyet P, Satoh H, Lundberg JM. Relative involvement of substance P and CGRP mechanisms in antidromic vasodilation in the rat skin. *Acta Physiol Scand* 1992;146:537–538.
33. Satoh H, Delay-Goyet P, Lundberg JM. Involvement of tachykinin and CGRP receptors in the cardio-pulmonary effects evoked by capsaicin in the guinea-pig. *Regul Pept* 1993;46:297–299.
34. Holzer P. Capsaicin: cellular targets, mechanisms of action and selectivity for thin sensory neurons. *Pharmacol Rev* 1991;43,4:143–201.
35. Lou Y-P. Regulation of neuropeptide release from pulmonary capsaicin-sensitive afferents in relation to bronchoconstriction. *Acta Physiol Scand* 1993;suppl. 612:1–88.
36. Lou Y-P, Lundberg JM. Experimental studies on sensory neuropeptide release in relation to postmortem bronchoconstriction in isolated rat and guinea-pig lung. *Respiration* 1994;submitted.
37. Lou Y-P, Franco-Cereceda A, Lundberg JM. Different ion channel mechanisms between low concentrations of capsaicin and high concentrations of capsaicin and nicotine regarding peptide release from pulmonary afferents. *Acta Physiol Scand* 1992;146:119–127.
38. Lou Y-P, Franco-Cereceda A, Lundberg JM. Variable alpha$_2$-adrenoceptor-mediated inhibition of bronchoconstriction and peptide release upon activation of pulmonary afferents. *Eur J Pharmacol* 1992;210:173–181.

39. Lou Y-P, Lundberg JM. Inhibition of low pH evoked activation of airway sensory nerves by capsazepine, a novel capsaicin-receptor antagonist. *Biochem Biophys Res Comm* 1992;189:537–544.
40. Franco-Cereceda A, Lundberg JM. Capsazepine inhibits low pH- and lactic acid-evoked release of calcitonin gene-related peptide from sensory nerves in guinea-pig heart. *Eur J Pharmacol* 1992; 221:183–184.
41. Haass M, Hock M, Richardt G, Schöming A. Neuropeptide Y differentiates between exocytotic and nonexocytotic noradrenaline release in guinea-pig heart. *Naunyn Schmiedebergs Arch Pharmacol* 1989;340:509–515.
42. Haass M, Förster C, Kranzhöfer R, Richardt G, Schöming A. Role of calcium channels and protein kinase C for release of norepinephrine and neuropeptide Y. *Am J Physiol* 1990;259:R925–R930.
43. Franco-Cereceda A, Saria A, Lundberg JM. Differential release of calcitonin gene-related peptide and neuropeptide Y from the isolated heart by capsaicin, ischaemia, nicotine, bradykinin and ouabain. *Acta Physiol Scand* 1989;135:173–187.
44. Lundberg JM, Rudehill A, Sollevi A, Hamberger B. Evidence for co-transmitter role of neuropeptide Y in pig spleen. *Br J Pharmacol* 1989;96:675–687.
45. Lundberg JM, Änggård A, Theodorsson-Norheim E, Pernow J. Guanethidine-sensitive release of neuropeptide Y-like immunoreactivity in the cat spleen by sympathetic nerve stimulation. *Neurosci Lett* 1984;52:175–180.
46. Lundberg JM, Rudehill A, Sollevi A. Pharmacological characterization of neuropeptide Y and noradrenaline mechanisms in sympathetic control of pig spleen. *Eur J Pharmacol* 1989;163;103–113.
47. Schoups A, Saxena VK, Tombeur K, De Potter WP. Facilitation of the release of noradrenaline and neuropeptide Y by the a_2-adrenoceptor blocking agents idazoxan and hydergine in the dog spleen. *Life Sci* 1988;42:517–523.
48. Haass M, Cheng B, Richardt G, Lang RE, Schömig A. Characterization and presynaptic modulation of stimulation-evoked exocytotic co-release of noradrenaline and neuropeptide Y in guinea pig heart. *Naunyn Schmiedebergs Arch Pharmacol* 1989;339:71–78.
49. Lundblad L, Änggård A, Saria A, Lundberg JM. Neuropeptide Y and nonadrenergic sympathetic vascular control of the cat nasal mucosa. *J Auton Nerv Syst* 1987;20:189–197.
50. Lundberg JM, Saria A, Franco-Cereceda A, Hökfelt T, Terenius L, Goldstein M. Differential effects of reserpine and 6-hydroxydopamine on neuropeptide Y and noradrenaline in peripheral neurons. *Naunyn Schmiedebergs Arch Pharmacol* 1985;328:331–340.
51. Schalling M, Franco-Cereceda A, Hemsén A, et al. Neuropeptide Y and catecholamine synthesizing enzymes and their mRNAs in rat sympathetic neurons and adrenal glands: studies on expression, synthesis and axonal transport after pharmacological and experimental manipulations using hybridization techniques and radioimmunoassay. *Neuroscience* 1991;41:753–766.
52. Pernow J, Thorén P, Millberg B-I, Lundberg JM. Renal sympathetic nerve activation in relation to reserpine-induced depletion of neuropeptide Y in the kidney of the rat. *Acta Physiol Scand* 1988; 134:53–59.
53. Lundberg JM, Al-Saffar A, Saria A, Theodorsson-Norheim E. Reserpine-induced depletion of neuropeptide Y from cardiovascular nerves and adrenal gland due to enhanced release. *Naunyn Schmiedebergs Arch Pharmacol* 1986;332:163–168.
54. Lundberg JM, Rudehill A, Sollevi A, Hamberger B. Frequency- and reserpine-dependent chemical coding of sympathetic transmission: differential release of noradrenaline and neuropeptide Y from pig spleen. *Neurosci Lett* 1986;63:96–100.
55. Pernow J, Kahan T, Lundberg JM. Neuropeptide Y and reserpine-resistant vasoconstriction evoked by sympathetic nerve stimulation in dog skeletal muscle. *Br J Pharmacol* 1988;94:952–960.
56. Lacroix JS, Stjärne P, Änggård A, Lundberg JM. Sympathetic vascular control of the pig nasal mucosa. (2): reserpine-resistant, non-adrenergic nervous responses in relation to neuropeptide Y and ATP. *Acta Physiol Scand* 1988;133:183–197.
57. Lundberg JM, Pernow J, Lacriox J-S. Neuropeptide Y: sympathetic cotransmitter or modulator? *NIPS* 1989;4:13–17.
58. Muramatsu I. The effect of reserpine on sympathetic, purinergic neurotransmission in the isolated mesenteric artery of the dog: a pharmacological study. *Br J Pharmacol* 1987;91:467–474.
59. Lundberg JM, Pernow J, Fried G, Änggård A. Neuropeptide Y and noradrenaline mechanisms in relation to reserpine induced impairment of sympathetic neurotransmission in the cat spleen. *Acta Physiol Scand* 1987;131:1–10.
60. Hall GT, Gardner TD, Potter EK. Attenuation of long-lasting effects of sympathetic stimulation after repeated stimulation. *Circ Res* 1990;67:193–198.

61. Modin A, Pernoe J, Lundberg JM. Maintained overflow of neuropeptide Y and noradrenaline upon repeated renal and splenic sympathetic nerve stimulation in controls but not after reserpine. *J Auton Nerv Syst* 1994; in press.
62. Brodin E, Gazelius B, Lundberg JM, Olgart L. Substance P in trigeminal nerve endings: Occurrence and release. *Acta Physiol Scand* 1981;111:501–503.
63. Lundberg JM. Evidence for coexistence of vasoactive intestinal polypeptide (VIP) and acetylcholine in neurons of cat exocrine glands. *Acta Physiol Scand* 1981;suppl496:1–57.
64. Öhlén A, Persson MG, Lindbom L, Gustafsson LE, Hedqvist P. Nerve-induced nonadrenergic vasoconstriction and vasodilatation in skeletal muscle. *Am J Physiol* 1990;258:H1334–H1338.
65. Morris JL. Roles of neuropeptide Y and noradrenaline in sympathetic neurotransmission to the thoracic vena cava and aorta of guinea-pigs. *Regul Pept* 1991;32:297–310.
66. Lundberg JM, Änggård A, Fahrenkrug J. Complementary role of vasoactive intestinal polypeptide (VIP) and acetylcholine for cat submandibular gland blood flow and secretion. II. Effects of cholinergic antagonists and VIP antiserum. *Acta Physiol Scand* 1981;113:329–336.
67. Holmquist F, Stief CG, Jonas U, Andersson K-E. Effects of the nitric oxide synthase inhibitor N^G-nitro-L-arginine on the erectile response to cavernous nerve stimulation in the rabbit. *Acta Physiol Scand* 1991;143:299–304.
68. Burnett AL, Lowenstein CJ, Bredt DS, Chang TSK, Snyder SH. Nitric oxide: a physiologic mediator of penile erection. *Science* 1992;257:401–403.
69. Lundberg JM, Modin A, Weitzberg E. Interactions between parasympathetic and local nitric oxide-mediated control of blood flow. *Pharmacol Toxicol* 1993;72supplII:S19.
70. Edwards AV, Garrett JR. Nitric oxide-related vasodilator responses to parasympathetic stimulation of the submandibular gland in the cat. *J Physiol (Lond)* 1993;464:379–392.
71. Furchgott RF, Zadawski JV. The obligatory role of endothelial cells in the relaxation of arterial smooth muscle by acetylcholine. *Nature* 1980;288:373–376.
72. Buxton ILO, Cheek DJ, Eckman D, Westfall DP, Sanders KM, Keef KD. N^G-nitro L-arginine methyl ester and other alkyl esters of arginine are muscarinic receptor antagonists. *Circ Res* 1992; 72:387–395.
73. Grider JR, Murthy KS, Jin J-G, Makhlouf GM. Stimulation of nitric oxide from muscle cells by VIP: prejunctional enhancement of VIP release. *Am J Physiol* 1992;262:G774–G778.
74. Thomas G, Cole EA, Ramwell PW. N^G-monomethyl L-arginine is a nonspecific inhibitor of vascular relaxation. *Eur J Pharmacol* 1989;170:123–124.
75. Gaw AJ, Aberdeen J, Humphrey PA, et al. Relaxation of sheep cerebral arteries by vasoactive intestinal polypeptide and neurogenic stimulation: inhibition by L-N^G-monomethyl arginine in endothelium-denuded vessels. *Br J Pharmacol* 1991;102:567–572.
76. McDonald DM, Mitchell RA, Gabella G, et al. Neurogenic inflammation in the rat trachea II. Identity and distribution of nerves mediating the increase in vascular permeability. *J Neurocytol* 1988;17:605–628.
77. Modin A, Weitzberg E, Lundberg JM. Nitric oxide regulates peptide release from parasympathetic nerves and vascular reactivity to VIP in vivo. *Eur J Pharmacol* 1994; in press.

Molecular and Cellular Mechanisms of Neurotransmitter Release, edited by Lennart Stjärne, Paul Greengard, Sten Grillner, Tomas Hökfelt, and David Ottoson, Raven Press, Ltd., New York © 1994.

16

Quantal Analysis of Excitatory Postsynaptic Currents at the Hippocampal Mossy Fiber-CA3 Pyramidal Cell Synapse

Eberhard von Kitzing, Peter Jonas, and Bert Sakmann

Department of Cell Physiology, Max-Planck-Institute for Medical Research, D-69120 Heidelberg, Germany

Excitatory synapses in the CNS show use-dependent and long-lasting changes in synaptic efficacy. To identify whether the underlying mechanisms occur pre- or postsynaptically (for a review see Stevens (1)), it is necessary to examine release at a single synapse. Ideally, one would like to simultaneously record electrical signals in the nerve terminals of the presynaptic neuron responsible for transmitter release as well as the postsynaptic signals that indicate the effect of the released transmitter. In rapidly transmitting synapses like excitatory synapses of the CNS it is, however, not yet possible to record from presynaptic terminals, to control their electrical activity, or to directly monitor the release process (e.g., by capacitance measurements). Therefore, recordings of fluctuations in amplitude of postsynaptic potentials or currents (EPSCs or EPSPs) have been used to estimate the size of quantal events, and derive the (average) number of released vesicles and the release probability by use of a statistical model of the release process (quantal analysis; see reviews (2–4)). To do this, a detailed knowledge of the postsynaptic action of the transmitter released from single vesicles is essential so that the characteristics of the release mechanism can be inferred, thus making it possible to distinguish presynaptic from postsynaptic changes in synaptic transmission.

In this article we review experimental results on stimulus evoked unitary EPSCs and on spontaneously occurring miniature EPSCs (mEPSCs) obtained at the excitatory synapse formed between terminal boutons of the mossy fibers of granule cells in the dentate gyrus and the apical dendrites of pyramidal cells in the CA3 subfield of the rat hippocampus (MF-CA3 synapse). Here, in a subset of synapses, unitary EPSC amplitude histograms were observed that showed multiple, evenly spaced peaks (5). In the first part the shape of these unitary EPSC and mEPSC amplitude histograms is analyzed to derive the average amplitude $\langle q \rangle$ and variability of the quantal events, and the statistical parameters of the release process (N_R and p_R) that determine the quantal content of EPSCs. This is done by comparing the predicted distributions of spatially and temporally homogeneous release (standard binomial release), as well as spatially nonuniform and time-dependent nonstationary release models with the measured amplitude histograms of unitary EPSCs. Subsequently, the possible contribution of postsynaptic sources to amplitude fluctuations of unitary EPSCs and multiple peaks in amplitude distributions is considered, and possible pitfalls of analyzing unitary EPSCs in terms of $\langle q \rangle$, N_S, and p_R without independent measurement of the amplitude distribution of quantal events at the same synapse are illustrated by simulations.

At many excitatory synapses, the nerve terminals of a single presynaptic axon form a distributed synapse with boutons located along the dendrites of the postsynaptic neuron (e.g., as discussed by Andersen (6)). This implies that transmitter is released from different boutons, resulting in production of unitary EPSCs by superposition of quantal events generated at the different active zones. The quantal events contributing to unitary EPSCs may therefore differ in time course and amplitude depending, for example, on the anatomical particulars of dendrites (e.g., basal versus apical dendrites) and on the size and geometry of spines and postsynaptic densities (PSDs). MF-CA3 synapses on the other hand are rather compact in structure since they are formed by a single large mossy fiber bouton that arises from a granule cell in the dentate gyrus (7). Because individual boutons arising from the same mossy fiber axon are far apart, it also seems unlikely that a granule cell axon branch contacts the same CA3 pyramidal cell more than once (8). Unlike many other excitatory CNS synapses, the unitary EPSC at MF-CA3 synapses thus seems to be generated at a single bouton-spine complex, implying that all PSDs of a synapse are located at a similar electrotonic distance from the somatic recording pipette and that the contributions of quantal events from different active zones to unitary EPSCs are attenuated to the same extent (9).

RESULTS

Unitary EPSCs, generated by an action potential propagating along a single granule cell axon, could be evoked by focal stimulation of a granule cell in the dentate gyrus using fine-tipped extracellular stimulation pipettes (Fig. 1A). Although CA3 pyramidal cells have extensive apical and basal dendrites, the MF boutons form synapses with large dendritic spines (or thorns, visible as thickenings of apical dendrites in Fig. 1B) that are located on apical dendrites relatively close to the soma. Tight-seal whole-cell recording with a somatic patch pipette (Fig. 1C) from CA3 pyramidal cells in brain slices under visual control (10) permits a resolution of EPSCs with amplitudes of a few picoamperes.

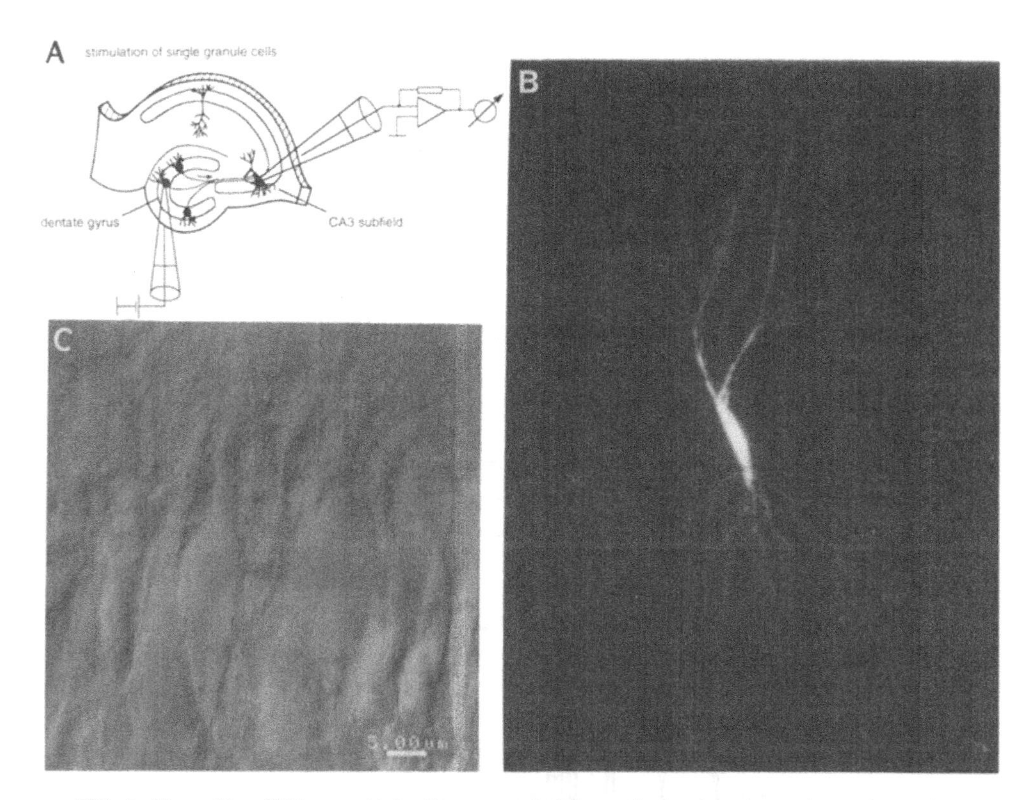

FIG. 1. Mossy fiber-CA3 pyramidal cell synapse. **A.** Schematic drawing of experimental setup in slice preparations for recording unitary excitatory postsynaptic currents (EPSCs). Recording from pyramidal neuron in CA3 subfield in whole-cell configuration, stimulation of a single granule cell in dentate gyrus with a 3 MΩ glass pipette as extracellular stimulation electrode and low stimulus intensity (≤15 V, corresponding to ≤5 μA, and 50–100 μs duration). The experiments were done at room temperature (22°C). Slices were obtained from 15–24-day-old rats. **B.** CA3 pyramidal neuron after intracellular filling with Lucifer Yellow. Thorny excrescences (large dendritic spines) located at the shaft of apical dendrites are visible as thickenings of dendrites. Thorny excrescences form the postsynaptic part of MF-CA3 synapses. **C.** Pyramidal neuron in the CA3 region of rat hippocampus with recording pipette (right) attached to cell body in a living slice preparation. Video image obtained by infrared-differential interference contrast microscopy. (From Stuart et al., ref. 10.) Image taken by N. Spruston (Heidelberg).

Distributions of Unitary EPSC and mEPSC Amplitudes at MF-CA3 Synapses

Unitary EPSC Amplitude Histograms With Multiple Peaks

If the quantal contributions of evoked unitary EPSCs were directly resolvable as distinct peaks in amplitude histograms, it might be possible to determine size and variability of quantal events as well as the release parameters from these distributions. Figure 2A illustrates examples of unitary EPSCs recorded at high extracellu-

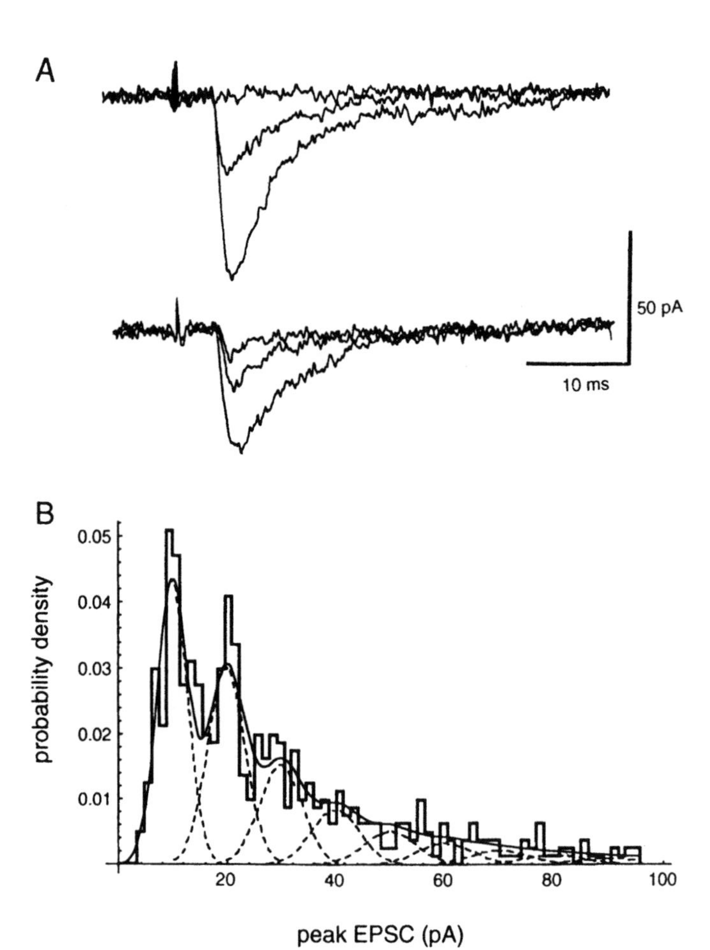
A
50 pA
10 ms
B
probability density
0.05
0.04
0.03
0.02
0.01
20
40
60
80
100
peak EPSC (pA)

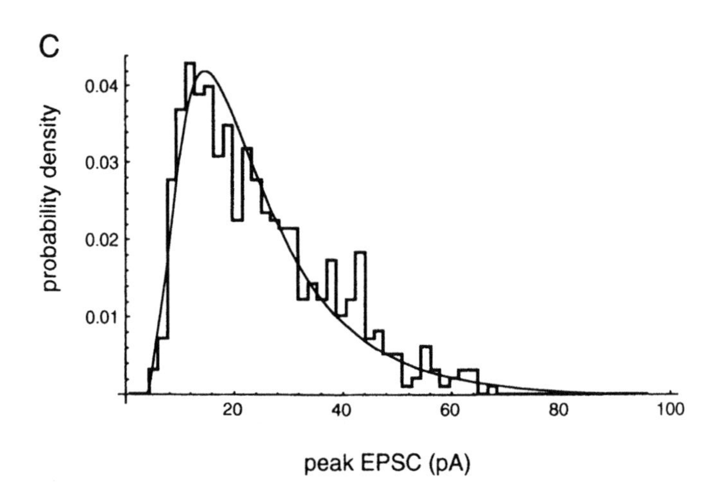
C
probability density
0.04
0.03
0.02
0.01
20
40
60
80
100
peak EPSC (pA)

lar Mg^{2+} and low extracellular Ca^{2+} concentrations (5). Under these conditions, which are known to reduce the quantal content of endplate currents at the neuromuscular junction (2) unitary EPSCs fluctuated in peak amplitude and almost equidistant peaks in the amplitude histogram were apparent (Fig. 2B) in a subset of our experiments (five of nine MF-CA3 synapses). In the remainder only a single broad peak was observed (Fig. 2C), and the EPSC histograms were skewed and well fitted with various empirical monotonous distributions. Several tests were performed that suggested statistical artifacts were not likely to be the source of the multiple peaks in unitary EPSC histograms: data were displayed at different bin widths, amplitude histograms from subsets of the total data set were analyzed, and the autocorrelation function was used to test for the presence of equidistant peaks (5). Unitary EPSC histograms that passed these tests were then fitted by sums of Gaussian functions (Fig. 2B) but no attempt was made to infer models of transmitter release from these fits (5). The apparent peak spacing obtained from the Gaussian fits was between 7 and 12 pA (at −70 mV to −90 mV membrane potential). The fit of peaked unitary EPSC histograms by sums of Gaussians suggested that unitary EPSCs are generated by statistical superposition of quantal events that have amplitudes described by a Gaussian distribution and a variability characterized by a coefficient of variation (CV_q) of 22% (5) (Fig. 2B).

Miniature EPSC Amplitude Histograms

If it were possible to obtain, from a single MF-CA3 synapse, both the amplitude distribution of spontaneously occurring mEPSCs and unitary EPSCs, one would be able to derive a rigorous statistical model of transmitter release, as at the neuromuscular junction (2). At the MF-CA3 synapse and at mammalian central synapses in general the situation is, however, more complicated. Figure 3A shows examples of mEPSCs recorded from a CA3-pyramidal cell at a membrane potential close to the resting membrane potential (−70 mV). All mEPSCs shown were selected on the basis of their fast rise times (Fig. 3B) in order to sample from synapses located

FIG. 2. Fluctuation in peak amplitude of unitary excitatory postsynaptic currents (EPSCs) in a MF-CA3 synapse elicited by electrical stimulation of a granule cell soma in the dentate gyrus. (From Jonas et al., ref. 5.) **A.** Examples of stimulus-evoked unitary EPSCs with low quantal content in extracellular solution containing 3 mM Mg^{2+} and 1 mM Ca^{2+}. Three successive unitary EPSCs in each family of traces. Brief upward deflection marks stimulus artifact. Membrane potential −90 mV. Note that in one trial no response was observed (failure). **B.** Histogram of unitary EPSC peak amplitudes. The histogram (593 events) is characterized by several well separated peaks and is fitted by the sum (continuous line) of eight equally spaced Gaussian distributions (broken lines). Assuming that the amplitude of quantal events follows a Gaussian distribution, the mean amplitude of quantal events is 10 pA and the width is 2.8 pA (CV_q of 28%). **C.** Histogram of unitary EPSC peak amplitudes lacking well separated peaks. The histogram is fitted by an empirical skewed function consisting of two exponential terms. The mean unitary EPSC amplitude is 24.5 pA, the width is 13.0 pA, and the third root of the skewness is 12.3 pA.

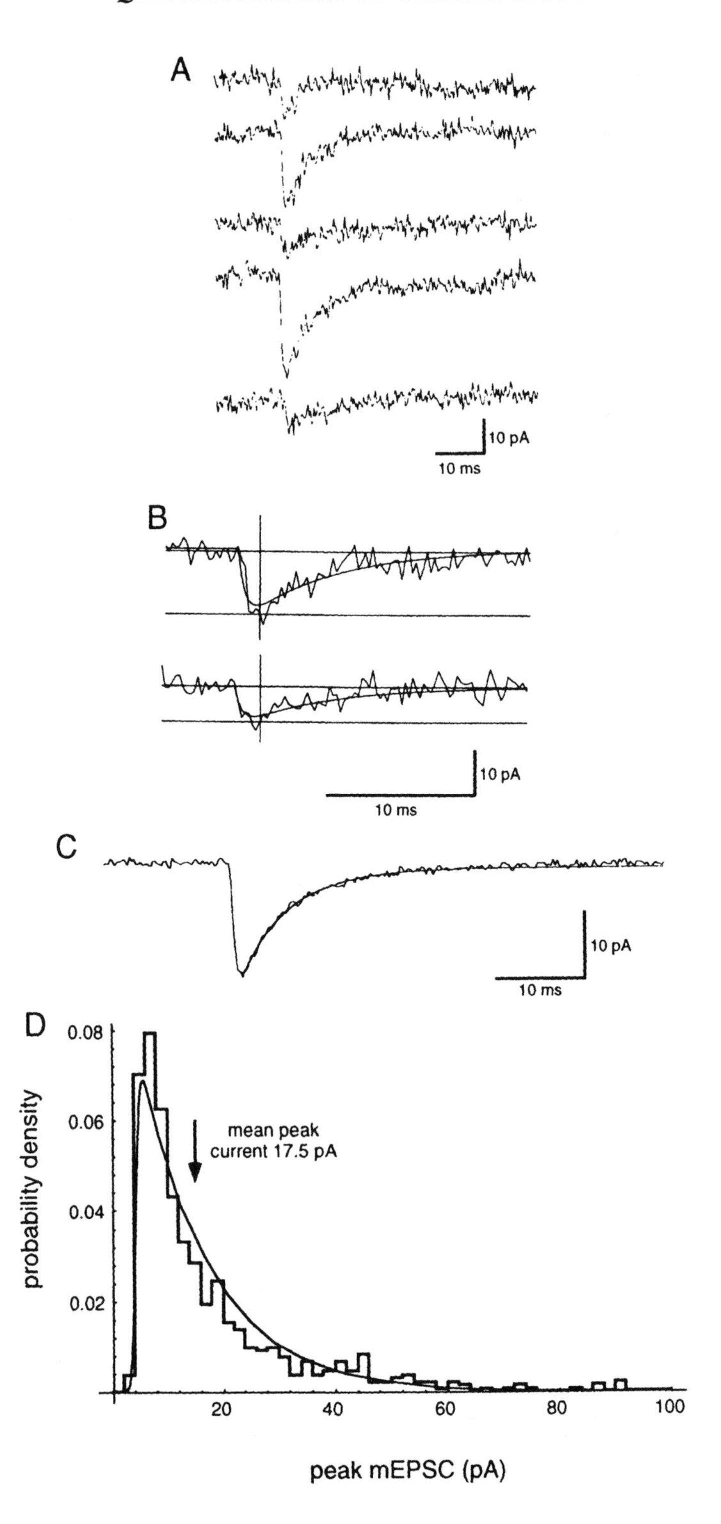
A
10 pA
10 ms
B
10 pA
10 ms
C
10 pA
10 ms
D
0.08
0.06
0.04
0.02
probability density
mean peak
current 17.5 pA
20
40
60
80
100
peak mEPSC (pA)

close to the soma (presumably MF-CA3 synapses). These mEPSCs had a small (10–20 pA) average peak amplitude (Fig. 3C), and their amplitudes showed a large variability. Amplitude histograms could not be described by a simple Gaussian function but were strongly skewed, as observed at other central synapses (e.g., as discussed by Edwards et al. (11), Ropert et al. (12), and Silver et al. (13)), with a median in the range between 6 pA and 10 pA (Fig. 3D). Thus the distributions of quantal events inferred from unitary EPSC fluctuations (Fig. 2B) and of mEPSCs (Fig. 3D) are comparable in their mean amplitudes but greatly differ in their variability and skewness.

Each CA3 pyramidal cell receives synaptic input from about 30 MF-boutons located at different regions of the apical dendrite (e.g., as discussed by Amaral et al., (8) and Gaiarsa et al. (14)), and each bouton faces separate PSDs located opposite to the respective active zones (on average 7 ± 4 PSDs per medium-sized thorn, with a range of 1–37 PSDs per thorn (7)). This suggests that mEPSCs recorded from CA3 pyramidal cells are sampled from about 30 MF-CA3 synapses containing a total of about 210 active zones and PSDs. The large variability of mEPSC amplitudes may be caused by inhomogeneity of space clamp conditions due to the distribution of MF-CA3 synapses along the apical dendrite, as well as by the inhomogeneity of quantal sizes between different MF-CA3 synapses and even between different PSDs of the same MF-CA3 synapse. Finally, mEPSCs generated at other excitatory synapses (e.g., those on basal dendrites) contribute to mEPSC histo-

FIG. 3. Properties of miniature excitatory postsynaptic currents (mEPSCs) recorded from CA3 pyramidal neurons (redrawn from Jonas et al., ref. 5). **A.** Examples of spontaneously occurring mEPSCs recorded at −70 mV membrane potential in the presence of 1 μM tetrodotoxin. Only mEPSCs with rise times of 0.8 ms or less were selected. **B.** A pattern recognition algorithm was used to detect spontaneously occurring miniature EPSCs. It exploited the fact that the time course of the miniature events was roughly known; specifically, that their rise was much faster than their decay. The pattern function chosen was of the form

$$p(t) = \{1-\exp[-(t-t_{start})/\tau_{onset}]\} \exp[-(t-t_{start})/\tau_{decay}] \text{ for } t \geq t_{start}$$
$$\text{and } p(t) = 0 \text{ for } t < t_{start}$$

normalized to a peak amplitude of 1. The t_{onset} was chosen as 0.5 milliseconds, t_{decay} as 6 milliseconds. For a given time window, the time course of the pattern p(t) was compared to the time course of the measured current I(t) by performing a linear regression analysis on p as a function of I for all values of t within the window. The correlation coefficient then provided a measure of how well the pattern fitted the data, and the steepness of the regression line corresponded to the amplitude of the event. The window was then shifted sample point by sample point over the whole set of recorded data until the values for the correlation coefficient and the amplitude exceeded critical values (0.66 and 3.4 pA, respectively). This algorithm turned out to be more robust than routines using simple threshold crossing criteria. In particular, brief transient events arising from baseline noise were ignored even when they had a relatively large amplitude. **C.** Mean time course of mEPSC obtained by averaging 20 individual mEPSCs selected for fast rise times. Decay time constant (represented by thin line superimposed on trace) is 5.6 milliseconds. **D.** Distribution of mEPSC amplitudes recorded from a CA3 pyramidal neuron with rise times of 0.8 milliseconds or less (622 events). The mean mEPSC peak amplitude (17.5 pA) is indicated by arrow. Width of distribution (standard deviation) is 15.8 pA and the third root of its skewness (third central moment) is 20.8 pA. The continuous line represents a maximum likelihood fit of a "skewed" Gaussian to the histogram.

grams. At present no method is available to identify those mEPSCs generated by glutamate release from a particular MF-bouton. Therefore it is not possible to fit statistical models of transmitter release to unitary EPSC distributions, assuming that the mEPSC distribution represents the amplitude distribution of quantal events of a particular MF-CA3 synapse.

Construction of a Passive Cable Model of a CA3 Pyramidal Cell

Despite the location of MF-CA3 synapses close to the cell soma, EPSCs generated at MF-CA3 synapses will be distorted in time course as well as attenuated in amplitude when recorded with a somatic pipette. To estimate the size of these errors and the number of open glutamate receptor (GluR) channels that mediate quantal events, the passive electrical properties of CA3 pyramidal cells were reconstructed (5) (Fig. 4A, B), and conductance changes at the most proximal and most distal MF-CA3 synapses were simulated. The recorded amplitude of EPSCs generated in distal thorns may be attenuated by as much as a factor of 3 to 4 (Fig. 4C, D). The effect of electrotonic attenuation on the shape of unitary EPSC distributions is illustrated by the simulation shown in Fig. 4E, F where it is assumed that the synapse is 0.2 length constants from the recording pipette. Clearly, both the size of individual peaks in the EPSC histogram and their spacing are drastically reduced during somatic recording of unitary EPSCs generated at a synapse located on the dendrite. It is therefore possible that the mean amplitudes of quantal events derived from unitary EPSC and mEPSC histograms (Fig. 2B, Fig. 3D) are substantially underestimated. For typical MF-CA3 synapse locations and typical series resistances of the recording pipette in the whole-cell configuration (5 MΩ), the amount of attenuation of EPSCs can be as much as fourfold. Taking into account voltage and space clamp imperfections, the simulations suggest that the 'real' amplitudes of quantal events underlying unitary EPSC fluctuations range between about 8 pA (negligible space clamp error) and 36 pA (large space clamp error). Based on estimates of single-channel conductance of glutamate receptor (GluR) channels of the AMPA receptor subtype, (5) it is estimated that the average number of postsynaptic GluR channels open during the peak of a quantal event would be between 15 (lower estimate) and 65 (upper estimate). These numbers are small compared to estimates of open acetylcholine receptor channels at the neuromuscular junction, but comparable to the estimates determined from quantal events at inhibitory CNS synapses (11) and at excitatory synapses at visual cortex stellate cells (15) and on cerebellar granule cells (13,16).

Models of Transmitter Release Derived from Amplitude Fluctuations of Unitary EPSCs

The shape of unitary EPSC histograms is determined on the one hand by the mean amplitude of quantal events $\langle q \rangle$ and its coefficient of variation (CV_q), and on the other hand by the statistical properties of the release mechanism. Histograms of unitary EPSCs with fluctuating amplitudes may be used to estimate the statistical parameters of transmitter release (i.e., number of releasable vesicles (N_S) and their release probability (p_R) following invasion of the terminal by an action potential). The number of releasable vesicles is thought to be identical to the number of release sites loaded with releasable vesicles. These two numbers are generally obtained by fitting model distributions to EPSC histograms (2–4). Since mEPSC histograms indicative of quantal events of a single synapse could not be obtained, unitary EPSC histograms were fitted simultaneously with different quantal event distributions and different release models. One problem with this analysis is that temporal fluctuations of quantal event amplitudes may contribute to fluctuations of unitary EPSC amplitudes thought to primarily reflect release statistics.

Presynaptic Sources of Variability

The most important presynaptic source of fluctuations in size of unitary EPSCs is the number of quanta of transmitter released (N_R) per trial. For a synapse with N_S release sites that all have a uniform p_R, the probability to release N_R quanta follows a binomial distribution, assuming statistical independence between release sites; here this is referred to as the standard model of release. However, there is no a priori reason for assuming that the p_Rs are identical at different release sites; rather, one would expect that p_R varies between spatially separate release sites. This leads to a nonuniform model of release (represented by a multinomial event distribution also called compound-binomial distribution (4)). The width of a multinomial distribution is considerably smaller than that of a binomial distribution, given that N_s is the same and the mean of individual release probabilities $\langle p_R \rangle$ equals that of the binomial distribution. A nonuniform distribution of individual release probabilities thus tends to narrow the amplitude histogram of unitary EPSCs for the same number of release sites. Moreover, there is no reason to assume that p_R at a given release site is always constant in time from trial to trial; it is possible that release at a release site follows a nonstationary distribution. Because unitary EPSCs collected for an amplitude histogram are measured at different times, one would expect a time averaging of bi- or multinomial distributions with varying mean release probabilities in a nonstationary model of release. In contrast to nonuniform distributions of release probabilities, the nonstationary distribution of release probabilities always results in a broadening of the amplitude histogram of unitary EPSCs as compared to histograms expected for a release process following standard binomial statistics (see also Faber and Korn (17) and Korn et al. (18)). For standard and nonuniform release models, the variance of the distribution of release probabilities must be less than its mean.

It is often assumed that one vesicle per presynaptic action potential is released from a single active zone (3,19,20) (see chapter 19), that all vesicles roughly contain the same number of transmitter molecules, and that the fusion of the vesicle occurs in the same position with respect to the PSD. It is not clear if these assump-

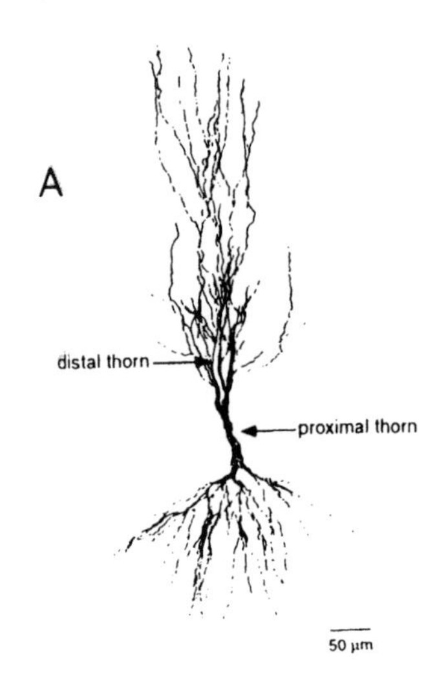
A
distal thorn
proximal thorn
50 μm

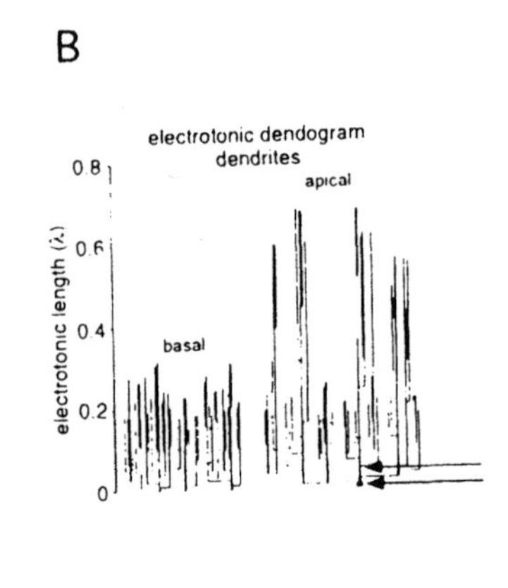
B
electrotonic dendogram
dendrites
apical
basal
electrotonic length (λ)
0.8
0.6
0.4
0.2
0

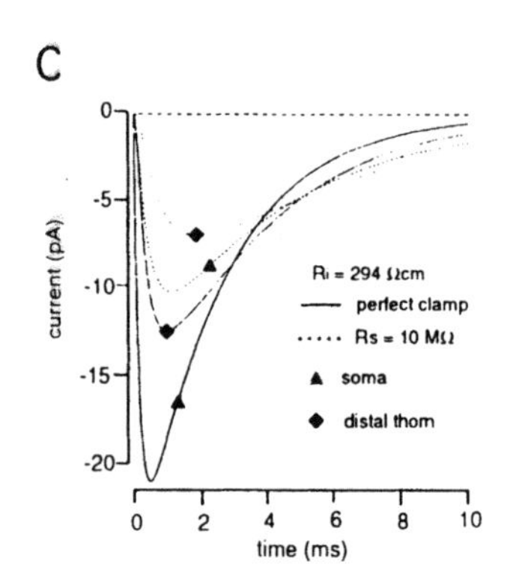
C
current (pA)
time (ms)
perfect clamp
Rs = 10 MΩ
soma
distal thorn

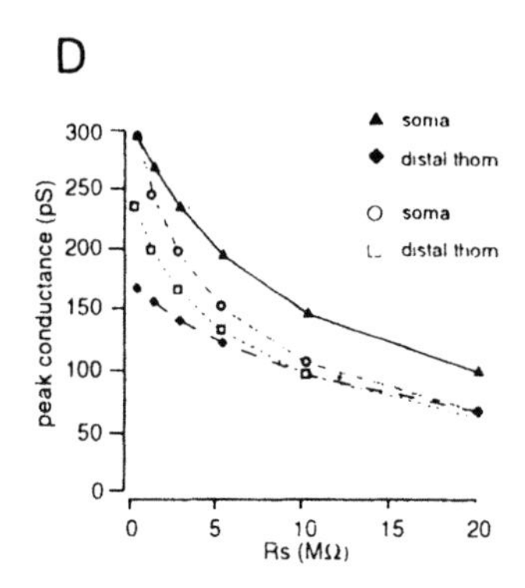
D
peak conductance (pS)
Rs (MΩ)
soma
distal thorn
soma
distal thorn

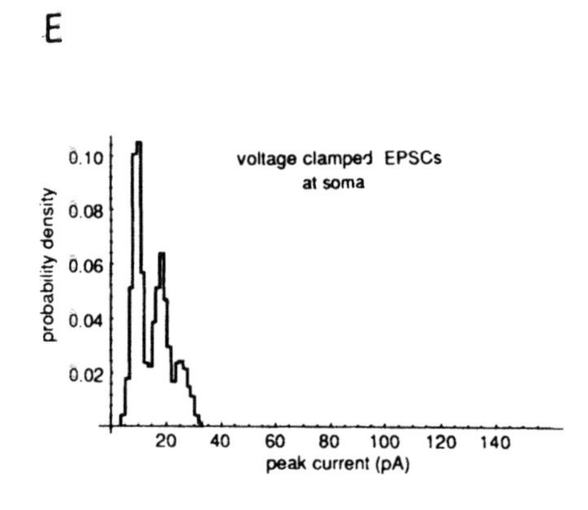
E
voltage clamped EPSCs
at soma
probability density
peak current (pA)

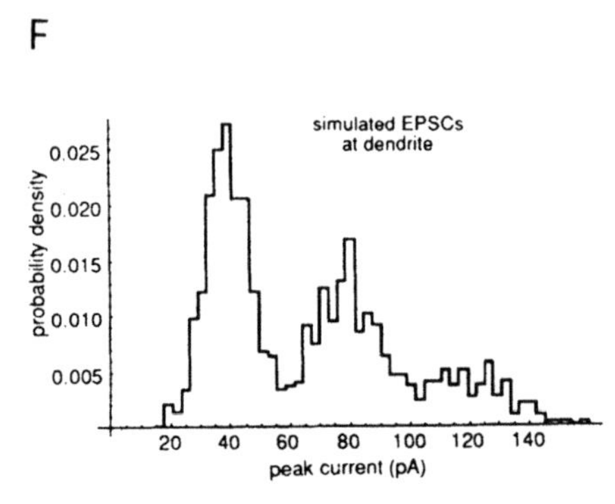
F
simulated EPSCs
at dendrite
probability density
peak current (pA)

tions hold at the MF-CA3 synapse. If the postsynaptic GluRs are saturated by glutamate released from one vesicle, these sources of variability would not be significant. On the other hand, if the postsynaptic GluR channels are not saturated and operate in the linear range of their activation curve, these sources of variability may become significant (21).

Postsynaptic Sources of Variability

Probably the most important source contributing to the variability of unitary EPSCs is the variation of the number of GluR channels in different PSDs facing different active zones. It is known from electron microscopic observations that the area of PSDs varies considerably at MF-CA3 synapses (7). The distribution of PSD area sizes could reflect the distribution of the number of GluR channels per PSD,

FIG. 4. Branching cable model of a reconstructed CA3 pyramidal neuron (redrawn from Jonas et al., ref. 5) to simulate effects of insufficient space clamp on unitary excitatory postsynaptic current (EPSC) amplitudes. **A.** Camera lucida drawing of CA3 pyramidal neuron filled with biocytin for 30 minutes in the whole-cell configuration and stained by avidin-horseradish peroxidase reaction. Location of most proximal and most distal thorny excrescences in this cell are indicated by arrows. **B.** Electrotonic "dendrogram" of the neuron shown in **A**. Location of thorns indicated by arrows. $C_m = 0.683\ \mu Fcm^{-2}$, $R_m = 164{,}000\ \Omega cm^2$, $R_i = 294\ \Omega cm$, and $G_{shunt} = 0$ nS. These parameters were obtained from fitting the voltage transient in response to a brief current pulse. **C.** Simulation of EPSCs at MF-CA3 synapses generated at proximal or distal thorn and recorded with whole-cell pipette at the soma. Note slowing of rise and decay time course and attenuation of amplitude for distal thorn and dependence on pipette access resistance. Current transient injected into the dendrite based on fast application experiments (rise time = 0.2 ms, decay time constant = 2.5 ms, 300 pS peak conductance. (From Colquhoun et al., ref. 28.) **D.** Attenuation of EPSC peak conductance generated at proximal or distal thorn and recorded with whole-cell pipette at the soma for two values of R_i (294 Ωcm, filled symbols; 70Ω cm, open symbols). **E,F.** Simulation of the effect of electrotonic attenuation by dendritic cable properties on EPSC amplitude distributions generated at MF-CA3 synapses located at a distal thorn on an apical dendrite (0.2 of a length constant from the soma). Three independent release sites opposite three separate postsynaptic densities (PSDs) with 100 ± 19 available glutamate receptor (GluR) channels in each PSD were assumed. The number of simulated amplitudes in each of the three peaks was taken to follow the standard release model with three release sites and a release probability of 0.47. In **E** the histogram of EPSC peak amplitudes recorded with a somatic "voltage clamp" pipette (5 MΩ series resistance) is shown using a bin width of 1.4 pA. The mean amplitudes of the three peaks in this histogram are 9.5 ± 1.7 pA, 18.1 ± 2.33 pA, and 25.7 ± 2.5 pA. The average number of open channels during a quantal event is close to 70. In **F** the histogram of the "real" EPSC peak amplitudes at the synapse are shown using a bin width of 2.9 pA. Mean peak amplitudes are 39.4 ± 7.5 pA, 79 ± 11 pA, and 118 ± 12 pA. The branching cable model illustrated in **B** was used for Monte Carlo simulations of EPSCs, using the kinetic scheme described in Jonas et al., (ref. 5), assuming that a 1 ms pulse of 3 mM glutamate represents the glutamate concentration transient in the synaptic cleft after release from one vesicle. Membrane potential is -70 mV, and the single channel current amplitude is 0.56 pA. Note that the mean peak amplitude of quantal events derived from "voltage clamped" EPSCs (shown in **E**) is only one fourth of the "real" amplitude (shown in **F**), and consequently the number of open channels during a quantal event calculated from the peak distance would be underestimated fourfold. (from Faber and Korn, ref. 17.)

assuming that the channel density per unit area is constant (22). Furthermore, even if all PSDs contained the same number of GluR channels, one would expect variability of the amplitude of the quantal event due to the stochastic channel gating (23). The magnitude of this variability will be largely determined by the maximum open probability p_O of the channels and the number of channels n_A present in a PSD. The number of open channels would then follow a binomial distribution with the parameters p_O and n_A. Finally, GluR channels show several different conductance states (24,25) suggesting that the single channel current amplitude at a given membrane potential will itself be described by a distribution with a nonzero width.

Uniform and Nonuniform Release

Figure 5A illustrates a unitary EPSC histogram from Jonas, et al. (5) that was fitted by maximizing the likelihood of the fit, assuming a "skewed Gaussian" distribution of the quantal events and the standard release model (Fig. 5B). The skewed Gaussian G_s is obtained by summing Gaussian functions G with decreasing amplitude:

$$G_s(x, x_0, \sigma, f) = (1\text{-}f)\ \Sigma\ f^n\ G(x, x_0 \pm n\,\sigma, \sigma)$$

where $0 \leq f < 1$ gives the degree of skewness. The sign in this equation is positive for positive skewness (the tail directed toward larger amplitudes). This skewed Gaus-

FIG. 5. Unitary excitatory postsynaptic current (EPSC) amplitude histogram fitted with sums of skewed Gaussians and assuming binomial release models. **A.** Histogram of measured unitary EPSC amplitudes at the MF-CA3 synapse with well separated peaks. The histogram is shown together with the approximate expected standard deviation of the probability for each bin (thin lines). **B.** Maximum likelihood fit (continuous line) to the above set of amplitudes allowing for skewed quantal event (EPSC) amplitude distribution. Uniform release probabilities that are distributed binomially are assumed (standard model of release). The derived number of release sites N_S is 8 (7.6) and the release probability p_R is 0.3. According to the fit the assumed quantal event (quantal EPSC, dotted line) is characterized by a peak amplitude at 7.0 pA, a mean amplitude of 8.3 pA, a standard deviation of 1.7 pA, and a third root of the skewness of 1.5 pA. Individual peaks are resolvable in the histogram. The derived quantal EPSC amplitude distribution in this experiment (dotted line) is by a factor of 9.3 narrower (standard deviation = 1.7 pA versus 15.8 pA) compared to measured mEPSC amplitude distributions (see Fig. 3D). **C.** Maximum likelihood fit to experimental amplitudes suggests a nonuniform release model. For fitting it was assumed that the amplitude distribution of the multiquantal events follows a skewed Gaussian. The quantal event is characterized by a peak amplitude of 6.9 pA and a mean amplitude of 9.7 pA (dotted line). The standard deviation has the value of 2.5 pA and the third root of the skewness is 2.5 pA. The quantal event distribution derived from this fit has a considerably larger width and is more skewed compared to that derived from the fit assuming a uniform release model as shown in **B**. The distribution of amplitudes of quantal peaks of first and higher order is narrower than the respective binomial one. This indicates that the release probabilities are nonuniform (i.e., differ for different release sites), with a mean release probability $\langle p_R \rangle$ of 0.27, a standard deviation of the individual release probabilities of 0.40, and a third root of the skewness of 0.54.

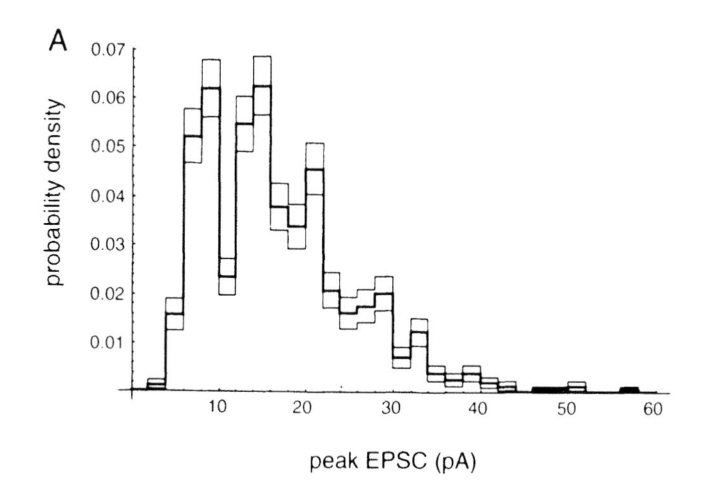
A
0.07
0.06
0.05
0.04
0.03
0.02
0.01
probability density
10
20
30
40
50
60
peak EPSC (pA)

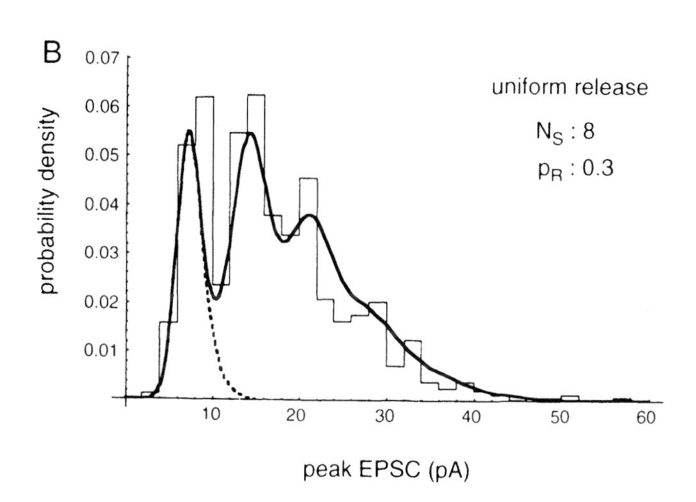
B
0.07
0.06
0.05
0.04
0.03
0.02
0.01
probability density
uniform release
N_S : 8
p_R : 0.3
10
20
30
40
50
60
peak EPSC (pA)

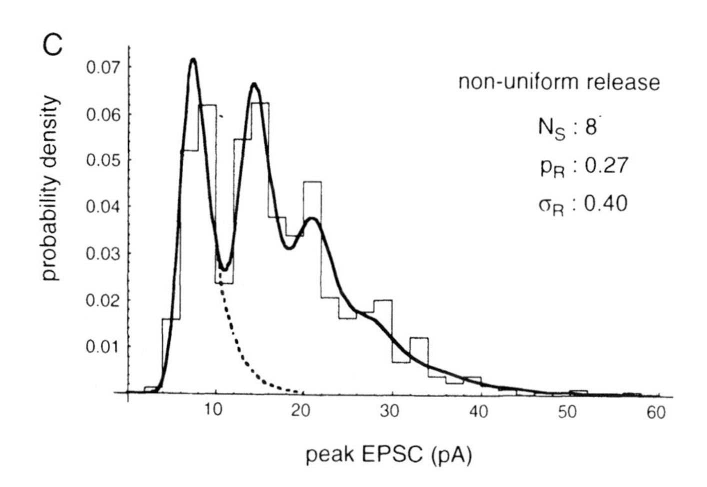
C
0.07
0.06
0.05
0.04
0.03
0.02
0.01
probability density
non-uniform release
N_S : 8
p_R : 0.27
σ_R : 0.40
10
20
30
40
50
60
peak EPSC (pA)

sian has its mean at x_s, a standard deviation of σ_s and a skewness S (third central moment) of

$$x_s = x_0 \pm f/(1\text{-}f)\ \sigma;\ \sigma_s = (1 - f + f^2)/(1 - f)\ \sigma \text{ and } S = f(1 + f)/(1\text{-}f)^3\ \sigma^3$$

The apparent CV_q of the quantal event amplitude in this fit is 20%. This could indicate that the successive peaks in this unitary EPSC histogram represent the mean amplitudes of unitary EPSCs with quantal contents of one, two, etc., and that each quantal event is mediated by a relatively constant number of open GluR channels. The derived number of releasable vesicles (or release sites) and the release probabilities in this particular experiment were estimated to be $N_S = 8$ and $p_R = 0.3$. The histogram was also fitted using a distribution of release probabilities not derived from specific release models. The fit indicated nonuniform release probabilities (Fig. 5C). The likelihood of this fit was better than that of the binomial fit; the log-likelihood ratio was -0.08. The apparent CV_q of the quantal event amplitude was 26%, and the number of release sites was eight as in the other fit. Thus, both the mean amplitude and variability of quantal events derived from the best fitting nonuniform release model are not largely different from one that assumes a Gaussian distribution of quantal events (5).

In four other experiments where the unitary EPSC amplitude distribution was judged to contain multiple peaks, the standard release model gave a rather poor fit, even when the amplitude distribution of quantal events was allowed to be skewed. One example is shown in Fig. 6. The standard release model can account either for the peaks (Fig. 6B, dashed line) or for the tail of the distribution (continuous line according to maximum likelihood fit), but not for both peaks and tail at the same time. Assuming a nonuniform release model by fitting multinomial distributions does not improve the quality of the fit, because this would make the distribution even narrower (see above). The observed tails in unitary EPSC distributions can be accounted for by assuming either nonstationary release probabilities or multiple vesicle release at each release site.

Nonstationary Release

There are several reasons why p_R could change from trial to trial. Two vesicles successively docked at a release site are not identical (e.g., they may contain slightly different amounts of membrane proteins necessary for docking and fusion). As intracellular Ca^{2+} is essential for transmitter release, oscillations of Ca^{2+} concentration in the nerve terminal (26,27) could generate fluctuations in the fusion probabilities of docked vesicles. A tailed distribution of release probabilities would satisfactorily account for the tail in the histogram of unitary EPSC amplitudes (Fig. 7A). The log-likelihood ratio of the fit with this model compared to a fit with the standard model is -0.07. The fits of four multipeak unitary EPSC histograms with this model gave mean amplitude estimates of quantal events that were comparable

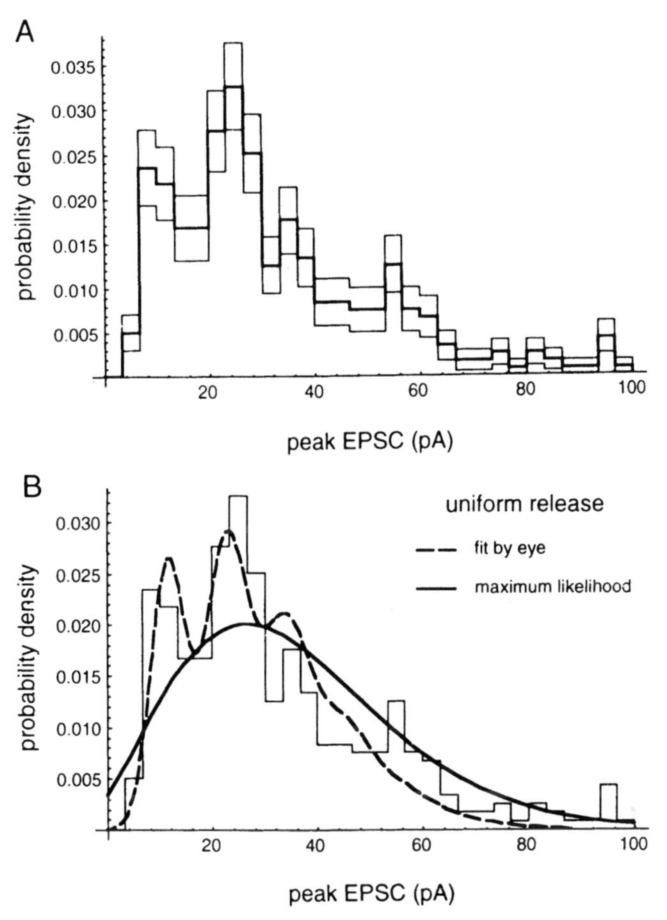

FIG. 6. Unitary excitatory postsynaptic current (EPSC) amplitude histogram with "tails." **A.** Histogram of measured unitary EPSC amplitudes with well separated peaks for which the standard release model assuming skewed Gaussians for the quantal event amplitudes did not give a satisfactory fit. Thin lines represent expected standard deviation for each bin. **B.** The standard uniform release model (binomially distributed release probabilities) can either account for the lower order peaks (dashed line, with parameters adjusted prior to optimization by eye to provide reasonable start values for the maximum likelihood fit) or for the long tail in the experimental amplitude histogram. In the resulting maximum likelihood fit, the peaks are sacrificed to describe the tail adequately (continuous line). The distribution of quantal event amplitudes became too broad to be resolved, and the average amplitude of quantal events is not related to the peak positions in the histogram.

to those obtained by fitting independent Gaussians (5) but had somewhat larger apparent CV_qs (30% on average). The number of release sites was estimated to vary between 8 and 21 for the different experiments, with respective mean release probabilities between 0.20 and 0.28.

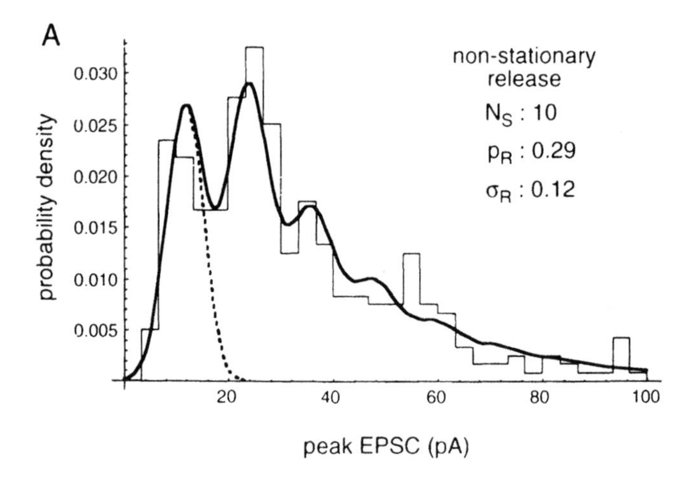

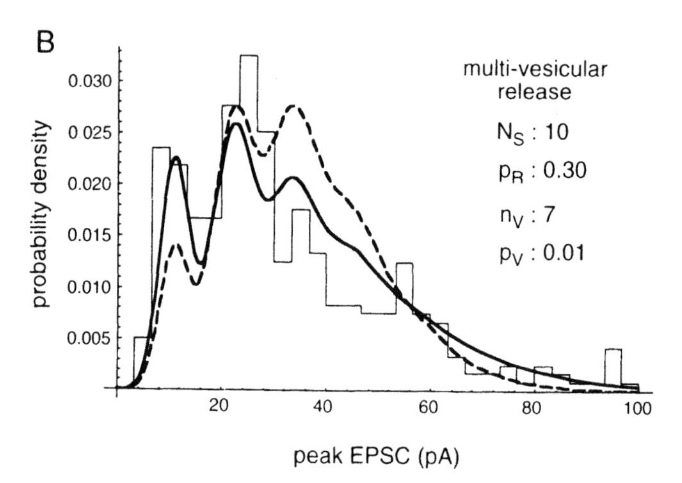

FIG. 7. Unitary excitatory postsynaptic current (EPSC) histograms with tail fitted by nonstationary release or multivesicular release models. **A.** A model-free maximum likelihood fit to experimental histogram shown in Fig. 6A resulted in a nonstationary release model with ten release sites. The mean release probability is 0.29. The distribution of amplitudes of peaks of first and higher order is too broad to be binomial; thus, nonstationary fluctuations in release probability at all release sites were assumed with a standard deviation of 0.12 and a third root of the skewness of 0.19. The derived mean amplitude of quantal events (dotted line) is 11.0 pA, with a standard deviation of 3.7 pA (CV_q of 30%) and a third root of the skewness of −2.7 pA (i.e., its tail is directed towards smaller amplitudes). **B.** Fit by standard release model allowing for several vesicles to be liberated at each release site (continuous line). The content of a single vesicle is assumed to activate only a small proportion of the available glutamate receptor (GluR) channels of a postsynaptic density (PSD). The assumed number of release sites N_S is ten and the release probability p_R for each release site is 0.30. The first quantal peak is determined by the number of releasable vesicles of $n_V = 7$ per release site, with the individual release probability p_V of 0.01. For comparison, a binomial distribution of Gaussian quantal peaks is fitted (dashed line) with the same number of release sites and the same mean unitary EPSC amplitude. It is considerably less "tailed" compared to the multivesicle release model.

Multiple Vesicle Release

Synchronized release of several vesicles at the same release site (i.e., from the active zone) may lead to linear superposition of their postsynaptic effects if the synapse operates in the linear range of the dose-response curve for activation of GluRs in the PSD. This model differs from the model of multiple vesicle release at different release sites (see chapter 19). If the number of active release sites and the number of vesicles released from an active site were binomially distributed, the resulting distribution would be more tailed compared to the respective binomial distribution (Fig. 7B). Because the EPSCs generated at a single release site have a peaky amplitude distribution, the maxima of first and higher orders superimpose coherently.

Postsynaptic Densities and Fluctuations of Unitary EPSCs

So far it has been assumed that the contribution of the size of the PSDs to the shape of multipeak unitary EPSC histograms is almost negligible, since the distributions of quantal events derived from fits to unitary EPSC histograms were assumed to be Gaussian or skewed Gaussian but always had a single peak. The peaks of EPSC histograms (or of the fitted distributions) thus reflected the means of unitary EPSCs of different quantal content. The MF-CA3 synapses have several spatially well separated presynaptic active zones and associated PSDs of different areas, presumably containing different numbers of GluR channels. In conditions where the GluR binding sites of a PSD are almost or completely saturated following release of glutamate from a single vesicle, this source of variability of quantal events could dominate the peaks of unitary EPSC histograms. The possible effect of PSD sizes on mEPSC and unitary EPSC histograms was therefore simulated based on electron microscopic reconstructions of MF-CA3 synapses (7).

Number of Receptors Clustered in a PSD

The lower limit of the number of available GluR channels in a PSD has been estimated from the average amplitude of quantal events as 20–100 (5). The average area of a PSD in CA3 pyramidal cell spines is 0.2 μm^2 (7). If the postsynaptic GluR channels are localized exclusively in the PSD, this suggests a lower limit of the channel density in a PSD of 100–500 channels μm^{-2} (i.e., higher than in the somatic membrane (28)).

Simulation of mEPSC Distributions

The PSD areas of a fully reconstructed MF-CA3 synapse determined by Chicurel and Harris (7) were used to simulate the expected distribution of mEPSCs generated at this synapse, assuming a channel density of 100 μm^{-2} (Fig. 8A). In the simulated mEPSC distribution, three separate peaks are visible that reflect the different areas of the PSDs, two of which might be resolved experimentally (Fig. 8B). The simulated mEPSC distribution arising from a total of three synapses reconstructed from Chicurel and Harris no longer shows clear peaks, but is characterized by a skewed shape. The experimentally determined mEPSC histograms (Fig. 8C) and simulated mEPSC distributions (Fig. 8D) superimpose for mEPSC amplitudes up to about 30 pA, but the larger amplitude events observed experimentally are not present in the simulated distribution (Fig. 8D). This might be due to the fact that the simulation was based only on a small sample of reconstructed synapses.

Simulation of Unitary EPSC Distributions

We have also simulated unitary EPSC amplitude distributions with the anatomical data used for the simulation of mEPSC distributions and assuming the standard release mechanism. Regularly spaced peaks are observed in the simulated unitary EPSC amplitude distribution (Fig. 9A, continuous line). The peaks do not, however, represent simply the means of simultaneous superpositions of different numbers of quantal events (i.e., unitary EPSCs of different quantal content) as can be seen by comparing these peaks (Fig. 9A, solid line) with the distributions of unitary EPSCs of different quantal content (Fig. 9A, broken lines, indicated by Q1 to Q5). The simulated unitary EPSC distribution can be fitted reasonably well with a sum of Gaussian functions (as is done often for experimental unitary EPSC distributions (5)), and the first four Gaussian components (G1 to G4) were equidistant (Fig. 9B,

FIG. 8. Simulated miniature excitatory postsynaptic current (mEPSC) amplitude distributions based on postsynaptic density (PSD) area measurements. **A.** Schematic representation of the reconstruction of six PSDs of a single MF-CA3 synapse from Chicurel and Harris (ref. 7). The lines represent the geometry of a single thorn composed of six spine heads each carrying a PSD (circles). The areas of the circles are proportional to the respective areas of the measured PSDs, with stippled circle representing apical dendrite (modified with permission from Chicurel and Harris, ref. 7). **B.** Simulated mEPSC amplitude distribution for the MF-CA3 synapse shown in **A.** It is assumed that at each presynaptic active zone opposite a PSD, one vesicle can be released, and that the release probabilities at all release sites are the same and no multiquantal events are occurring. A channel density of 100 glutamate receptors (GluRs) μm^{-2} for the PSD is assumed. The single channel current is 0.62 pA and its open probability during the EPSC peak is 0.7. An instrumental noise of 0.62 pA is assumed. The amplitude distribution of quantal events (mEPSCs) is represented by the continuous line, and the distributions of the events generated at different PSDs are shown by broken lines and are numbered according to **A. C.** Experimental mEPSC histogram as published by Jonas (ref. 33). The histogram is given together with the approximate expected standard deviation of the probability for each bin (thin line). **D.** Simulation of mEPSC amplitude distribution (continuous line) expected for mEPSCs generated at three synapses reconstructed by Chicurel and Harris (ref. 7). The mEPSC distributions generated by the three individual synapses are indicated by long dashes (twelve PSDs), short dashes (six PSDs), and dots (one PSD). To simulate mEPSC distributions, the same parameters as detailed in **B** were used. Experimental mEPSC histogram as in Fig. 8 is shown for comparison.

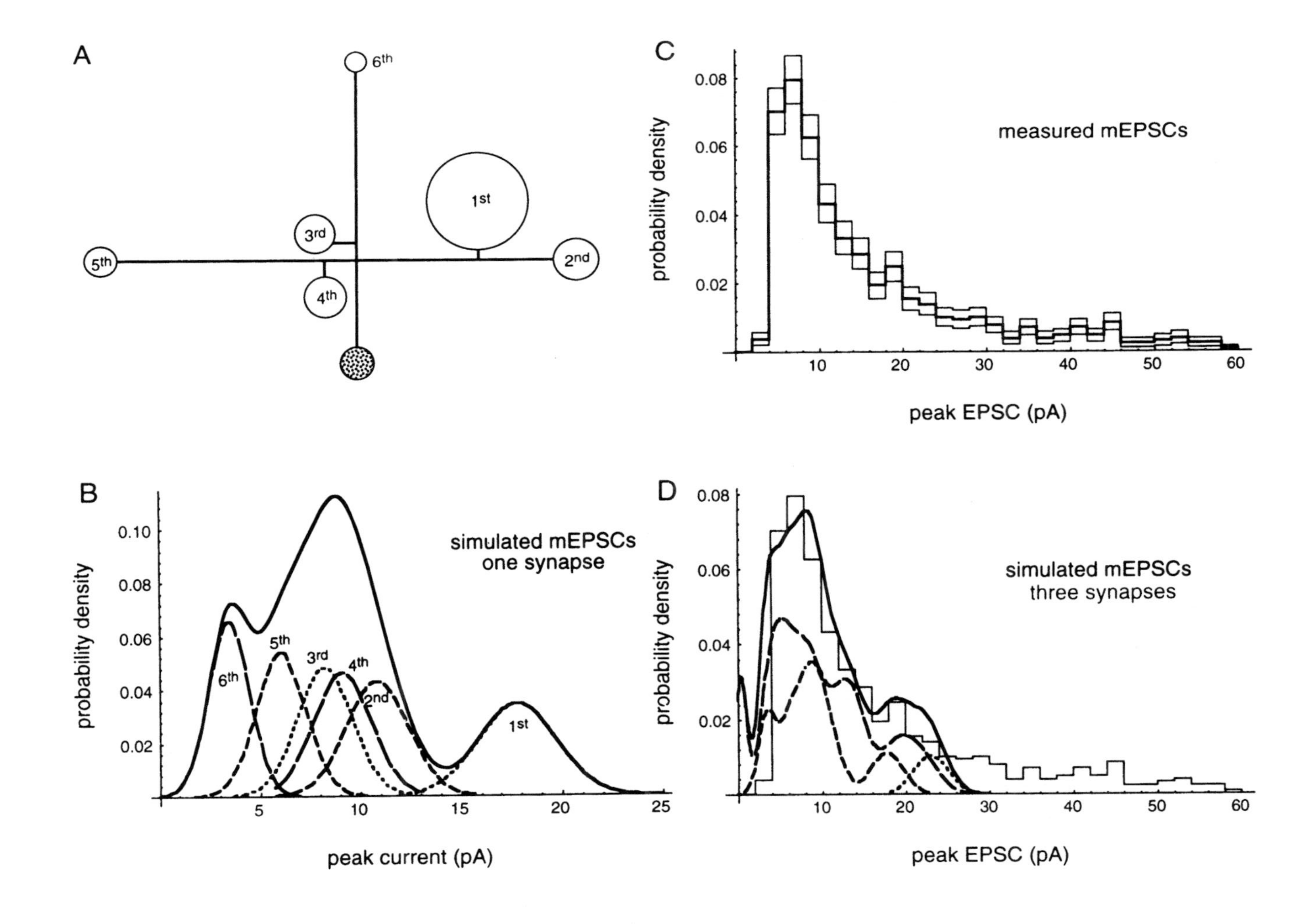
A
6th
1st
3rd
5th
2nd
4th
B
probability density
simulated mEPSCs
one synapse
6th
5th
3rd
4th
2nd
1st
peak current (pA)
C
probability density
measured mEPSCs
peak EPSC (pA)
D
probability density
simulated mEPSCs
three synapses
peak EPSC (pA)

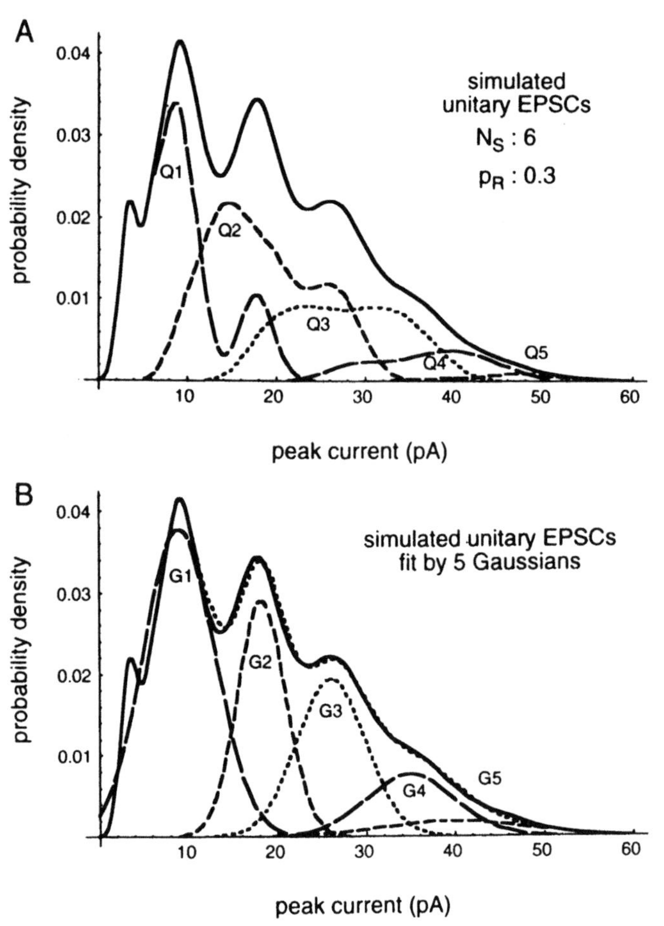

FIG. 9. Simulated unitary excitatory postsynaptic current (EPSC) amplitude distribution based on postsynaptic density (PSD) area measurements. **A.** Simulated unitary EPSC distribution (solid line) expected for a MF-CA3 synapse with six active zones facing PSDs as illustrated in **Fig. 8A.** The standard binomial release model was used, with a release probability of 0.3 at each release site. It is assumed that each active zone has one release site. Alternatively, if each active zone has several release sites the content of one vesicle is assumed to saturate all glutamate receptor binding sites of a PSD. The amplitude distribution of quantal events is shown by broken line labeled Q1. The distributions of unitary EPSC amplitudes expected from the superposition of two to five quantal events (i.e., unitary EPSCs with quantal contents of two to five) are indicated by broken lines labeled Q2 to Q5. Unitary EPSCs with a quantal content of six were very rare and are not shown. **B.** Simulated unitary EPSC distribution shown in **A** (solid line) fitted by a sum (dotted line close to the continuous line) of five Gaussian distributions (broken lines, labeled G1 to G5). The mutual nearest neighbor distances of the first four Gaussians are almost equal, with a mean value of 8.6 pA. The normalized CV_qs for the Gaussians fitted to the first four peaks are 43%, 21%, 25%, and 28%.

broken lines). On the other hand, the variance for the Gaussian G_n did not follow the expected relation n $V_q + V_B$, where V_q is the variance of the quantal event and V_B the variance of the background noise. A comparison between the fitted Gaussian distributions in Fig. 9B and the underlying distributions of unitary EPSCs of different quantal content in Fig. 9A reveals that the two are largely unrelated (the width of Q_n is in most cases larger than the width of G_n). This suggests that the interpretation of the unitary EPSC histograms in terms of sums of Gaussian components may be misleading in synapses where the PSDs opposite different active zones contain different numbers of GluR channels.

It has been found that the width of Gaussians in fitted EPSC amplitude histograms remains virtually constant with increasing peak order (29) or even decreases as the release probability increases (e.g., see Jack et al., Chapter 18). The variation of the number of GluR channels in different PSDs of a synapse could provide an explanation for this observation. Each PSD contributes to the distribution of quantal events. The contribution of each PSD is characterized by a mean current $\langle c_n \rangle$ and a variance V_{cn}. The variance V_1 of the quantal events depends on the average $\langle V_q \rangle$ of the individual variances V_{ci} and the variance $V_{\langle q \rangle}$ of the individual mean currents $\langle c_n \rangle$:

$$V_1 = \langle V_q \rangle + V_{\langle q \rangle}$$

If the synapse contains N_D postsynaptic densities, the amplitude distribution of N_D simultaneous quantal events has the variance V_N of

$$V_N = N_D \cdot \langle V_q \rangle$$

For large variations in the sizes of the PSD areas, $N_D \cdot \langle V_q \rangle$ may be considerably smaller than $V_{\langle q \rangle}$, leading to $V_1 > V_N$. Such behavior is simulated for the unitary EPSC distribution shown in Fig. 10. It is obvious that the width of the distribution of EPSCs with a quantal content of six (broken line labeled by Q6 in Fig. 10) is smaller than the width of the distribution of monoquantal events (broken line labeled by Q1). The variances of the simulated multiquantal EPSCs relative to the monoquantal events are 1.63 for the EPSCs generated by double events, and 1.91, 1.83, 1.38, and 0.57 for the EPSCs due to release of three to six quanta, respectively.

DISCUSSION

We have attempted to derive the parameters of quantal release at the MF-CA3 synapse, the mean amplitude of quantal events $\langle q \rangle$ the quantal variability CV_q, the number of release sites N_S and the release probability p_R from measurements of evoked unitary EPSCs at the MF-CA3 synapse. Because the spontaneous mEPSCs from a single MF-bouton could not be identified, we have attempted to derive both the average amplitude and variability of quantal events as well as the release parameters from peaky histograms of unitary EPSC amplitudes.

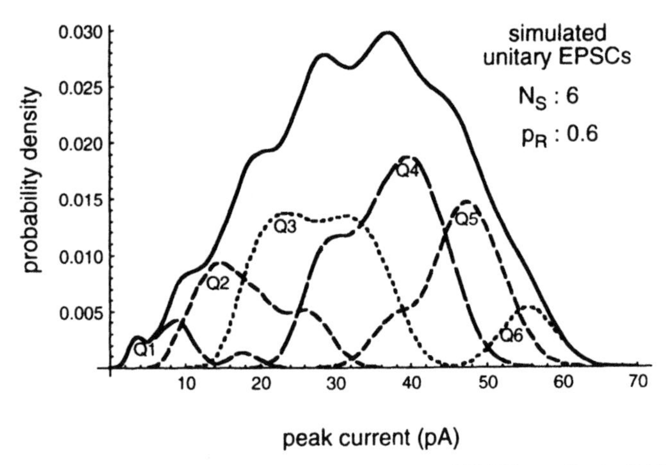

FIG. 10. Simulated unitary excitatory postsynaptic current (EPSC) amplitude distribution based on postsynaptic density (PSD) area measurements assuming high release probability (continuous line). The standard release model was assumed ($p_R = 0.6$). The other parameters are the same as in Fig. 8B and Fig. 9A. Broken lines represent amplitude distribution of quantal events (line labeled Q1) and distributions of unitary EPSCs generated by superposition of two to six quantal events (lines labeled Q2 to Q6). The higher p_R was chosen to make the distribution generated by superposition of six quantal events visible.

Quantal Events at MF-CA3 Synapses

The amplitude distribution of mEPSCs is strongly skewed, which probably reflects biological sources more than inadequate space-clamp conditions. The skewness most likely reflects sampling of mEPSCs from many excitatory synapses on CA3 pyramidal cells, as well as intrinsic variability of mEPSCs generated at individual synapses. Nevertheless, the mean amplitude of mEPSCs selected for short rise times indicates that the conductance increase underlying a quantal event at the MF-CA3 cell synapse is small, on the order of a few hundred picosiemens. This is comparable to the quantal conductance increase derived from the peak separation of unitary EPSC histograms and reflects opening of relatively few GluR channels. The variability in amplitude of quantal events derived from unitary EPSCs is much smaller than that of mEPSCs, at least in the subset of synapses analyzed here. As expected, the best fits of skewed quantal distributions and release models to multipeak EPSC histograms yielded quantal event distributions with larger variability than those derived by fitting sums of independent Gaussians (mean $CV_q = 30\%$ for skewed Gaussian distributions versus 22% for independent Gaussians). If saturation of the binding sites of GluR channels in the PSD is assumed to occur during the peak of EPSCs, then these relatively small CV_qs of quantal event amplitudes could be due primarily to different numbers of GluRs in the PSDs opposite different active zones of a single MF-CA3 synapse.

Release Mechanisms at MF-CA3 Synapses

The shape of the multipeak EPSC histograms in one MF-CA3 synapse could be adequately described by the standard model or a nonuniform release model. In this experiment the estimated number of release sites was eight. The multipeak EPSC histograms from four other synapses showed a pronounced tail at larger amplitudes that is not accounted for by these models. The assumption of nonstationary release, where the release probabilities at the release sites of a single bouton vary in time, however, accounts for these EPSC histograms. The derived number of release sites N_S in the 5 MF-CA3 synapses studied varied between 8 and 21 (mean = 12). This is in the range of the number of presynaptic active zones and PSDs measured morphologically (range between 1 and 37; mean = 7 ± 4 PSDs in synapses with intermediate size thorns (7)). Therefore, the assumption that the release parameter N_S derived from quantal analysis of unitary EPSCs reflects the number of presynaptic active zones of a single MF-bouton could hold for MF-CA3 synapses, implying the validity of the "one vesicle hypothesis " (3,19,20). On the other hand, N_S could also reflect the number of separate PSDs of a single thorn. Saturation of GluR binding sites following glutamate release from a single vesicle would imply that an individual active zone of an MF-bouton and the corresponding PSD of the thorn act as an "all or nothing" unit during synaptic transmission. Even if several vesicles could be liberated from a single active zone, only a single quantal event would result. The functional consequences of this assumption (e.g., with respect to the mechanisms that change the quantal content of evoked EPSCs) would be similar to those of the "one vesicle hypothesis " (3,19,20).

In the quantal analysis just described, it was assumed that multiple peaks in histograms of unitary EPSC amplitudes are the result of statistical superposition of quantal events of rather uniform size. On the other hand, a nonuniformity in the amplitude distribution of quantal events might be expected from the anatomy of PSDs, and this nonuniformity may contribute significantly to both the position and width of individual peaks in unitary EPSC distributions. Therefore, the usefulness of multipeak unitary EPSC histograms for deriving the amplitude and variability of quantal events for derivation of release parameters may be rather limited if the distribution of quantal events of the same bouton is not known from independent measurements.

Given these uncertainties in the interpretation of multipeak unitary EPSC histograms, EPSC histograms that do not show several peaks cannot be assumed a priori to be generated by any particular release mechanism (e.g., following standard (binomial) or Poisson release statistics) (26,30,31). The indirect method of quantal analysis (CV^{-2} versus mean current analysis) based on such histograms may therefore be incorrect (17) and thus not very useful for locating unequivocally the pre- or postsynaptic site of the change in synaptic transmission that underlies, for example, long-term potentiation (LTP).

Outlook

The conclusion from the attempted quantal analysis of unitary EPSCs at MF-CA3 synapses is that, for a more rigorous characterization of the quantal event and for the derivation of the factors determining the quantal content of EPSCs, additional experiments will be required. Simultaneous recordings from the apical dendrite and the soma of the same CA3 pyramidal cell (32) could be useful to obtain an accurate estimate of space-clamp errors and thus estimates of the number of open channels that generate a quantal event in each particular experiment. Electrical depolarization and stimulation of single MF-boutons would facilitate determination of the amplitude distribution of mEPSCs and unitary EPSCs generated by the same bouton thorn complex. This may be possible using patch pipette recordings from MF-boutons, which has recently been achieved (N. Spruston, unpublished). Finally, accurate estimates of the number of glutamate molecules contained in synaptic vesicles and of the density of GluR channels in PSDs could resolve the issue of whether or not glutamate released from a single vesicle saturates the binding sites of the different GluR channel subtypes present in PSDs.

SUMMARY

Unitary excitatory postsynaptic current (EPSC) amplitude histograms obtained from synapses between mossy fibers (MF) and CA3 pyramidal cells in hippocampus were analyzed with the aim of deriving the mean peak current $\langle q \rangle$ of quantal events, its coefficient of variation CV_q, the number of release sites N_S, and the release probability p_R. Unitary EPSC histograms with multiple peaks were fitted satisfactorily under the assumption that quantal events have a slightly skewed amplitude distribution and that the release mechanism is described by models with either binomial (standard), nonuniform, or nonstationary statistics. The average peak current of the quantal event derived from these fits reflected the opening of between 15 and 65 glutamate receptor (GluR) channels of the AMPAR subtype. The variability in the amplitude of quantal events is characterized by a CV_q of 25% to 30%. The best fits to multiple peak EPSC histograms are obtained if it is assumed that the number of release sites contained within a single MF-bouton is between 8 and 21, a value comparable to the published number of morphologically measured presynpatic active zones and postsynaptic densities (PSDs) of individual MF-CA3 synapses. This suggests that at the MF-CA3 synapse each presynaptic zone and its associated PSD act as independent units, contributing one quantal event to unitary EPSCs. Simulations of unitary EPSC distributions, based on measurements of PSD size could, however, indicate that the interpretation of multipeak EPSC amplitude distributions at MF-CA3 synapses in terms of fluctuating release probabilities might be an oversimplification, and that postsynaptic factors may contribute significantly to fluctuations of unitary EPSC amplitudes.

REFERENCES

1. Stevens CF. Quantal release of neurotransmitter and long-term potentiation. *Cell* 1993;72(suppl): 55–63.
2. Katz B. *The release of neural transmitter substances*. Liverpool: Liverpool University Press; 1969.
3. Korn H, Faber DS. Regulation and significance of probabilistic release mechanisms at central synapses. In: Edelmann GM, Gall WE, Cowan WM, eds. *Synaptic function*. Chichester: Wiley, 1987;57–108.
4. Redman S. Quantal analysis of synaptic potentials in neurons of the central nervous system. *Physiol Rev* 1990;70:165–198.
5. Jonas P, Major G, Sakmann B. Quantal components of unitary EPSCs at the mossy fibre synapse on CA3 pyramidal cells of rat hippocampus. *J Physiol* 1993;472:615–663.
6. Andersen P. Synaptic integration in hippocampal CA1 pyramides. *Prog Brain Res* 1990;83:215–222.
7. Chicurel ME, Harris KM. Three-dimensional analysis of the structure and composition of CA3 branched dendritic spines and their synaptic relationships with mossy fiber boutons in the rat hippocampus. *J Comp Neurol* 1992;325:169–182.
8. Amaral DG, Ishizuka N, Claiborne B. Neurons, numbers and the hippocampal network. *Prog Brain Res* 1990;83:1–11.
9. Johnston D, Brown TH. Interpretation of voltage-clamp measurements in hippocampal neurons. *J Neurophysiol* 1983;50:464–486.
10. Stuart GJ, Dodt H-U, Sakmann B. Patch-clamp recordings from the soma and dendrites of neurons in brain slices using infrared video microscopy. *Pflügers Archiv* 1993;423:511–518.
11. Edwards FA, Konnerth A, Sakmann B. Quantal analysis of inhibitory synaptic transmission in the dentate gyrus of rat hippocampal slices: a patch-clamp study. *J Physiol* 1990;430:213–249.
12. Ropert N, Miles R, Korn H. Characteristics of miniature inhibitory postsynaptic currents in CA1 pyramidal neurones of rat hippocampus. *J Physiol* 1990;428:707–722.
13. Silver RA, Traynelis SF, Cull-Candy SG. Rapid-time-course miniature and evoked excitatory currents at cerebellar synapses *in situ*. *Nature* 1992;355:163–166.
14. Gaiarsa JL, Beaudoin M, Ben-Ari Y. Effect of neonatal degranulation on the morphological development of rat CA3 pyramidal neurons: inductive role of mossy fibers on the formation of thorny excrescences. *J Comp Neurol* 1992;321:612–625.
15. Stern P, Edwards FA, Sakmann B. Fast and slow components of unitary EPSCs on stellate cells elicited by focal stimulation in slices of rat visual cortex. *J Physiol* 1992;449:247–278.
16. Traynelis SF, Silver RA, Cull-Candy SG. Estimated conductance of glutamate receptor channels activated during EPSCs at the cerebellar mossy fiber-granule cell synapse. *Neuron* 1993;11:279–289.
17. Faber DS, Korn H. Applicability of the coefficient of variation method for analyzing synaptic plasticity. *Biophys J* 1991;60:1288–1294.
18. Korn H, Fassnacht C, Faber DS. Is maintenance of LTP presynaptic? *Nature* 1991;350:282.
19. Faber DS, Korn H. Binary mode of transmitter release at central synapses. *Trends Neurosci* 1982;5: 157–159.
20. Korn H, Mallet A, Triller A, Faber DS. Transmission at a central inhibitory synapse. (2) Quantal description of release with a physical correlate for binomial n. *J Neurophysiol* 1982;48:679–707.
21. Bekkers JM, Richerson GB, Stevens CF. Origin of variability in quantal size in cultured hippocampal neurons and hippocampal slices. *Proc Natl Acad Sci USA* 1990;87:5359–5362.
22. Lisman JE, Harris KM. Quantal analysis and synaptic anatomy—integrating two views of hippocampal plasticity. *Trends Neurosci* 1993;16:141–147.
23. Faber DS, Young WS, Legendre P, Korn H. Intrinsic quantal variability due to stochastic properties of receptor-transmitter interactions. *Science* 1992;258:1494–1498.
24. Cull-Candy SG, Usowicz MM. Multiple-conductance channels activated by excitatory amino acids in cerebellar neurons. *Nature* 1987;325:525–528.
25. Jahr CE, Stevens CF. Glutamate activates multiple single channel conductances in hippocampal neurons. *Nature* 1987;325:522–525.
26. Malinow R. Transmission between pairs of hippocampal slice neurons: quantal levels, oscillations, and LTP. *Science* 1991;252:722–724.

27. Melamed N, Helm PJ, Rahamimoff R. Confocal microscopy reveals coordinated calcium fluctuations and oscillations in synaptic boutons. *J Neurosci* 1993;13:632–649.
28. Colquhoun D, Jonas P, Sakmann B. Action of brief pulses of glutamate on AMPA/kainate receptors in patches from different neurones of rat hippocampal slices. *J Physiol* 1992;458:261–287.
29. Larkman A, Stratford K, Jack J. Quantal analysis of excitatory synaptic action and depression in hippocampal slices. *Nature* 1991;350:344–347.
30. Bekkers JM, Stevens CF. Presynaptic mechanism for long-term potentiation in the hippocampus. *Nature* 1990;346:724–729.
31. Malinow R, Tsien RW. Presynaptic enhancement shown by whole-cell recordings of long-term potentiation in hippocampal slices. *Nature* 1990;346:177–180.
32. Stuart GJ, Sakmann B. Active propagation of somatic action potentials into neocortical pyramidal cell dendrites. *Nature* 1994;367:69–72.
33. Jonas P, Sakmann B. Glutamate receptor channels in isolated patches from CA1 and CA3 pyramidal cells of rat hippocampal slices. *J Physiol* 1992;455:143–171.

Molecular and Cellular Mechanisms of Neurotransmitter Release, edited by
Lennart Stjärne, Paul Greengard, Sten Grillner,
Tomas Hökfelt, and David Ottoson,
Raven Press, Ltd., New York © 1994.

17

The Nature of Quantal Transmission at Central Excitatory Synapses

John M. Bekkers and *Charles F. Stevens

*Department of Neuroscience, John Curtin School of Medical Research, Canberra ACT 2601 Australia; and *Department of Molecular Neurobiology, The Salk Institute, La Jolla, California 92037*

Although the quantal hypothesis (1) is well established for synaptic transmission at the neuromuscular junction, the applicability of the Katz statistical theory for transmitter release to central neurons remains controversial (2). We have been attempting to test the extent to which the Katz formalism is suitable for describing synaptic transmission at excitatory synapses on hippocampal neurons.

The following presentation considers properties of quanta recorded from central neurons, then examines the adequacy of the Katz theory for central synaptic transmission. We shall argue that quanta at individual synapses vary markedly in size, and that the Katz theory, usually in the Poisson limit, provides an accurate description of transmission, but only if the quantal variability is taken into account.

CHARACTERIZATION OF CENTRAL QUANTA

We feel that any quantal analysis attempted without specific information about the nature of the quantum is likely to be misleading. Because a detailed characterization of quantal properties—that is, a description of the elementary synaptic current here called the "mini"—is, of course, crucial for testing the adequacy of any theory of synaptic transmission, we have measured directly the size and time course of minis for excitatory central synapses (3,4). Minis are defined as the spontaneous and randomly occurring synaptic currents that are observed after action potential production has been blocked.

Minis are difficult to study in central neurons because they arise at random times and at various dendritic locations. The problem is that minis produced at synapses distant from the somatic recording site are attenuated by cable properties of the dendritic tree, so that any inherent mini size variations are difficult to disentangle from size fluctuations due to site of origin. Because spontaneous synaptic releases occur at locations throughout the dendritic tree, one would always record minis of different sizes with a somatic electrode. But without independent knowledge of the underlying time course, it is impossible to decide if a particular mini is small because it originated at a relatively distant site or small because the conductance change responsible is small (and slow). The study of spontaneous minis, then, can by itself never define the time course and inherent size distribution. To circumvent this difficulty, we have used a method (see description that follows) that increases the rate of mini release locally so that the effects of cable filtering are not a factor in determining the variability in mini size and shape. We find that mini amplitude is quite variable even when the source is at a constant electronic distance from the recording site, and that the ratio of the smallest mini to the largest in a sample of 100 is likely to be 10:1. An example of a mini size distribution recorded from a single anatomically identified synaptic contact is illustrated in Fig. 1A, and the distribution from a population of synaptic contact in Fig. 2. Thus, if one carried out a quantal analysis with the arbitrary assumptions that minis have a constant size, and that all of the size variations in evoked synaptic currents result from probabilistic release mechanisms, the conclusions could be seriously in error.

In the following, we outline our characterization of mini size variations. Because cortical synaptic density is about $10^9/\text{mm}^3$ (see ref. 5), the study of synapses at specific dendritic locations in brain is difficult. We therefore turned initially to an investigation of synapses in culture where synaptic sites are arrayed at low density in two dimensions. We are aware, of course, that any conclusions derived from study of cultured neurons must be confirmed in brain slices.

The method we have used to increase mini frequency at a defined position in the dendritic tree is the local application of hypertonic solution. Fatt and Katz (6) noted that hypertonic solution produces an increase in the rate of mini production, and Hartline (7) demonstrated that the effect is, at the frog neuromuscular junction, a local one. That is, local application of hypertonic solution increases mini frequency only very near the point at which the hypertonic solution is applied. The mechanism of this effect is still not understood (Chap. 24; Brosius et al., ref. 8) but we have found that it works for central synapses as well as at the neuromuscular junction. When a region of dendrite is locally bathed with hypertonic solution, the rate of mini production increases only if the region of microperfusion possesses synaptic contacts (3).This method has been described in detail in Bekkers and Stevens (9).

The experiments just cited demonstrate that mini release occurs only if the microperfusion area includes at least one synaptic bouton; the method has a spatial resolution, in culture, of 10 to 15 μm. Hartline (7) showed that release is restricted to the immediate area at which hypertonic solution has been applied, and we have repeated his experiments for central neurons. The idea is to evoke release at different distances from the recording site and then to examine the extent to which cable attenuation (using the observed morphological properties of the neuron under study), can account for the changes in shape and size of the recorded mini currents.

We applied hypertonic solution to the cell body and to a dendrite at distances up

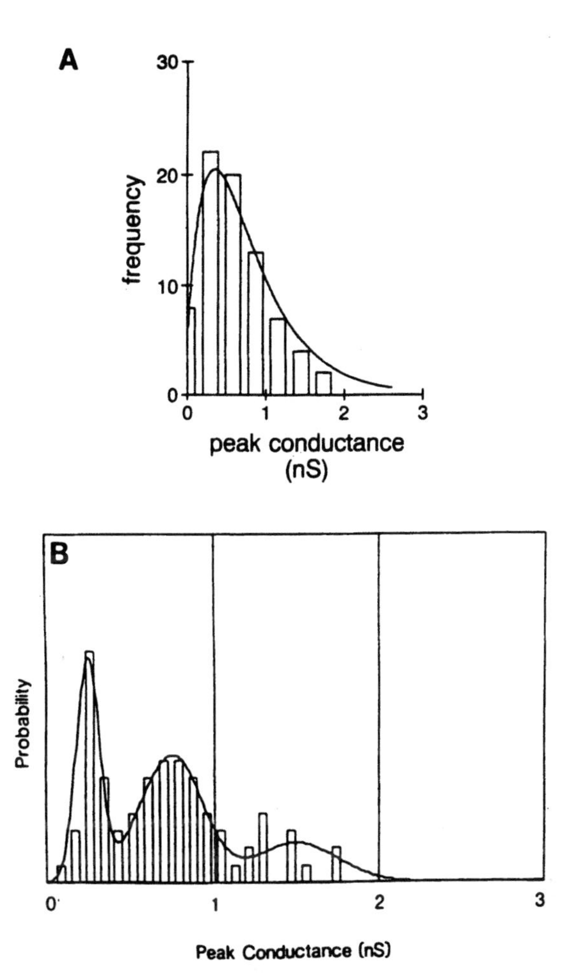

FIG. 1. Histograms estimating the probability density of peak mini conductances. Ordinate is relative frequency and abscissa is mini peak conductance. Data from a single histochemically identified synapse in hippocampal culture (Fig. 2 of Bekkers et al., ref. 4). Number of entries in histogram is 91. **A.** Histogram constructed with bin width of .25 ns **B.** Same data with bin width of .087 ns. The smooth curve is the sum of three Gaussian functions fitted by eye.

to 240 μm, and collected minis evoked from each microperfusion site (10).Further, we measured the dendritic diameter and the passive electrical properties of the neuron and found that the cable equation accurately predicted the observed currents at each location from the waveform of synaptic currents observed with application of hypertonic solution to the soma. If the effects of hypertonicity at one site had spread as much as 50 μm in a typical dendrite, which corresponds to about a 20% change in peak mini amplitude, we would have easily detected the effect with these experiments. Thus we conclude that the minis collected with application of hypertonic

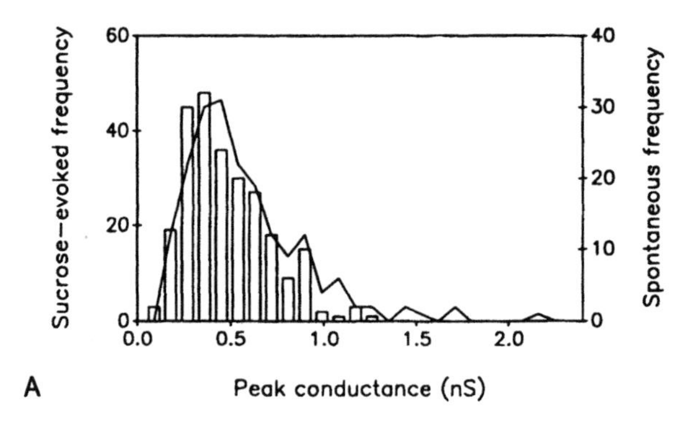

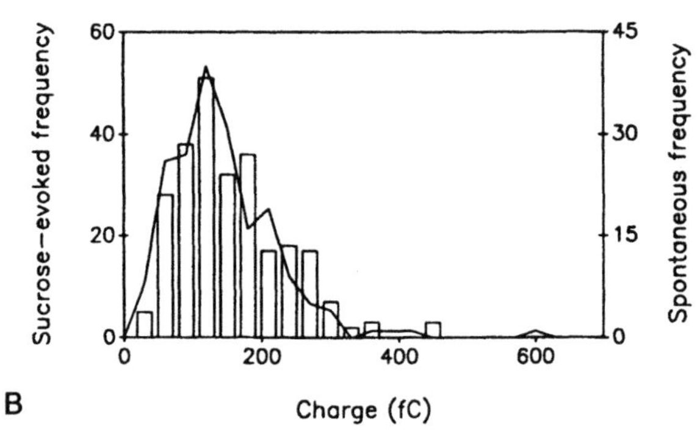

FIG. 2. Histograms estimating the probability density of mini peak amplitude taken from a population of synapses on a neuron in hippocampal culture. This neuron had been in culture for 13 days. External solution contained 10 mM Mg, 100 μM picrotoxin and 1 μM tetrodotoxin. The holding potential was −60 mV. The bar graphs present data from minis whose release was evoked by the application of hypertonic solution (.5 M sucrose in saline), and the line from spontaneous minis collected from the same neuron. **A.** Frequency as a function of peak conductance. The left ordinate is for minis whose release was evoked by application of hypertonic solution and the right ordinate for spontaneously occurring minis. **B.** The same data replotted for mini size measured by charge transfer (the integral of the mini current) rather than peak conductance.

solution at one site suffer essentially no size and shape variation that can be attributed to cable filtering in these experiments.

Minis that arise at one location exhibit marked size variations when the mini frequency is increased by focal application of hypertonic solution. In culture, for example, the average peak conductance of minis recorded without dendritic cable attenuation is about 1 ns, and the distribution of mini sizes is always skewed with a long tail of higher values and a coefficient of variation between .5 and .6 (4). Are these minis typical of the "normal" ones? Although minis evoked in this way at the frog neuromuscular junction do seem typical (7,11) other authors have reported that quantal size is altered by prolonged application of hypertonic solutions (e.g.,

Van der Kloot, 12). We therefore have examined this issue for our central synapses by comparing hypertonic solution-evoked minis with those that occurred spontaneously. As illustrated in Figure 2A, spontaneously occurring and hypertonic solution evoked minis have essentially the same amplitude distributions. Because minis that occur at remote locations are decreased in size when recorded by a somatic electrode, this agreement between evoked and spontaneously occurring minis might be fortuitous. We have therefore compared the size of the charge transferred per mini (e.g., the mini's integral). The idea is that, although the peak amplitude falls e-fold in about 170 μm (in a 1 μm diameter dendrite), the charge transfer recorded decays according to the DC length constant of the dendritic tree (see Bekkers and Stevens 10). In our cultured neurons, with dendritic trees in the range of 300 to 600 μm long, the DC length constant is 1000 μm (for a 1 μm diameter dendrite), so the dendritic trees should be reasonably compact electrically for slowly varying signals. Thus, we would expect the charge transfer per mini to be much less distorted by the cable properties of the dendritic tree than the peak current. When the charge transfer for spontaneous minis and those evoked by hypertonic solution are compared, the result again is that the distributions are essentially indistinguishable, as illustrated in Figure 2B. Thus, minis whose release is evoked by hypertonic solution are, under the conditions of our experiments, typical.

What is the source of the mini variability? One possibility is that each synapse has its own characteristic mini amplitude—this might vary little in size—but that mini size varies from synapse to synapse in a way that produces the mini amplitude distributions we have reported. The alternative to this population explanation of mini size variations is that mini amplitude fluctuates from one release to the next at individual synapses. Of course these possibilities are not mutually exclusive; the mini size variations we report could result from any mixture of these two extreme alternatives. To examine this question, we have recorded the distribution of mini sizes from small populations of boutons. The hypertonic solution was applied to specific locations, minis were collected, and then the boutons present in the area of application were stained with antibodies to synapsin and counted (4).Only locations at which discrete effects occurred were used for this analysis; that is, we required that the mini frequency decreased to essentially background levels when the hypertonic solution was applied to adjacent dendritic regions where no synapses were present.

The result of these experiments (4) was that we found no significant correlation between the variance of the mini size distributions and the number of boutons to which hypertonic solution was applied. If the mini sizes at an individual bouton had little variance, then we should have found histograms with discrete peaks, one for each bouton present. Furthermore, the specific shape of the mini size distribution should have varied from site to site; with only a few boutons present, minis could not have been represented with the same relative frequencies for each size category, as would have been the case for a large population. Although some synapse-to-synapse variation in mini size must occur, we conclude that the major source of variability is present at the individual synapse.

The central quantum, then, varies considerably in size from one release to the next at individual synapses. One can envision several mechanisms through which the size of one mini, or the time at which it occurred, might influence the size of the next mini. Suppose, for example, that larger, more "mature" quanta are released preferentially. Then large minis would tend to occur together, and smaller minis would be seen later in the series of quantal releases after the "mature" quanta had been exhausted. Another possibility is that significant desensitization is produced by each mini. Then, on average, the size of one mini would alter the size of successive ones because of accumulated desensitization. We have checked for independence in the size of successive minis by computing serial correlograms, and find no significant serial correlations. The correlogram in Fig. 3 presents data from the single, histochemically identified bouton that gave rise to the mini amplitude distribution presented in Fig. 1. Clearly, no serial correlations in mini size are evident that could account for the observed size distribution. We conclude, therefore, that the mini size variation is not the effect of one mini on the next.

For our present purposes, the precise source of mini variability is not important. We note, however, that the variability probably arises from a presynaptic mechanism because the response to focal iontophoretic application of glutamate does not fluctuate (Trussell, personal communication; see Trussell et al., 13). A clue to the possible mechanism is provided by the histogram presented in Fig. 1B that was derived from minis that were produced at a single bouton; the Fig. 1B histogram is based on the same data, with different bin size, that produced the Fig. 1A histogram. Clear peaks are evident in this histogram with a smaller bin size, and these peaks could be explained if the single quantum consists of various numbers of vesicular releases. For example, the first peak would correspond to one vesicle, the second to two, etc. Monte Carlo simulations have revealed, however, that the peaks evident in Fig. 1B could have arisen by chance and thus might represent a statistical artifact. Because we have not been able to study single boutons at will and find histograms with peaks like the one illustrated whenever we wish, we are uncertain if the peaks are real or merely represent a statistical artifact. Only when this phenomenon can be repeatedly produced and experimentally manipulated will the explanation for the mini size variability emerge.

To summarize: minis at individual synapses in culture vary considerably in size, probably because of differences in the quantity of neurotransmitter released, and the amount of variability at one synapse seems to be greater than the contributions between synapses. Tantalizing peaks appear sometimes in the mini size distributions, and these peaks could arise from a mechanism in which the mini represents multivesicular releases with variable numbers of vesicles from mini to mini.

Do minis have the same properties in the brain, or is the mini variability described above just an artifact of tissue culture? Unfortunately, repeating the experiments just described above seems impossible for brain slices because stimulation of defined numbers of synapses is so difficult when they are present in such high densities. Indeed, the difficulty of studying defined populations of synapses is what drove us to carrying out experiments in culture. Even though the precise experiments that we described cannot be carried out in brain slices, the conclusions we reached can be verified for brain by comparing predictions based on the culture-derived properties of minis with observations made in slices. Because we find that most of the size distribution of minis arises at individual synapses, and that the contributions of synapse-to-synapse size variation is generally small, then mini size distributions from a population of synapses in slices should have about the same size and shape as the distributions we have found in culture. Bekkers et al. (4) checked this prediction and found that it is accurate; that is, local application of hypertonic solution to the specific sites on the surface of brain slices produced minis with roughly the same time course and amplitude distribution as those observed in culture. Although we could not determine the size of the population of synapses that contributed to our mini sample in slices, the number was probably not very large. Raastad et al. (14) have reached a similar conclusion about the variability of central quanta based on analysis of synaptic currents evoked by "minimal" stimulation that activates only one or a few synapses.

Although roughly similar in time course and amplitude, minis recorded from neurons in slices are systematically slower (43%) and smaller (mean peak conductance of .21 rather than .81 ns) than those recorded from cultured neurons. At least part of this difference is the result of cable filtering, because application of hypertonic solution to cell bodies and the first 100 μm of apical dendrite did not produce minis in slices, so that the population of minis studied in slices would have arisen a distance of more than 100 μm from the recording site. Although cable filtering could account for the culture/slice differences we have found, we cannot exclude

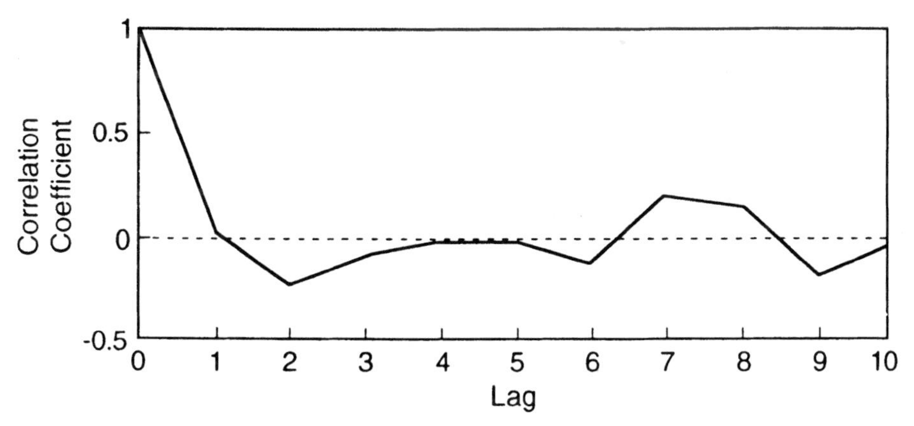

FIG. 3. Serial correlogram computed for the 91 minis from a single synapse (the size distributions of these minis are presented in Fig. 1). The ordinate is the serial correlation coefficient and the abscissa is the lag (i.e., number of intervening minis). The serial covariance function is computed from the equation

$$C(k) = \frac{1}{N} \sum_{j=1}^{N} g_j g_{j+k}$$

where $C(k)$ is the serial covariance with lag k, N is the number of minis in the sample, and g_j is the deviation of the peak conductance of the jth mini in the sample from the mean; the subscripts indicate the order in which the minis occurred. The serial correlation coefficient plotted on the ordinate is the ratio $C(k)/C(0)$.

the possibility that synapses in culture are, in some way (e.g., have higher receptor densities) systematically different from those in brain. The major conclusion, that mini size is quite variable, seems to hold for both culture and brain slices.

THE KATZ THEORY FOR CENTRAL SYNAPSES

The Katz theory for quantal release holds that evoked synaptic currents are sums of minis that occur at random according to coin flipping theory. That is, N release sites are available and a quantum of neurotransmitter (= a mini) is released at each site with a probability p. N is generally considered to be fixed over time, and p is assumed to have the same value for each release site. Although neither of these assumptions is probably true (15–17) the resulting formalism usually provides an accurate description of synaptic transmission at the neuromuscular junction. We have tested the Katz formalism, using the observed mini size distributions, for both cultured neurons and in slices and have found it to be satisfactory in the sense that the observed distributions of evoked synaptic current amplitudes is not significantly different from the predictions of the theory. Specifically, we have supposed that the probability p_k of k quanta being released is (in the Poisson limit)

$$p_k = \frac{m^k}{k!} e^{-m}$$

where m is the mean number of quanta released for each synaptic activation. If the Poisson limit is inadequate, we use the binomial distribution for p_k. Note that, in the Poisson limit, no assumption is needed about the uniformity of release probabilities at different synapses. If $w_k(g)$ is the probability of finding a peak conductance g when k quanta are released, then

$$w_{k+1}(g) = \int_0^\infty w_k(g - g')w_1(g')dg'$$

where $w_l(g)$ is the observed probability density function for peak mini amplitude. Thus these functions are found by the appropriate n-fold convolution of the observed mini size histogram. The observed amplitude fluctuations $f(g)$ are then, according to this extension of the Katz theory, described by the equation

$$f(g) = \sum_{k=0}^{\infty} p_k w_k(g)$$

A relatively stringent test of the adequacy of this theory is illustrated in Fig. 4. Here we used the spontaneous mini distribution (the distribution of charge transfers—that is, the integral of synaptic currents—was used to minimize cable filtering effects) and collected synaptic currents at three different calcium concentrations (0.5, 0.75, and 1.0 mM). We have used the Dodge and Rahamimov (18) theory to determine how the quantal content varies with calcium concentration. Specifically, quantal content $m(c)$ is given by the Dodge-Rahamimov equation

$$m(c) = 3.6\left(\frac{c}{1 + \frac{r}{K_r} + \frac{c}{K_c}}\right)^4$$

where c is the calcium concentration (mM) and r is the magnesium concentration ($r = 10$ mM for all experiments); the magnesium dissociation constant K_r is 3 mM and the calcium dissociation constant K_c is .5 mM.

The fits illustrated—none are significantly different from the observed amplitude distributions—used the Poisson approximation just given without free parameters beyond the constants, whose values are indicated, estimated for the Dodge-Rahamimov equation. The dotted line for the 1 mM calcium concentration data in Fig. 4 is the best binomial fit with a number of release sites taken to be ten; the Poisson approximation was always superior to the binomial fit until the number of release sites was increased to the point that the two distributions effectively coincided. Clearly, the Katz theory is satisfactory for these synapses. Equally clearly, the broad distribution of mini sizes would preclude an accurate test of the Katz theory if the properties of minis were not known independently, especially for the lower calcium concentrations.

In summary, the Katz theory is, without exception for the synapses we have studied, adequate for describing central synaptic transmission. This is not to say that the theory is "correct," but merely that it provides an accurate quantitative characterization of the experimental data, usually with only a single parameter (because we generally found synapses had low enough release probabilities to be effectively at the Poisson limit). Our most detailed and complete analyses have been carried out in culture, so the extent to which the Katz theory provides an adequate description for slices requires further investigation. However adequate the Katz theory is at the level of precision usually achieved in electrophysiological experiments, we anticipate that a more complicated theory will ultimately be needed when more is understood about synaptic transmission; we expect that the present Katz theory will be seen to be an approximation, valid under certain circumstances, to the detailed theory that finally emerges. Indeed, Barrett and Stevens (19,20) have proposed extensions of the Katz theory that account for more aspects of synaptic transmission, but these extensions are certainly only crude approximations to the actual situation.

Although we are not confident that the peaks in the mini histogram are other than statistical artifacts, it is interesting to investigate the effect they would have on the evoked histogram were they actually present. If transmission at multiple synapses were studied simultaneously, and if no mechanism were present to ensure that the peaks occurred at precisely the same values of synaptic current for every synapse (the most common situation), then the peaks of one synapse's mini distribution would usually fall in the troughs of another's, so that no peaks would appear in the overall histogram of evoked release amplitudes. But if only one synapse were being stimulated, or if the population of synapses that were contributing to the synaptic currents happened to have the peaks of their distributions aligned, then peaks would

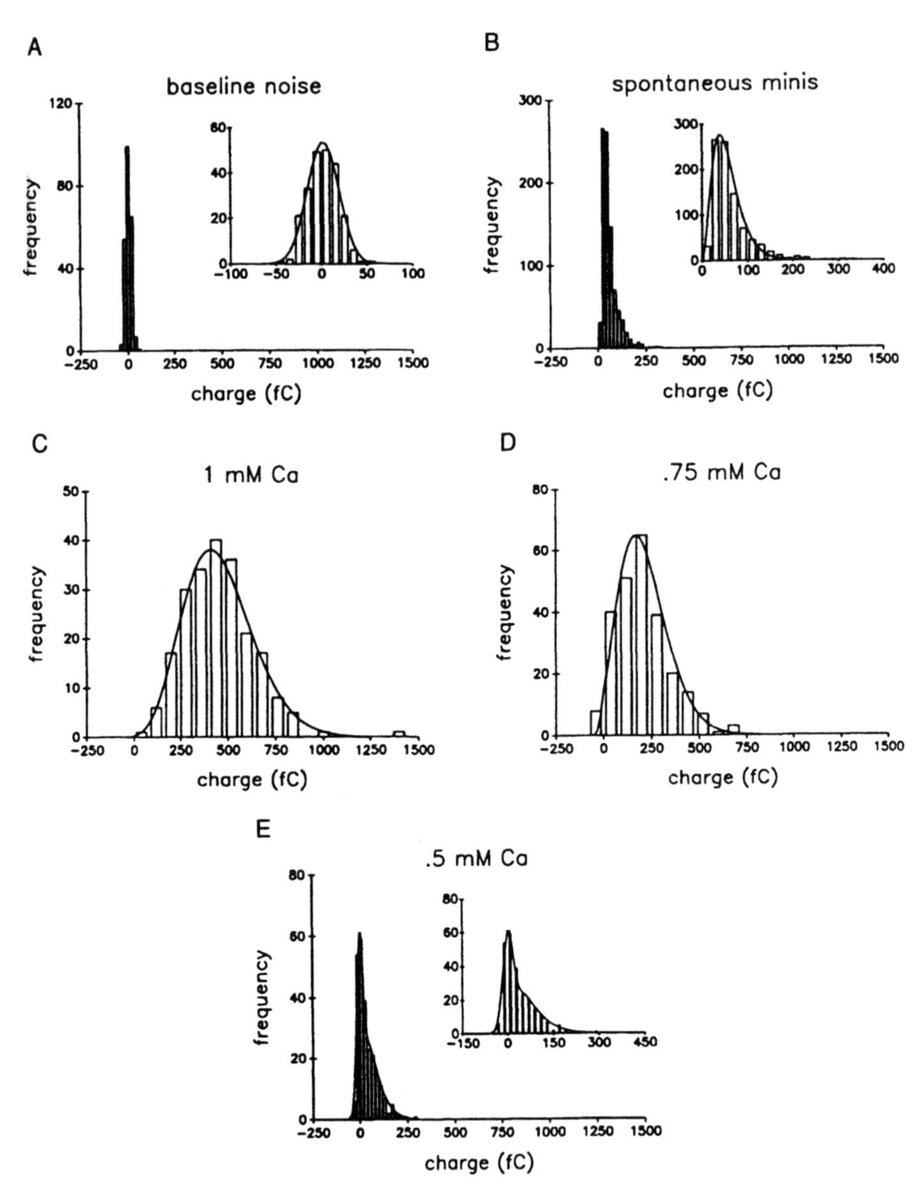

FIG. 4. Observed (bar graphs) and predicted (smooth lines) amplitude distributions for evoked excitatory synaptic currents. The ordinate specifies frequency and the abscissa mini size measured by charge transfer (integral of the mini currents). The upper pair of graphs illustrate the amplitude of the recording noise and the amplitude distribution of minis that occurred spontaneously. The bottom three histograms present evoked epsc data at three different calcium concentrations as indicated in the figure. The theoretical curves are obtained from the equations presented in the text using the illustrated mini size distribution.

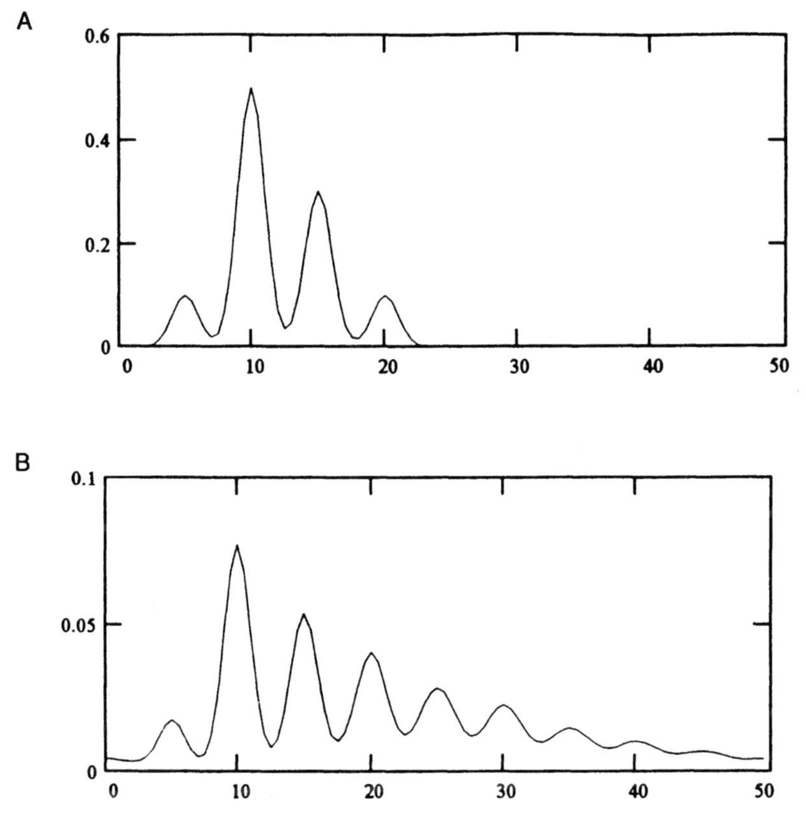

FIG. 5. Hypothetical mini distribution with four discrete subquantal peaks and the histogram of evoked peak synaptic conductance changes that would result from the Katz theory (using the equations presented in the text) from a synapse that has the illustrated hypothetical mini size distribution. Ordinate is relative frequency and abscissa is peak conductance measured in arbitrary units.

appear in the distribution of evoked synaptic current amplitudes (see Chap. 16). These peaks would not correspond to the usual quanta, however, but instead to the subquanta that make up the mini distribution. To appreciate this, suppose for the sake of illustration that the mini distribution had, as illustrated in Fig. 5A, four peaks, and further suppose that release occurred according to a Poisson process with a mean quantal content of $m = 3$. Then, assuming the Katz theory as described by the equations just mentioned, we find that the evoked distribution would be the one illustrated in Fig. 5B. Here, by hypotheses, Katz's theory is correct and the mini distribution has peaks like those described by Edwards et al. (21), Silver et al. (22), or illustrated in Fig. 1A. But if one were to take the peaks in the evoked release histogram as a measure of quantal size (and assume that release at a single synapse exhibits no variability), then Katz's theory would appear not to apply.

Several authors (refs this book) have found peaks in their histograms of evoked release amplitudes and have assumed that these peaks correspond to the release of individual quanta. If our analysis is accurate, then the incorrect identification of the quantum would make the Katz theory appear to be inapplicable under these circumstances even if it were, in fact, adequate with the correct identification of the quantum.

ACKNOWLEDGMENTS

This work was supported by National Institutes of Health Grant NS12961 and the Howard Hughes Medical Institute.

REFERENCES

1. Katz B. *The release of neural transmitter substances*. Liverpool: Liverpool University Press, 1969.
2. Edwards F. LTP is a long term problem. *Nature* 1991;350:271–272.
3. Bekkers JM, Stevens CF. NMDA and non-NMDA receptors are co-localized at individual excitatory synapses in cultured rat hippocampus. *Nature* 1989;341:230–233.
4. Bekkers JM, Richerson GB, Stevens CF. Origin of variability in quantal size in cultured hippocampal neurons and hippocampal slices. *Proc Natl Acad Sci USA* 1990;87:5359–5362.
5. Stevens CF. How cortical interconnectedness varies with network size. *Neural Computation* 1989;1: 473–479.
6. Fatt P, Katz B. Spontaneous subthreshold activity at motor nerve endings. *J Physiol Lond* 1952; 117:109–128.
7. Hartline FF. *Cable filtering and waveform variability of miniature end-plate currents at the frog neuromuscular junction*. Doctoral dissertation, University of Washington Medical School, 1978.
8. Brosius DC, Hackett JT, Tuttle JB. Ca^{2+}-independent and Ca^{2+}-dependent stimulation of neurosecretion in avian ciliary ganglion neurons. *J Neurophysiol* 1992;68:1229–1234.
9. Bekkers JM, Stevens CF. Osmotic stimulation of presynaptic terminals. In: Kettenmann H, Grantyn R, eds. *Practical electrophysiological methods*. New York: John Wiley & Sons: 1992:150–154.
10. Bekkers JM, Stevens CF. Two different ways evolution makes neurons larger. In: Storm-Mathisen J, Zimmer J, Ottersen OP, eds. *Understanding the brain through the hippocampus: the hippocampal region as a model for studying brain structure and function. Prog Brain Res*. Elsevier Science Publ, Norway. 1990;83:37–45.
11. Dionne VE, Stevens CF. Voltage dependence of agonist effectiveness at the frog neuromuscular junction: Resolution of a paradox. *J Physiol Lond* 1975;251:245–270.
12. Van der Kloot W. Pretreatment with hypertonic solutions increases quantal size at the frog neuromuscular junction. *J Neurophysiol* 1987;57:1536–1554.
13. Trussell LO, Thio LL, Zorumski CF, Fischbach GD. Rapid desensitization of glutamate receptors in vertebrate central neurons. *Proc Natl Acad Sci USA* 1988;85:2834–2838.
14. Raastad M, Storm JF, Andersen P. Putative single quantum and single fibre excitatory postsynaptic currents show similar amplitude range and variability in rat hippocampal slices. *Europ J Neurosci* 1992;4:113–117.
15. Ryan TA, Reuter H, Wendland B, Schweizer FE, Tsien RW, Smith SJ. The kinetics of synaptic vesicle recycling measured at single presynaptic boutons. *Neuron* 1993;11:713–724.
16. Rosenmund C, Clements JD, Westbrook GL. Nonuniform probability of glutamate release at a hippocampal synapse. *Science* 1993;262:754–757.
17. Hessler NA, Shirke AM, Malinow R. The probability of transmitter release at a mammalian central synapse. *Nature* 1993;366:569–572.
18. Dodge FA Jr, Rahaminoff R. Co-operative action of calcium ions in transmitter release at the neuromuscular junction. *J Physiol Lond* 1967;193:419–432.

19. Barrett EF, Stevens CF. The kinetics of transmitter release at the frog neuromuscular junction. *J. Physiol* 1972;277:691–708.
20. Barrett EF, Stevens CF. Quantal independence and uniformity of presynaptic release kinetics at the frog neuromuscular junction. *J. Physiol* 1972;227:665–689.
21. Edwards FA, Konnerth A, Sakmann B. Quantal analysis of inhibitory synaptic transmission in the dentate gyrus of rat hippocampal slices: a patch-clamp study. *J Physiol* 1990;430:213–249.
22. Silver RA, Traynelis SF, Cull-Candy SG. Rapid-time-course miniature and evoked excitatory currents at cerebellar synapses *in situ*. *Nature* 1992;355:163–166.

Molecular and Cellular Mechanisms of Neurotransmitter Release, edited by Lennart Stjärne, Paul Greengard, Sten Grillner, Tomas Hökfelt, and David Ottoson, Raven Press, Ltd., New York © 1994.

18

Quantal Analysis of the Synaptic Excitation of CA1 Hippocampal Pyramidal Cells

Julian J. B. Jack, Alan U. Larkman, Guy Major, and Ken J. Stratford

University Laboratory of Physiology, Oxford University, Oxford OX1 3PT, England

Although it would be a general assumption that synaptic excitation in the central nervous system is achieved by a quantal mechanism analogous to that described at the vertebrate neuromuscular junction, direct evidence has been of very recent origin. The key feature of a quantal process, as observed at the neuromuscular junction, is that an individual evoked response is made up of an integral number of units, with the exact number of units varying from trial to trial. Thus, the most convincing evidence for such a process is that an amplitude histogram, made up of a series of stimulus trials, shows clear, equally spaced peaks. In earlier work on the spinal cord, the histograms were not usually "peaky" (1,2) with the notable exception of synaptic excitation of dorsal spinocerebellar cells (3,4). The reason for this was thought to be the poor signal-to-noise ratio occurring with conventional intracellular recording in an in vivo preparation. The in vitro brain slice preparation offered the possibility of greatly reducing the background synaptic "noise," but early studies with intracellular recording in the hippocampal slice failed to reveal peaky histograms (5,6). The potential improvement of recording with the whole-cell patch method, which Edwards et al. (7) used to show "quantal" behavior for synaptic inhibition in the hippocampus, did not reveal equivalent peaks in the histograms for synaptic excitation (8–10). The study of presumed single quantal events evoked from a restricted region of the hippocampal cell (in culture and slice) by Bekkers and Stevens (9,11) led to the conclusion that there was substantial inherent variability in the size of quanta evoked successively from a single release site and thus to the expectation that evoked histograms would not show peaks.

With this background, it was a great surprise to us when we discovered that we could regularly find peaky histograms for synaptic excitation of CA1 pyramidal cells (2,12) particularly as we were using conventional sharp electrode recording. This led us to claim that we had demonstrated the quantal nature of excitatory transmission in this region of the hippocampus—a claim promptly challenged by

Clements (13). Since both the original publication (12) and the ensuing correspondence (13,14) were severely curtailed in length by the journal in which they appeared, it may be helpful if we set out our own forms of skepticism when first confronted with the experimental observations, including, of course, the grounds on which we overcame them.

ARE PEAKY HISTOGRAMS RELIABLE?

There are at least three reasons for being suspicious about the peaky histograms we reported. The first centers around the issue of reliability: are the observed peaks spurious, arising for example from finite sampling from an underlying smooth distribution? The second reinforces the first, because we used conventional intracellular electrode voltage recording rather than the currently fashionable whole cell-patch electrode in voltage clamp mode; the recording noise is expected to be greater with the former method and thus more likely to blur any peaks. The third reason is that we reported in 1991 that the best fits (with a maximum likelihood optimizing procedure) to our data histograms were obtained using zero or very low levels of quantal variance. This seemed completely at odds with the reports of substantial quantal variance measured from spontaneous events (9,11,15).

Finite Data Sampling

The key objection to the interpretation of our data is that it is possible, when sampling from a smooth distribution, to obtain histograms with spurious, roughly equally spaced peaks. This was exactly the point that Clements (13) raised and that we acknowledged to be correct (14). We had already been aware of this problem and were developing an extension to the autocorrelation method of Magleby and Miller (16) in order to check that this was not the explanation for the peaks we

FIG. 1. Reliability of peaky amplitude histograms. **A.** Histogram for all 250 recorded trials for an excitatory postsynaptic potential (EPSP) using a stimulation rate of 0.1 Hz. The histogram is binned conventionally using a bin width of 40 μV. This excitatory postsynaptic potential (EPSP) was used for Fig. 1 (Larkman et al., ref. 17). **B.** Same data as in **A,** but binned finely (bin width, 5 μV) and smoothed using a digital Gaussian filter. Using either display method, the histograms shown in **A** and **B** both show clear peaks, separated by about 280 μV. **C.** Result of autocorrelation procedure used to assess the likelihood that the peaks in **A** and **B** could have arisen by random sampling from a smooth distribution. We calculated the "autocorrelation score" for the histogram, which is a measure of its degree of peakiness and equality of peak spacing (Larkman et al., ref. 17; Stratford, ref. 24). We then drew a series of random samples, each of 250 trials, from a smooth Gaussian distribution with the same mean and standard deviation as the data histogram. We calculated the autocorrelation score for each of these samples. The curve shows the distribution of autocorrelation scores for 50,000 samples, and the vertical line indicates the score for the recorded data. Only 48 of the random samples gave scores as high as or higher than the recorded data. This indicates that the likelihood of a histogram as peaky as the data arising by sampling artifact from a smooth Gaussian distribution is only about 1 in 1,000.

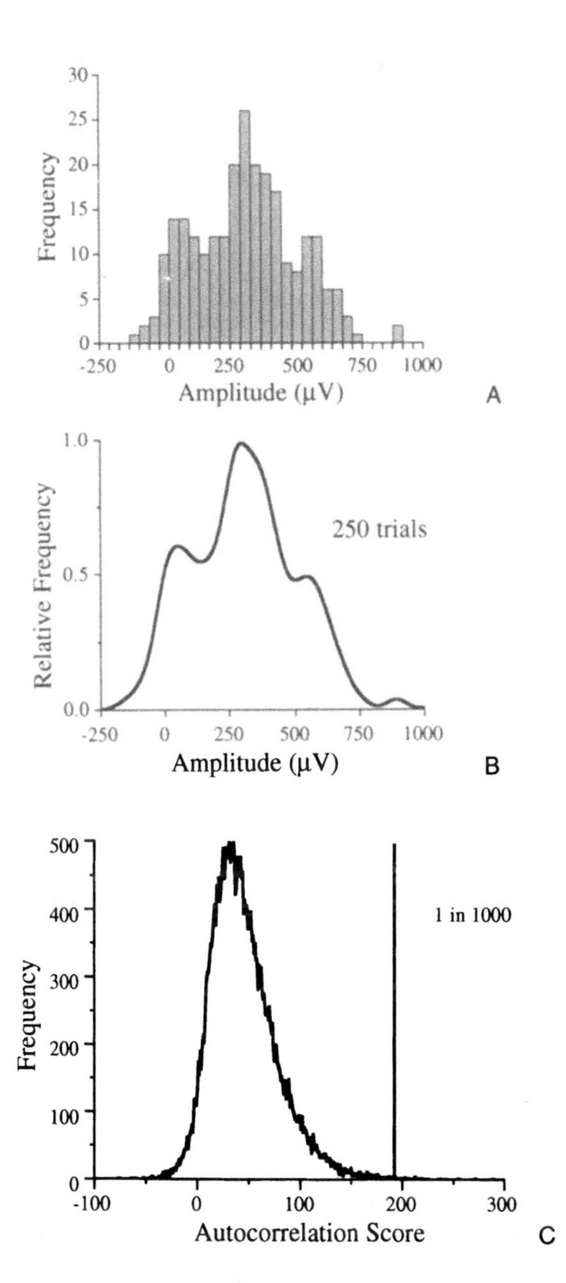

30
25
20
15
10
5
0
Frequency
-250
0
250
500
750
1000
Amplitude (μV)
A
1.0
0.5
0.0
Relative Frequency
250 trials
-250
0
250
500
750
1000
Amplitude (μV)
B
500
400
300
200
100
0
Frequency
1 in 1000
-100
0
100
200
300
Autocorrelation Score
C

observed (17,18). By using a Monte Carlo simulation technique we could give a numerical estimate of the likelihood that peaks arose as a finite sampling artifact. For some of our histograms this could be quantified as $p<0.001$, so that we were very confident that the observation of regularly spaced peaks was reliable (see Fig. 1).

We have been gratified that subsequently several groups, using their own methods of validation, have confirmed that it is possible to obtain peaky histograms for synaptic excitation in CA1 hippocampal pyramidal cells (19–22).

Whole-Cell Patch Voltage Clamp Recording Versus Intracellular Electrode Voltage Recording

The resistances of the electrodes we used for intracellular recording were of the order of 50–70 MΩ, whereas a typical whole-cell patch electrode has a resistance about one order of magnitude less. Thus, the noise contributed by the electrode would be expected to be comparably higher with the high resistance electrode. Nevertheless, this does not mean that the signal-to-noise ratio would necessarily be an order of magnitude lower. In our experience, the problematic noise arises not just from the electrode but from the preparation; the power spectrum of this latter noise has a similar frequency distribution to that of synaptic potentials and thus there are no obvious ways of filtering it out without degrading the signal. Thus the signal-to-noise ratio with whole-cell recording should be better but not necessarily by an order of magnitude.

Two factors can offset the advantage of the whole cell patch technique, both of which are a consequence of the conventional choice with this method of voltage clamp mode rather than voltage recording ("current clamp"). The first is that excitatory inputs, evoked by fibers in the stratum radiatum, on CA1 hippocampal pyramidal cells are at some electrotonic distance away from the somatic electrode. Under voltage clamp, the peak value of the current recorded suffers much more severe attenuation than the peak value of recorded voltage in current clamp mode, for the same electrotonic distance. Thus, for a given electrode resistance the signal-to-electrode noise ratio is more favorable with voltage recording. (For further discussion, see Major (23).)

The second factor concerns stability of recording. It is well known that electrode resistance can drift with time. The great advantage of the current clamp mode is that the voltage recorded is unaffected by such electrode resistance drifts, unlike the current recorded under voltage clamp. Unless the changing "series resistance" is exactly compensated, the magnitude of the peak current can be severely affected. This is illustrated in Fig. 2, which uses data presented in Fig. 10B of Major (23). Note that an uncompensated change in series resistance from 5 MΩ to 20 MΩ can roughly halve the peak amplitude of the recorded synaptic current. Smaller, undetected resistance changes could significantly affect the peak amplitude, tending to

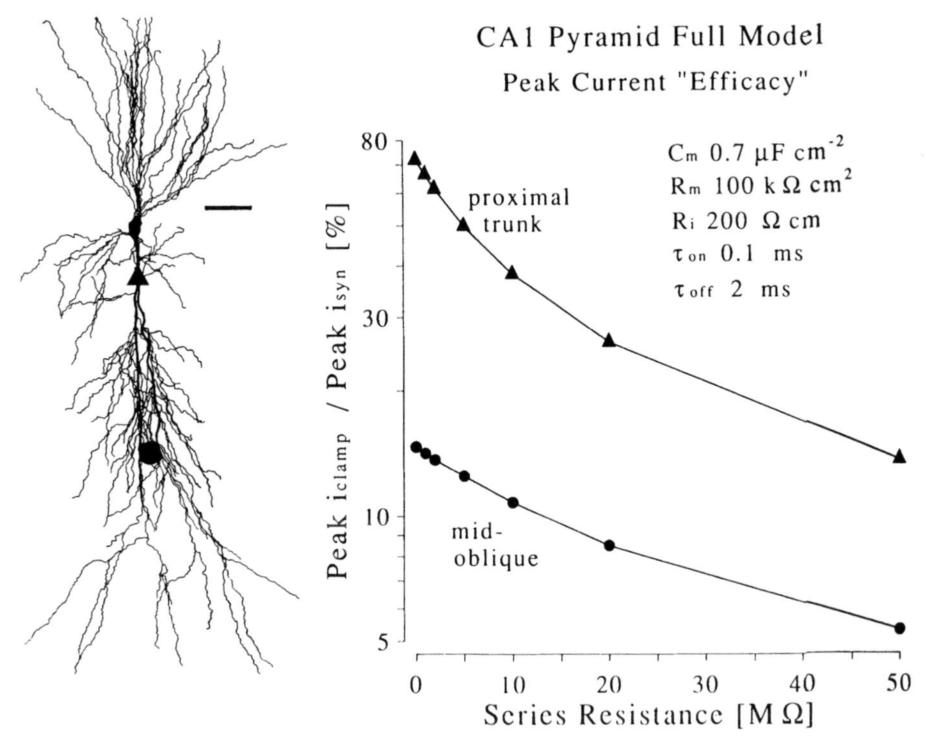

FIG. 2. Modeling study to predict the attenuation of peak synaptic current under somatic voltage clamp. The left panel shows a drawing of a CA1 pyramidal cell that was injected with biocytin, reconstructed, measured (scale, 50 μm), and then used to construct a passive electrical model (Major, ref 46; Major et al., ref. 47; Stratford et al., ref. 73). The electrical parameters were estimated using the cell's response to a brief pulse of injected current. The right panel shows a graph of the predicted effect of series resistance on the peak amplitude of the synaptic current that would be recorded at the soma (expressed as a percentage of the peak at the synaptic site). Two different synaptic locations are considered, one on a proximal part of the apical trunk (at the site indicated by a triangle on the drawing), the other midway along a terminal segment of an apical oblique dendrite (indicated by a circle). Under perfect voltage clamp, with zero series resistance, the peak current recorded at the soma is about 80% of its value at the synaptic site for the proximal location (triangles) and only 15% for the midoblique location (circles). Any series resistance reduces the peak current still further, and the curve is particularly steep for low series resistance values and the proximal input (note logarithmic y-axis). Small drifts in series resistance can therefore make a substantial difference to the attenuation of peak synaptic currents, and may contribute to blurring of amplitude histograms. See inset on right panel for the values of the passive electrical parameters; the time course of the synaptic current was modeled using a double exponent with $\tau_{on} = 0.1$ milliseconds, $\tau_{off} = 2$ milliseconds.

blur peaks in a histogram and cause an apparent "quantal variance" that would be an experimental artifact.

For these reasons, the apparent greater advantage of the whole-cell patch clamp technique may be rather less than customarily assumed, if the goal is to make stable records of the amplitude fluctuations of a synaptic process.

Apparent Zero Quantal Variance and "Noise Reduction"

One of the most puzzling features of the histograms we obtained was that, unlike the classical histograms obtained at the vertebrate neuromuscular junction, the peaks did not get more blurred for larger amplitude as compared to smaller amplitude entries. If there is any quantal variance then, on the classical account, the magnitude of this variance will increase linearly with the number of quanta released. Yet in our 1991 paper we found that, when we performed conventional optimizing routines, the best fits to our histograms were most commonly given by a model that assumed zero quantal variance. In subsequent data-fitting studies we have found that the same data are generally even better fitted if the model allows that successively larger amplitude peaks have successively **less** variance; in other words, the apparent variance attributable to quanta is negative rather than positive. Of 23 histograms, 18 were best fitted on the assumption of apparent negative quantal variance and the rest were fitted best by the assumption of zero or minimal quantal variance (24). To express this result in qualitative terms, the peaks associated with progressively larger numbers of quanta tend to get sharper rather than more blurred. This was, to us, a startling result. We have considered two possible mechanisms that might contribute to the origin of this phenomenon.

The first mechanism starts from the observation, first reported by us (2,12) and subsequently also noted by Liao et al. (20), Voronin et al. (21), and Redman (22), that the independently recorded noise had a variance that was greater than the variance associated with each peak entry that the MLE procedure found to be optimum. The basic observation was that the peaks in the amplitude histogram are sharper than those expected from independent noise measurements.

We presumed that there had to be some difference between noise arising spontaneously and the level of noise associated with an evoked synaptic potential. One possible mechanism by which this could arise is outlined by Jack et al. (2). The basic assumptions of the model are that there is a "refractory" period following release at each site and that a significant proportion of the spontaneous synaptic noise arises from the release sites on the stimulated fiber. This combination, in modeling studies, produces a reduction in the variance contributing to each peak entry compared with the background noise variance, with somewhat greater variance reductions for the large amplitude peaks (24). Inadequate experimental information means that it is difficult to quantify how significant this possible mechanism is, but it is otherwise difficult to explain the degree of difference between the variance of the evoked amplitude peaks and of the noise (12). As pointed out by Jack et al. (2), a further factor is that the histograms we examined were selected, from larger data sets, partly on the basis of the clarity of the peaks, so that an element of the "reduction" was due to "biased" sampling; thus an unknown combination of these two mechanisms has to be considered.

The discussion in the last paragraph offers an explanation for two different phenomena; 1) the observation that the peaks in the evoked histogram are "sharper" (i.e., less variance) than expected if noise and signal are added independently, and

2) the phenomenon of "apparent negative quantal variance." While both data selection and signal-noise interaction may contribute to the first phenomenon, only signal-noise interaction would be expected to generate the second. Since the degree of "apparent negative quantal variance" was so striking, we remained uneasy that signal-noise interaction was the full explanation. Furthermore, in two histograms we obtained (at much lower stimulus rates than our customary 5 Hz), there was no convincing evidence of "noise reduction," but there was clear evidence of apparent negative quantal variance (see Fig. 5). In considering a second possible explanation, we need to introduce a general discussion of quantal variance.

QUANTAL VARIANCE

It may be helpful to consider first, in principle, some of the factors that may contribute to quantal variance. For the type of rapid transmitter release process we are considering, there is now excellent evidence that release occurs at sites with a very sophisticated structural organization (e.g., see Landis et al. (25)), including a specialized postsynaptic area, the postsynaptic density (26). Generally there is an unknown number of these specialized sites associated with each presynaptic fiber; furthermore, recent structural evidence suggests that in hippocampal area CA1 there may be multiple sites between single axons and a CA1 pyramidal cell and that some of these synapses may occur at different electrotonic distances (27). For these reasons it is important to distinguish between two kinds of quantal variance: 1) that occurring at a single release site (from trial to trial), and 2) that which represents the variability in effects produced by different sites (on average), at the recording electrode. Following Walmsley (28) we will call these two kinds of quantal variance Type I and Type II.

Type I: Variance at a Single Release Site

In the past, two quite distinct mechanisms have been suggested for the action of a quantum of transmitter. At the neuromuscular junction, where some 1,000–2,000 channels are opened as the result of the release of roughly 10,000 molecules of vesicular acetylcholine, it is usually assumed that the dominant factor in setting the variability in response from trial to trial is the number of molecules of transmitter contained in different vesicles. Individual vesicles vary in diameter (29) and may also vary in the concentration of transmitter accumulated, so that there is a general expectation that the number of molecules per vesicle will vary. A larger number of molecules released would be expected to open more postsynaptic ionophores (30). In contrast to this mechanism, Redman and his colleagues (31–33) suggested that for excitatory action on the spinal motoneuron the number of transmitter molecules released far exceeded the number of postsynaptic ionophores, so that all the receptor sites would be occupied. If the kinetic behavior of the occupied receptor were such that the probability of opening was unity, then no variability in the peak conductance would be expected, since the smallest number of molecules in a vesicle would still be adequate to occupy all sites. Alternatively, the kinetics of receptor behavior may lead to a probability of less than one that the channel opens when occupied by transmitter, a much more realistic possibility given recent experimental measures. (34–37) For the glutamatergic AMPA channels, Jonas, Major, and Sakmann (36) have shown experimentally that the coefficient of variation of the peak conductance was well fitted by the simple binomial expression

$$QCV = \sqrt{\frac{(1 - p)}{np}}$$

where QCV denotes quantal coefficient of variation, p = probability of channel opening, and n = number of channels (see Jonas et al., Fig. 13D (36)).

Which of these two mechanisms operates at CA1 hippocampal excitatory synapses? They do not need to be mutually exclusive; some variance might be contributed by variation in the number of transmitter molecules released (and hence in the number of receptors occupied) and further variance contributed by the probabilistic nature of the ionophore opening. These two variances, brought about by presumed independent processes, would be expected to add linearly.

Several groups have modeled various aspects of these two processes (36,38–43) and Fig. 3 offers a schematic summary of these studies. The abscissa plots the number of responsive channels (some receptors may be desensitized, for example, by ambient levels of glutamate—see Colquhoun et al. (44)). The curve on the left plots the expected quantal variance on the assumption that all channels are doubly liganded and the opening probability of the channel p is 0.67. The curve with a maximum to the right is a very rough estimate of the amount of variation contributed by differing numbers of transmitter molecules in different vesicles. It has been assumed, on the basis of variation in vesicular diameter, that the coefficient of variation in number of transmitter molecules is of the order of 30%. When the number of responsive receptors matches the average number of transmitter molecules, there will be a roughly linear relationship between the two (30,39) but as the number of transmitter molecules exceeds the number of receptors there is very little change in the number of channels opening, for different amounts of transmitter released (see Fig. 4 in Faber et al. (42)). The curve is very schematic and only correct in a qualitative sense, since an adequate range of simulations has not been performed. Furthermore, the form of the curve will, to some extent, depend on factors such as diffusion and removal of transmitter from the subsynaptic space, the binding affinity and kinetics of the receptor, etc. The dashed line represents the total variation contributed by the two sources, and closely follows the form of the mechanism contributing greater variation except in the region centered at about 500 channels, where the two mechanisms contribute approximately equally to the total variation. Note that this relationship is roughly U-shaped, with a minimum in the region of 200–500 channels when the expected Type I quantal coefficient of variation is of the order of 4%.

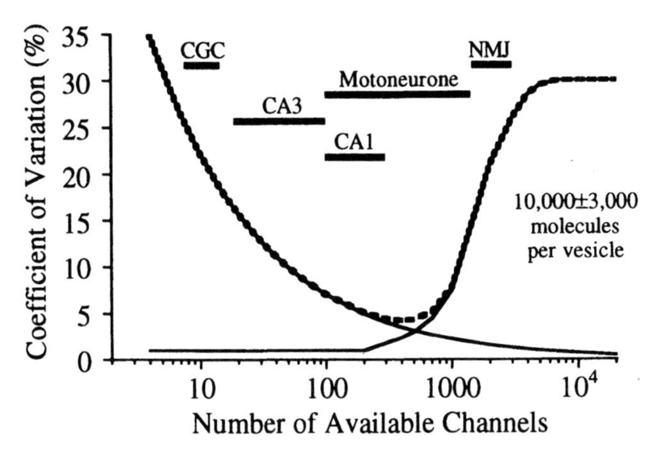

FIG. 3. Graph showing likely levels of Type I quantal variance (expressed as coefficient of variation of the mean quantal size) for synapses with different numbers of functional synaptic channels. The solid exponential line with its maximum to the left shows the expected variance caused by the probabilistic opening of fully liganded channels, assuming an open probability of 0.67 (though this could vary for different synapses). Variance from this source decreases progressively as the number of functional channels increases. The solid sigmoid line with its maximum to the right shows the expected variance of the synaptic event due to variations in number of transmitter molecules released, assuming that a vesicle releases 10,000 ± 3000 (SD) molecules. This curve is a very rough approximation derived from published data (Jonas et al., ref. 36; Land et al., ref. 39; Bartol et al., ref. 41; Faber et al., ref. 42). An assumption of a smaller number of molecules released (e.g., 4,000 ± 1,300, (see Larkman et al., ref. 12)), would shift this curve to the left. The dashed line shows the net coefficient of variation for both sources. For most numbers of channels, the total variance is dominated by one or the other effect; only in the range of 200–500 functional channels do both sources of variance contribute significantly. The horizontal bars indicate the approximate number of channels that need to be available (at $p = 0.67$) to obtain the number of channels actually opening during the peak of a single quantal event at various excitatory synapses.

CGC—cerebellar granule cell, about 10 channels open (Traynellis et al., ref. 37).
CA3-mossy fiber synapses on CA3 pyramidal cells, 14–65 channels open (Jonas et al., ref. 36); CA1—90–200 channels open (present study); Motoneuron—Ia afferent synapses on spinal motoneurons, 75–1,000 channels open, depending on synaptic location (Jack et al., ref. 2); NMJ—vertebrate neuromuscular junction, 1,000–2,000 channels open.

How does the analysis just stated relate to the expected Type I quantal coefficient of variation at CA1 excitatory synapses? As Fig. 3 illustrates, the crucial variable to determine is the average number of channels available to be opened by the action of one quantum. This requires the measurement or inference of three quantities: 1) the average peak conductance at the synaptic site; 2) the conductance of an open channel; and 3) the probability of opening of a channel at the time of peak conductance. Measurements of 2) have been made by Jonas and Sakmann (45) for CA1 and CA3 AMPA receptors, although they apply to extrasynaptic receptors. Both CA3 and CA1 channels had a conductance of 8 picosiemens. Similarly, Jonas, Major, and Sakmann (36) have estimated, for extrasynaptic receptors, that the open probability is of the order of 0.78. A value, $p = 0.67$, slightly below this has been chosen for illustration in Fig. 3 because the measurements we made were of peak voltage rather than peak current and thus represented a partial "integration" of the current, including times when the probability of opening is lower. As pointed out by Kullmann, (43) there are factors that tend to both increase and decrease the coefficient of variation when sampling from a wider time interval, but the overall effect is likely to slightly increase the measure, which is equivalent to a lowering of the effective probability.

The most problematic measure is the estimate of the average peak conductance at the synaptic site. This has to be deduced from a measure of the peak voltage (or peak current, if using whole-cell patch clamp) recorded at the soma. In extensive modeling of CA1 hippocampal pyramidal cells (see Major (46) and Major, Evans, et al. (47)) we have calculated that, for the excitatory postsynaptic potential (EPSP) time courses we obtained, our average quantal peak voltage of 130 μV would be produced by the opening of about 90 channels. This is likely to be an underestimate because the time course of synaptic current assumed had a decay time constant of 2 milliseconds, that observed at room temperature (44,48) whereas our observations were made on slices at approximately 35°C. According to the temperature dependence studies on AMPA currents made by Feldmeyer, Silver, and Cull-Candy (48) the time course at this higher temperature would have a decay time constant of slightly less than 1 millisecond, meaning that only about 60% of the charge calculated for room temperature would be entering (allowing a Q_{10} of 1.3 for electrodiffusional charge movement through the channel). Thus, an estimate of 90 channels opening is very conservative, the more likely figure after temperature correction being of the order of 150 channels.

This figure is much greater than the estimate made by Kullmann (43). In the example of his Fig. 1, the quantal amplitude at the soma under whole-cell patch clamp had a peak of 5 pA, which for a somatic input would be the equivalent of eight channels opening. Kullmann makes a correction for the dendritic location and uncompensated series resistance, which led him to estimate that as few as 20 channels could be opening. From the time course of the current (Kullmann's Fig. 1), we would estimate from our modeling studies that a midapical oblique dendritic input, approximately 0.3 λ from the soma, and with an uncompensated series resistance of about 10 mΩ, would generate this current (assuming a time course of synaptic conductance appropriate to the temperature, 23°C, at which the recordings were made). The amount of attenuation prevailing in this circumstance would be a factor of 10 (see Fig. 2, midoblique input and, for further details, Fig. 10 of Major (46)). Thus we would regard an estimate of 80 channels opening as more appropriate, bringing Kullmann's data into reasonable alignment with ours. Similar reasoning applies to other data measuring quantal size (e.g., see Liao et al.(20)), and it is of interest that Bekkers and Stevens (9) made an estimate, from their study of spontaneous miniature excitatory postsynaptic currents (EPSCs) in culture at room temperature, that an average of approximately 110 AMPA channels were opened; this suggests that the quantal mechanism in their cultured cells may be similar, in at least one respect, to those observed in slices.

The conclusion of this discussion is that about 100 channels are opened with an average sized quantum. From Fig. 3, this would imply that the quantal coefficient of variation would be about 6%. Even if there was some inaccuracy in the assumed peak probability of opening, the estimate will not vary greatly. Within the range $p = 0.5$ to $p = 0.9$, the expected coefficient of variation would be roughly 7% to 3%. Alternatively, it may be that the synaptic ionophores are different from the extrasynaptic ionophores studied by Sakmann and colleagues. For example, they may have a larger unit conductance, like those in the cerebellum, where Traynellis, Silver, and Cull-Candy (37) have estimated a value of 20 picosiemens. For our data, this would imply about 40 channels opening and a coefficient of variation of roughly 9% (11% to 5% for the range of probability of opening between 0.5 and 0.9).

What is most striking about these tentative conclusions is that we have estimated that one form of quantal variance (Type I) should lie somewhere in the range of coefficient of variation of 3% to 11%, with a favored estimate of 6%. This is in striking contrast to the two most discordant estimates in the literature of roughly 0% (12) and the much larger values of 42% (slices) or 55% (cultured cells) measured by Bekkers, Richerson, and Stevens (11) for putative single-site miniature EPSCs. Can these large discrepancies be resolved? In order to gain further insight into our estimate of zero quantal variance (or less!), we need next to consider Type II quantal variance.

Type II: Variance Between Sites

Although there are examples in the central nervous system where a presynaptic axon supplies a single bouton, with a single release site, to a postsynaptic cell (see Gulyas et al. (49)), the more common organization may be that reported by Sorra and Harris (27) in which many single axons supply multiple boutons, some with more than one release site, to a single postsynaptic cell (see also Walmsley (50), and Pierce and Mendell (51)). If there are multiple release sites, which could be more or less electrotonically remote from the recording electrode, it is very likely that quanta evoked at one release site will have an average amplitude different from those at another site because of different degrees of electrotonic attenuation. Furthermore, there is no a priori reason why the number of postsynaptic receptor channels should be exactly the same at two separate release sites (e.g., see Jonas et al., (36) and Busch and Sakmann (40)). Both of these factors could contribute to Type II quantal variance.

Does Type II variance contribute to the variance associated with amplitude entries in an evoked histogram in the same way as Type I variance? As Walmsley (28) has pointed out, the qualitative expectation is that Type II variance will not generate amplitude histograms with a linear increase in the variance associated with the number of quanta released, as for Type I variance. Consider a simple example, where a presynaptic fiber has two release sites that produce peak EPSP voltages of 90 μV and 110 μV at the recording electrode. If there is no Type I variance, the amplitude when both sites release is always 200 μV but the amplitude entry for release from a single site will be a mixture of 90 and 110 μV. Thus there will be more total variance associated with the single quantum amplitude entry than for the case when two quanta are released. It will be obvious that this is a possible basis for "apparent negative quantal variance," since the entry associated with two quanta has **less** variance than that associated with one, whereas if there were only Type I quantal variance, the variance with two quanta would be double that associated with one quantum.

We have explored this further in modeling studies and some of our preliminary results are illustrated in Fig. 4. The left half of Fig. 4 illustrates the results for a varying number of release sites, assuming an equal probability of release at each release site, and that there is no Type I variance. The entry number on the abscissa refers to the number of quanta released and the ordinate plots the magnitude of the variance (normalized in units). The diagonal line shows the behavior of Type I variance (increasing linearly), whereas the curves for Type II variance are convex upwards with a maximum when the number of quanta released is half the number of

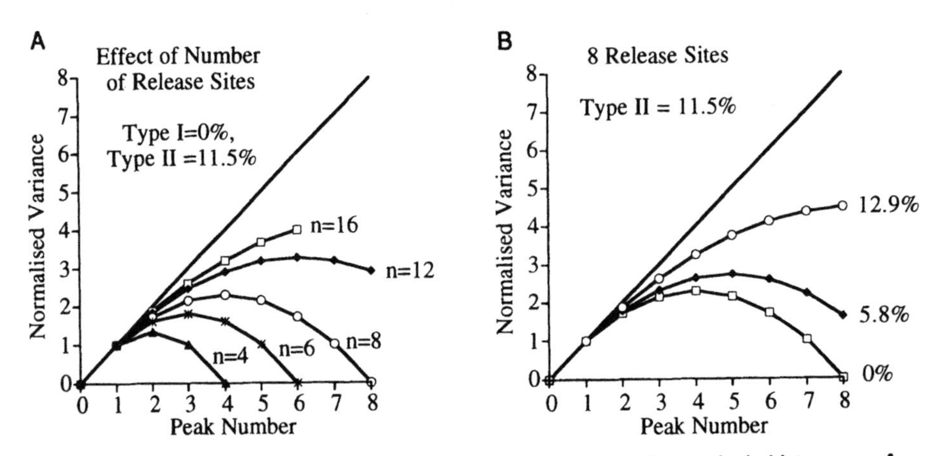

FIG. 4. Effects of Type I and Type II quantal variance on successive peaks in histograms. **A.** Graph showing the variance of successive peaks in an amplitude histogram, normalized to the variance of the first nonfailures peak (designated peak number 1), for different numbers of release sites. The diagonal shows the behavior of Type I variance, which increases linearly with entry number. The curves show the effect of 11.5% Type II variance, with zero Type I variance, for different numbers of release sites (assuming the probability of release is the same at all sites). Note that the variance is zero for peaks corresponding to release from all available sites, and is maximal when half the available sites release. **B.** The effect of combining Type I and Type II quantal variance. The case for eight release sites and 11.5% Type II variance is illustrated for different levels of Type I variance (Upper trace, 12.9%, equivalent to 30 active channels at $p = 0.67$; middle trace, 5.8%, equivalent to 150 active channels at $p = 0.67$; lower trace, 0%). When the level of Type I variance is greater than that for Type II, the variance of successive peaks increases monotonically (open circles). If Type II variance dominates, then the variance of the high peak numbers will be less than those in the center of the histogram. This effect will be revealed only when the probability of release is high, and will give rise to peaks on the right side of the histogram that are sharper than those in the center.

release sites (if this is an odd number, n, then the variance is maximal and equal at $(n-1)/2$ and $(n+1)/2$. Notice that the variance is zero when all sites release.

The right half of Fig. 4 illustrates the effect of combining Type I and Type II variance. When Type I variance of about 6% (coefficient of variation) is combined with Type II variance with about double the coefficient of variation (four times the variance), the curve still shows a maximum, but now when five (out of eight possible) quanta are released. If there is more Type I than Type II variance, then the variance monotonically increases, although the maximum variance, for eight quanta, reflects the Type I variance only (open circles). In general, there will be only "apparent negative quantal variance" if the Type II variance substantially exceeds the Type I variance. Furthermore, in order to detect the phenomenon, the probability of release at the sites has to be fairly high so that the entries with larger numbers of quanta are well represented in the histogram (thus the procedure makes estimates of the variance associated with these entries). This certainly holds for our experimental preparations, since the estimates of the average probability of release that we have made quite commonly exceed $p=0.5$ (see Larkman et al., (17) and Hannay et al., (52)).

We can conclude from the above discussion that a possible contributing factor to our observation of apparent zero or negative quantal variance arises from the technique we used to make the estimate. The optimizing procedure was given the task of determining a linear slope for the amount of variance associated with each peak entry, and commonly the optimum solution was a negative slope. It is now clear that such a finding is not incompatible with significant quantal variance, provided that Type II quantal variance is greater than Type I. For most of our data, it is not possible to quantify the ratio of Type I to Type II variance because of the compounding phenomenon of noise reduction. Figure 5 illustrates more recent data where there is no evidence of noise reduction, but clear evidence that Type II variance considerably exceeds Type I.

The method we have used to illustrate the ratio of Type I and Type II variance is not suited to all experimental conditions, particularly if the release probability is relatively low, and it is worth considering whether some other approach can be adopted in this circumstance. Jonas, Major, and Sakmann (36) addressed this issue by carefully measuring the variance associated with the first entry (i.e., after subtraction of noise variance). The remaining coefficient variation would reflect the sum of the two quantal variances; they found a value of 22%. If an independent estimate can then be made of the Type I variance, the remaining variance would be attributable to Type II. They calculated a range of values for the number of channels opened, on average, for a single quantum (14–71 channels) and concluded that Type I variance was dominated by probabilistic behavior of the channels (rather than by fluctuations in amount of transmitter released). They deduced that the coefficient of variation associated with Type II variance was between four times (71 open channels) and two times (14 open channels) that associated with Type I variance. Thus, although our approach has been quite different, these two estimates are in approximate accord; one qualification to make to this apposition is that Jonas et

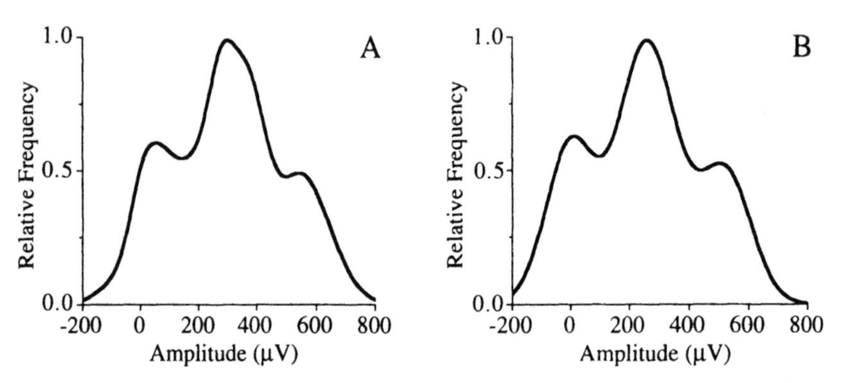

FIG. 5. Simulation of an experimental histogram. **A,B.** Experimental (**A**) and simulated (**B**) amplitude histograms for the excitatory postsynaptic potential (EPSP) shown in Fig. 1. The right peak shown on **A** and **B** in **Fig. 1** has been removed (this peak consisted of only two trials, inspection of which revealed that they were both contaminated by spontaneous EPSPs close to the peak of the evoked EPSP). A feature of the experimental histogram (**A**) is that deconvolution shows that the first and third peaks can be well matched using the experimental noise (standard deviation, 70 μV), while the central peak (corresponding to the release of one quantum) requires a higher standard deviation (95 μV). The shape of this histogram could be approximately matched by simulating 5,000 trials using a simple binomial generator with $n=2$, $p=0.48$, Type I variance of up to 6% (coefficient of variation) and quantal sizes of 220 μV and 300 μV (**B**). This is equivalent to Type II variance of 22% (coefficient of variation). Both histograms are shown after smoothing using a digital Gaussian filter.

al. were studying excitatory synaptic action on CA3 hippocampal pyramidal cells, and it is known that the (extrasynaptic) channels on these cells are subtly different in properties from those of CA1 (44). Furthermore, Jonas et al. were recording in voltage clamp mode, so minor changes in series resistance might have contributed some "blurring" to their measurements. Using the logic they followed, this might reduce their estimate of the magnitude of Type II variance.

Despite this general agreement, there remains the problem of understanding the data of Bekkers and Stevens (9,11) since they had carefully designed experiments that sought to estimate the quantal variance associated with a single release site (or very few) and had found very large values (~50%) for the coefficient of variation of, putatively, only Type I quantal variance. Is there any way of explaining this gross discrepancy? Before considering some possibilities, it is necessary to review current evidence about the nature of spontaneous "miniature" synaptic events.

SPONTANEOUS SYNAPTIC EVENTS

In the classical account arising from the studies of Katz and others at the vertebrate neuromuscular junction, the spontaneously occurring endplate potentials were found to be of the same amplitude and variance as the first entry in the evoked amplitude histogram. Thus, the spontaneous events came to be a calibration measure for the size and variability of a single quantum. At the discretely localized neuromuscular junction there was no major degree of electrotonic attenuation and, since there appears to be a substantial number of receptors within diffusional reach of the transmitter, it is generally assumed that the main component of the coefficient of variation (~30%) comes from variations in the number of transmitter molecules in each vesicle (see Fig. 3).

By contrast, spontaneous synaptic events recorded in central nerve cells are more complicated. They can arise from any release site on the cell, so they can suffer widely differing degrees of electrotonic attenuation. Consonant with this, there is a very wide distribution in their amplitudes; the full range is often difficult to quantify because the smallest events may not be detectable above baseline noise, but spreads of up to a factor of 10 have been reported for CA1 hippocampal cells in slice (e.g., see Manabe et al. (53)) and even greater ranges—up to a hundredfold—for cells in culture (15). An obvious move would be to attribute most of this amplitude range to varying degrees of electrotonic attenuation; but the balance of evidence is against this being more than a partial contributing factor. For example, Finch et al. (15) found no consistent correlation between the rise time (a guide to electrotonic location) and the amplitude of their spontaneous events. In addition, both Bekkers et al. (11) and Raastad et al. (54) evoked spontaneous events from a restricted region of CA1 cells in slices and still found a wide spread of amplitudes. It is evident, therefore, that the variance of spontaneous synaptic events arising from a restricted region of the cell can be considerable. There are two questions that need to be asked: 1) are spontaneous synaptic events always monoquantal, or can they sometimes represent multivesicular release?; and 2) is the variance associated with monoquantal spontaneous events attributable primarily to Type II variance, or is there substantial Type I variance? A supplementary question that then arises, if the answer to the second question is that it is primarily Type II variance, is whether the magnitude of Type II variance associated with all sites (spontaneous events) is significantly greater than the Type II variance of the release sites belonging to a single presynaptic fiber.

These questions will be addressed in turn, but because the evidence for excitatory action on hippocampal pyramidal cells is fragmentary, examples will be drawn more widely.

Are Spontaneous Events Always Monoquantal?

There is a long history, going back to Liley (55) that some spontaneous synaptic events may be "multiquantal." The basic piece of evidence is that sometimes the amplitude histogram of these spontaneous events is made up of a variable number of equally spaced peaks (see for example, 7,56–60). In most of these examples there were good grounds for assuming that the spontaneous events arose from a relatively restricted region (in electrotonic terms). In an elegant study, Ulrich and Luscher (61) recorded spontaneous synaptic currents from organotypic cultures of spinal cord motoneurons and used their time course as a guide to their electrotonic location; in consequence, they were able to correct the amplitudes for relative electrotonic attenuation and then reported that their data was best fitted by a sum of two Gaussian curves of single and double mean amplitude and variance. They interpreted these results as reflecting a combination of either one or two quanta, released spontaneously.

Two quite different interpretations have been offered for these data. The more common is that the different peaks each represent the release of an integral number of vesicles and that the mechanism by which "spontaneous" release occurs can lead either a single or several vesicles to be triggered to discharge their contents simultaneously. In keeping with this suggestion are the observations made by Melamed et al. (62) at the lizard neuromuscular junction that there can be spontaneous fluctuations in intracellular calcium concentration coordinated within a synaptic bouton (involving distances of the order of 5 μm). If this general interpretation is correct, it is also of interest to consider whether multivesicular release occurs over a spatial domain that would allow the contents of each vesicle to mix diffusionally, to act on the same group of receptors. This is important because, if so, it would require that there was a linear relationship between total number of transmitter molecules and postsynaptic response in order that the peaks were equally spaced in amplitude. In the case of the transmitter glutamate acting on AMPA receptors, this is only likely to occur if all the vesicles were released within a domain where the maximum distance apart was less than 1 μm, since the peak time of the conductance (maximal occupation of the receptors) is of the order of 0.5 milliseconds. Alternatively, each vesicle could be released from more separated sites and act on separate sets of postsynaptic receptors. Korn et al. (60) present a correlation between the proportions of single or multiple spontaneous quanta and the distribution of the number of separate release sites in a single bouton, evidence strongly favoring the latter interpretation (see Chapter 19).

The second type of explanation offered for multiple peaks in the histogram of spontaneous synaptic amplitudes is that they all represent the release of the contents of a single vesicle, but the number of responsive postsynaptic receptors is "quantized" at different release sites (7,64). This explanation would require that the total number of postsynaptic receptors was always considerably less than the average number of molecules in a vesicle. The only piece of independent evidence at all suggestive of this possibility is that, as Sakmann (65) has pointed out, the distribution of the areas of postsynaptic densities at excitatory synapses in CA1 hippocampal pyramidal cells is roughly bimodal (see Harris et al. (66)). There is some danger in correlating the area of a postsynaptic density with the number of available postsynaptic receptors, since we know that AMPA receptors are not the principal protein content of the density (26). If this second type of explanation were correct, it would mean that the dominant type of quantal variance was of Type II, with the added complication that the distribution was multimodal rather than unimodal.

The answer to the question posed in this section is that there is clear evidence that spontaneous events are not always unimodal, but that the degree to which they display multimodality may vary between different preparations. The simplest interpretation of the multimodality is that it represents multivesicular release and some direct evidence consonant with this has been offered by Korn et al. (60) for inhibitory synaptic events. On the other hand, it is possible that the multipeaked histograms of spontaneous events represent differences between release sites, with the number of responsive postsynaptic receptors being distributed in a multimodal manner. It then becomes an awkward terminological problem whether "quantum" should refer to the effects of a single vesicle, to the multivesicular effect (a quantum composed of subunits (67)), or to the "irreducible minimum" in terms of the postsynaptic response. In order to proceed to the second question posed, we will for convenience assume that "monoquantal" refers to the first peak of the postsynaptic amplitude histogram.

Is the Variance of a Monoquantal Spontaneous Event Primarily Due to Type I or to Type II Variance?

There are two possibilities concerning the main factor that sets Type I variance: 1) the variance is dominated by the stochastic opening behavior of the postsynaptic ionophores, or 2) it is related to the variability in the number of transmitter molecules released by a single vesicle (see Fig. 3). By contrast, Type II variance can be a consequence of two quite independent mechanisms: 1) the average number of postsynaptic channels opening at different sites, and 2) the degree of electrotonic attenuation suffered by the current spreading from each release site to the recording electrode. The two factors could compound; or, as has been suggested for the adult motoneuron, the first factor could compensate for the latter when recording from the soma (2,33). In this section we would like to restrict the question to the first kind of Type II variance, and decide for the spontaneous quantum, as well as for the evoked quantum, whether the contribution of variation in the number of receptors between sites is more important than Type I variance.

The experiments of Bekkers and Stevens (9,11) were explicitly designed by them to address this question. Because these experiments are so important, we would like to discuss them separately (see information that follows). In this section we will briefly review other evidence pertaining to glutamatergic AMPA receptors, where there are good grounds for assuming that electrotonic attenuation does not make a major contribution to the Type II variance. Two sets of data are available where it is possible to make approximate guesses. In the experiments of Silver et al. (58) and Traynelis et al. (37) on cerebellar granule cells, the amplitude of the first spontaneous peak was equivalent to a conductance, on average of 117 picosiemens. They calculated that the (average) amplitude of the channels contributing to spontaneous potentials was 15 picosiemens, so that on average about eight channels open. Thus, a substantial amount of Type I variance would be expected: e.g., if the channel opening probability at the peak is 0.67, the coefficient of variation due to Type I would be ~20%. This could explain a substantial amount of the total variance observed (value not given, but of the order of 25%; see Fig. 2D of Silver et al.

(58)). It is possible that there is a higher peak opening probability ($p>0.8$), suggesting a Type I variance of the order of 10% or less. In that case, Type II variance is likely to be greater than Type I.

It is also unlikely that the data of Ulrich and Luscher (61) can be attributed primarily to Type I variance. They calculated that their single quantum represented an average of 54 channels opening and that the coefficient of variation of the quantum was 28%. The maximum Type I variance that could be expected from the probabilistic opening of channels, for this number of channels opening, would be a coefficient of variation of less than 14% (with a very low probability of channel opening) that still leaves the bulk of the variance (equivalent to a coefficient of variation of 24%) to be explained. Thus, if it is Type I variance, it would have to be attributed to variation in the number of transmitter molecules released, with a large number of receptors available each having a very low probability (~0.01) of opening when transmitter binds. The alternative is that these spinal cord AMPA receptors are similar to those in the hippocampus, with a reasonably high probability of opening; if so, the Type I variance would generate a coefficient of variation of 8%, so that the total variance could only be explained by assuming a large Type II variance (with a coefficient of variation of ~27%). This is equivalent to suggesting that the number of available receptors is 80 ± 20 at the different release sites.

The answer to the question raised in this section is not definitive. There is only fragmentary evidence pertaining to whether, after selection (or correction) so that the effect of electrotonic attenuation is minimal, Type I or Type II variance is more significant in contributing to the coefficient of variation of spontaneous monoquantal events. Nevertheless, it is clear that electrotonic attenuation can contribute significantly to the total Type II variance. For example, the modeling of the CA1 pyramidal cell by Major (46) indicates that the peak synaptic effect can have a range of at least sixfold for EPSCs and at least threefold for EPSPs (see Table 4, Major (46)), assuming a fixed synaptic current at all release sites. It is unlikely, therefore, that, when considering all release sites, the total Type II variance does not, in general, exceed Type I variance.

Thus, it is also very likely that the Type II variance associated with quanta arising from the release sites of a single presynaptic fiber is less than the total Type II variance associated with all sites. The fact that it is (often) possible to detect peaks in the amplitude histograms of putative single-fiber inputs means that the total coefficient of variation associated with each peak entry has to be a maximum of the order of 30%, implying at most a roughly threefold range of quantal amplitudes. This is the minimal contribution of electrotonic attenuation to Type II variance (for all release sites) and, as discussed, there is also likely to be variation in the number of receptors at different sites. Another piece of evidence is that the average quantal size measured for different presynaptic excitatory fibers in CA1 cells shows a wide range. For example, Liao et al. (20) normalized their quantal sizes (before induction of long-term potentiation) to allow for different electrotonic position and series resistance of the recording electrode, then found a tenfold range (see their Fig. 10b).

This is suggestive, for excitation of CA1 cells, that there can be substantial Type II variance even after correcting for electrotonic attenuation, but that the amount is much less for the sites of a single presynaptic fiber.

Although there are qualifications about using the area of a postsynaptic density as a guide to the number of functional channels (see last paragraph), the data of Sorra and Harris (27) are also very suggestive of the same conclusion. They found that the postsynaptic densities at different release sites of the same fiber, on the same cell, differed by only 35% on average (not more than 81%), whereas the same fiber could innervate postsynaptic densities on *different* postsynaptic cells that differed on average by 100%, ranging up to differences of 650%. Given the wide range in postsynaptic densities on a single postsynaptic cell (66) this is in accord with the idea that Type II variance associated with different numbers of postsynaptic receptors will be greater for all sites (at one electrotonic location) than for those associated with a single presynaptic fiber.

The Experiments of Bekkers and Stevens

Bekkers and Stevens were well aware of the necessity to distinguish between Type I and Type II variance, and consequently designed some apparently clear-cut experiments whose aim was to minimize the contribution of Type II variance to their estimates. They reached the conclusion that in both conventional slice preparations and in cultured neurons, the Type I variance was very considerable. They pointed out that, because of the magnitude of the quantal variance (coefficient of variation of the order of 50%), it would be very unlikely that peaks would be evident in evoked single-fiber amplitude histograms (9,11). Since there is now extremely good evidence that peaks can be detected, which are not sampling artifacts, it is important to reconsider their data. It is possible that the experimental design tends to generate a greater Type I variance than that prevailing in other experiments.

The reason for this suggestion (which may also apply to the study of spontaneous potentials by Raastad et al. (54)) is that "spontaneous" miniature potentials were "provoked" from a restricted region of the neuron by hypertonic solutions. In the most definitive experiments, involving cultured neurons, Bekkers et al. (11) recorded miniature potentials that were apparently arising from a single synaptic bouton. In order to ensure that the experimental sample was predominantly from this restricted site, it was important to check that the "contamination" from "unprovoked" spontaneous potentials was minimal. Bekkers et al. quote values of a frequency of <0.2 Hz for "unprovoked" miniatures, compared with a value of >5 Hz for the provoked miniatures. In Fig. 2 of Bekkers and Stevens (9) it is possible to count 15–20 provoked miniatures as a result of a 300-millisecond burst of hypertonic solution—in other words, a frequency of the order of 50 Hz.

It is still uncertain how many separate release sites there may be on a single synaptic bouton in culture, but Sorra and Harris (27) have noted that boutons with more than one release site are relatively common (one quarter of their serial section sample) in adult CA1 hippocampus. Thus, if boutons in culture are comparable, there may be two or more release sites, each of which could either separately or simultaneously release the contents of vesicles. Bekkers et al. (11) note that some of their histograms had discrete peaks and therefore make the suggestion that some of the total variance they recorded could arise from concerted release of the contents of various numbers of vesicles. This is the explanation recently proposed by Stevens (68) but he favors regarding the effects of a single vesicle as being a subquantum (67) because he assumes the data arise from a single release site.

There appear to us to be some real difficulties with this explanation. The first is that, if there is only a single release site, there will have to be a fairly linear relationship between amount of transmitter released and the number of channels opened. This seems unlikely, as just discussed. A further problem arises, however, from the frequency at which the release process occurs. If the cultured CA1 cells of Bekkers and Stevens have AMPA receptors with the same properties as those studied by Colquhoun et al. (44) then, after activation by the brief pulse of glutamate released from the vesicle(s), they are substantially desensitized. The rapid desensitization process has a slow recovery time; for CA1 receptors, the time constant for recovery is roughly 60 milliseconds, with about half the receptors desensitized by a pulse of glutamate concentration comparable to that occurring in the synaptic cleft (69) Thus, provoking vesicular release at average frequencies as high as 50 Hz (but with wide variations in the intervals between each release—see Fig. 2 of Bekkers and Stevens, (9)) will mean that the number of undesensitized receptors available to react to transmitter could fluctuate widely. This is in sharp contrast to the experimental techniques used to produce single-fiber histograms where the frequency at which synaptic events are evoked is a maximum of 5 Hz (12) or, more commonly, much lower frequencies (e.g., 1 Hz or less). In all of these experiments, recovery from rapid desensitization should be complete. Thus, even if Stevens is correct that the Bekkers and Stevens data do arise from a single release site, the Type I quantal variance would be greatly overestimated because the desensitization process could vary the number of available receptors by a factor of nearly 2 compared with the experiments stimulating presynaptic fibers. If, on the other hand, there is more than one release site per bouton in the Bekkers and Stevens preparation, the blurring by desensitization may be less, but the total variance they have measured would be attributable to an unknown mixture of Type I variance, Type II variance, and variations in the number of vesicles simultaneously released. The only safe conclusion that we feel can be reached is that, although these experiments are incisive in conception, the possibilities we have raised mean that they cannot be regarded as a definitive measure of Type I variance. In the other experiments reported by Bekkers and Stevens (on adult slice preparations), the hypertonic solutions had a spatial spread of at least 15 μm, and would therefore provoke release from a large number of sites. Since there are of the order of two dendritic spines per μm of dendritic length in CA1, the total number of activated release sites could be considerable: hence the possibility of substantial Type II variance contributing to the measurements. This point also applies to the study of Raastad et al. (54).

CONCLUDING REMARKS

In this account we have tried to illustrate how some surprising data that we obtained, and the criticism it attracted on publication (12,13), have led us to explore the mechanisms underlying quantal behavior of putative single-fiber and spontaneous excitatory synaptic events in the CA1 region of the hippocampus. Our analysis and review is far from complete, but we are reasonably confident about the following conclusions:

1. At a single release site, a limited number of postsynaptic channels open (~100). It is likely that virtually all available postsynaptic receptors are occupied by transmitter and that the only major factor contributing to variability from trial to trial is the probabilistic opening of the channels. It is likely that the coefficient of variation for this form of quantal variance, which we call Type I, is about 6%. Thus, qualitatively, a single site behaves in an "all-or-nothing" manner, depending on whether or not a quantum is released presynaptically. This latter conclusion echoes the suggestion, first made by Redman and his colleagues (31,32) for excitatory action on the motoneuron.
2. A single presynaptic fiber can make several synaptic connections with a postsynaptic cell. Each of these release sites can contribute to an evoked single-fiber histogram. There exists significant variability in the number of postsynaptic channels at each release site, and this variability contributes to Type II quantal variance. In addition, differing degrees of electrotonic attenuation in the propagation of the synaptic event to the recording electrode can contribute to the total Type II variance.
3. Although we have not yet quantified the magnitude of Type II quantal variance, the form of our amplitude histograms—which show "apparent negative quantal variance"—has led us to conclude qualitatively that Type II quantal variance can be greater than Type I quantal variance.
4. Spontaneous miniature synaptic events are not a reliable "calibration" for the expected quantal size of a single-fiber histogram, even if the spontaneous potentials arise predominantly from the same restricted region (electrotonically) of the postsynaptic cell. There are two reasons for this conclusion: a) spontaneous potentials may be multiquantal, and b) the Type II variance between postsynaptic sites in one region of a cell may be greater than the Type II variance associated with the sites belonging to a single fiber. This latter conclusion should hardly be surprising given the profusion of mechanisms in the hippocampus for increasing or decreasing the synaptic efficacy of a single fiber input, without necessarily affecting another synaptic input whose release sites could be adjacent (e.g., see examples (70,71)).

It is of interest to compare these conclusions with the classical account given by Katz and his colleagues for the vertebrate neuromuscular junction. No distinction was made between Type I and Type II variance but the evidence, as far as it goes, suggests that Type I variance is probably larger than Type II. In any event, the distinction is not important in analyzing their evoked histograms because the difference in contribution of Type II from Type I variance is only evident when the number of quanta released is 10% or more of the total number of release sites (see Fig. 4, left graph). The classical studies were made on junctions with a large number of release sites, but only a small number of quanta were released. In this circumstance, both Type I and Type II quantal variance contribute to the total variance of an entry in the same manner, thus making the analysis more straightforward.

Another convenient feature of the frog neuromuscular junction is that multiquantal spontaneous synaptic events appear to be uncommon; furthermore, the regions from which the spontaneous events arise appear to be those regions from which evoked events are also released (72), thus tending to minimize any possible difference between the amount of Type II variance associated with each type of synaptic event.

We do not regard our results as showing that the account given by Katz of quantal release is invalid (cf, Stevens (68)). The account we favor for a description of quantal mechanisms of excitation in CA1 only differs from it in two principal respects. The first is that a single quantum involves the opening of a small number of postsynaptic channels, and thus it is unlikely that variation in the number of transmitter molecules in a vesicle is reflected in quantal variance. The second respect is that the experimental circumstance prevailing in the classical analysis meant that it was unnecessary to distinguish between Type I and Type II quantal variance.

ACKNOWLEDGMENTS

This work has been supported by grants from the Wellcome Trust and the Royal Society. We would like to thank Lindi Wahl for help with the simulations for Fig. 5.

REFERENCES

1. Redman S. Quantal analysis of synaptic potentials in neurons of the central nervous system. *Physiol Rev* 1990;70(1):165–198.
2. Jack JJB, Kullmann DM, Larkman AU, Major G, Stratford KJ. Quantal analysis of excitatory synaptic mechanisms in the mammalian central nervous system. In: *Cold Spring Harbor Symposia on Quantitative Biology LV*, Cold Spring Harbor Laboratory Press, 1990;57–67.
3. Walmsley B, Edwards FR, Tracey DJ. The probabilistic nature of synaptic transmission at a mammalian excitatory central synapse. *J Neurosci* 1987;7(4):1037–1046.
4. Walmsley B, Edwards FR, Tracey DJ. Nonuniform release probabilities underlie quantal synaptic transmission at a mammalian excitatory central synapse. *J Neurophysiol* 1988;60(3):889–908.
5. Sayer RJ, Redman SJ, Andersen P. Amplitude fluctuations in small EPSPs recorded from CA1 pyramidal cells in the guinea pig hippocampal slice. *J Neurosci* 1989;9(3):840–850.
6. Sayer RJ, Friedlander MJ, Redman SJ. The time course and amplitude of EPSPs evoked at synapses between pairs of CA3/CA1 neurons in the hippocampal slice. *J Neurosci* 1990;10(3):826–836.
7. Edwards FA, Konnerth A, Sakmann B. Quantal analysis of inhibitory synaptic transmission in the dentate gyrus of rat hippocampal slices: a patch-clamp study. *J Physiol* 1990;430:213–249.
8. Malinow R, Tsien RW. Presynaptic enhancement shown by whole-cell recordings of long-term potentiation in hippocampal slices. *Nature* 1990;346:177–180.
9. Bekkers JM, Stevens CF. NMDA and non-NMDA receptors are co-localised at individual excitatory synapses in cultured rat hippocampus. *Nature* 1989;341:230–232.

10. Bekkers JM, Stevens CF. Presynaptic mechanism for long-term potentiation in the hippocampus. *Nature* 1990;346:724–729.
11. Bekkers JM, Richerson GB, Stevens CF. Origin of variability in quantal size in cultured hippocampal neurons and hippocampal slices. *Proc Natl Acad Sci* 1990;87:5359–5362.
12. Larkman A, Stratford K, Jack J. Quantal analysis of excitatory synaptic action and depression in hippocampal slices. *Nature* 1991;350:344–347.
13. Clements J. Quantal synaptic transmission? *Nature (correspondence)* 1991;353:396.
14. Larkman A, Stratford K, Jack J. Quantal synaptic transmission? Reply. *Nature (correspondence)* 1991b;353:396.
15. Finch DM, Fisher RS, Jackson MB. Miniature excitatory synaptic currents in cultured hippocampal neurons. *Brain Res* 1990;518:257–268.
16. Magleby KL, Miller DC. Is the quantum of transmitter release composed of subunits? a critical analysis in the mouse and frog. *J Physiol* 1981;311:267–287.
17. Larkman A, Hannay T, Stratford K, Jack J. Presynaptic release probability influences the locus of long-term potentiation. *Nature* 1992;360:70–73.
18. Stratford K, Larkman A, Jack J. Is excitatory synaptic transmission quantal in rat hippocampal slices in vitro? *J Physiol* 1993;459:157P.
19. Kullmann DM, Nicoll RA. Long-term potentiation is associated with increases in both quantal content and quantal amplitude. *Nature* 1992;357:240–244.
20. Liao D, Jones A, Malinow R. Direct measurement of quantal changes underlying long-term potentiation in CA1 hippocampus. *Neuron* 1992;9:1089–1097.
21. Voronin LL, Kuhnt U, Hess G, Gusev AG, Roschin V. Quantal parameters of "minimal" excitatory postsynaptic potentials in guinea pig hippocampal slices: binomial approach. *Exp Brain Res* 1992; 89:248–264.
22. Redman S. 1994;(Chap. 20)
23. Major G. Solutions for transients in arbitrarily branching cables: III. Voltage clamp problems. *Biophys J* 1993;65:469–491.
24. Stratford KJ. Quantal analysis of excitatory synaptic transmission. *D.Phil Thesis. Oxford University* 1992.
25. Landis DMD, Hall AK, Weinstein LA, Reese TS. The organisation of cytoplasm at the presynaptic active zone of a central nervous system synapse. *Neuron* 1988;1:201–209.
26. Kennedy MB. The postsynaptic density. *Current Opinion in Neurobiol* 1993;3:732–737.
27. Sorra KE, Harris KM. Occurrence and three-dimensional structure of multiple synapses between individual radiatum axons and their target pyramidal cells in hippocampal area CA1. *J Neurosci* 1993;13(9):3736–3748.
28. Walmsley B. Quantal analysis of synaptic transmission. In: Wallis DI, ed. *Electrophysiology: A practical approach.* Oxford: IRL Press, 1993;109–141.
29. Tatsuoaka H, Reese TS. New structural features of synapses in the anteroventral cochlear nucleus prepared by direct freezing and freeze-substitution. *J Comp Neurol* 1989;290:343–357.
30. Hartzell HC, Kuffler SW, Yoshikami D. Postsynaptic potentiation: interaction between quanta of acetylcholine at the skeletal neuromuscular synapse. *J Physiol* 1975;251:427–463.
31. Edwards FR, Redman SJ, Walmsley B. Statistical fluctuations in charge transfer at Ia synapses on spinal motoneurones. *J Physiol* 1976;259:665–688.
32. Edwards FR, Redman SJ, Walmsley B. Non-quantal fluctuations and transmission failures in charge transfer at Ia synapses on spinal motoneurones. *J Physiol* 1976;259:689–704.
33. Jack JJB, Redman SJ, Wong K. The components of synaptic potentials evoked in cat spinal motoneurones by impulses in single group Ia afferents. *J Physiol* 1981;321:65–96.
34. Dilger JP, Brett RS. Direct measurement of the concentration- and time-dependent open probability of the nicotinic acetylcholine receptor channel. *Biophys J* 1990;57:723–731.
35. Jahr CE. High opening probability of NMDA receptor channels by L-glutamate. *Science* 1992;255: 470–472.
36. Jonas P, Major G, Sakmann B. Quantal components of unitary EPSCs at the mossy fibre synapse on CA3 pyramidal cells of rat hippocampus. *J Physiol* 1993;472:615–663.
37. Traynelis SF, Silver RA, Cull-Candy SG. Estimated conductance of glutamate receptor channels activated during EPSCs at the cerebellar mossy fibre-granule cell synapse. *Neuron* 1993;11:279–289.
38. Wathey JC, Nass MM, Lester HA. Numerical reconstruction of the quantal event at nicotinic synapses. *Biophys J* 1979;27:145–164.

39. Land BR, Salpeter EE, Salpeter MM. Kinetic parameters of acetylcholine interaction in intact neuromuscular junction. *Proc Natl Acad Sci* 1981;78,7200–7204.
40. Busch C, Sakmann B. Synaptic transmission in hippocampal neurons: numerical reconstruction of quantal IPSCs. In: *Cold Spring Harbor Symposia on Quantitative Biology LV,* Cold Spring Harbor Laboratory Press, 1990;69–80.
41. Bartol TM Jr, Land BR, Salpeter EE, Salpeter MM. Monte Carlo simulation of miniature endplate current generation in the vertebrate neuromuscular junction. *Biophys J* 1991;59:1290–1307.
42. Faber DS, Young WS, Legendre P, Korn H. Intrinsic quantal variability due to stochastic properties of receptor-transmitter interactions. *Science* 1992;258:1494–1498.
43. Kullmann DM. Quantal variability of excitatory transmission in the hippocampus: implications for the opening probability of fast glutamate-gated channels. *Proc R Soc Lond B* 1993;253:107–116.
44. Colquhoun D, Jonas P, Sakmann B. Action of brief pulses of glutamate on AMPA/kainate receptors in patches from different neurones of rat hippocampal slices. *J Physiol* 1992;458:261–287.
45. Jonas P, Sakmann B. Glutamate receptor channels in isolated patches from CA1 and CA3 pyramidal cells of rat hippocampal slices. *J Physiol* 1992;455:143–171.
46. Major G. The physiology, morphology and modelling of cortical pyramidal neurones. *D.Phil Thesis. Oxford University* 1992.
47. Major G, Evans JD, Jack JJB. Solutions for transients in arbitrarily branching cables: II. voltage recording with somatic shunt. *Biophys J* 1993;65:423–449.
48. Feldmeyer D, Silver RA, Cull-Candy SG. Temperature dependence of EPSC time course at the rat mossy fibre-granule cell synapse in thin cerebellar slices. *J Physiol* 1993;459;481P.
49. Gulyas AI, Miles R, Sik A, Toth K, Tamamaki N, Freund TF. Hippocampal pyramidal cells excite inhibitory neurons through a single release site. *Nature* 1993;366:683–687.
50. Walmsley B. Central synaptic transmission: studies at the connection between primary afferent fibres and dorsal spinocerebellar tract (DSCT) neurones in Clarke's column of the spinal cord. *Prog Neurobiol* 1991;36:391–423.
51. Pierce JP, Mendell LM. Quantitative ultrastructure of Ia boutons in the ventral horn: scaling and positional relationships. *J Neurosci* 1993;13(11):4748–4763.
52. Hannay T, Larkman AU, Stratford KJ, Jack JJB. A common rule governs the locus of both short and long-term potentiation. *Current Biol* 1993;3(12):832–841.
53. Manabe T, Renner P, Nicoll RA. Postsynaptic contribution to long-term potentiation revealed by the analysis of miniature synaptic currents. *Nature* 1992;355:50–55.
54. Raastad M, Storm JF, Andersen P. Putative single quantum and single fibre excitatory postsynaptic currents show similar amplitude range and variability in rat hippocampal slices. *Eur J Neurosci* 1992;4:113–117.
55. Liley AW. Spontaneous release of transmitter substance in multiquantal units. *J Physiol* 1957;136: 595–605.
56. Bornstein JC. Spontaneous multiquantal release at synapses in guinea-pig hypogastric ganglia: evidence that release can occur in bursts. *J Physiol* 1978;282:375–398.
57. Ropert N, Miles R, Korn H. Characteristics of miniature inhibitory postsynaptic currents in CA1 pyramidal neurones of rat hippocampus. *J Physiol* 1990;428:707–722.
58. Silver RA, Traynelis SF, Cull-Candy SG. Rapid-time-course miniature and evoked excitatory currents at cerebellar synapses *in situ. Nature* 1992;355:163–166.
59. Liu G, Feldman JL. Quantal synaptic transmission in phrenic motor nucleus. *J Neurophysiol* 1992; 68(4):1469–1471.
60. Korn H, Bausela F, Charpier S, Faber DS. Synaptic noise and multiquantal release at dendritic spines. *J Neurophysiol* 1993;70(3):1249–1254.
61. Ulrich D, Luscher H-R. Miniature excitatory synaptic currents corrected for dendritic cable properties reveal quantal size and variance. *J Neurophysiol* 1993;69(5):1769–1773.
62. Melamed N, Helm PJ, Rahamimoff R. Confocal microscopy reveals coordinated calcium fluctuations and oscillations in synaptic boutons. *J Neurosci* 1993;13(2):632–649.
63. Korn H, et al., 1994;(Chap. 19).
64. Edwards FA, LTP is a long term problem. *Nature* 1991;350:271–272.
65. Kitzing E. von, Jones P, and Sakmann B. 1994; (Chap. 16).
66. Harris KM, Stevens JK. Dendritic spines of CA1 pyramidal cells in the rat hippocampus: serial electron microscopy with reference to their biophysical characteristics. *J Neurosci* 1989;9(8):2982–2997.

67. Kriebel ME, Llados F, Matteson DR. Spontaneous subminiature end-plate potentials in mouse diaphragm muscle: evidence for synchronous release. *J Physiol* 1966;262:553–581.
68. Stevens CF. Quantal release of neurotransmitter and long-term potentiation. *Cell* 1993;72(Suppl.): 55–63.
69. Clements JD, Lester RAJ, Tong G, Jahr CE, Westbrook GL. The time course of glutamate in the synaptic cleft. *Science* 1992;258:1498–1501.
70. Bliss TVP, Collingridge GL. A synaptic model of memory: long-term potentiation in the hippocampus. *Nature* 1992;361:31–39.
71. Malenka RC, Nicoll RA. NMDA-receptor-dependent synaptic plasticity: multiple forms and mechanisms. *TINS* 1993;16(12):521–527.
72. Barrett EF, Barrett JN, Martin AR, Rahamimoff R. A note on the interaction of spontaneous and evoked release at the frog neuromuscular junction. *J Physiol* 1974;237:453–463.
73. Stratford K, Mason A, Larkman A, Major G, Jack JJB. The modelling of pyramidal neurones in the visual cortex. In Durbin R, Miall C, Mitchison G, eds. *The computing neurone*, New York: Addison-Wesley; 1989;296–312.

Molecular and Cellular Mechanisms of Neurotransmitter Release, edited by Lennart Stjärne, Paul Greengard, Sten Grillner, Tomas Hökfelt, and David Ottoson, Raven Press, Ltd., New York © 1994.

19

The One-Vesicle Hypothesis and Multivesicular Release

Henri Korn, Cyrille Sur, Stephane Charpier, Pascal Legendre, and *Donald S. Faber

*Department of Cellular Neurobiology, INSERM U261, Institut Pasteur, 75724 Paris Cedex 15, France; and *Department of Anatomy and Neurobiology, The Medical College of Pennsylvania, Philadelphia, Pennsylvania 19129*

The one-vesicle hypothesis signifies that following an impulse an active zone releases the contents of only one synaptic vesicle, with a probability, p (1,2). This proposal is now taken as being "generally accepted" for central junctions and for peripheral ones as well (3,4) but without a clear perspective on its evolution and implications. For example, it raises mechanistic questions that remain unsolved (5). Some of them are related to a possible refractory period of release following an exocytotic event. Furthermore, it is important to recognize that, lacking a final verification with high resolution imaging, this concept relies primarily on the linkage, albeit often unexpressed, of the vesicular theory and its morphological correlates to the animodel distribution of quantal events originating from a simple release site.

Originally, with Poisson fits, where n is so large as to be undefined, this term was thought to represent the number of units available for release at a single junction (6) and was further equated with the size of the vesicular pool (see Katz (7)). As the superiority of the binomial model emerged, n became smaller and an identifiable entity, and there was an "inclination to assign this parameter to a finite number of release sites . . . based more on intuition than on experimental evidence" (8). At that time, the arguments were speculative. The major point was that by restricting the extent of either the releasing region or the "area" seen by the recording electrode, at frog and crayfish neuromuscular junctions, binomial n was in the range of the number of presynaptic release sites thought to contribute to the responses (9,10,11). While this congruence implies the release of no more than one quantum per active zone, the authors carefully avoided going as far as defining a vesicle as a quantal unit, with the possible exception of Wernig (12) who also considered the alternative that a quantum is due to the simultaneous discharge of several packets. In retrospect, it is surprising that this connection was not made, given the cultural dominance of the vesicular hypothesis and evidence that a vesicle could accommodate the amount of transmitter required to produce a quantal conductance at the frog and snake neuromuscular junctions (13).

During the same period the focus in the central nervous system was on the question of whether transmission is quantal (see refs. 14,15). Resolution of this with its associated uncertainties due to difficulties in assessing amplitudes of unitary events, required that statistical analysis of fluctuating responses be supported by morphological information. This information was obtained by staining the presynaptic cells after collecting the physiological data. In contrast to the more extensive peripheral junctions, these methods could be better combined in central neurons, since most of their afferents have a restricted set of terminals on a given target and can be stained with functionally inert dyes. While they enabled the formulation of the one-vesicle hypothesis, they have only been successful when the quanta could be clearly resolved (as discussed by Korn and Faber (16)).

Initial Studies of Inhibitory Connections

Figure 1 summarizes the main conclusion obtained with identified inhibitory interneurons presynaptic to the goldfish Mauthner (M-) cell, *at the level of its soma*. This preparation proved to be fortunate since most presynaptic glycinergic cells have a few terminals, which are localized to the region of the soma and axon hillock (17,18). Also, the quantal increments are due to the opening of a large number of Cl^- channels and their size can be raised well above the noise level by increasing the driving force for this ion (19,20). Equally important, and unlike the dendritic afferents (see next paragraph), there is only one active zone per terminal bouton (18).

The scheme presented in Fig. 1 seemed to be more parsimonious than others that incorporated multivesicular release since it was assumed, by analogy with the neuromuscular junction, that: 1) there would be enough receptor sites to detect additional exocytotic events, and 2) the contents of one vesicle could account for the quantal conductance. It provided an internally consistent framework for interpreting experimental data on both evoked and spontaneous release.

Simultaneous intracellular recordings were used to study evoked release in vivo and to analyze response fluctuations according to a binomial model. The main results of these investigations can be summarized as follows:

1. Amplitude histograms of evoked inhibitory postsynaptic currents (IPSCs) could exhibit several distinct classes that were equally spaced and were, therefore, in agreement with the quantal theory, although they were masked in a number of experiments by the instrumental noise.
2. For many neurons the optimal binomial parameter n was equivalent or close to the total number of boutons that each one established on the M-cell. Such is the case in Fig. 2A1, where this term was equal to ten, which was also the number of endings exhibited by the stimulated presynaptic cell (Figure 2A2).
3. After an impulse, each terminal thus appeared to release in an all-or-none man-

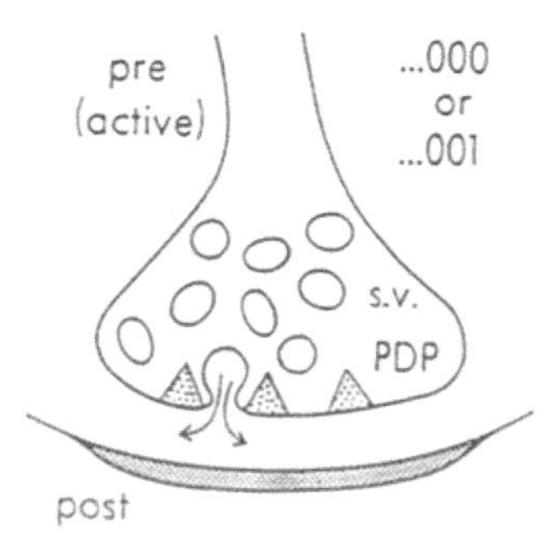

FIG. 1. The one vesicle hypothesis. Diagram of a synaptic bouton having only one active zone with its associated presynaptic dense projections (PDP) and a postsynaptic (post) membrane receptor matrix (shaded). The binary notation to the right signifies that at most one synaptic vesicle (s.v.) releases its contents into the cleft at a given time, either at rest or when the presynaptic (pre) ending is depolarized (active) by an action potential. This secretory process is stochastic and evokes a quantal response (q) in the target cell. Modified from Faber and Korn, ref. 2.

ner, and it represents an independent functional unit. Histological and ultrastructural studies of HRP-labeled boutons have shown (18) that most of them contain only one active zone (Fig. 2B–C). The quantum defined indirectly by the statistical analysis would then reflect the postsynaptic effect of the amount of transmitter issued by each release site.

It should be stressed here that, as for all quantal studies, results are model dependent (16). In this study, a simple binomial, which assumes that the release probability is the same for all terminals of a given afferent neuron, was used. Although not all fits were statistically significant (1,21) its attractiveness appeared to be the correlation between statistical and structural parameters. Also, optimal values of n for a given connection could sometimes differ by one or two units, depending upon the estimate of background noise used by the deconvolution algorithm (i.e., during extraction of binomial parameters from "noisy" histograms), and confidence limits could not be placed on these estimates.

A strong support for the results obtained with the binomial statistics is the similarity between the quantal size that they predicted and the spontaneously occurring miniature inhibitory currents recorded in voltage clamp. This agreement is shown in Fig. 3. The analog diagram in Fig. 3A provided the basis for normalizing inhibitory postsynaptic potentials IPSPs recorded in different cells and for converting quantal size in mV to its underlying conductance, g_q (20). This calculation was possible because E, the driving force for Cl^-, is, on average, twice the magnitude of the full-sized collateral IPSP, V_{coll} (see also Faber and Korn (2,22)), and it yielded a value in the range of 35 nS. This estimate implicates the opening of at least 1,000 Cl^--channels per q. Subsequently, an independent analysis of inhibitory quanta, which were isolated by blocking presynaptic impulses with tetrodotoxin (TTX) and could be distinguished from background noise (Fig. 3B1), gave the same results.

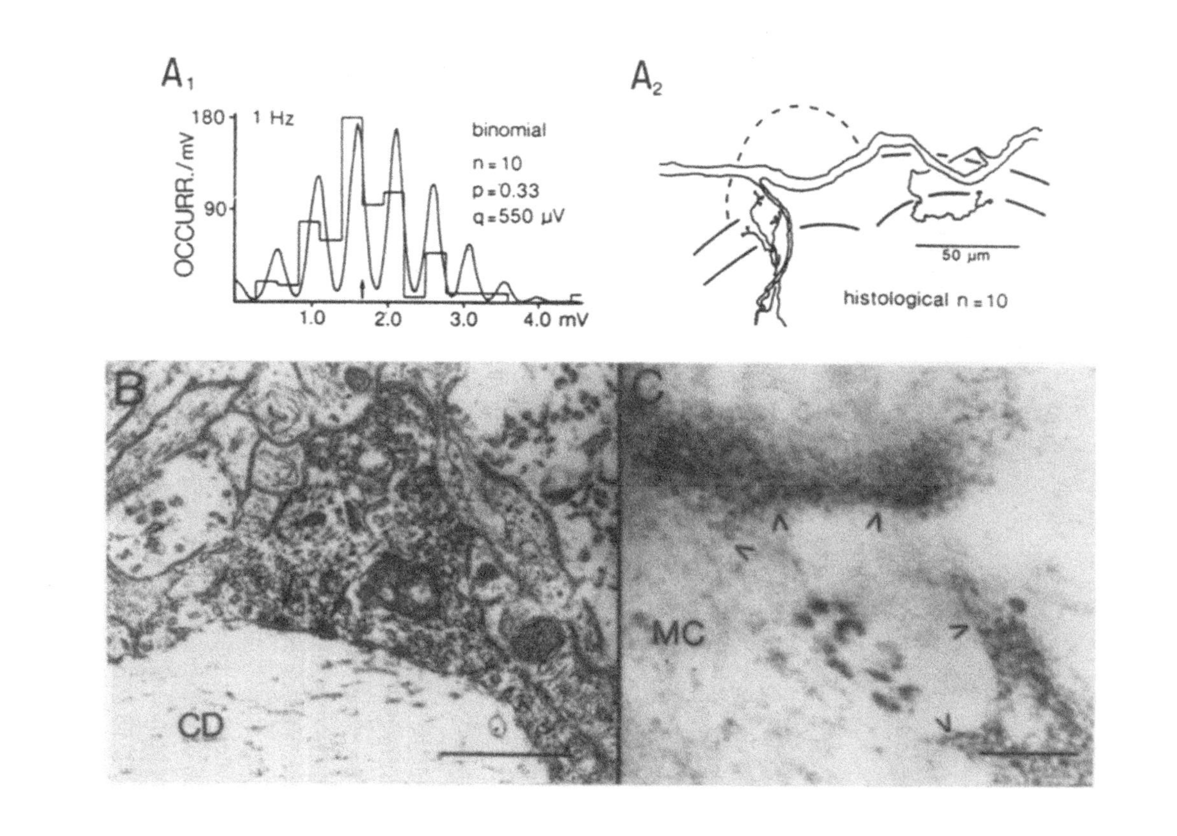

A1
180
90
OCCURR./mV
1 Hz
binomial
n = 10
p = 0.33
q = 550 µV
1.0
2.0
3.0
4.0 mV
A2
50 µm
histological n = 10
B
CD
C
MC

Amplitude histograms of these spontaneous events were unimodal, and Gaussians (Fig. 3B2) had means in the predicted range (23,24).

Quantal Variability and Postsynaptic Saturation

As seen in Fig. 3B2, measured quantal amplitudes were dispersed around a mean, and this variability exceeded that of the background noise, σ_n, with a net biological coefficient of variation (CV) of ~15% (23,24) similar to that found in hippocampal slices (25). This needed to be tested, particularly since it is in contrast with the conclusion obtained when the deconvolution method was applied without constraints to the excitatory Ia to motoneuron synapses, that quantal size is invariant, a stability taken as reflecting saturation of postsynaptic receptors (15,26). The latter would render it difficult to detect and compare the postsynaptic effects of one vesicle versus several, regardless of whether the term "saturation" refers to binding of all receptor sites, opening of all channels, or even a constant quantal waveform.

This question was considered by using Monte Carlo simulations of quantal currents for central junctions with a wide range of receptor complements (27). Structural and biophysical constraints of the synapse were included in the algorithm (Fig. 4A) and series of runs were conducted with a fixed set of parameters. The states of individual receptor-channel complexes following an "exocytosis" could be distinguished (Fig. 4B), and it appeared that: 1) at no time were all channels opened, and 2) a given channel behaved probabilistically (Fig. 4C), even though all binding sites could be occupied. Consequently, the peak amplitude and time course of a "quantum" (i.e., population response) reflects this stochastic property, which was confirmed in experiments with adult and embryonic M-cell preparations. Modeling one synapse alone yielded a quantal CV of about 10% to 20%, and this value should be even larger in neurons with postsynaptic matrices of different sizes.

The simulations also showed that quantal size would be relatively insensitive to the amount of transmitter in a vesicle, unless the ratio of molecules released to the number of available channels (with two binding sites each) was less than 5:1 (see also Busch and Sakmann, ref. 28). Thus, this approach cannot resolve the question posed here: namely, whether one or several vesicles are released by a given active

FIG. 2. Correspondence between the binomial parameter n and the number of active zones established by inhibitory interneuron on the Mauthner cell. **A1.** Amplitude histogram of 165 unitary inhibitory postsynaptic potentials (IPSPs) evoked by stimulations of a presynaptic neuron at a low rate (1 Hz) and a superimposed computer-determined best fit (curved line) obtained assuming a binomial process of quantal release, with the indicated parameters. **A2.** Camera lucida reconstruction of the HRP-filled interneuron and its relationship with the Mauthner cell (thick line) and its axon cap (dashed). Ten terminal boutons were identified, and this complement corresponds to the statistical n in A1 (see also text; from Korn and Faber, ref. 14). **B.** Ultrastructural characteristics of a stained unmyelinated club ending synapsing on a M-cell cap dendrite (CD). The presynaptic dense projections are indicated by arrows. **C.** En face view of an ethanolic phosphotungstic acid (EPTA)-stained terminal in contact with the Mauthner cell (MC) illustrating that each synaptic bouton contains only one presynaptic vesicular grid. The periphery of the bouton is outlined by open arrowheads. CA1.: 0.2 μm in **B** and **C.** From Korn and Triller, ref. 60.

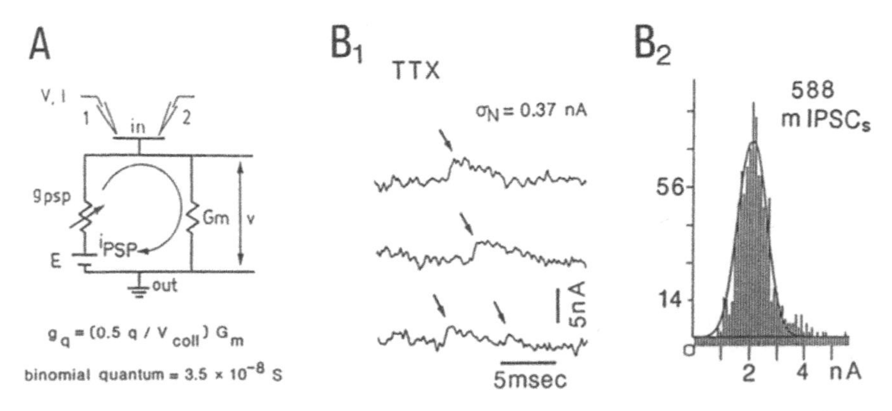

FIG. 3. Experimental verification of the quantal size predicted from binomial analysis. **A.** Calculations of quantal conductance (g_q). Equivalent circuit of the postsynaptic membrane, indicating that the inhibitory response is due to an increased conductance (g_{PSP}) that is variable and in series with a battery (E) representing the chloride driving force. The current (i_{PSP}) resulting from the opening of the corresponding channels flows across the input conductance (G_m) of the cell and produces a voltage drop (v) across the membrane. Two intracellular microelectrodes (V,I, 1 and 2) can be used to determine Gm and E, as well as the amplitude of the full-sized collateral inhibitory postsynaptic potential (IPSP), V_{coll}. Below is the equation for estimating g_q from the quantal size, q, derived from statistical analysis of fluctuating unitary IPSPs and its average value (modified from Faber and Korn, ref. 20). **B1,B2.** Unimodal distribution of spontaneous miniature currents recorded in voltage clamp after blocking presynaptic impulses with tetrodotoxin (TTX). **B1.** Examples of quanta (arrows), the smallest of which is almost obscured by instrumental noise (σ_N). **B2.** Amplitude histogram of a population of miniature inhibitory postsynaptic currents (IPSCs) in the same experiment, with a Gaussian fit of the 588 smallest ones. The mean quantal size was 2 nA, with $g_q \approx 30$ nS (i.e., close to the predicted value). The quantal variance was larger (see text) than σ_N^2 (modified from Korn et al., ref. 23, with remeasurements of the data for **Fig. 3B2**).

zone following an impulse, since the amount of transmitter packaged in a vesicle is not known for central synapses. Yet, it does indicate that multiple exocytoses would be effective only if this ratio was no more than 3:1. This could very well be the case, given that an enhancement of postsynaptic responses is observed when adjacent synapses are coactivated (29).

Morphological Correlates of Multivesicular Release

In contrast to observations in the M-cell soma and the neuromuscular junction, histograms of miniature events recorded in central neurons are often skewed (25,30,31), and this creates uncertainties for quantal analysis (16,32). The various explanations include the release of multiple vesicles (25,33) or distortions due to dendritic cable properties (34,35,36). In the case of the M-cell lateral dendrite, where cable properties are not a critical issue (19) histograms may be skewed, and evidence has been obtained that multivesicular released supports the one-vesicle hypothesis rather than contradicting it (37,38). Specifically, by again combining an

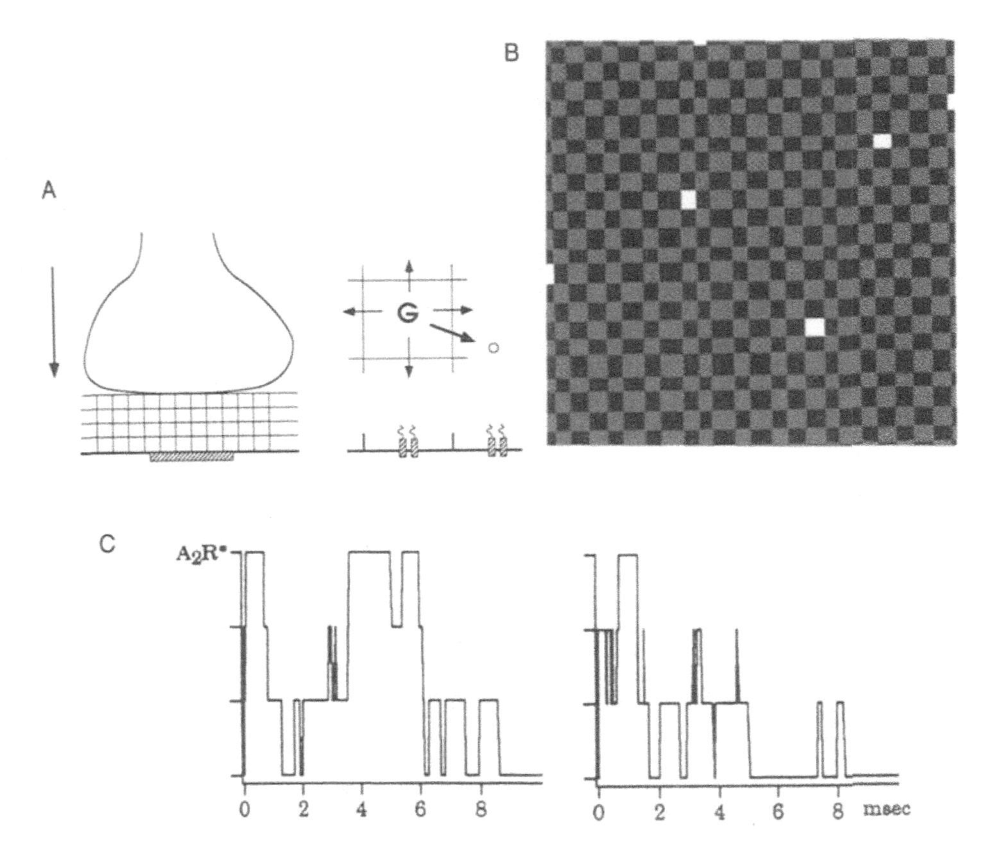

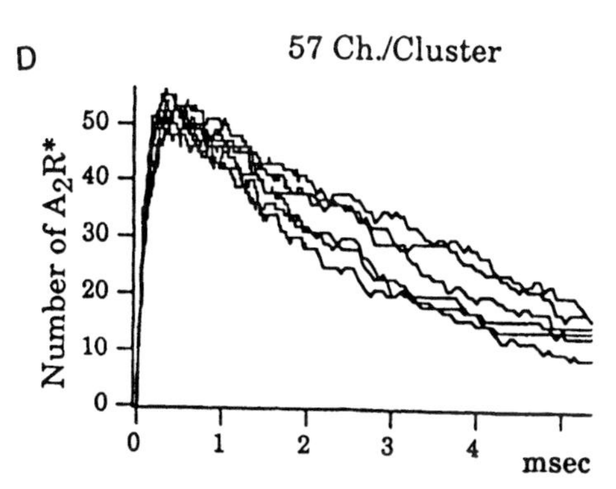

FIG. 4. Variability of quanta simulated with fixed parameters **A. Left:** transverse representation of a presynaptic bouton. The extracellular space is divided into four layers containing "cells" of 6×6×5 nm, and the postsynaptic membrane is associated with receptor-operated channels (hatched). **Right:** expanded view of one such compartment, containing a molecule of agonist (here glycine, G) free to diffuse in all directions (arrows); the circle to the right indicates its next location. Both receptor binding sites must be occupied for a chloride channel opening (below). **B.** Absence of functional "saturation" following an exocytotic event. The different states of individual receptors (squares) within an aggregate at the peak of the quantal response (0.5 mS) are color coded. Each complex can be bound to two transmitter molecules and be opened (red) or closed (green), or it can be single-bound (blue) or unoccupied (white). **C.** State history of a channel chosen randomly from the matrix, during two successive runs of the program. Openings occur only at the highest level of the ordinate (A_2R^*) and are both stochastic and repetitive. **D.** Superimposed simulated quanta (n = 6) obtained with a cluster containing 57 channels, after the "release" of 10,000 molecules of glycine. Note the fluctuations of the peak amplitude and of the decay phase, which are accounted for by the unpredictable behavior of the channel population, similar to that shown in C (modified from Faber et al., ref. 27).

analytical algorithm with ultrastructural methods, we found that in the presence of TTX asymmetrical histograms are multiquantal, the number of components being equivalent to the maximum complement of release sites in individual afferent boutons. Thus, large "miniature" events isolated by TTX are presumably due to synchronous exocytosis at adjacent active zones.

Recordings of synaptic noise were made in current clamp from the lateral dendrite about 300 μm from the soma. As in the latter, spontaneous synaptic noise is predominantly inhibitory and presynaptic impulses can be blocked by TTX (Fig. 5A). Long segments of such activity were collected and numerous individual components were recognized and measured with an automatic program described and assessed elsewhere (Ankri et al., 39). Visual inspection of sample traces already revealed a wide range of event amplitudes, despite the presence of TTX (Fig. 5B).

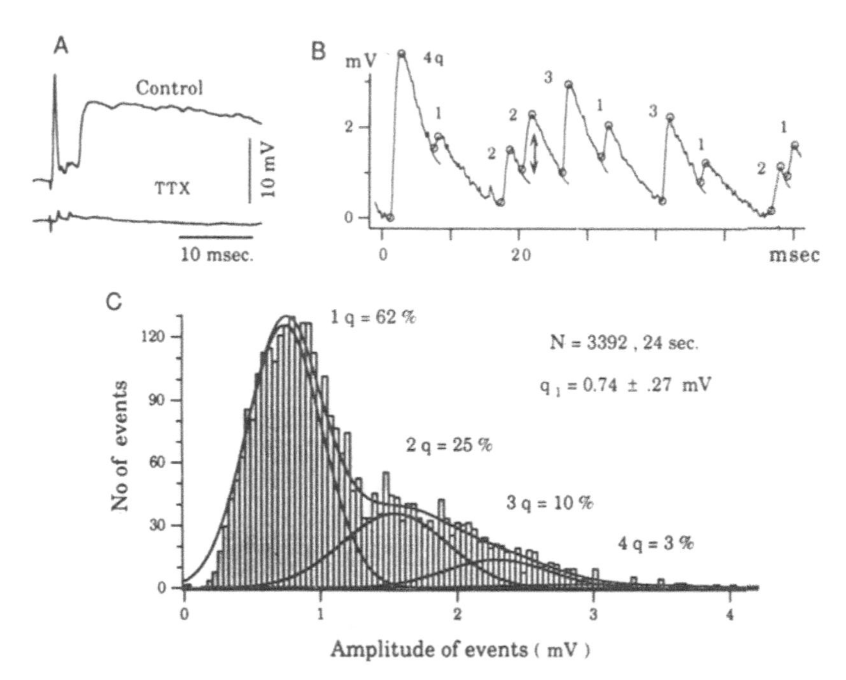

FIG. 5. Discrete size classes of spontaneous depolarizing dendritic inhibitory postsynaptic potentials (IPSPs) recorded in absence of presynaptic impulses. **A.** Antidromic spike and recurrent collateral IPSP before (above) and during (below) superfusion with tetrodotoxin (TTX)-containing saline. **B.** Typical segment of depolarizing inhibitory synaptic noise, showing detection of individual events (pairs of open circles) and measurements of their amplitudes (quantal contents indicated next to each response). The vertical line with arrowheads illustrates that amplitudes of responses within bursts were measured after extrapolating the decay of the preceding component. **C.** Amplitude distribution of 3,392 events detected over a period of 24 seconds, after complete block of evoked synaptic responses (**A,** below). Four Gaussian subdistributions (solid curves) were extracted by the stochastic-approximation expectation maximization (SAEM) algorithm; their means are integer multiples of that of the first one (q_1, as specified), and their relative weights (in percentage) decrease progressively (from Korn et al., 1993, unpublished).

Corresponding histograms were skewed to the right (Fig. 5C), with a predominance of smaller sizes, reminiscent of those attributed to single quanta in the case of compound somatic control recordings (24). On the other hand, putative multiquantal peaks were far less apparent than when amplitude distributions of somatic unitary events had been assessed visually (23,24). Such distributions of dendritic responses might actually represent mixtures of Gaussians, suggestive of multivesicular release if separated by quantal increments, and this turned out to be the case when the histograms were processed as described in the next paragraph.

A specific algorithm was used to identify combinations of an unknown number, k, of Gaussians within the histograms, with the hypothesis that a variable number of quanta occurred simultaneously, but without any assumptions about their sizes or separations and standard deviations. This method, called the stochastic-approximation version of the expectation-maximization procedure (40) randomly selects the starting Gaussian parameters. The problem was to: 1) determine k, and 2) estimate their individual means (μ_j), standard deviations (σ_j), and weights (w_j) on the basis of a maximum likelihood criterion (L). For this purpose we used a modified version of the expectation-maximization (EM) algorithm (41) that is designated (42) as the stochastic-approximation EM (SAEM) and initially incorporates a random, or simulation, procedure (39,43) between the E and the M steps. It has similarities with simulated annealing since it avoids local minima. Starting with a distribution of N events x_i ($i=1,N$), the program randomly selects the parameters $\theta_j=(\mu_j, \sigma_j, w_j)$ defining k Gaussian functions (e.g., G_j). It then follows three steps. The first is estimation (E), in which the probability

$$P(G_j / x_i; \theta_j) = w_j \cdot \text{Gauss}(x_i, \theta_j) / \sum_{e=1,k} w_e \cdot \text{Gauss}(x_i, \theta_e)$$

is calculated, where Gauss (x_i, θ_j) is the probability value expected for x_i in θ_j. The second is simulation (S) to generate a complete data set by randomly selecting for each point a first order variable e_s defining a partition H_l, H_k of the population such that $P(H_j / e_s ; \theta_j) = 1$. In the third, the parameters are maximized (M) by $n+1$ successive iterations for each Gaussian according to

$$\mu_j^{n+1} = (P(H_j / x_i; \theta_j^n))^{-1} \cdot \sum_{i=1,N} P(G_e / x_i; \theta_j^n) \cdot x_i$$

$$\sigma_j^{n+1} = (P(H_j / x_i; \theta_j^n))^{-1} \sum_{i=1,N} P(G_e / x_i; \theta_j^n) \cdot [x_i - \mu_j^{n+1}] \cdot [x_i - \mu_j^{n+1}]^T$$

$$w_j^{n+1} = N^{-1} \cdot [\sum_{i=1,N} x_i / (P(G_j / x_i; \theta_j) = 1)]$$

Finally, the likelihood criterion is

$$\ln [L(x, \theta^n)] = \sum_{i=1,N} \ln [\sum_{j=1,k} P_j (\text{gauss}(x_i; \theta_j))].$$

Convergence was assessed with the Wilks (44) test. The solution including k Gaussians was preferred to that with $k+1$ when $2[\ln(L_{k+1}) - \ln(L_k)] < \chi^2$ for three degrees of freedom, accounting for the extra μ, σ, and w in θ_{k+1} ($p>0.05$).

The fits of Fig. 5C are representative of results obtained in this manner. In the control, there were five to six Gaussian classes; the first (postulated q_1) had a mean and standard deviation of 0.74 ± 0.22 mV, and the subsequent increments averaged 0.69 mV (not shown). After TTX, with the number of classes reduced to four; their magnitudes were 0.74 ± 0.28, 1.54 ± 0.39, 2.28 ± 0.34, and 3.01 ± 0.66 mV, respectively, and their relative weights diminished progressively, as indicated. A total of 11 data segments with 1,500 to 5,000 events in each, obtained from the 7 best experiments were pooled; the average normalized sizes were $q_2 = 1.89 \pm 0.20\ q_1$, $q_3 = 3.04 \pm 0.37\ q_1$, and $q_4 = 4.44 \pm 0.62\ q_1$.

The corresponding weights of these classes were 54.5 ± 15.7, 32.1 ± 10.2, 10.8 ± 6.6, and 3.6 ± 2.1%, respectively (see also Korn et al. (38)). Given the high mean frequency at which the responses occurred (141 Hz in Fig. 5 and 112 Hz overall), it was necessary to determine whether the compound events were due to collisions of single quanta at short intervals, such that they were not distinguished as being discrete. Tests on the resolving power of the detection program were conducted on simulated data. Single quanta with constant amplitudes and kinetic parameters were mimicked in the absence and presence of instrumental noise, at mean frequencies ranging from 80–320 Hz. At the higher frequencies there were bursts of events in rapid succession. The sample traces in Fig. 6A provide examples where the computer could not discriminate all the underlying components (arrows), particularly when there were no clear inflections on the rising phases. This occurred when the interval between successive events was less than their time to peak. Hence, as shown in Fig. 6B, there was a small but significant fraction of apparent doublets and triplets even in the absence of background noise, and their proportions increased as a function of frequency.

The example of Fig. 6 confirms that the merging of independent "quanta" does lead to a substantial misallocation by the detection algorithm. Yet, it is possible to predict the number of collisions, assuming an exponential distribution of intervals and an inability to discriminate transients separated by less than their time to peak, ΔT, as detailed elsewhere (39). Briefly, the probability density function of inter-event intervals is $f(\Delta t) = \lambda e^{-\lambda \Delta t}$, where λ is the mean frequency. Then, the probability that n independent events cannot be distinguished but appear as one is $p_n E = e^{-\lambda \Delta T}(1 - e^{-\lambda \Delta T})^{n-1}$. In the case of Fig. 5C, the expected relative weights of q_1 to q_4 would have been 86.8%, 11.5%, 1.5% and 0.2%, respectively, for $\Delta T = 1.0$ msec, these last three values being appreciably smaller than the observed ones; that is, compound responses cannot be attributed to collisions alone.

While addressing this statistical aspect of the problem, we also questioned whether the structure of dendritic inputs is the same, or different, from those on the soma. By combining ethanolic phosphotungstic acid (EPTA) staining (18) with postembedding immunocytochemistry, it is possible to identify the contours of terminal boutons and determine if they were glycinergic or γ-aminobutyric acid-ergic (GABAergic). Active zones were recognized as distinct entities on the basis of: 1) a continuous stain associated with the presynaptic dense projections (PDPs), probably facing the postsynaptic differentiation, and 2) a distance between adjacent release sites of at least twice the average separation of PSDs within a grid. Figure 7A1–A3,

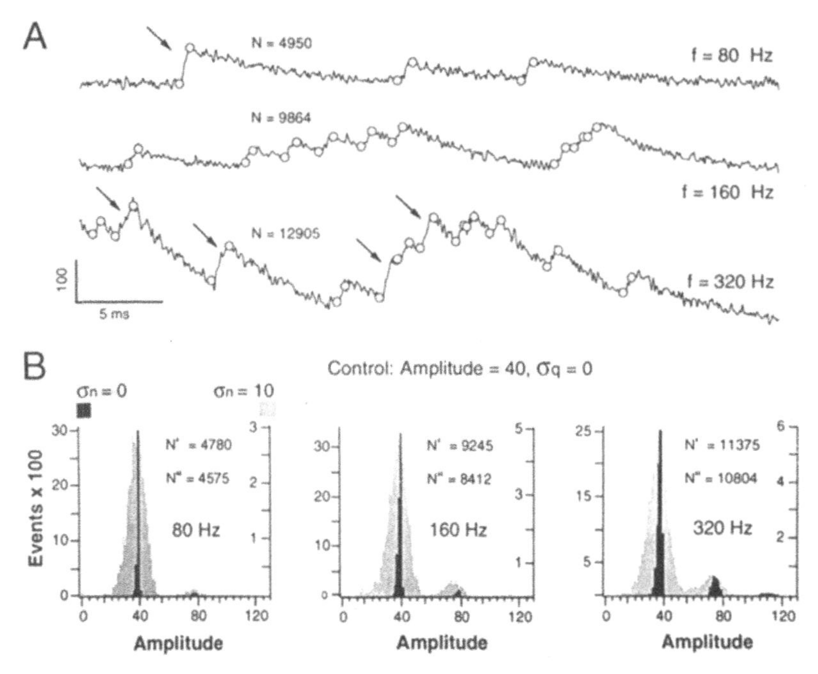

FIG. 6. Automatic detection and measurement of simulated events at different frequencies. Simulated quanta had the same amplitude ($q = 40$, $\sigma_q = 0$) and invariant times to peak (0.8 msec) and decay times (5 msec), and were convolved with white noise. They occurred randomly, at the indicated average frequencies (f). **A.** Sample sweeps at 80, 160, and 320 Hz (from top to bottom), with a background noise of 10; the total number (N) of responses for each corresponding data set is indicated above. A subroutine for detecting inflections was activated. Superimposed events produced by collisions and that still cannot be distinguished from each other by the program are designated by arrows. **B.** Amplitude histograms of the observed components, in absence (black, N') and presence (gray, N") of noise. Note that as the mean rate increases, and noise is added, there is 1) an apparent loss of single quanta (compare N, N', and N") that can be explained by 2) a progressively larger number of doublets, and the appearance of triplets at 320 Hz, due to collisions, and 3) a greater variance of each peak, even in the absence of background noise (see also test—from Ankri et al., 39).

FIG. 7. Morphological substrate for multiple release sites in dendritic inhibitory bouton. A_1–A_3. En face views of active zones in three terminal boutons (outline by arrowheads) stained with ethanolic phosphotungstic acid (EPTA) and having—from left to right—one, two, and three release sites (arrows—same calibration in the three panels). **B.** Transverse views of three other endings (labeled b_1–b_3) with one or more active zones. The illustrated terminals were glycinergic, as revealed with gold particles. **C.** Plot of active zone (A.z.) area (ordinate) versus the size, in μm^2, of the corresponding boutons (abscissa) for 79 histochemically identified inhibitory endings randomly sampled along the lateral dendrite of three preparations. The two parameters are not correlated ($r = 0.17$). (Sur, Triller, and Korn, unpublished.)

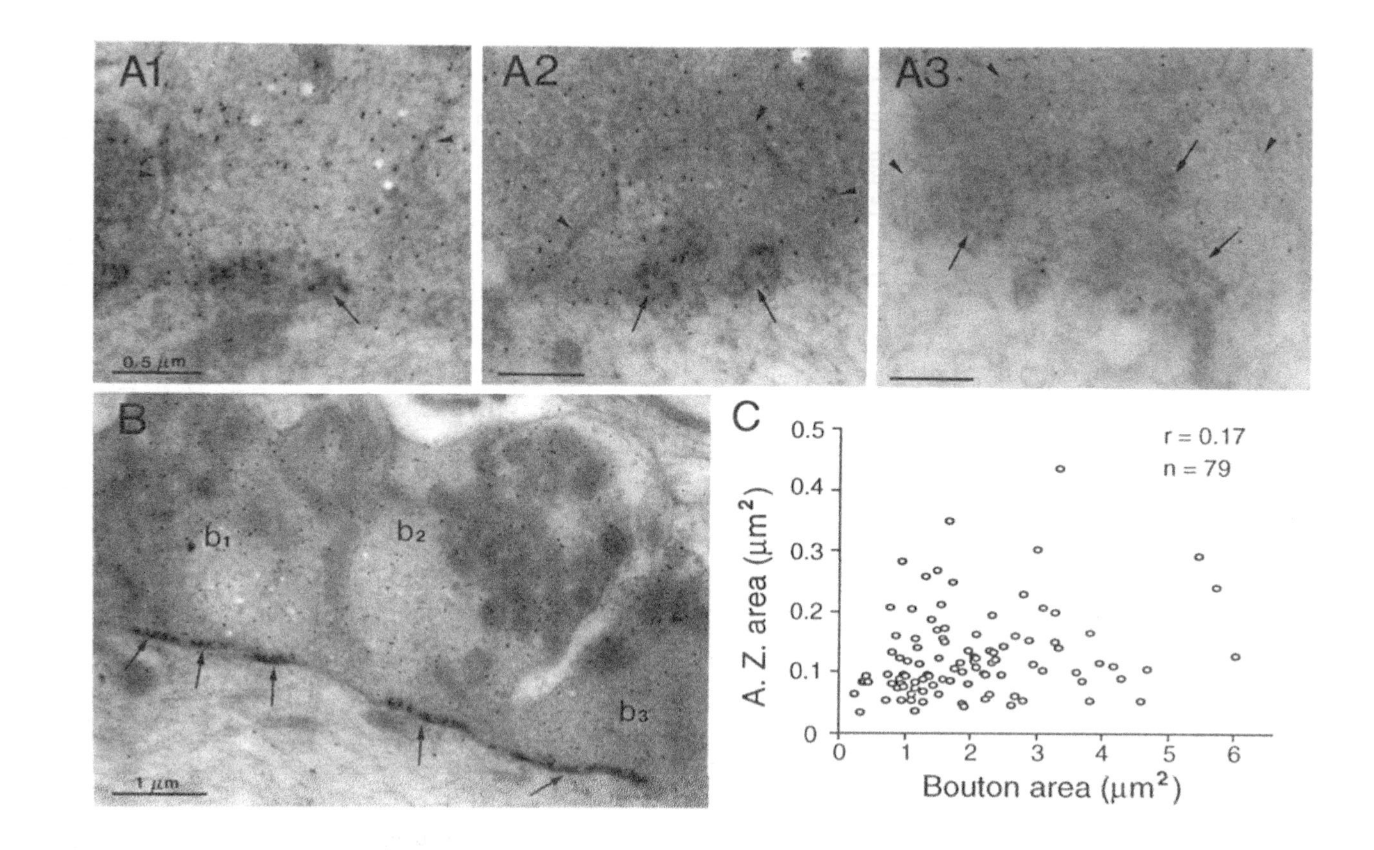
A1
0.5 μm
A2
A3
B
b1
b2
b3
1 μm
C
r = 0.17
n = 79
0.5
0.4
0.3
0.2
0.1
0
A. Z. area (μm²)
0
1
2
3
4
5
6
Bouton area (μm²)

and B show that the terminals, which correspond to the small vesicle boutons (SVBs) of Nakajima, (45) can have more than one active zone (see also Tuttle et al. (46)). In 87 glycinergic and 96 GABAergic profiles, the proportions having one (63.2% and 67.7%), two (29.8% and 30.2%), three (6.9% and 1%), and four (0% and 1%) active zones were statistically the same (47). In another 115 boutons whose borders were not as clearly defined, the distribution was similar (i.e., 61.7%, 33%, 4.3%, and 0.8%, respectively, when grouping observations for the two transmitters). The number of grids was not related to the size of the terminals (Fig. 7C), a finding that may be relevant to the mechanisms underlying multiquantal release. Specifically, the areas of glycinergic boutons with one and two active zones were $3.04 \pm 2.15\ \mu m^2$ (n=43) and $3.22 \pm 1.56\ \mu m^2$ (n=24), respectively; at GABAergic ones they were $2.48 \pm 1.29\ \mu m^2$ (n=36) and $2.71 \pm 2.12\ \mu m^2$ (n=18). In addition, the surface areas of the release sites were independent of the transmitter in the endings and their dendritic location.

A comparison between the physiological and morphological data was made by pooling the results of the TTX experiments conducted on the dendrite. It is presented in Fig. 8A, which shows that, theoretically, collisions of single quanta would be too infrequent to provide the basis for the composite events. In contrast, the proportions of IPSPs representing one to four postulated quanta are quite similar to the relative weights of profiles having the equivalent amount of release sites. This correspondence suggests that each terminal on the M-cell dendrite can release synchronously as many vesicles as the number of active zones it contains (Fig. 8B). Consequently, even in the presence of TTX, spontaneous responses are not necessarily monoquantal, a phenomenon that might underlie the skewed configuration of distributions, not only in the M-cell but also in other preparations. But, as shown below, this does not rule out the "one vesicle hypothesis," first advanced after using the binomial model to analyze evoked responses.

Evidence for the Synchronization Model

The scheme of Fig. 8B is supported by the results of experiments where the progressive effect of low Ca^{2+} and high Mg^{2+} on synaptic noise was studied after the disappearance of the collateral IPSP and of large events. At the onset of this block of synaptic transmission there were three Gaussians (Fig. 9A) separated by constant increments, as in TTX. Specifically, q_1 equalled 0.67 ± 0.27 mV, while q_2 and q_3 equalled 1.40 ± 0.43 and 2.23 ± 0.25 mV, respectively; thus, they were 2.1 and 3.3 q_1 (a larger component did not have enough events to be fit). These values are comparable to those for the first three peaks extracted by the SAEM algorithm from control data where, for example, the size of the postulated q_1 was 0.66 ± 0.21 mV (Fig. 9B). Furthermore, as the probability of release was further diminished, the number of classes in the dendrite was reduced to one (i.e., to singlets (Fig. 9C)); its mean and variance (0.64 ± 0.23 mV) were close to those of the earlier composite histograms, thus providing an independent validation of this protocol. The antidromically evoked responses in Fig. 9D demonstrate that the block of evoked re-

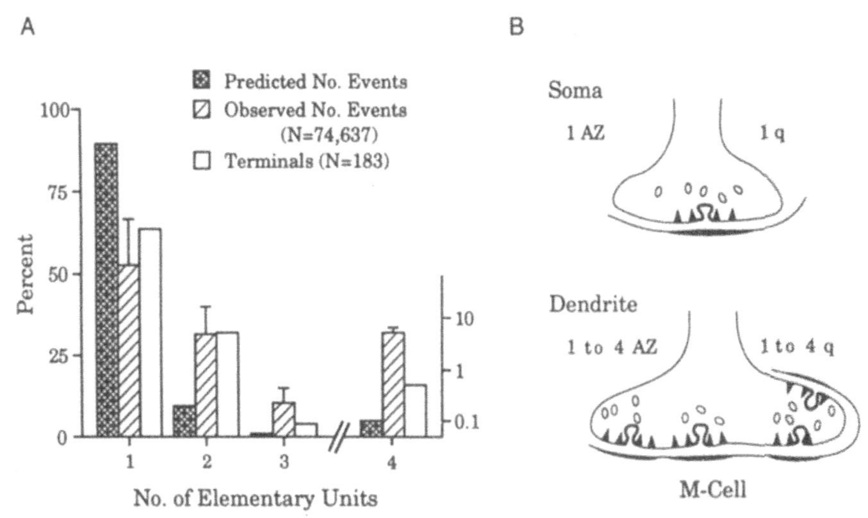

FIG. 8. Morphofunctional correlation favoring the concept of synchronized release. **A.** Histograms comparing the releative weights of the predicted (gray) and observed (empty) number of events with one to four quanta in the presence of tetrodotoxin (TTX) with the distribution of identified terminals (hatched) having the corresponding complement of active zones. The size (N) of the experimental data sets are as indicated. The physiological results are means and standard deviations from 16 pooled experiments; the overall frequency, 112 Hz, was used to calculate the predicted values, assuming that only single quanta were issued independently (i.e., that larger components would be due to collisions). The ordinate scale is linear, except for the fourth class of elementary units, for which it is logarithmic. Note that: 1) this assumption alone cannot account for the observed distributions of multiquantal responses, and 2) there is better agreement between the actual quantal contents and the corresponding numbers of release sites in the presynaptic population. **B.** Extension of the one-vesicle hypothesis, as formulated for the soma (above) incorporating synchronized multivesicular release in the case of the more complex terminals on the dendrite (below) (from Korn et al., unpublished).

lease was already effective at a time when the histogram of spontaneous events was still skewed (Fig. 9A). It should be stressed that the action of the modified divalent cation solution was variable in time and effectiveness, probably due to the deep location of the M-cell and its afferent synapses within the brain. Nevertheless, in three out of eight experiments a single Gaussian could be isolated, and in another two there were only singlets and doublets, with a clear predominance of the former (i.e., $w_1 \geq 80\%$ (38)). As with the example of Fig. 9, this simplification of the distributions was associated with a decrease in the overall event frequency, indicative of the lowered probability of release.

Previously undescribed oscillations of spontaneous inhibitory noise provide additional support for synchronized release and hints about its underlying mechanism. Long-lasting segments of control recordings from the dendrite displayed a striking periodicity in the range of 50 Hz that was obvious visually. More remarkably, this regularity persisted in the absence of presynaptic impulses (Fig. 10) and when the medulla was exposed to the low Ca^{++} solution. It could be detected by calculating

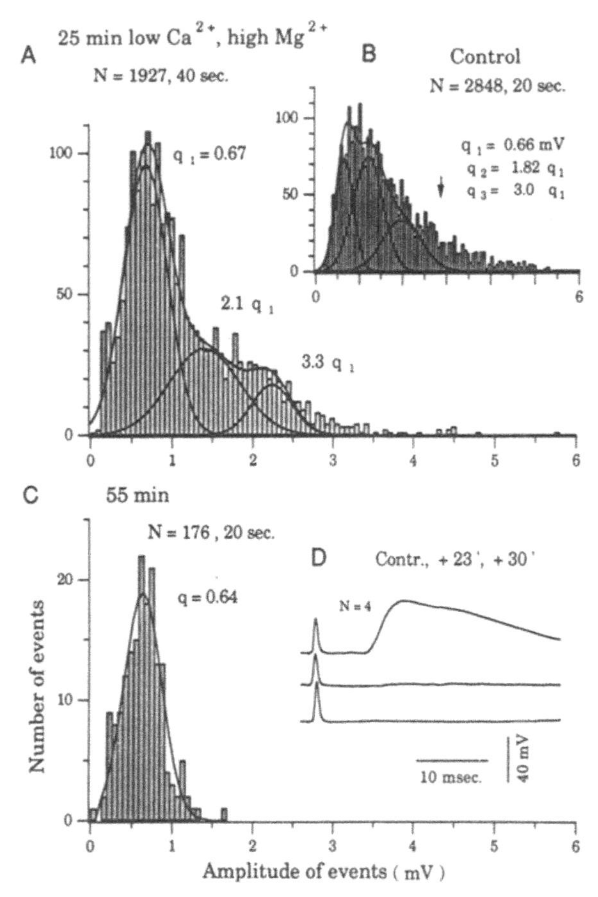

FIG. 9. Isolation of single quanta in the dendrite in conditions of reduced probability of release. **A.** Histogram of inhibitory postsynaptic potential (IPSP) amplitudes recorded in the dendrite 25 min after the onset of superfusion with a modified divalent cation solution (50 μM CNQX and 100 μM APV were used throughout this experiment). The stochastic-approximation expectation maximization (SAEM)-fitting algorithm extracted three amplitude classes, suggesting that there was still multivesicular release at this stage. **B.** Control distribution with Gaussians for the three smallest components (values left of arrow), although the whole range of IPSPs was fit. The indicated increments of the class means are compatible with a multiquantal process. **C.** After a total of 55 min in the low Ca^{++} solution, only the smallest quantal component (q_1) occurring at a low rate remained, and its distribution was comparable to those in **A** and **B**, with similar means and standard deviations (see text). **D.** Averaged antidromic spikes and full-sized collateral IPSPs at the indicated times, showing persistent block of impulse-dependent synaptic transmission shortly after exposure of the brain to lowered Ca^{++} and elevated Mg^{++} concentrations (from Korn et al., unpublished).

the autocorrelation function, even when not prominent in the voltage traces (Fig. 10A). The origin of this phenomenon has been puzzling since, if TTX indeed blocks the impulse activity in the deepest afferents, it is difficult to invoke a presynaptic mechanism that would require the concerted discharge from a population of independent afferents. A more plausible explanation would involve individual active

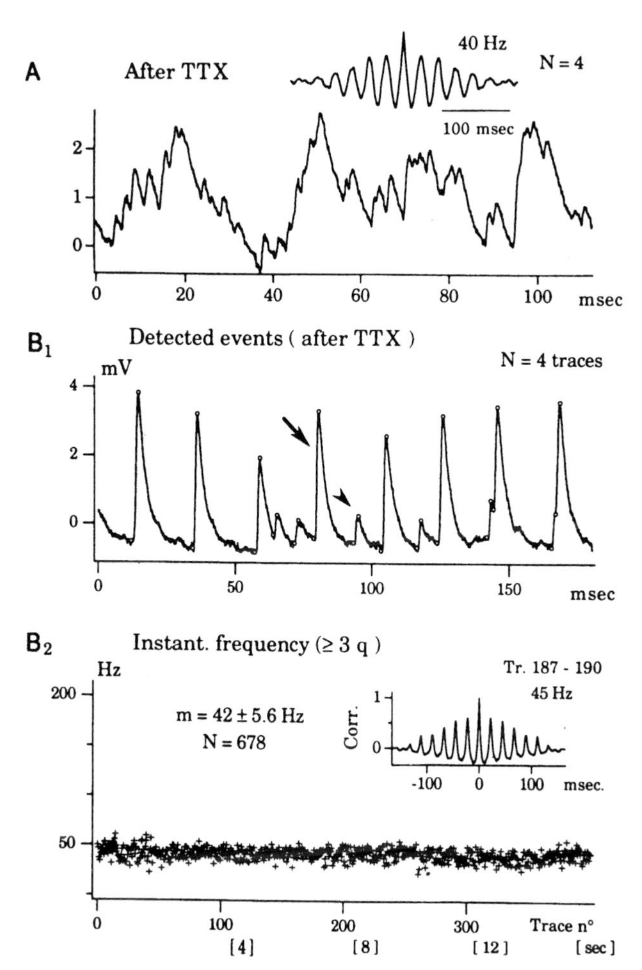

FIG. 10. Recurrent 40 Hz oscillations of multiquantal inhibitory postsynaptic potentials (IPSPs). **A.** Sample segment of synaptic noise recorded in the M-cell dendrite during exposure to tetrodotoxin (TTX), which led to complete abolition of the recurrent collateral IPSP and isolated the axon spike; the amplitude histogram had four size classes (not shown), as in the case of **Fig. 5. Inset.** Autocorrelation function, computed over 245 msec, showed a dominant 40 Hz rhythm. $\mathbf{B_1}$, $\mathbf{B_2}$: Evidence from the same experiment, suggesting that multivesicular release from individual terminals can undergo the same periodicity. $\mathbf{B_1}$. Sample data set, with clock-like appearance of multiquantal events (arrow), whereas singlets (arrowhead) seem more irregular. $\mathbf{B_2}$. Plot of instantaneous frequency versus time, of all events whose amplitudes were equal to or greater than three quanta (q) during more than 15 sec, the average rate being as signified. **Inset.** Autocorrelation function for a 245-msec epoch (corresponding to the four indicated traces) representative of this phenomenon (Korn et al., unpublished).

zones or boutons releasing at the observed frequency. Instantaneous frequency plots for all recorded events did not exhibit this trend, whereas it became particularly apparent when the frequency of only the large responses, corresponding to at least three quanta in the amplitude histograms (e.g., Fig. 5, 5C, 9A), dominated. A dramatic example of this selectivity is in Fig. 10B1, B2 where large IPSPs were predominant and had a clock-like periodicity for more than 10 sec. The average frequency, 42 ± 5.6 Hz, corresponded to the first harmonic in the autocorrelation function computed on all events over a shorter epoch. The most reasonable explanation is that this is an intrinsic rhythmicity reflecting repetitive release from individual boutons and being more apparent when occurring at endings that have multiple release sites. A similar observation at a tonic crayfish giant synapse has been attributed to vesicle cycling time (48). More generally, it presumably is indicative of a maximal release rate at the "activated" sites (e.g., in the face of a maintained increase in intracellular calcium concentration).

DISCUSSION

There are two novel findings here. One is that, in the presence of TTX, histograms of the remaining dendritic IPSPs are skewed and multiquantal. Thus, this toxin, which has yielded similar distributions in other studies (25,30,49,50,51) may not always be the ultimate tool for isolating true miniatures. Second, 40–50 Hz oscillations of spontaneous synaptic events can be observed in this reticulospinal neuron, not obviously to afferent firing.

Results described here were obtained using an automated procedure. The detection and measurement steps, which filtered high frequency instrumental noise and minimized operator control (39), did not always yield peaks in the dendritic distribution histograms as clear as those previously reported for compound somatic noise (23,24). However, comparable multiquantal classes were identified with the SAEM algorithm, which does not have an inherent tendency to converge to such solutions, as tested with simulated data. Also, it can distinguish mixtures of two Gaussians having the same means but different σ's, and it would not force Gaussian solutions to exponential distributions.

Adequacy of Dendritic Recordings

Several potential sources of skewness have to be considered. Since the program may not distinguish events starting within 0.8–1.0 msec of each other (i.e., one time to peak), the largest ones might have been composites. Corrections made for such collisions indicated, however, that even at high frequencies of occurrence this does not explain the observed proportions of the different size classes, which rather match those of active zones in terminals. Distortions due to cable properties along the dendrite and/or contamination of dendritic recordings by distant somatic events were also, if any, secondary. It has been shown that the dendritic length constant is at least 250 μm (19) and in some experiments Cl^--loading was restricted to the recording zone, such as in Fig. 9D where the control collateral IPSP triggered by antidromic stimulation did not contain the shorter latency somatic component. This, and occasional paired somatic and dendritic recordings, suggest that the two compartments of the M-cell may be more isolated than suspected thus far, probably related to a proximal dendritic constriction (52). In some experiments with TTX, however, when both regions were purposely loaded with Cl^-, the SAEM recognized an additional peak slightly smaller than the dendritic quantum (not shown). Finally, since glycinergic and GABAergic synapses are both found on the lateral dendrite (53) one could question whether the two types of junctions could account for distinct size classes in the histograms. But blockers to their receptors did not selectively affect any component (unpublished), and there were no subunits in the single Gaussians remaining in low Ca^{++} when assessing either amplitudes or shape indices.

Data from the M-cell of the Zebrafish Larvae

For some unspecified reasons, results obtained with the M-cell are at times taken as originating from a rather unusual system. They differ somewhat from those of some higher vertebrate preparations, particularly excitatory synapses in mammalian spinal cord, hippocampus, and neocortex where, in TTX, histograms tend to be skewed and less peaky and the number of opened channels per quantum is small. Given that most of the latter were obtained with either embryonic or young systems, it is appropriate to propose here that in some cases these differences may be age-related, in view of recent investigations with the embryonic zebrafish M-cell (54,55).

Patch clamp techniques have been successfully applied to the isolated whole brain of 52-hour-old zebrafish (*Brachydanio rerio*) where the M-cell can be recognized with light microscopy. Whole-cell somatic recordings disclosed an intense spontaneous synaptic activity that is primarily inhibitory and almost completely blocked by 0.5–1 μM strychnine. Although 1 μM TTX reduced the frequency of the responses, the remaining IPSCs had a wide range of amplitudes, even when 10 mM Mg^{++} was added to the superfusate (Fig. 11A). The miniature IPSCs had a mean time to peak and a decay time constant of 0.8 msec and 5.3 msec, respectively, similar to those found in the adult goldfish. However, the composite histograms obtained with the programs described hereon had a more complex envelope; they exhibited two initial peaks separated by an increment equal to the size of the first, followed by a skewed and extended tail (Fig. 11B). The first two components were well fitted by Gaussians, giving a mean quantal amplitude of about 35 pA (at a holding potential of −50 mV) with a coefficient of variation of the first one ranging between 0.2 and 0.3. Outside-out recordings showed two major classes of glycine-activated receptor channels, one with multiple conductance states (81–86, 59–69, and 40–43 pS), the other, which was dominant, having one conductance state of 40–43 pS. Since a putative conductance of 25 pS was used to calculate the number of channels in the "adult" quantum (1,23,24), the estimate of ~1000 should now be revised accordingly. Assuming that the first Gaussian in the composite histograms is a quantum, there would be 14–22 channels opened at its peak, and the correspondence between the mean decay time of miniature IPSCs and the mean open time of the channels (4–5 msec) suggests that they open once after exocytosis.

These data suggest that, in the CNS of the fish, a developmental process may

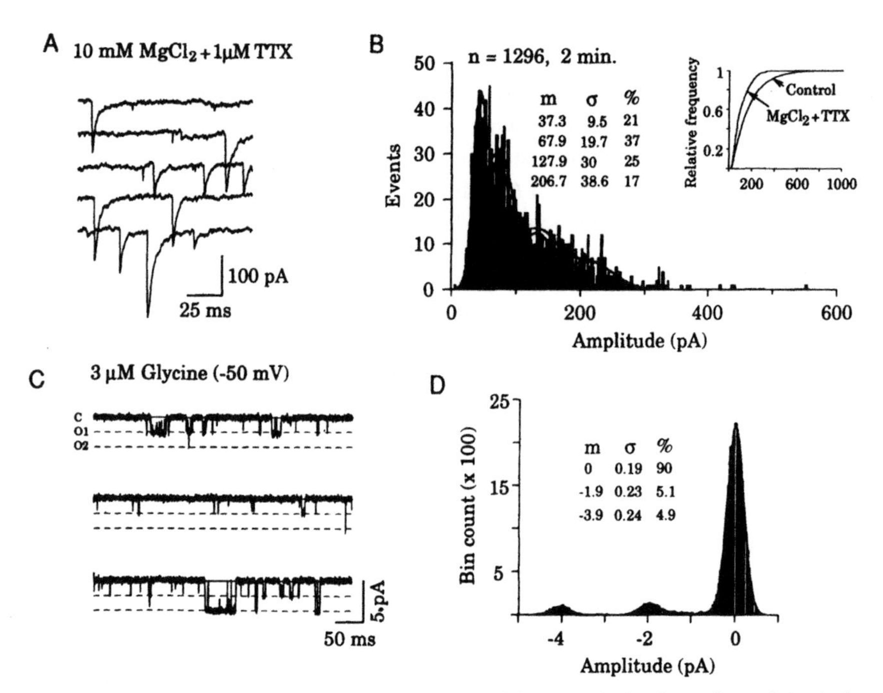

FIG. 11. Miniature inhibitory postsynaptic currents (IPSCs) and single-channel correlates in the embryonic zebrafish M-cell. Patch-clamp recordings at the somatic level, with a holding potential of −50 mV from cells 52 hours after fertilization. **A.** Whole-cell configuration, during continuous bath application of a solution containing 1 μM TTX, 10 mM $MgCl_2$, and 50 μM $CaCl_2$, when the synaptic noise was dominated by IPSCs. **B.** Amplitude distribution of events recorded in the same conditions; the stochastic-approximation expectation maximization (SAEM) algorithm revealed four amplitude classes with the indicated means (m), standard deviations (σ), and relative weights (%). Note that only the second is an integer multiple of the postulated "quantal" unit and that there is a wide range of amplitudes. **Inset.** Cumulative amplitude histograms for events recorded before (control, N = 3,861, in 20 sec) and during superfusion with the Na^+ and Ca^{2+} blockers (N = 1,296, in 2 min) showing only a partial effect of these compounds on the IPSCs amplitudes. **C.** Single-channel currents evoked by application of 3 μM glycine to an outside-out patch. In this example, in addition to the closed (c) state there are two open levels (01,02—horizontal dashed lines). **D.** Point-per-point amplitude histogram of the single-channel activity, with Gaussian fits (data segment = 15 sec) and indicated parameters. The mean currents for the two open states correspond to conductances of 38 and 78 pS, respectively. Assuming that the smaller conductance arises from channels which are activated synaptically, the first peak in **B** represents the opening of about 20 channels during a quantum (from 55).

account for at least some of the different interpretations and uncertainties related to quantal analysis. The complex distribution of IPSC amplitudes of the 52 hours M-cell could be due to: 1) multivesicular release (although, as in the adult, most somatic presynaptic endings only have one active zone) (P. Rostaing and A. Triller, personal communication); other hypothesis being that, 2) as suggested by outside-out recordings (55), two classes of glycine receptors having different main conductance states can be grouped in postsynaptic clusters; or 3) multiple receptor aggregates either facing individual release sites (3,25) or activated by transmitter diffusion from adjacent junctions (29). In any case, the approximately fifty-fold difference in quantal conductance from the embryonic to adult M-cell confirms proposals that this parameter is correlated with the input resistance of the M-cell (22), which drops from >50 MΩ (55) to ≤1 MΩ in the adult goldfish.

Overall Perspective

Despite all the evidence in its favor, final confirmation of the one-vesicle hypothesis will only be obtained with knowledge of the transmitter contents of single packets and with imaging of vesicles during release. The most commonly evoked alternative that of, postsynaptic quantization due to saturation, cannot be ruled out, but this view is less satisfactory, particularly since morphofunctional correlations point to a presynpatic locus. This is especially evident in the case of the adult M-cell dendrite, where our findings are easily explained without the need to implicate multiple receptor clusters, the responses of which should have remained synchronized in low Ca^{++}. Also, the CV of monoquantal events (0.31 ± 0.1, $n = 33$ from 16 experiments), extracted with minimal error by the SAEM, is greater than the variability of the apparent size of postsynaptic receptor matrices, as judged by immunocytochemistry (56). Finally, observations of a nonzero variance waveform of a quantum (27) and facilitated synaptic conductances in the presence of an elevated background level of transmitter (29) are incompatible with saturation. In any case, all postsynaptic binding sites being occupied at the peak of a quantum, due to the release of a large number of transmitter molecules, is not incompatible with the one-vesicle hypothesis.

The equivalence found in earlier studies between the relative numbers of active zones and recorded quanta does not necessarily imply that the probability of spontaneous exocytosis per site, p_s, is the same at all inhibitory boutons. Even if synchrony dominates, it most likely is not the rule, and may even wax and wane, as with the oscillations. If so, p_s may be larger at boutons having three and four active zones (4.7%) than suggested by the fraction of IPSPs with three and four quanta (~10% in TTX), since they also would contribute smaller events. This equivalence does not imply that release is synchronous in all terminals having multiple active zones. Rather, synchronicity could be specific to endings with closely spaced release sites, as is the case with the inhibitory boutons on the M-cell dendrite, which have similar sizes. Regardless, it is difficult to predict if spike-triggered exocytotic events are independent in such boutons, as they appear to be in two examples of large terminals having numerous synaptic contacts (57,58,59) or synchronized, which could produce quantal distributions with unusual statistical properties.

In a recent paper (A.I. Gulyas et al., Nature, 1993, 366:683–687) it was shown that EPSPs are evoked in hippocampal interneurons by single release sites; their distribution is compatible with the notion that transmission is monoquantal at this type of connection - see also O. Arancio, H. Korn, A. Fulyas, T. Freand and R. Miles. *J Physiol.* (Lond), in press.

ACKNOWLEDGMENTS

We thank Dr. A. Triller for supervising the morphological studies and N. Ankri for computer programming. This work was supported in part by DRET (92/058) and NIH (NS-15335) grants.

REFERENCES

1. Korn H, Mallet A, Triller A, Faber DS. Transmission at a central inhibitory synapse. II. Quantal description of release with a physical correlate for binomial n. *J Neurophysiol* 1982;48:679–707.
2. Faber DS, Korn H. Binary mode of release at central synapses. *Trends Neurosci* 1982a;5:157–159.
3. Edwards F. LTP is a long term problem. *Nature* 1991;350:271–272.
4. Stevens CF. Quantal release of neurotransmitter and long term potentiation. *Cell 72/Neuron 10* 1993;55–63.
5. Triller A, Korn H. Activity-dependent deformations of presynaptic grids at central synapses. *J Neurocytol* 1985;14:177–192.
6. del Castillo J, Katz B. Quantal components of the end-plate potential. *J Physiol (Lond)* 1954;124: 560–573.
7. Katz B. *The release of neural transmitter substances*, Springfield, Illinois: Charles C Thomas; 1969.
8. Martin AR. Junctional transmission. II. Presynaptic mechanisms. In: Brookhart JM, Mountcastle VB, Sec. eds. *Handbook of Physiology, The Nervous System. Vol I. Cellular Biology of Neurons, Part I. Americ Physiol Soc* 1977;329–355.
9. Atwood HL, Johnston HS. Neuromuscular synapses of a crab motor axon. *J Exp Zool* 1968;167: 457–470.
10. Zucker RS. Changes in the statistics of transmitter release during facilitation. *J Physiol (Lond)* 1973; 229:787–810.
11. Wernig A. Estimates of statistical release parameters from crayfish and frog neuromuscular junctions. *J Physiol (Lond)* 1975;244:207–221.
12. Wernig A. Localization of active sites in the neuromuscular junction. *Brain Res* 1976;118:63–72.
13. Kuffler SW, Yoshikami D. The number of transmitter molecules in a quantum: an estimate from iontophoretic application of acetylcholine at the neuromuscular junction. *J Physiol* 1975;251:465–482.
14. Korn H, Faber DS. Regulation and significance of probabilistic release at central synapses. In: Edelman GM, Gall WE, Cowan WM, eds. *Synaptic Function*, New York: Wiley, 1987;57–108.
15. Redman S. Quantal analysis of synaptic potentials in neurons of the central nervous system. *Physiol Rev* 1990;70:165–198.
16. Korn H, Faber DS. Quantal analysis and synaptic efficacy in the CNS. *Trends Neurosci* 1991;14: 439–445.
17. Triller A, Korn H. Morphologically distinct classes of inhibitory synapses arise from the same neurons: ultrastructural identification from crossed vestibular interneurons intracellularly stained with HRP. *J Comp Neurol* 1981;203:131–155.
18. Triller A, Korn H. Transmission at a central inhibitory synapse. III. Ultrastructure of physiologically identified terminals. *J Neurophysiol* 1982;48:708–736.
19. Faber DS, Korn H. Electrophysiology of the Mauthner cell: basic properties, synaptic mechanisms, and associated networks. In: Faber DS, Korn H, eds. *Neurobiology of the Mauthner cell*. New York: Raven Press; 1978;47–131.
20. Faber DS, Korn H. Transmission at a central inhibitory synapse: I. Magnitude of unitary postsynaptic conductance change and kinetics of channel activation. *J Neurophysiol* 1982b;48:654–678.
21. Korn H, Mallet A. Transformation of binomial input by the postsynaptic membrane at a central synapse. *Science* 1984;225:1157–1159.
22. Faber DS, Korn H. Unitary conductance changes at a teleost Mauthner cell inhibitory synapses: a voltage clamp and pharmacological analysis. *J Neurophysiol* 1988b;60:1982–1991.
23. Korn H, Burnod Y, Faber DS. Spontaneous quantal currents in a central neuron match predictions from binomial analysis of evoked responses. *Proc Natl Acad Sci USA* 1987;84:5981–5985.

24. Korn H, Faber DS. Transmission at a central inhibitory synapse. IV. Quantal structure of synaptic noise. *J Neurophysiol* 1990;63:198–222.
25. Edwards FA, Konnerth A, Sakmann B. Quantal analysis of inhibitory synaptic transmission in the dendate gyrus of rat hippocampal slices: a patch-clamp study. *J Physiol (Lond)* 1990;430:213–249.
26. Jack JJB, Redman SJ, Wong K. The components of synaptic potentials evoked in cat spinal motoneurones by impulses in single group Ia afferents. *J Physiol (Lond)* 1981;321:65–96.
27. Faber DS, Young WS, Legendre P, Korn H. Intrinsic quantal variability due to stochastic properties of receptor-transmitter interactions. *Science* 1992;258:1494–1498.
28. Busch C, Sakmann B. Synaptic transmission in hippocampal neurons: numerical reconstruction of quantal IPSCs. *The Brain*. Cold Spring Harbor Symposium on Quantitative Biology. Vol. LV 1990; 69–80.
29. Faber DS, Korn H. Synergism at central synapses due to lateral diffusion of transmitter. *Proc Natl Acad Sci USA* 1988a;85:8708–8712.
30. Ropert N, Miles R, Korn H. Characteristic of miniature inhibitory postsynaptic currents in CA1 pyramidal neurones of rat hippocampus. *J Physiol (Lond)* 1990;428:707–722.
31. Silver RA, Traynelis SF, Cull-Candy SG. Rapid-time-course of miniature and evoked excitatory currents at cerebellar synapses *in situ*. *Nature* 1992;355:163–166.
32. Larkman A, Stratford K, Jack JJB. Quantal analysis of excitatory synaptic action and depression in hippocampal slices. *Nature* 1991;350:344–347.
33. Bekkers JM, Richerson GB, Stevens CF. Origin of variability in quantal size in cultured hippocampal neurons and hippocampal slices. *Proc Natl Acad Sci USA* 1990;87:5359–5362.
34. Johnston D, Brown TH. Interpretation of voltage-clamp measurements in hippocampal neurons. *J Neurophysiol* 1983;50:464–486.
35. Hestrin S, Nicoll RA, Perkel DJ, Sah P. Analysis of excitatory synaptic action in pyramidal cells using whole-cell recording from rat hippocampal slices. *J Physiol (Lond)* 1990;422:203–225.
36. Bekkers JM, Stevens CF. Presynaptic mechanism for long-term potentiation in the hippocampus. *Nature* 1990;346:724–729.
37. Korn H, Ankri N, Faber DS. Synchronization of quantal inhibitory dendritic responses. *Soc Neurosci Abstr* 1992;18:790.
38. Korn H, Bausela F, Charpier S, Faber DS. Synaptic noise and multiquantal release at dendritic synapses. *J Neurophysiol* 1993;70:1249–1254.
39. Ankri N, Legendre P, Faber D S, Korn H. Automatic detection of spontaneous synaptic responses in central neurons. *J Neurosci Methods* 52(1):87–100, 1994.
40. Celeux G, Diebolt J. The EM and the SEM algorithms for mixtures: statistical and numerical aspects. *Cahiers du CERO* 1990;32:135–151.
41. Dempster AP, Laird NM, Rubin DB. Maximum likelihood from incomplete data via the EM algorithm. *J Res Statist Soc B* 1977;39:1–21.
42. Celeux G, Diebolt J. A stochastic approximation type EM algorithm for the mixture problem. *Inst Nat Rech Inf Autom* (INRIA), Research Report, 1991;1383:1–31.
43. Celeux G, Diebolt J. The SEM algorithm: a probabilistic teacher algorithm derived from the EM algorithm for the mixture problem. *Comput Stat Quart* 1985;2:73–82.
44. Wilks SS. The large sample distribution of the likelihood ratio for testing composite hypotheses. *Ann Meth Stat* 1938;9:60–62.
45. Nakajima Y. Fine structure of the synaptic endings on the Mauthner cell of the goldfish. *J Comp Neurol* 1974;156:375–402.
46. Tuttle R, Masuko S, Nakajima Y. Small vesicle bouton synapses on the distal half of the lateral dendrite of the goldfish Mauthner cell: freeze fracture and thin section study. *J Comp Neurol* 1987; 265:254–274.
47. Sur C, Korn H, Triller A. Size and number of active zones in Mauthner cell inhibitory afferents increase from soma to dendrite. *Soc Neurosci Abstr* 1992;18:1339.
48. Lin JW, Llinas R. The role of vesicular mobilization in tonic transmitter release and potentiation in the crab T-fiber giant synapse. *Soc Neurosci Abstr* 1992;18:576.
49. Cohen I, Kita H, Van Der Kloot W. The intervals between miniature end-plate potentials in the frog are unlikely to be independently or exponentially distributed. *J Physiol (Lond)* 1974;236:327–339.
50. Dennis MJ, Harris AJ, Kuffler SW. Synaptic transmission and its duplication by focally applied acetylcholine in parasympathetic neurons in the heart of the frog. *Proc Roy Soc London B Biol Sci* 1971;177:509–539.

51. Liu G, Feldman JL. Quantal synaptic transmission in phrenic motor nucleus. *J Neurophysiol* 1992; 68:1468–1471.
52. Zottoli S. Comparative morphology of the Mauthner cell in fish and amphibians. In: Faber DS, Korn H, eds. *Neurobiology of the Mauthner cell.* New York: Raven Press; 1978;13–45.
53. Petrov T, Seitanidou T, Triller A, Korn H. Differential distribution of transmitter defined afferents on an identified neuron. *Brain Res* 1991;559:75–81.
54. Legendre P, Korn H. Inhibitory synaptic currents in the immature Zebrafish brain. *Soc Neurosci Abstr* 1992;18:569.16.
55. Legendre P, Korn H. Glycinergic inhibitory synaptic currents and related receptor-channels in the Zebrafish brain. *The Europ J Neurosci*, in press.
56. Triller A, Seitanidou T, Franksson O, Korn H. Size and shape of neurotransmitter receptor clusters in a central neuron exhibit a somatodendritic gradient. *The New Biologist* 1990;2:637–641.
57. Lin JW, Faber DS. Synaptic transmission mediated by single club endings on the goldfish Mauthner cell. II. Plasticity of excitatory postsynaptic potentials. *J Neurosci* 1988;8:1313–1325.
58. Walmsley B, Wienawa-Narkiewicz E, Nichol M. The ultrastructural basis for synaptic transmission between primary muscle afferents and neurons in Clarke's column of the cat. *J Neurosci* 1985;8: 2095–2106.
59. Walmsley B, Edwards FR, Tracey DJ. Non uniform release probabilities underlie quantal synaptic transmission at a mammalian excitatory central synapse. *J Neurophysiol* 1988;60:889–902.
60. Korn H, Triller A. Synaptic mechanisms in the light of structural insights: quantal release in the central nervous system and its physical constraints. *Proc 45th Ann Meet EMSA* 1987;688–689.

Molecular and Cellular Mechanisms of Neurotransmitter Release, edited by Lennart Stjärne, Paul Greengard, Sten Grillner, Tomas Hökfelt, and David Ottoson, Raven Press, Ltd., New York © 1994.

20

Probabilistic Secretion of Quanta at Excitatory Synapses on CA1 Pyramidal Neurons

Christian Stricker, Anne C. Field, and Stephen Redman

Division of Neuroscience, John Curtin School of Medical Research, Australian National University, Canberra ACT 0200, Australia

Since the first use of statistical analysis of synaptic transmission by Del Castillo and Katz (1) to analyze transmitter release at the skeletal neuromuscular junction, the technique has become an important tool in studying junctional mechanisms at both central and peripheral synapses (2,3,4,5). The application of statistical techniques to synapses on central neurons is complicated by many factors. One problem is that the amplitude distribution of spontaneous miniature potentials originating from the activated synapses is rarely available. This means that an independent measure of the quantal amplitude and its variability is not available. Another problem is that release may occur from numerous release sites associated with the activated axon(s) and the amplitudes of the responses from each of these sites may not be identical, resulting in unequal increments in the evoked response. A third difficulty is that the probability of release may vary for the different release sites. Since it is not possible to obtain direct measurements of these unknowns, different models can be formulated to provide quantitative descriptions of transmitter release and its postsynaptic effect. A comparison of these different models on a sound statistical basis can provide more detailed insight into how transmitter is liberated and acts on the postsynaptic cell. The aim is to choose with confidence the best model describing the evoked responses, under circumstances where the signal-to-noise conditions and limited sample size are usually not helpful in resolving the parameter values of the model.

The main reason for pursuing these statistical approaches to synaptic transmission is that they have the potential to separate presynaptic changes from postsynaptic changes when synaptic transmission is altered. The current interest in the mechanisms of long-term potentiation (LTP) has given new impetus to refining these statistical methods. The analyses used have lacked a sound statistical basis for model comparison. Once the appropriate model parameters are obtained, confidence limits are needed to determine the reliability of release statistics. They can then be used to determine the significance of apparent changes in quantal size and release probabilities following modifications of synaptic strength.

In this chapter we have analyzed the synaptic currents evoked in a CA1 pyramidal neuron by minimal stimulation of stratum radiatum. Probability density functions of the peak amplitudes of the currents have been used to obtain the parameters for different models of transmission. The parameters were adjusted to maximize the likelihood using an expectation-maximization (EM) algorithm. The statistical reliability of each model was assessed and confidence limits were obtained for the parameters of the most reliable model using resampling and bootstrap techniques (6,7). In the particular example analyzed in this chapter, we show that the most compelling statistical model of transmission is one where the synaptic currents generated at each release site have similar amplitudes with negligible variance. Alternative models in which transmission is quantized, but where the variance associated with transmission at each site is sufficiently large to obscure a multimodal structure in the density of the evoked current, can be rejected on a sound statistical basis. All models that constrained the release probabilities to be either uniform or different at each release site could be rejected. However, the data was most closely matched when the release probabilities were uniform.

METHODS

Preparation of Hippocampal Slices

Brains were removed rapidly from 17–28-day-old rats decapitated on a guillotine, and placed into ice cold artificial cerebrospinal fluid (ACSF) containing (in mM): NaCl 124; KCl 3; $NaHCO_3$ 26; NaH_2PO_4 2.5; $CaCl_2$ 2.5; $MgSO_4$ 1.3; glucose 10, gassed with 95% O_2/5% CO_2. The brain was hemisected along the midline, and the cut surface glued to the stage of a vibratome (Camden) with cyanoacrylic glue. The stage was surrounded with ice cold, gassed ACSF, and slices were cut at a thickness of 400 μm. The overlying cortex was removed from transverse hippocampal sections, which were then placed in a holding chamber and incubated at 34°C for 1 hour. They were then maintained at room temperature (20–22°C) until recordings were made.

Electrophysiology

The slices were transferred to a recording chamber, where they were superfused with gassed ACSF containing 10 μM bicuculline methiodide (Sigma) at 30°C. A cutting tool was made that contained a 50 μm gap in a razor blade, and this was used to make a transverse cut across the entire CA1 field, leaving only a 50 μm tissue bridge in stratum radiatum between CA3 and the region in CA1 from which recordings were made (8). Whole-cell voltage clamp recordings were obtained from CA1 neurons using the "blind" patch clamp technique (9). Pipettes (3–5 MΩ) contained (in mM): Cs gluconate 130; CsCl 9.5; ethyleneglycol-bis-(β-amino-ethyl ether) N, N′-tetra-acetic acid (EGTA) 0.2; N-tris[hydroxymethyl]methyl-2-amino-ethanesulphonic acid (TES buffer) 10; adenosine 5′-triphosphate (ATP) 3; guanosine 5′-triphosphate (GTP) 0.3; $CaCl_2$ 0.8; $MgCl_2$ 3; pH 7.3; osmolarity 270 mOsm/l. Excitatory postsynaptic currents (EPSCs) (average amplitude <5 pA) were evoked at 1 Hz by low-intensity, constant-current stimulation of stratum radiatum, using very fine tipped tungsten electrodes or NaCl-filled patch electrodes placed on the distal side of the tissue bridge.

The position of the stimulating electrode was altered, and the stimulus current was varied, until a stimulus-response result such as that shown in Fig. 1B was achieved. As the stimulus current was increased, a range of current was found for which the EPSC amplitude remained constant. Ideally, this range should be at least 10 μA. The stimulus current was then adjusted to the center of this plateau range (as shown by the arrow in Fig. 1B), and recordings of the EPSC were taken at 1 Hz for as long as low noise conditions could be maintained. Access resistance was continuously monitored by recording the current response to a .5 mV depolarizing voltage step.

Data Acquisition

Individual records were filtered at 1 kHz, using a 4-pole Bessel filter, digitized at 5 kHz and stored on disk. The currents were analyzed offline. The peak current was determined by taking the difference of the average currents in two time windows: one placed in the baseline region between the stimulus artifact and the current, and the other placed over the peak of the response (10). The two time windows were visualized on a monitor by two pairs of cursors that followed the time course of each record and that could be moved forwards or backwards in time. The noise current was calculated by placing the two time windows in the baseline region (before the stimulus artifact), preserving the same temporal relationships as was used for calculating the peak current, and again calculating the difference between the averages of the current in each window. A noise current and a peak current were obtained for each record, unless spontaneous currents created obvious distortions of the baseline region or the evoked current, in which case the record was rejected. Figure 1C shows the peak currents as a function of time, and Fig. 1D illustrates the average of successive groups of 60 peak amplitudes, indicating a stable response amplitude. Data were used only from a continuous period of recording in which the mean remained within one standard deviation of its initial value. It was further required that the standard deviation of the evoked response, and the access resistance, remained constant throughout this period.

Probability Density Function for the Peak Current and the Baseline Noise

Each peak measurement (x_i) was convolved with a normal distribution $G(0,\sigma)$, where σ depended on the standard deviation of the baseline noise (σ_n), the sample

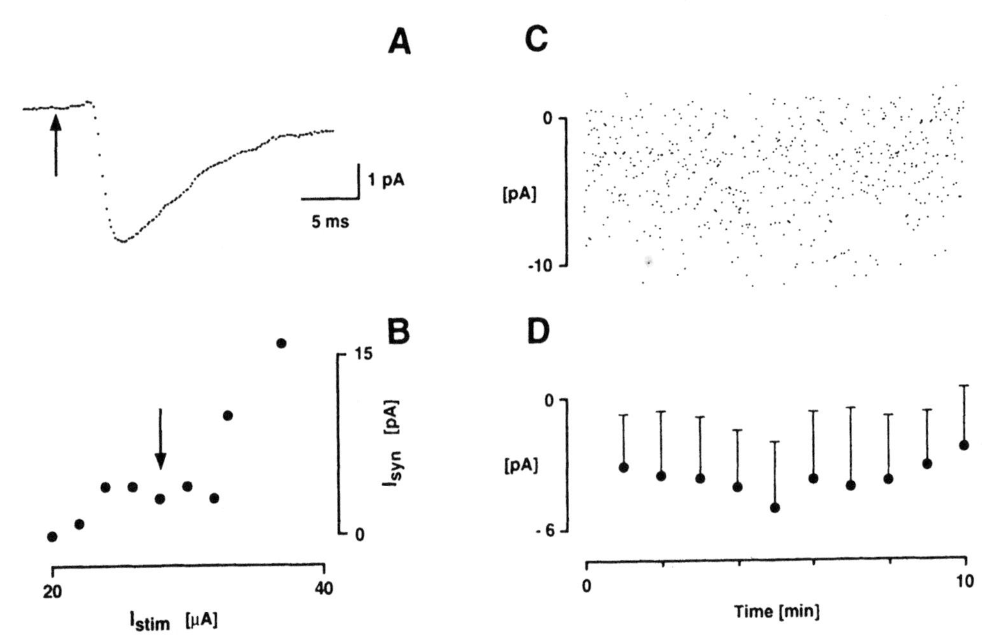

FIG. 1. A. Average time course of excitatory postsynaptic current (EPSC). The stimulus artifact has been subtracted. The stimulus was delivered at the time indicated by the arrow. **B.** The stimulus strength was adjusted between 20 μA and 37 μA, and the synaptic current at each response was averaged over 30 responses. The average for each stimulus current is shown. The stimulus current used to obtain the data in C and D is indicated by the arrow. **C.** The peak amplitude of sequentially evoked synaptic currents (N = 526). **D.** Averages of 60 consecutive peak currents with one-sided standard deviation.

size (N), and a preliminary estimate of the quantal current (11). Typically, with N = 500 and quantal current in the range 2–5 pA, $\sigma = 0.25\sigma_n$. The probability density of the peak EPSC (P(x)) was obtained from

$$P(x) = \frac{1}{N}\sum_{i=1}^{N} G(x_i,\sigma)$$

This procedure for creating a density function avoids the nonlinear bias introduced by binning procedures. A density function for the baseline noise was calculated in the same manner.

Statistical Models of Synaptic Transmission

The expectation-maximization (EM) algorithm was used to calculate the best fit of P(x) to different statistical models of synaptic transmission, using a maximum likelihood criterion (12). The models to be considered here are all based on the premise that the EPSC occurs as one of an unknown number of discrete amplitudes, and the variability about each amplitude depends on the baseline noise and intrinsic sources. The variability is sufficiently small for individual peaks associated with each amplitude to be detected in the density function. The most general model is one where the discrete amplitudes and their probabilities are unconstrained. Variations of this model are: 1) a quantal model where adjacent discrete amplitudes are separated by equal amounts, and where the variance associated with each amplitude can increase in a quantal manner; 2) a binomial model, where the probabilities in 1) are constrained to satisfy the binomial distribution; and 3) a compound binomial model, where the probabilities in 1) must satisfy the constraint that the release probabilities are all real, positive, and possibly different. The equations used in the EM algorithms for 1), 2), and 3) have been derived by Stricker and Redman (13).

The probability density of the baseline noise was represented by a mixture of two normal distributions. The EM algorithm was used to determine the parameters of this mixture.

Model Comparison

The models just described constitute a nested or hierarchical set. Any two models can be selected in which one is a smoothly parameterized subhypothesis of the other. Procedures for comparing the relative adequacy of nested models to fit data have been discussed in Horn (14). The maximum log-likelihood for the fit to each of the two models is calculated. Twice the ratio of these log-likelihoods asymptotes to a χ^2 distribution. This statistic and the difference in the number of degrees of freedom between the two models were used to determine the level of significance for rejection of one of the models.

When the models are not nested (e.g., the two models in Fig. 2B having K discrete amplitudes for one case and $K + 1$ amplitudes for another), the ratio of the maximum log-likelihoods does not asymptote to a χ^2 distribution and the Wilks test just described cannot be used. A method for discriminating non-nested models is described in Stricker et al. (15).

Confidence Limits for Model Parameters

Resampling techniques have been used to calculate the standard deviation of each parameter required to describe a particular model of transmission. The original data set was resampled randomly, with replacement, to obtain a sample of the same size as the original. This procedure was repeated 250 times. Efron (16) has calculated that 50 to 200 resamplings are needed to estimate the moments of a statistic. To prevent the possibility of bias, a balanced resampling scheme is used (17). In this scheme, the distribution formed by combining all the resampled sets is identical to the distribution of the original sample.

The EM algorithm was applied to the probability density formed by each resampling, assuming the same model of transmitter release as for the original sample.

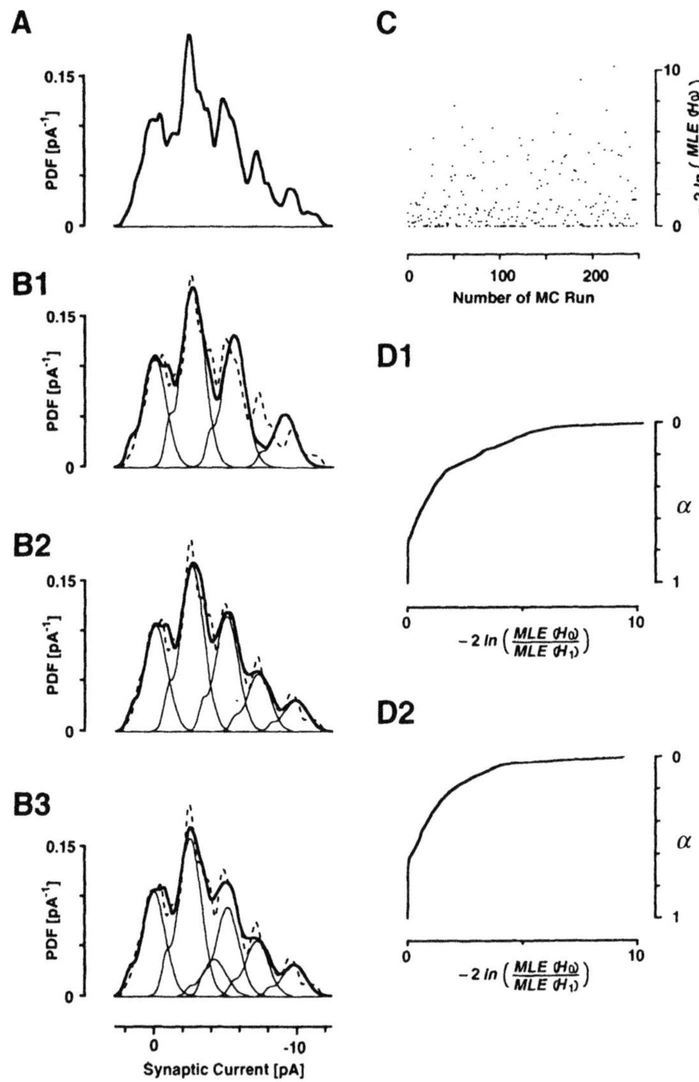

FIG. 2. A. The probability density of the peak amplitudes shown in Fig. 1C. B1, B2, B3: The best-fitting mixture distributions for K=4, K=5 and K=6, respectively. The details on the component amplitudes and probabilities are given in Table 1. The heavy line is the mixture distribution, which is the sum of the component distributions, each indicated by a light continuous line. The dashed line in each case is the probability density for the data, as shown in **A. C.** The mixture model in B1 was generated, and Monte Carlo sampling of this model was used to make 250 new density functions, each with sample size 526. The best fitting mixture model with K=4 (H_o) and with K=5 (H_1) was obtained for each pair of densities using a maximum likelihood criterion. The Wilks statistic (−2 times the difference in the logarithm of the maximum likelihoods for H_o and H_1) was calculated and its sequential values for the 250 calculations are shown. **D1.** The values in C have been rank-ordered, and formed into a cumulative density function with the ordinate $\alpha = 1\text{-p}$, where p is the cumulative probability. **D2.** This cumulative probability was calculated as for D1, except that the mixture model in B2 with K=5 was H_o, and the mixture model with K=6 was H_1.

Each probability density provides a parameter set, from which the mean and standard deviation of each parameter can be calculated.

RESULTS

The EPSC to be analyzed in detail was introduced in Fig. 1. The probability density formed from the records is shown in Fig. 2A. This density function was formed by convolving each peak amplitude with a normal distribution having zero mean and $0.2\sigma_n$ standard deviation. (For this EPSC, $\sigma_n = 0.87$ pA). All the normal distributions were added and the area of the summed distributions was normalized to one. Unconstrained models were fitted to this density function using the EM algorithm, and different numbers (K) of discrete amplitudes in the mixture were assumed. The only likely values of K were 4, 5, and 6, as judged by the maximum likelihood estimate (MLE) for different values of K, and the fit by eye. The discrete amplitudes, and the probabilities associated with these amplitudes, for the models with K = 4, 5, and 6 are given in Table 1.

The unconstrained models for K = 4 and 5 are illustrated in Fig. 2B1, 2B2. The log-likelihoods (LL) for these models fitting the data were −1299.64 and −1276.18 for K = 4 and 5, respectively. To determine the most reliable model for the observed density function, the Wilks statistic was used; this is −2(LL(K = 4) − LL(K = 5)) and equals 46.92. This statistic is not distributed as a χ^2 distribution, as the two models with K = 4 and K = 5 cannot be formed by a smooth parametric transition from one to the other. Instead, the required distribution of the statistic was calculated by making the model with K = 4 the null hypothesis (H_o), generating the mixture distribution corresponding to K = 4, and using Monte Carlo sampling to generate 250 new data sets, each having the same number of samples (526) as the original sample. Each data set was formed into a probability density as in Fig. 2A, and for each density the EM algorithm was used to calculate the best-fitting model for K = 4 and K = 5. The Wilks statistic was calculated for each of these pairs of

TABLE 1. *The parameter values for optimal fits of unconstrained mixture models to the data, with 4, 5, and 6 components in the mixture*

	K = 4		K = 5		K = 6	
Component	*P*	μ	*P*	μ	*P*	μ
0	0.23	−0.01	0.22	0.03	0.22	0.04
1	0.38	−2.71	0.35	−2.54	0.33	−2.47
2	0.28	−5.56	0.24	−4.93	0.08	−4.19
3	0.11	−9.12	0.12	−7.17	0.19	−5.11
4			0.07	−9.78	0.12	−7.19
5					0.07	−9.78
LL	−1299.64		−1276.18		−1275.83	

P, probability of each component in the mixture; μ, amplitude in pA; LL, maximum log-likelihood obtained for the fit of each model to the data.

mixture models, and its sequential values are given in Fig. 2C. These values were rank ordered along a normalized ordinate (representing cumulative probability), as shown in Fig. 2D1. This is the statistic required for the null hypothesis described, and it indicates that the mixture model with $K=4$ can be rejected with $\alpha<0.004$.

A similar procedure was used in an attempt to discriminate between the models with $K=5$ and $K=6$. Notice from Table 1 and Fig. 2B3 that the result for $K=6$ contains a component (μ_2) that occurs with a probability smaller than the probabilities of occurrence of its neighbors (μ_1 and μ_3); this can only occur in a release process when different release sites associated with the same axon do not act independently. Using the model with $K=5$ as H_o, Monte Carlo sampling was used to generate 250 data sets. Each of these was used to find the best fitting models for $K=5$ and $K=6$. The distribution of the Wilks statistic was calculated, and is shown in Fig. 2D2. The model with $K=5$ was a better fit in about half of the trials. The model with $K=5$ cannot be rejected, and we accepted the mixture with $K=5$ on the grounds of parsimony (a smaller number of degrees of freedom) and the need for nonindependence in the release process when $K=6$.

The next step was to examine the hypothesis that the presence of approximately regularly occurring peaks in Fig. 2A is a sampling artifact, and that the probability distribution in Fig. 2A was drawn from a unimodal distribution. A mixture of two normal distributions that represent the baseline noise must be added to the unimodal distribution to account for any failures in transmission. If the variance associated with the currents arising from individual active sites was large in comparison with their amplitudes, regularly occurring peaks would not be present in the probability distribution of the evoked current when it was reliably sampled. In Fig. 3, the null hypothesis is that the unimodal distribution describing the transmission process is a gamma distribution. The best fit of the combination of a gamma distribution and the noise distribution to the data was obtained using an EM algorithm, and it is shown in Fig. 3B. The optimal value of α was 4.64, and the cumulative probability for the gamma distribution was 0.76. The LL was -1289.62, while the LL for the model with $K=5$ in Fig. 3A (as in Fig. 2C) was -1276.18. The Wilks statistic is 26.88. As the gamma distribution and the mixture distribution are not nested, the Wilks statistic is not distributed as χ^2. The distribution of this statistic was calculated in the same way as the statistic to distinguish between $K=4$ and $K=5$. Figure 3C shows the Wilks statistic ($-2(LL(GAMMA)-LL(K=5))$ as it was calculated for each of the 250 sample sets and arranged in rank order as a cumulative distribution. The gamma distribution can be rejected with $\alpha<.004$, which means that it is very unlikely that the peaks in the observed distribution could arise by chance from finite sampling of the gamma distribution. A similar result was obtained when other unimodal distributions were used in place of a gamma distribution. These distributions were the Weibull distribution, the cubic transform of a normal variable (18), and a normal distribution.

The mixture distribution in Fig. 3A with $K=5$ was obtained without placing constraints on the amplitudes of the discrete components or their probabilities. The variance of each component was restricted to be the variance of the noise, and a

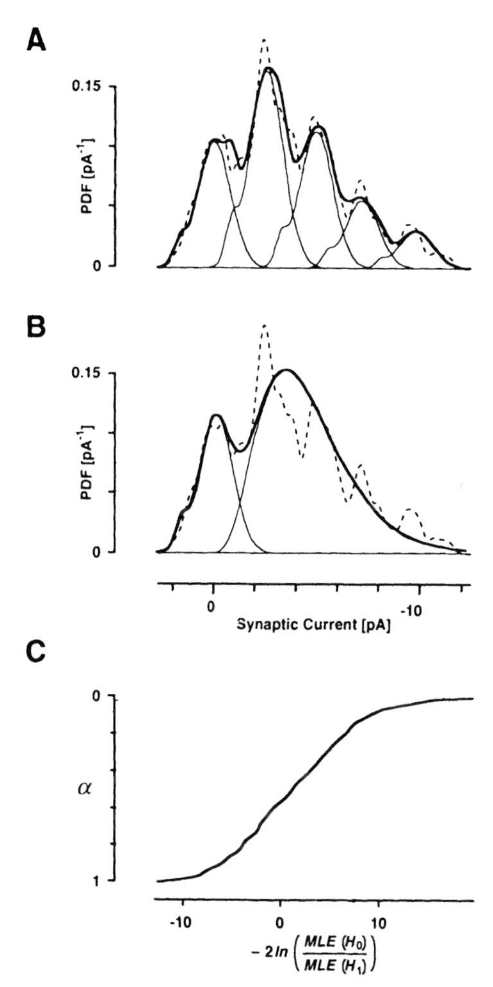

FIG. 3. A. The unconstrained mixture model from Fig. 2B2 ($K = 5$) is repeated here for comparison. **B.** The heavy line is the sum of two density functions. One is the density representing the baseline noise (a mixture of two normal distributions) centered on 0.07 pA and with a cumulative probability of 0.24. The other density is a gamma distribution, with $\alpha = 4.64$, and a cumulative probability of 0.76. The cumulative probabilities, α, and the offset for the noise density were obtained using an expectation-maximization (EM) algorithm. The light lines indicate the cross-over of the noise and gamma distributions. The dashed lines indicate the original data. **C.** The Wilks statistic calculated from 250 Monte Carlo samplings from the mixture distribution shown in B as H_o, and with the mixture distribution shown in A as H_1.

cursory inspection of the fit to the observed distribution with $K = 5$ indicates that little additional variance can he accommodated. The addition of a quantal variance is considered in the next paragraphs.

A more appropriate model than the mixture distribution in Fig. 3A may be one where the amplitudes are separated by quantal increments. The probabilities associated with each amplitude may also be restricted to satisfy a binomial release process where the release probabilities are all equal, or a compound binomial process where the release probabilities may all be different. These latter models are all nested versions of the unconstrained model, and in each case a null hypothesis can be tested using the Wilks statistic distributed as a χ^2 distribution. This is done in Fig 4.

The unconstrained mixture model with $K=5$ is shown in Fig. 4A. The optimized mixture model with the amplitudes constrained to have quantal separation, but with no constraints on probability, is shown in Fig 4B. Only minor changes occurred in the discrete amplitudes and their probabilities (Table 2). The mean separation in the unconstrained model was 2.44 pA, and the separation in the quantal model was 2.45 pA. The Wilks statistic for the quantal model tested against the unconstrained mixture model (with $K=5$) was 2.02. There are 9 degrees of freedom in the unconstrained model, and 6 in the quantal model, giving a difference of 3 degrees of freedom for the χ^2 distribution. On this basis, quantal separation cannot be rejected and on the basis of parsimony, quantal separation has to be accepted.

When quantal variance was added to the model, the changes were negligible (Table 2). The quantal amplitude was 2.44 pA and there was little change in the probabilities for each component. The quantal variance was 0.03 pA^2, which corresponds to a coefficient of variation for the quantal current of 0.07. The Wilks statistic for this model, when tested against the quantal model without quantal variance, is 0.62 (with one degree of freedom), which is not significant (i.e., we reject the mixture with quantal variance on the grounds that the extra degree of freedom has not made a significant difference in the quality of the fit between the model and the data).

When the uniform binomial constraint was applied to the release probabilities, combined with a quantal separation of discrete amplitudes and $K=5$, the result is shown in Fig. 4C. The quantal separation was 2.45 pA, and the release probability

TABLE 2. *The parameter values for the optimal fit of three different models (as indicated)*

	Unconstrained Quantal Model	Unconstrained Quantal Model with Quantal Variance	Binomial Model with Quantal Separation
Component			
0	0.22	0.22	
1	0.34	0.34	
2	0.26	0.26	$p_r=0.365$
3	0.12	0.12	
4	0.06	0.06	
Quantal Size	−2.45	−2.44	−2.45
Offset	0	0.02	0
CV	–	0.07	–
LL	−1277.19	−1276.88	−1299.24

The quantal amplitude is given in pA. P_r, release probability for the binomial model; CV, coefficient of variation of the quantal current. The offset [pA] allows for the failure peak (and all successive peaks) to be offset from zero if a stimulus artifact of synaptic field current is mixed with the evoked response.

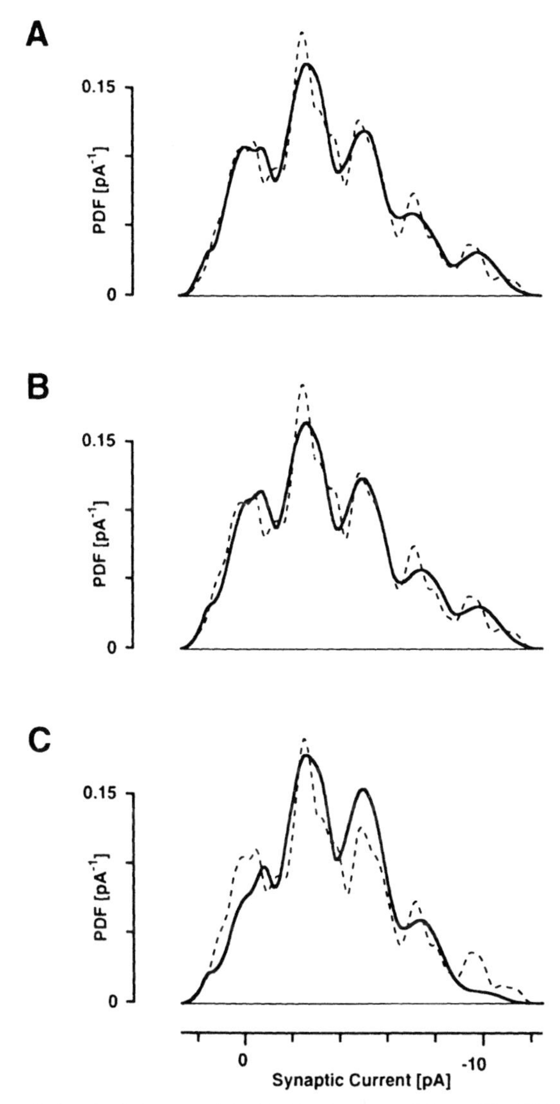

FIG. 4. A. The unconstrained mixture model with K=5 repeated from Fig. 2B2. **B.** The heavy line is the mixture model with K=5, but constrained to have quantal separation between adjacent peak amplitudes. The dashed line is the density function of the original data. The quantal amplitude was 2.54 pA, and the probabilities associated with each amplitude are given in Table 2. **C.** The heavy line is the optimal fit to the data (dashed line) when the mixture model was constrained to have quantal separation of amplitudes and uniform binomial release probabilities. The parameters for this model are given in Table 2.

was 0.37 (Table 2). The Wilks statistic for the fit to this model compared with the quantal model without probability constraints was 42.6 with 3 (6 for quantal, 3 for binomial) degrees of freedom. The binomial model can be rejected with $\alpha < .005$. However, we could do no better when a non-uniform binomial model with different release probabilities at each of the four release sites was assumed. The best solution was almost identical to that shown in Fig. 4C, but it involved more degrees of freedom.

Although the binomial and compound binomial models can be rejected, it is useful to see how close the amplitude probabilities required by the binomial model are to the amplitude probabilities of the best model, in relation to the quantal model with unconstrained probabilities. An indication of this is given in Fig. 5. Here, the confidence intervals for the quantal model without probability constraints are indicated. The data in Fig. 1C were resampled, with replacement, in a balanced way. Each sample ($N = 526$) was used to make a probability density, and the EM algorithm for the quantal model was used to calculate the model parameters (as in Fig. 4B for the original data). This process was repeated 250 times. The discrete amplitudes and their probabilities for each density were used to calculate the standard deviations for the probability and amplitude of each component in the mixture. The result is shown in Fig. 5, where the error bars correspond to ± 1 standard deviation (SD), and the mean of each probability and amplitude is the same as that obtained for the original data (as in Fig. 4B). The amplitudes and probabilities for the binomial model (Fig. 4C), and the mixture model without constraints on amplitude separation and probabilities (Fig. 4A), are also indicated.

DISCUSSION

The purpose of this article was to demonstrate rigorous statistical procedures for choosing between different models of synaptic transmission that are all biologically plausible. These procedures have been illustrated with one EPSC. To date, we have fully analyzed four EPSCs using the methods illustrated in this paper, and more lessons may be learned as more are analyzed.

The Quantal Nature of Transmission

The concept of quantal transmission at excitatory synapses in the mammalian central nervous system has been difficult to establish. Attempts to elucidate the nature of transmission rely on the observation of distributions of currents which have clearly defined peaks. Some skepticism has been raised about apparent peaks in the distributions of evoked currents (19). It has been suggested that the relatively small number of experimental observations (finite sampling) may not represent the underlying statistical process, and that these peaks have arisen by chance. Another problem with the analysis procedures commonly used is a tendency to overfit the data. The statistical procedures presented in this chapter aim to overcome both of these problems by using classical statistical tests. The only implied assumption in the statistical interpretation is that, of all the models that describe the data adequately, the one having the least degrees of freedom provides the best explanation of the data (on the grounds of parsimony).

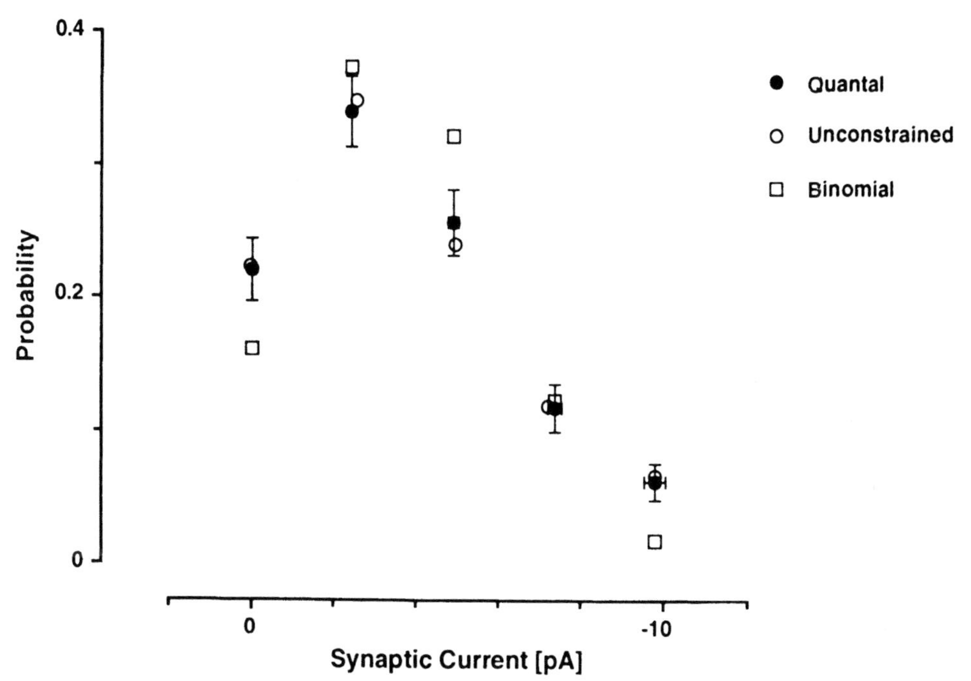

FIG. 5. The filled circles indicate the amplitude and probabilities of the mixture components for the quantal model, as shown in **Fig. 4B** and Table 2. The error bars indicate ± 1 standard deviation (SD) for the probabilities and amplitudes. The open circles and open squares indicate the amplitudes and probabilities for the unconstrained mixture model and the binomial model, respectively.

With four EPSCs analyzed to date, we have been unable to reject the description that, for each EPSC, the evoked responses consist of discrete amplitudes separated by quantal increments, with negligible quantal variance. The statistical differences between the fits to the data when no quantal separation and no quantal variance are imposed, compared with models where quantal separation with and without quantal variance is assumed, are not significant. That is, after having made allowance for various other models of transmission, we have to accept the quantal model of transmission at this CA3–CA1 synapse. It is somewhat surprising to find that such a description still holds given the experimental conditions, where possibly several Schaffer collaterals make contact at numerous release sites located at different regions of the dendrites, where the responses from each of these sites may vary at the soma, and where inadequate space clamp conditions prevail.

The Absence of Quantal Variability

The lack of quantal variance in the statistical description of EPSCs has also been observed by others (20,21). These descriptions are inconsistent with an alternative quantal model that assumes that the currents evoked at each active site have such large variability that the density of the evoked EPSC contains no apparent peaks (18,22). Under these conditions a unimodal distribution such as a gamma distribution is an appropriate statistical description. Such a description was rejected with a high reliability for the EPSC analyzed in this chapter, and for the other EPSCs.

The discrete model has certain biological implications. It requires the EPSCs arising at the different active sites to have very similar amplitudes at the soma. If there are differences in the electrotonic locations of these sites with respect to the soma, the synaptic currents at each site would need to be adjusted to compensate for synaptic location. There is evidence from the studies of transmission between group Ia synapses and motoneurons that such a compensation occurs at those synapses (23). If no differences in electrotonic locations exist, then similar conductance changes must occur at each active site. The negligible variability in the quantal amplitude at the soma can only occur if the synaptic currents at each active site satisfy the requirements just mentioned. It also implies that the stochastic nature of channel opening must contribute negligible variability to the synaptic current from each site. This latter source of variability would be expected to be significant if channel opening probabilities are not close to 1.0, and saturating concentrations of glutamate are not present at the receptors in the synaptic cleft (24). Hestrin (25) reported a channel opening probability of 0.7 for AMPA/kainate-gated channels in somatic membrane patches of visual cortical neurons. He calculated that on average only one out of three of the synaptic channels at an active site are opened when transmitter is released. On this basis, a coefficient of variation (CV) of 0.2 in the quantal current would be expected. Similarly, Jonas et al. (26) estimated a CV of 0.22 for the quantal current at mossy fiber synapses on CA3 neurons, and Kullmann (27) gave one example of an EPSC evoked in a CA1 neuron that required a CV of 0.15.

Quantal variability with a CV of 0.15 or greater is incompatible with the EPSC analyzed in the results, and with the three other EPSCs not illustrated. It should be pointed out that the quantal variance model used in this analysis, and used routinely in quantal analysis, may not be appropriate when transmission occurs at a small number of active sites distributed to different regions of a dendritic tree. The variability arising from stochastic channel activity is additive for all the quanta that contribute to a response. This is the classical quantal variance model which is traditionally used. Intersite variability due to different quantal currents, and electrotonic factors, will not add in the same way. For instance, if inter-site variability was the only source of variability, the responses resulting from transmission at all the active sites associated with the stimulated axon would have no variance. Responses resulting from transmission at different subsets of these active sites, with each subset having the same number of sites, will show variability. The quantal variance model does not take this into account. However, when no quantal variance of any kind can be accommodated, this awkward issue does not arise.

The Release Process at Each Active Site

The question of whether transmitter is released with the same release probability at different release sites associated with a single axon, or whether different release probabilities apply at different release sites, is problematic. In the example illustrated in this chapter, the best fit to the data was obtained for uniform release probabilities, but this model could be rejected compared with one where no constraints were placed on the release process at each site. The best fits for the three EPSCs not illustrated were obtained with different release probabilities at each site, but these fits were still inferior to those obtained for the unconstrained probabilities. All could be rejected with $\alpha < .05$. Nonuniform release probabilities best describe the release processes at synapses formed between primary afferents and spinal cord neurons, (23,28,29) and evidence for nonuniform release at synapses formed between cultured hippocampal neurons has recently appeared (30) The likelihood ratio test used in this analysis is a more powerful test than that used in the studies on transmission at synapses in the spinal cord. The question arises as to whether finite sampling misrepresents the underlying statistical model to the extent that the correct model cannot be successfully fitted to the sampled data.

The confidence limits on the probabilities associated with each amplitude in the quantal model, shown in Fig. 5, indicate that the best fit of a binomial process results in probabilities that are significantly different from those obtained for the unconstrained quantal model. Three of the five probabilities associated with the binomial model are two or more standard deviations away from the corresponding probabilities for the unconstrained probability model. Before attaching any significance to this mismatch, it is important to interpret these confidence limits correctly. The confidence limits are a measure of the uncertainty in the estimates of the model parameters arising from the sample size from which the density function was constructed, the signal-to-noise conditions (as measured by quantal current/noise standard deviation), the number of components in the underlying mixture and the distribution of the probabilities of these components. In previous analyses, the effect of these factors has been guessed or, at best, studied with Monte Carlo simulations. The provision of confidence limits on the model parameters is an important step in introducing rigor to the analysis. It is thus very unlikely that the binomial release process was rejected because of uncertainties introduced by the experimental conditions and the analysis procedure. However, the confidence limits convey no information on the reliability of the original sample. With only one sample from the experiment, there is no way that information on its reliability can be obtained. Thus it remains possible that the failure of the binomial and compound binomial distributions to account for the release probabilities in the four EPSCs analyzed to date is because of a combination of finite sampling of the underlying process and the introduction of a more powerful statistical test than the χ^2 test that was used previously (e.g., see Walmsley et al., ref. 28).

An alternative explanation for our inability to match binomial statistics with the data may be that the basic assumptions that only a single quantal EPSC can be generated at a single release site, and that release from adjacent release sites in the same nerve terminal occurs independently, may be incorrect. Two recent studies on spontaneous miniature EPSCs recorded in neurons have reported multiple peaks in amplitude distributions of the mEPSCs (31,32). However, there have been other reports of spontaneous mEPSCs and inhibitory postsynaptic currents (IPSCs) where multiquantal peaks were not apparent (26,33,34,35). If multiquantal release can occur at some active sites, or if release from adjacent sites in the same nerve terminal does not occur independently, a completely different interpretation of the probabilities associated with different discrete amplitudes will be needed.

The Quantal Current at CA3-CA1 Synapses

The number of axons excited by the stimulation scheme is unknown. The first obvious plateau in the stimulus strength-response curve was used to set the stimulus strength, but if several axons had excitation thresholds that were very similar, the first obvious plateau may correspond to the excitation of several axons. Whether one or more than one axon was excited, the main requirement was that the same number of axons were excited with each stimulus. Even this requirement cannot be guaranteed if the stimulus-response curve shifted along the stimulus strength axis. There was clear evidence that for some of the EPSCs analyzed (but not shown in this paper), intermittent excitation of the axon(s) occurred. The appearance of discrete amplitudes in the EPSC implied that if more than one axon was activated, the quantal amplitudes for the release sites associated with different axons were similar.

CA1 neurons (Stricker, Field, and Redman, unpublished), the quantal amplitude was $2.12 \pm .06$ ($n=2$) compared with 2.2 ± 0.1 for the EPSCs discussed in this paper. Paired recordings provide the greatest security that only one presynaptic neuron is reliably activated. As the quantal amplitude obtained by this method was very similar to the value obtained using extracellular stimulation, this suggests that the EPSCs evoked by extracellular stimulation can be used to determine the quantal amplitude.

These measures of quantal current at the soma are serious underestimates of the actual currents at the active sites. A somatic voltage clamp has little voltage control over a synaptic conductance located more than 0.1 space constants (λ) from the soma (36,37,38,39). The distortion and attenuation of the synaptic current depends upon the electrotonic location of the synapse, the time course of the synaptic conductance, and the series resistance present with the whole-cell recording technique (36,37,38,39). If the series resistance is 20 MΩ, and the electrotonic location is 0.5λ, a twenty fold attenuation of the peak current is possible. This suggests that the actual quantal current may be approximately 40 pA, corresponding to a quantal conductance of about 600 pS. Assuming AMPA/kainate channels with 10 psiemens conductance, this corresponds to the opening of about 60 channels.

The Need for Morphological Data on CA3-CA1 Connectivity

Statistical analysis of synaptic transmission has provided much insight on the details of release processes, and synaptic organization. The real insights have come where details of the termination pattern of single axons with target neurons have been available, and especially when the termination pattern at the synaptic connection that was used to evoke the recorded responses was reconstructed (40,41). At present there exists no detailed information on the number of contacts made by a single CA3 pyramidal cell with a single CA1 pyramidal cell, how widespread these contacts might be over the dendrites, and how the preterminal collateral system is structured. It is essential that this information becomes available if analyses such as the one reported in this chapter are to be correctly interpreted.

ACKNOWLEDGMENTS

This work was supported by fellowships of the Swiss National Science Foundation and the Schweizerische Stiftung für Medizinisch-Biologische Stipendien to C.S. We are grateful to John Bekkers for his comments on this paper.

REFERENCES

1. Del Castillo J, Katz B. Quantal components of the end-plate potential. *J Physiol (Lond)* 1954; 124:560–573.
2. Martin AR. Quantal nature of synaptic transmission. *Physiol Rev* 1966;45:51–66.
3. Kuno M. Quantum aspects of central and ganglionic synaptic transmission in vertebrates. *Physiol Rev* 1971;51:647–678.
4. McLachlan EM. The statistics of transmitter release at chemical synapses. In: Porter, R, ed. *International review of physiology. neurophysiology III*. Baltimore, Maryland: University Park, 1978;17; 49–117.
5. Redman S. Quantal analysis of synaptic potentials in neurons of the central nervous system. *Physiol Rev* 1990;70:165–198.
6. Efron B. Bootstrap methods: another look at the jack-knife. *Ann of Stat* 1979;7:1–26.
7. Efron B, Tibshirani R. Bootstrap methods for standard errors, confidence intervals, and other measures of statistical accuracy. *Stat Sci* 1986;1:54–77.
8. Andersen P, Silfvenius H, Sundberg SH, Sveen O. A comparison of distal and proximal dendritic synapses on CA1 pyramids in guinea-pig hippocampal slices *in vitro*. *J Physiol* 1980;307:273–299.
9. Blanton MG, LoTurco JJ, Kriegstein AR. Whole cell recording from neurons in slices of reptilian and mammalian cerebral cortex. *J Neurosci Meth* 1989;30:203–210.
10. Sayer RJ, Redman SJ, Andersen P. Amplitude fluctuations in small EPSPs recorded from CA1 pyramidal cells in the guinea pig hippocampal slice. *J Neurosc* 1989;9:840–850.
11. Silverman BW. *Density estimation for statistics and data analysis*. London: Chapman and Hall; 1986.
12. Kullmann DM. Applications of the expectation-maximization algorithm to quantal analysis of post-synaptic potentials. *J Neurosci Meth* 1989;30:231–245.
13. Stricker C, Redman SJ. Statistical models of synaptic transmission evaluated using the expectation-maximisation algorithm. *Biophys* 1994 (in press).
14. Hom R. Statistical methods for model discrimination. *Biophys J* 1987;51:255–263.
15. Stricker C, Daley D, Redman SJ. Statistical analysis of synaptic transmission: model discrimination and confidence limits. *Biophys* 1994 (in press).

16. Efron B. More efficient bootstrap computations. *J Am Stat Assoc* 1990;85:79–89.
17. Davison A, Hinkley D, Schechtman E. Efficient bootstrap simulation. *Biometrika* 1986;73:555–556.
18. Bekkers JM, Stevens CF. Presynaptic mechanism for long-term potentiation in the hippocampus. *Nature* 1990;346:724–729.
19. Clements JD. Quantal synaptic transmission? *Nature* 1991;353:396.
20. Larkman A, Hannay T, Stratford K, Jack J. Presynaptic release probability influences the locus of long-term potentiation. *Nature* 1992;360:79–83.
21. Liao D, Jones A, Malinow R. Direct measurement of quantal changes underlying long-term potentiation in CA1 hippocampus. *Neuron* 1992;9:1089–1097.
22. Raastad M, Storm JF, Andersen P. Putative single quantum and single fiber excitatory postsynaptic currents show similar amplitude range and variability in rat hippocampal slices. *Eur J Neurosci* 1992;4:113–117.
23. Jack JJB, Redman SJ, Wong K. The components of synaptic potentials evoked in cat spinal motoneurones by impulses in single group Ia afferents. *J Physiol* 1981;321:65–96.
24. Faber DS, Young WS, Legendre P, Korn H. Intrinsic quantal variability due to stochastic properties of receptor-transmitter interactions. *Science* 1992;258:1494–1498.
25. Hestrin S. Activation and desensitization of glutamate-activated channels mediating fast excitatory synaptic currents in the visual cortex. *Neuron* 1992;9:991–999.
26. Jonas P, Major G, Sakmann B. Quantal analysis of unitary EPSCS at the mossy fibre synapse on CA3 pyramidal cells of rat hippocampus. *J Physiol* 1993;72:615–633.
27. Kullmann DM. Low quantal variability implies that fast glutamate-gated channels open with a high probability. *Proc R Soc London, Ser B* 1993;253:107–116.
28. Walmsley B, Edwards FR, Tracey DJ. Nonuniform release probabilities underlie quantal synaptic transmission at a mammalian excitatory central synapse. *J Neurophysiol* 1988;60:889–908.
29. Clamann HP, Mathis J, Lüscher H-R. Variance analysis of excitatory postsynaptic potentials in cat spinal motoneurons during posttetanic potentiation. *J Neurophysiol* 1989;62:403–416.
30. Rosenmund C, Clements JD, Westbrook GL. Nonuniform probability of glutamate release at a hippocampal synapse. *Science* 1994;262:754–757.
31. Liu G, Feldman JL. Quantal synaptic transmission in phrenic motor nucleus. *J Neurophysiol* 1992; 68:1468–1471.
32. Ulrich D, Lüscher H-R. Miniature excitatory synaptic currents corrected for dendritic cable properties reveal quantal size and variance. *J Neurophysiol* 1993;69:1769–1773.
33. Korn H, Faber DS. Transmission at a central inhibitory synapse IV. Quantal structure of synaptic noise. *J Neurophys* 1990;63:198–222.
34. Kraszewski K, Grantyn R. Unitary, quantal and miniature GABA-activated synaptic chloride currents in cultured neurons from the rat superior colliculus. *Neuroscience* 1992;47:555–570.
35. Gleason E, Borges S, Wilson M. Synaptic transmission between pairs of retinal amacrine cells in culture. *J Neurosci* 1993;13:2359–2370.
36. Rall W, Segev I. Space-clamp problems when voltage clamping branched neurons with intracellular microelectrodes. In: Smith Jr TG, Lecar H, Redman SJ, Gage PW, eds. *Voltage and patch clamping with microelectrodes*. American Physiological Society. 1993;191–215.
37. Spruston N, Jaffe DB, Williams SH, Johnston D. Voltage- and space clamp errors associated with the measurement of electrotonically remote synaptic events. *J Neurophysiol* 1993;70:781–802.
38. Major G, Evans JD, Jack JJB. Solutions for transients in arbitrarily branching cables: II. Voltage clamp theory. *Biophys J* 1993;65:450–468.
39. Thurbon D, Field AC, Redman RJ. Electrotonic profiles of interneurons in stratum pyramidale of the CA1 region of rat hippocampus. 1994 *J Neurophysiol* 1994;71:1948–1958.
40. Redman S, Walmsley B. Amplitude fluctuations in synaptic potentials evoked in cat spinal motoneurones at identified group Ia synapses. *J Physiol (Lond)* 1983;343:135–145.
41. Korn H, Mallet A, Triller A, Faber DS. Transmission at a central inhibitory synapses. II. Quantal description of release, with a physical correlate for binomial *n*. *J Neurophysiol* 1982;48:679–707.

Molecular and Cellular Mechanisms of Neurotransmitter Release, edited by
Lennart Stjärne, Paul Greengard, Sten Grillner,
Tomas Hökfelt, and David Ottoson,
Raven Press, Ltd., New York © 1994.

21

Low Synaptic Convergence of CA3 Collaterals on CA1 Pyramidal Cells Suggests Few Release Sites

Per Andersen, Mari Trommald, and Vidar Jensen

Department of Neurophysiology, University of Oslo, 0317 Oslo, Norway

MICROCONNECTIVITY DATA ARE NECESSARY FOR RELEASE MODELS

Synapses on hippocampal pyramidal cells are probably representative for many synapses in cortical tissue. The release probability of individual hippocampal synapses is similar to those of autonomic synapses. Virtually all hippocampal synapses are of the en passage type and are distributed along the largely unmyelinated axons in rats (1). When there is stimulation of a small bundle of afferent fibers, or an individual axon, it is likely that the release occurs at a low probability from a single site per bouton. Comparing the release at individual CA1 synapses caused by either puffs of hyperosmolar solutions, spontaneous CA3 activity, or stimulation of a small number of afferent fibers, all gave the same overall magnitude of synaptic currents (2,3,4,5). In contrast to the situation in autonomic synapses, however, the size of these individual synaptic currents are highly variable. Among the different models devised to explain this phenomenon, three major proposals are: 1) variable sized quanta released from an individual bouton; 2) a single axon is connected to the same target cell by several boutons, each of which with individual and variable release probability; and 3) an axon may have several boutons connected to the same postsynaptic cell, but with different release probabilities.

In order to choose the best model to explain the release behavior at single CA3-derived synapses on CA1 pyramidal cells, it is essential to know the connectivity between the afferent fibers and individual target cells. Although the hippocampal formation has been intensely studied, only the overall connectivity is known. Such data indicate which areas of the hippocampal formation are connected to groups of cells or fields in other parts of the system. What is needed for the release studies is more detailed information. We need to know how many boutons are carried by a single fiber, and how many synaptic contacts a single fiber makes with various target cells. Furthermore, can a single axon make contact with several spines of the same target cell or even neighboring spines on the same dendritic branch? Or, alternatively, does the effect of a single axon distribute itself to a restricted number of cells in its area of distribution? For a precise answer to these questions, one would like to have a high recording resolution, with both pre- and postsynaptic cells marked. Only electron microscopy offers sufficient resolution to prove that a possible contact is a real synapse. As an initial step one can stain both a group of presynaptic axons and a single postsynaptic cell with fluorescent dyes. In such preparations, one can look for possible appositions that are candidates for connections.

Here we make such a survey of one particular synaptic connection in the hippocampus: the Schaffer collateral system terminating on CA1 pyramidal cells. We have determined the bouton density and the axonal branching by staining afferent fibers with rhodamine-coupled dextran. We have also measured the pattern of dendritic branching of individual CA1 cells filled with Lucifer yellow and reconstructed the dendritic tree in a confocal laser-scanning microscope. Furthermore, with the same technique, the spine density along the dendrites of basal and secondary apical dendrites have been estimated. Although the final microconnectivity pattern is not possible with this technique, the method can give an outline of the types of connectivity we can expect to find.

SEVERAL METHODS WERE USED TO ELUCIDATE THE MICROCONNECTIVITY

Adult male Wistar rats (150–250 g) where anesthetized with halothane. After the brain was taken out and cooled in a calcium-containing artificial cerebrospinal fluid (ACSF) at 2°C, it was divided in two, and one hippocampus was gently dissected out. A portion of the middle part of the hippocampus was placed on the vibratome table and cut while submerged in cold (4°C) oxygenated ACSF. Some slices were cut on a tissue chopper. Transverse or longitudinal slices of 400 μm thickness were cut and transferred to a chamber where the slices rested on a net in the interphase between ACSF below and humidified CO_2/O_2 gas above. The composition of the ACSF was as follows (in mM): NaCl, 124; KCl, 3; $NaHPO_4$, 1.25; $CaCl_2$, 2; $MgCl_2$, 2; $NaHCO_3$, 26; glucose, 10. The gas was bubbled with 5% CO_2 in O_2 to maintain a pH of 7.4. The temperature in the surrounding water jacket was regulated at 30–32°C, giving a temperature of the slices themselves of 28–30°C.

After stabilization for about 1 hour, a stimulation electrode consisting of sharpened tungsten needle insulated with varnish was placed in the stratum radiatum, delivering 50 μsecond cathodal pulses at 30–150 μA. Recording micropipettes were pulled from borosilicate glass and filled with 4% Lucifer yellow (LY) in 0.2 M LiCl. Once placed intracellularly in a CA1 pyramidal cell, the dye was either allowed to diffuse out passively from the recording electrodes or ejected with hyperpolarizing current pulses. Only cells with well-maintained membrane potentials (< -65mV) and action potential consistently above 85 mV were found to fill sufficiently well to allow a full analysis of the dendritic structure. After filling, the slices were left for 1–4 hours before they were taken out of the chamber, fixed in 4% formaldehyde (10–30 minutes), and dehydrated in an alcohol series of increasing strength. Finally, the tissue was immersed in methyl salicylate to clear the tissue for the optical analysis, then mounted in DPX medium under a cover slip. Once the sections were cleared, they were controlled in a fluorescent microscope to see whether individual cells were filled. In the tightly packed hippocampal formation, the penetrating microelectrode can sometimes lodge dye in more than one cell; this can confuse the analysis, since the fluorescent dendrites may belong to different cells. Only preparations in which there was clearly a single cell injected were accepted for further analysis.

In the laser-scanning confocal microscope, the cells were scanned with 40x or 100x times objectives, depending upon whether an overview of the dendritic structure or spine measurement was the aim. The configuration of the dendritic tree was assessed by digitizing strategic points along the dendritic tree, using the monitor to assist the measurements in the X–Y plane. The depth was given by the depth value of the different optical sections. With a special program (provided by T. Blackstad, Univ. Oslo, unpublished), these data were reconstructed to give the total length of the dendritic tree. The high resolution scans were used to estimate the distribution of spines along the different dendritic segments. The sections were partly visualized in a projection form with several optical sections superimposed in a single plane, and partly by scanning individual sections to see whether the spine candidates satisfied the identification criteria.

Axons were stained in slices by delivering a small crystal of rhodamine-coupled dextran from a tungsten needle dipped repeatedly in a solution of dye in water and dried. These deposits were delivered on the CA3 side of sections along the CA3/CA1 border, made with a knife made by two razor blade chips with a small gap between. In this way, nearly all afferents to the CA1 region were severed, except a 50 μm thick bridge over which rhodamine stained axons could be seen to traverse from the CA3 region towards the CA1 target area. Care had to be taken to deliver the dye to circumscribed points in the tissue. In order to prevent the surface film of fluid to dilute and distribute the dye over a wider area, the surface was flushed with a large amount of ACSF immediately after dye delivery, and the excess was quickly sucked off. In this way, only a small region (20–100 μm diameter, Fig. 2) was exposed to the dextran. Later, the slices were dehydrated as just described and examined in the confocal microscope. The axons could be followed for a considerable distance into the CA1 region by scanning at different depths through the tissue. The boutons appeared as swellings with a diameter of 0.6–0.9 μm.

PATTERNS OF DIVERGING AND CONVERGING OF HIPPOCAMPAL CONNECTIONS

A discussion on the overall hippocampal connectivity must start with the number of cells in the various fields and the mainly unidirectional flow of impulses through the hippocampal formation (Fig. 1). Here, we give numbers based on studies on

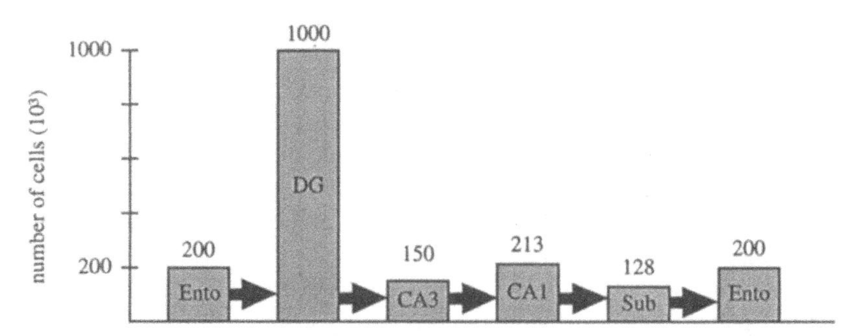

FIG. 1. The estimated number of cells in different areas of the hippocampus in adult Wistar rats. Heavy arrows indicate the unidirectional flow of information.

Wistar rats. Starting with the entorhinal cells, these number about 200,000 in the rat (6). The entorhinal axons form the perforant path that feeds on to the dentate granule cells, whose number has been estimated to between 635,000 (7) and 2,170,000 (8) on one side of the rat hippocampus. We have assumed a number of 1,000,000 granule cells (Fig. 1).

The CA3 cells number is taken as about 150,000, being a mean value between the estimates of 143,000 (9) and 157,000 (10) whereas there are 213,000 CA1 cells (10). The smallest amount of cells are found in the subicular region, where only 128,000 cells are taken as an estimate (10). The height of the different columns in Fig. 1 indicates the degrees of divergence and convergence that exist between cells in the different areas.

THE SCHAFFER AXONAL SYSTEM

The CA3 neurons have a set of long and heavily branching collaterals. One of these branches is called the Schaffer collateral after its discoverer, Karl Schaffer (11). Figure 2 shows an overview of some of the largest axonal branches. One axon collateral runs into the fimbria and further towards the lateral septal nucleus, accumbens nucleus, and anterior part of the hypothalamus. This collateral also sends off a commissural branch to the contralateral side, where it branches widely, coursing roughly through the CA1 area as on the ipsilateral side. Ipsilaterally, there are multiple collaterals running in the longitudinal direction (the longitudinal association path (lap)), both in the septal and in the temporal directions that connect a CA3 neuron to other CA3 cells. There is also a collateral that goes into the hilus of the dentate region, where it branches profusely to distribute along a considerable stretch of the hippocampal formation (not drawn). Finally, the Schaffer collateral system bends off and divides to make a bush-shaped figure stretching through a large part of the CA1 region. Along all the branches, boutons are found at regular intervals. In addition, there are a number of shorter collaterals in the vicinity of the cell of origin.

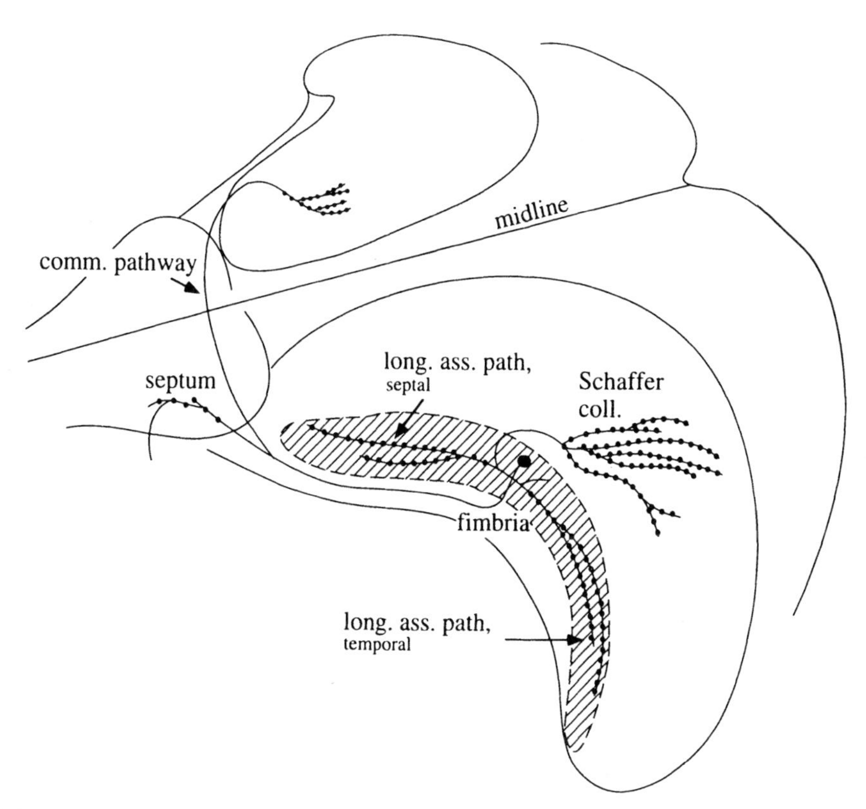

FIG. 2. Schematic drawing of the connections of a CA3 cell. The axon divides in a number of branches. Ipsilaterally, the longitudinal association pathway (long. ass. path.) runs both temporally and septally. In addition, the Schaffer collateral forms a bush-like pathway in a roughly transverse direction across the CA1 field.

We have concentrated on the Schaffer collateral system for three reasons: 1) the system is relatively homogenous because virtually all afferent fibers to the stratum oriens and radiatum of CA1 are derived from the ipsi- and contralateral CA3 neurons (12); 2) it has received a lot of attention in physiological studies; and 3) it shows greater divergence along the septotemporal dimension than other hippocampal pathways.

Total Length and Number of Boutons of the Schaffer Collaterals from One CA3 Neuron

After staining a small number of CA3 neurons with rhodamine-conjugated dextran, the fluorescence emitted from various parts of the CA3 axonal system signals the large number of collaterals (Fig. 3). The three-dimensional reconstruction of the total axonal tree of single CA3 neurons injected with HRP also emphazises the large

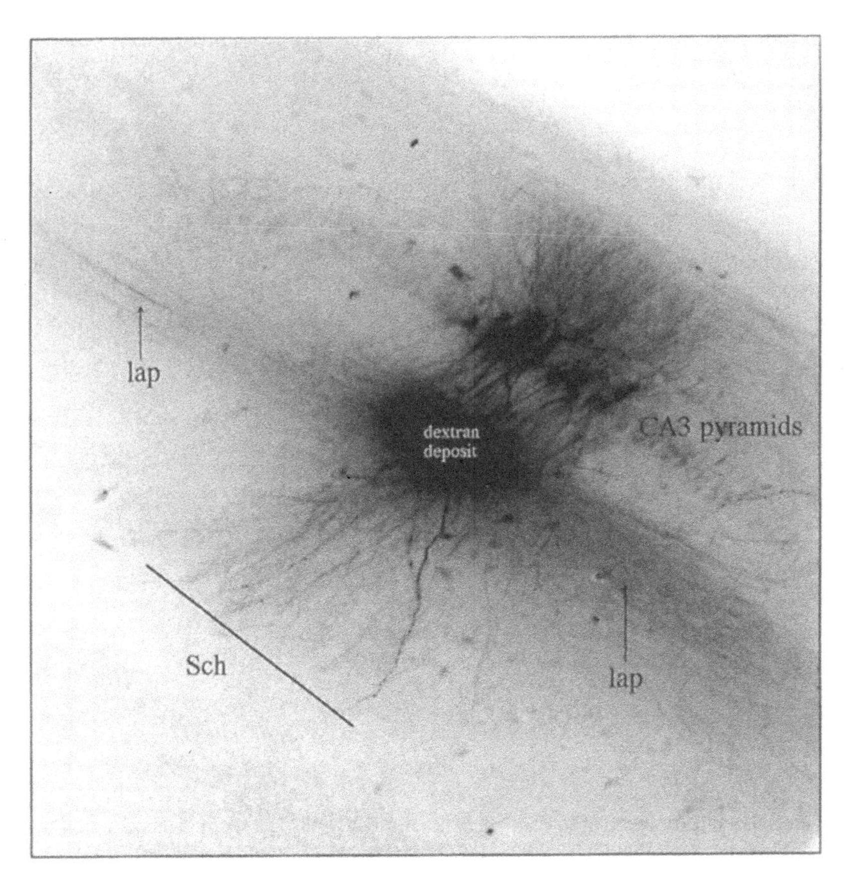

FIG. 3. Single confocal laser scan in a horizontal direction through areas CA3 and CA1 in a hippocampal formation that is unfolded in a single plane. Cells with dendrites and axonal branches are stained by deposition of a few crystals of rhodamine-conjugated dextran in str. radiatum of CA3. The septal pole is up to the left.

Sch, Schaffer collaterals; lap, longitudinal association path.

fraction of the collaterals in the Schaffer system (13). Using biocytin injections and subsequent tracing of the axonal branches, the total length of the entire axonal tree was measured to between 150 and 300 mm (14). Taking an average value of 200 mm as the total length, and using the amount of fluorescence emitted from various sectors of the collaterals, we estimated the Schaffer collateral bush to comprise 40% of the total, or 80 mm (Fig. 3). With an interbouton distance of 4 μm, the number of boutons in a unilateral Schaffer tree is about 20,000. If the CA1 cells receive only one of these boutons each, a single CA3 cell would distribute its Schaffer collateral synapses to 20,000 of the 213,000 cells, giving a connectivity figure of 1:10. With increasing concentration of synaptic contacts per CA1 cell, the proportion of CA1 target cells contacted by a single CA3 cell will shrink accordingly.

Interbouton Distance

Depending upon the method of estimation, somewhat different values have been found for the interbouton distance along intrahippocampal axons. After staining cells with intracellular injection of horseradish peroxidase, Ishizuka et al. (15) found a mean value of 7 μm between boutons along fibers in the stratum radiatum and stratum oriens in the rat. A similar figure, 7.8 μm, was found by Sayer and Andersen (unpublished observation) for individual Schaffer collateral fibers following HRP injections in CA3 cells. However, with biocytin injection Li et al. (14) found a higher value of 4.7 ± 0.5 μm for rat axons in the Schaffer collateral system. With dextran conjugates, we now have found similar values, in that the proximal portion of the Schaffer collaterals (near the CA3 border) showed 4.5 ± 0.8 μm between each bouton, whereas the distance at the other end of CA1 near the subicular border measured 3.6 ± 0.6 μm. Based upon the dextran conjugate studies, 4 μm can be taken as a mean value for the interbouton distance of the Schaffer collateral system.

How Many Pyramidal Cells are Contacted by an Individual Schaffer Collateral Axon?

The form of the Schaffer collaterals is a bush with repeatedly branching structure. For purposes of calculation, here we convert this progressively branching bush to a set of equally long branches. With a mean CA3-subiculum distance of 2.8 mm, the length of one Schaffer collateral bush is equivalent to 28 full-length branches between the CA3 and subicular borders. If they are spread evenly over a septotemporal distance of 4 mm, and within both stratum oriens and radiatum (total depth of about 600 μm), the cross-sectional tissue area per fiber is $(4{,}000 \times 600)/28 = 8.5 \times 10^4$ μm^2, which means that the average distance between these 40 equivalent branches will be about 290 μm. In order to find how many such branches can be found per target cell, we calculate the tissue area enclosing the basal and apical dendrites of a single CA1 pyramidal cell in the septotemporal plane (i.e., in a plane at right angles to the penetrating Schaffer collaterals). Based upon LY-filled cells, the basal and apical dendrites are found inside areas of 4.3×10^4 μm^2 and 7.0×10^4 μm^2, respectively, or 11.3×10^4 μm^2 total. On average, therefore, only slightly more than one branch from a single Schaffer collateral bush will be found within the borders of the dendritic tree of a given CA1 target cell.

Spine Numbers of CA1 Target Cells

The detailed connectivity obviously depends upon the number of contact sites between afferent fibers and an individual target cell. Virtually all excitatory synapses terminate on spines (16). According to Westrum and Blackstad, (1) each CA1 radiatum bouton contacts "several spines," while Sorra and Harris (17) suggest that 25% of these boutons have two spines. Here, we estimate that each bouton contacts a mean value of 1.5 spines. On the other hand, because each spine is served by one bouton only, the number of spines can be used as a good estimate for the number of synapses on an individual cell.

We have used two methods for estimating the total number of dendritic spines: 1) intracellular staining with Lucifer yellow and detailed study of the cell in a confocal laser-scanning microscope, and 2) determination of average density of asymmetric boutons from electron microscopical sections and calculation of the number of boutons belonging to an individual cell in areas contacted by the Schaffer collateral system. With the first technique (Fig. 4) one can see individual spines at a rather high density. With well-stained cells, it is evident that virtually all portions of the basal and secondary apical dendrites are covered with spines at a higher density than that normally seen in Golgi preparations. By two different investigators, we found an average density of 1.9 spines/μm length, both in the basal dendritic region and along the secondary apical dendrites. These measurements are shrinkage-corrected from direct measurements of tissue dimensions before and after dehydration (18). In two fully identified cells the total dendritic length was 13.2 and 10.2 mm. Using a mean value of 12 mm and a spine density of 1.9 spine/μm length, the total spine number on an individual CA1 cell is estimated to about 23,000. Since about one fifth of these are in the stratum lacunosum-moleculare, the spine number per CA1 neuron in regions where the Schaffer collaterals terminate is about 20,000.

Another way to calculate the number of spines is to start with electron microscopical sections and count the synapse density. From three-dimensional series reconstructions, a 100 μm^2 CA1 radiatum tissue contained an average of 42 asymmetric synapses (19). This gives an average head center distance of 1.54 μm. In order to find the tissue volume occupied by an individual CA1 cell, we have taken all dendritic trees and collapsed them into a series of parallel and equally thick cylinders with lengths corresponding to the distance from the alvear border to the hippocampal fissure. Taking the packing of cell bodies in the pyramidal layer to be four cells high, and assuming an average transverse diameter of CA1 pyramidal cells of 20 μm, the diameter of the equivalent cylinder is $(20\ \mu m / \sqrt{4}) = 10\ \mu m$. Assuming 250 μm for the thickness of stratum oriens and 350 μm for the stratum radiatum, the volume of the part of such an equivalent cylinder that receives Schaffer synapses is 47,124 μm^3. The tissue volume surrounding an individual spine is a cube with sides of 1.54 μm, equalling 3.65 μm^3. The total number of spines in the Schaffer-influenced part of the CA1 equivalent cylinder volume is then (47,124/3.65) = 12,910, a number just over one half of the 20,000 acquired through Lucifer yellow staining. Due to its direct approach, the latter method is likely to give the most accurate result.

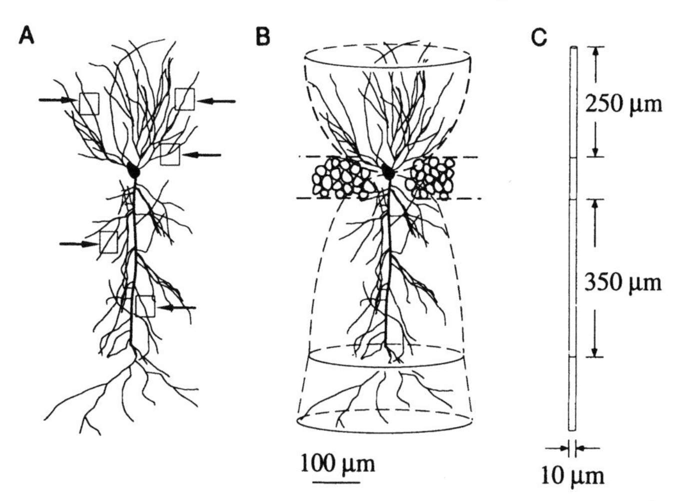

FIG. 4. A. Projection of a Lucifer yellow-filled CA1 pyramidal cell as analyzed by a confocal laser-scanning microscope. Selected dendrites (arrows) were analyzed to measure the spine density. **B.** Envelope of the tissue volume enscribing the basal and apical dendrites of the cell in **A. C.** An explanation of the method of calculation of the equivalent cylinder, and the volume of tissue exclusively occupied by one CA1 cell and its dendrites.

Estimate of the Average Connectivity Between Individual CA3 Cells and CA1 Neurons

If each of 150,000 CA3 cells has 20,000 boutons in their Schaffer collateral bush on one side, this system contains a total of 3×10^9 boutons. Assuming that each bouton serves a mean value of 1.5 spines, and neglecting contacts on nonpyramidal cells, a total of 4.5×10^9 CA1 spines can be contacted. Taking the CA1 cell number as 213,000, each being covered with 20,000 spines on the dendritic parts contacted by Schaffer afferents, the total number of relevant CA1 spines is about 4×10^9, a figure of the same order of magnitude as the number of boutons. In addition, commissural fibers from the contralateral CA3 will also occupy a number of the CA1 spines. Although the uncertainties involved in the estimated distributions of both axonal branches and spines are appreciable, these figures should be taken as an index that the number of contacts from a single CA3 cell to a single CA1 neuron is likely to be quite low.

CONCLUSIONS

In general, these calculations suggest that the number of contacts between an individual CA3 cell and a target CA1 cell are most likely quite small. Each of the 150,000 CA3 cells has 20,000 boutons in its Schaffer collateral bush, by which it can influence about 30,000 CA1 spines. If this influence is to be distributed over half the hippocampus, this means that every third of the approximately 100,000 CA1 neurons would have one spine each in contact with a single CA3 cell. These data are compatible with models explaining release data from patch-clamp experiments on the basis of a single (or very few) release sites per afferent fibers.

In case there should be a concentration of CA3 efferent axons on a smaller number of CA1 cells, some other CA1 cells must then be devoid of such contacts. In order for such an arrangement to exist, there also must be guiding principles for the growing axons so that a single parent axon may connect repeatedly to a particular target cell among a large number of seemingly identical neurons. Whether an average distribution of low density contacts, as suggested here, or a peaky distribution of more intense innervation governs the CA3/CA1 connectivity, remains unsettled. A solution to this problem demands adequate marking of both pre- and postsynaptic elements, and remains a challenging task for future neurobiology.

REFERENCES

1. Westrum LE, Blackstad TW. An electron microscopic study of the stratum radiatum of the rat hippocampus (regio superior, CA1) with particular emphasis on synaptology. *J Comp Neurol* 1962; 119:281–292.
2. Bekkers JM, Stevens CF. Presynaptic mechanism for long-term potentiation in the hippocampus. *Nature* 1990;346:724–729.
3. Malinow R, Tsien RW. Presynaptic enhancement shown by whole-cell recordings of long-term potentiation in hippocampal slices. *Nature* 1990;346:177–180.
4. Redman S. Quantal analysis of synaptic potentials in neurons of the central nervous system. *Physiol Rev* 1990;70:165–198.
5. Raastad M, Storm JF, Andersen P. Putative single quantum and single fibre excitatory postsynaptic currents show similar amplitude range and variability in rat hippocampal slices. *Eur J Neurosci* 1992;4:113–117.
6. Amaral DG, Ishizuka N, Claiborne B. Neurons, numbers and the hippocampal network. *Prog Brain Res* 1990;83:1–11.
7. Seress L, Pokorny J. Structure of the granular layer of the rat dentate gyrus: a light microscopic and golgi study. *J Anat (Lond)* 1991;133:181–195.
8. West MJ, Andersen AH. An allometric study of the area dentata in the rat and mouse. *Brain Res Rev* 1980;2:317–348.
9. Gaarskjaer FB. Organisation of the mossy fiber system of the rat studied in extended hippocampi. I. Terminal area related to number of granule and pyramidal cells. *J Comp Neurol* 1978;178:49–72.
10. Cassell MD. The numbers of cells in stratum pyramidal of the rat and human hippocampal formation. Ph.D. Thesis, University of Bristol, UK, 1980.
11. Schaffer K. Beitrag Zur Histologie der Ammonshornformation. *Arch mikr Anat* 1892;39:611–632.
12. Hjorth-Simonsen A. Some intrinsic connections of the hippocampus in the rat: an experimental analysis. *J Comp Neurol* 1973;147:145–162.
13. Tamamaki N, Nojyo Y. Disposition of the slab-like modules formed by axon branches originating from single CA1 pyramidal neurons in the rat hippocampus. *J Comp Neurol* 1990;291:509–519.
14. Li X-G, Somogyi P, Ylinen A, Buzsaki G. The hippocampal CA3 network: an in vivo intracellular labelling study. *J Comp Neurol* 1994;339:181–208.
15. Ishizuka N, Weber J, Amaral DG. Organization of intrahippocampal projections originating from CA3 pyramidal cells in the rat. *J Comp Neurol* 1990;295:580–623.
16. Andersen P, Blackstad TW, Lømo T. Location and identification of excitatory synapses on hippocampal pyramidal cells. *Exp Brain Res* 1966;1:236–248.

17. Sorra KE, Harris KM. Occurrence and three-dimensional structure of multiple synapses between individual radiatum axons and their target pyramidal cells in hippocampal area CA1. *J Neurosci* 1993;13:3736–3748.
18. Trommald M, Jensen V, Andersen P. Analysis of dendritic spines in rat CA1 pyramidal cells stained with a fluorescent dye. *J Comp Neurol* 1994; (in press).
19. Andersen P, Silfvenius H, Sundberg SH, Sveen O. A comparison of distal and proximal dendritic synapses on CA1 pyramids in hippocampal slices in vitro. *J Physiol* 1980;307:273–299.

Molecular and Cellular Mechanisms of Neurotransmitter Release, edited by Lennart Stjärne, Paul Greengard, Sten Grillner, Tomas Hökfelt, and David Ottoson, Raven Press, Ltd., New York © 1994.

22

Depression and Augmentation of Quantal Release in Adrenal Chromaffin Cells

Erwin Neher and *Ludolf von Rüden

*Department of Membrane Biophysics, Max-Planck-Institute for Biophysics Chemistry, Am Faßberg, D-37077 Göttingen, Germany; and *Department of Molecular and Cellular Physiology, Howard Hughes Medical Institute, Stanford University, Stanford, California 94305*

It has been known for some time that release of neurotransmitters at the nerve terminal is controlled by calcium (1) and other regulators (2). However, elucidation of the molecular mechanisms of secretion control has been difficult because, with the exception of the squid giant synapse, nerve terminals are not very accessible to experimental manipulation. Therefore, neurosecretory cells such as chromaffin cells of the adrenal medulla or pituitary cells have been used as welcome "model nerve terminals" (3,4). These model systems, however, have inherent limitations such as:

1. Unlike neuronal synapses, neuroendocrine cells do not have a built-in sensor of released substances. Conventional chemical detection has very limited time resolution such that many kinetic details, important for an understanding of mechanisms, are concealed. Capacitance measurements (5) and, more recently, electrochemical detection (6,7) have partially overcome this limitation.
2. Although secretion from neuroendocrine cells is not structured around active zones, it has become clear recently that there is a small pool of granules ($\approx$1% of all granules of a cell) that can be released much more rapidly than the rest. It turns out that many experimental manipulations, commonly used for studying the secretory process, readily deplete this pool (see information that follows), such that the measured response reflects both the secretory mechanism and steps involved in vesicle supply. Our recent finding (8,9) that both vesicle supply and secretion are Ca dependent further complicates matters, since Ca sensitivity cannot be used to distinguish the process of vesicle supply from that of secretion itself.
3. Gradients in Ca concentration $[Ca^{++}]$ exist inside the cell when Ca channels are open; therefore, the exact value of Ca concentration at the release site is not

known. This is true in spite of many investigations using advanced Ca-imaging techniques (10). Such techniques demonstrate large scale Ca-gradients, but cannot resolve microdomains around single channels, which have nanometer dimensions and build up and collapse within microseconds (11,12).

As a result of these complications we are just starting to answer the most basic questions concerning the relationship between the rate of exocytosis and calcium concentration, and of the average Ca concentration reached at the release site during electrical activity. This information would be very important in evaluating Ca-dependent vesicular proteins (13) as candidates for the Ca-dependent regulator (s) of exocytosis. Such paucity of information persists despite the fact that in the whole-cell configuration we can perfectly clamp the membrane potential, can control free calcium by Ca buffers and caged Ca compounds, and can readily introduce Ca probes such as fura-2.

In the next sections we will summarize recent advances in answering some of the questions just raised. We will show how experimental manipulations change the size of a release-ready pool of granules, and discuss a recently proposed model (14) that, in the simplest possible form, describes the situation of a release-ready pool in which vesicles are both consumed and replenished by Ca-dependent reactions. We point out some interesting consequences of the specific functional forms of the Ca dependencies, which confer properties to the model reminiscent of "short-term plasticity" at neuronal synapses.

SECRETORY DEPRESSION DUE TO A LIMITED POOL OF RELEASE-READY GRANULES

Indications for a limited pool of release-ready granules have come from many recent studies both on chromaffin cells (15,8) and on pituitary cells (16–20) (also see chapters 6 and 13). We have studied the degree of depression induced by trains of depolarizing pulses and by caffeine-induced Ca-transients in bovine adrenal chromaffin cells under whole-cell conditions. Figure 1A shows capacitance changes during a train of five depolarizing pulses, each 150 milliseconds in duration. Unfortunately, capacitance measurements are not valid during depolarization, since voltage-dependent conductances contribute nonlinear components to the complex admittance of the cell. Comparing capacitance values before and after depolarization, however, it is seen that the first pulse induces a sizable response, and that subsequent responses are markedly reduced. The reduction might be due to a reduced size of the release-ready pool (depression) or to a reduced stimulus strength, such as decreasing Ca current (as a consequence of inactivation of Ca channels). In order to show that the reduction is indeed depression we applied pulse protocols that, within a pulse train, provided for monotonously increasing Ca currents. In this case, responses still showed depression (9). Similar depression has been shown before in pituitary cells (18) and pancreatic beta cells (21). In the case of pituitary cells depression was preceded by some kind of facilitation during the first two or three pulses in a train. We sometimes see facilitation, predominantly in cells with small

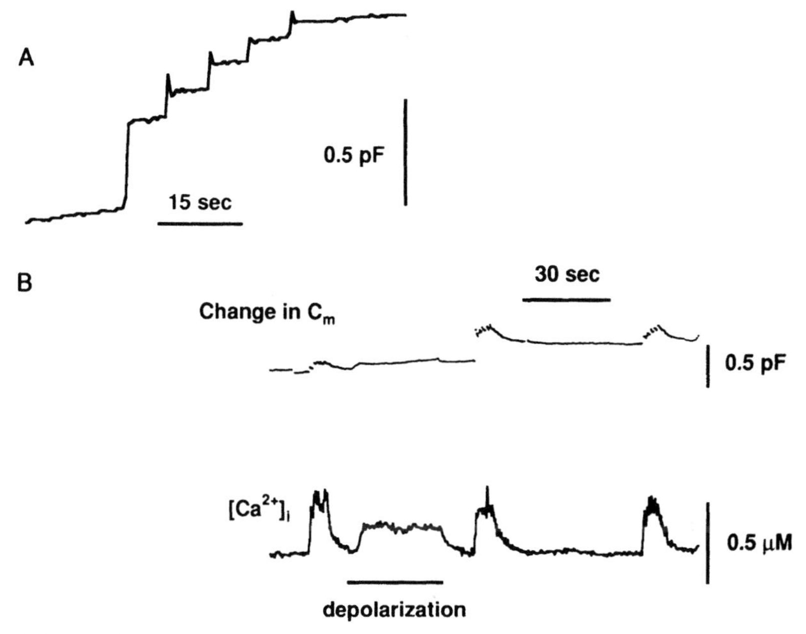

FIG. 1. Depression and augmentation of secretory responses. **A.** This panel shows the depression of capacitance increases evoked by a train of five depolarizing pulses. Pulses of 150-millisecond duration from −70 to +10 mV were applied with 7-second intervals. The responses to individual pulses, at the compressed time scale of the record, appear as step-like increases. The overshoot, apparent in some of the responses, is artifactual (see text). **B.** Depolarizing trains (5 pulses each to 0 mV, 200-millisecond duration, 1-second interval) were applied repetitively at 60-second intervals. The lower trace ($[Ca^{++}]_i$) shows the average calcium concentration within the cell; the upper trace shows the cell capacitance (change in C_m). Pulse trains show up as transient increases in both $[Ca^{++}]_i$ and capacitance. Capacitance changes are transient due to partial membrane retrieval following exocytosis. Membrane retrieval is quite variable from cell to cell (compare parts **A** and **B**). In the intervals between trains the cell was either held at −70 mV or was slightly depolarized (to around −35 mV) to enhance basal calcium. This enhancement of basal calcium caused augmentation of subsequent secretory responses. When the experiments were performed using perforated patch recordings (as in the given example), both secretory depression and augmentation were more pronounced than in whole-cell recordings. With Nystatin, repetitive cycles of augmented and control responses could be observed over tens of minutes, while in whole-cell recordings augmented responses and overall secretion were lost after a few stimuli, suggesting the washout of important components of the exocytotic apparatus. Methods: bovine adrenal chromaffin cells were isolated as described by Fenwick et al., 1982 (25). Electrophysiological experiments were carried out using either the whole-cell or the slow whole-cell mode of the patch-clamp technique (26,27). For fluorescence measurements the calcium indicator fura-2 (28) was used. Catecholamine release was measured by monitoring membrane capacitance as described by Neher and Marty (5). The extracellular solution contained in mM: NaCl, 120; $CaCl_2$, 2; $MgCl_2$, 2; glucose, 50; pH, 7.2. The pipette solution in mM: Cs glutamate, 145; NaCl, 8; $MgCl_2$, 1; Mg-ATP, 2; GTP, 0.3; NaHEPES, 10; fura-2, 0.1. Experiments were performed at room temperature (≈25°C).

Ca currents. Depression, however, is much more common, using the aforementioned pulse protocol.

We also could induce depression by release of calcium from intracellular stores. Figure 2 shows Ca transients induced by caffeine application. Short depolarizations were applied before and after three bouts of caffeine stimulation. The first caffeine transient led to a capacitance increase of 380 fF ($\approx$152 vesicles released), which we believe depleted the pool of release-ready granules. Evidence for depletion in this trace is twofold:

1. The depolarizing pulse after the bouts of caffeine stimulation produced a response much smaller than those recorded before caffeine application.
2. The capacitance responses became successively smaller with each application of caffeine. Certainly this is partly due to the decreasing Ca responses. However, even within the first stimulus the capacitance response ceased while $[Ca^{++}]$ was high. This is seen more clearly in Fig. 2B where the rate of capacitance increase (the time derivative of the capacitance signal) is plotted against $[Ca^{++}]$. Such a plot reveals a pronounced hysteresis with points originating from the rising phase of the signal lying well above corresponding points of similar $[Ca^{++}]$ from the falling phase.

Ca responses to agents that release calcium from intracellular stores (caffeine, ionomycin, bradykinin) vary widely from cell to cell (see also (15)). It is consistently observed that "large" responses lead to depression and hysteresis, whereas "small" responses do not (see next paragraph).

The fact that two different ways of increasing intracellular Ca concentration led to similar depression of the secretory response suggests that it is actually $[Ca^{++}]$ and not depolarization or nonspecific effects of caffeine that causes depression. In addition, the kinetic features observed during step-like increases in $[Ca^{++}]$ using caged Ca compounds (8,17,22) point towards $[Ca^{++}]$ as the direct cause for both eliciting secretion and depression.

RECOVERY FROM DEPRESSION IS ACCELERATED BY ELEVATED CALCIUM

When enough time is allowed after a "depressing" train of pulses, subsequent pulse trains again elicit secretory responses. Figure 1B shows three pulse trains with 60-second intervals between trains. During the pauses between trains the cell was either held at a resting potential of -70 mV or else slightly depolarized. Depolarization was such that Ca channels were partially activated, leading to a small elevation in $[Ca^{++}]$. Increased $[Ca^{++}]$ caused some degree of secretion, as evidenced by an upward slope in the Cm trace during the depolarizing episode. Surprisingly, the secretory response to the pulse train following the depolarizing episode was larger than that following the rest episode in spite of the fact that during the former episode there was some "consumption" of release-ready granules. The rate of vesicle delivery, therefore, must have been appreciably higher at elevated $[Ca^{++}]$ than at rest.

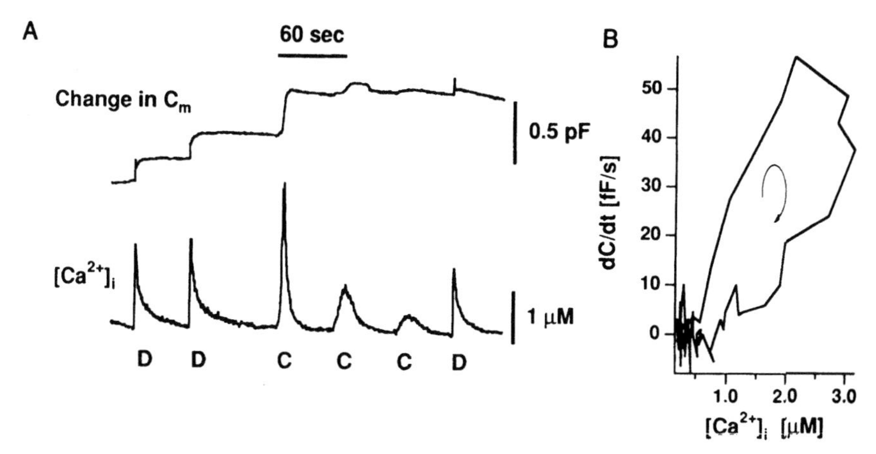

FIG. 2. Depression after a large release transient. **A.** Changes in the cytosolic calcium concentration ($[Ca^{++}]_i$) and membrane capacitance (change in C_m) in response to either short depolarizations (200 milliseconds from −70 to +0 mV; D) or applications of caffeine (C; 10 mM) are shown. **B.** A short segment of the capacitance trace from part **A** (around the first caffeine-induced calcium transient) was differentiated, and the resulting dC/dt was plotted versus the calcium concentration for each data point. The arrow indicates the progression of time. The experiment was performed in the whole-cell mode.

A similar phenomenon was observed when $[Ca^{++}]$ was slightly buffered to elevated values by appropriate Ca/EGTA mixtures. This led to slowly increasing capacitance, as shown by Augustine and Neher (15). Depolarizing voltage pulses then led to responses superimposed onto the slow trend. We performed a series of experiments in which average $[Ca^{++}]$ varied from 200–800 nM but all other conditions were identical. When sizes of individual responses to depolarizing stimuli were plotted against either preceding $[Ca]_i$ or against the slope of the preceding capacitance, a positive correlation was observed between these quantities. This, again, is a trend against the expectation that an ongoing secretion should deplete the pool of release-ready granules. It is explained by assuming that elevated $[Ca^{++}]$ augments vesicle delivery, thus overcompensating for consumption. This confirms previous evidence for Ca dependence of a priming process obtained in studies on permeabilized cells (23) and in kinetic studies using caged calcium (6).

THE TWO-STEP MODEL OF SECRETION CONTROL

The simplest possible model describing the dynamics of a pool of release-ready granules involves two steps: granules migrate from a reserve pool (termed A) to a pool of release-ready granules (B) from which they are exocytosed. This two-pool model may well be too simplistic. In fact, there is evidence from caged calcium measurements (8,17,22) that the response to a Ca step has more than two components. Nevertheless, analysis of the simplest case shows some interesting features, as demonstrated by Heinemann et al. (14). In their model both vesicle supply rate

(k_1) and the vesicle consumption rate (k_2) are Ca dependent (see Fig. 3). A Ca-independent backward rate (k_{-1}) was introduced to limit the size of pool B.

In order to measure the Ca dependence of rate k_2, one would like to change Ca in a known fashion and measure rate of release (or capacitance increase) at fixed pool B size. Thus, the Ca stimulus should be small and brief enough not to deplete pool B. On the other hand, the Ca change should be slow enough to allow for spatial equilibration of $[Ca^{++}]$ so that the fura-2 signal will faithfully report $[Ca^{++}]$ at release sites. Depolarizing stimuli are inappropriate, because they are known to cause marked Ca gradients (15). Both requirements are met, however, for caffeine-induced Ca transients if cells with a relatively small response are selected. The $[Ca^{++}]$ rises and falls within seconds, such that spatial equilibrium should be reached (15). During small responses, unlike the case of Fig. 2B, no hysteresis is seen in plots of dCap/dt versus $[Ca^{++}]$. A fit to the data points then suggests a third-power relationship between secretion and $[Ca^{++}]$ (14). This power dependence is compatible with secretion data from the neuromuscular junction and from the squid giant axon (see Augustine et al. (24) for review). Using reasonable assumptions about the size of pool B, Heinemann et al. (14) then arrived at the expression for k_2 given in Fig. 3. Given k_1, k_2 of the reaction scheme can be obtained from the "steady-state" secretion data of Augustine and Neher (15). It turns out that a reasonable fit can be obtained by assuming a Michaelis-Menten-type Ca dependence of the vesicle supply rate k_1 and a fixed backward rate k_{-1}, as listed in Fig. 3. Simulations using this model give results consistent with the results on the Ca dependence of the recovery process from depression just presented. In fact, the latter experiments were prompted by the expectations of the two-step model.

The model is also consistent with experiments using caged-calcium compounds. (8) Recent experiments using DM-nitrophen to elicit step-like changes in $[Ca^{++}]$ both in pituitary cells (22) and in chromaffin cells (unpublished observations) show that the fastest kinetic component of secretory responses has a rate that, like the rate of k_2 of our model, increases with a high power of $[Ca^{++}]$ ($n = 2.7$ in Thomas et al. (22)). For $[Ca^{++}]$ above 20 μM, however, the experimental rate, unlike k_2 in our simple model, saturates. Nevertheless, there is rough agreement between our model predictions on k_2 and measured rates in the range 1–10 μM.

The particular forms of the vesicle supply rate (k_1) and the vesicle consumption rate (k_2) imply some interesting features, as illustrated in Fig. 3. It is seen that, in a certain range of $[Ca^{++}]$, the vesicle supply rate (calculated for a fixed size of pool A) exceeds the vesicle consumption rate, which is proportional to the sum of k_2 and k_{-1} (calculated for fixed size B), such that within this range the number of release-ready granules increases beyond the value at resting $[Ca^{++}]$. Only when $[Ca^{++}]$ rises above ≈ 1.1 μM does the consumption rate exceed the supply rate, resulting in depression of pool B. The steady-state size of pool B (under the assumption of fixed pool A size) is displayed in Fig. 3, which shows a bell-shaped dependence. Also shown in Fig. 3B is the time constant, predicted from the model, with which pool B approaches steady state as a function of $[Ca^{++}]$, again assuming fixed pool A size. This time constant varies between about 3 minutes at low $[Ca^{++}]$ and 10–20 sec-

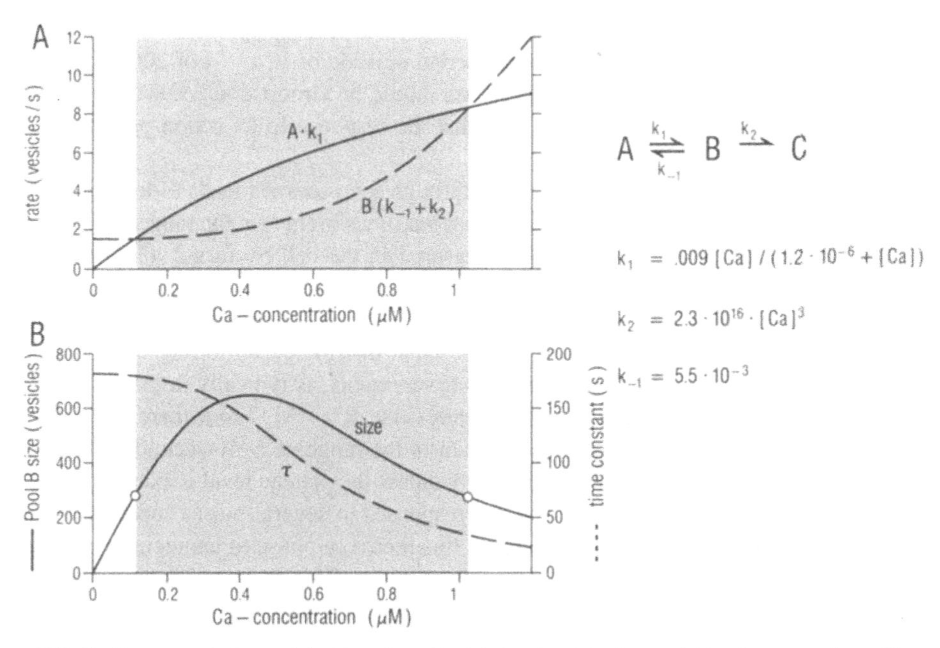

FIG. 3. Some predictions of the Two-Step Model. On the right the model is given, together with rate constants (in sec^{-1}); concentrations should be entered in mol/l. In part **A** rates are calculated for fixed pool sizes: A = 2000 vesicles; B = 272 vesicles. The value chosen for pool **B** is the steady state value at a resting $[Ca^{++}]$ of ≈100 nM. The curves represent the product Ak_1, which is the rate of vesicle delivery to pool B, and the product $B(k_{-1}+k_2)$, which is the total rate of vesicle consumption from pool B. It is seen that the two curves intersect at ≈100 nM. Between 100 nM and approximately 1 μM, vesicle delivery rate is higher than consumption rate, which implies that pool B size will increase above the basal value if $[Ca^{++}]$ is suddenly increased into that range. The new "steady state" as a function of [Ca] is given in part **B**, together with the time constant with which it is reached. Actually, B is not in a steady state, because pool A depletes. However, this is a very slow process due to the large size of pool A.

onds at high $[Ca^{++}]$. In summary, the model predicts that the number of release-ready granules should be increased following episodes of slightly elevated [Ca] and should be depressed following short episodes of high [Ca]. It should also be depressed after prolonged episodes at reduced $[Ca^{++}]$. The time scale over which a cell "integrates" its Ca history is between minutes and several tens of seconds. Strong stimuli can depress the cell very rapidly due to the high power relationship of k_2. The model qualitatively describes all the experimental findings just given.

MODEL PREDICTION FOR ACTION POTENTIALS AND CA-RELEASE TRANSIENTS

Given the features of the model previously discussed, it is expected that short pulses of Ca increase (such as occur during action potentials) should be particularly effective in releasing catecholamines, if they occur during episodes of modestly elevated calcium. At the end of a 200-second episode of [Ca^{++}] at 200 nM, for instance, the pool of release-ready granules should be almost doubled with respect to its steady state size at 100 nM, such that the response to an action potential is correspondingly larger.

During action potentials, influx of Ca through ion channels leads to locally elevated [Ca^{++}] beneath the membrane, about ten times higher (in the range 2–5 μM) than the average [Ca^{++}] that can be measured in the cell by fura-2 (15). In this range, the rate of secretion (Bk_2) is high and the number of vesicles in pool B rapidly decays. Following a depolarization, [Ca^{++}] at the membrane drops back within 100–200 milliseconds to a plateau value distributed uniformly throughout the cell (15) which, according to fura-2 measurements, is typically in the range of 50 nM to several hundred nM above steady state [Ca^{++}]. From there, [Ca^{++}] decays to a basal value with a time constant in the range of 5–10 seconds. During the latter episode very little secretion occurs unless the plateau level is exceptionally high due to large Ca^{++} influx, or when responses to several pulses superimpose. However, [Ca^{++}] is high enough during this recovery phase to appreciably speed up the refilling of the pool of release-ready granules. This may lead to either overfilling or incomplete recovery depending on the exact relationship between peak Ca and average Ca, and also on the time constant of Ca decay. Thus, slight modifications of model assumptions and stimulation frequency will either produce overall augmentation or depression in the response to repetitive pulse-like stimulations (see Heinemann et al. (14) for more detail).

A type of [Ca^{++}] signal very appropriate for augmentation of action-potential responses occurs after hormone-induced release of calcium from intracellular stores. Characteristic Ca transients followed by a plateau phase of elevated calcium can be observed in response to such stimuli in many cell types. According to our model, the plateau phase should be particularly effective in filling the pool of release-ready granules, since both amplitude (200–500 nM) and duration (tens of seconds) favor filling of pool B. Unless the peak phase of the Ca-release transient is very large, little secretion occurs, such that the net effect is overfilling (see also Heinemann et al. (14)). Subsequent action potentials should then show an augmented secretory response.

In bovine chromaffin cells, histamine, bradykinin, and several other agents lead to Ca release from intracellular stores. We tested the above model prediction using histamine. Figure 4 shows a series of pulse trains similar to that of Fig. 2. At alternate pauses between trains, histamine was applied elevating [Ca^{++}] for 40 to 60 seconds. As in the case of Fig. 2, responses following Ca elevation were consistently higher than those following rest episodes. Augmentation was between 50% and 100% (see von Rüden and Neher (9) for more detail). This effect may confer functional significance to the particular type of signal commonly observed during Ca release from intracellular stores.

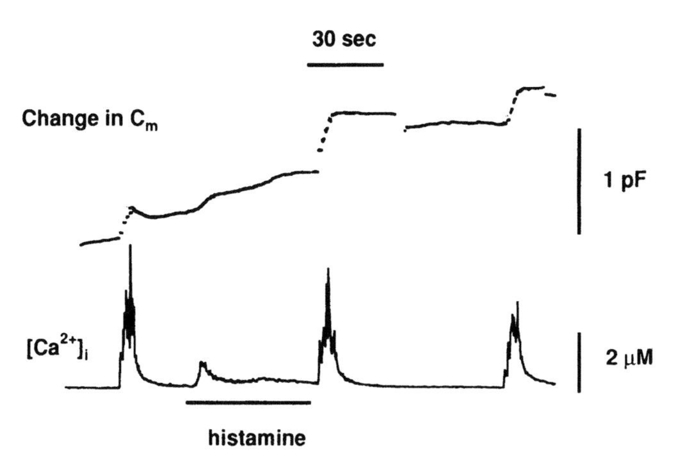

FIG. 4. Augmentation after a spike and plateau in calcium. Changes in the cell capacitance and the corresponding calcium are shown for trains of depolarizing pulses (like those in Fig. 1). Between the first two trains histamine (20 μM) was applied, resulting in a calcium release transient followed by a plateau phase of elevated calcium. This plateau phase evoked an augmented secretory response to the subsequent stimulus.

ACKNOWLEDGMENT

We thank Michael Pilot for the preparation of chromaffin cells and Dr. Kevin Gillis for suggestions on the manuscript.

REFERENCES

1. Katz B. *The release of neural transmitter substances*. Liverpool, England: Liverpool University Press; 1969.
2. Greengard P, Valtorta F, Czernik AJ, Benfenati F. Synaptic vesicle phosphoproteins and regulation of synaptic function. *Science* 1993;259:780–785.
3. Burgoyne RD, Morgan A. Regulated exocytosis. *Biochem J* 1993;293:305–316.
4. Zorec R. Exocytosis in pituitary cells. *Acta Pharmacol* 1992;42:281–286.
5. Neher E, Marty A. Discrete changes of cell membrane capacitance observed under conditions of enhanced secretion in bovine adrenal chromaffin cells. *Proc Natl Acad Sci USA* 1982;79:6712–6716.
6. Jankowski JA, Schroeder TJ, Holz RW, Wightman RM. Quantal secretion of catecholamines measured from individual bovine adrenal medullary cells permeabilized with digitonin. *J Biol Chem* 1992;267:18329–18335.
7. Chow RH, von Rüden L, Neher E. Delay in vesicle fusion revealed by electrochemical monitoring of single secretory events in adrenal chromaffin cells. *Nature* 1992;356:60–63.
8. Neher E, Zucker RS. Multiple calcium-dependent processes related to secretion in bovine chromaffin cells. *Neuron* 1993;10:21–30.
9. von Rüden L, Neher E. A Ca-dependent early step in the release of catecholamines from adrenal chromaffin cells. *Science* 1993;262:1061–1065.
10. Tsien R. Fluorescent probes of cell signaling. *Ann Rev Neurosci* 1989;12:227–253.

11. Simon S, Llinas R. Compartmentalization of the submembrane calcium activity during calcium influx and its significance in transmitter release. *Biophys J* 1985;48:485–498.
12. Chad J, Eckert R. Calcium domains associated with individual channels can account for anomalous voltage relations of Ca-dependent responses. *Biophys J* 1984;45:993–999.
13. Jahn R. Suedhof TC. Synaptic vesicles and exocytosis. *Ann Rev Neuroscience* 1993;17:219–246.
14. Heinemann C, von Rüden L, Chow RH, Neher E. A two-step model of secretion control in neuroendocrine cells. *Pflügers Arch* 1993;424:105–112.
15. Augustine GJ, Neher E. Calcium requirements for secretion in bovine chromaffin cells. *J Physiol* 1992;450:247–271.
16. Thomas P, Surprenant A, Almers W. Cytosolic Ca^{2+}, exocytosis, and endocytosis in single melanotrophs of the rat pituitary. *Neuron* 1990;5(723):723–733.
17. Thomas P, Wong JG, Almers W. Millisecond studies of secretion in single rat pituitary cells stimulated by flash photolysis of caged Ca^{2+}. *Embo J* 1992;12:303–306.
18. Bookman RJ, Lim NF, Schweizer FE, Nowycky M. Single cell assays of excitation-secretion coupling. *Annals NY Acad Sci* 1992;635:352–364.
19. Lindau M, Stuenkel EL, Nordmann JJ. Depolarization, intracellular calcium and exocytosis in single vertebrate nerve endings. *Biophys J* 1992;61:19–30.
20. Lim NF, Nowycky MC, Bookman RJ. Direct measurement of exocytosis and calcium currents in single vertebrate nerve terminals. *Nature* 1990;344:449–451.
21. Gillis KD. Misler S. Enhancers of cytosolic cAMP augment depolarization-induced exocytosis from pancreatic B-cells: evidence for effects distal to Ca^{2+} entry. *Pflügers Arch* 1993;424:195–197.
22. Thomas P, Wong JG, Lee AK, Almers W. A low affinity Ca^{2+} receptor controls the final steps in peptide secretion from pituitary melanotrophs. *Neuron* 1993;11:93–104.
23. Bittner MA. Holz RW. Kinetic analysis of secretion from permeabilized adrenal chromaffin cells reveals distinct components. *J Biol Chem* 1992;267:16219–16225.
24. Augustine GJ, Charlton MP, Smith SJ. Calcium action in synaptic transmitter release. *Ann Rev Neurosci* 1987;10(633):633–693.
25. Fenwick EM, Marty A, Neher E. Sodium and calcium channels in bovine chromaffin cells. *J Physiol* 1982;331:599–635.
26. Hamill OP, Marty A, Neher E, Sakmann B, Sigworth FJ. Improved patch-clamp techniques for high-resolution current recording from cells and cell-free membrane patches. *Pflügers Arch* 1981; 391:85–100.
27. Rae JL. Fernandez J. Perforated patch recordings in physiology. *NIPS* 1991;6:273–277.
28. Grynkiewicz G, Poenie M, Tsien RY. A new generation of Ca^{2+} indicators with greatly improved fluorescence properties. *J Biol Chem* 1985;260:3440–3450.

Molecular and Cellular Mechanisms of Neurotransmitter Release, edited by Lennart Stjärne, Paul Greengard, Sten Grillner, Tomas Hökfelt, and David Ottoson, Raven Press, Ltd., New York © 1994.

23

Nonuniformity and Plasticity of Quantal Release at Crustacean Motor Nerve Terminals

Harold L. Atwood, Robin L. Cooper, and J. Martin Wojtowicz

Department of Physiology, University of Toronto, Toronto, Ontario M5S 1A8, Canada

Synapses differ in their physiological properties. In addition to the many differences associated with the varieties of neurotransmitters and cotransmitters, and different species of receptor, there is variability in the amount of transmitter released from the presynaptic terminal and in temporal variation of release during imposed patterns of stimulation. Manifestations of synaptic plasticity related to presynaptic mechanisms have been known for some time. Prominent among these are short-term homosynaptic facilitation and depression (1,2); longer-lasting changes such as augmentation, potentiation, and long-term facilitation (3); and semipermanent or permanent changes such as neuronal adaptation (4), long-term potentiation (5), kindling (6), long-term depression (7), and long-term sensitization (8). Some of the latter changes are complex, and require participation of more than one pre- or postsynaptic element. In general, our current frame of reference for nervous system function relates short-term synaptic plasticity to rapid modification of response, and the longer-lasting synaptic effects to neural pathway modification and to aspects of learning and memory. It must be admitted, however, that the behavioral relevance of some of the synaptic plasticity phenomena uncovered by various experimental manipulations remains to be unambiguously established.

For the study of synaptic efficacy and plasticity, several different preparations have proven suitable for particular phenomena. Crustacean neuromuscular preparations have been important for investigation of homosynaptic plasticity and certain types of synaptic interaction (e.g., the presynaptic mechanisms of short-term facilitation (9) and presynaptic inhibition (10) were initially investigated in a crayfish limb muscle). Since then, the range of synaptic phenomena amenable to investigation in crustacean motor systems has expanded to include long-term facilitation, neuronal adaptation to activity, and different forms of synaptic depression (2). The advantages of crustacean motor neurons for such studies lie in their large size, their accessibility for experimental investigation, and their possession of a range of synaptic types that appear to imitate central synapses more closely than do other neuromuscular junctions.

Crustacean motor neurons often form synapses with different properties of transmission on different members of their postsynaptic target populations. This affords the opportunity to study factors that regulate synaptic "strength," (i.e., the amount of transmitter released per synapse) at individual synapses of a single neuron. Furthermore, the ability of these neurons to undergo either short- or long-lasting activity-dependent alterations of synaptic performance opens up a model for investigating well-defined plastic changes.

We present here observations on the structural and physiological correlates of synaptic strength, and on long-term facilitation, a simple form of activity-dependent synaptic modification. Two questions will be addressed: 1) what factors contribute to synaptic strength on terminals of a single neuron?; and 2) how is synaptic strength modified in response to neuronal activity?

MOTOR NEURON TYPES

Crustacean motor neurons are diverse in their physiology and morphology. Most muscles receive both excitatory and inhibitory innervation. Excitatory motor neurons range in type from phasic to tonic (2,11). The phasic neurons are characterized physiologically by a low level of impulse production (often limited to rapid escape responses), and by a large release of transmitter for the first few impulses in a series. Rapidly developing synaptic depression becomes apparent with maintained stimulation. In contrast, tonic neurons display ongoing low levels of impulse activity, and are involved in postural responses and most locomotory activities. Synaptic transmission is typically resistant to depression and exhibits short-term facilitation.

In limb muscles, branches of "tonic" or "slow" axons are known to provide physiologically distinct synapses to different muscle fibers (12). This produces a favorable situation for investigation of synaptic differences in a single motor neuron, and will be discussed further in the next section.

Although phasic motor axons are typically larger than tonic motor axons, their synaptic terminations on the muscle fibers are characteristically very slender and threadlike. The synaptic endings of tonic motor neurons are quite different; they are comprised of many varicosities connected by intervening bottlenecks. The thin endings of phasic motor neurons are very profuse, accounting in part for the large synaptic potentials that they generate. Mitochondrial content of the tonic endings is higher, and this likely explains the greater fatigue resistance.

The physiological and morphological phenotypes of phasic motor neurons can be shifted towards the tonic phenotype by increasing the impulse activity or the sensory input (13). This conversion process (termed "long-term adaptation") takes several hours to a day to start and, once established, persists for weeks. It requires an intact neuron, involves protein synthesis, and is known to involve the mitochondrial system as one component (14). However, the conversion to the tonic phenotype is never complete; thus, genetic or tropic regulation sets limits to what can be achieved in adaptation to altered activity.

SINGLE-NEURON DIFFERENCE IN TRANSMISSION

High-output and low-output synapses of a single motor neuron of tonic type have been found in many crustacean limb muscles, including the "opener" muscle of the crayfish claw and leg (15,16). This preparation has been used more than any other for physiological investigations of crustacean synapses, and will be the "type specimen" of the present discussion. The utilization of functionally different synapses appears to be an important part of the mechanism by which a single excitatory motor axon can grade muscle contractions. High-output synapses recruit their target muscle fibers at relatively low frequencies of impulse firing. Low-output synapses, which facilitate greatly as the frequency of firing increases, add increasing numbers of recruited muscle fibers, and increased depolarization and tension in each fiber, at high frequencies of impulse firing (15).

Innervation of the crayfish leg's opener muscle is illustrated in Figs. 1 and 2. The two major efferent axons, opener excitor (OE) and specific opener inhibitor (OI), course over the inner surface of the muscle and provide several concurrent branches to each muscle fiber. A small branch of a second "common" inhibitor innervates a few fibers in the proximal region of the muscle (17) but is often hard to see in fluorescently stained preparations. The proximal region of the muscle's inner surface is a location for high-output synapses; most of the central region is supplied by low-output synapses (18).

The individual branches and varicosities (boutons) of the two regions do not differ greatly in appearance (Fig. 2). However, in the proximal region, small clusters of boutons on short branches close to the main axon are evident; these are not seen in the central region. Quantitatively, there is a difference in the extent of axonal branching in the two regions: the "branching factor" (number of times each primary branch divides) and terminal length are often two to three times greater in the central (low-output) region (19). The number of synaptic boutons per muscle fiber is quite variable in fluorescently stained preparations, but in general lower in the proximal region.

These observations indicate a less profuse innervation of muscle fibers that exhibit large excitatory postsynaptic potentials (EPSPs) at low frequency. The same situation, but more extreme, has been described in a crab leg muscle (20). The larger EPSP amplitude in the proximal region cannot be explained by a larger number of boutons on proximal fibers.

Muscle fibers with large EPSPs have higher input resistances than those with smaller EPSPs but, in the crayfish leg opener muscle, this factor accounts for only about 25% of the overall difference in EPSP amplitude (21). The major determinant of EPSP differences is the evoked quantal release of transmitter at individual boutons. Evoked release of transmitter can be investigated with macropatch electrodes placed over synaptic boutons (22). Quantal content at representative boutons typically is higher in the proximal region (Table 1). Thus, average currents are invariably larger in this region (Fig. 3). These observations suggest that the properties of individual synapses on boutons of the two regions differ, since the number of boutons is in general smaller in the proximal region.

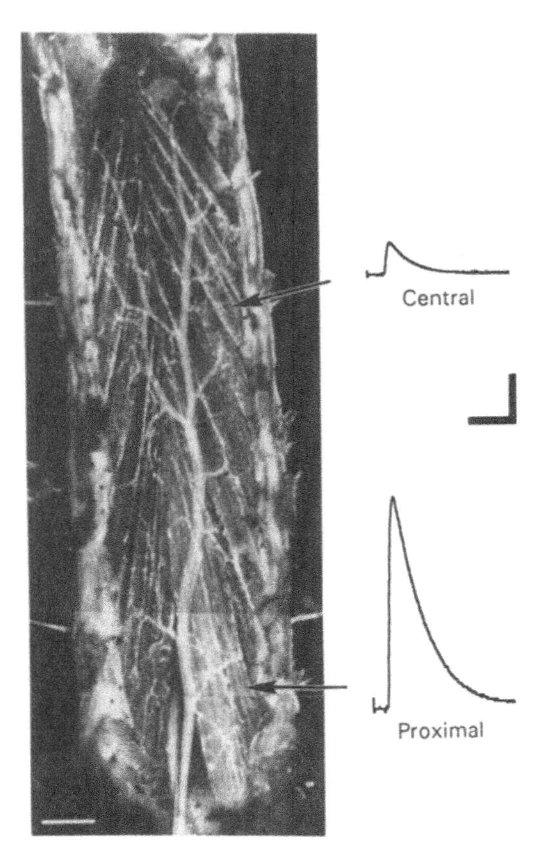

FIG. 1. Innervation of the crayfish leg opener muscle: overview of the innervation on the inner surface of the muscle, as observed by confocal microscopy. The preparation was stained with the fluorescent dye, 4-Di-2-Asp. Averaged ($n = 500$ sweeps) excitatory postsynaptic potentials (EPSPs) recorded with intracellular electrodes during stimulation of the excitatory motor axon at 1 Hz are shown for representative proximal and central muscle fibers. (Scale bar, 250 μm; electrical calibration bars, 0.4 mV and 50 milliseconds)

Since the mechanisms of differential synaptic performance in the crayfish opener muscle reside primarily in the individual synapses and boutons, it is pertinent to examine the structural features of synapses in the regions of high and low output. Correlation of structural differences at the synaptic level with properties of quantal release can be achieved through use of vital fluorescent dyes, such as 4-Di-2-Asp, (23) which permit visualization of the recording site for the macropatch electrode (Fig. 4A). A marker for subsequent electron microscopy can be deposited by the recording electrode during or after recording; we have found that fluorescent polystyrene beads (Duke Scientific Company), 0.5 μm in diameter, are suitable, since they can be observed on the embedded specimen and also in sections cut for electron microscopy. If a recording site on a specimen can be successfully located, serial

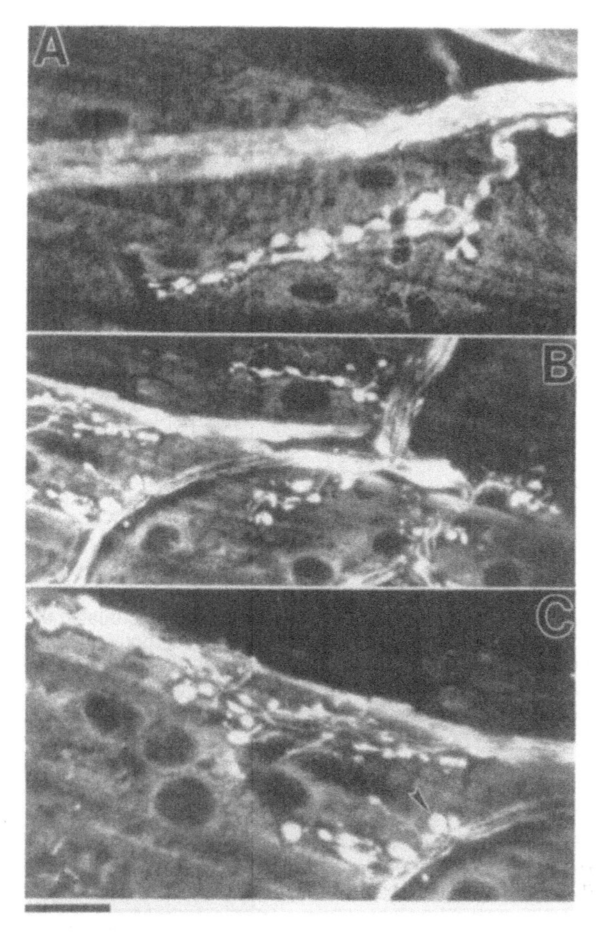

FIG. 2. Terminal branches and synaptic boutons of central and proximal regions observed in a preparation stained with 4-Di-2-Asp. **A.** Examples from the central (low-output) region. Extended branches with many boutons are prevalent. **B, C.** Examples from the proximal (high-output) region. Note clustered boutons close to the main axon. Example of a bouton close to the axon is marked in **C.** (Scale bar, **A** and **C,** 25 μm; **B,** 10 μm)

sectioning followed by electron microscopy and subsequent reconstruction (Fig. 4B) can be employed to determine the number of synapses at the recording site, and their fine structure.

Results of serial reconstructions have shown that the number of individual synapses at a 20 μm diameter recording site is of the order of 30 to 40. Binomial statistical analysis (24) allowing for nonuniform probabilities of release for individual response elements (tentatively taken to be active synapses), generally yields much smaller values for the number of response elements n, at low frequencies of stimulation (Table 1). This general result has also been found in a crab leg muscle.

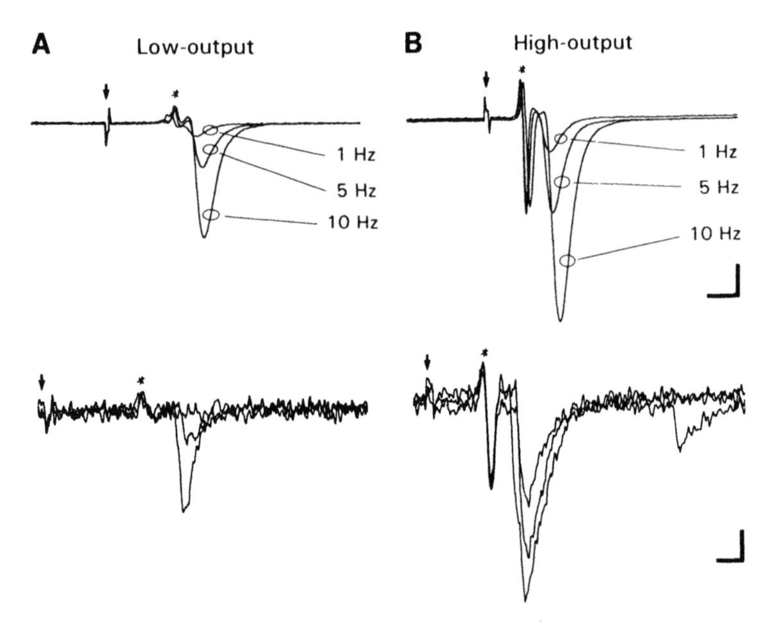

FIG. 3. Synaptic events observed at visualized boutons with macropatch electrodes. **A.** Average (1,000 sweeps) synaptic currents from central (low-output) bouton at different frequencies, with three superimposed individual sweeps shown below. **B.** Average synaptic currents and individual sweeps from proximal (high-output) bouton. (Calibration bars: vertical, 0.05 nA; horizontal, 0.15 milliseconds for upper records, 0.5 milliseconds for lower records)

TABLE 1. *Quantal release of transmitter and frequency facilitation at representative low-output and high-output synaptic boutons (Data for low-output example after Wojtowicz et al., 1994: (52))*

	Stimulation (Hz)	Quantal content	Parameters
Low-output recording	2	0.051	$\bar{p} = 0.05$ $n = 1$
	4	0.150	$\bar{p} = 0.05$ $n = 3$
	10	0.56	$\bar{p} = 0.08$ $n = 7$
High-output recording	1	2.14	$\bar{p} = 0.53$ $n = 4$
	5	5.77	$\bar{p} = 0.83$ $n = 7$
	10	12.48	$\bar{p} = 0.92$ $n = 14$

$\bar{p}$, mean release probability; n, number of response elements

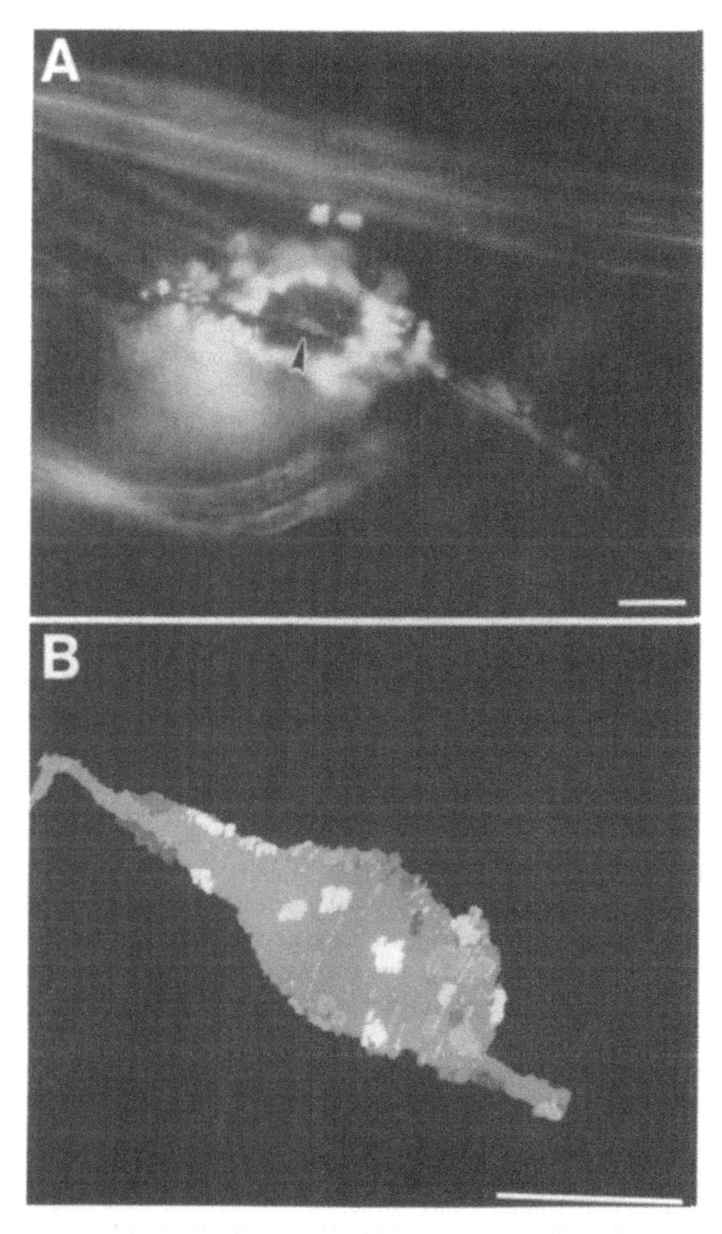

FIG. 4. Labeling and reconstruction of an identified recording site. **A.** Fluorescence microscopy of a terminal bouton stained with 4-Di-2-Asp and deposited fluorescent polystyrene beads, arranged in a ring around the bouton. (Scale bar, 20 μm) **B.** Bouton with synapses reconstructed from serial electron micrographs taken from the site shown in **A.** (Scale bar, 5 μm)

(25) At face value, the observations suggest that only a few of the synapses seen with electron microscopy at a recording site are physiologically active at low frequencies.

The structural complexity of synapses on a single bouton is known to differ in crustaceans (26) In the crayfish opener muscle, most have a single small presynaptic dense bar (see Fig. 5) thought to be associated with the active zone for transmitter release (27) Some, however, have either no dense bar, or more than one ("complex synapses"). The number of complex synapses corresponds more closely to the values of *n* obtained with binomial analysis at low frequencies of stimulation, than does the total number of synapses. For example, in the low-output of Table 1, there were 10 synapses with 2 or 3 dense bodies out of the 43 present in the recording site, and at 10 Hz, the value of *n* was 7. From this, the hypothesis arises that the complex synapses have a higher probability for transmission, and that the less complex synapses (which form the majority) contribute very little at low frequencies of stimulation.

Comparison of samples of high- and low-output synapses at the structural level, using the serial reconstruction approach, indicates a difference in both number and length of presynaptic dense bars associated with their active zones (Table 2). At high-output synapses, the dense bars are about one and a half times longer in both crayfish (Table 2) and lobster (28,29), and the number per synapse is on average larger (i.e., there are more complex synapses). The absolute number of synapses per unit length of terminal (and thus for a recording site) is about the same for high- and low-output terminals in crayfish (Table 2). For a crab muscle, the number of synapses per length of terminal was found to be greater for low-output terminals. (30) Previous studies have consistently shown fewer synapses per length of nerve terminal for inhibitory axons in crustaceans (25); yet, the quantal output at low frequencies is estimated to be greater (16). Some of the inhibitory synapses are particularly large and complex in crabs and crayfish. The occurrence of complex synapses, and longer presynaptic dense bars, appear to be correlated with high-output transmission, whereas the number of synapses per length of nerve terminal is not. From this alone, it is clear that not all synapses of a single neuron are of equal effectiveness.

A hypothesis relating the occurrence of longer or more numerous dense bars to higher probability of transmitter release arises from the observation that large intra-

FIG. 5. Synapses of the crayfish opener muscle: fine structure of the active zone (after Govind et al., 1994: (66)). **A.** Representative presynaptic dense bar with clustered synaptic vesicles in a nerve terminal of the proximal region. **B.** Representative presynaptic dense body from a synapse of the central region; this structure is shorter than its counterpart in **A. C.** Freeze-fracture preparation of the active zone, P-face view, to illustrate a cluster of large intramembranous particles representing putative calcium channels, together with circular shallow depressions adjacent to these particles (fused synaptic vesicles). **D.** Freeze-fracture preparation of the active zone, E-Face view, showing a circular depression with surrounding protrusions representing synaptic vesicles. **E.** View of the active zone in a grazing thin section showing a presynaptic body surrounded by synaptic vesicles. (Scale bar, 0.25 μm)

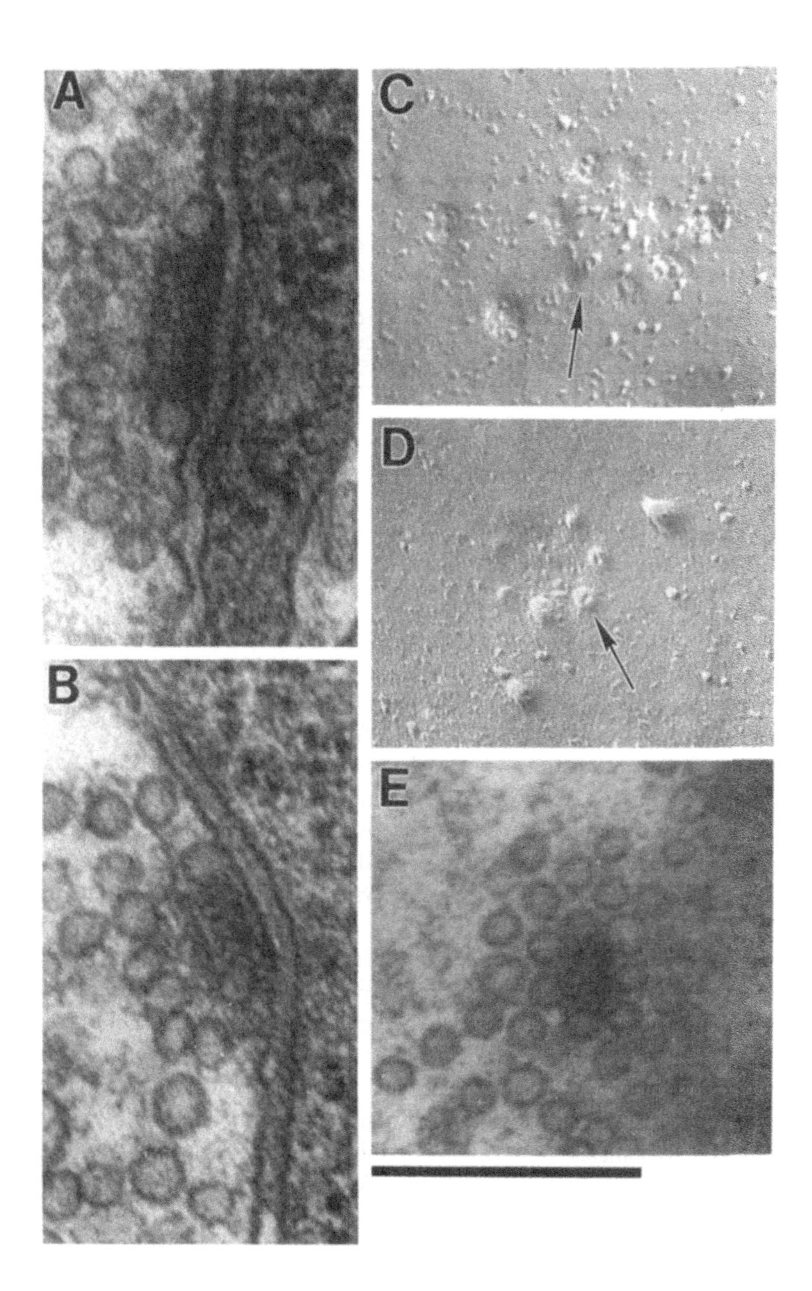
A
B
C
D
E

TABLE 2. *Structural features of synapses in serially sectioned high-output and low-output terminals (after Govind et al., 1994: (66))*

	Proximal (High-output)	Central (Low-output)	p
Synapses			
Number per μm	3.9	3.5	
Mean surface area (μm²)	0.36 ± 0.04	0.30 ± 0.05	ns
Contact area per terminal length	1.38 ± 0.14	1.20 ± 0.15	ns
Dense bars			
Mean number per synapse	1.39 ± 0.16	0.96 ± 0.1	<0.001
Mean length (μm)	0.10 ± 0.006	0.085 ± 0.006	<0.001
Length per μm of fiber	0.51 ± 0.03	0.25 ± 0.02	<0.001

membranous particles observed with freeze-fracture techniques (27) are located at the active zone (Fig. 5). The packing density of these particles is similar at high- and low-output synapses in both crayfish and lobster: one particle per 300–400 nm^2 active zone area. Work on several other systems has led to the proposal that these particles are synaptic calcium channels (31) or more likely a mixture of calcium channels and calcium-activated potassium channels (32). Given similar packing densities, more of these channels occur in synapses with longer or multiple dense bars. A nonlinear relationship between the concentration of calcium at the active zone and release of transmitter (33–35) would favor the occurrence of a higher probability of transmitter release at such synapses, if in fact the entry of calcium during impulse activity is related to the number of "channel particles" seen at the active zone. In some complex synapses, dense bars occur 100–200 μm apart (Fig. 6); thus, lateral interaction between calcium domains of adjacent active zones could occur, since the calcium domains at individual channels may extend that far (36,37).

A prediction of the hypothesis is that a greater change in calcium concentration will occur with impulse activity in boutons of the high-output region. Direct measurements to test this have now been made using an injected calcium indicator, Calcium Green 2 (obtained from Molecular Probes, Eugene, Oregon). Images of similar-sized boutons obtained with confocal microscopy show a greater increase in calcium at high-output boutons with low frequencies of stimulation (Figs. 7,8). This confirms the hypothesis for crayfish terminals. A similar situation is thought to occur in frog neuromuscular junctions with different transmission properties (38,39).

FREQUENCY FACILITATION

If synapses differ in their ability to release transmitter vesicles, and if the majority have only a small or negligible probability for transmission at low frequencies, there would be a large pool of inactive synapses on a typical nerve terminal. Synaptic transmission improves greatly as the frequency is raised; proportionately, this effect

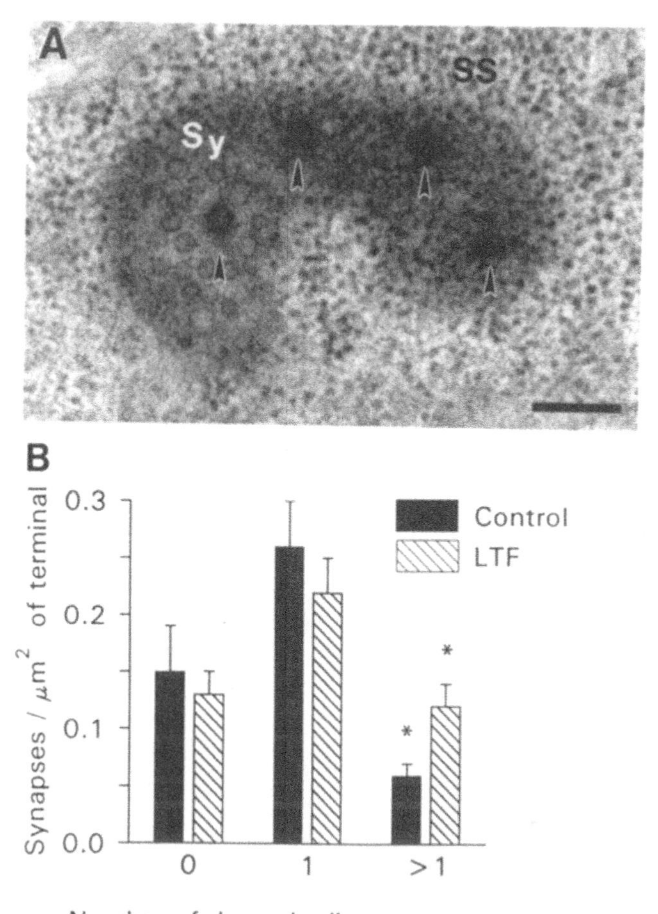

FIG. 6. Evidence for synaptic structural changes with induction of the long-lasting phase of long-term facilitation (after Wojtowicz et al., 1994: (52)). **A.** Face view in grazing section of a synapse with multiple dense bars, from a terminal in which long-term facilitation had been induced. (Scale bar, 0.2 μm) **B.** Histogram to show overall increase in complex synapses in serially sectioned terminals after induction of long-term facilitation.

is more pronounced at low-output synapses (15). A possible mechanism for frequency-dependent enhancement of transmission is recruitment of inactive synapses.

The improvement of transmission is linked in part to the frequency-dependent calcium influx observable (Fig. 7). With many impulses, calcium builds up and persists in the terminal, and this has been correlated with post-tetanic enhancement of transmission (40). However, available evidence suggests that after a single impulse, rapid reactions leading to short-term facilitation of transmission outlast the elevation in free calcium in the terminal (41–43). Thus, short-term facilitation and

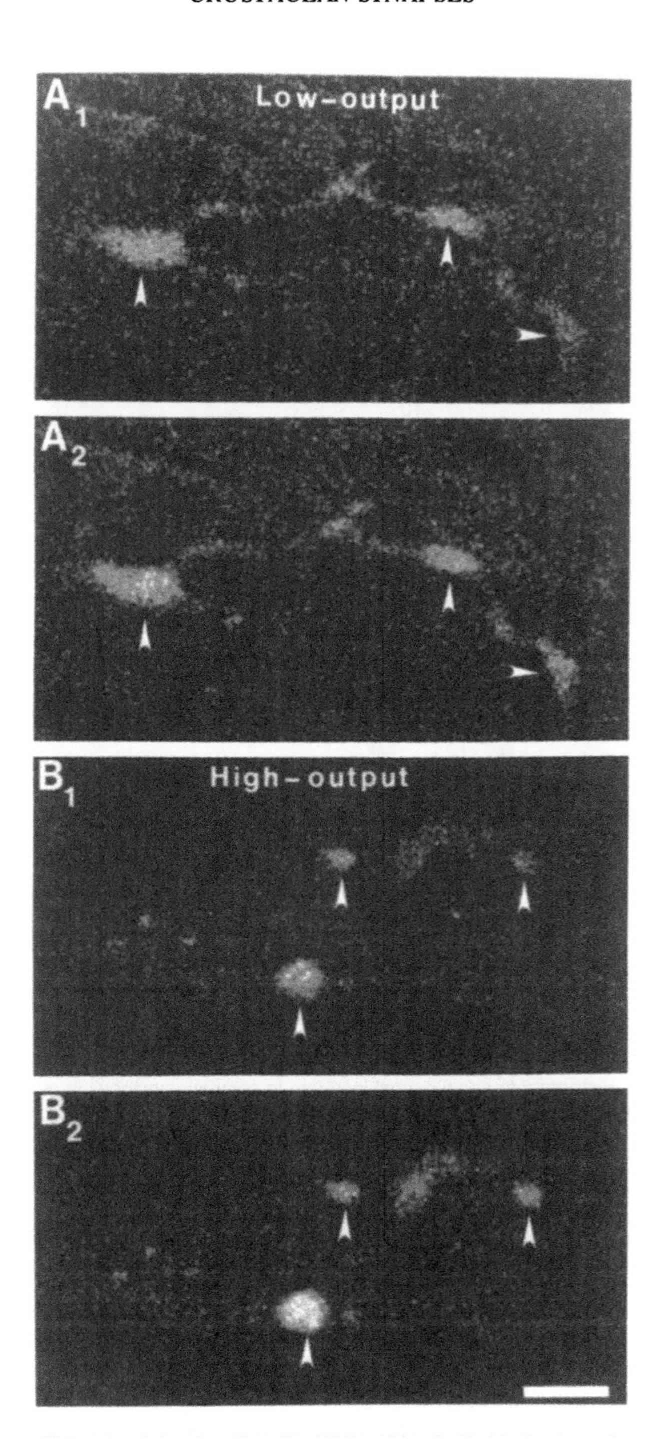
A1
Low-output
A2
B1
High-output
B2

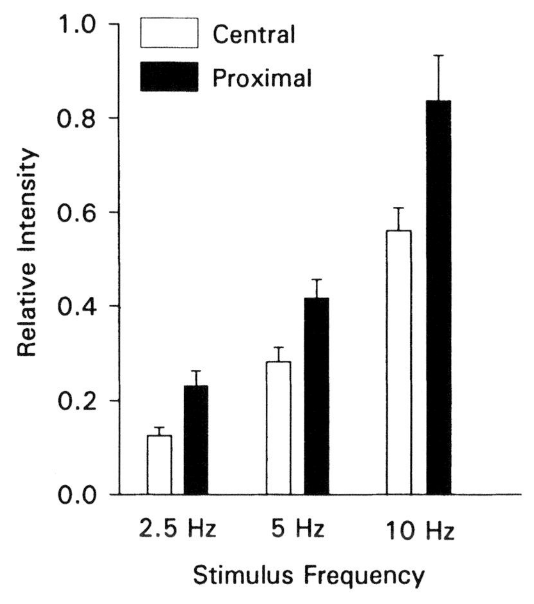

FIG. 8. Relative changes in fluorescence $[(F_{stim} - F_{rest})/(F_{rest})]$ at low- and high-output boutons at increasing frequencies of stimulation, in several preparations loaded with Calcium Green 2. Calcium entry is significantly greater in proximal boutons. Means (±SE) of 21 central and 14 proximal boutons are shown.

longer-term potentiation of transmission appear to have different mechanisms, but both are ultimately linked to calcium influx at or near the synapse.

Reactions associated with calcium influx could lead to recruitment of synapses to the active state as the frequency of stimulation is raised. This could account in part for the large increase in transmission. If recruitment of synapses occurs, quantal analysis should show an increase in *n* at higher frequencies of stimulation. At the same time, probability of transmission should increase overall in the binomial model, because synapses with initially low probabilities would participate more frequently, and because those already active would have improved rates of transmission.

Results available from previous work (44–46) and current results (Table 1) are generally consistent with this interpretation. Thus, both *n* and *p* increase overall at many sites. At 10 Hz, *n* is still much less than the number of morphologically defined synapses known to occur at the recording site. At high-output terminals,

FIG. 7. Changes in calcium concentration at different frequencies of stimulation in boutons bearing high-output and low-output synapses, observed with confocal microscopy. Signals are stronger in high-output boutons. **A.** Images of low-output boutons loaded with Calcium Green 2 at rest (**A1**) and during 10 Hz stimulation (**A2**). **B.** Images of high-output boutons loaded with Calcium Green 2 at rest Hz (**B1**) and during 10 Hz stimulation (**B2**). (Scale bar, 5 μm)

both in crabs (25) and crayfish (Table 1), very high probabilities of release are frequently seen in binomial results at low frequencies. This implies the occurrence of a few highly reliable (complex) synapses on high-output terminals. At the same time, the appearance at some proximal recording sites of additional synapses (larger *n*) at higher frequencies, with persistently high overall probability, is suggestive of a threshold effect for synaptic recruitment: certain synapses may pass abruptly from an inactive state to a fully active state at a threshold condition.

The picture that emerges from this work is one of flexibility within the nerve terminal: synaptic recruitment and synaptic probability can both change over the short term to provide sufficient transmitter release to meet functional demands.

LONG-TERM FACILITATION

Following a maintained train of stimuli at frequencies above 5 Hz, synaptic transmission at low frequencies is persistently enhanced (47). This phenomenon is referred to as the long-lasting phase of long-term facilitation (48). Participation of the cyclic adenosine monophosphate (AMP) second messenger system is necessary for this effect to occur (49). Once induced, low-frequency transmission is elevated with little decrement for hours (Fig. 9). The persistent aftereffect far outlasts detectable changes in free intraterminal calcium (40) and may in fact not require much, if any, calcium for its induction (48).

Several studies based upon quantal analysis and variants of the binomial model (24,50,51) have indicated that the elevated transmitter release is associated with an increase in *n*, with little change in overall probability, *p* (Fig. 9). This result contrasts with that for short-term facilitation, in which both *n* and *p* increase (Table 1).

Selective increase in *n* implies semipermanent recruitment of synapses to the active low-frequency pool. On the basis of the observations on high-output and low-output synapses, it would be reasonable to predict structural alteration of a few synapses to more complex form. Synapses with longer or more numerous presynaptic dense bodies would be one possibility. The number of modified synapses would not be large, since low-frequency transmission increases during the long-lasting phase of long-term facilitation by 50% to 100% (compared with increases of up to 2,000% during tetanic stimulation).

Structural features of synapses on low-output terminals of the opener muscle's central region were assayed by serial section electron microscopy after they had been subjected to inducing tetanic stimulation. There was no evidence of an increase in the total number of synapses, nor of an increase in the population of "readily releasable" synaptic vesicles (52). However, the incidence of complex synapses was higher in the facilitated samples than in controls (Fig. 6). Also, in specimens fixed during high-frequency stimulation, a few synapses were found in which presynaptic dense bodies appeared to be splitting, or recently separated; such events may signify an increase in the number of presynaptic dense bodies. Such occurrences were not seen in unstimulated or in poststimulation samples. Collectively, the results suggest that one manifestation of homosynaptic plasticity in decentral-

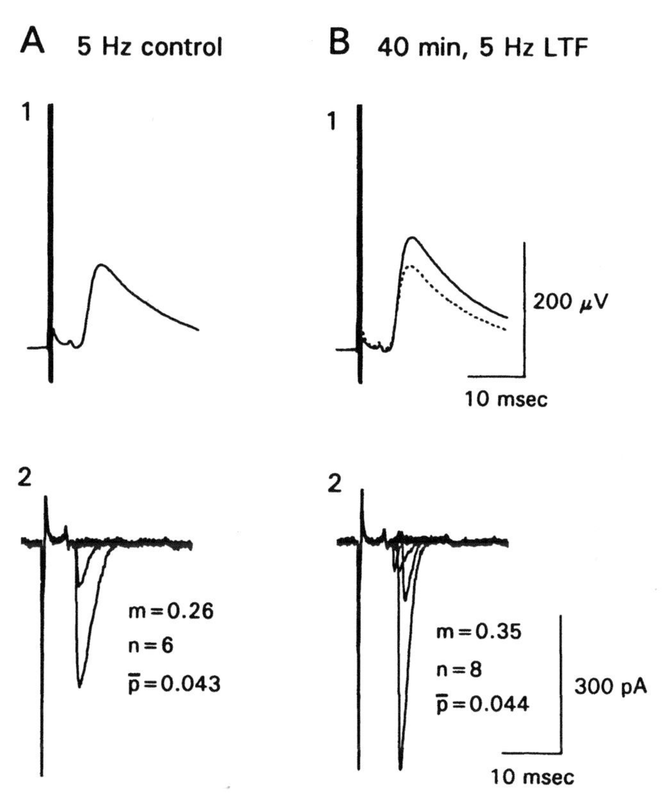

FIG. 9. Long-term facilitation. Records taken before (**A**) and after (**B**) tetanic stimulation to illustrate long-term facilitation in a single muscle fiber and at an innervation site on the same fiber (after Wojtowicz et al., 1994: (52)). Quantal content is increased, mainly due to an increase in *n*. Calculated values for binomial statistical parameters (nonuniform probability model) are shown for the extracellular records.

ized crayfish axons is generation of additional complex synapses during and after stimulation that induces long-term facilitation. Since complex synapses are thought to be more likely to release transmitter at low frequencies of stimulation, a structural basis for the long-lasting phase of long-term facilitation can be put forward.

CONCLUSION

The results presented for crustacean synapses, taken in conjunction with work on other neurons, lead to the general hypothesis that neurons produce synapses with diverse physiological performance correlated with recognizable structural specialization. Structural analysis provides a framework for visualizing synaptic changes underlying both short-term and long-term alterations in performance. Crustacean motor neurons, which regulate muscle activity in part through frequency-dependent recruitment of muscle fibers, provide good experimental models for examining the processes utilized by neurons with thousands of individual synapses in ongoing activity and in adjustments to altered conditions.

Differences in synaptic strength at individual terminals of crustacean motor axons have been hypothesized to result from differences in terminal excitability (12,53) from differences in intraterminal calcium level (54) or from differences in calcium cooperativity (54). The present observations favor a "structural" hypothesis: high-output synapses have larger or more numerous active zones and, as a consequence, a larger number of associated calcium channels. Structural evidence for identification of calcium channels as large intramembranous particles is not complete but is strongly supported by observations on a variety of synapses. Although in crayfish synapses the difference in putative calcium channels is not great, nonlinear relationships between intracellular calcium and transmitter release translate into substantial differences in output of transmitter for a single impulse. The present work confirms the hypothesis through observations of relative calcium entry in synaptic boutons: high-output boutons have a significantly larger calcium signal at low frequencies of impulse activity.

The structural hypothesis for crustacean synapses is in accord with similar observations of other systems, in particular frog neuromuscular junctions. Output of transmitter is different for junctions on different muscles (38) and also is known to vary along the length of an individual junction (55). The variations in transmitter output are related to differences in calcium entry (39) and correlated with the length of active zones and thus with the number of putative calcium channels (56,57). Crustacean junctions differ from those of the frog in synaptic organization, since they possess more numerous smaller synapses and active zones; but the theme of more calcium entry for high-output synapses appears to be shared.

The phenomenon of frequency facilitation is particularly well expressed at synapses of crustacean motor neurons. The structural observations provide a possible explanation. Synapses are imagined to differ in their recruitability, with more complex synapses (those with more numerous or larger presynaptic dense bars) being the easiest to recruit. As frequency of stimulation rises, the less complex synapses are recruited. This explanation is consistent with the structural information, and with statistical analysis.

Crustacean long-term facilitation is a simple form of activity-dependent synaptic modification, requiring only activity in the terminals of a single motor neuron, without mediation of the cell body. Establishment of long-lasting synaptic enhancement can take place without any transmitter release, and thus probably does not require feedback from the postsynaptic (muscle) cell. Interactions with other neurons are not required; thus, the process differs from long-term potentiation and other long-lasting changes requiring actions of neurotransmitters or neurohormones for initiation. The synaptic modifications underlying crustacean long-term facilitation are thought to involve structural modification of existing synapses. Electron microscopy of serially sectioned terminals yields a higher count of complex synapses after induction of long-term facilitation, with no change in total number of synapses.

Thus, the changes induced locally in the terminal by extra impulse activity push a few synapses into a high-output condition. Quantal analysis is consistent with this interpretation, since the number of active units (n) increases, while average probability of release changes little.

The synaptic changes discussed here are the simplest ones known for the crustacean motor neurons. Other changes, requiring protein synthesis in the cell body, can be induced over a longer time period by altering neural activity or sensory input (4). Thus, the neuron is equipped with a wide range of activity-dependent responses that seem adaptive for altered demands.

Speculations about the mechanism for initial establishment and subsequent maintenance of the different types of synapse in crustacean muscles have, in the past, invoked either retrograde trophic signals from muscle fiber to nerve terminal (58) or timing of synapse formation as a factor in establishing the final physiological properties of the nerve terminal (59). The more restricted growth pattern of terminals in the proximal region of the muscle suggests the possibility of an interaction between nerve terminal and muscle fiber to regulate synapse characteristics. Neuromuscular junctions of frogs appear to be regulated in position and strength in a manner consistent with the properties of the muscle fibers they innervate (60). There is good experimental evidence that crustacean synapses adjust to the electrical properties of the muscle fibers as they grow (61,62) or atrophy (63). In other neuromuscular systems, endplate morphology can be manipulated by altering the electrical effects of synapses on their target muscle fibers (64,65). Since proximal fibers have higher input resistance than central fibers, and thus generate a larger electrical signal for a given synaptic current, any interactions between muscle fiber and nerve terminal dependent upon the magnitude of electrical signal in the muscle fiber would be stronger in the proximal fibers. If nerve terminal growth is responsive to such an interaction, it would be inhibited in the proximal region more than in the central region, producing a higher incidence of short nerve terminals and stubby "cluster" synapses (Fig. 2). A further outcome of such a process could be a slower growth rate of proximal terminals, which may entail a slower turnover of resident synapses and a higher incidence of complex (possibly long-lived) synapses. This scheme, though speculative, could account for some features of the characteristic matching of synapse and muscle fiber properties in the crayfish and other crustaceans.

ACKNOWLEDGMENTS

We thank Marianne Hegström-Wojtowicz, Jonathan Wasserman, and Al Shayan for help with data analysis and illustrations. Electron micrographs were taken by Leo Marin and by Joanne Pearce. C. K. Govind contributed to work on high- and low-output synapses. The manuscript was typed by Julie Weedmark. Support for the work came from research grants from NSERC, Canada; MRC, Canada; and the Network on Neural Regeneration and Functional Recovery, Canada (Fellowship to Robin Cooper).

REFERENCES

1. Zucker RS. Short-term synaptic plasticity. *Annu Rev Neurosci* 1989;12:13–31.
2. Atwood HL, Wojtowicz JM. Short-term and long-term plasticity and physiological differentiation of crustacean motor synapses. *Int Rev Neurobiol* 1986;28:275–362.
3. Atwood HL. Organization and synaptic physiology of crustacean neuromuscular systems. *Prog Neurobiol* 1976;7:291–391.
4. Atwood HL, Lnenicka GA. Role of activity in determining properties of the neuromuscular system in crustaceans. *Am Zool* 1987;27:977–989.
5. McNaughton BL. The mechanism of expression of long-term enhancement of hippocampal synapses: current issues and theoretical implications. *Annu Rev Physiol* 1993;55:375–396.
6. Dragunow M, Currie RW, Faull RLM, Robertson HA, Jansen K. Immediate-early genes, kindling and long-term potentiation. *Neurosci Biobehav Rev* 1989;13:301–313.
7. Ito M. Long-term depression. *Annu Rev Neurosci* 1989;12:85–102.
8. Bailey CH, Kandel ER. Structural changes accompanying memory storage. *Annu Rev Physiol* 1993;55:397–426.
9. Dudel J, Kuffler SW. Mechanism of facilitation at the crayfish neuromuscular junction. *J Physiol* 1961;155:530–542.
10. Dudel J, Kuffler SW. Presynaptic inhibition at the crayfish neuromuscular junction. *J Physiol* 1961;155:543–562.
11. Sherman RG, Atwood HL. Correlated electrophysiological and ultrastructural studies of a crustacean motor unit. *J Gen Physiol* 1972;59:586–615.
12. Atwood HL. Variation in physiological properties of crustacean motor synapses. *Nature* 1967; 215:57–58.
13. Lnenicka GA. The role of activity in the development of phasic and tonic synaptic terminals. *Ann NY Acad Sci* 1991;627:197–211.
14. Lnenicka GA, Atwood HL, Marin L. Morphological transformation of synaptic terminals of a phasic motoneuron by long-term tonic stimulation. *J Neurosci* 1986;6:2252–2258.
15. Bittner GD. Differentiation of nerve terminals in the crayfish opener muscle and its functional significance. *J Gen Physiol* 1968;51:731–758.
16. Atwood HL, Bittner GD. Matching of excitatory and inhibitory inputs to crustacean muscle fibers. *J Neurophysiol* 1971;34:157–170.
17. Wiens TJ. Triple innervation of the crayfish opener muscle: the astacuran common inhibitor. *J Neurobiol* 1985;16:183–191.
18. Thompson CS, Atwood HL. Synaptic strength and horseradish peroxidase uptake in crayfish nerve terminals. *J Neurocytol* 1984;13:267–280.
19. Harrington CC. An investigation of structural relationships of the crayfish neuromuscular junction. M.Sc. Thesis, University of Toronto. 1993.
20. Tse FW, Marin L, Jahromi SS, Atwood HL. Variation in terminal morphology and presynaptic inhibition at crustacean neuromuscular junctions. *J Comp Neurol* 1991;304:135–146.
21. Cooper R, Wojtowicz JM, Atwood HL. Characterization of high- and low-output synapses from a single motor neuron. *Soc Neurosci Abstr* 1993;19.
22. Dudel J. The effect of reduced calcium on quantal unit current and release at the crayfish neuromuscular junction. *Pflügers Arch* 1981;391:35–40.
23. Magrassi L, Purves D, Lichtman JW. Fluorescent probes that stain living nerve terminals. *J Neurosci* 1987;7:1207–1214.
24. Wojtowicz JM, Smith BR, Atwood HL. Activity-dependent recruitment of silent synapses. *Ann NY Acad Sci* 1991;627:169–179.
25. Atwood HL, Tse FW. Changes in binomial parameters of quantal release at crustacean motor axon terminals during presynaptic inhibiton. *J Physiol* 1988;402:177–193.
26. Jahromi SS, Atwood HL. Three-dimensional ultrastructure of the crayfish neuromuscular apparatus. *J Cell Biol* 1974;63:599–613.
27. Pearce J, Govind C, Shivers RR. Intramembranous organization of lobster excitatory neuromuscular synapses. *J Neurocytol* 1986;15:241–252.
28. Walrond JP, Govind CK. Active zone structure at lobster high- and low-output synapses. *Soc Neurosci Abstr* 1987;13:317.
29. Walrond JP, Govind CK, Heustis S. Two structural adaptations for regulating transmitter release at lobster neuromuscular synapses. *J Neurosci* 1993;13:4831–4845.

30. Atwood HL, Marin L. Ultrastructure of synapses with different transmitter-releasing characteristics on motor axon terminals of a crab, *Hyas areneas*. *Cell Tissue Res* 1983;231:103–115.
31. Pumplin DW, Reese TS, Llinas R. Are the presynaptic membrane particles the calcium channels? *Proc Natl Acad Sci USA* 1981;78:7210–7213.
32. Roberts WM, Jacobs RA, Hudspeth AJ. Colocalization of ion channels involved in frequency selectivity and synaptic transmission at presynaptic active zones of hair cells. *J Neurosci* 1990;10:3664–3684.
33. Augustine GJ, Charlton MP, Smith SJ. Calcium entry into voltage-clamped presynaptic terminals of squid. *J Physiol* 1985;367:143–162.
34. Augustine GJ, Charlton MP, Smith SJ. Calcium entry and transmitter release at voltage-clamped nerve terminals of squid. *J Physiol* 1985;367:163–181.
35. Smith SJ, Augustine GJ, Charlton MP. Transmission at voltage-clamped giant synapse of the squid: evidence for cooperativity of presynaptic calcium action. *Proc Natl Acad Sci USA* 1985;82:622–625.
36. Zucker RS, Fogelson AL. Relationship between transmitter release and presynaptic calcium influx when calcium enters through discrete channels. *Proc Natl Acad Sci USA* 1986;83:3032–3036.
37. Simon SM, Llinas RR. Compartmentalization of the submembrane calcium activity during calcium influx and its significance in transmitter release. *Biophys J* 1985;48:485–498.
38. Pawson PA, Grinnell AD. Postetanic potentiation in strong and weak neuromuscular junctions: physiological differences caused by a differential Ca^{2+}-influx. *Brain Res* 1984;323:311–315.
39. Pawson PA, Grinnell AD. Physiological differences between strong and weak frog neuromuscular junctions: a study involving tetanic and posttetanic potentiation. *J Neurosci* 1990;10:1769–1778.
40. Delaney KR, Zucker RS, Tank DW. Calcium in motor nerve terminals associated with posttetanic potentiation. *J Neurosci* 1989;9:3558–3567.
41. Robitaille R, Charlton MP. Frequency facilitation is not caused by residual ionized calcium at the frog neuromuscular junction. *Ann NY Acad Sci* 1991;635:492–494.
42. Sivaramakrishanan S, Brodwick MS, Bittner GD. Presynaptic facilitation at the crayfish neuromuscular junction: Role of calcium-activated potassium conductance. *J Gen Physiol* 1992;98:1181–1196.
43. Bain AI, Quastel DMJ. Multiplicative and additive Ca^{2+}-dependent components of facilitation at mouse endplates. *J Physiol* 1992;455:383–405.
44. Hatt H, Smith DO. Non-uniform probabilities of quantal release at the crayfish neuromuscular junction. *J Physiol* 1976;259:395–404.
45. Wernig A. Changes in statistical parameters during facilitation at the crayfish neuromuscular junction. *J Physiol* 1972;226:751–759.
46. Zucker RS. Changes in the statistics of transmitter release during facilitation. *J Physiol* 1973;229: 787–810.
47. Sherman RG, Atwood HL. Synaptic facilitation: Long-term neuromuscular facilitation in crustaceans. *Science* 1971;171:1248–1250.
48. Wojtowicz JM, Atwood HL. Presynaptic long-term facilitation at the crayfish neuromuscular junction: voltage-dependent and ion-dependent phases. *J Neurosci* 1988;8:4667–4674.
49. Dixon D, Atwood HL. Adenylate cyclase system is essential for long-term facilitation at the crayfish neuromuscular junction. *J Neurosci* 1989;9:4246–4252.
50. Wojtowicz JM, Atwood HL. Long-term facilitation alters transmitter releasing properties at the crayfish neuromuscular junction. *J Neurophysiol* 1986;55:484–498.
51. Wojtowicz JM, Parnas I, Parnas H, Atwood HL. Long-term facilitation of synaptic transmission demonstrated with macro-patch recording at the crayfish neuromuscular junction. *Neurosci Lett* 1988;90:152–158.
52. Wojtowicz JM, Marin L, Atwood HL. Activity-induced changes in synaptic release sites at the crayfish neuromuscular junction. *J Neurosci* 1994;14:3688–3702.
53. Dudel J. Graded or all-or-nothing release of transmitter quanta by local depolarizations of nerve terminals on crayfish muscle. *Pflügers Arch* 1983;398:155–164.
54. Parnas I, Parnas H, Dudel J. Neurotransmitter release and its facilitation in crayfish muscle. V. Basis for synapse differentiation of the fast and slow type in one axon. *Pflügers Arch* 1982;395:261–270.
55. Bennett MR, Jones P, Lavidis NA. The probability of quantal secretion along visualized terminal branches at amphibian (*Bufo marinus*) neuromuscular synapses. *J Physiol* 1986;379:257–274.
56. Propst JW, Herrera AA, Ko C-P. A comparison of active zone structure in frog neuromuscular junctions from two fast muscles with different synaptic efficacy. *J Neurocytol* 1986;15:525–534.

57. Propst JW, Ko CP. Correlations between active zone ultrastructure and synaptic function studied with freeze-fracture of physiologically identified neuromuscular junctions. *J Neurosci* 1987;7:3654–3664.
58. Frank E. Matching of facilitation at the neuromuscular junction of the lobster: a possible case for influence of muscle on nerve. *J Physiol* 1973;233:635–658.
59. Govind CK, Atwood HL, Lang F. Synaptic differentiation in a regenerating crab-limb muscle. *Proc Natl Acad Sci USA* 1973;70:822–826.
60. Nudell BM, Grinnell AD. Regulation of synaptic position, size and strength in anuran skeletal muscle. *J Neurosci* 1983;3:161–176.
61. Lnenicka GA, Mellon J. Changes in electrical properties and quantal current during growth of identified muscle fibres in the crayfish. *J Physiol* 1983;345:261–284.
62. DeRosa RA, Govind CK. Transmitter output increases in an identifiable lobster motoneurone with growth of its muscle fibers. *Nature* 1978;273:676–678.
63. Lnenicka GA, Mellon J. Transmitter release during normal and altered growth of identified muscle fibres in the crayfish. *J Physiol* 1983;345:285–296.
64. Brown MC, Ironton R. Motor neurone sprouting induced by prolonged tetrodotoxin block of nerve action potentials. *Nature* 1977;265:459–461.
65. Wines MM, Letinsky MS. Inactivity-induced motor nerve terminal sprouting in amphibian skeletal muscles chronically blocked by alpha-bungarotoxin. *Exp Neurol* 1991;111:115–122.
66. Govind CK, Pearce J, Wojtowicz JM, Atwood HL. "Strong" and "weak" synaptic differentiation in the crayfish opener muscle: structural correlates. *Synapse* 1994;16:45–58.

Molecular and Cellular Mechanisms of Neurotransmitter Release, edited by Lennart Stjärne, Paul Greengard, Sten Grillner, Tomas Hökfelt, and David Ottoson, Raven Press, Ltd., New York © 1994.

24

Regulation of Transmitter Release by Muscle Length in Frog Motor Nerve Terminals

Dynamics of the Effect and the Role of Integrin-ECM Interactions

Bo-Ming Chen and Alan D. Grinnell

Department of Physiology, Jerry Lewis Neuromuscular Research Center, University of California, Los Angeles, Los Angeles, California 90024

INTRODUCTION: EFFECT OF MUSCLE STRETCH ON RELEASE FROM MOTOR NERVE TERMINALS

Muscle length exerts a powerful influence on transmitter release from frog motor terminals. In their paper first describing miniature endplate potentials (mEPPs), Fatt and Katz (1) noted that a 10% to 15% stretch of frog sartorius muscle from its rest length reversibly increased the frequency of mEPPs by two-and-a-half- to threefold, as did "pulling on the motor nerve." Soon thereafter, Hutter and Trautwein (2) found that the amplitude of EPPs of partially curarized sartorius fibers increased at a rate of 7%/mm stretch from rest to 1.25 × the rest length. At reduced quantal content, in a low Ca^{++} Ringer, both evoked quantal content and mEPP frequency increased by 25%-to-30%/mm stretch. In contrast, mEPP amplitude remained essentially constant. Hutter and Trautwein ruled out the hypothesis that stretch acts by depolarizing the nerve terminal membrane potential, since the mEPP frequency increase due to K^{+} depolarization was blocked by Mg^{++}, which was not the case for the acceleration due to stretch. Turkanis further noted an inverse relationship between resting mEPP frequency and the magnitude of the stretch effect (3). The stretch effects were unaffected by tetrodotoxin (TTX) or by increasing the Mg^{++}/Ca^{++} ratio to as high a value as 10/0.1. Together these findings suggest that Ca^{++} influx through voltage-sensitive channels does not alter release probability. This work, however, predated our knowledge of stretch-activated channels, some of which are nonselective cation channels permeant to Ca^{++} (4,5). Consequently, a role for Ca^{++} influx in this stretch effect cannot be discounted. Furthermore, both Hutter and Trautwein, and Turkanis recorded before and after the stretch, in most cases withdrawing the electrode during the stretch; hence the time course of the change in release properties was not resolved.

These studies leave two major questions unanswered. First, what is the functional significance, if any, of the stretch effect? Is the naturally occurring stretch in a given muscle sufficient to alter release properties significantly, and does the change in release occur rapidly enough to be functionally important? Second, what mechanism(s) underlie the stretch effect? Does it occur independent of extracellular Ca^{++}? If so, what explains the altered release probability?

DYNAMICS OF THE STRETCH EFFECT AND ITS FUNCTIONAL SIGNIFICANCE

The frog sartorius, on which most of the work on the stretch effect has been done, is a long (approximately 30 mm), thin muscle attaching proximally to the pelvic midline and distally to the bones of the lower leg just below and medial to the knee. It principally serves to adduct the femur, but also helps to flex the knee. Hence, it is stretched by abduction and extension of the lower leg. In our experiments, we define "rest length" as that measured when the frog is on its back, its femur extended at about 45° from the body axis and on the plane of the body, with the distal hindlimb slightly flexed. In this position, the sarcomere spacing is approximately 2.3 μm, and passive movement of the limb can increase the length of the muscle by about 2 mm (6%) at one extreme or decrease it by about 2 mm (6%) at the other. These changes are smaller than the 20% stretch used routinely by Hutter and Trautwein (2) and Turkanis (3) but still large enough to affect release probability. The questions are how much, and how rapidly?

To answer these questions, we recorded EPPs and mEPPs during changes in length, using "floating" electrodes, consisting of just the terminal shank and tip of a microelectrode filled with 0.6 M K^{+} acetate and suspended from a delicate coiled electrode wire (Fig. 1). A muscle was attached at either end to rigid rods that were free to rotate around axes halfway along their length. The other sartorius muscle from the same animal was attached at the other ends of the rigid rods. This muscle could be stimulated directly by a pair of silver wires placed on opposite surfaces of the muscle. Contraction of the second muscle stretched the first, from which recordings were being made. A nerve stimulus could be delivered to the first muscle to generate an EPP at any time during the stretch. An isometric tension transducer (Statham Instruments) attached at one axis measured the time course and magnitude of tension change. The other axis was fixed. The absolute magnitude of the stretch was measured by direct observation of the positions of both ends of the muscle along a calibrated grid.

Contraction typically began about 10–12 milliseconds after the beginning of a stimulus train and reached a peak in 80–100 milliseconds. In the example shown in

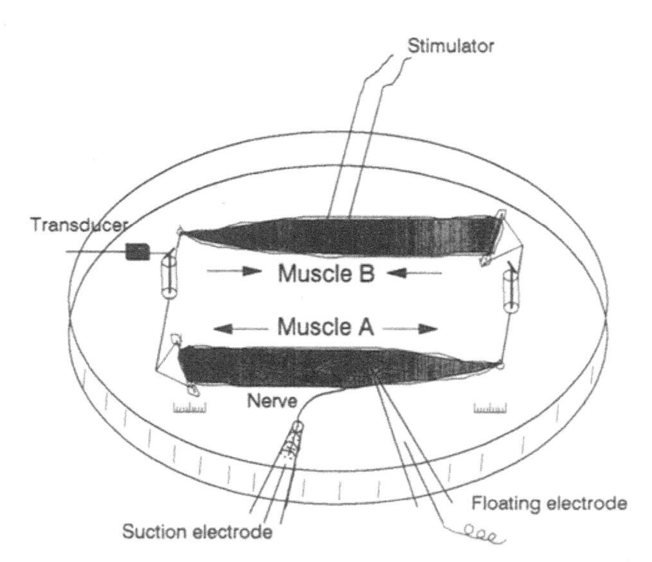

FIG. 1. A diagram of the preparation used to study the dynamics of length modulation of release. A floating electrode monitored endplate potential (EPP) amplitude or miniature endplate potential (mEPP) frequency at junctions in muscle A as it was stretched by tetanic contraction of muscle B. The rods to which the muscles were attached rotated around a fixed axis at one end, while the other end was attached to an isometric tension transducer. Length changes of muscle A were measured by eye through the dissecting microscope using calibrated grids at each end.

Fig. 2, the amplitude of the EPP was increased by about 58% at 40 milliseconds and 93% at 80 milliseconds during the onset of stretch. Figure 3 shows a more extensive record of EPPs recorded at one junction at different times during a series of such stretches. The EPP amplitudes characteristically increased sharply over the first 2–3 mm of stretch from rest (about 6% to 9% stretch, the maximum change the muscle might normally undergo). With passive shortening from rest length, the EPP amplitude and mEPP release probability decreased over the first millimeter (about 3%) of shortening, but then tended to approach a minimum value. The effects on EPPs and mEPPs of shortening during active contraction were not determined.

Clearly, the enhancement of release occurs rapidly. To measure accurately the time course of the phenomenon, it would be ideal to be able to change instantaneously from one length to another. Unfortunately, even a floating electrode cannot be held in a fiber during such a stretch. A realistic compromise is to compare the enhancement at any given length during the development of tension, with the enhancement seen at the same length when held statically. If the enhancement is instantaneous, there will be no difference in these measurements. If the enhancement develops gradually, it will be less at any given length during a stretch (depending on the rate of stretch) than when the muscle is held for some time at the stretched length.

Stretch driven by tetanic muscle contraction has the advantage that it is quasinatural in its magnitude and time course. It has the disadvantage, apparent in Fig. 3,

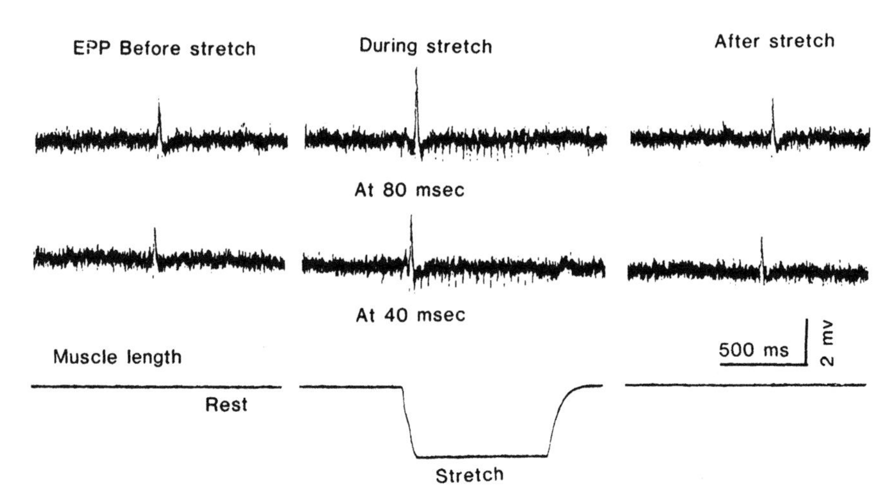

FIG. 2. Sample records of EPPs before, during, and after a stretch produced by 50 Hz direct stimulation of muscle B, with stimuli to the nerve in muscle A timed to occur 40 and 80 milliseconds after the beginning of muscle stimulation (i.e., during the development of tension). Both muscles were in Ringer containing 4.5/ μM d-tubocurarine chloride (dTC).

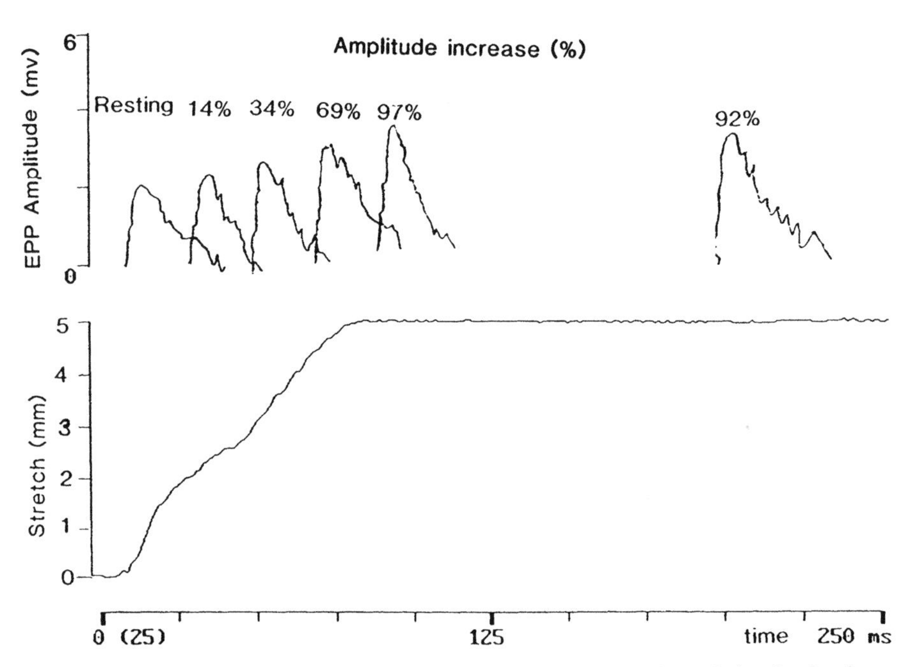

FIG. 3. Tracings of endplate potentials (EPPs) at rest and at different times during the development of a 5-mm stretch driven by tetanic contraction of the other muscle. The EPPs shown were timed to occur at 20-millisecond steps during different stretches, which might explain some of the variability in the percent enhancement compared with rest amplitude (indicated above each tracing). The degree of EPP enhancement at any given length during the onset of stretch can be compared with the degree of enhancement seen when the muscle was maintained at the same length (Fig. 4).

that it is not a constant, uniform rate of change. Nevertheless, experiments of this type reveal that a delay exists from the time the muscle is stretched to a given length to the time the enhancement reaches the level appropriate to that length in a maintained stretch. Figure 4 shows a comparison of the percentage increase in EPP during stretch ("dynamic") and at different maintained lengths ("static"). The displacement of these lines on the X axis shows that there is a delay in full development of the enhancement, which can be determined, given a knowledge of the contraction/stretch speed of the muscle. In this case, a displacement of about 0.5 mm on the X axis corresponded to a delay of approximately 7–10 milliseconds. This onset rate sets constraints on the types of mechanisms proposed to explain the phenomenon.

Another important property of the enhancement is that it is maintained essentially constant at any given length (Fig. 5). There is little or no "sag" with time, and the new level of synaptic efficacy is maintained essentially indefinitely. Indeed, in studying frog neuromuscular physiology, it is imperative that the length be known and controlled (e.g., see Grinnell and Herrera (6)).

At the termination of a tetanus, as the muscle returns toward its rest length, EPP amplitude declines with approximately the same time course of the development of enhancement (data not shown).

This phenomenon is robust. Stretching the sartorius by 1–2 mm (3% to 6%) can increase the EPP amplitude in some fibers by as much as 80% to 100%. The mean

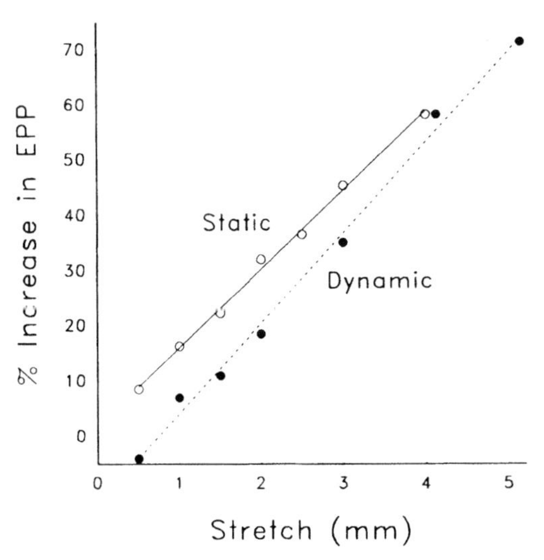

FIG. 4. Comparison at one junction of the percentage increase in endplate potential (EPP) amplitude at different lengths during the onset of stretch ("dynamic" records) with the enhancement at different maintained lengths ("static"). The displacement of the two lines represents delay in the buildup of stretch-induced enhancement of release. In this case, the dynamic record shows enhancement approximately equivalent to that at static lengths 0.5 mm shorter. Given the speed of the muscle stretch, that corresponds to a 7–10-millisecond delay in development of the phenomenon.

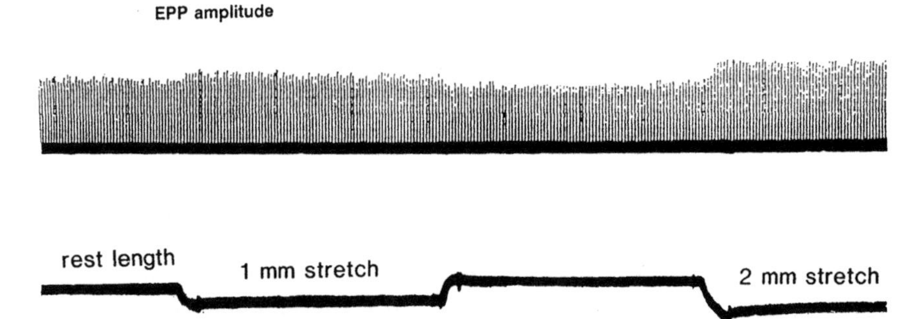

FIG. 5. Reproduction of chart recorder (Graphtec Linearcorder VII, WR 3101) tracings of end-plate potentials (EPPs) evoked by successive stimuli at 1 Hz at rest length and when the muscle was stretched by 1 mm and 2 mm (approximately 3% and 6%). Although EPP amplitudes varied at any given length, the average amplitude was well maintained at each length.

effect, measured in 26 curarized junctions in 6 muscles, was a $39.8 \pm 3.8\%$ (SEM) increase for a 2 mm (about 6%) stretch ($p<.01$). If all junctions in unblocked preparations at rest were already suprathreshold, such an increase in release would be functionally irrelevant. However, even when stretched to 1.05 times rest length, many junctions in a sartorius muscle are subthreshold. In a muscle stretched to this extent, increasing the Ca^{++} concentration in the Ringer from 1.8 to 3.0 mM increases the twitch tension by approximately 20%, indicating that all of the junctions on approximately that fraction of the fibers in the muscle are subthreshold for muscle contraction in normal frog Ringer (NFR) (6). An even larger percentage of the fibers contract when the muscle is given brief tetanic stimulation (7).

A phenomenon that increases efficacy by as much as 40% for a 6% stretch must constitute a powerful peripheral amplification of the spinal stretch reflex. Since the smallest motor units have the weakest mean EPP amplitudes (7,8) they are likely to be the most strongly affected by stretch. As was just noted, shortening of a muscle reduces the efficacy of its synapses. In a frog sitting in its normal alert posture, the sartorius is close to its minimum length. Leg extensors, on the other hand, would be stretched, with their synapses proportionately strengthened. With hindlimb extension, as in a jump, the sartorius would be passively stretched by as much as 4 mm, and its synapses strengthened correspondingly at a time when it is called on to help return the legs to their sitting position.

MECHANISM OF STRETCH ENHANCEMENT OF RELEASE

Lack of Dependence on External Ca^{++}

Since Ca^{++} plays such a prominent role in transmitter release, an obvious initial hypothesis would be that muscle stretch somehow causes a Ca^{++} influx that facilitates release. That a high Mg^{++}/low Ca^{++} Ringer does not eliminate the phenomenon (2) suggests otherwise. Still, stretch might open Ca^{++}-permeant channels that are not blocked by Mg^{++} or other conventional blocking agents. To exclude this possibility, the effects of stretch must be tested in the absence of external Ca^{++}. Such an experiment is difficult because, even in the presence of Mg^{++}, frog muscle tends to show widespread spontaneous twitching in a 0 Ca^{++} solution. However, the introduction of μ-conotoxin (μ-CgTX), which blocks Na^{+} channels in muscle but not nerve (9,10) makes such experiments feasible. A muscle exposed to 25 μM μ-CgTX for 30 minutes remains quiescent and unexcitable for several hours with apparently normal synaptic potentials. In preparations treated in this manner and then placed for a minimum of 1 hour in Ringer containing 0 Ca^{++}, 2 mM Mg^{++}, and 1 mM EGTA (0 Ca^{++} Ringer), the EPP was eliminated and the mEPP frequency fell significantly, from a mean of 1.27 ± 0.3/s (n = 68) in normal frog Ringer (NFR) to $0.52 \pm .09$/s (n = 22) in 0 Ca^{++} Ringer. In general, there is a close correlation between mEPP frequency and EPP quantal content in a Ca^{++} containing Ringer (6,11).

Even with no Ca^{++} in the external solution, mEPP frequency increased by 14.4 ± 2.7 (n = 34) with a 1 mm (3%) stretch—about half the change that was observed in NFR (Fig. 6). The effect of 0 Ca^{++} treatment is only partially reversed 1 hour after return to NFR, when measurements of the stretch effect showed a 23.5% mean increase. The frequency of spontaneous quantal release at any given junction oscillates slowly (12,13). At the low mEPP frequencies characteristic of 0 Ca^{++} Ringer, these oscillations can obscure the effects of stretch. However, when such a preparation is tetanized for 1 minute at 100 Hz, the mEPP frequency rises markedly, possibly due to Na^{+} displacement of bound internal Ca^{++} (14,11). As the elevated frequency declines toward the resting value after a tetanus, stretch clearly increases mEPP frequency. Figure 7 shows characteristic results, in which the enhancement averaged about 15% (range 7% to 23%) when tested at 1-minute intervals during the 7-minute period after the tetanus. The increase in mEPP frequency with stretch in 0 Ca^{++} Ringer cannot be attributed to Ca^{++} influx through stretch-activated or any other kind of channel. The rapidity of the change (7–10 milliseconds to full development) argues against second messenger involvement or changes in pump activity that could affect ionic gradients. The sustained nature of the effect and its immediate reversibility are inconsistent with an increase in cytoplasmic free Ca^{++} due to displacement of bound internal Ca^{++} by Na^{+} entering through voltage or stretch-activated channels, as in the case of tetanic stimulation. Moreover, TTX does not block the effect (3) and, based on the time and number of terminal action potentials required to significantly increase mEPP frequency (11) it is unlikely that sufficient Na^{+} could enter in a few milliseconds and have such an effect. The rapidity and magnitude of the effect imply that the changes in tension or length associated with terminal stretch are acting close to release sites.

Morphological Evidence for ECM-terminal Plasma Membrane Connections at the Active Zones

Candidate connections between the ECM and active zone regions of the nerve terminal can be demonstrated. Deeply etched EM transverse sections through frog

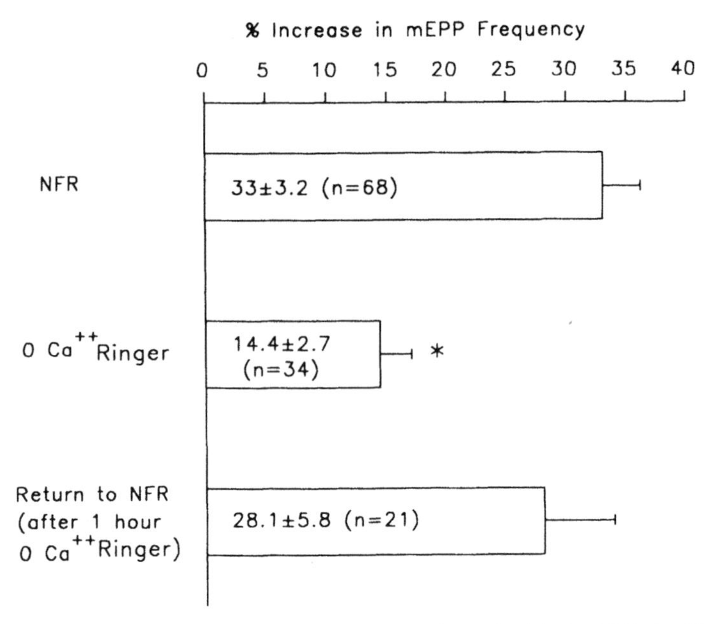

FIG. 6. Persistence of stretch-enhancement of release in a 0 Ca^{++}, 2 mM Mg^{++} Ringer. Bars show mean (±SEM) increase in miniature endplate potential (mEPP) frequency with a 1 mm stretch in NFR, in the 0 Ca^{++} Ringer, and after return to normal frog Ringer (NFR). The enhancement in 0 Ca^{++} 2 mM Mg^{++} Ringer was significantly less than in original NFR ($p<.001$) or after return to NFR ($p<.02$). The values in NFR before and after treatment were not significantly different. (Composition of NFR in mM: NaCl, 116; KCl, 2; $CaCl_2$, 1.8; $MgCl_2$, 1; $NaHCO_3$, 1; dextrose, 3; Hepes, 5) (pH, 7.4))

neuromuscular junctions show strands connecting the basal lamina and the presynaptic terminal membrane. These connections are present throughout the synaptic cleft, but their density is greatest at active zones (see Heuser (15) Fig. 7). Figure 8 shows a fortuitous example of a deeply etched freeze fracture obtained in our laboratory that shows these connections clearly. In this fracture, which jumped from presynaptic terminal bouton to postsynaptic membrane at an active zone, strands of ECM material appear to attach selectively at high density in the region of the intramembranous particles of the presynaptic active zone. It is not easy to interpret such micrographs, or to know the appearance of the matrix before the etching process. Such images, however, suggest a structure whereby stretch of the muscle and ECM could exert tension effectively near the sites of release in the nerve terminal.

ECM and Plasma Membrane Molecules Mediating Mechanical Transduction

Mechanical connections of cells to the ECM or surrounding substrate are critical to functions of specialized cells (e.g., serving as receptors for vibration, angular acceleration, pressure, and gravity). Likewise, most living cells depend on such

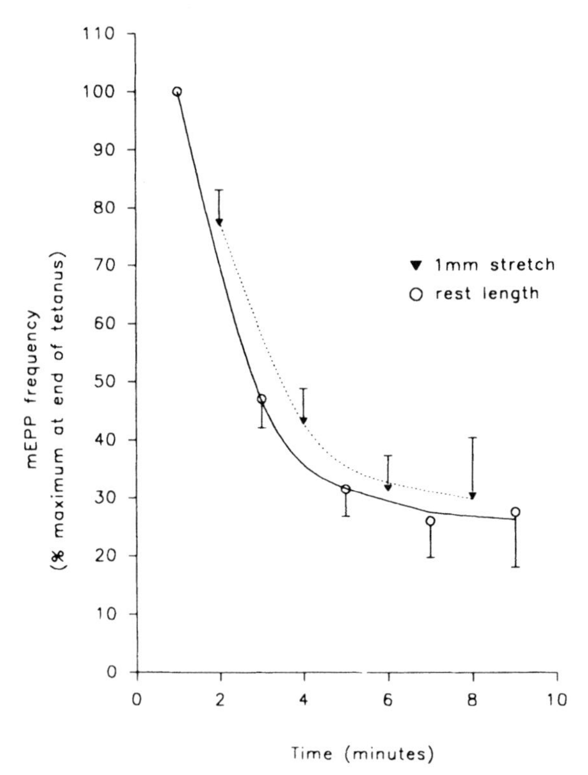

FIG. 7. Curves derived by averaging results of 12 experiments in which a 60-second, 100-Hz tetanus in 0 Ca^{++}, 2 mM Mg^{++} Ringer was used to increase the miniature endplate potential (mEPP) frequency to a mean of 4.42/s (from a mean pretetanus value of 0.61), following which the frequency was measured at rest and 1 mm stretched length at 1-minute intervals during recovery. The dashed line connects the mEPP frequency values in the stretched condition; the solid line is based on measurements at rest length. The displacement of the two indicates a stretch effect, at elevated mEPP frequency, in 0 Ca^{++} Ringer, of 10% to 20% (mean $14.7 \pm 1.5\%$ at 1-minute intervals for 7 minutes). Thus the percentage increase in mEPP frequency with stretch was not significantly different with or without tetanic elevation of intracellular Ca^{++}.

interactions for migration, process extension, and maintenance of their shape. In some cases, the ECM contributes to regulation of gene expression, protein synthesis, secretion, and ion transport (16). Mechanical stimuli are amplified by the structure of the ECM and transduced into both mechanical and electrochemical responses inside cells. In turn, the internal cytoskeletal forces can be transmitted externally to act on the surrounding substrate (16). Although several types of transmembrane receptors interact with the ECM, the most widespread and versatile appear to be the integrins (17,18,19). The integrins are a large family of heterodimeric transmembrane receptors, each composed of an α and β subunit. On the cytoplasmic side, the subunits are linked to actin through accessory proteins such as talin, vinculin, and α-actinin. Extracellularly, the subunits interact with ECM mole-

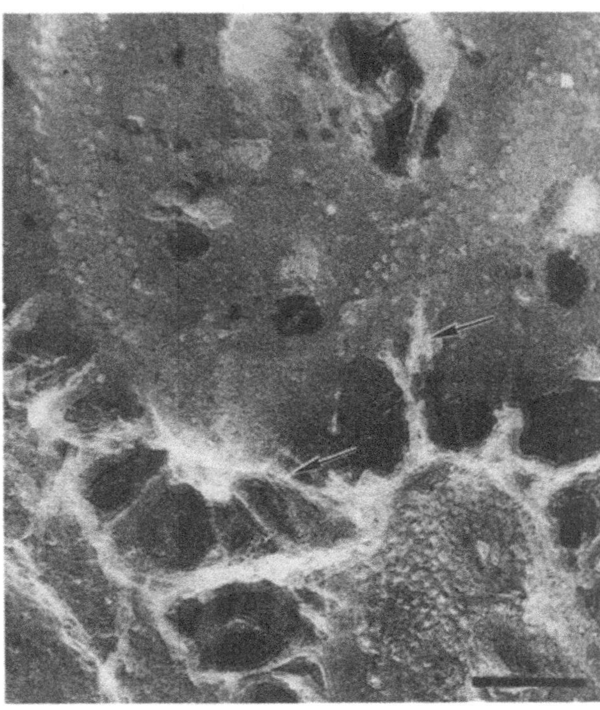

FIG. 8. Deep-etched freeze-fracture replicas of portions of two synaptic boutons in which the fracture jumped from the terminal membrane through the ECM to the muscle fiber membrane at active zones. The double rows of intramembranous particles (IMPs) are evident, with strands of ECM prominently connecting near the IMPs (arrows). These strands are candidates for the connections that mediate the effects of stretch. ACh receptors are evident as prominent particles in the lower right region of the fracture on the right. (Calibration, 200 nm)

cules or adhesion receptors on other cells. At least 14 α and 8 β subunits have been described (19) These fit together noncovalently in numerous combinations, some almost unique in their binding specificities. In general, the α-β interaction that forms the extracellular binding site depends on the presence of divalent cations, especially Ca^{++} (20,21) That the length-dependent modulation of release, although still present, was reduced in magnitude in a 0 Ca^{++}, 2 mM Mg^{++} Ringer (Fig. 6, 7) is consistent with the integrins being involved in the stretch effect.

Our initial test of this hypothesis has been guided by another feature common to many integrin-ligand binding interactions: the integrins bind to sites that contain a sequence of three amino acids—arginine-glycine-aspartic acid (RGD). Fibronectin (FN), vitronectin (VN), tenascin, thrombospondin, collagen, and laminin are among the ECM molecules with such RGD sites. Depending on the specific integrin and ECM molecule, this binding can be disrupted by swamping the preparation with short peptides containing the RGD sequence (18). Particularly effective in many instances is the sixmer GRGDSP (hereafter referred to as RGD). The control peptide argine-glycine-glutamic acid (RGE) has no effect (22).

To test the hypothesis that the RGD-specific binding between integrins and ECM ligands might mediate the stretch effect, we first treated a muscle with EGTA-buffered 0 Ca^{++}, 2 mM Mg^{++} Ringer for 1 hour, expecting that many of the postulated integrin-ligand bonds would be disrupted. This solution was replaced either by NFR or NFR containing a low concentration of either RGD or RGE. At 0.1 or 0.2 mM concentration, RGD sharply reduced the enhancement of mEPP frequency due to 1 mm stretch; 0.2 mM RGE had no greater effect than return to NFR alone (Fig. 9). After return either to NFR or NFR + RGE, the stretch effect was slightly less than before the zero Ca^{++} treatment (mean $33 \pm 3.2\%$), perhaps because integrin-ligand bonds had been broken and not completely reestablished after return to a normal Ca^{++} concentration (see also Fig. 6).

Peptide concentrations as low as 0.1 mM sharply reduce the stretch effect. This order of sensitivity to RGD interference is characteristic primarily of vitronectin receptors, some of which can, however, also bind fibronectin with almost equal affinity (23). Vitronectin (VN)-receptor binding can be stabilized by either Mg^{++} or Ca^{++}, and may require physiological levels of Ca^{++} (21). We therefore tested

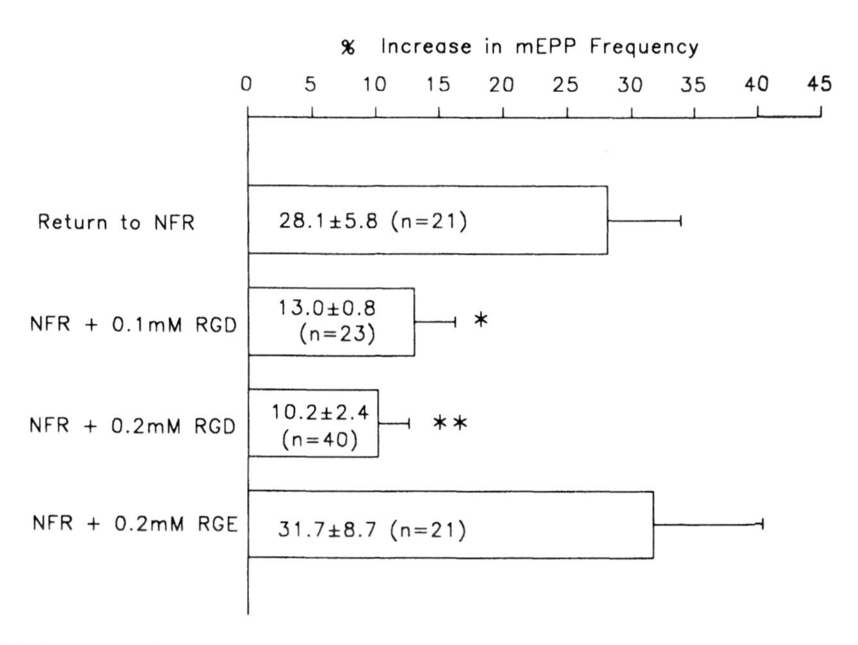

FIG. 9. Effects of arginine-glycine-aspartic acid (RGD) and the control arginine-glycine-glutamic acid (RGE) peptides on the magnitude of the stretch effect after pretreatment for 1 hour in 0 Ca^{++}, 2 mM Mg^{++} Ringer. The suppression in the effect in 0.1 mM RGD, compared with return to normal frog Ringer (NFR) and to NFR+0.2 mM RGE, was significant at the $p<.05$ level. Suppression by 0.2 mM RGD was significant at $p<.005$.

an EGTA-buffered "low-divalent Ringer" (50 μM Ca^{++}, 50 μM Mg^{++}) that we thought might be more effective than 0 Ca^{++}, 2 mM Mg^{++} Ringer in dissociating receptor binding. In the low-divalent Ringer, 1 mm stretch still caused a large enhancement of mEPP frequency (Fig. 10). When 0.2 mM RGD was added to the low divalent Ringer, however, the stretch effect was sharply reduced (by 67% at 1 mm stretch, $p<.001$; by 47% at 2 mm stretch, $p<.003$). By itself the low-divalent Ringer was not as effective as 0 Ca^{++}, 2 mM Mg^{++} Ringer in reducing the effects of stretch. In the low-divalent Ringer, however, the RGD peptide was highly effective in reducing enhancement.

Without pretreatment with 0 Ca^{++} or low-divalent Ringer, 0.2 mM RGD in NFR had comparatively little effect on the stretch enhancement (23±6.1% enhancement of mEPP frequency for a 1 mm stretch; 44.7±24, n=7, for 2 mm stretch—neither were significantly different than in NFR alone, given the small amount of data). Indeed, even in NFR containing 2 mM RGD, without prior zero Ca^{++} treatment, there was still a 27.3±13% (n=12) enhancement with 2 mm stretch, compared with 54.6±6.7% in NFR controls (n=23). The RGE (2 mM) had no effect at all on the stretch-induced increase in mEPP frequency (enhancement with 2 mm stretch=54.3±6.2%, n=12) (see Fig. 9).

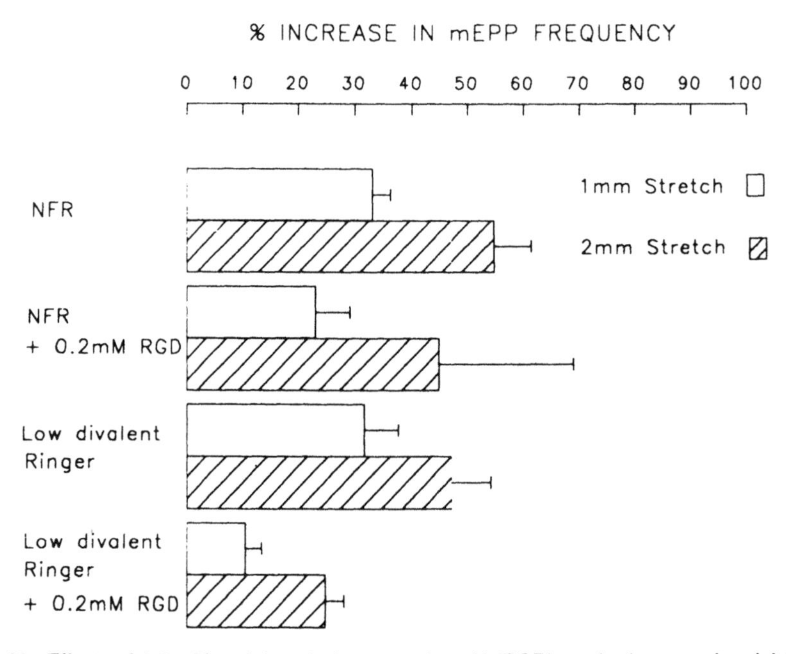

FIG. 10. Effects of 0.2 mM arginine-glycine-aspartic acid (RGD) on the increase in miniature endplate potential (mEPP) frequency with 1 and 2 mm stretch and added to normal frog Ringer (NFR) or a low-divalent Ringer containing only 50 μM Ca^{++} and 50 μM Mg^{++}. The RGD had a small but not statistically significant effect in NFR, but a large effect in the low-divalent Ringer ($p<.001$ at 1 mm, $p<.003$ at 2 mm). The effect of stretch in the low-divalent Ringer, by itself, was not significantly different from the effect in NFR.

INTERRELATIONSHIP BETWEEN SYNAPTIC STRENGTH AND THE STRETCH EFFECT

An important clue to the mechanism of stretch-induced enhancement of release almost certainly lies in the observation that there is an inverse relationship between synaptic strength (EPP amplitude) and the magnitude of enhancement with any given stretch (Fig. 11A). A similar inverse relationship exists for mEPP frequency in NFR. Interestingly, in preparations first treated with 0 Ca^{++}, 2 mM Mg^{++} Ringer and then returned to NFR plus 0.2 mM RGD, the mean enhancement with 2 mm stretch was reduced ($26.6 \pm 3\%$, $n=42$ versus $39.8 \pm 3.8\%$, $n=26$, in the NFR alone; $p<.01$) and there was little or no evidence of the inverse relationship (Fig. 11B). Terminals with larger EPPs and mEPP frequencies, which probably exhibit a larger resting Ca^{++} influx and resting Ca^{++} concentration inside the terminal, (24,11) exhibit reduced responses to stretch. This is not because the stronger terminals are already under greater tension. When partially depolarized by elevating the K^{+} concentration in the Ringer, the same population of junctions showed a reduced stretch effect proportional to the increase in mEPP frequency. Apparently, as the internal Ca^{++} concentration rises, it masks or inhibits the effects of stretch.

CONCLUSIONS

These findings indicate that divalent cation-sensitive integrin-ECM binding interactions involving the RGD sequence of amino acids in the ligand are responsible for at least part of the length-dependent modulation of release. The identity of the ECM molecule(s) involved in these bonds is unknown, nor has the integrin been characterized. However, the integrin binds the RGD sequence of one or more ECM ligands with high affinity and requires divalent cations, with Ca^{++} favored over Mg^{++}. Even in the presence of 2 mM Mg^{++}, a 0 Ca^{++} Ringer strongly reduced the stretch effect and made the preparation susceptible to more complete block by low concentrations of the peptide. A Ringer containing only 50 μM each of Ca^{++} and Mg^{++} did not reduce the effect of stretch, but did make the junctions susceptible to block by the RGD peptide.

Of great interest is the mechanism whereby tension on the integrin-ECM bonds alters release probability. The cytoplasmic domains of integrins are associated with cytoskeletal elements, and in some cases are responsible for maintaining the stability of cytoskeletal organization and cell shape (16). Furthermore, integrin-ligand binding or mechanical stimulation may: a) elevate Ca^{++} influx (e.g., see Schwartz (25)); b) release chemical second messengers that can cause the release of intracellular Ca^{++} (e.g., see Wirtz and Dobbs (26), Ng-Sikorski et al. (27), and Jaconi et al. (28)); c) activate the Na^{+}/H^{+} antiporter and cause cytoplasmic alkalinization; (29) and d) alter phosphatidylinositol metabolism through phospholipase C and 1,2-diacylglycerol (30). The cytoplasmic domains of some integrins activate tyrosine or serine-threonine protein kinases (31,32). However, where binding has been shown

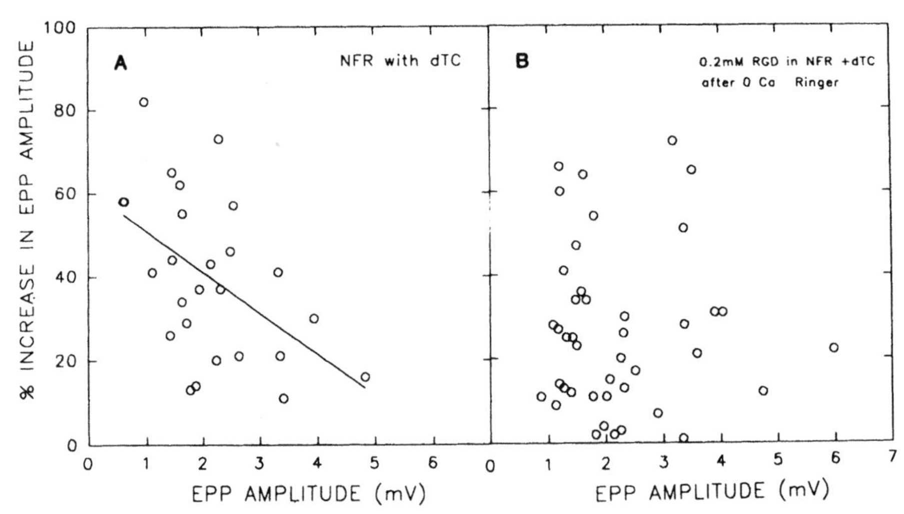

FIG. 11. Relationship between the magnitude of the stretch-enhancement of EPPs in normal frog Ringer (NFR) (**A**) and in NFR containing 0.2 mM arginine-glycine-aspartic acid (RGD) after treatment in the zero Ca^{++}, 2 mM Mg^{++} Ringer (**B**). In both cases the NFR contained 5 μM d-tubocurarine chloride dTC. Endplate potential (EPP) amplitude and degree of enhancement were inversely correlated in NFR, but there was no evident relationship in the presence of RGD. (Data in **A** from six muscles, in **B** from five muscles.)

to affect intracellular Ca^{++} concentration or second messenger activity, the changes take minutes or longer to become significant and last in some cases for hours. The speed of development and reversal of stretch-related modulation of release probability is difficult to reconcile with second-messenger involvement or Ca^{++} buildup from intracellular sources. More likely, tension on the integrins somehow mechanically alters the conformation or location of some component(s) of the release apparatus, increasing its sensitivity to Ca^{++} or to some other trigger of release.

SUMMARY

Changes in muscle length cause large changes in the probability of transmitter release from frog motor nerve terminals. A 5% to 10% stretch from rest length can increase EPP amplitude or mEPP frequency by more than 100%. The phenomenon is fully reversible and extremely rapid. Within 7–10 milliseconds of the stretch, the enhancement is complete, and it is maintained essentially constant at the new level for as long as the stretch is sustained. Given these properties, the length modulation of release is unquestionably of functional importance, strongly amplifying the spinal stretch reflex.

The stretch-induced enhancement of transmitter release persists at a reduced level in a 0 Ca^{++}, 2 mM Mg^{++} Ringer. This finding indicates a lack of dependence on

Ca^{++} influx from outside the terminal. Release of Ca^{++} from intracellular stores close to release sites cannot be ruled out as a contributing factor. Our results, however, suggest a mechanism involving physical connections between the extracellular matrix and the nerve terminal that can alter release probability directly. Morphological evidence for connections that might be responsible can be demonstrated in micrographs of deep-etched freeze fractures through neuromuscular junctions.

Hypothesizing that the ECM-nerve terminal connections responsible for the stretch effect involve proteins from the integrin family and knowing that many of the integrin-ECM binding interactions occur at sites on the ECM proteins containing the amino acid sequence RGD, we treated preparations with 0 Ca^{++}, 2 mM Mg^{++} Ringer to reduce integrin binding and then returned the muscle to normal Ringer containing 0.1–0.2 mM of a six-amino-acid peptide containing the RGD sequence. This peptide strongly suppressed the stretch effect, while a control peptide (RGE) had no effect. A 50 μM Ca^{++}/50 μM Mg^{++} Ringer had little effect on stretch enhancement but permitted a strong inhibition of enhancement when RGD was present.

The identity of the ECM molecule(s), the integrin(s), and the mechanism of enhancement of release are unknown. However, our findings imply that much or all of the length-dependent modulation of release probability is mediated by an RGD-sensitive integrin-ECM interaction that depends more on external Ca^{++} than on Mg^{++}

ACKNOWLEDGMENTS

We wish to thank Susan Hulsizer for help during the early stages of this research, and Steve Meriney, Steve Young, Bruce Yazejian, Richard Weiss, and Amir Kashani for advice and assistance at many points during this work and preparation of the manuscript. S.A. Kick provided valuable advice on the manuscript. Bibbi Wolowski's skill and patience made the freeze-fracture deep etch micrographs possible. Supported by NIH grant NS06232.

REFERENCES

1. Fatt P, Katz B. Spontaneous subthreshold activity at motor nerve endings. *J Physiol* 1952;117:109–128.
2. Hutter OF, Trautwein W. Neuromuscular facilitation by stretch of motor nerve endings. *J Physiol* 1956;133:610–625.
3. Turkanis SA. Effects of muscle stretch on transmitter release at end-plates of rat diaphragm and frog sartorius muscle. *J Physiol* 1973;230:391–403.
4. Sachs F. Mechanical transduction by membrane ion channels: a mini review. *Mol Cell Biochem* 1991;104:57–60.
5. Davis MJ, Donovitz JA, Hood JD. Stretch-activated single-channel and whole cell currents in vascular smooth muscle cells. *Am J Physiol* 1992;262:C1083–8.
6. Grinnell AD, Herrera AA. Physiological regulation of synaptic effectiveness at frog neuromuscular junctions. *J Physiol* 1980;307:301–317.

7. Grinnell AD, Trussell L. Synaptic strength as a function of motor unit size in the normal frog sartorius. *J Physiol* 1983;338:221–243.
8. Trussell LO, Grinnell AD. The regulation of synaptic strength within motor units of the frog cutaneous pectoris muscle. *J Neurosci* 1985;5:243–254.
9. Cruz LJ, Gray WR, Olivera BM, Zeikus RD, Kerr L, Yoshikami D, Moczydlowski E. *Conus geographus* toxins that discriminate between neuronal and muscle sodium channels. *J. Biol Chem* 1985;260:9280–9288.
10. Robitaille R, Charlton MP. Presynaptic calcium signals and transmitter release are modulated by calcium-activated potassium channels. *J Neurosci* 1992;12:297–305.
11. Pawson PA, Grinnell AD. Physiological differences between strong and weak neuromuscular junctions: a study involving tetanic and posttetanic potentiation. *J Neurosci* 1990;10:1769–1778.
12. Meiri H, Rahamimoff R. Clumping and oscillations in evoked transmitter release at the frog neuromuscular junction. *J Physiol Lond* 1978;278:513–523.
13. Pawson PA, Grinnell AD. Oscillation period of MEPP frequency at frog neuromuscular junctions is inversely correlated with release efficacy and independent of acute Ca^{2+} loading. *Proc R Soc Lond B* 1989;237:489–499.
14. Lev-Tov A, Rahamimoff R. A study of tetanic and post-tetanic potentiation of miniature end-plate potentials at the frog neuromuscular junction. *J Physiol Lond* 1980;309:247–273.
15. Heuser J. 3-D visualization of membrane and cytoplasmic specializations at the frog neuromuscular junction. In: Taxi J, ed. *Ontogenesis and functional mechanisms of peripheral synapses*. Inserm Symposium 13. Elsevier/North-Holland Biomedical Press, 1980;139–155.
16. Ingber D. Integrins as mechanochemical transducers. *Curr Opin Cell Biol* 1991;3:841–848.
17. Albelda SM, Buck CA. Integrins and other cell adhesion molecules. *FASEB J* 1990;11:2868–2880.
18. Reichardt LF, Tomaselli K. Extracellular matrix molecules and their receptors: functions in neural development. *Annu Rev Neurosci* 1991;14:531–570.
19. Hynes RO. Integrins: versatility, modulation, and signaling in cell adhesion. *Cell* 1992;69:11–25.
20. Gailit J, Ruoslahti E. Regulation of the fibronectin receptor affinity by divalent cations. *J Biol Chem* 1988;263:12927–12933.
21. Kirchhofer D, Grzesiak J, Pierschbacher MD. Calcium as a potential physiological regulator of integrin-mediated cell adhesion. *J Biol Chem* 1991;266:4471–4477.
22. Pierschbacher MD, Ruoslahti E. Influence of stereochemistry of the sequence Arg-Gly-Asp-Xaa on binding specificity in cell adhesion. *J Biol Chem* 1987;36:17294–17298.
23. Orlando RA, Cheresh DA. Arginine-glycine-aspartic acid binding leading to molecular stabilization between integrin alpha v beta 3 and its ligand. *J Biol Chem* 1991;266:19543–19550.
24. Grinnell AD, Pawson PA. Dependence of spontaneous release at frog junctions on synaptic strength, external calcium, and terminal length. *J Physiol* 1989;418:397–410.
25. Schwartz MA. Spreading of human endothelial cells on fibronectin or vitronectin triggers elevation of intracellular free calcium. *J Cell Biol* 1993;120:1003–1010.
26. Wirtz W, Dobbs LG. Calcium mobilization and exocytosis after one mechanical stretch of lung epithelial cells. *Science* 1990;250:1266–1269.
27. Ng-Sikorski J, Andersson R, Patarroyo M, Andersson T. Calcium signaling capacity of the CD11b/CD18 integrin on human neutrophils. *Exp Cell Res* 1991;195:504–508.
28. Jaconi MEE, Theler JM, Schlegel W, Appel RD, Wright SD, Lew PD. Multiple elevations of cytosolic-free Ca^{2+} in human neutrophils: initiation by adherence receptors of the integrin family. *J Cell Biol* 1991;112:1249–1257.
29. Schwartz MA, Ingber DE, Lawrence M, Springer TA, Lechene C. Multiple integrins share the ability to induce elevation of intracellular pH. *Exper Cell Res* 1991;195:533–535.
30. Cybulsky AV, Carbonetto S, Cyr MD, McTavish AJ, Huang Q. Extracellular matrix-stimulated phospholipase activation is mediated by beta 1-integrin. *Am J Physiol* 1993;264:C323–C332.
31. Guan JL, Trevithick JE, Hynes RO. Fibronectin/integrin interaction induces tyrosine phosphorylation of a 120-kDa protein. *Cell Regul* 1991;2:951–964.
32. Kornberg L, Juliano RL. Signal transduction from the extracellular matrix: the integrin-tyrosine kinase connection. *TIPS* 1992;13:93–95.

Molecular and Cellular Mechanisms of Neurotransmitter Release, edited by Lennart Stjärne, Paul Greengard, Sten Grillner, Tomas Hökfelt, and David Ottoson, Raven Press, Ltd., New York © 1994.

25

Quantal Secretion from Single Visualized Synaptic Varicosities of Sympathetic Nerve Terminals

Max R. Bennett

Department of Physiology, University of Sydney, Sydney, N.S.W. 2006, Australia

The analysis of quantal secretion of transmitter at nerve terminals in the peripheral and central nervous system has used the amphibian neuromuscular junction as the paradigm nerve terminal (1,2). Following the first description of quantal secretion at this terminal (3), it was shown that the size of the spontaneous potentials or quanta, which follow a Gaussian distribution, is determined by the amount of transmitter secreted and not by saturation of a limited number of receptors at each synapse (4,5). The possibility that quantal size is determined by a limited number of receptors at central synapses (e.g., see Redman (1), Edwards et al. (6), and Faber et al. (7)), together with the apparently non-Gaussian distribution of the spontaneous potentials that determine the quantal size, indicates fundamental differences between central synapses and the amphibian neuromuscular junction. Indeed, all the peripheral terminals so far investigated possess non-Gaussian distributions of spontaneous potentials so that the Gaussian description of the quantal size is not applicable for these nerve terminals. In this work, the results of recording quantal secretion from single visualized synaptic varicosities of sympathetic nerve terminals are presented. These observations suggest that the mammalian sympathetic nerve terminal may be a more suitable preparation to act as a paradigm for central synaptic transmission than the motor-nerve terminal.

STRUCTURE OF THE SYNAPTIC VARICOSITY

Each sympathetic nerve terminal in smooth muscle or cardiac muscle consists of several collateral sprays of hundreds of synaptic varicosities (Fig. 1). These collateral sprays typically cover a territory of several hundred microns (Fig. 1). Individual sympathetic axons course over the surface of a muscle in bundles of several

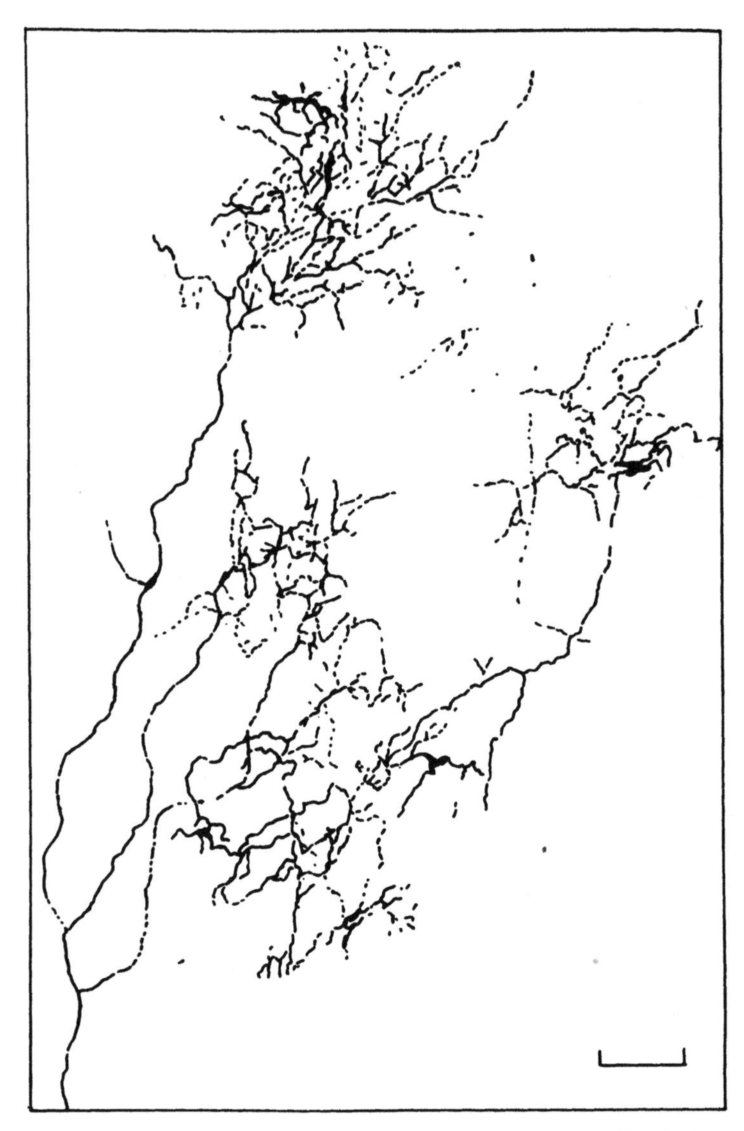

FIG. 1. The termination of a single sympathetic nerve in a smooth muscle. The diagram shows one axon (lower left corner) branching twice to give rise to three or four systems of varicose nerve terminals. Each varicosity constitutes a single synapse. (Calibration, 100 μm) (From Bennett, ref. 37, with permission.)

axons before they break up into collateral sprays of varicosities. The compound 3-3 diethyloxardicarbocyanine iodide ($DiOC_2(5)$) can be used as a vital stain to fluoresce the varicosities and so show their distribution over the surface of a muscle (8). This occurs as the $DiOC_2(5)$ attaches to mitochondria, which are in such high density in the varicosities compared with the muscle that differential fluorescence of the varicosities is possible. In the mouse vas deferens the fluorescent collateral branches of sympathetic axons can be observed to leave small axon bundles; they then run freely over the surface of the muscle as single collaterals forming synaptic varicosities (Fig. 2 (9)). The size and distribution of synaptic varicosities along individual terminal branches is about the same when these are fluoresced with $DiOC_2(5)$ as when they are fluoresced for catecholamines (Fig. 3B, mean length 1.1 μm; mean diameter 0.9 μm; mean intervaricosity distance 4.6 μm). The $DiOC_2(5)$ fluorescent technique has a resolution down to at least 0.8 μm. This is shown by the fact that occasionally a spermatozoa passes into the field of view (Fig. 4). As its tail is 0.8 μm diameter and this is easily identified, the resolution of this $DiOC_2(5)$ technique is certainly sufficient to give the dimensions of the varicosities as well as their spatial distribution.

We have recently carried out an ultrastructural reconstruction of the relationship between varicosities and smooth-muscle cells on the surface of the mouse vas def-

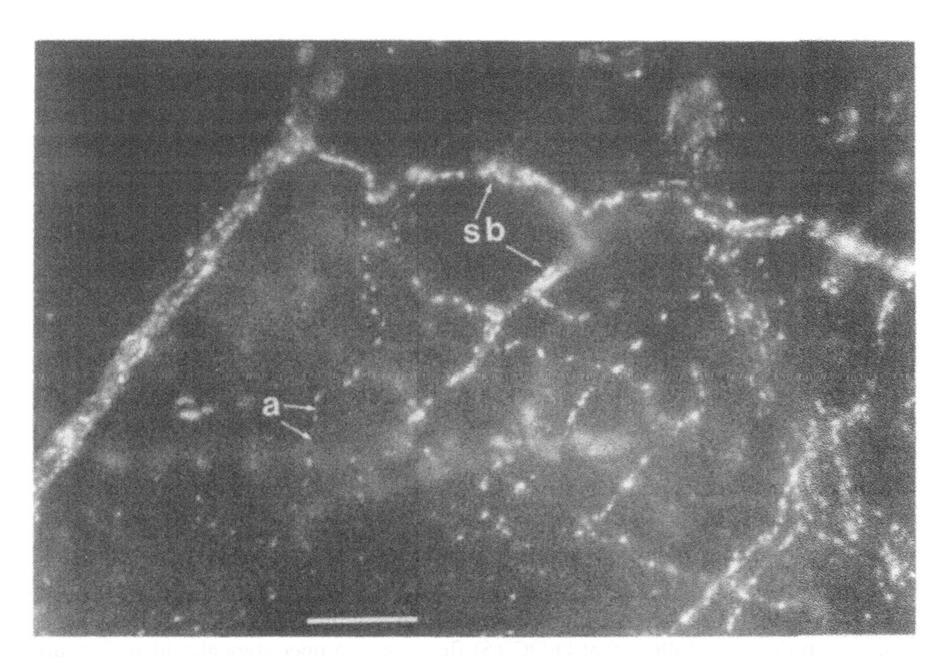

FIG. 2. The distribution of axons on the surface of the smooth muscle of the mouse vas deferens. This low-power magnification shows $DiOC_2(5)$ fluorescent axons; the beaded appearance of the axons indicates the distribution of varicosities in small axon bundles (sb) and individual axons (a). (Calibration, 1 μm) (From Lavidis and Bennett, ref. 38, with permission.)

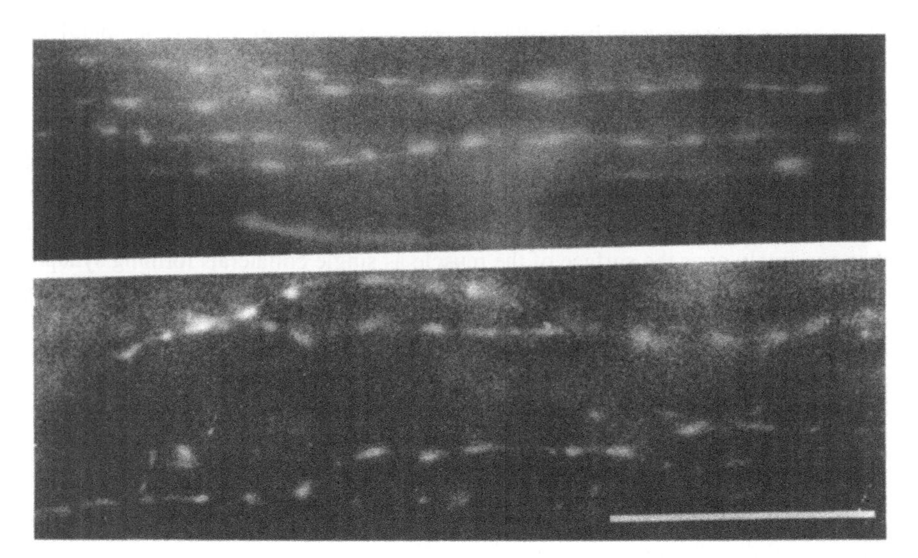

FIG. 3. *Upper panel* shows catecholamine-fluorescent axons following use of the FAGLU method; although the focus is on the surface axons, background fluorescence from axons a smooth-muscle cell deeper in the tissue is also evident in these whole mounts. *Lower panel* shows $DiOC_2(5)$ fluorescent axons; the background fluorescence is less than in the upper panel as the tissue was only exposed to $DiOC_2(5)$ for sufficient time to stain the surface axons in these whole mounts. (Calibration, 20 μm) (From Lavidis and Bennett, ref. 38, with permission.)

erens (Cottee, Lavidis, and Bennett, unpublished observation). Small axon bundles leave the large, Schwann cell enclosed axon bundles and pass over the surface of the muscle partially enclosed in Schwann, but at least 200 nm from the muscle. Individual varicose collaterals leave the small axon bundle and come into close contact with the muscle cells, so that the basal lamina of the varicosities and of the muscle cells fuse, indicating a separation between the presynaptic and postsynaptic membranes of about 50 nm (Fig. 5). Serial ultrastructural reconstruction of individual varicosities shows that they are about 1 μm long, which agrees with the estimates from the $DiOC_2(5)$ fluorescence; the region of close contact extends for a considerable part of the length and width of the region of close apposition of the varicosities (Fig. 5); this is of the order of 0.5 μm^2. Synaptic vesicles are in highest density at the presynaptic membrane over much of the area of the close contact (Fig. 5). It is doubtful whether any equivalent to an active-zone structure can be identified at these synapses; no postsynaptic specialization is evident either (Fig. 5). Although there is agreement between the dimensions of the synaptic varicosities and the spacing between them according to both ultrastructural and $DiOC_2(5)$ fluorescence techniques, it does not follow that $DiOC_2(5)$ fluorescence unequivocally identifies single varicosities. Until a single string of $DiOC_2(5)$ fluoresced synaptic varicosities is reconstructed at the ultrastructural level, the possibility remains that each $DiOC_2(5)$ fluorescent synaptic varicosity consists of more than one ultrastructurally identified

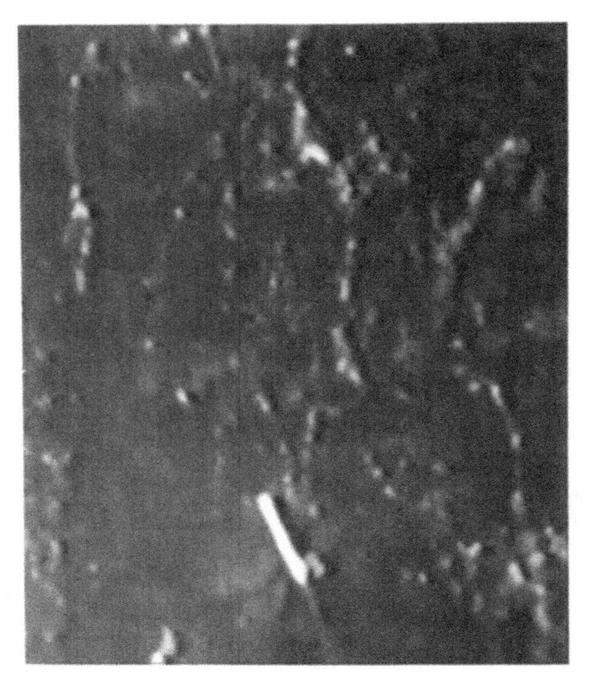

FIG. 4. A $DiOC_2(5)$ fluorescent field of varicosities on the surface of the vas deferens. At the lower edge of the field is a spermatozoa. The tail of the spermatozoa is about 0.8 μm in diameter. This gives a "natural" calibration for the figure.

synaptic varicosity; the electrophysiological recordings from $DiOC_2(5)$ fluorescent varicosities are described in the next section with this caveat in mind.

QUANTAL SECRETION FROM VISUALIZED SYNAPTIC VARICOSITIES

Microelectrodes, filled with bath solution and of 4–25 μm diameter, may be placed over one to five synaptic varicosities visualized with $DiOC_2(5)$ fluorescence. Negative pressure is not applied to the microelectrode, so as to avoid the formation of a seal between the tip of the microelectrode and the varicosities. Spontaneous excitatory junction potentials (SEJPs) are then observed, with amplitudes in the range from 10 μV up to about 250 μV; these most often form an amplitude-frequency distribution that could be reasonably well described by a gamma distribution (10,11). Stimulation of the sympathetic nerves at 0.1 Hz gives rise to evoked excitatory junction potentials (EJPs; Fig. 6). Figure 6 shows a recording made by an electrode placed over a few varicosities; approximately four different size classes of EJP's can be observed during this recording, together with several occasions in which no EJP occurred at all in response to nerve stimulation. Amplitude-frequency

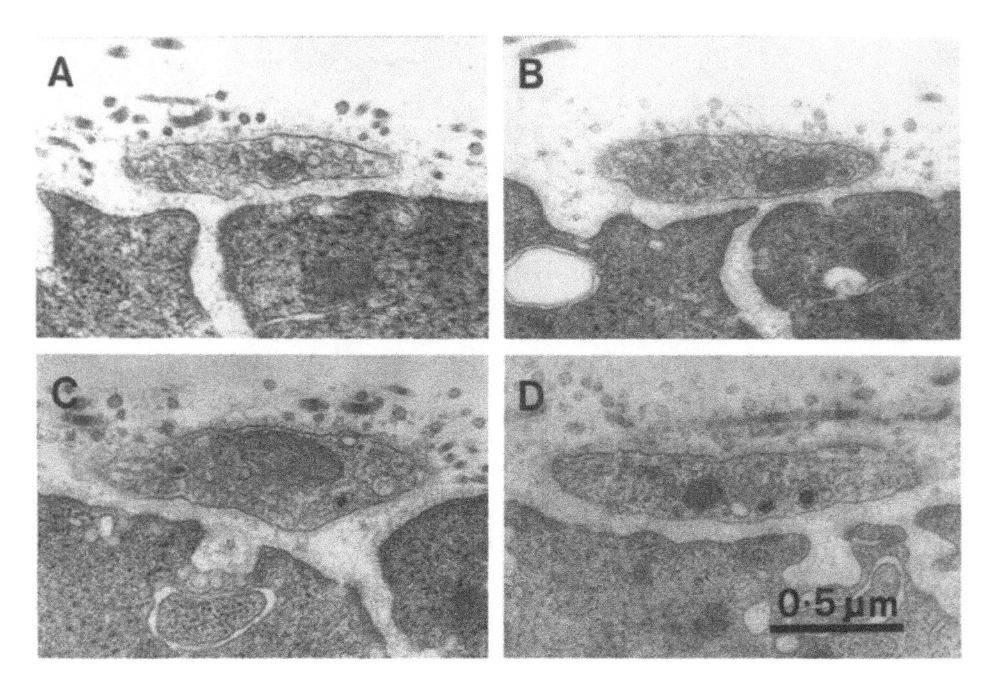

FIG. 5. Electron micrographs chosen from serial sections through a single varicosity on the surface of the smooth muscle of the vas deferens.

distributions of both spontaneous and evoked EJPs were constructed for about 100 stimuli after the number of varicosities recorded from was ascertained using the $DiOC_2(5)$ technique. Results for a recording from five varicosities with a large-diameter microelectrode are shown in Fig. 7 (12). The upper panel gives the position of the recording electrode placed over the five synaptic varicosities; the lower panel gives the amplitude-frequency distribution of both SEJPs and EJPs, together with a curve fitted according to binomial statistics in which the unit size is given by the gamma distribution of SEJPs (see Bennett and Lavidis (13)). The histogram of EJPs possesses three peaks and the binomial fit gave an $n = 6.8 \pm 1.4$. Figure 8 presents another result using a large-diameter microelectrode, recording this time from four varicosities; the amplitude-frequency distribution of SEJPs extends out to 130 μV and was again treated as a gamma distribution in order to fit the binomial curve to the histogram of EJPs. In this case no clear peaks were observed in the EJP histogram, and the binomial analysis has an $n = 5.0 \pm 0.6$.

Smaller-diameter microelectrodes were used to record from two varicosities on each of two different sympathetic nerve terminals (Figs. 9, 10). These gave spontaneous SEJP distributions that spread out as far as 130 μV, and gamma curves were again fitted to these in order to obtain the unit size for the binomial curves that were used to describe the EJP histograms. These binomial curves gave an $n = 1.88 \pm 0.38$ and $n = 1.65 \pm 0.25$ for the results in Fig. 9 and Fig. 10, respectively. There is then a correlation between the number of varicosities recorded from and the value of the

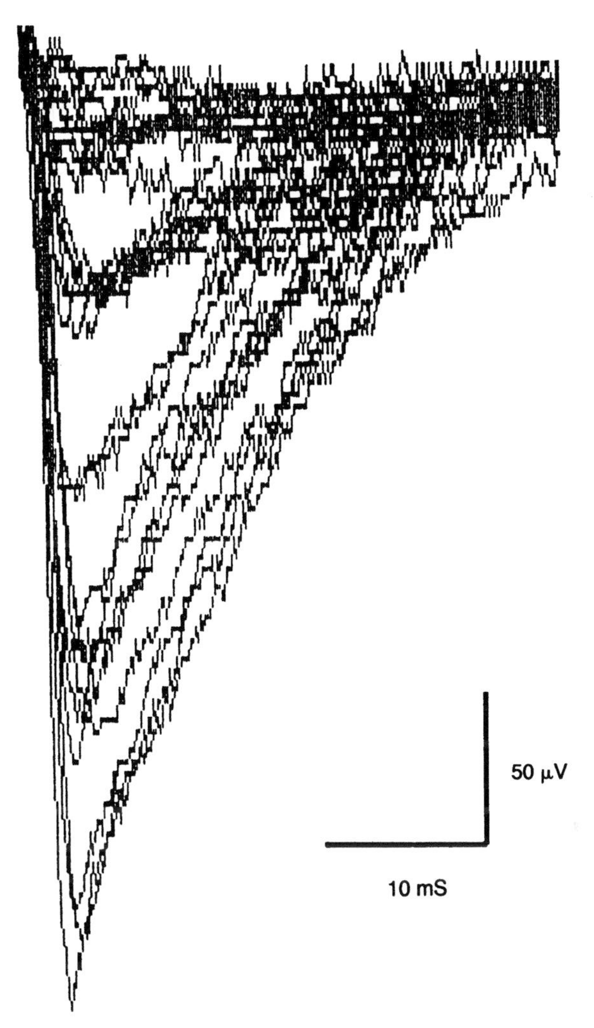

FIG. 6. Recordings of excitatory junction potentials (EJPs) from visualized varicosities (using $DiOC_2(5)$ staining). The results are shown for 30 successive impulses. The baseline is due to the number of failures of secretion.

binomial parameters n. Figure 11 shows that n increases on average by 0.8 for each additional varicosity recorded from. (Compare with the hippocampus, in which n increases at a greater rate than the number of synaptic boutons recorded from; see Bekkers et al.) (14)). These results indicate that each synaptic varicosity secretes at most one quantum, whose amplitude is given by a gamma-like distribution (15,16).

Small diameter microelectrodes were used to measure quantal secretion from one or two varicosities at different positions along single terminal branches. The average quantal secretion over 100 impulses varied greatly between sets of one or two varicosities along the same terminal branch (see Figs. 12, 13). Some varicosities

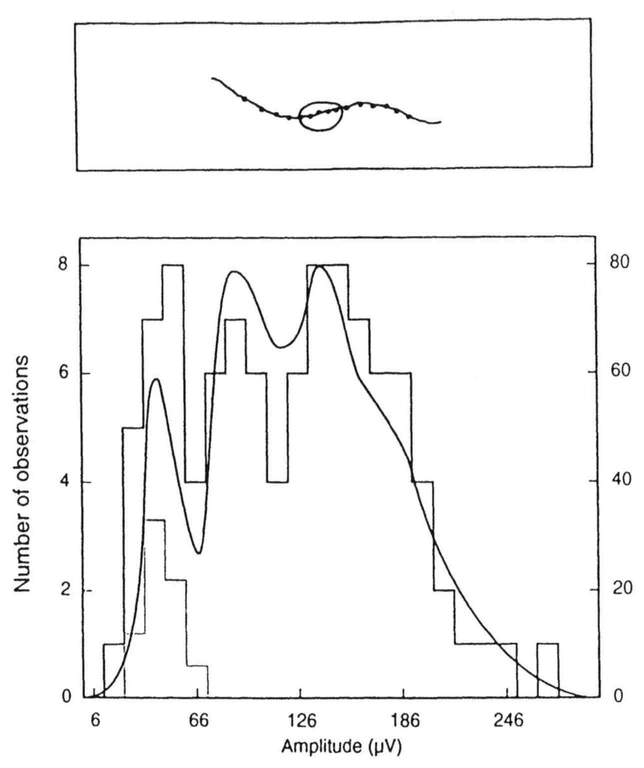

FIG. 7. The amplitude-frequency distribution of excitatory junction potentials (EJPs) (continuous line histogram) and spontaneous EJPs (broken line histogram) recorded with a 20 μm microelectrode placed over 5 varicosities (shown in the *upper panel*). The frequency of the spontaneous EJPs is given by the left-hand ordinate and of the EJPs by the right-hand ordinate. The EJP distribution is well described by binomial statistics (continuous curve; using a gamma fit to the spontaneous EJPs), in which the value of the parameters (mean ± S.E.M.) are $m = 2.8 \pm 0.2$, $p = 0.41 \pm 0.09$, and $n = 6.8 \pm 1.4$; the predicted number of failures was 3 compared with the observed number of 2. Note that the binomial parameter n is about the same value as the number of varicosities recorded from. (From Karunanithi et al., ref. 12, with permission.

had more than four times the probability for secretion than others (Fig. 13). In one case the secretion probability was about 0.1 in an external calcium concentration ($[Ca^{2+}]_o$) of 4 mM for most varicosities, and then increased to 0.2 for another set of varicosities along the same branch (Fig. 12). This nonuniformity in the probability for quantal secretion from different synaptic varicosities along single terminal branches was observed in all cases studied.

CALCIUM DEPENDENCE OF QUANTAL SECRETION FROM VARICOSITY SYNAPSES

In order to determine the calcium dependence of secretion from synaptic varicosities, the calcium concentration in the recording microelectrode was lowered to

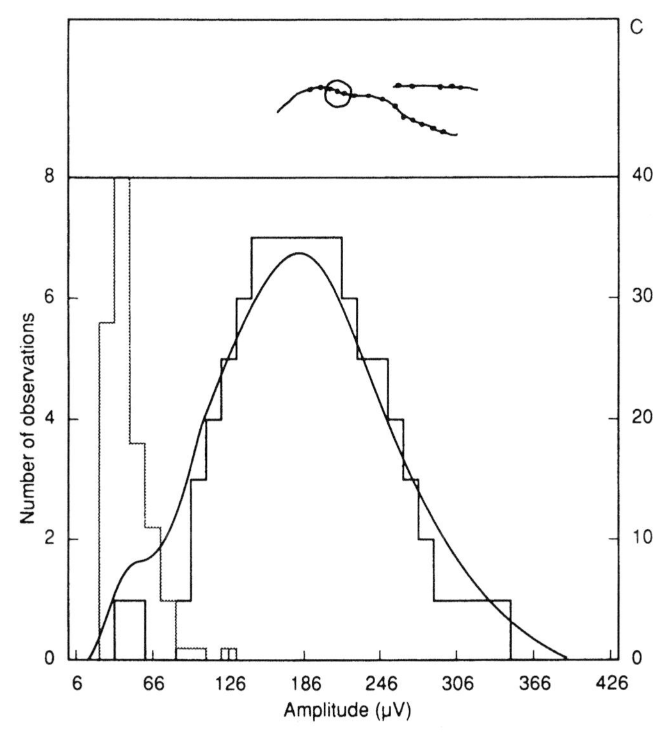

FIG. 8. The amplitude-frequency distribution of excitatory junction potentials (EJPs) (continuous line histogram) and spontaneous EJPs (broken line histogram) recorded with a microelectrode placed over four varicosities (shown in the *upper panel*). The frequency of the spontaneous EJPs is given by the left-hand ordinate and of the EJPs by the right-hand ordinate. The EJP distribution is described by binomial statistics (continuous curve; using a gamma fit to the spontaneous EJPs) in which the value of the parameters (mean ± S.E.M.) are $m = 3.2 \pm 0.2$, $p = 0.63 \pm 0.06$, $n = 5.0 \pm 0.6$; the predicted number of failures was 2 compared with 6 observed. (From Karunanithi et al., ref 12, with permission.)

0.5 mM and the $[Ca^{2+}]_0$ in the bath varied between 1.0 mM and 4.0 mM. It was anticipated that the $[Ca^{2+}]_0$ in the bath would determine the Ca^{2+} at the varicosity recorded from if a low Ca^{2+} was used in the recording microelectrode and this did not seal over the varicosity. This was shown to be the case, as sometimes during recording of EJPs there was a sudden precipitous decrease in the rate of quantal secretion, indicating that the microelectrode with the low Ca^{2+} had sealed around the varicosity, lowering its probability for secretion. Small diameter microelectrodes (4 μm) were used to record quantal secretion from each of six different synaptic varicosities in $[Ca^{2+}]_0$ ranging from 1.0 mM to 2.0 mM. The results indicated that the probability of quantal secretion is steeply dependent on $[Ca^{2+}]_0$. In all six synaptic varicosities the probability for secretion increased as the fourth power of $[Ca^{2+}]_0$, independent of the initial probability in 1 mM $[Ca^{2+}]_0$.

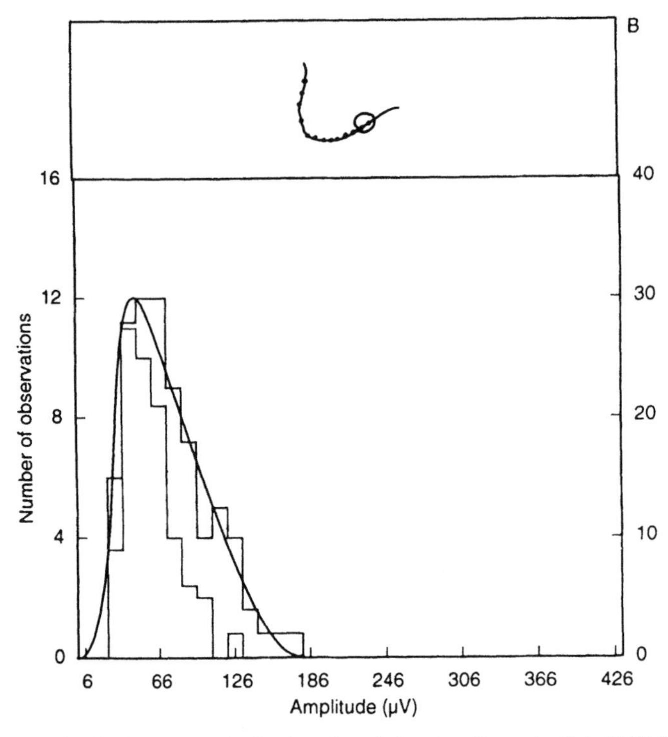

FIG. 9. The amplitude-frequency distribution of excitatory junction potentials (EJPs) (continuous line histogram) and spontaneous EJPs (broken line histogram) recorded with a microelectrode placed over two varicosities (shown in *upper panel*). The frequency of the spontaneous EJPs is given by the left-hand ordinate and of the EJPs by the right-hand ordinate. The EJP distribution is well described by binomial statistics (continuous line; using a gamma fit to the spontaneous EJPs) in which the value of the parameters (mean ± S.E.M.) are $m = 0.98 \pm 0.08$, $p = 0.59 \pm 0.08$, $n = 1.65 \pm 0.25$, the predicted number of failures was 22 compared with 24 observed. (From Karunanithi et al., ref. 12, with permission.)

A POISSON MIXTURE OF GAUSSIANS DESCRIBES SPONTANEOUS QUANTAL SECRETIONS AT SYNAPTIC VARICOSITIES

The existence of amplitude-frequency distributions of spontaneous EJPs that were much better approximated by gamma distributions than by Gaussian distributions (e.g., see Figs. 8, 10) prompted a more detailed study of spontaneous EJPs from single synaptic varicosities recorded with 4 μm-diameter electrodes. Amplitude-frequency histograms were collected from synaptic varicosities for which sufficient numbers of SEJPs were recorded and the smallest SEJPs could be detected free of the noise level. Many of the histograms showed peaks that tended by eye to be integer multiples of the first peak size (upper panel in Fig. 14). Peaks in such amplitude-frequency histograms can arise from sampling errors so, if the sample size is increased, the peaks disappear (17,18,19). It has not been possible to collect

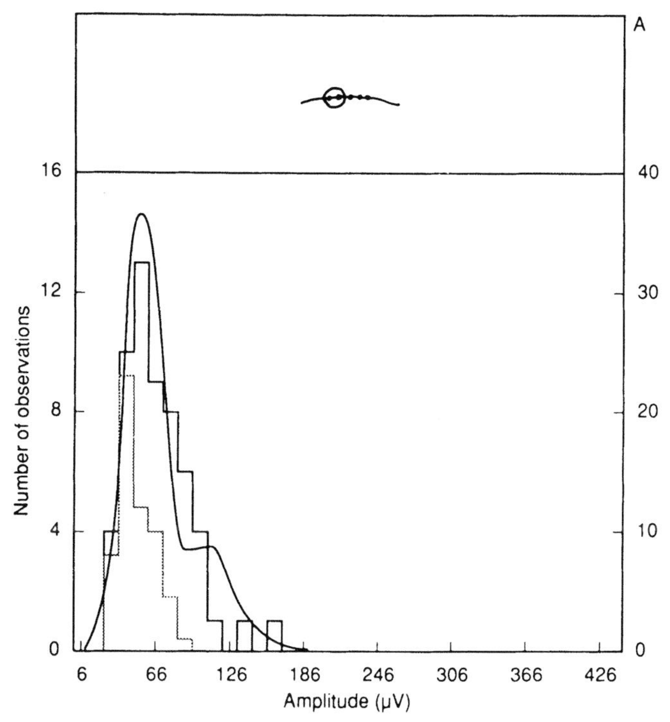

FIG. 10. The amplitude-frequency distribution of excitatory junction potentials (EJPs) (continuous line histograms) and spontaneous EJPs (broken line histogram) recorded with a microelectrode placed over two varicosities (shown in *upper panel*). The frequency of the spontaneous EJPs is given by the left-hand ordinate and of the EJPs by the right-hand ordinate. The EJP distribution is well described by binomial statistics (continuous line; using a gamma fit to the spontaneous EJPs), in which the value of the parameters (mean ± S.E.M.) are $m = 0.70 \pm 0.07$, $p = 0.38 \pm 0.08$, and $n = 1.88 \pm 0.38$. The predicted number of failures is 40 compared with 42 observed. (From Karunanithi et al., ref. 12, with permission.)

more than about 200 spontaneous potentials from single synaptic varicosities to this time, so it is uncertain whether these peaks are genuine or a statistical artifact (see also Bekkers et al. (14) and Stevens (20)) for a similar problem when interpreting spontaneous potential distributions from small numbers of synaptic boutons). Nevertheless, the gamma distribution curve was frequently inadequate for describing these histograms, as there were no values at the origin of the distribution as required by a gamma distribution. We therefore examined the possibility that the amplitude-frequency histograms could be best described by a Poisson mixture of Gaussians, as first introduced into synaptic physiology by del Castillo and Katz (3) to analyze evoked transmitter release at the amphibian neuromuscular junction. There are many reasonable models that could give rise to a Poisson distribution of SEJPs. For example, consider a process in which clusters of different numbers of synaptic vesicles (one, two, etc.) surround each voltage-dependent calcium channel in the

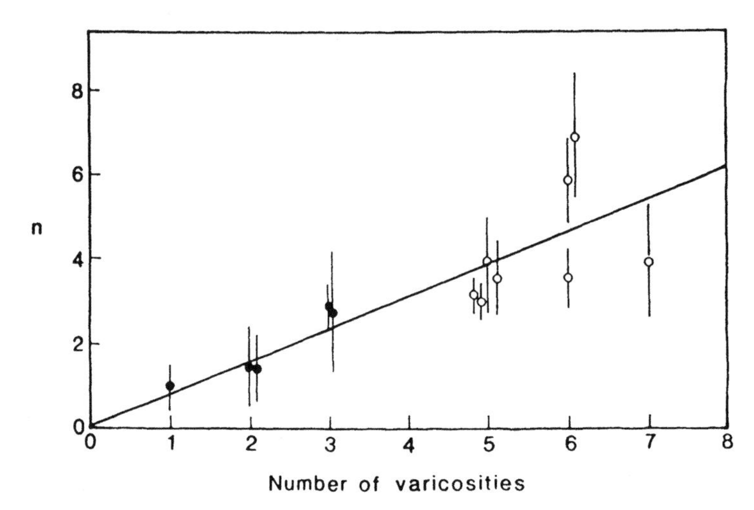

FIG. 11. Changes in the binomial parameter "n" (ordinate) with the number of varicosities recorded from. Filled symbols give observations from small-diameter electrodes and open symbols give observations from large-diameter electrodes. Vertical bars give ±S.E.M. There is a correlation between binomial parameter "n" and the number of varicosities (regression line has a correlation coefficient of 0.74). (From Lavidis and Bennett, ref. 8, with permission.)

synaptic varicosity; an SEJP then involves one of these calcium channels opening spontaneously, allowing for a calcium influx that triggers the exocytosis of the vesicles in the cluster. If λ is the average number of vesicles in a cluster and the secretion of one vesicle gives rise to an SEJP distributed as a Gaussian variate with mean γ and variance σ^2, then the amplitude-frequency distribution of SEJPs will follow a Poisson random sum of Gaussians such that

$$f(x) = \sum_{k=1}^{\infty} \frac{\lambda^k}{k!} \frac{e^{-\lambda}}{(1-e^{-\lambda})} \frac{1}{\sigma\sqrt{k}} \varphi\left(\frac{x-k\gamma}{\sigma\sqrt{k}}\right) \qquad [1]$$

in which $f(x)$ is the frequency of observing an SEJC of amplitude xmV, k is the number of vesicles released, and φ indicates the normal density function. I am grateful to Professor John Robinson of the School of Mathematics at Sydney University for this derivation. The upper panel in Fig. 14 shows the fit of equation 1 to the SEJP histogram, in which the values of γ, σ, and λ used were determined by maximum likelihood, giving 19.1, 4.2, and 1.8, respectively. A bootstrap method (21) was used, under the assumption that the model is correct, in order to determine that unbiased estimates of γ, σ, and λ had been obtained and to determine the true variability of these estimates. To this end the values of γ, σ, and λ obtained from the data were used to generate simulated data with a particular random number seed; examples of such simulated data are shown in the lower panel of Fig. 14 for the data in the upper panels. Maximum likelihood procedures were then applied to estimate γ, σ, and λ from this simulated data. This procedure was repeated 100 times, using a different random number seed on each occasion. The distribution of γ, σ, and λ

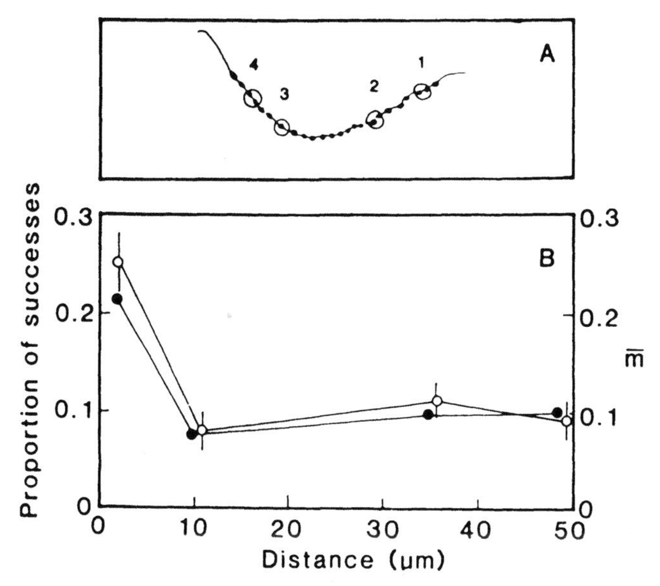

FIG. 12. Changes in the secretion of transmitter and m at different positions along a varicose axon recorded with a small-diameter microelectrode of 4 μm. The *upper panel* gives a drawing of the varicose terminal, derived from the $DiOC_2(5)$-fluoresced nerve; ellipses give the outline of the recording electrode tip and the numbers indicate the sequence in which recordings were made. The lower panel gives the proportion of stimuli that gave an excitatory junction potential (EJP) at different positions along the length of the terminal axon (filled circles) as well as m (open circles; bars give S.E.M.); the origin of the distance is taken as the extreme left-handed or right-handed recording site on the axon. All graphs of the proportions of successes and of m have been plotted in the direction along the axon branch, which gives a declining proportion of successes. (From Lavidis and Bennett, ref. 8, with permission.)

from one set of 100 simulations (that for the Fig. in the upper panel of Fig. 14) are shown in Fig. 15. Here it can be seen that the means of the estimates of γ, σ, and λ from the simulations are almost the same as that from the data (19.1/19.2, 4.2/4.1, and 1.8/1.77, respectively), indicating that the estimates are unbiased. Furthermore, the variability of the estimates of γ, σ, and λ (the standard error of the estimates) is less than 10% of their means. The goodness of fit of the curve to the data in Fig. 14 (upper panel) was not tested (e.g., by the x^2-test). It is possible then that each synaptic varicosity can secrete more than the contents of a single vesicle, if the vesicle is taken as the basis for the subquanta.

The average values of γ and σ for seven different SEJP distributions were 23.6 uV and 5.8 uV, respectively. As the standard deviation of the electrical noise (σ_n) was about 5 uV, the standard deviation of the subquanta composing γ (σ_γ) is 3 uV (from $\sigma = (\sigma_n^2 + \sigma_\gamma^2)^{0.5}$). The coefficient of variation of the noise (σ_n/γ) is then 0.20 and, of the subquanta (σ_γ/γ), 0.13. These values may be compared with those for spontaneous potentials at inhibitory synapses on granule cells in the hippocampus $\sigma_n/\gamma = 0.20$ and $\sigma_\gamma/\gamma = 0.22$ (6) and for excitatory transmission to stellate

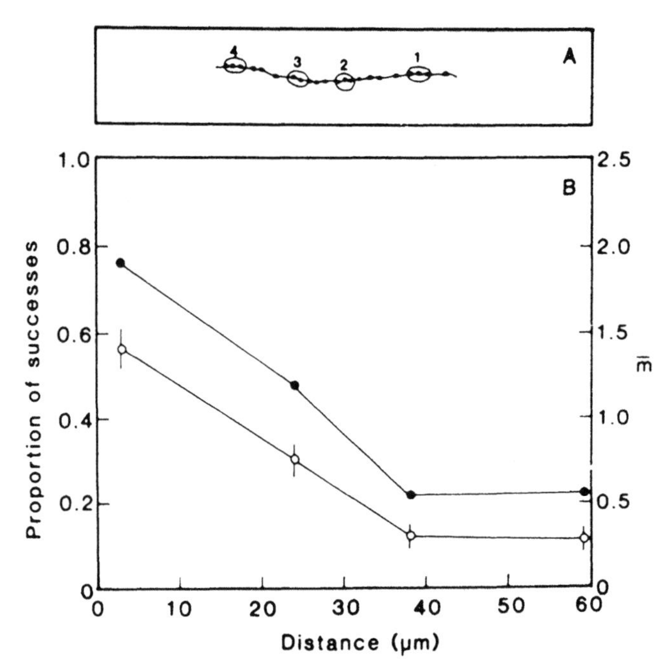

FIG. 13. Changes in the proportion of stimuli that gave an excitatory junction potential (EJP) at different positions along a varicose axon recorded with a small-diameter electrode of 6 μm as well as changes in m along the varicose axon. The *upper panel* (**A**) gives a drawing of the varicose terminal, derived from the DiOC$_2$(5)-fluoresced nerve; the ellipses give the outline of the recording electrode tip and the numbers indicate the sequence in which recordings were made. The *lower panel* (**B**) gives the proportion of stimuli that gave an EJP at different positions along the length of the terminal axon (filled circles), as well as m (open circles); the origin of the distance is determined as in Fig. 12. The vertical bars on the symbols give ±S.E.M. (From Lavidis and Bennett, ref. 9, with permission.)

cells ($\sigma_n/\gamma = 0.14$) (22). At the mammalian motor endplate, σ_γ/γ is about 0.22 for the spontaneous synaptic potentials that are distributed as a single Gaussian (23). The coefficient of variation for the entire spontaneous distribution (i.e., including all subquanta) for the seven recordings was 0.47; this may be compared with 0.51 for spontaneous excitatory postsynaptic currents in visual cortex (24) and 0.42 for spontaneous excitatory currents in hippocampal neurons (14). The question arises as to how much of the coefficient of variation of the spontaneous distributions for the subquantal size may be attributed to variance arising from the activation of receptors (7). For the spontaneous excitatory synaptic currents in the visual cortex this has been estimated as about 0.2 (24). The coefficient of variation may be estimated from evaluation of $[(1\text{-}p_0)/(N.p_0)]^{0.5}$ where p_0 is the open probability of receptor channels at the peak of the spontaneous current and N is the number of available receptors (25). The conductance underlying the spontaneous potentials at synaptic varicosities secreting adenosine 5′-triphosphate (ATP) is about 2 nanoSiemens (26) and the unitary conductance change is about 5 picoSiemens (27,28); it follows that

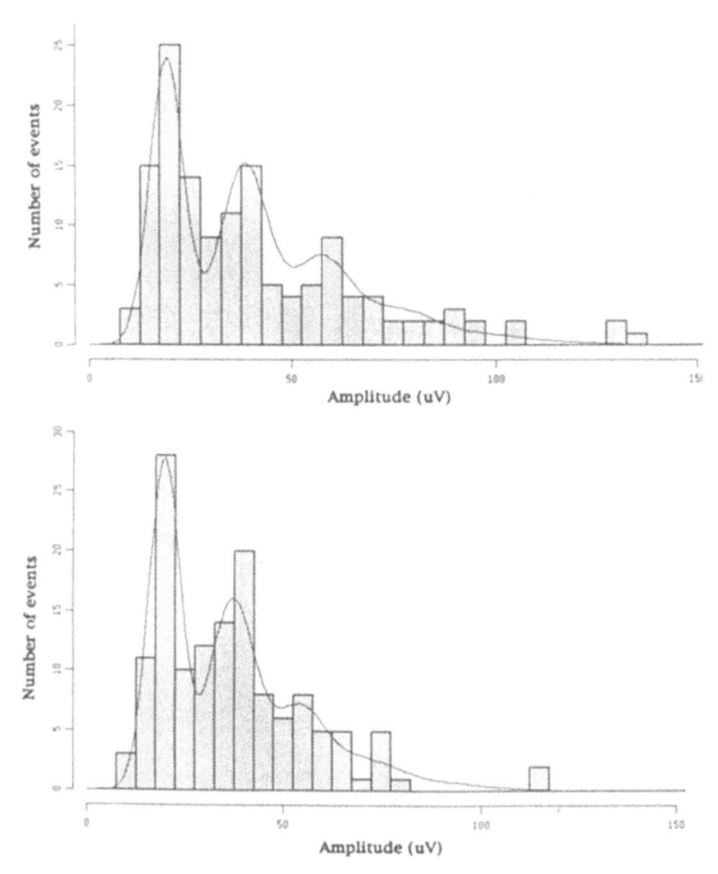

FIG. 14. Amplitude-frequency histograms of spontaneous excitatory junction potentials (EJPs) recorded from a single varicosity with a 4 μm diameter microelectrode. *Upper panel* gives observed data and lower panel gives simulations (see text). The continuous line for both histograms gives the Poisson predictions in which γ, σ, and λ had values of 19.1, 4.2, and 1.8, respectively, for the data; the values of γ, σ, and λ for the simulation were 17.2, 4.0, and 1.6, respectively. For the standard deviations on the data estimates for γ, σ, and λ, see Fig. 15.

about 400 purinergic receptors are activated at the peak of the spontaneous potentials. The value of p_0 is not known, but taking it as about 0.5 in order to produce the largest coefficient of variation gives a value for this of 0.05. Only a relatively small amount of the coefficient of variation of the quanta or subquanta can then be attributed to variability in transmitter–receptor kinetics. Most of the variability in the subquantal size most likely arises from variation in the amount of transmitter in the subquanta. It seems likely that there are sufficient numbers of purinergic receptors at a synaptic varicosity to accommodate multisubquantal releases. This situation is quite different then from that at excitatory synapses on stellate cells in the visual cortex (24) or for inhibitory synapses on granule cells in the hippocampus (6): at

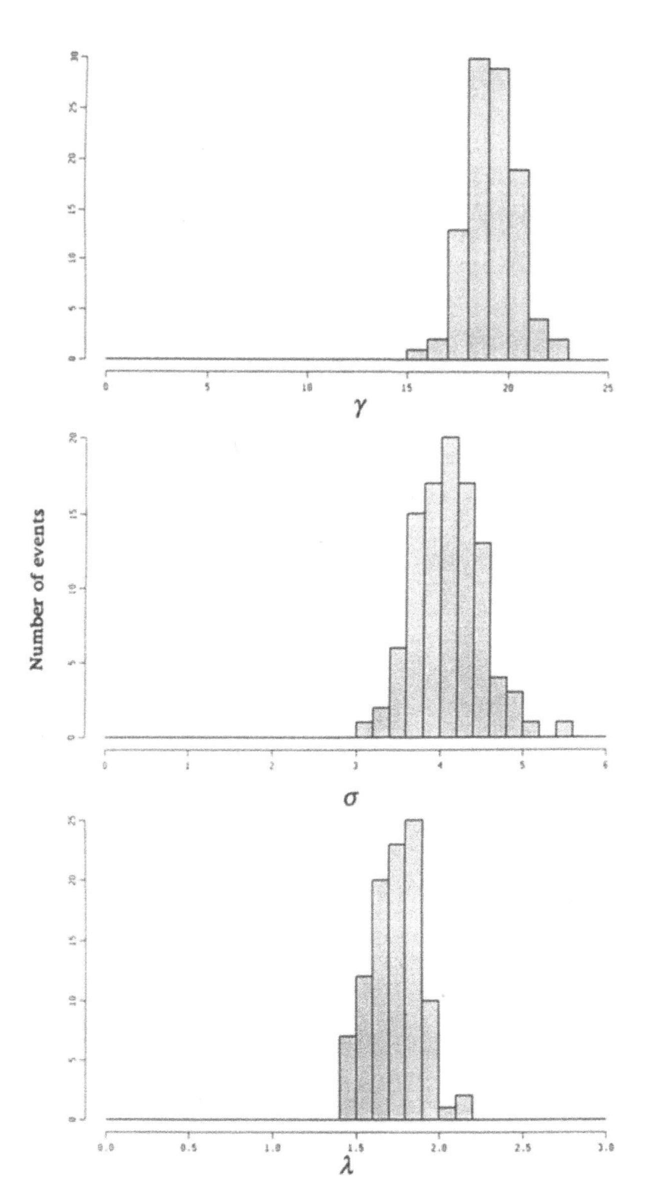

FIG. 15. Frequency distributions for the estimate of γ (*upper panel*), σ (*middle panel*), and μ (*lower panel*) for the 100 simulations of the distribution given in Fig. 14. This bootstrap procedure gave standard deviations for the parameter in Fig. 14 of $\gamma = 19.1 \pm 1.2$, $\sigma = 4.2 \pm 0.4$, and $\lambda = 1.8 \pm 0.15$; thus the standard deviations are less than 10% of the parameter values. The bootstrap means for γ, σ, and λ were 19.2, 4.1, and 1.77, respectively. Since these means are almost the same as the estimates from the data, they are unbiased.

the former synapses, about half of the variability in quantal secretion arises from transmitter receptor interaction; in the latter case, there is very little variability, and given the small number of γ-aminobutyric acid (GABA) receptors involved (about 10), this has been attributed to saturation of the receptors by an excess of transmitter (see also Faber et al. (7)). It has been argued that unless the coefficient of variation of the subquanta is less than about 0.15, single peaks in the amplitude histogram will not be evident (6,29). Furthermore, the coefficient of variation of the noise should be less than 0.5 (2). At the synaptic varicosities, both of these criteria have been met; the coefficient of variation of the subquanta is typically 0.12 and the coefficient of variation of the noise is about 0.2.

It is interesting to note that not only the synaptic varicosities of sympathetic nerve terminals generate Poisson-like amplitude-frequency distributions of spontaneous potentials, but so do developing motor-nerve terminals and mature synaptic boutons of central terminals. During development of motor-nerve terminals, the amplitude-frequency distribution of spontaneous endplate potentials is at first skewed; the distribution then becomes multimodal, with the modes tending to occur at integer multiples of the first mode. Finally, the distribution assumes a Gaussian form (30). Recently Lo, Wang, and Poo (31) have shown that motor-nerve terminals developing in tissue culture clearly show a period during which the spontaneous endplate potentials do not follow a Gaussian distribution (Fig. 16). These can be fitted with a Poisson mixture of Gaussians, as shown in Fig. 16. An amplitude-frequency distributions of spontaneous excitatory postsynaptic potentials has been determined for mossy-fiber synapses on granule cells in the cerebellum (32). This does not follow a Gaussian distribution, nor can it be fitted by a gamma distribution as the smallest spontaneous potentials are well clear of zero, at which a gamma distribution must begin (Fig. 17). The distribution has been fitted by a Poisson mixture of Gaussians as shown in Fig. 17. Poisson-like distributions of spontaneous postsynaptic currents have been observed at inhibitory synapses on the somas of neurons in the hippocampus (Fig. 17). Poisson-like distributions of spontaneous inhibitory postsynaptic potentials in granule cells in the dentate gyrus that have had excitatory activity blocked with tetrodotoxin (TTX) and added Mn^{2+} have been observed (6) (Fig. 18, upper panel; lower panel gives simulations). The curve in Fig. 18 gives the Poisson predictions according to equation 1 in which γ, σ, and λ had values of 10.4, 2.4, and 1.4, respectively. Edwards et al. (6) fitted Gaussian distribution to the first peak at 10.7, which agrees well with the maximum likelihood fit to this peak in the Poisson of 10.4.

A POISSON MIXTURE OF POISSONS MAY DESCRIBE EVOKED QUANTAL SECRETIONS

The question arises as to the form that amplitude-frequency distributions of evoked synaptic potentials are likely to take if the quantal size is distributed as a Poisson mixture of Gaussians. If the evoked quantal secretion itself follows a Poisson process, then transmitter release consists of a Poisson random sum of Poissons,

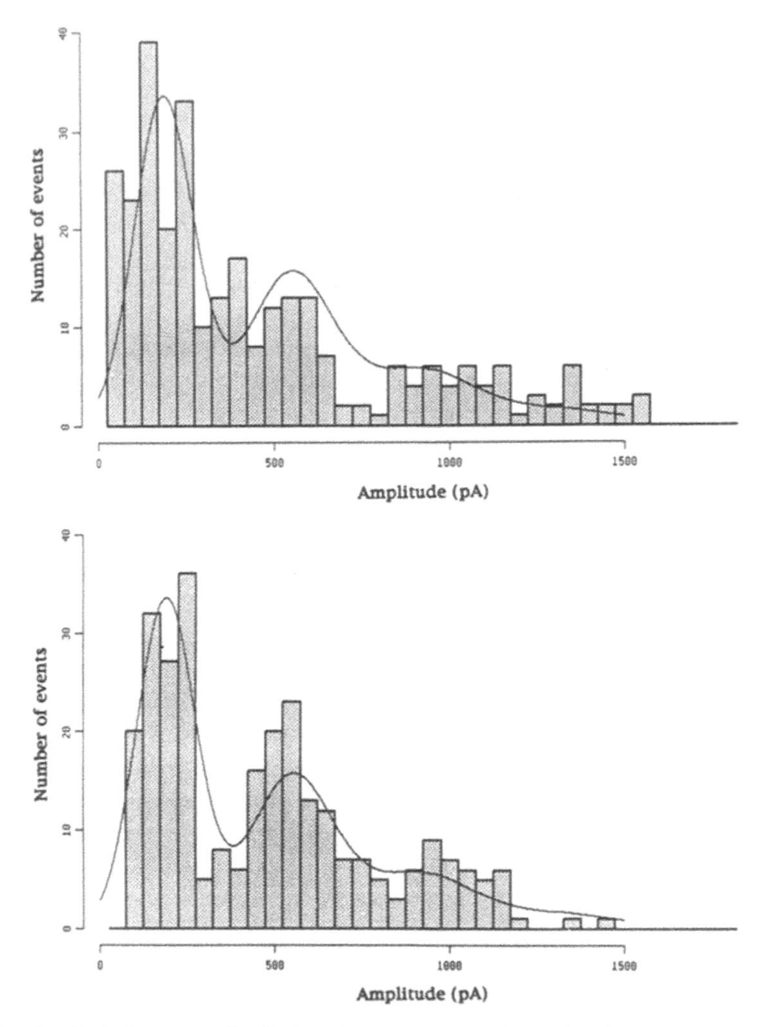

FIG. 16. Amplitude-frequency distribution of spontaneous excitatory junctional currents at newly formed motor-nerve terminals on striated muscle cells in tissue culture. The upper panel gives the data (data from ref. 31); the curve gives a Poisson fit to the data in which γ, σ, and λ had the values of 362, 85, and 1.3, respectively. The lower panel gives a Poisson curve fit to a simulated set of data in the bootstrap.

as the quantal size is a Poisson sum of Gaussians. One way in which this could arise is if the average number of voltage-dependent calcium channels that open is increased to μ during the nerve terminal action potential. In this case the evoked secretions follow a Poisson (μ) random sum of Poisson (λ) according to

$$f(y) = \sum_{n=1}^{\infty} p_n \frac{1}{\sigma\sqrt{n}} \varphi\left(\frac{y_i - ny}{\sigma\sqrt{n}}\right) \qquad [2]$$

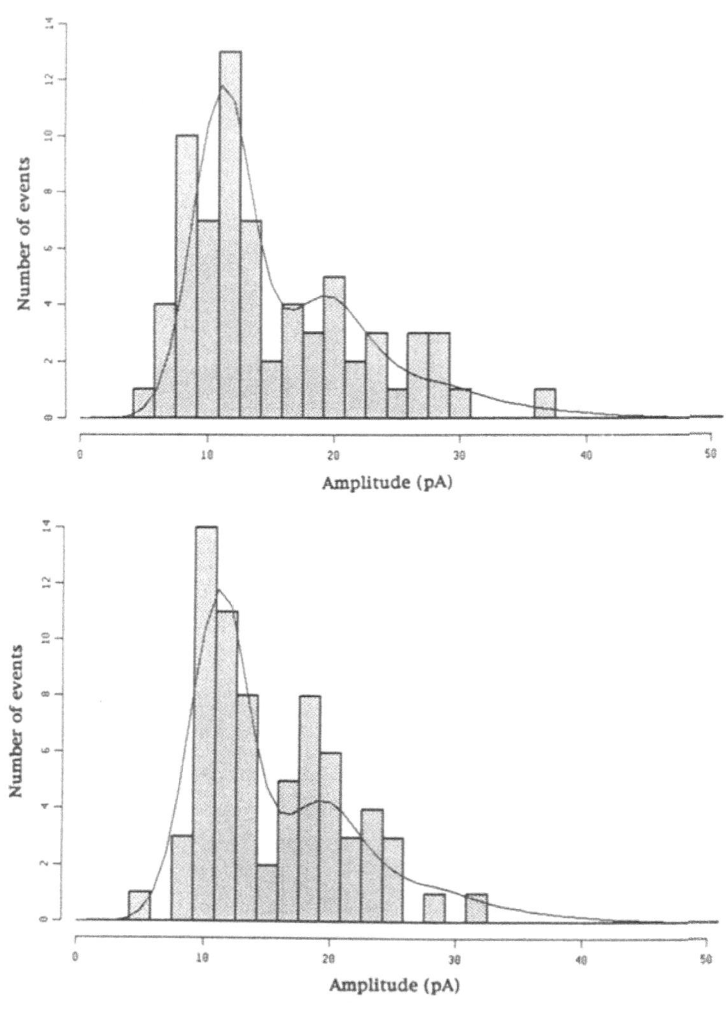

FIG. 17. Amplitude-frequency distribution of spontaneous excitatory postsynaptic currents recorded from granule cells in slices of cerebellum. The *upper panel* gives the data from ref. 32 in which the Poisson curve is fitted with values for γ, σ, and λ, of 8.08, 2.32, and 1.01, respectively. The *lower panel* gives a Poisson curve fit to a simulated set of data in the bootstrap.

where

$$p_n = \frac{\lambda^n \exp[\mu(e^{\lambda} - 1)]}{1 - \exp[\mu(e^{-\lambda} - 1)]} \frac{\mu_n'}{n!} \qquad [3]$$

and where μ_n' is the *nth* moment of $P_0(\mu e^{-\lambda})$. I am again indebted to Professor John Robinson for this derivation.

The upper panel of Fig. 19 shows another amplitude-frequency histogram of

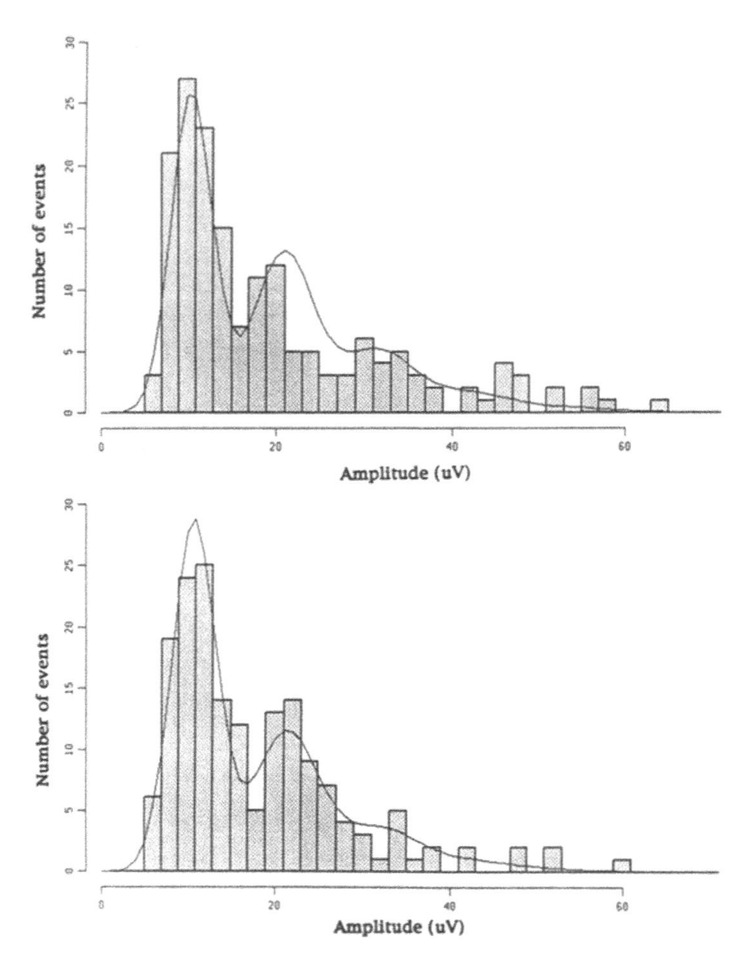

FIG. 18. Amplitude-frequency histogram of spontaneous miniature inhibitory postsynaptic current amplitudes recorded from a rat dentate gyrus granule cell. The bath solution contained tetrodotoxin (TTX) (1 μM) to block sodium-dependent action potentials and Mn^{2+} (2 mM), with no added Ca^{2+}, to block calcium-dependent action potentials. The data is from ref. 6. The *upper panel* gives this data, which was fitted by a Poisson with γ, σ, and λ values of 10.4 ± 0.7, 24 ± 0.20, and 1.4 ± 0.1, respectively. (Edwards, Konnerth, and Sakmann fitted Gaussians to the three peaks with mean and standard deviations (from left to right) of 10.7 ± 2.5, 19.2 ± 2.4, and 31.0 ± 4 pA. The *lower panel* gives a simulation in which γ, σ, and λ were 10.6, 2.6, and 1.12, respectively.

spontaneous inhibitory postsynaptic currents recorded from granule cells in the hippocampus by Edwards, Konnerth, and Sakmann (6), together with the Poisson prediction curve. The lower panel of Fig. 19 shows the simulated amplitude-frequency distribution of evoked inhibitory postsynaptic currents, according to equation 2 for a μ of 5.29. Note that peaks occur in this amplitude-frequency histogram, and these are multiples of the first peak in the spontaneous histogram given in the upper panel.

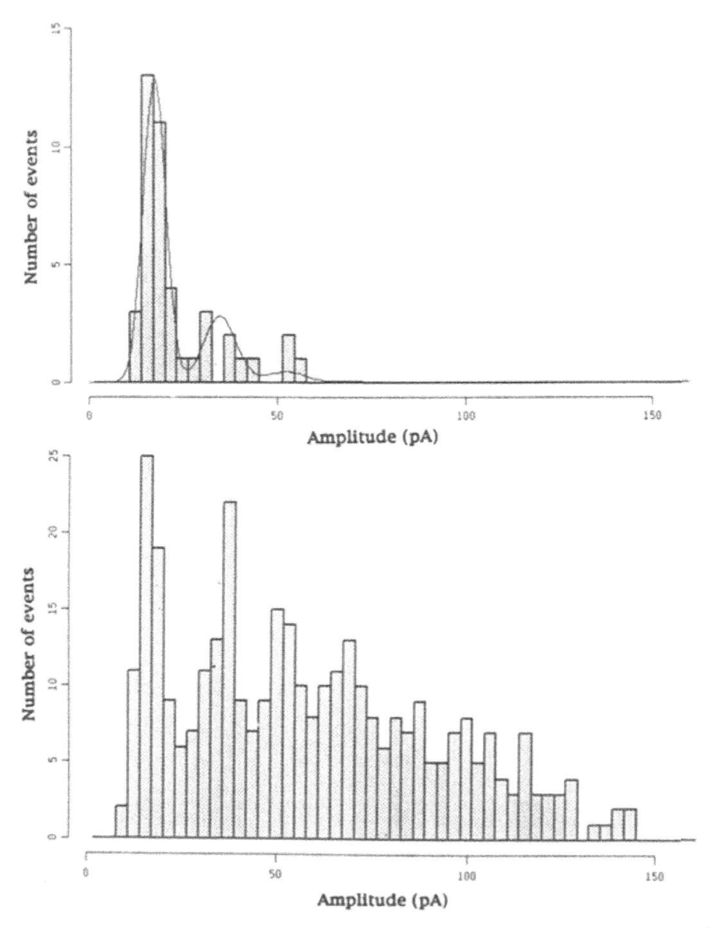

FIG. 19. *Upper panel* gives amplitude-frequency histogram of spontaneous miniature inhibitory postsynaptic current amplitudes recorded from a rat dentate gyrus granule cell in the presence of 0.5 μM of tetrodotoxin (TTX). (Data is from ref. 6.) The continuous curve gives the Poisson fit to this data with γ, σ, and λ having values of 17.5 ± 1.8, 2.98 ± 0.32, and 0.62 ± 0.17, respectively. The *lower panel* gives a simulation of the evoked inhibitory postsynaptic current amplitude-frequency histogram for a Poisson evoked process with $\mu = 5.29$.

Evoked amplitude-frequency distributions of this kind have been observed by Edwards, Konnerth, and Sakmann (6), as Fig. 20 shows. The upper panel in Fig. 20 gives both the spontaneous (filled histogram) and evoked (open histogram) distributions recorded by these authors. The lower panel shows a simulated evoked distribution (open histograms; $\mu = 5.3$) using the quantal sizes given by the filled histograms reproduced from the upper panel. The histogram of simulated evoked synaptic currents (lower panel) is similar to that of the histograms of observed evoked synaptic currents (upper panel); peaks in both evoked histograms occur at about the same positions, although the tail of the simulated histogram is longer than

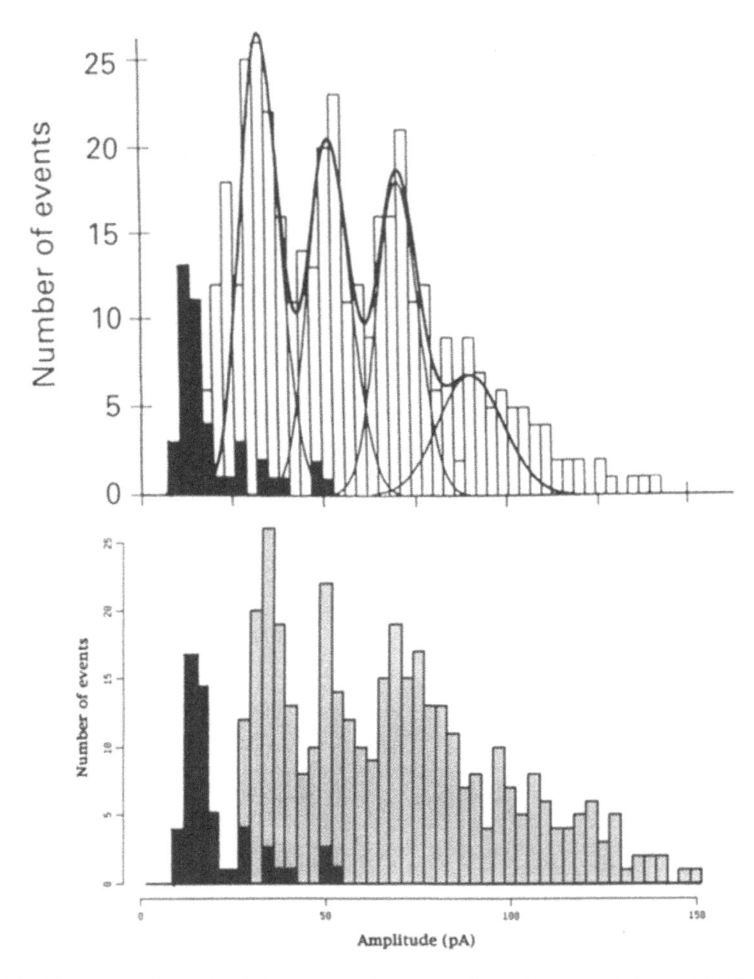

FIG. 20. *Upper panel,* amplitude-frequency histogram of spontaneous miniature inhibitory postsynaptic current (IPSC) amplitudes recorded from a rat denate gyrus granule cell in the presence of 0.5 μM tetrodotoxin (TTX) (filled histogram) and of evoked IPSCs in the absence of TTX (open histogram). The thick smooth lines superimposed on the histogram represent the sum of Gaussian distributions that best fit the data. The thin smooth lines represent the individual Gaussian of this fit. The peak of the spontaneous miniature IPSC amplitude distribution is 13.8 pA; the separation between the peaks of the evoked distribution is 17.4 pA. (From Edwards et al., ref. 6, with permission.) *Lower panel,* amplitude-frequency histogram of simulated evoked IPSCs (open histogram) based on the Poisson fit to the spontaneous IPSCs (filled histogram in *upper* and *lower panels*) with values of γ, σ, and λ of 17.5 ± 1.8, 2.97 ± 0.32, and 0.62 ± 0.17, respectively, and a rate for the evoked Poisson of $\mu = 5.29$. Note that the peaks in the evoked and simulated histogram occur at multiples of 17.5, as do the data in the upper panel. The first peak in the evoked histogram is masked by the spontaneous distribution (black histogram) and so has been entirely removed.

that of the observed data. These results suggest the possibility that synaptic boutons use a quantal size that is distributed as a Poisson mixture of Gaussians and, in the case illustrated, that the average number of voltage-dependent calcium channels that open during an impulse is about five. However, there are important reasons related to the very small number of channels that seem to be present beneath a synaptic bouton compared with the large estimates for the amount of transmitter secreted by a vesicle for rejecting this suggestion (e.g., Edwards et al. (6)). In the peripheral nervous system, we have not yet been able to collect sufficient numbers of both evoked and spontaneous quantal secretions from a single synaptic varicosities to carry out an inquiry as to whether evoked release follows a Poisson mixture of Poissons.

GAUSSIAN VERSUS POISSON DISTRIBUTIONS OF QUANTAL AMPLITUDES

The amphibian neuromuscular junction has been the paradigm synapse for studying the biophysics of secretion since it was first intensively studied by Eccles, Katz, and Kuffler (33) in Sydney 50 years ago. The concept of quantal secretion was developed from an analysis of spontaneous endplate potentials generated at this junction (3), which follow a Gaussian distribution. However, we have seen that during development of this synapse, quantal secretion does not at first follow Gaussian laws, but rather the quanta fall into discrete size categories giving a multimodal amplitude-frequency distribution (30).Only later during development does this give way to a Gaussian distribution. Particular ion or drug treatments of the mature amphibian endplate can even break up the Gaussian distributions into subquantal units as they occur during normal development (34,35). It might be possible then that the mature motor-nerve terminal is not the paradigm synapse, as it is highly specialized to ensure a high safety factor for transmission to skeletal muscles. Indeed, the motor-nerve terminal does not even appear to be structurally similar to any other peripheral synaptic nerve terminal in the Vertebrata, which consist of chains of synaptic varicosities or boutons. The appropriate paradigm synapse may then be provided by the synaptic varicosities of sympathetic nerve terminals, which after all gave us the idea of chemical transmission in the first place (36).

ACKNOWLEDGMENTS

I am most grateful to my colleagues for their contribution to the work described above. To Professor John Robinson for theoretical analysis, Nick Lavidis for the experimental work on the nonuniformity of evoked quantal secretion, Greg McLeod for the experimental work on spontaneous quantal secretions, Shanka Karunanithi for experimental work on calcium dependence of evoked secretion, and Lynn Cottee for electron microscope work.

REFERENCES

1. Redman S. Quantal analysis of synaptic potentials in neurones of the central nervous system. *Physiol Rev* 1990;70:165–198.
2. Korn H, Faber DS. Quantal analysis and synaptic efficacy in the CNS. *TINS* 1991;14:439–445.
3. del Castillo J, Katz B. Quantal components of the end-plate potential. *J Physiol* 1954;124:560–573.
4. Kuffler SW, Yoshikami D. The distribution of acetylcholine sensitivity at the postsynaptic membrane of vertebrate skeletal twitch muscles: iontophoretic mapping in the micron range. *J Physiol* 1975a;244:703.
5. Kuffler SW, Yoshikami D. The number of transmitter molecules in a quantum: an estimate of iontophoretic application of acetylcholine at the vertebrate neuromuscular junction. *J Physiol* 1975b;251:465.
6. Edwards FA, Konnerth A, Sakmann B. Quantal analysis of inhibitory synaptic transmission in the dentate gyrus of rat hippocampal slices: a patch-clamp study. *J Physiol* 1990;430:213–249.
7. Faber DS, Young WS, Legendre P, Korn H. Intrinsic quantal variability due to stochastic properties of receptor-transmitter interactions. *Science* 1992;258:1494–1501.
8. Lavidis NA, Bennett MR. Probabilistic secretion of quanta from visualized sympathetic nerve varicosities in mouse vas deferens. *J Physiol* 1992;454:9–26.
9. Lavidis NA, Bennett MR. Probabilistic secretion of quanta from visualized varicosities along single sympathetic nerve terminals. *J Autonom Nervous Syst* 1993a;43:41–50.
10. McLachlan E. An analysis of the facilitated release of acetylcholine from preganglionic nerve terminals. *J Physiol* 1975;245:447–466.
11. Bennett MR, Florin T, Pettigrew AG. The effect of calcium ions on the binomial statistic parameters that control acetylcholine release at preganglionic nerve terminals. *J Physiol* 1976;257:597–620.
12. Karunanithi S, Lavidis NA, Bennett MR. Evidence that each varicosity on the surface of the mouse vas deferens secretes ATP. *Neurosci Letts* 1993;161:157–160.
13. Bennett MR, Lavidis LA. The effect of calcium ions on the secretion of quanta evoked by an impulse at nerve terminal release sites. *J Gen Physiol* 1979;74:429–456.
14. Bekkers JM, Richerson GB, Stevens CF. Origin of variability in quantal size in cultured hippocampal neurones and hippocampal slices. *Proc Nat Acad Sci USA* 1990;87:5359–5362.
15. Bennett MR. The probability of quantal secretion at visualized varicosities of sympathetic nerve terminals. *News in Physiological Sciences* 1993a;8:199–201.
16. Bennett MR. Probability of neurotransmitter secretions at individual release sites. In: D. Powis, ed. *Neurotransmitter release and modulation*. Cambridge University Press, 1993b; (*In press*).
17. Magleby KL, Miller DC. Is the quantum of transmitter release composed of subunits? *J Physiol* 1981;311:267–287.
18. Clements, J. Quantal synaptic transmission. *Nature* 1991;353:396.
19. Kullman DM. Quantal analysis using maximum entropy noise deconvolution. *J Neurosci Methods* 1992;44:47–57.
20. Stevens CF. Quantal release of transmitter and long-term potentiation. *Cell* 72/Neuron 10 (Suppl) 1993;55–63.
21. Hall P. *The Bootstrap and Edgeworth Expansion*. New York: Springer-Verlag; 1992.
22. Stern P, Edwards FA, Sakmann B. Fast and low components of unitary EPSCs on stellate cells elicited by focal stimulation in slices of rat visual cortex. *J Physiol* 1992;449:247–278.
23. Boyd IA, Martin AR. The end-plate potential in mammalian muscle. *J Physiol* 1956;132:74–91.
24. Hestrin S. Activation and desensitization of glutamate-activated channels mediating fast excitatory synaptic currents in the visual cortex. *Neuron* 1992;9:991–999.
25. Sigworth FJ. The variance of sodium current fluctuations at the node of Ranvier. *J Physiol* 1980; 307:97–129.
26. Hirst GDS, Neild TO. Some properties of spontaneous excitatory junction potentials recorded from arterioles of guinea pig. *J Physiol* 1980;303:43–60.
27. Benham CD, Tsien RW. A novel receptor-operated CA^{2+}-permeable channel activated by ATP in smooth muscle. *Nature* 1987;328:275–278.
28. Friel DD. An ATP-sensitive conductance in single smooth muscle cells from the rat vas deferens. *J Physiol* 1988;401:361–380.
29. Clements, J. A statistical test for demonstrating a presynaptic site of action for a modulator of synaptic amplitude. *J Neurosci Methods* 1990;31:75–88.

30. Bennett MR, Pettigrew AG. The formation of synapses in amphibian striated muscle during development. *J Physiol* 1975;252:203–239.
31. Lo Y, Wang T, Poo M. Repetitive impulse activity potentiates spontaneous acetylcholine secretion at developing neuromuscular synapses. *J Physiol Paris* 1991;85:71–78.
32. Silver RA, Traynelis SF, Cull-Candy SG. Rapid-time-course miniature and evoked excitatory currents at cerebellar synapses in situ. *Nature* 1992;355:163–166.
33. Eccles JC, Katz B, Kuffler SW. Effect of eserine on neuromuscular transmission. *J Neurophysiol* 1943;5:211–230.
34. Kriebel ME, Gross CE. Multimodal distribution of frog miniature endplate potentials in adult, denervated and tadpole leg muscle. *Prog Neurobiol* 1974;64:85–103.
35. Tremblay JP, Laurie RE, Colonnier M. Is the mepp due to the release of one vesicle or to the simultaneous release of several vesicles at one active zone? *Brain Res Reviews* 1983;299–314.
36. Elliott TR. The action of adrenaline. *J Physiol* 1905;732:401–467.
37. Bennett MR. Autonomic Neuromuscular Transmission. *Monograph of the Physiological Society No. 30*. Cambridge: Cambridge University Press; 1972.
38. Lavidis NA, Bennett MR. Sympathetic axon: varicosity distributions on the surface of the mouse vas deferens. *J Auton Nerv Syst* 1993b;43:41–50.

Molecular and Cellular Mechanisms of Neurotransmitter Release, edited by Lennart Stjärne, Paul Greengard, Sten Grillner, Tomas Hökfelt, and David Ottoson, Raven Press, Ltd., New York © 1994.

26

Neurotransmitter Release Mechanisms in Autonomic Nerve Terminals

Thomas C. Cunnane and Tim J. Searl

University Department of Pharmacology, Mansfield Road, Oxford OX1 3QT, United Kingdom

The purpose of this chapter is to explain how electrophysiological techniques have helped us understand the mechanisms involved in the release of neurotransmitters from postganglionic autonomic nerve terminals. As a starting point, we will briefly discuss neurotransmitter release mechanisms at the skeletal neuromuscular junction and then describe the characteristic features of release from varicose autonomic nerve terminals on an impulse-to-impulse basis. Particular emphasis will be placed on neurotransmitter release mechanisms in postganglionic sympathetic nerve terminals where noradrenaline and adenosine 5′-triphosphate (ATP) are thought to be co-stored and co-released. We will also briefly discuss how we can set about studying the release of acetylcholine at "muscarinic synapses."

NEUROTRANSMITTER RELEASE MECHANISMS AT THE SKELETAL NEUROMUSCULAR JUNCTION

Our present understanding of the mechanisms underlying neurotransmitter release are based on the pioneering electrophysiological studies of Katz and his colleagues in the early 1950s. Fatt and Katz (1) carried out their experiments on amphibian nerve-skeletal muscle preparations and used intracellular recording techniques to measure the postjunctional response to acetylcholine released from the somatic motor nerves at the motor endplate. These preparations were used because they had two unique advantages: 1) individual muscle fibers are usually innervated by a single nerve terminal at the motor endplate (i.e., the specialized region of the muscle fiber in close apposition to the nerve terminal), and 2) muscle fibers are electrically isolated from one another, so all of the electrical activity recorded by an intracellular microelectrode positioned in the endplate region can be attributed to acetylcholine released from a single nerve terminal.

Spontaneous Electrical Activity at the Motor Endplate

When a skeletal muscle fiber is impaled with a microelectrode at the endplate region, small random spontaneous depolarizations are recorded. These spontaneous potentials were first discovered in 1952 by Fatt and Katz, who termed them miniature endplate potentials (MEPPs). The MEPP arises from the random binding of acetylcholine molecules, released as a multimolecular packet of some 3,000–10,000 molecules, to a patch of nicotinic receptors located on the surface of the muscle fiber at the endplate region. The acetylcholine receptors are mainly located on the lips of the postsynaptic membrane folds. These potentials are potentiated by anticholinesterases, reduced in a graded manner by the nicotinic receptor antagonist (+)-tubocurarine, and are absent in muscle that had been previously denervated. On the basis of these observations it was suggested that MEPPs resulted from the release of packets of acetylcholine from motor nerve terminals (2).

Evoked Electrical Activity at the Motor Endplate

When the motor nerve innervating the muscle is stimulated, a large endplate potential (EPP), which has a similar time course to the MEPP, is recorded in the impaled cell. The EPP normally triggers a muscle action potential and contraction but, in the presence of (+)-tubocurarine or when preparations are bathed in physiological saline containing high Mg^{2+} and low Ca^{2+} concentrations, the EPP is reduced below threshold.

The Relationship Between the MEPP and the EPP

The exact relationship between the spontaneous MEPP and the nerve-evoked EPP was elucidated by del Castillo and Katz (3). When the amplitude of the EPP was reduced by increasing the Mg^{2+} and reducing the Ca^{2+} concentration, individual EPPs fluctuated in amplitude in a random stepwise manner, the step unit having a similar amplitude to the MEPP. In addition, a proportion of the stimuli failed to elicit EPPs. Based on these observations it was suggested that the EPP was composed of an integral number of all-or-none units (quanta), identical in size to the MEPP. These ideas are encompassed in the *quantal hypothesis,* the fundamental idea that evoked neurotransmitter release occurs in standard multimolecular "packets" (identical to the spontaneously released units), the size of which is independent of the event that caused the release (2).

Anatomical Basis of the Quantum

The quantal hypothesis of chemical transmission was widely accepted in the early 1950s, and about this time subcellular organelles termed vesicles were discovered in the presynaptic terminals of neurons (4,5,6,7). The question naturally arose whether a single quantum could be equated with the amount of acetylcholine sequestered within a single vesicle. This concept became known as the *vesicle hypothesis* (3). Further supporting evidence included the fact that synaptic vesicles prepared from several different tissues by homogenization and centrifugation have been shown to contain neurotransmitters (8). Finally, at the frog neuromuscular junction the best morphological evidence for the vesicular release of acetylcholine came from the studies of Heuser et al. (9) and Heuser and Reese (10) who developed a technique in which the morphological changes associated with acetylcholine release could be studied with a high degree of temporal resolution. Nerve-skeletal muscle preparations were rapidly frozen shortly after stimulation and morphological changes corresponding to exocytotic events were detected at active zones. The vesicle hypothesis is believed to hold true for chemical transmission at practically all chemical synapses and neuroeffector junctions (11).

However, the quantal and vesicular hypothesis has not been accepted unequivocally at all cholinergic nerves. It has been suggested that, in the *Torpedo* electric organ, a major fraction of the acetylcholine released by the nerve action potential originates from the cytoplasm rather than the vesicular store. Biochemical studies suggest that a large proportion of the acetylcholine released in response to a nerve impulse has a cytosolic rather than a vesicular origin (12,13). Recent functional studies with vesamicol, an inhibitor of the vesicular acetylcholine carrier, suggest that cytoplasmic release of acetylcholine occurs through vesicular transporters that have become incorporated into the nerve terminal membrane following exocytotic release. (14) These transporters now pump acetylcholine out of the nerve terminal (leakage) rather than into vesicles. Vesamicol has no effect on the quantal size of previously filled vesicles but reduces the acetylcholine content of newly formed vesicles. (15,16) These studies strongly support the idea that the EPP results from the vesicular release of acetylcholine.

Quantal Content of the Endplate Potential

Under conditions of increased Mg^{2+} and low Ca^{2+} the distribution of amplitudes of evoked EPPs can be fitted by Poisson statistics (3,17). However, under normal physiological conditions at many neuroeffector junctions and synapses, the release process appeared often to be better fitted by a simple binomial model (18,19,20) where there is a population of n units with a uniform probability of releasing a single quantum p. This type of analysis is of particular interest in that both n and p may have physical correlates—n may represent the number of quanta available for release or the number of active release sites and p may be a measure of the cytosolic concentration of Ca^{2+}. It has been suggested that n may reflect the number of "active zones," (i.e., the regions of the nerve terminal specialized for the release of neurotransmitter packaged in vesicles). Such regions have a characteristic morphology, and studies with labeled ω-conotoxin and α-bungarotoxin have shown that the prejunctional calcium channels are aggregated here in discrete clumps opposite discrete patches of postjunctional nicotinic receptors (21).

Nonuniform Release of Neurotransmitter

At the amphibian neuromuscular junction, release of quanta has been shown to be intermittent and spatially nonuniform under certain experimental conditions. (22,23) The sites of neurotransmitter release where the motor nerve axon first makes contact with the muscle-fiber membrane have a higher average probability of quantal release than those from more distal release sites (24,25). It has also been demonstrated that very low probability release sites are intermingled with sites of neurotransmitter release with relatively high probabilities (22).

Factors Affecting Neurotransmitter Release

The general idea therefore emerges that, under normal conditions, all potential release sites do not have the same probability of discharging a quantum of neurotransmitter. At the amphibian neuromuscular junction, a considerable nonuniformity in the probability of release at different sites along the nerve terminal has been demonstrated using focal extracellular recording techniques when the external Ca^{2+} concentration was reduced (24). The neurotransmitter release process is in fact better fitted by a compound binomial model in which the probability of neurotransmitter release varies at different release sites along the same nerve terminal. One possible explanation is that the number and therefore density of voltage-gated Ca^{2+} channels aggregated at a particular release site determines the probability of secretion; further experimentation will be required to confirm this explanation.

The Calcium Hypothesis

The demonstration by Katz and colleagues that the number of quanta released in response to nerve stimulation is dependent on the extracellular Ca^{2+} concentration (28) indicated the pivotal role of calcium ions in neurotransmitter release. In addition, Katz and Miledi (27,28,29) found that in preparations where propagation of the nerve impulse had been blocked by tetrodotoxin (TTX), acetylcholine release could still be evoked by focal depolarization of the nerve terminal, but only if Ca^{2+} ions were present in the extracellular fluid. These early studies led to the *calcium hypothesis,* the idea that the release of neurotransmitter is dependent on the entry of Ca^{2+} into the nerve terminal. The calcium hypothesis is now widely believed to hold for all nerves, but it is fair to say that the fundamental biochemical events that activate the release mechanisms subsequent to calcium entry remain largely unknown. While a great deal of progress has been made in characterizing the properties of the various neuronal vesicular and cytoplasmic proteins thought to be involved in neurotransmitter release, the overall picture is still far from clear. The next decade should see the triumph of molecular biology over this problem and herald in a new era in our understanding of secretory mechanisms in a wide variety of cell types. Even at this early stage, it is clear that the biochemical machinery regulating neurotransmitter release will differ across species and from tissue to tissue. Different neurons that nevertheless release the same neurotransmitter have different biochemical machinery involved in release (30,31).

Calcium Channels Involved in Neurotransmitter Release

The entry of calcium ions into nerve terminals occurs through voltage-activated calcium channels. There appear to be several types of calcium channels involved in neurotransmitter release depending on the species. Experiments with two toxins—ω-conotoxin, which blocks calcium influx through channels which are termed N-type, and funnel web spider toxin, which blocks calcium influx through P-type channels—were used to study nerve-evoked transmission at the skeletal neuromuscular junction. In the frog it is clear that N-type channels are involved in evoked release, whereas at mammalian skeletal neuromuscular junctions P-type, channels are thought to be involved (32). However, the Q and R-type calcium channels have recently been described and are abundant in the central nervous system (see chapter 11). At other mammalian cholinergic junctions, in both the peripheral and central nervous systems, evoked neurotransmitter release is blocked by ω-conotoxin, which suggests that in some nerve terminals different populations of calcium channels may be present.

Summary

The analysis of the vertebrate skeletal neuromuscular junction has provided the most detailed evidence to date for the processes involved in neurotransmitter release. The key ideas that have emerged are that, following nerve stimulation, about 100 or so quanta are secreted from each nerve terminal by an action potential. Single quanta are released in an all-or-none manner at specialized release sites known as *active zones,* and each active zone probably behaves as a discrete unit. There is a high safety factor for the transmission of excitation from nerve to skeletal muscle, and individual skeletal muscle fibers contract when endplate regions are depolarized to the threshold for the initiation of muscle action potentials.

AUTONOMIC NEUROEFFECTOR TRANSMISSION

What then are the comparable neurotransmitter release mechanisms in the complex nerve terminal networks of postganglionic autonomic nerves? At first sight it might seem that there are enormous differences between release mechanisms at the skeletal and the autonomic neuroeffector junctions, but what we will now highlight is the remarkable similarity.

Postganglionic Autonomic Nerve Terminals

Until relatively recently, surprisingly little was known about the characteristic features of neurotransmitter release from individual varicosities on postganglionic autonomic nerve terminals on an impulse-to-impulse basis. In the last decade a bewildering variety of chemicals have been shown to be stored in, and released from, autonomic nerve terminals either as a primary neurotransmitter or as a modulator. Perhaps the most surprising finding is that ATP is released with noradrenaline and acts in its own right as a primary neurotransmitter in many sympathetically innervated vascular and nonvascular smooth muscles (31,33,34,35). In this chapter we will confine much of our discussion to electrophysiological studies of neurotransmitter release mechanisms in postganglionic sympathetic (noradrenaline and ATP) and parasympathetic (acetylcholine; muscarinic receptors) nerve terminals.

Issues to Resolve at the Autonomic Neuroeffector Junction

Some of the issues that will need to be clarified in the future include the following questions:

- Are all autonomic neurotransmitters released in quantal packets?
- What is the probability of release at the level of the individual varicosity?
- Is the nerve terminal actively or passively invaded by the action potential?
- Do neuronal K^+-channels normally limit release?
- Are there specialized neuroeffector junctions?
- What are the mechanisms underlying frequency-dependent facilitation?
- What are the mechanisms underlying the local regulation of neurotransmitter release (e.g., α-autoinhibition)?
- Is more than one neurotransmitter released from the same vesicle within the same varicosity, or are they stored in separate vesicles; definitions of co-release versus co-transmission can often be confusing in the literature.

We will now address some of these questions, but clearly many experiments will need to be carried out in the future to unequivocally answer all of the current questions.

METHODS TO STUDY NEUROTRANSMITTER RELEASE IN THE AUTONOMIC NERVOUS SYSTEM (ANS)

Much of our early knowledge of the mechanism of storage and release of the sympathetic neurotransmitter(s) is based to a large extent on biochemical measurements of the release of tritium-labeled catecholamines from electrically stimulated preparations and on morphological observations (36).

Biochemical Studies of Noradrenaline Release

Most of the noradrenaline in sympathetic neurons is synthesized and located in the varicosities of the terminal axon, although synthesis also occurs throughout the cell body and axon. The noradrenaline is largely contained in a heterogeneous store comprising both small (25–60 nm) and large (70–160 nm) dense-cored, granular vesicles, the percentage varying enormously from tissue to tissue and between species. In the rodent vas deferens, the large dense-cored vesicles account for approximately 5% of the vesicle population, whereas in the ox spleen they account for as much as 50% (37). There is convincing evidence that nerve impulses release only noradrenaline that is bound or stored in vesicles, and not extravesicular neurotransmitter (11,36). The study of the release mechanism has been facilitated by the discovery of drugs that can interfere specifically with the neurotransmitter release process. Most of the evidence suggests that the noradrenaline which is "free" within the cytosol cannot be released by electrical stimulation (38). Thus it appears that only noradrenaline stored in vesicles can be released by nerve action potentials.

Exocytotic Release Mechanism

Early experiments utilizing the adrenal medulla as a model system demonstrated that catecholamines (noradrenaline, adrenaline) are secreted from chromaffin cells together with all the soluble components of the storage granule but not the cytoplasmic proteins such as lactate dehydrogenase or phenylethanolamine-N-methyl transferase (39,40). These observations were primarily responsible for the proposal that the release mechanism for catecholamines from chromaffin cells of the adrenal medulla is exocytotic. Similar studies were carried out on sympathetic nerves and virtually identical conclusions drawn. The enzyme dopamine-β-hydroxylase and the protein chromogranin A, which are found only within vesicles in sympathetic nerve terminals, are released together with noradrenaline following sympathetic nerve stimulation. The fact that the unique constituents of the vesicle are released during sympathetic nerve activity demonstrates unequivocally that it is the noradrenaline sequestered within vesicles that is released by nerve action potentials rather than noradrenaline "free" within the cytosol. Electron microscopy studies have also shown vesicles fusing with the axoterminal membrane in varicose regions of noradrenergic fibers (41,42,43). Thus there is now convincing biochemical, morphological, and electrophysiological evidence that sympathetic nerves release neurotransmitter in multimolecular packets by an exocytotic mechanism. The question as to whether the packets of neurotransmitter are of uniform size is at present unresolved, especially given the "apparent" variation of vesicle sizes reported in sympathetic nerve terminals (see factors just discussed).

Neurotransmitter Release at the Level of the Individual Varicosity

A single sympathetic neuron can give rise to 10,000–30,000 varicosities that course throughout tissues weaving a complex three-dimensional path. What then are the characteristic features of neurotransmitter release at the level of the individual varicosity? Several hypotheses have been put forward based on the key issue of whether catecholamine secretion from sympathetic nerve terminals is graded or quantal in nature (i.e., whether only a small fraction or the entire neurotransmitter content of a vesicle is released by the invading nerve action potential or whether the entire vesicle content is released). At one time, it was usual to make the assumption that all varicosities were equally capable of discharging neurotransmitter on depolarization. Viveros and his colleagues concluded that, as far as the adrenal medulla was concerned, secretion was quantal and that the size of the quantum was determined by the entire catecholamine content of the individual vesicle (44). Studies by Neher and his colleagues using patch-clamp and electrochemical techniques have resolved this issue in chromaffin cells (45). The mechanism of neurotransmitter release from sympathetic nerve terminals, on the other hand, was more uncertain. Clues to the mechanisms involved in noradrenaline release came from biochemical studies and the key findings are outlined in the next section.

Evidence That Only a Fraction of Vesicle Content Is Released

Folkow and his colleagues measured the release of noradrenaline from the vasoconstrictor nerves of cat calf muscle. They found that the fraction of the total noradrenaline released by a nerve action potential was about 1/50,000 of the total noradrenaline content of the vasoconstrictor nerve endings. They made two assumptions: 1) most of the noradrenaline in the tissue is contained in vesicles in the varicosities, and 2) the majority of the varicosities release neurotransmitter when excited by a nerve action potential. Estimates were made of the number of vesicles in a varicosity and the number of molecules of noradrenaline contained in a vesicle, and the fractional release of the noradrenaline store per nerve impulse was determined. The results suggested that only 1% to 3% of the neurotransmitter content of a vesicle is released per nerve impulse (46). An alternative explanation proposed by the same research group was that not all varicosities are capable of discharging neurotransmitter when depolarized by an action potential. This hypothesis requires that a noradrenergic varicosity discharges neurotransmitter only once in response to every 30–100 nerve impulses (i.e., release is intermittent). This latter hypothesis was rejected, since there was no evidence then that evoked release from sympathetic nerves occurred intermittently.

Evidence That the Entire Vesicle Content Is Released

Bevan and co-workers applied another approach to this problem that did not involve any assumptions about the amount of noradrenaline contained in a varicosity. They estimated the amount of noradrenaline released per varicosity in the rabbit pulmonary artery by counting the number of varicosities per unit of tissue, and determined the amount of noradrenaline released per unit. The results suggested that the contents of one vesicle may be released on average every seven to eight impulses and are consistent with the hypothesis that the neurotransmitter content of one vesicle is released from each varicosity intermittently (47). Biochemical techniques lack the time resolution required to resolve these issues, so we must turn to electrophysiological studies to see how further progress has been made.

Electrophysiological Studies of Neurotransmitter Release

Novel electrophysiological approaches have been developed that have led to a new understanding of neurotransmitter release mechanisms at the level of the individual varicosity in sympathetic nerve terminals (48,49,50,51,52).

The electrophysiological experiments have been carried out on sympathetically innervated smooth-muscle preparations (vas deferens, arteries, and arterioles). It is important to consider first those aspects of postjunctional mechanisms in smooth muscle that have a bearing on the interpretation of electrophysiological records, namely:

- the relationship between individual smooth-muscle cells
- the relationship between individual smooth-muscle cells and the varicosities from which neurotransmitter is released
- the electrical activity recorded from smooth muscle cells following nerve stimulation

Postjunctional Mechanisms—Basic Considerations

These mechanisms are equally applicable to sympathetic and parasympathetic neuroeffector junctions. Individual smooth-muscle cells are electrically coupled to their neighbors, therefore, the electrical activity recorded with an intracellular microelectrode results not only from the impaled cell but also from cells remote from the point of impalement. Furthermore, smooth muscle receives a multiple innervation. In some smooth muscles a single stimulus evokes a transient depolarization, the excitatory junction potential (EJP) whose amplitude is graded with stimulus strength (i.e., as the stimulus intensity is raised, EJP amplitude increases until all contributing nerve fibers are excited). During a train of stimuli, successive EJPs increase in amplitude from the first in a train and at 1 Hz reach a steady state in about six to eight stimuli (Fig. 1A). This progressive increase in EJP amplitude is termed facilitation and we will return to this phenomenon later. In the absence of stimulation, spontaneous excitatory junction potentials (SEJPs) are recorded (Fig. 1B).

Absence of Postjunctional Specialization

At many synapses and neuroeffector junctions, specific receptors for the neurotransmitter are concentrated in the regions of the postjunctional membrane in direct

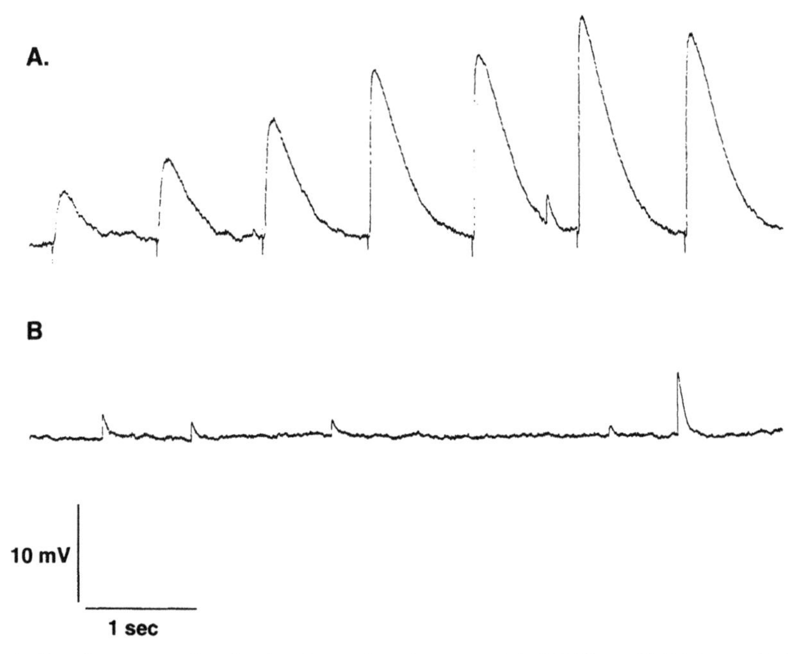

FIG. 1. Spontaneous and evoked excitatory junction potentials (SEJPs and EJPs) recorded with intracellular microelectrodes in the guinea-pig vas deferens. **A.** EJPs evoked by a train of seven stimuli at 1 Hz. **B.** SEJPs. Note the growth in EJP amplitude during the train (facilitation).

apposition to the nerve terminals, but at the neuroeffector junction in smooth muscle there is little anatomical evidence for postjunctional specialization. (53) The apparent absence of postjunctional specialization may reflect a lack of suitable electron-dense probes to identify specific junctional receptors. We await the development of suitable gold-labeled antibody probes directed at specific receptors. Electrophysiological data suggest that sympathetic neuroeffector junctions behave to all intents and purposes like "classical synapses " (11,31).

Nature of the Sympathetic Neurotransmitter(s)

Noradrenaline has been the most intensively studied neurotransmitter in the sympathetic nervous system and, until relatively recently, was generally regarded as the neurotransmitter mediating electrical events at practically all sympathetic neuroeffector junctions. However, now a considerable body of evidence suggests that sympathetic nerves secrete not only noradrenaline but also ATP and neuropeptide Y (35,11). Perhaps the most surprising recent finding is that ATP acts as a rapid neurotransmitter in many sympathetically innervated tissues. Interestingly, it has recently been reported that ATP acts as a fast neurotransmitter in the brain (54).

METHODS TO STUDY NEUROTRANSMITTER RELEASE

There are a number of ways to study neurotransmitter release mechanisms in sympathetic nerve terminals: 1) classical pharmacological methods in which the mechanical response of the end-organ in response to nerve stimulation is recorded; 2) biochemical measurements of the overflow of neurotransmitter from innervated preparations as discussed earlier; 3) morphological methods; and 4) electrical recording techniques. From this point on we shall be discussing results obtained using electrophysiological recording techniques.

Electrophysiological Studies of Neurotransmitter Release

Burnstock and Holman (55,56) were the first to record EJPs in the guinea-pig vas deferens using intracellular recording techniques. When a single stimulus was applied to the hypogastric nerve trunk, an EJP that lasted for about a second was evoked. In the absence of nerve stimulation SEJPs, which are unaffected by tetrodotoxin, were recorded every few seconds; this was the first evidence for the packeted release of neurotransmitter from sympathetic nerve terminals. Examples of EJPs and SEJPs in the guinea-pig vas deferens are shown in Fig. 1. Note that the time course of the EJP is several orders of magnitude slower (about 1,000 milliseconds) than the equivalent EPP (2–20 milliseconds) at the skeletal neuromuscular junction. Unlike the skeletal neuromuscular junction, where fast-twitch muscle fibers are normally innervated by a single nerve terminal, it has been difficult to establish the exact relationship between the spontaneous and evoked EJP in smooth muscle.

Amplitude Distributions of SEJPs

It was previously noted that the amplitude distribution of spontaneous MEPPs can be used to determine the average size of the quantum of neurotransmitter secreted from a single nerve terminal at the skeletal neuromuscular junction. Unfortunately, the amplitude distribution of SEJPs at the autonomic neuroeffector junction cannot be used in this way. It has not therefore been possible to determine statistically the quantal size and quantal content of the EJP with any degree of confidence, for three main reasons: 1) the difference in time course of the SEJP and EJP means that they are not directly comparable; 2) the amplitude distribution of the SEJPs is not normally unimodal but skewed towards the noise level of the recording system, making it virtually impossible to determine the mean quantal size; and 3) statistical tests are only valid if the activity of a single junction can be identified or if the innervation is limited to a specific area of the muscle fiber. Smooth-muscle cells receive a multiple innervation and, because of electrical coupling between the cells, the electrical activity recorded in any one cell reflects the activity of neurotransmitter released both locally and at sites remote from the recording electrode.

These properties of smooth muscle explain, at least in part, why the SEJP amplitude distribution is skewed; neurotransmitter action close to the recording electrode produces a larger SEJP than that occurring at distant release sites due to electrotonic attenuation of the signals. However, variable numbers of neurotransmitter molecules in the packets, the width of the junctional cleft, variation in input impedance of the different smooth-muscle cells in the tissue, and variations in the distribution and numbers of postjunctional receptors may all contribute to the shape of the SEJP amplitude distribution. To address these problems, novel electrophysiological approaches have been developed to study neurotransmitter release mechanisms in postganglionic sympathetic nerve terminals.

Electrophysiological Method to Study Neurotransmitter Release from a Single Varicosity

A number of years ago an electrophysiological method was introduced that allowed us to study neurotransmitter release from one or a small population of varicosities on an impulse-to-impulse basis (48,49). It is useful to give a brief description of the method and to summarize the main results obtained. Essentially, the membrane potential was recorded from smooth-muscle cells with an intracellular microelectrode, and EJPs were evoked by stimulating the excitatory innervation with trains of stimuli at low frequencies (0.1–2 Hz). Close examination of the rising phases of a series of EJPs revealed discontinuities that were highlighted by electronically differentiating each EJP (i.e., the rate of rise of each EJP was measured). These discontinuities now appeared as transient peaks in the rate of depolarization which were termed discrete events (48,49). A prominent discrete event was taken as a measure of the postjunctional response of a single smooth-muscle cell to one or a few quanta of neurotransmitter secreted from varicosities in close apposition to the impaled cell. Signals from more distant varicosities are attenuated and appear as a slow nonintermittent component that represents the summed activity of many release sites (Fig. 2). By examining the rising phase of the EJP in this way we can break down individual EJPs into their component parts. The special relationship of the nerve terminals to the smooth-muscle cell and the intermittent nature of the neurotransmitter release mechanism have allowed the events associated with the activation of single (or small groups of) varicosities to be described with a degree of resolution previously unobtainable.

Characteristic Features of Release from Individual Varicosities

Action potential evoked neurotransmitter release from an individual varicosity has the following characteristics:

Release is highly intermittent—individual varicosities do not secrete a quantum of neurotransmitter in response to every nerve action potential, but only at about one to three times per 100 stimuli at 0.1–1 Hz).

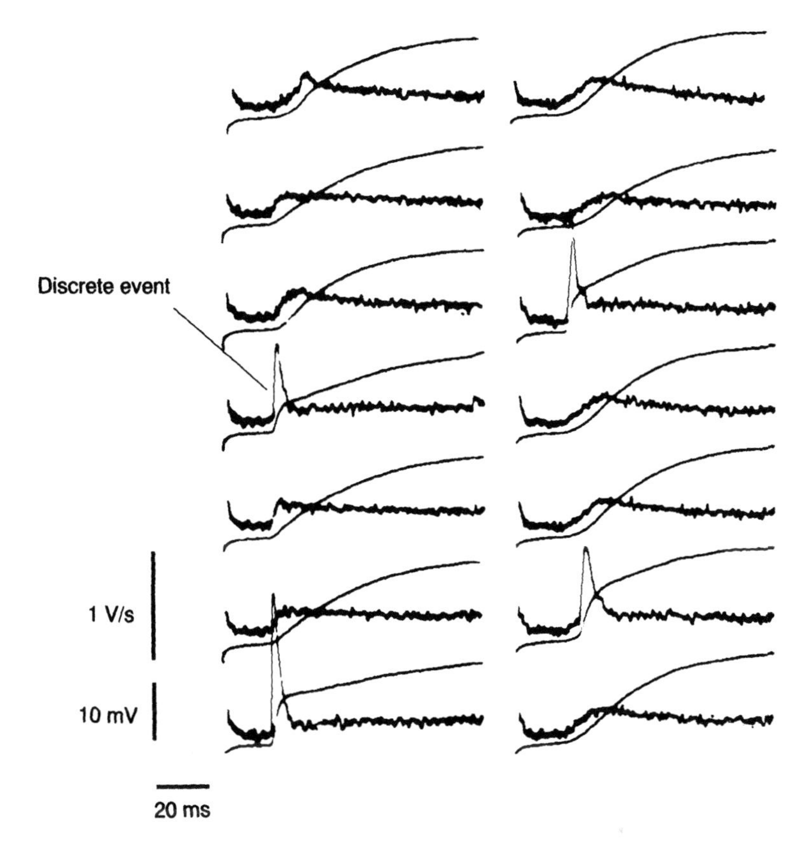

FIG. 2. Excitatory junction potentials (EJPs) and their first-time derivatives (discrete events) in the guinea-pig vas deferens. EJPs were evoked by trains of stimuli at 0.91 Hz. Note the rapid, intermittently occurring rates of rise (discrete events) and the slower, nonintermittent component representing the summed activity of many release sites.

Release is monoquantal—when a release site on a varicosity is activated by the nerve action potential, only a single packet of neurotransmitter is secreted.

The quantum released by the nerve impulse is the same as that secreted spontaneously—the evoked "unit" of neurotransmitter action is equivalent to the "unit" released spontaneously in the absence of stimulation.

These characteristics are best illustrated by reference to Fig. 3, which shows the responses of a cell presumably innervated by only a single or at most two varicosities. Discrete events were evoked in an all-or-none manner, intermittently, even though the stimulus strength applied to the innervating nerve fiber was supra threshold. The lower record shows an evoked discrete event superimposed on a spontaneous discrete event recorded in the same cell, showing them to be virtually identical.

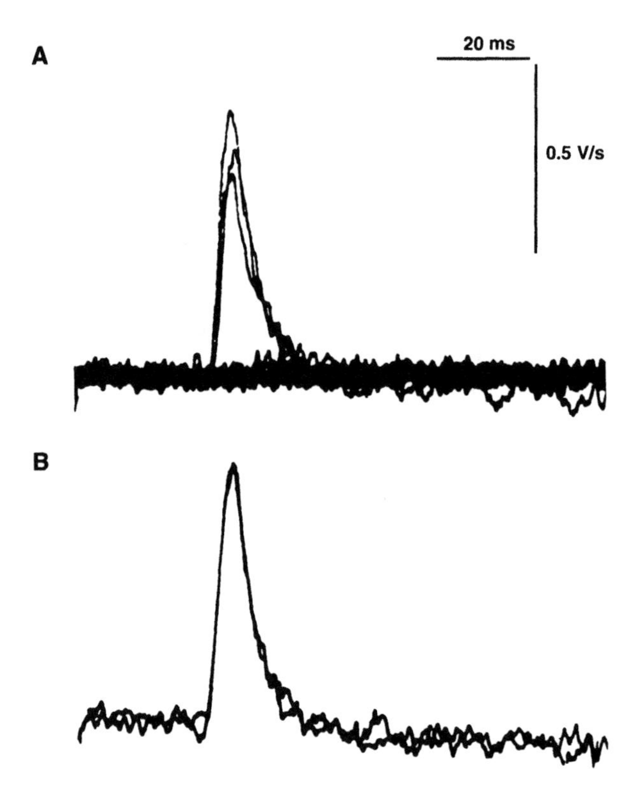

FIG. 3. Characteristic features of discrete events recorded in a single cell of the guinea-pig vas deferens which was presumably innervated by only one or a few varicosities. **A.** Only three discrete events were evoked by a total of 450 stimuli delivered at 1 Hz. **B.** An evoked discrete event from **A** photographically superimposed on a spontaneous discrete event (differentiated spontaneous excitatory junction potential (SEJP)).

Taking the simplest interpretation of the vesicle hypothesis—that one quantum can be equated with the entire neurotransmitter contents of one vesicle—this result provides good evidence for a monoquantal release mechanism.

Lack of Synaptic Variance

One characteristic feature of release from the same or closely related release site(s) is the appearance of virtually identical events within a few stimuli of each other during long trains of stimuli. This finding suggests that there can be a striking lack of variance in the postjunctional response produced by neurotransmitter secreted from the same or closely related site (see chapter 18). In a low-probability system, we think it highly unlikely that several multiquantal releases would occur from the same or closely related group of varicosities. For similar reasons we do not believe that the normal spontaneous event is made up of several quanta released almost synchronously in the absence of a nerve action potential.

It is likely, although we have no evidence for this statement yet in the sympathetic nervous system, that postjunctional receptors opposite the point of neurotransmitter release are saturated. There seems little point in having a mechanism in which multiple quanta are secreted from a release site but produce only a response equivalent to a single quantum. Therefore, varicosities function in a digital all-or-none manner; graded activity in populations of neurons is accomplished by the progressive recruitment of previously silent release sites (see latter part of this chapter). One important aim of future research will be to determine whether single varicosities have more than one release site, a question currently being addressed by Bennett and Lavidis (see Bennett's Chapters 1 and 25) and in our laboratory.

Studies in Arterioles

Hirst and Neild (57,58) studied neurotransmitter release mechanisms in postganglionic sympathetic nerve terminals innervating the guinea-pig submucosal arterioles and also concluded that neurotransmitter release occurs intermittently. A short segment of arteriole was isolated from the branching arteriolar arcade by cutting each end. In these short segments, which are only one or two smooth-muscle cells thick, SEJPs and EJPs recorded with two electrodes impaled in cells at either end of the preparation had virtually identical amplitudes (i.e., there was little spatial attenuation of the signals). Furthermore, SEJPs and EJPs had similar time courses. This similarity between the time course of the SEJP and the EJP is because the cut ends of the arteriolar segment seal to form a high resistance between the cytoplasm and the extracellular fluid. Therefore, current can escape only locally, and the time course of both the SEJP and EJP is determined by the rate of charge loss across the arteriolar membrane. Thus, using this preparation, it was possible to directly determine the relationship between the SEJP and the EJP.

One of the most interesting findings was that the amplitude distribution of SEJPs was unimodal (and not skewed as in most smooth-muscle preparations) and that the amplitudes corresponded closely to the smallest recorded EJPs. An analysis of the EJP amplitude distributions suggested that the majority of EJPs had a quantal content of only 2 or 3. In these preparations the number of varicosities was estimated from fluorescent microscopy studies to be 112 and 224. Luff et al. (59) later demonstrated that fluorescent microscopy underestimates the number of varicosities in this preparation by two or threefold. Taken together, these results provide convincing evidence that the average probability of release of a single quantum from an individual varicosity was less than 0.01 (60).

Summary

Both the biochemical and the electrophysiological studies of neurotransmitter release from single or small groups of varicosities (i.e., release sites) suggest that action-potential-evoked neurotransmitter release occurs intermittently at the level of the individual varicosity (61). The electrophysiological studies also confirm that evoked neurotransmitter release is quantal, and that the quantum probably represents the whole neurotransmitter contents of a single vesicle. It should be remembered that these data do not necessarily relate to the release of noradrenaline per se, as the EJP and SEJP are generated by the release of ATP or a related purine nucleotide (11) (see later in this chapter). However, the bulk of evidence to date suggests that both noradrenaline and ATP are co-stored and co-released, and, since ATP is known to be present in the same vesicles as noradrenaline (36) it follows that the EJP and SEJP will measure indirectly the packeted release of noradrenaline.

Possible Causes of Intermittence

Physiologists often use isolated preparations bathed in physiological solutions. The question can be raised whether subtle changes in the ionic environment of small varicose fibers would compromise action potential conduction in complex nerve terminal networks in vitro. One hypothesis to explain intermittence of neurotransmitter release would be failure of the full action potential to propagate to all potential sites of neurotransmitter release in the highly branched and varicose nerve terminal network.

It has been argued that action potentials would have difficulty passing along fibers in which there was more than a fivefold increase in diameter. Such a situation occurs in the terminal regions of autonomic nerves in which varicosities alternate with stretches of thin nonvaricose axon (0.1 μm diameter), producing the characteristic "string of beads" appearance (approximately 1 μm diameter). It has also been suggested that an action potential in a preterminal axon, encountering an abrupt increase in impedance upon reaching the first varicosity, may fail to invade subsequent varicosities along the chain. If the action potential could be modified by appropriate conductances in this region, then an attractive mechanism for modifying neurotransmitter release would be the result. In addition to reducing release from the first varicosity, the action potential would fail to propagate to other varicosities along the chain. More distant varicosities would be depolarized only to the extent to which they are depolarized by electrotonic conduction from the point of propagation failure. This type of mechanism has been proposed as the means of providing an effective means of regulating neurotransmitter release locally.

The other possible explanation of intermittence would be that the probability of depolarization-secretion coupling in the invaded varicosity is low and much less than one. To distinguish between these two possibilities at the sympathetic neuroeffector junction, we have developed a novel focal extracellular recording technique (50). This technique has shed further light on neurotransmitter release mechanisms in varicose nerve terminals.

Focal Extracellular Recording

The essential idea is to record the nerve terminal impulse in varicosities coursing along the surface of a sympathetically innervated tissue using a small suction electrode. By gently drawing into the electrode the innervated smooth-muscle cells, it is also possible to monitor neurotransmitter release from the invaded varicosities. The technique can then be used to establish the relationship between the arrival of the nerve terminal impulse and neurotransmitter release on an impulse-to-impulse basis. The principles involved in the use of a suction electrode have been described elsewhere (31). Briefly, a resistance is created at the junction between the edge of the suction electrode and the smooth-muscle surface. A potential change proportional to the postjunctional current is created over this "seal" resistance. The polarity of the potential is determined by the direction of current flow over the seal resistance. A negative-going potential reflects postjunctional current originating inside the suction electrode, while neurotransmitter action outside the electrode generates a positive-going signal. The use of suction electrodes therefore allows the electrical activity originating from a small population of varicosities and smooth-muscle cells located under the suction electrode to be investigated.

This method was first applied to the surface varicosities innervating the guinea-pig vas deferens. The extracellular signals recorded with suction electrodes reflect the time course of the conductance changes underlying the intracellularly recorded junction potentials (51).

Spontaneous Excitatory Junction Currents (SEJCs)

The SEJCs recorded from the surface of the vas deferens in the absence of stimulation are produced by the spontaneous release of neurotransmitter. These SEJCs have characteristics similar to intracellularly recorded SEJPs (51). The overall time courses of the SEJC and the SEJP are similar and there is no obvious relationship between the amplitude and the time constant of decay. These findings support the view that the time course of the SEJP is determined by the time course of neurotransmitter action rather than passive membrane properties.

Evoked Excitatory Junction Currents (EJCs)

The pharmacological profile of the evoked EJCs in the guinea-pig vas deferens was largely the same as that of the SEJCs (51). The configuration of the EJCs recorded from the guinea-pig vas deferens depends on the intensity of stimulation and the site of recording. When only the effects of neurotransmitter from the varicosities located under the electrode are recorded, EJCs are similar in both amplitude and time course to the SEJCs recorded in the same attachment. Activity outside the electrode produces a positive-going signal that in some attachments distorts the negative-going EJC. Great care must be taken in the interpretation of records under these circumstances, particularly with regard to the amplitude and the time course of individual events. If an SEJC represents the release of a single quantum, then these findings show that EJC results from the evoked quantal release of neurotransmitter from sympathetic nerve terminals. Negative-going EJCs recorded at a particular latency occur intermittently (Fig. 4), supporting the previous observation that neu-

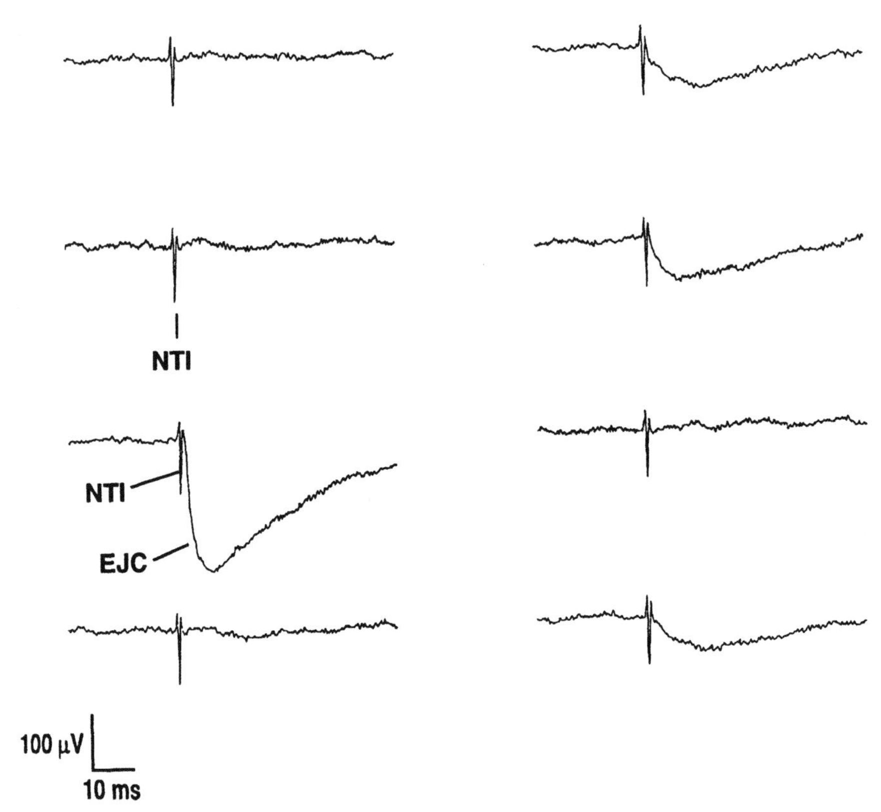

FIG. 4. A series of eight traces showing the relationship of the nerve terminal impulse (NTI) to transmitter release in the guinea-pig vas deferens. The NTIs and excitatory junction currents (EJCs) were evoked during a train of suprathreshold stimuli at 1 Hz. Although the NTI arrived on every occasion, the transmitter release process was not always activated.

rotransmitter release from individual varicosities occurs intermittently during trains of low-frequency stimuli (48,49,58,62).

Monoquantal Secretion

The evidence obtained with this extracellular technique supports the view that only a single quantum is secreted by a varicosity (62). The key evidence can be briefly summarized:

The amplitude distributions of SEJCs and EJCs recorded in the same attachment are similar; multimodal distributions of the amplitudes of evoked events are not observed (Fig. 5).

During trains of stimuli, EJCs are recorded with identical amplitude and time courses; indeed, pairs of identical EJCs are often recorded within a few stimuli of

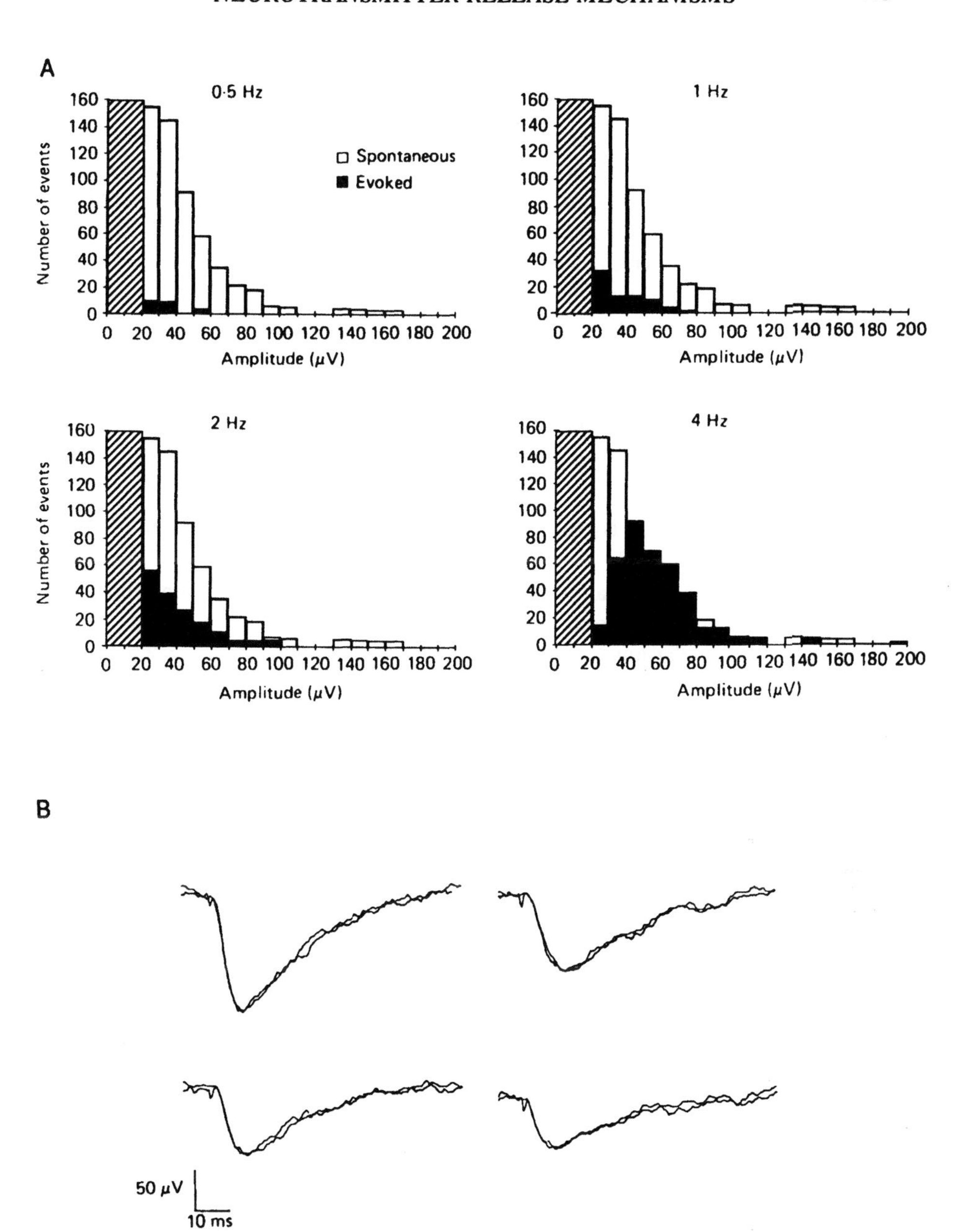

FIG. 5. Comparison of amplitude distributions of spontaneous excitatory junction currents (SEJCs) and EJCs recorded in the same attachment. **A.** EJCs were evoked by trains of 500 stimuli at 0.5, 1, 2, and 4 Hz. The amplitude distribution of SEJCs is superimposed on that of the EJCs at each frequency for ease of comparison. The hatched area represents twice background noise level and is the level where the presence or absence of events could not be stated with any degree of confidence. **B.** Examples of SEJCs and EJCs of similar amplitude and time course recorded in the same attachment. Individual records are superimposed for ease of comparison.

each other, suggesting, as with the discrete event, that previous release facilitates subsequent release from the same site. Similar observations have been made elsewhere and may be a characteristic feature of neurotransmitter release from closely related release sites (63). The importance of these findings (as discussed earlier for discrete events) is that quanta released from the same or closely related release site may produce postjunctional responses with little or no variance within the noise level of the recording system.

Individual EJCs, even though there is a wide variation in their amplitudes, can often be matched with individual SEJCs in the same attachment (see Fig. 5B). Taking the simplest interpretation of the quantal hypothesis that the SEJC represents the postjunctional response to the release of a single quantum, then varicosities normally secrete only one quantum, intermittently, when the neurotransmitter release mechanism is activated by long trains of low-frequency stimuli.

Is an SEJC Caused by the Synchronous Release of Several Quanta?

The question has been raised several times during this symposium whether a spontaneous event results from the synchronous release of several quanta. At the skeletal neuromuscular junction it is known that MEPPs that are two or three times the amplitude of the mean MEPP are occasionally recorded. In this case it seems likely that two or three individual MEPPs released almost synchronously summate to produce the larger spontaneous events. In the case of the sympathetic neuroeffector junction we think it unlikely that, in a low probability system, evoked EJCs of identical shape, occurring within a few stimuli of each other, could represent multiquantal releases from the same varicosity; if this were so then the matching SEJC would also have to be the result of a multiquantal release. At present, we cannot rule out the possibility that there are several release sites on some varicosities, each capable of occasionally releasing one or more quanta almost synchronously. Indeed, it seems reasonable, given the variable degree of shapes and sizes of individual varicosities, that this will indeed turn out to be the case. To resolve the problem of the uni- versus multiquantal release capability of a single varicosity, it will be necessary to develop methods of recording the secretion of a single packet of neurotransmitter from a single identified varicosity or secretory cell. Considerable progress to this end has been made by several research groups: release of single quantum from chromaffin cell (45): focal recording from visualized living varicosities (52).

Relationship of the Nerve Action Potential to Neurotransmitter Release

It is of more than passing interest to study the relationship between action potential propagation in, and neurotransmitter release from, the same string of varicosities. In all experiments, EJCs were always preceded by a nonintermittent nerve impulse. This signal is believed to be the nerve terminal impulse, since the occurrence of the EJC in a narrow latency band is strictly dependent on its arrival (Fig.

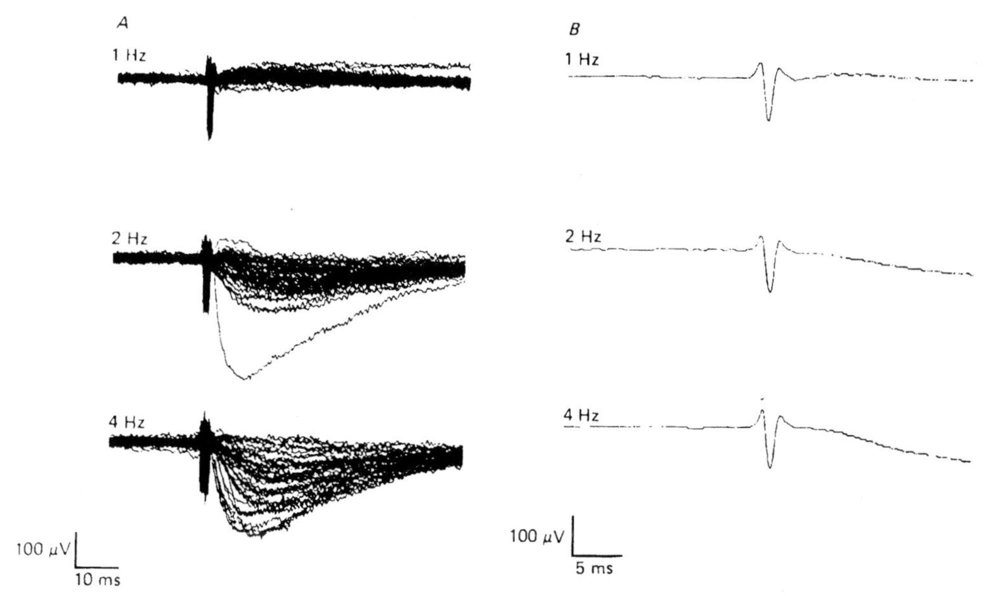

FIG. 6. Frequency-dependent facilitation of transmitter release in the guinea-pig vas deferens. **A.** Extracellular recording of nerve terminal impulses and excitatory junction currents (EJCs) evoked by stimulation of the hypogastric nerve with trains of 100 suprathreshold stimuli at 0.5, 1, and 2 Hz. **B.** Averages of 20 nerve terminal impulses and associated EJCs in the same attachment, showing that the facilitation is not associated with a detectable alteration in the configuration of the nerve terminal impulse.

6). In addition, the sensitivity of the impulse to the adrenergic neuron-blocking agents bretylium and guanethidine, which are taken up into sympathetic nerves by uptake$_1$, supports the view that it is the impulse propagating in a sympathetic nerve terminal (51).

The nerve impulses were evoked in an all-or-none manner about the stimulus threshold and retained their shape during repetitive stimulation, suggesting that they were single-unit recordings. An important observation was that there was no detectable variation in the size of the nerve impulse when EJCs occurred and when they did not, which could have accounted for the intermittence of neurotransmitter release (Fig. 6). Thus intermittence cannot be accounted for by fluctuations in the amplitude of the nerve terminal impulse or by failure of the nerve impulse to propagate in the terminals; action potentials invade sympathetic nerve terminals, faithfully, in an unvarying manner when the action potential is initiated by stimulation of the postganglionic axon (62).

Focal extracellular recording studies of both impulse propagation and neurotransmitter release have been carried out in the mouse vas deferens (64) and various arteries (65,66,52). Essentially similar results have been obtained, namely that neurotransmitter release is intermittent and that intermittence is not due to failure of impulse propagation in sympathetic nerve terminals.

Visualization of Living Varicosities Using Fluorescent Probes

The ability to place small-diameter electrodes (4 μm) next to or over single varicosities has been greatly facilitated by the visualization of strings of living varicosities (52). The dye 3-3-diethyloxardicarbocyanine iodide binds to the membranes of neuronal mitochondria. Since the nerve terminals contain a relatively high density of mitochondria, the varicosities fluoresce more intensely than the underlying smooth-muscle cells. In low concentrations, this dye does not produce noticeable changes in either spontaneous or evoked neurotransmitter release, or in the postjunctional sensitivity of the smooth muscle to the released neurotransmitter (52).

Active Or Passive Invasion of Varicose Nerve Terminals

Focal extracellular recording techniques have also shed light on the relationship of the nerve terminal action potential to neurotransmitter release, specifically on whether there is active or passive invasion of nerve terminals. One question that has received much attention is whether the entire length of a varicose nerve terminal is excitable. In order to answer this question, studies of the nerve terminal impulse in several preparations have been carried out. Normally when studying the effects of a drug, it is necessary to apply the drug to the solution bathing the whole preparation and not just to the region of recording. The great advantage of the extracellular suction-electrode technique is that drugs can be applied internally within the electrode to perfuse the region of recording only. Following internal perfusion of an extracellular recording electrode with TTX, active invasion of the varicosities within the patch is blocked but electrotonic invasion from the point of block (just outside the electrode) still occurs (Fig. 7). Interestingly, local application of TTX totally inhibits the occurrence of EJCs, showing that active invasion by a TTX-sensitive nerve terminal impulse is a fundamental requirement for evoked neurotransmitter release to occur (51). It is of some interest that the passive depolarization fails to activate sufficient calcium channels to trigger neurotransmitter release under these conditions. Perhaps TTX exerts other effects in sympathetic nerve terminals, even modifying calcium-entry directly (a very controversial statement if true!). Studies of focal depolarization of nerve terminals with pulses of different durations should help to solve this issue. Katz and Miledi (29) also concluded that active invasion of somatic motor nerve terminals is necessary for neurotransmitter release to occur. When the suction electrode is made very small (~5 μm) it is noteworthy that locally applied TTX now cannot abolish impulse propagation, as the area of membrane sodium channels blocked by TTX is now too small to prevent the active region ahead of the nerve impulse from exciting the membrane beyond the patch (Lavidis, personal communication).

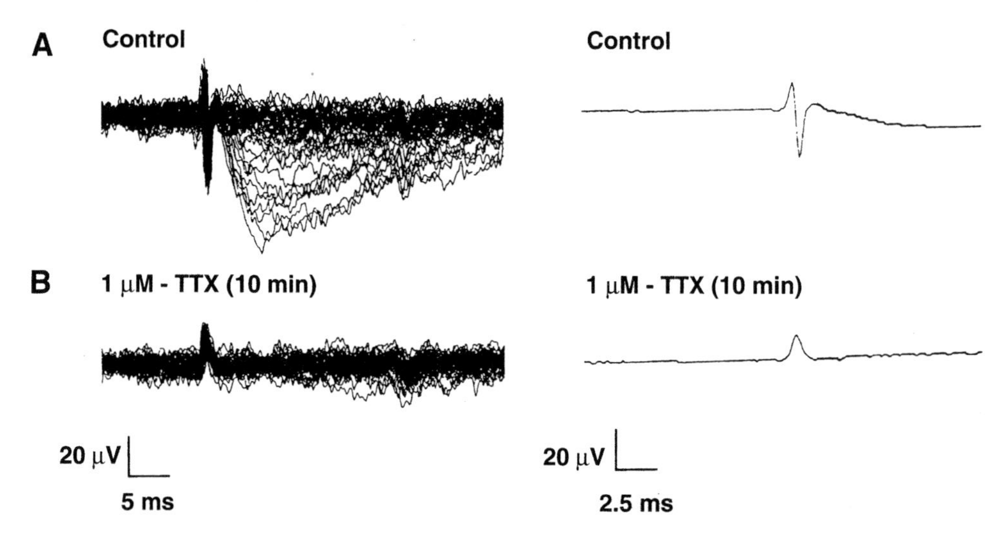

FIG. 7. Effects of local application of tetrodotoxin (TTX) inside the suction electrode on impulse propagation in sympathetic nerve terminals. **A.** Nerve impulses and excitatory junction currents (EJCs) evoked by a train of 50 stimuli at 1 Hz before and following perfusion of the suction electrode with 1 μM TTX for 10 minutes. **B.** Averages of 80 nerve impulses recorded in the same attachment before and during addition of TTX. Although the nerve impulse still electrotonically invaded the terminals, the residual depolarization was insufficient to activate the transmitter release mechanism.

N-type Calcium Channels Control Evoked Release in Sympathetic Nerve Terminals

What is the nature of the calcium channel controlling neurotransmitter release in sympathetic nerve terminals? One indirect way to address this question is to study the effects of L- and N-type calcium channel blockers. The N-type calcium channel blocker ω-conotoxin in low concentrations (0.01–0.1 μM) powerfully inhibits neurotransmitter release (Fig. 8). In contrast, nifedipine (1–100 μM) and other L-type calcium channel blockers have no effect on evoked neurotransmitter release, except at concentrations where they exert significant local anesthetic effects on nerve terminal impulses (67).

Effects of Potassium Channel Blockers

Drugs that prolong the duration of the nerve terminal action potential powerfully increase neurotransmitter release (i.e., tetra ethylammonium and 4-aminopyridine (Fig. 9; studies carried out with James Brock)). One factor setting the degree of intermittence is therefore the duration of the nerve terminal action potential. Presumably, the degree of activation and nature of the potassium conductances in sym-

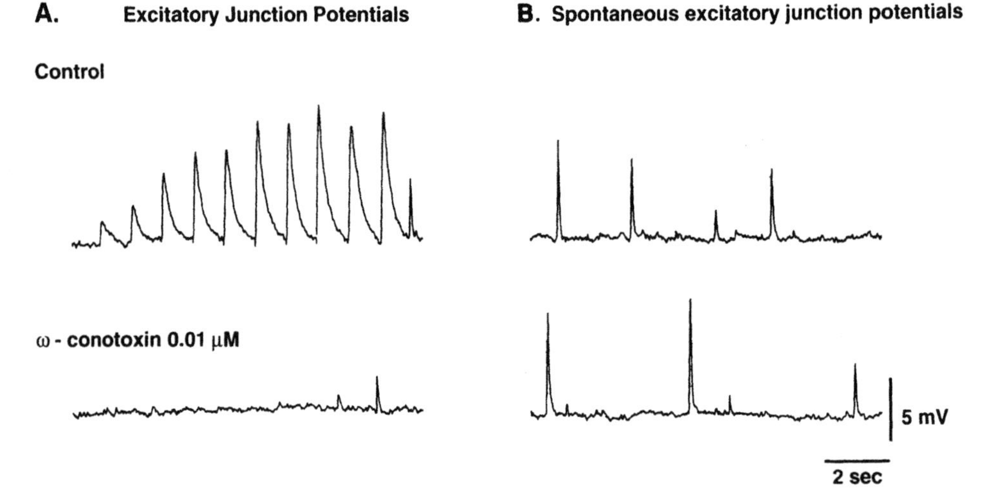

FIG. 8. Effects of ω-conotoxin GVIA (0.01 μM) on intracellularly recorded electrical activity in the guinea-pig vas deferens. **A.** Inhibition by ω-conotoxin of evoked excitatory junction potentials (EJPs) (10 pulses, 1 Hz) in the guinea-pig vas deferens. The results shown are from a maintained impalement in a single cell before (control) and 10 min after ω-conotoxin. **B.** SEJPs recorded from the same cell as in Fig. 8A, before and 10 minutes after the application of ω-conotoxin.

pathetic nerve terminals set the duration of the nerve terminal impulse and, therefore, the amount of calcium that enters per impulse.

Frequency-dependent Facilitation

A key question is whether the duration of the nerve terminal impulse varies with the frequency of nerve stimulation. In all experiments to date, facilitation has been observed as a change in the number of EJCs evoked per train of stimuli (Fig. 6). As the frequency of stimulation is increased (constant number of pulses in the train), the number of EJCs evoked per train increases (Fig. 6). The degree of facilitation can vary markedly from attachment to attachment and in some cases is rather weak. Thus it seems likely that there is a variation in the degree of facilitation produced in different regions of the same nerve terminal. In attachments where marked facilitation was observed, no detectable change in the shape or size of the nerve terminal impulse was detected ($\leq$ 4 Hz) that might have accounted for facilitation (Fig. 6). It should be remembered that focal extracellular recording techniques will tend to emphasize the high-frequency components in the nerve terminal impulse, and important slower conductances may not be detectable by this method. Nevertheless, it seems likely that the increased probability of release associated with facilitation is due to mechanisms operating at the level of depolarization-secretion coupling. The impression gained to date is that facilitation involves either the recruitment of previ-

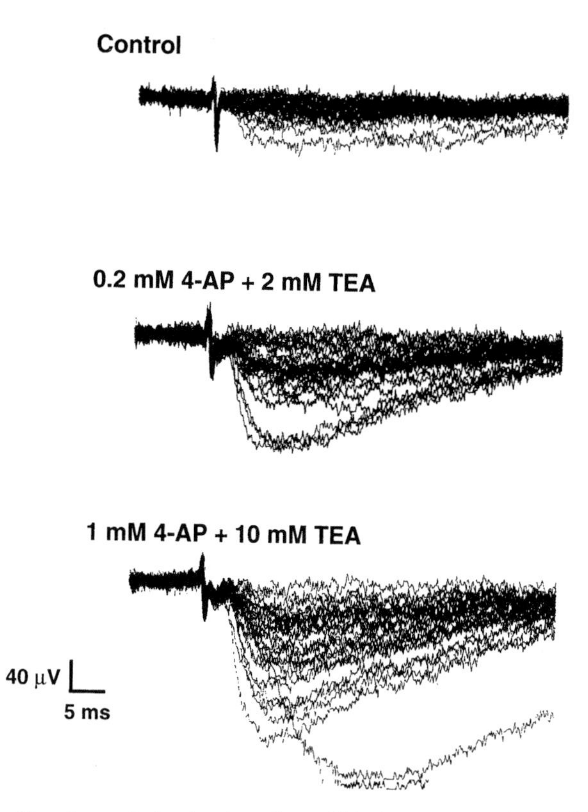

FIG. 9. Effects of locally applied potassium channel blockers on neurotransmitter release in the guinia-pig vas deferens. Tetraethylammonium and 4-aminopyridine were applied inside the recording suction electrode by means of a perfusion system. Electrical activity was evoked by trains of 25 stimuli at 1 Hz. Neurotransmitter release is greatly enhanced by increasing concentrations of potassium channel blockers. One factor setting the probability of release at a constant frequency is the degree of activation of potassium channels and therefore duration of the nerve terminal impulse.

ously silent varicosities or more release sites on the same varicosity, or perhaps a combination of these mechanisms.

Biochemical Machinery Underlying Facilitation

During trains of stimulation the amount of neurotransmitter released per action potential increases (i.e., frequency-dependent facilitation). What then are the biochemical mechanisms underlying facilitation? Much work has been carried out on the mechanisms involved in the release of neurotransmitter from sympathetic nerves, but the molecular mechanisms linking calcium entry into the nerve terminal to the subsequent fusion of vesicles with the terminal membrane remain unclear. It is likely that the frequency-dependent increase in neurotransmitter release is dependent on the levels of residual calcium in the nerve terminals between stimuli that activate as yet uncharacterized biochemical mechanisms.

The possible role of second messengers in neurotransmitter release has been aided by the identification within nervous tissue of the group of enzymes that couples receptor activation to cellular events. Of particular interest are the adenosine 3′,5′-cyclic phosphate (cAMP)-dependent protein kinases (PKA) and the calcium/phospholipid-dependent protein kinases (PKC). These have been found in many nerve cells and are concentrated in the nerve terminal regions (see Walaas and Greengard, 1991) (68).The discovery of multiple forms of both PKA and PKC with a variable neuronal distribution suggests that the biochemical machinery of neurotransmitter release may be highly neuron specific.

There is much evidence to show a role for PKC in modulating the release of noradrenaline. Recently it has been demonstrated that PKC activators and inhibitors have powerful effects on the EJP, which is an indirect measure of the release of the co-transmitter ATP (78). Thus, the phorbol ester, phorbol 12, 13-dibutyrate (PDBu), a PKC activator, increases the amplitude of the first EJP in a train and markedly alters the pattern of facilitation (69,78). Furthermore, the PKC inhibitors staurosporine and Ro-31, 8220 prevent the effects of PDBu. Perhaps the most interesting finding is that PKC inhibitors when given alone can greatly reduce or abolish facilitation at 1 Hz with little effect on release to the first stimulus in a train (31). The observations with PKC antagonists suggest that PKC plays an important role in facilitation in sympathetic nerve terminals. The analogous process in the squid giant axon is post-tetanic potentiation, where similar mechanisms operate (see chapter 10).

Prejunctional Receptors

Another factor to be taken into account when neurotransmitter release mechanisms are investigated is the degree of activation of prejunctional (presynaptic) receptors. Perhaps the most widely studied of these is the α_2-adrenoceptor, which is thought to play a major role in regulating neurotransmitter release from sympathetic nerve terminals (i.e., α-autoinhibition (70,71)). We will now consider some electrophysiological evidence that has a bearing on the role of α-autoinhibition under the conditions of in vitro experiments.

The cornerstone of autoinhibition theory is the fundamental observation that α-adrenoceptor antagonists (e.g., yohimbine) increase electrically-evoked neurotransmitter release during a train of stimuli without affecting release to the first stimulus. It is assumed that yohimbine potentiates neurotransmitter release by removal of an ongoing α-autoinhibition, but others have suggested this is the sole mechanism (72).This question can be investigated on an impulse-to-impulse basis by using intracellular and focal extracellular recording techniques.

Yohimbine increases the amplitude of all but the first few EJPs in a train evoked at 1 Hz in the guinea-pig vas deferens (Fig. 10A), but has no effect on EJPs recorded in tissues taken from animals pretreated with reserpine to deplete neuronal noradrenaline stores (Fig. 10B) (73,74,75).These results show that yohimbine in-

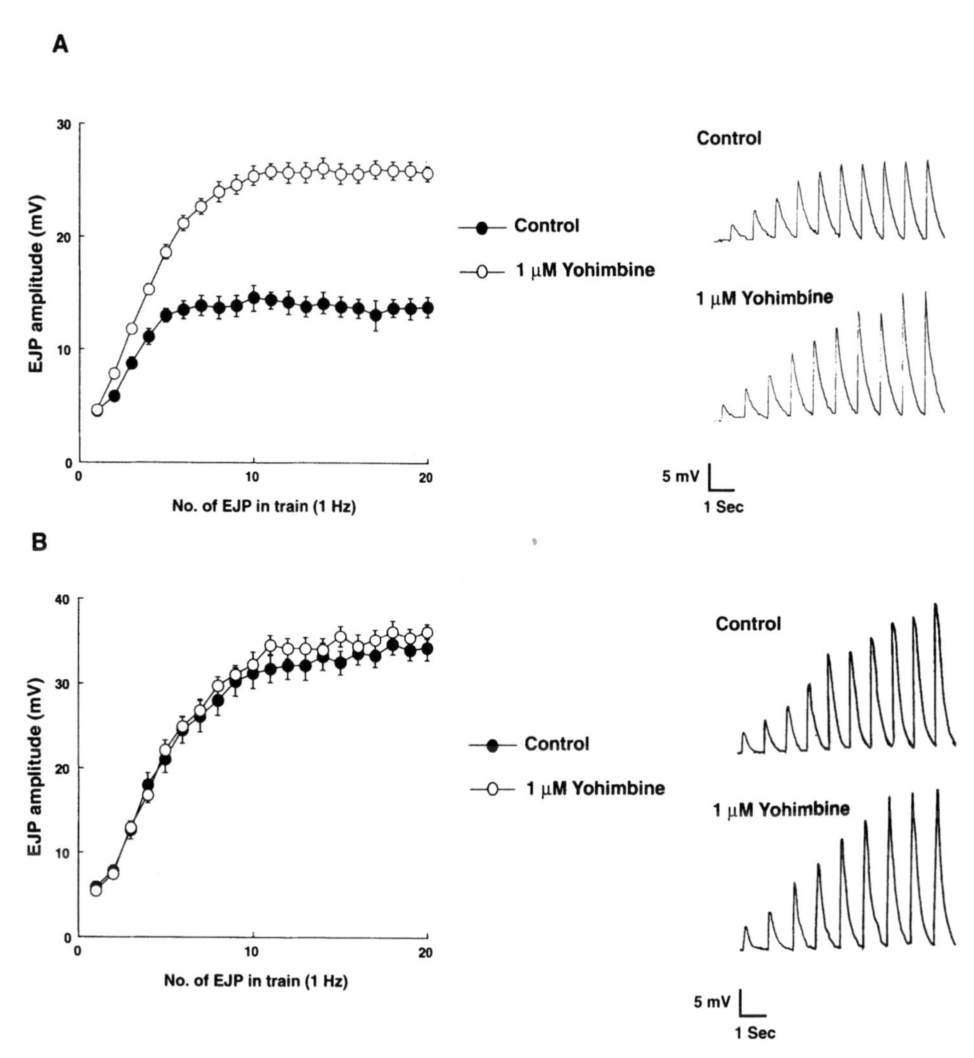

FIG. 10. Effects of yohimbine (1 μM) on excitatory junction potential (EJP) amplitude in control and reserpinized guinea-pig vas deferentia. **A.** In control tissues EJPs facilitated during the first five to eight stimuli. **B.** In reserpinized tissues, facilitation continues for 10 to 15 stimuli. Yohimbine markedly increased the amplitude of the fully facilitated EJP in control tissues but had no effect on EJPs in reserpinized tissues. Noradrenaline released by nerve action potentials therefore acts through prejunctional α-adrenoceptors to limit the magnitude of facilitation and therefore ATP release. The data represent the mean amplitude ± S.E.M. of eight control cells in the presence of yohimbine in the same tissue. Trains of 10 EJPs at 1 Hz after each procedure are shown inset in the panels (from Brock et al., ref. 74).

creases neurotransmitter release by interrupting α-autoinhibition and not by a nonselective action (e.g., K^+-channel block in nerve terminals). Further studies have shown that yohimbine has no effect on the nerve terminal impulse at concentrations that powerfully increase neurotransmitter release. These data also strongly support the view that the EJP is not generated by the release of noradrenaline. Indeed, one function of noradrenaline is to modulate the release of ATP (see next section). It is noteworthy that in reserpinized- and yohimbine-treated control tissues the development of full facilitation of EJPs requires more stimuli, suggesting that the functional role of the prejunctional α-adrenoceptors is to limit the extent of facilitation.

Co-transmission

If the neurotransmitter generating the EJP is noradrenaline it would be expected that EJPs would be greatly reduced in amplitude or abolished in tissues taken from reserpinized animals (when noradrenaline levels cannot be detected by HPLC and in addition noradrenaline synthesis has been inhibited by α-methyl para tyrosine). However, EJPs can easily be recorded under these circumstances and indeed EJPs facilitate to even greater amplitudes than in control tissues (Fig. 10). The neurotransmitter generating the EJP is unlikely to be noradrenaline. The bulk of the available evidence suggests that ATP is the transmitter responsible for the generation of EJPs in many vascular and nonvascular smooth muscles (11). Therefore, the apparent good correlation between biochemical studies of noradrenaline overflow and electrophysiological studies of neurotransmitter release (which demonstrate intermittence) should be questioned. Changes in membrane potential produced by nerve stimulation cannot be attributed to the release of noradrenaline but rather to ATP (11). We therefore know a great deal from electrophysiological studies about the characteristic features of ATP release from sympathetic nerve terminals, but surprisingly little about the basic mechanisms by which noradrenaline is released.

Intermittent Release of Endogenous Noradrenaline

The problem of endogenous noradrenaline release has recently been addressed by Stjärne and co-workers using small carbon-fiber electrodes and continuous amperommetry. Using this elegant and sensitive technique, they have been able to measure the release of endogenous noradrenaline from the sympathetic nerves innervating the rat tail artery. They have demonstrated directly that noradrenaline is released intermittently when evoked by trains of electrical stimuli at 0.1 Hz (76) (see Chap. 27).

Summary

Evidence obtained with focal extracellular electrodes has complemented and extended the data obtained from intracellular studies on neurotransmitter release and unequivocally demonstrated that, at the level of the individual varicosity, action potential-evoked neurotransmitter release occurs intermittently. Studies with this technique have further shown that intermittence is not due to failure of the nerve impulse to invade the secretory terminals. Active invasion of the terminals by a sodium-dependent nerve action potential is an obligatory requirement for neurotransmitter release to occur. Electrotonic invasion of varicosities from a point of nerve impulse blockade (focal TTX) does not activate the neurotransmitter release mechanism (Fig. 7). The action potential appears to arrive in the nerve terminal in an unvarying manner at frequencies of stimulation up to 4 Hz. The main factor that determines the amount of neurotransmitter released is the rate of arrival of action potentials in the secretory terminals, and this leads to an increase in the number of active varicosities. Thus it is possible to achieve graded responses in innervated tissues by varying the centrally determined pattern of impulse traffic arriving in the periphery. The progressive recruitment of previously silent varicosities would be a nice mechanism to produce both graded contractions of different regions of smooth muscle and integrated responses that would also be dependent on the degree of overlap of the innervation.

PARASYMPATHETIC NERVOUS SYSTEM

In comparison with the sympathetic nervous system, we know relatively little about acetylcholine release from postganglionic parasympathetic nerve terminals, and there are a number of reasons why this is so. First, acetylcholine in postganglionic parasympathetic neuroeffector junctions acts on muscarinic receptors. Muscarinic receptors are linked to G-proteins and second messenger systems, and therefore muscarinic EJPs tend to be much slower and more diffuse than the responses recorded in preparations where the neurotransmitter acts on directly-coupled ligand-gated channels. Second, it has not been possible to record spontaneous quantal responses in parasympathetic preparations with any degree of certainty. This has meant that it has been impossible to determine, using electrophysiological techniques the characteristic features of acetylcholine release. Recently, we have begun to address this issue by adopting two separate approaches. First, we have utilized the drug vesamicol, which inhibits acetylcholine uptake into synaptic vesicles. In twitch-tension studies of the guinea-pig ileum and chicken esophagus, stimulation at 0.1 Hz in the presence of (−)-vesamicol produces a rapid reduction in twitch amplitudes (75–100 stimuli). In contrast, the inactive (+)-isomer of vesamicol has no effect on electrically evoked contractions and in both cases responses to exogenously applied acetylcholine are unaffected (Fig. 11). These findings suggest that only a small pool of acetylcholine containing vesicles can be released by nerve action potentials, and that these vesicles are normally rapidly recycled and reused.

Our second approach has been to investigate acetylcholine release in a preparation previously shown to have fast muscarinic responses by the use of the sucrose gap technique (77). We decided to investigate this preparation in more detail using conventional intracellular recording techniques. Field stimulation of the preparation elicited EJPs with time courses similar to purinergic EJPs in the vas deferens of the guinea-pig. However, the EJPs had a characteristically long latency, and were unaffected by purinoceptor antagonists and abolished by low concentrations of the muscarinic receptor antagonist atropine (0.1 μM) and are therefore muscarinic in

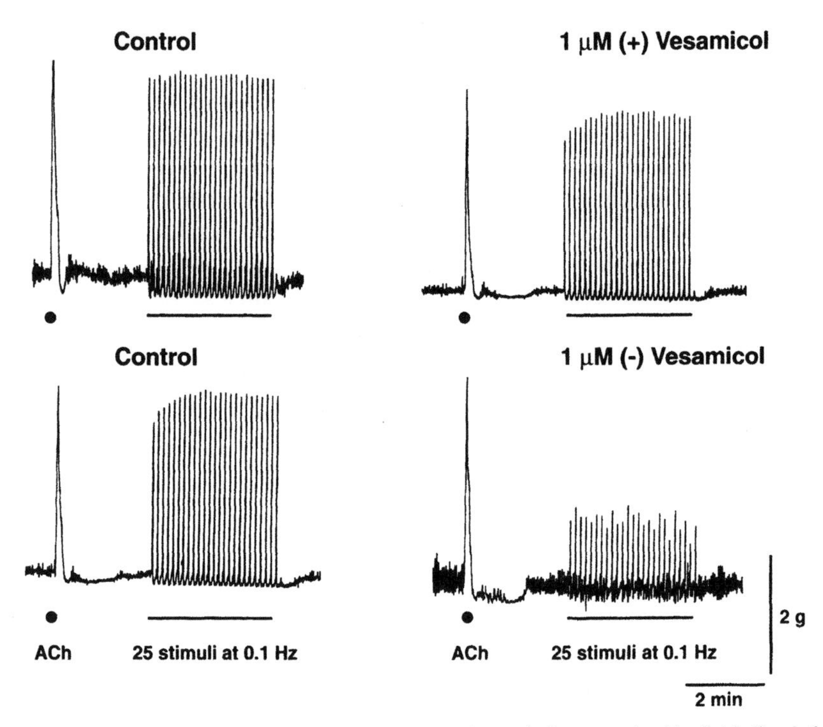

FIG. 11. Effects of vesamicol on contractions of the guinea-pig ileum evoked by field stimulation and exogenously applied acetylcholine (ACh). Intrinsic nerves were field stimulated with trains of 25 stimuli at 0.1 Hz (p.w. 0.5 mV, supramaximal voltage) delivered at 5-minute intervals. In the absence of vesamicol, contractions were well maintained. After 30 minutes exposure to 1 μM (−)vesamicol neuronally evoked contractions were markedly inhibited. In contrast, contractions elicited by exogenously applied acetylcholine or by nerve stimulation in the presence of (+)vesamicol were unaffected. Contractions were unaffected by the ganglion blocker hexamethonium (100 μM) but abolished by the muscarinic antagonist cyclopentolate (0.3 μM) (not shown). The effect of vesamicol is therefore prejunctional and stereospecific. These findings suggest that: 1) acetylcholine is released from vesicles, and 2) only a small pool of vesicles is available for rapid recycling at this muscarinic neuroeffector junction. • 1 μM acetylcholine bolus added to 20 ml bath. The bar denotes the period of electrical stimulation.

nature. The most significant finding is the presence of SEJPs, suggesting that acetylcholine release occurs in packets. However, this preparation is complex, containing intramural ganglia and other neurotransmitters apart from acetylcholine. At present it seems that three different types of SEJPs are present:

those resulting from the release of acetylcholine
those resulting from the spontaneous firing of nerves
those resulting from calcium transients in smooth muscle

The effects of TTX on EJPs and SEJPs are shown in Fig. 12.

Despite this obvious complexity, our early studies suggest that there are many similarities between neurotransmitter release in postganglionic parasympathetic and sympathetic nerves. Thus the quantal content of the EJP appears to be low, there is

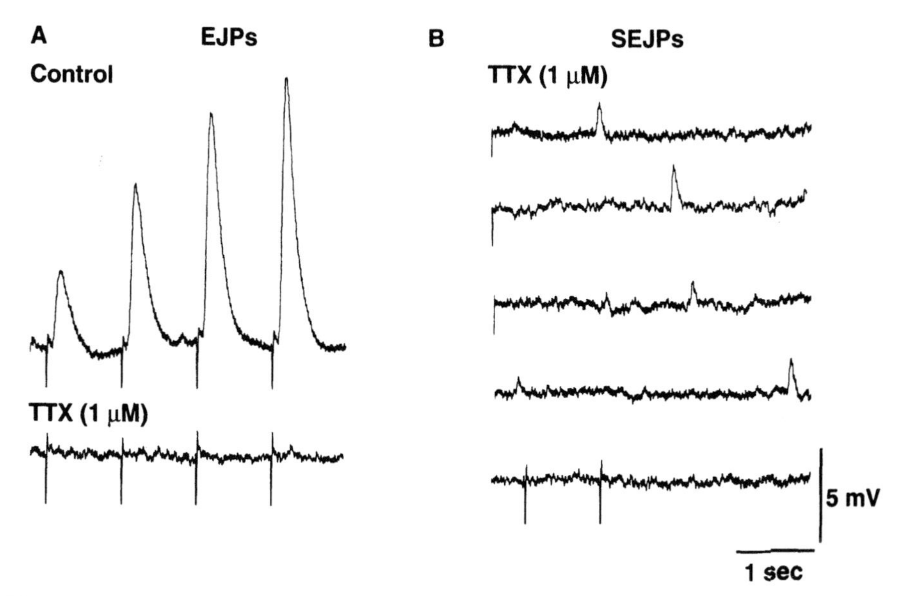

FIG. 12. Effects of tetrodotoxin (TTX) on excitatory junction potentials (EJPs) and spontaneous excitatory junction potentials (SEJPs) in the chick esophagus. **A.** EJPs evoked by trains of four stimuli at 1 Hz before and after TTX (1 μM). **B.** SEJPs recorded after the tissue had been exposed to TTX (1 μM) for 1 hour.

marked facilitation of EJPs, and protein kinase C appear to play a major role in qauntitation of acetylcholine release (30). The release of acetycholine is regulated through N-type calcium channels since low concentrations of ω-conotoxin (0.01-0.1 μM) reduce or abolish electrically evoked release. It is interesting that when the onset of the effects of ω-conotoxin (0.1 μM) are studied in a single cell (Fig. 13), EJPs fluctuate in amplitude, failures occur, and components of the EJP are revealed that are similar to individual SEJPs in the same cell.

In the future we hope to apply focal extracellular recording techniques to resolve the obvious questions regarding the relationship between the site of acetylcholine release and the postjunctional response, and the possible postjunctional interactions between different neurotransmitter release sites.

Preliminary electronmicroscopy studies of ChAT stained nerves show that some varicosities make close contact with individual smooth-muscle cells and contain vesicles (Somogyi and Cunnane, unpublished). It therefore seems probable that muscarinic synapses release packets of acetylcholine from a small pool of vesicles in a conventional manner.

Key Unresolved Issues at Sympathetic Neuroeffector Junction

While much progress has been made in our understanding of neurotransmitter release mechanisms in varicose nerve terminals, it is clear that there are many unresolved issues, including:

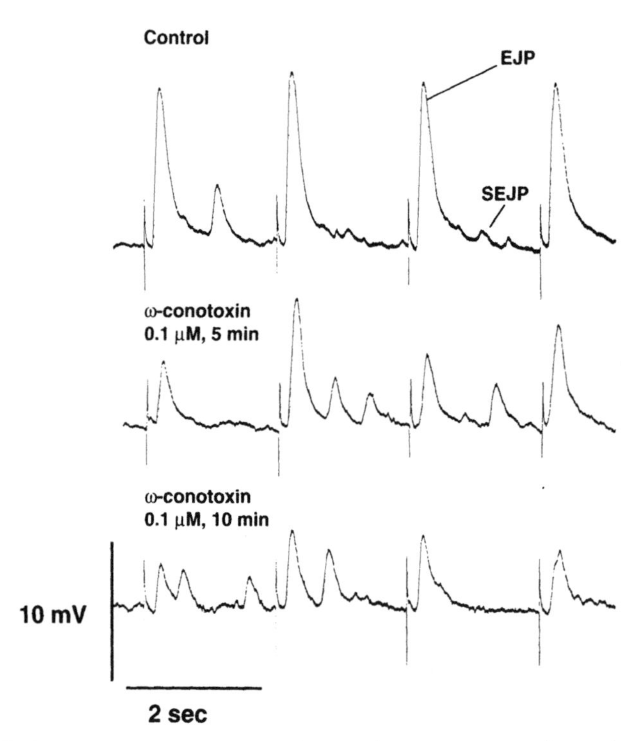

FIG. 13. Excitatory junction potentials (EJPs) and spontaneous excitatory junction potentials (SEJPs) recorded at a muscarinic neuroeffector junction. EJPs were evoked by a train of stimuli at 0.5 Hz. Upper panel, control; middle panel, 5 minutes after ω-conotoxin; lower panel, 10 minutes after ω-conotoxin. (From Searl et al., ref. 30.)

Does a varicosity secrete more than one quantum of neurotransmitter or have more than one release site?

Does a varicosity secrete several vesicles synchronously, and is this the quantum?

What are the postjunctional specializations and locations of noradrenergic/purinergic receptors?

What is the distribution and location of calcium channels on the presynaptic membranes of varicosities—are they arranged in discrete rows or distributed randomly over the entire varicosity?

What are the step-by-step biochemical mechanisms activated by a rise in intraneuronal calcium concentration that lead to exocytosis?

We look forward with excitement to the results of future experiments in many multidisciplinary fields that will tackle these problems.

REFERENCES

1. Fatt P, Katz B. Spontaneous subthreshold activity at motor nerve endings. *J Physiol* 1952;117:109–128.
2. Katz, B. *Nerve, muscle and synapse*. New York: McGraw-Hill, 1966.
3. del Castillo J, Katz B. Quantal components of the end-plate potential. *J Physiol* 1954;124:560–573.

4. Palade GE. Electron microscope observations of interneuronal and neuromuscular synapses. *Anat Rec* 1954;118:335–336.
5. Palay SL. Electron microscope study of the cytoplasm of neurons. *Anat Rec* 1954;118:336.
6. De Robertis E, Bennett HS. Some features of the submicroscopic morphology of synapses in frog and earthworm. *J Biophys Biochem Cytol* 1955;1:47–58.
7. Robertson JD. The ultrastructure of a reptilian myoneural junction. *J Biophys Biochem Cytol* 1956; 2:381.
8. De Robertis E. Ultrastructure and cytochemistry of the synaptic region. *Science* 1967;156:907–914.
9. Heuser JE, Reese TS, Dennis MJ, Jan Y, Jan L, Evans L. Synaptic vesicle exocytosis captured by quick freezing and correlated with quantal transmitter release. *J Cell Biol* 1979;81:275–300.
10. Heuser JE, Reese TS. Structural changes after transmitter release at the frog neuromuscular junction. *J Cell Biol* 1981;88:564–580.
11. Brock JA, Cunnane TC. Electrophysiology of neuroeffector transmission in smooth muscle. In: Burnstock G, Hoyle CHV, eds. *Autonomic neuroeffector mechanisms*. Chur: Harwood Academic Publishers; 1992;121–213.
12. Tauc L. Non-vesicular release of neurotransmitter. *Phys Rev* 1982;62:857–893.
13. Dunant Y, Israël M. The release of acetylcholine. *Sci Amer* 1985;252:58–66.
14. Zemkova H, Vyskocil F, Edwards C. The effects of nerve terminal activity on non-quantal release of acetylcholine at the mouse neuromuscular junction. *J Physiol* 1990;423:631–640.
15. Searl T, Prior C, Marshall IG. Acetylcholine recycling and release at rat motor nerve terminals studied using (−)-vesamicol and troxypyrrolium. *J Physiol* 1992;444:99–116.
16. Prior C, Marshall IG, Parsons SM. The pharmacology of vesamicol: an inhibitor of the vesicular acetylcholine transporter. *Gen Pharmacol* 1992;23:1017–1022.
17. Boyd IA, Martin AR. The end-plate potential in mammalian muscle. *J Physiol* 1956;132:74–91.
18. Blackman JG, Purves RD. Intracellular recordings from ganglia of the thoracic sympathetic chain of the guinea-pig. *J Physiol* 1969;203:173–198.
19. Korn H, Faber DS. Regulation and significance of the probabilistic release mechanisms at central synapses. In: Edelman GM, Gall WE, Cowan M, eds. *Synaptic Function*. New York: John Wiley & Sons; 1987:57–108.
20. Wernig A. Estimates of statistical release parameters from crayfish and frog neuromuscular junctions. *J Physiol* 1975;244:207–221.
21. Robitaille R, Adler EM, Charlton MP. Strategic location of calcium channels at transmitter release sites of frog neuromuscular synapses. *Neuron* 1990;5:773–779.
22. Bennett MR, Lavidis NA. The effect of calcium ions on the secretion of quanta evoked by an impulse at nerve terminal release sites. *J Gen Physiol* 1979;74:429–456.
23. D'Alonzo AJ, Grinnell AD. Profiles of evoked release along the length of frog motor nerve terminals. *J Physiol* 1985;359:233–258.
24. Bennett MR, Lavidis NA. Variation in quantal secretion at different release sites along developing and mature motor nerve terminal branches. *Dev Brain Res* 1982;5:1–9.
25. Bennett MR, Jones P, Lavidis NA. The probability of quantal secretion along visualised terminal branches at amphibian (*Bufo marinus*) neuromuscular synapses. *J Physiol* 1986;379:257–274.
26. Dodge FA, Rahaminiff R. Cooperative action of calcium ions in transmitter release at the neuromuscular junction. *J Physiol* 1967;193:419–432.
27. Katz B, Miledi R. The measurement of synaptic delay, and the time course of acetylcholine release at the neuromuscular junction. *Proc R Soc Lond [Biol]* 1965;161:483–495.
28. Katz B, Miledi R. The timing of calcium action during neuromuscular transmission. *J Physiol* 1967a;189:535–544.
29. Katz B, Miledi R. The release of acetylcholine from nerve endings by graded electrical pulses. *Proc R Soc Lond [Biol]* 1967b;167:23–38.
30. Searl T, Wardell CF, Smith A, Cunnane TC. Is the biochemical machinery regulating neuro-transmitter release neurone-specific? *Br J Pharmacol* 1993;108:180P.
31. Brock JA, Cunnane TC. Neurotransmitter release mechanisms at the sympathetic neuroeffector junction. *Exp Physiol* 1993;78:591–614.
32. Protti DA, Sanchez VA, Cherksey BD, Sugimori M, Llinas R, Uchitel, OD. Mammalian neuromuscular transmission blocked by funnel web toxin *Ann NY Acad Sci* 1993;681:405–407.
33. Sneddon P, Westfall DP. Pharmacological evidence that adenosine triphosphate and noradrenaline are co-transmitters in the guinea-pig vas deferens. *J Physiol* 1984;347:561–580.
34. Sneddon P, Westfall DP, Fedan JS. Cotransmitters in the motor nerves of the guinea-pig vas deferens: electrophysiological evidence. *Science* 1982;218:693–695.

35. Burnstock G. Noradrenaline and ATP as cotransmitters in sympathetic nerves. *Neurochem Int* 1990; 17:357–368.
36. Smith AD, Winkler H. Fundamental mechanisms of the release of catecholamines. In: Blaschko H, Muscholl E, eds. *Handbook of experimental pharmacology* vol. 33: Catecholamines. Berlin: Springer-Verlag; 1972:538–617.
37. Klein RL, Lagercrantz H. Noradrenergic vesicles: composition and function. In: Stjärne L, Hedqvist P, Lagercrantz H, Wennmalm Å, eds. *Chemical neurotransmission 75 Years*. London: Academic Press, 1981:69–83.
38. Potter LT. Role of intraneuronal vesicles in the synthesis, storage and release of noradrenaline. *Circ Res* 1967;21Suppl.3:13–24.
39. De Potter WP, De Schaepdryver AF, Moerman EJ, Smith AD. Evidence for the release of vesicle-proteins together with noradrenaline upon stimulation of the splenic nerve. *J Physiol* 1969;204:102–104P.
40. Smith AD, De Potter WP, Moerman EJ, De Schaepdryver AF. Release of dopamine β-hydroxylase and chromogranin A upon stimulation of the splenic nerve. *Tissue Cell* 1970;2:547–568.
41. Thureson-Klein Å. Exocytosis from large and small dense cored vesicles in noradrenergic nerve terminals. *Neuroscience* 1983;10:245–252.
42. Thureson-Klein Å, Klein RL Exocytosis from neuronal large dense-cored vesicles. *Int Rev Cytol* 1990;121:67–126.
43. Fillenz M. Transmission: noradrenaline. In: Burnstock G., Hoyle C.H.V., eds. *Autonomic neuro-effector mechanisms*. Chur: Harwood Academic Publishers; 1992:323–365.
44. Viveros OH, Arqueros L, Kirshner N. Quantal secretion from adrenal medulla: all-or-none release of storage vesicle content. *Science* 1969;165:911–913.
45. Chow RH, Ludolf von R, Neher E. Delay in vesicle fusion revealed by electrochemical monitoring of single secretory events in adrenal chromaffin cells. *Nature* 1992;356:60–63.
46. Folkow B, Häggendal J. Some aspects of the quantal release of the adrenergic transmitter. *Bayer-Symposium.*, Berlin: Springer-Verlag; 1970; vol II:91–97.
47. Bevan JA, Chesher GB, Su C. Release of adrenergic transmitter from terminal nerve plexus in artery. *Agents Actions* 1969;1:20–26.
48. Blakeley AGH, Cunnane TC. Is the vesicle the quantum of sympathetic transmission? *J Physiol* 1978;280:30–31P.
49. Blakeley AGH, Cunnane TC. The packeted release of transmitter from the sympathetic nerves of the guinea-pig vas deferens: an electrophysiological study. *J Physiol* 1979;296:85–96.
50. Brock JA, Cunnane TC. Relationship between the nerve action potential and transmitter release from sympathetic postganglionic nerve terminals. *Nature* 1987;326:605–607.
51. Brock JA, Cunnane TC. Electrical activity at the sympathetic neuroeffector junction in the guinea-pig vas deferens. *J Physiol* 1988;399:607–632.
52. Lavidis NA, Bennett MR. Probabilistic secretion of quanta from visualized sympathetic nerve varicosities in mouse vas deferens. *J Physiol* 1992;454:9–26.
53. Gabella G. Fine structure of post-ganglionic nerve fibres and autonomic neuroeffector junctions. In: Burnstock G, Hoyle CHV, ed. *Autonomic neuroeffector mechanisms*. Chur: Harwood Academic Publishers; 1992;1–31.
54. Edwards FA, Gibb AJ, Colquhoun D. ATP receptor-mediated synaptic currents in the central nervous system. *Nature* 1992;359:103–104.
55. Burnstock G, Holman ME. The transmission of excitation from autonomic nerve to smooth muscle. *J Physiol* 1961;155:115–133.
56. Burnstock G, Holman ME. Spontaneous potentials at sympathetic nerve endings in smooth muscle. *J Physiol* 1962;160:446–460.
57. Hirst GDS, Neild TO. Evidence for two populations of excitatory receptors for noradrenaline on arteriolar smooth muscle. *Nature* 1980a;283:767–768.
58. Hirst GDS, Neild TO. Some properties of spontaneous excitatory junction potentials recorded from arterioles of guinea-pigs. *J Physiol* 1980b;303:43–60.
59. Luff SE, McLachlan EM, Hirst GDS. An ultrastructural analysis of the sympathetic neuromuscular junctions on arterioles of the submucosa of the guinea pig ileum. *J Comp Neurol* 1987;257:578–594.
60. Hirst GD, Edwards FW. Sympathetic neuroeffector transmission in arteries and arterioles. *Physiol Rev* 1989;69(2):546–604.
61. Cunnane TC. The mechanism of neurotransmitter release from sympathetic nerves. *Trends Neurosci* 1984;7:248–253.

62. Cunnane TC, Stjärne L. Transmitter secretion from individual varicosities of guinea-pig and mouse vas deferens: highly intermittent and monoquantal. *Neuroscience* 1984;13:1–20.
63. Robitaille R, Tremblay JP. Non-uniform release at the frog neuromuscular junction: evidence of morphological and physiological plasticity. *Brain Res Revs* 1987;12:95–116.
64. Åstrand P, Brock JA, Cunnane TC. Time course of transmitter action at the sympathetic neuroeffector junction in vascular and non vascular smooth muscle. *J Physiol* 1988;401:657–670.
65. Åstrand P, Stjärne L. On the secretory activity of single varicosities in the sympathetic nerves innervating the rat tail artery. *J Physiol* 1989a;409:207–220.
66. Åstrand P, Stjärne L. ATP as a sympathetic co-transmitter in rat vasomotor nerves—further evidence that individual release sites respond to nerve impulses by intermittent release of single quanta. *Acta Physiol Scand* 1989b;136:355–365.
67. Beattie DT, Cunnane TC, Muir TC. Effects of calcium channel antagonists on action potential conduction and transmitter release in the guinea-pig vas deferens. *Br J Pharmacol* 1986;89:235–244.
68. Walaas SI, Greengard P. Protein phosphorylation and neuronal function. *Pharmacol Rev* 1991; 43(3):299–349.
69. Wardell CF, Cunnane TC. Involvement of protein kinase C in transmitter release and α_2-autoinhibition in guinea-pig vas deferens. *Fund Clin Pharmacol* 1991;5:425.
70. Starke K. Presynaptic α_2-adrenoceptors. *Rev Physiol Biochem Pharmacol* 1987;107:73–146.
71. Gillespie JS. Presynaptic receptors in the autonomic nervous system. In: Sezekeres L, ed. *Handbook of experimental pharmacology*. vol 54: Adrenergic Activators and Inactivators. Part I. Berlin: Springer-Verlag, 1980;352–425.
72. Kalsner S, Quillan M. A hypothesis to explain the presynaptic effects of adrenoceptor antagonists. *Br J Pharmacol* 1984;82:515–522.
73. Cunnane TC, Manchanda R. Effects of reserpine pre-treatment on neuroeffector transmission in the vas deferens. *Clin Exp Physiol Pharmacol* 1989;16:451–455.
74. Brock JA, Cunnane TC, Starke K, Wardell CF. α_2-adrenoceptor mediated autoinhibition of sympathetic transmitter release in guinea-pig vas deferens studied by intracellular and focal extracellular recording of junction potentials and currents. *Naunyn Schmiedebergs Arch Pharmacol* 1990;324: 45–52.
75. Illes P, Starke K. An electrophysiological study of presynaptic α-adrenoceptor in the vas deferens of the mouse. *Br J Pharmacol* 1983;78:365–373.
76. Msghina M, Gonon F, Stjärne L. Intermittent release of noradrenaline by single pulses and release during short trains at high frequencies from sympathetic nerves in rat tail artery. *Neuroscience* 1993; 57:4.
77. Ohashi H, Ohga A. Transmission of excitation from the parasympathetic nerve to the smooth muscle. *Nature* 1967;216:291–292.
78. Wardell CF, Cunnane TC. Biochemical machinery involved in the release of ATP from sympathetic nerve terminals. *Br J Pharmacol* 1994;111:975–977.

Molecular and Cellular Mechanisms of Neurotransmitter Release, edited by
Lennart Stjärne, Paul Greengard, Sten Grillner,
Tomas Hökfelt, and David Ottoson,
Raven Press, Ltd., New York © 1994.

27

Spatiotemporal Pattern of Quantal Release of ATP and Noradrenaline from Sympathetic Nerves: Consequences for Neuromuscular Transmission

Lennart Stjärne, Per Åstrand, Jian-Xin Bao, *François Gonon, Mussie Msghina, and Eivor Stjärne

*Department of Physiology and Pharmacology, Karolinska Institutet, S-17177 Stockholm, Sweden; and *CNRS URA 1195, University of Lyon 1, 69373 Lyon (Cedex 08), France*

In this paper, the last in a sequence on quantal release in this volume, we will address four fundamental questions: (i) what difference does it make that the transmitter is released as a small, massive packet (the "quantum") into a restricted "junctional space" where the distance between nerve and muscle is in the order of the diameter of a synaptic vesicle (1), rather than in a more diffuse fashion? (ii) given that individual active zones operate in a binary fashion, (i.e., either ignore the nerve impulse or respond by releasing a single quantum (2)), and that they differ widely in probability of monoquantal release (3), how functionally important is the spatiotemporal pattern in which individual active zones release quanta (4)? (iii) is monoquantal release from individual active zones purely probabilistic, (5) or is the probability of probabilistic release regulated (6)? and (iv) are regulation of release and the responsiveness of the effector the only means by which the efficacy of neurotransmission is controlled, or does regulation of clearance of released transmitters play a role (7)?

BASIC PROPERTIES OF THE PREPARATION: THE EXPERIMENTAL APPROACH

The answers to these questions clearly vary with the tissue. We have addressed them in rat tail artery, a purely sympathetically innervated vascular model (8). A simplified diagram of this elegantly assymetrical, "mixed chemical-electrical synapse" is shown in Fig. 1A. The electrically coupled smooth-muscle cells in the media are lined on the intimal side by the endothelium (which here does not markedly influence the neurogenic contraction; Bao & Stjärne, unpublished) and on the adventitial side by a dense, two-dimensional plexus of nerve terminals. Each mm^2 of this surface is innervated by 40,000 varicosities, most of which establish a junctional relationship with a smooth-muscle cell in sites where the basal laminae of varicosity and muscle are fused, the neuromuscular separation 50–100 nm and "small dense-cored" synaptic vesicles aggregated near the occasionally visibly thickened, varicosity membrane (9). If each nerve impulse releases a single quantum from approximately 1% of the varicosities, as currently thought (10), then it explodes a small but massive packet into 400 junctional clefts per mm^2.

At least three putative transmitters are present in these sympathetic nerves: noradrenaline (NA), adenosine 5′-triphosphate (ATP) and neuropeptide Y (NPY). The electrical (11) and mechanical (12) responses of the smooth muscle are mediated

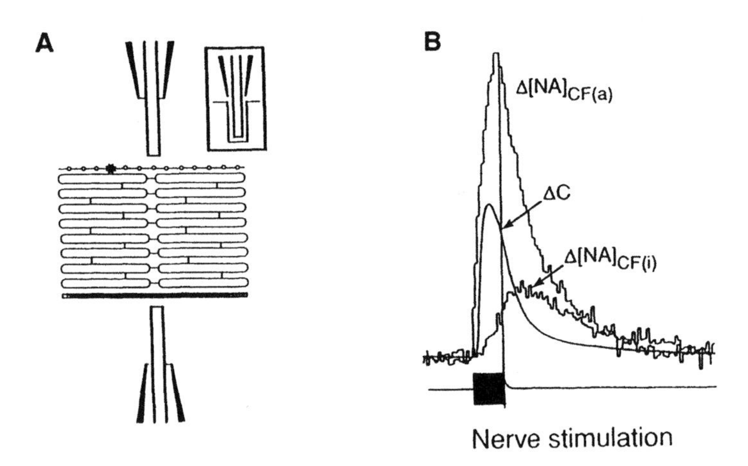

FIG. 1. A. Cartoon of a section through the wall of rat tail artery, showing the media with (in reality) 12–15 layers of electrically coupled smooth-muscle cells, the dense plexus of sympathetic nerve terminals junctionally innervating cells at the adventitial surface, the endothelium on the intimal side (8). Unfilled varicosities denote those in which the nerve impulse failed to trigger release, the star one from which a quantum was released. Also shown is the carbon-fiber (CF) electrode (active part: 8 × 25–100 μm), applied with slight pressure to the adventitial or intimal surfaces to monitor the concentration of released endogenous noradrenaline (NA) in the surrounding tissue pocket (see inset). **B.** The NA oxidation current recorded by differential pulse amperometry (DPA), with the electrode at the adventitial or intimal surface ($\Delta[NA]_{CF(a)}$, $\Delta[NA]_{CF(i)}$, respectively) and the contractile response (ΔC), evoked by electrical nerve stimulation with 400 pulses at 40 Hz in the presence of 3 μM cocaine and 30 μM corticosterone, added to block neuronal and extraneuronal uptake of released NA, respectively. The vertical line marks the end of nerve stimulation. Note that: 1) the rising phase of $\Delta NA]_{CF(a)}$ preceded that of the contractile response; 2) $\Delta NA]_{CF(a)}$ continued to rise while the tension declined somewhat, during ongoing stimulation; 3) tension had declined by 50% after the end of nerve stimulation, before $\Delta[NA]_{CF(i)}$ reached peak; and 4) $\Delta[NA]_{CF(i)}$ (the "sink") peaked later and at each point in time was smaller than $\Delta[NA]_{CF(a)}$ (the "source"). (From Stjärne et al., ref. 7, with permission.)

mainly by ATP and NA, respectively. Both occur in small as well as large dense-cored vesicles, in the large ones often together with NPY (13). The average varicosity contains some 500 small and 25 large vesicles, both with an electron-dense core (14). Both release their contents by exocytosis, the small vesicles from the active zone (15), the large ones probably from ectopic sites (16). Little is known about the release or functional role in this vessel of transmitters in large vesicles. We have therefore focused on the release and roles of NA and ATP in small vesicles.

Electrophysiological extracellular recording was used to detect the nerve impulse in the terminals, the release of quanta of ATP, and the lifetime of released ATP at the receptors (17). The NA oxidation current measured by differential pulse amperometry (DPA) during 40 milliseconds each second (18) or by continuous amperometry (19), was used to detect the net increase in the concentration of released endogenous NA at the carbon-fiber microelectrode (CF); the signal is therefore termed $\Delta[NA]_{CF}$ (Fig. 1A).

The usefulness of combining electrochemical and mechanical recording is illustrated in Fig. 1B. In this case, pharmacological tools and comparison between the kinetics of the $\Delta[NA]_{CF}$ responses at the adventitial and intimal surfaces with those of the contractile response were used to study the roles of neuronal and extraneuronal uptake of NA, and to find out if the wave of NA diffusing through the muscle wall triggers part of the neurogenic contraction (7). The blockers of neuronal uptake of NA, cocaine (3 μM), or desipramine (1 μM) enhanced the amplitude and duration both of the $\Delta[NA]_{CF}$ responses at the adventitial and intimal surfaces (7) and of the contractile response (20). Further addition of corticosterone (30 μM) to block extraneuronal uptake of NA was without effect (7). It can be seen that the rising phase of $\Delta[NA]_{CF}$ at the innervated adventitial surface preceded, but that at the intimal side followed after the mechanical response to nerve stimulation. These findings imply that: (i) extraneuronal uptake is negligible in this preparation (i.e., clearance occurs by neuronal uptake and diffusion (7)); (ii) the NA-induced component of the neurogenic contraction is likely to be mediated by α-adrenoceptors near the sites of neuronal uptake (i.e., in the plane of the nerve plexus, perhaps on directly innervated key cells (21)); and (iii) it is therefore the spread of excitation from these cells throughout the functional syncytium, not diffusion of released NA, that activates cells in deeper layers of the muscle wall (7).

RELEASE OF ATP AND NA IN RAT TAIL ARTERY

Release at Low Frequency: Monoquantal and Intermittent

When using microelectrodes with tip diameters of ≤50 μm, the pattern of extracellularly recorded electrical activity in rat tail (10) or femoral artery (22) during nerve stimulation at 0.1–1 Hz was basically similar to that observed in guinea-pig (17) or mouse (3) vas deferens. In the example in Fig. 2A,B it can be seen that the

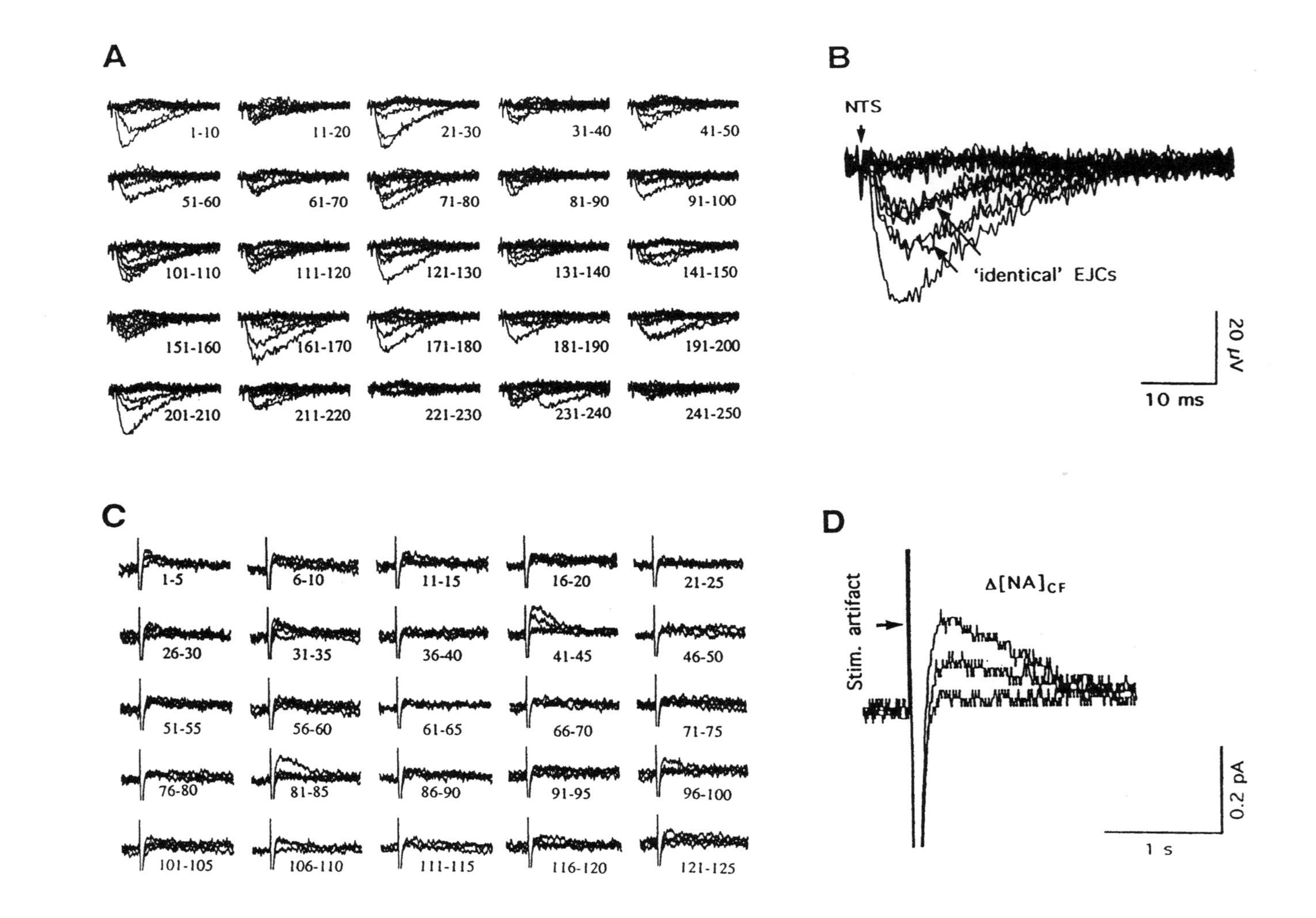
A
1-10
11-20
21-30
31-40
41-50
51-60
61-70
71-80
81-90
91-100
101-110
111-120
121-130
131-140
141-150
151-160
161-170
171-180
181-190
191-200
201-210
211-220
221-230
231-240
241-250
B
NTS
'identical' EJCs
20 μV
10 ms
C
1-5
6-10
11-15
16-20
21-25
26-30
31-35
36-40
41-45
46-50
51-55
56-60
61-65
66-70
71-75
76-80
81-85
86-90
91-95
96-100
101-105
106-110
111-115
116-120
121-125
D
Stim. artifact
Δ[NA]CF
0.2 pA
1 s

nerve terminal spike (NTS), the extracellular image of the nerve terminal action potential, was nonintermittent but frequently failed to trigger an excitatory junction current (EJC) (i.e., failed to release ATP), and that evoked EJCs fluctuated between a number of preferred amplitudes and time courses. The amplitude distribution of the evoked EJCs was closely similar to that of the spontaneous EJCs in the same patch, and individual evoked EJCs were often matched in amplitude and time course by individual spontaneous EJCs (10). Given that spontaneous EJCs are caused by single ATP quanta (17), we conclude that: (i) each evoked EJC in Fig. 2A,B reflects release of a single ATP quantum; (ii) EJCs that differ in amplitude and time course were caused by quanta of the same size but released from different sites; (iii) conversely, "identical" EJCs may represent "fingerprints" (23) of quanta released from the same site (24); and (iv) from the frequency of occurrence of prominent identical EJCs, the release probability in these "recognized" individual sites was less than 0.02 and could be as low as 0.002 (10). In other experiments, especially when stimulating the nerve fibers via a suction electrode and using a recording electrode with a larger tip, the pattern was different. The NTS was broader and often multiphasic, release failures rare, and the amplitude distribution of evoked EJCs dominated by an excess of large amplitude EJCs when compared with the spontaneous EJCs recorded in the same patch. This "low-resolution" pattern probably reflects activity in several nerve fibers; many evoked EJCs were probably caused by near-synchronous release of quanta from more than one site (22).

Continuous amperometry was employed to resolve the release of NA by single pulses (25). An example of the activity in rat tail artery recorded by this method during stimulation at 0.1 Hz (i.e., single pulses at 10-second intervals) is shown in Fig. 2C,D. The pattern was dominated by numerous apparent failures and marked amplitude fluctuation of individual evoked responses; no spontaneous NA oxidation current peaks were detected. These results provide the first direct evidence that the per pulse release of NA from sympathetic nerve varicosities may be intermittent during trains at low frequency and that NA, similarly to ATP in these nerves (10) and to catecholamines in chromaffin cells (26,27), may be released in quanta. The fact that no spontaneously released "NA quanta" were detected suggests that the

FIG. 2. A,B. Extracellular recordings in rat femoral artery (microelectrode tip diameter: 40 μm). Electrical activity evoked by nerve stimulation via a focal microelectrode with 250 pulses (100 μV, 50 μseconds) at 0.5 Hz, shown in groups of ten. Note the non-intermittency of the nerve terminal spike (NTS), the amplitude fluctuation of the excitatory junction current (EJC), and the frequent failures. **B.** Ten selected sweeps from **A** showing more clearly four failures and three amplitude classes of EJCs, with 3, 2, or 1 member(s), respectively. (10) **C.** Continuous amperometric recordings of the NA oxidation current in rat tail artery evoked by 125 pulses (100 μV, 50 μseconds) at 0.1 Hz, shown in groups of five. Note the large stimulus artifact, the occasional apparent failures, and the amplitude fluctuation of the responses to single pulses. **D.** Averages of six large and six intermediate-sized responses, and six apparent failures from **C,** to show more clearly the fluctuation of these responses. Note: 1) the similarity in intermittency and amplitude fluctuation of the EJC and the noradrenaline (NA) oxidation current, and 2) the thirtyfold difference in the duration of these responses. (From Msghina et al., ref. 25, with permission.)

evoked NA oxidation current peaks may not represent single quanta but instead, similarly to the evoked EJCs in low-resolution recordings (22), near-synchronous release of single quanta from more than one site.

Clearance of ATP and NA in Single Quanta

The duration of EJCs shows that the lifetime of released ATP at the P_{2x}-purinoceptors (28) was 50 milliseconds (Fig. 2B). Estimation of the lifetime or released NA at the carbon-fiber electrode is more problematic, as the rates of decay of the NA oxidation current measured by continuous amperometry (i.e., during maintained exposure of the electrode to +300 mV (19)) differ from those measured by DPA (i.e., intermittent exposure to +120 mV for 40 milliseconds every 500 or 1,000 milliseconds (18)). In the former method the total duration of this response to a single pulse is 1,500 milliseconds (Fig. 2B), in the latter 5,000 milliseconds (18). As amperometry is destructive (29), the milder method (DPA) may give a more correct estimate of the true lifetime of released NA at the carbon-fiber electrode, and therefore as well at the α-adrenoceptors driving the contraction. The issue remains open but it appears safe to conclude that clearance of NA in single quanta by reuptake and diffusion is from fifteen to a hundredfold slower than that of ATP by ecto-ATPase (28).

Per Pulse Release of ATP and NA During a Tetanus

Most electrophysiological studies of the release of sympathetic transmitters have been performed by intracellular recording of excitatory junction potentials (EJPs) during nerve stimulation at low frequencies. However, in one preparation, mouse vas deferens, the per pulse release of transmitter (now known to be ATP) during a high frequency train ("tetanus") has been examined by this technique (30). Two important observations were made: (i) that the per pulse release during trains at 10 Hz was first strongly facilitated, then moderately depressed, and (ii) that α-adrenoceptor blockers enhanced and prolonged the facilitation, and delayed and reduced the depression of release. We have instead used extracellular recording, in order to be able to study both the nerve impulse and the per pulse release of ATP, during a tetanus (700 pulses at 20 Hz) before and after adding the α_2-adrenoceptor blocker, yohimbine (1 μM). The experiments were performed in rat tail artery, and for comparison with the data described above, in mouse vas deferens as well (31). Examples of results in the two preparations, obtained under identical conditions, are shown in Fig. 3A,B.

In mouse vas deferens (Fig. 3A) the EJCs were first strongly facilitated during the tetanus, then declined rapidly towards the unfacilitated level; yohimbine enhanced the facilitation and delayed, but did not prevent, the depression. The results confirm the aforementioned findings by intracellular recording in this preparation (30). In rat tail artery under the same conditions (Fig. 3B), a very different picture emerged.

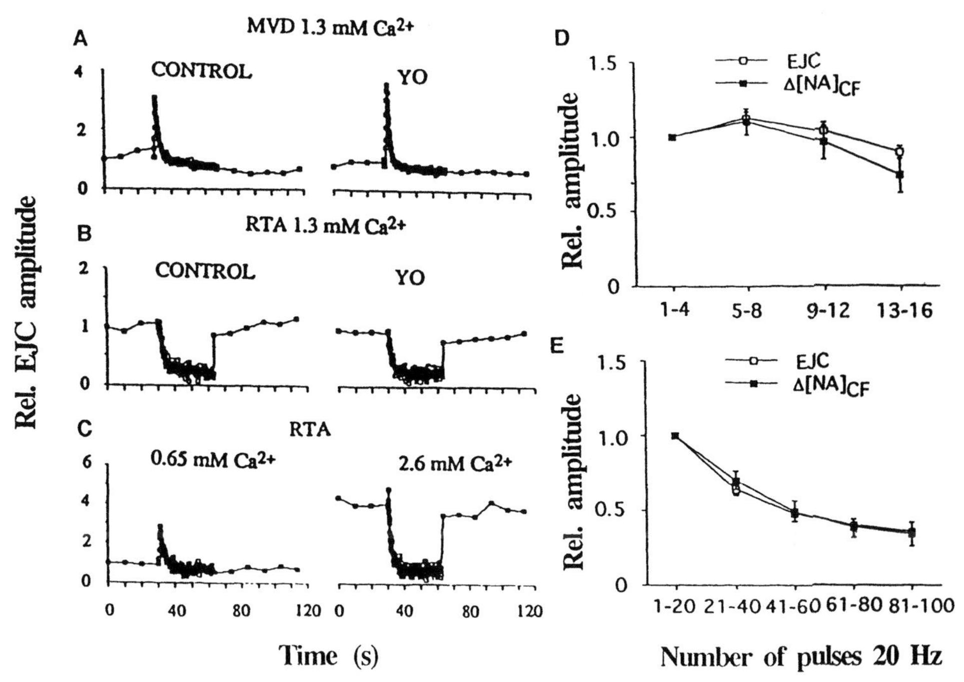

FIG. 3. A–C. Representative single experiments in mouse vas deferens (MVD) and rat tail artery (RTA), showing the amplitudes of excitatory junction currents (EJCs) first at 0.1 Hz, then during a tetanus (700 pulses at 20 Hz), then again at 0.1 Hz immediately after the tetanus in controls (**A,B**) and in the presence of yohimbine (YO), or (**C**) at low and high external Ca^{2+}. The EJCs in each experiment are normalized to the first control EJC at 0.1 Hz. **D,E.** Rat tail artery: comparison between the changes in the peak amplitudes of EJCs and the increments in the NA oxidation current measured by differential pulse amperometry (DPA) ($\Delta[NA]_{CF}$) caused by 4–16 pulses (**D**) or 20–100 pulses at 20 Hz (**E**). (From Msghina and Stjärne, ref. 31, with permission.)

The early facilitation of the EJCs during the tetanus was minimal and followed by a profound depression that lasted throughout the tetanus. The EJCs recovered rapidly (by 70% within 10 seconds) when switching back from 20 Hz to 0.1 Hz. The nerve terminal spike (NTS) changed moderately in size and shape, but these changes showed no temporal correlation with those in the EJCs. Nor did the tetanus change the current response to exogenous ATP, focally applied by pressure ejection. The depression of the EJCs during the tetanus was thus not due to conduction failure or desensitization of P_{2x}-purinoceptors, but more probably to reduced per pulse release of ATP. The fact that the presence of yohimbine did not affect the facilitation, depression, or recovery of the EJCs shows that these changes in the per pulse release of ATP were "intrinsic," not expressions of α_2-adrenoceptor-mediated control of the release machinery (31).

These conclusions are supported, indirectly, by similar findings in a range of other preparations, from neuromuscular junction in rat diaphragm to squid giant synapse (32). In these systems, the evoked current or potential response is progressively depressed during a tetanus and afterwards recovers within seconds, just as in rat tail artery. As in those preparations (33), the relative proportions of facilitation and depression of the EJCs in rat tail artery varied with the starting level of release (31). A 50% reduction in external Ca^{2+} to 0.65 mM reduced the amplitude of the first EJC and caused the pattern of subsequent EJCs (Fig. 3C) to be dominated by strong early facilitation followed by more moderate depression, a pattern similar to that in mouse vas deferens at 1.3 mM Ca^{2+} (Fig. 3A). The opposite effect was seen when, in the same experiment, an increase in external Ca^{2+} to 2.6 mM had amplified the EJCs at 0.1 Hz by fourfold. The EJCs now exhibited no facilitation during the tetanus and the depression developed even faster than at 1.3 mM Ca^{2+}. However, the time course of recovery of the EJCs after the tetanus was independent of the external Ca^{2+} level (in the range 0.65–2.6 mM) (31).

The EJC data reflect the per pulse release of ATP. The corresponding release of NA in rat tail artery during a tetanus at 1.3 mM external Ca^{2+} was assessed from the increments in the DPA response measured once per second ($\Delta[NA]_{CF}$), caused by 1, 2, 4, 8, 12, and 16 pulses, or 20, 40, 60, 80, and 100 pulses at 20 Hz. A comparison between NA release as assessed from these $\Delta[NA]_{CF}$ data and ATP release as reflected directly in the averaged EJCs caused by the corresponding stimulus trains is shown in Fig. 3D,E. Both releases were minimally facilitated during the first 16 pulses, and both subsequently declined progressively such that the release during the last 20 pulses was 25% to 30% of that during the first 20 pulses of the 100 pulse train. The release of NA thus appeared to parallel that of ATP, under these conditions (31).

Repetitive Release of Quanta from a Small Readily Releasable Pool

To clarify the implications of these results for the spatiotemporal pattern of quantal release of NA and ATP in rat tail artery, the depression of EJCs (Fig. 3B) was further analyzed by a method used to characterize the depression of acetylcholine (ACh) release during a tetanus in motor nerve terminals in rat diaphragm (34,35). According to the criteria in those studies, the EJC plot shown in Fig. 4A would show that: (i) the rapid decline in the EJC amplitudes (i.e., in the per pulse release of ATP quanta) was mainly due to depletion of a small readily releasable pool; (ii) the initial probability of monoquantal release from these sites was approximately 0.1; and (iii) the releaseable pool (35) in each initially contained, on average, 10 quanta (5) (i.e., made up ≤2% of the total pool).

As shown in other studies, the average probability of monoquantal release of ATP (17,3) and NA (36) in sympathetic nerve varicosities is approximately 0.01. The results just described thus imply that sympathetic nerves in rat tail artery utilized only 10% of their varicosities, each with an average release probability of 0.1 and a releasable pool of 10 quanta. During 20 Hz trains, the average varicosity within this subpopulation initially released two quanta per second. Its releasable pool would thus be completely depleted within 5 seconds, unless release was bal-

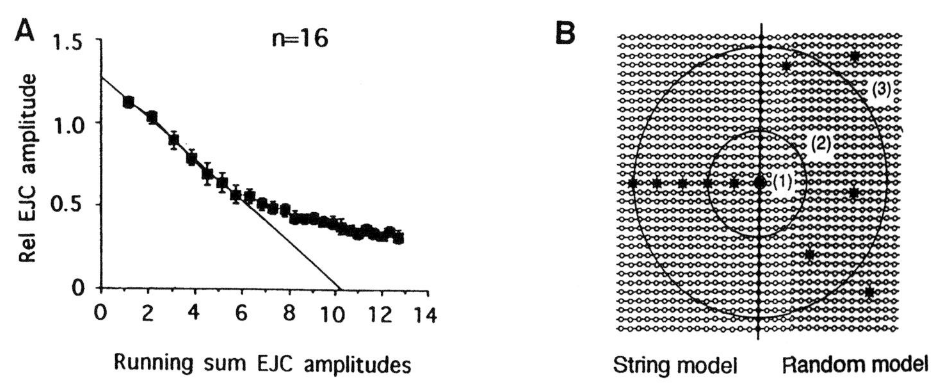

FIG. 4. A. The amplitude of each excitatory junction current (EJC) during 20 Hz trains at 1.3 mM Ca^{2+} (in 16 preparations of rat tail artery) was normalized to that of the first EJC and plotted against the sum of the amplitudes of all previous EJCs. The intercept of the regression line with the abscissa is assumed to give the initial number of releasable quanta, and the reciprocal of this value is assumed to give the initial release probability in the sites from which release had occurred. **B.** Two extreme models of the spatiotemporal pattern of monoquantal release in sympathetic nerve varicosities. The 33 varicosities in 33 strings are all potential release sites. At any given time only 1% of them respond to a nerve impulse, releasing a single quantum (average $p = 0.01$). Impulses during high-frequency bursts release quanta repeatedly from a subpopulation (≤10%) within which the average probability of the varicosities is tenfold higher ($p = 0.1$), while in that moment ≥90% of the varicosities are silent ($p = 0$). In the "random model," the distribution of active varicosities is probabilistic; in the "string model," they occur along the same terminal branches. A choice between the two models cannot be made by blind analysis with electrodes 2,3 (tip diameters 50–200 μm). It requires analysis of the quantal release of adenosine 5′-triphosphate (ATP) from visualized varicosities by electrode 1 (tip diameter 4 μm). (From Stjärne, ref. 37, with permission.)

anced by replenishment (32). Figure 4B shows two extreme spatial models (37), both of which satisfy these criteria: one in which the active varicosities are randomly distributed and one in which they occur in strings (4). A choice between these models requires focal recording of the activity of single visualized varicosities. As yet such analysis has been achieved only in mouse vas deferens (3). During nerve stimulation at 0.17 Hz (with 4 mM Ca^{2+} in the external medium), release from individual varicosities was observed to be monoquantal; the release probability was extremely nonuniform. Active varicosities (average $p = 0.1$) occurred in strings within which strong units ($p \geq 0.2$) alternated with weak ones ($p \leq 0.05$). If applicable to rat tail artery, these observations in mouse vas deferens clearly support the string model.

The string model is also supported by the effects of nerve stimulation for 40 minutes at 20 Hz on the fluorescence histochemical appearance of sympathetic nerve terminals in rat iris (38). In the absence of drugs, the morphology of the terminals was essentially unaffected by this intense nerve stimulation, indicating either that little release had occurred or else that it was balanced by increased NA synthesis. In support of the latter possibility, similar stimulation after blocking NA synthesis at the tyrosine hydroxylation step led to a substantial, spotted decline in the NA fluorescence; some branches were depleted while others were unchanged in appearance. Ten minutes after adding reserpine to block the catecholamine uptake transporter of the vesicles, similar nerve stimulation caused complete loss of fluorescence from all varicosities in all branches. In unstimulated controls, however, reserpine alone caused little or no decrease in the fluorescence within this time period (50 minutes). These results suggest that nerve impulses may eventually release NA from most or all sympathetic varicosities in the rat, and that individual branches may be activated in rotation (i.e., oscillate between an active and a silent state).

Conclusions

As shown in the previous paragraph, several lines of evidence support the string model for transmitter release from sympathetic nerves in rat tail artery. Two fundamental questions arise: (i) what mechanisms underlie this strange spatio-temporal pattern of release? (6), and (ii) how can such parsimonious release trigger and maintain an effective contraction? (7).

HOW CAN WE EXPLAIN RELEASE ACCORDING TO THE STRING MODEL?

The string model as defined in Fig. 4B requires that a regulatory factor causes terminal branches to be activated in rotation by nerve impulses (i.e., to oscillate between an active state in which the varicosities have a finite release probability ($0<p<1$) and a silent state in which the probability is zero ($p=0$)). Intermittent conduction failure in branch points, whenever occurring, clearly causes exactly this effect. In this case the regulatory factor is the nerve terminal action potential itself. But conduction failure is almost certainly not the main cause of the secretory intermittency in sympathetic nerve varicosities. This is indicated by several reports according to which the extracellularly recorded nerve terminal spike (NTS) is nonintermittent and constant in size and shape when the EJCs are highly intermittent. (10,22,17) That this is the case in rat tail artery is shown in Fig. 2A,B; note that the NTS followed by an EJC was indistinguishable from that which was not. It is important to note that the NTS represents the second time derivative of the intracellular nerve terminal action potential and hence, mainly reflects its rising phase, not its duration (17). It seems unlikely, however, that subtle variations in the duration of the action potential could cause 1% of the varicosities to release a quantum and 99% of them to ignore the nerve impulse. A more attractive alternative is that the action potential is a necessary but not sufficient condition for release (i.e., that a quantum is released only from active zones in which permissive factor(s) have made a vesicle "exocytosis-competent" (39). In the string model it is hypothesized that: (i) an unidentified mobile permissive factor "X" travels along nerve terminal branches; (ii) the presence of this factor or its local effects is required in order for the action potential to release a quantum; and (iii) the release probability is subject both to "local" and "remote" control by changes in the action potential-induced Ca^{2+} current near the active zone, and in the Ca^{2+}-dependent transport or properties of "X" respectively (40,41).

Experiments to Study Remote Control of Release

We have examined the feasibility of remote release control by extracellular recording in guinea-pig or mouse vas deferens or rat tail artery (40,41), using large internally perfused recording electrodes (internal tip diameter, 150–400 μm; Fig. 5A). Spontaneous EJCs still represent single quanta, but many individual evoked EJCs the synchronous release of quanta from many varicosities (17). The evoked EJCs were either negative- or positive-going; the former were caused by release of ATP inside, the latter by release outside the patch (42). Negative-going EJCs were not affected by addition of the P_{2x}-purinoceptor desensitizing agent (43), α,β-methylene ATP (10 μM), to the external medium but were blocked by addition of this agent inside the electrode. This agent was therefore present in the external medium throughout these experiments, both in order to block positive-going EJCs and to test for leakage across the seal. Only attachments in which EJCs were reasonably stable in spite of the presence of α,β-methylene ATP outside were accepted for further analysis. The results in the three examined tissues were basically similar (40,41); here we focus on those in rat tail artery.

Examples of the effects on the negative-going EJCs (i.e., release of ATP from sites in the patch), caused by: (i) removing Ca^{2+} or adding Ca^{2+} channel blockers in the medium inside or outside the electrode; (ii) adding the K^+ channel blocker, tetraethylammonium (TEA) outside; or (iii) adding agents activating α_2-adrenoceptors either inside or outside the patch, are shown in Figs. 5–7. Such changes in the medium inside the electrode were of course expected to strongly modulate the action potential-induced release from sites within the patch (local control). The question asked was whether the evoked release from these sites could also be modulated by similar changes in the medium outside the electrode (remote control).

Local and Remote Ca^{2+}-dependence of Transmitter Release

Removal of Ca^{2+} from the medium perfusing the electrode (i.e., changing from "normal Tyrode's solution" (1.3 mM Ca^{2+}, 0.5 mM Mg^{2+}) to "0 Ca^{2+}" (no added Ca^{2+}, 1.8 mM Mg^{2+}, 1 mM EGTA) abolished the evoked EJCs. This effect was expected, but not that 0 Ca^{2+} outside would also profoundly depress the evoked EJCs (i.e., the action potential-induced release from sites inside the patch (Fig. 5C)). The local inhibitory effect (0 Ca_i) was fully developed within 10 seconds and equally rapidly reversed when Ca^{2+} was readmitted inside; the remote effect (0 Ca_o) started developing after a lag time of 5 minutes, reached maximum after 10–15 minutes and was almost fully reversed within 15 minutes after restoring the Ca^{2+}

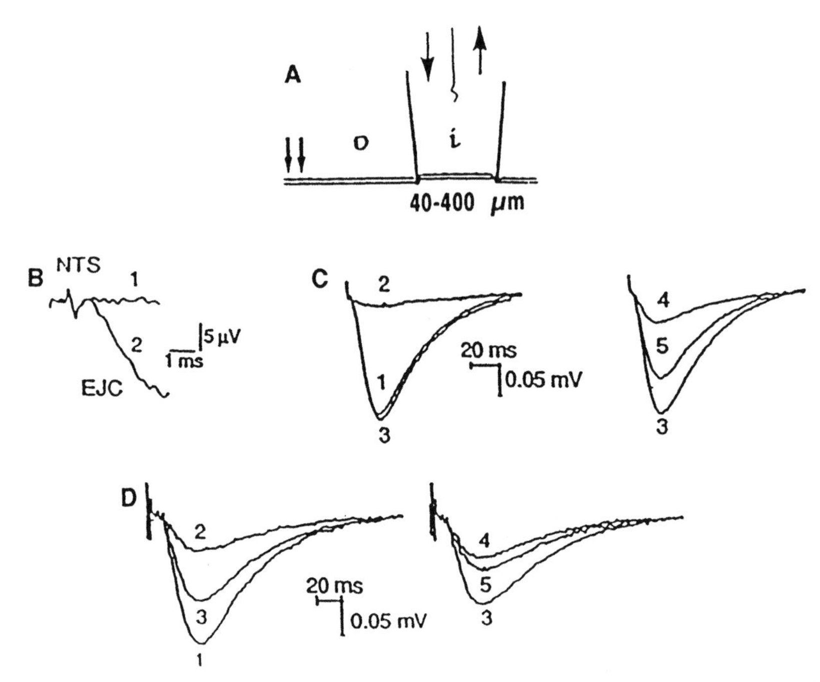

FIG. 5. A. The setup: varicose nerve terminals coursing over smooth muscle, the stimulating suction electrode (double arrow) and the internally perfused glass micropipette used as extracellular recording electrode. In **B** the tip diameter was 40 μm; in **C,D** it was 150–200 μm. In **C,D** the electrode and the bath were perfused separately; **i** denotes inside, **o** outside the electrode. **B.** Nerve stimulation at 1 Hz with a focal microelectrode. Averages of 50 sweeps in which the nerve terminal spike (NTS) was followed (*1*) by "failure" (*2*) by an excitatory junction current (EJC) (i.e., adenosine 5′-triphosphate (ATP) release. Note that the NTS was identical in the two traces. **C,D.** Nerve stimulation at 0.1 Hz. Each trace averages 10–20 consecutive sweeps. The P_{2x}-purinoceptor desensitizing agent, α,β-methylene ATP (10 μM) was present outside; addition of this agent inside the electrode had no effect on the NTS but rapidly abolished the EJCs. **C.** Effects of 0 Ca^{2+} inside or outside the electrode. (*1*) Control, Tyrode's solution (1.3 mM Ca^{2+}) inside. (*2*) 0 Ca^{2+} inside immediately (within 10 seconds) abolished the EJCs. (*3*) Equally rapid, complete reversal by 1.3 mM Ca^{2+} inside. (*4*) 0 Ca^{2+} outside also strongly depressed the EJCs; the effect was detectable after 5 minutes and maximal after 10–15 minutes. (*5*) Partial reversal by changing back to 1.3 mM Ca^{2+} outside. **D.** Effects of α_2-adrenoceptor activation inside or outside the electrode. (*1*) Control. (*2*) Addition of tyramine 100 μM inside profoundly depressed the EJCs. (*3*) Partial reversal by adding 1 μM yohimbine inside. (*4*) Further addition of oxymetheazoline 1 μM outside (with tyramine and yohimbine still present inside) also depressed the EJCs. (*5*) Partial reversal by 1 μM yohimbine outside. (From Stjärne et al., ref. 6, with permission).

concentration in the external medium. The effects of both 0 Ca_i and 0 Ca_o could be repeated several times during the same experiment (Fig. 6A). The effects of removing Ca^{2+} inside were independent of the caliber of the electrode, within the range examined and equally apparent at 23°C and 36°C (Fig. 6C,D). In contrast, the inhibition of EJCs by removing Ca^{2+} outside was inversely related to the electrode tip area and not detectable at 23°C (Figs. 6C,D; 7A).

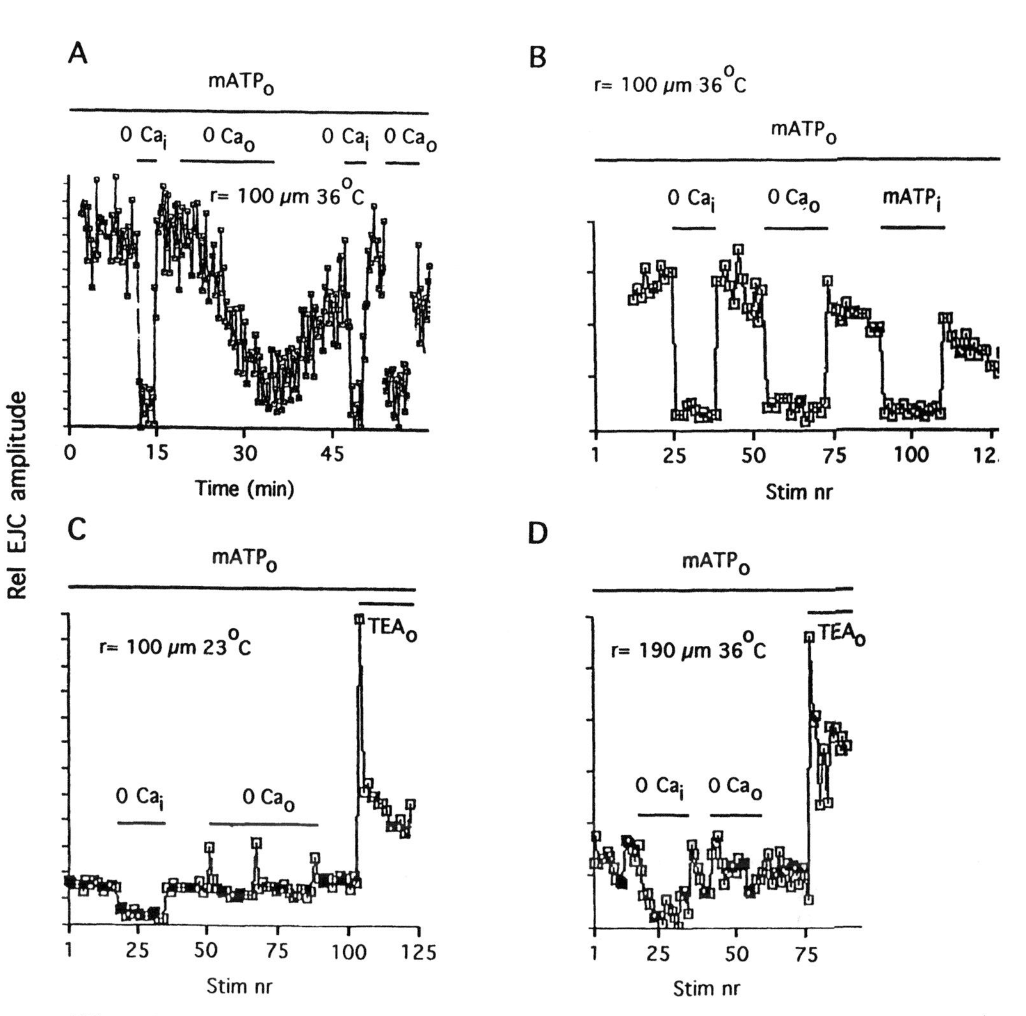

FIG. 6. Four single experiments in rat tail artery to show the effects on the peak amplitude of the excitatory junction currents (EJCs) at 0.1 Hz caused by "0 Ca^{2+}" inside or outside (0 Ca_i, 0 Ca_o), α,β-methylene adenosine 5′-triphosphate (ATP) (10 μM) outside and inside ($mATP_o$, $mATP_i$), and tetraethyl-ammonium (TEA, 20 mM) outside (TEA_o) the recording electrode (internal radius, r = 100 or 190 μm) at 36°C or 23°C, as indicated. Note that α,β-methylene ATP was present outside throughout. Except for the first 0 Ca_o in **A,** each change in the medium outside was followed by a 10-minute stimulus-free interval. **A.** Both 0 Ca_i and 0 Ca_o inhibited the EJCs reversibly; note the differences in time course. **B.** Similarly to 0 Ca_i and 0 Ca_o, $mATP_i$ reversibly blocked the EJCs. **C.** AT 23°C, 0 Ca_i rapidly inhibited the EJCs but 0 Ca_o had no effect after 10, 20, or 30 minutes. In contrast, TEA_o strongly amplified the EJCs (the rapid EJC rundown is typical for rat tail artery). **D.** AT 36°C, but with a larger pipette (radius 190 μm), the EJCs were inhibited by 0 Ca_i and enhanced by TEA_o, but 0 Ca_o had little or no effect.

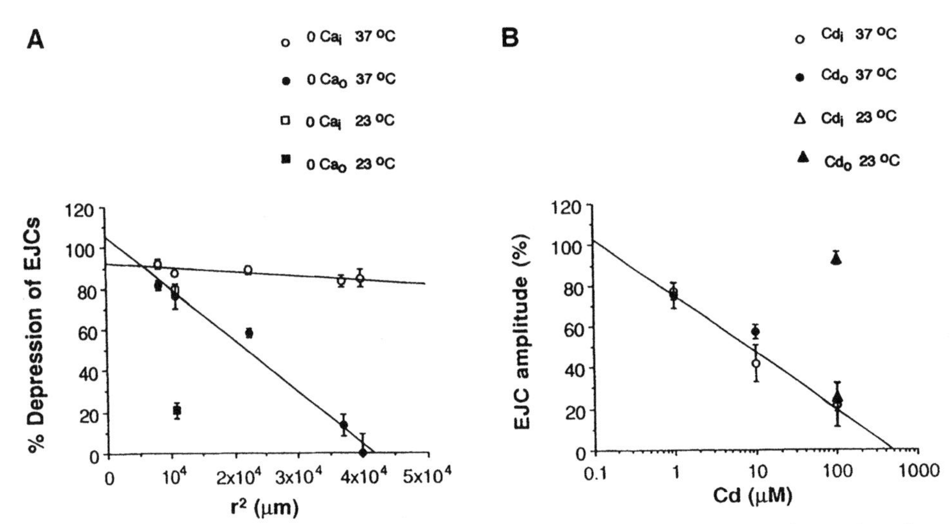

FIG. 7. Same experimental approach as in Figs. 5C,D, and 6. **A.** Percentage depression of excitatory junction currents (EJCs) by "0 Ca^{2+}" inside or outside ("0 Ca_i" or "0 Ca_o," respectively) at 37°C or 23°C, as a function of the area of the patch (given by r^2, the square of the internal radius of the tip of the recording electrode). **B.** Concentration-dependent depression of EJCs by cadmium (Cd^{2+}) inside or outside the recording electrode ("Cd_i" or "Cd_o," respectively) at 37°C or 23°C; the medium inside and outside contained the L-type Ca^{2+}-channel blocker, nifedipine (10 μM). Internal radius of electrode tip: 100 μm. The external medium contained 10 μM α,β-methylene ATP. Each point is mean ± S.E.M. (n = 4).

The EJCs were unaffected by the L-type Ca^{2+} channel blocker, nifedipine (10 μM), but as shown in earlier studies, blocked by further addition of cadmium (Cd^{2+}) (i.e., by block of N-type Ca^{2+} channels as well) inside or outside the electrode (40,41). Examples of these effects in rat tail artery are shown in Fig. 7B. The local and remote inhibitory effects of Cd^{2+} were similar in concentration-dependence (Fig. 7B) but differed greatly in time course. The local inhibition (i.e., by Cd^{2+} inside) developed rapidly and was fully reversed by washing; the remote effect (i.e., by Cd^{2+} outside) started after 5 minutes, reached maximum after 15 minutes, and was not reversed by wash for 30 minutes. The inhibitory effect of 100 μM Cd^{2+} inside was equally strong at 23°C and 36°C, while the same concentration of Cd^{2+} outside was almost without effect at 23°C.

Local and Remote K^+-Channel Dependence of Transmitter Release

Addition of TEA (≥1 mM) and/or 4-aminopyridine (4-AP, ≥0.25 mM) inside the recording electrode to block local K^+ channels caused the nerve terminal spike to acquire a late negative component (44), reflecting the broadening of the intracellular action potential, and greatly increased the peak amplitude and duration of evoked EJCs. These local effects were amplified when the drugs were added together inside the electrode at higher concentrations (20 mM TEA + 1 mM 4-AP).

The nerve terminals now fired spontaneous action potentials, in turn triggering "giant," tetrodotoxin-sensitive spontaneous EJCs, but the frequency, amplitude, and duration of the tetrodotoxin-insensitive, "true" spontaneous EJCs were not affected. Surprisingly, addition of TEA and 4-AP outside the electrode also amplified the evoked EJCs. In this case this effect was apparently exerted without prolonging the nerve terminal action potential in the patch (40,41).

Examples of the amplification of the evoked EJCs in rat tail artery by addition of TEA outside the electrode (TEA_o) are shown in Fig. 6C,D. This remote effect was slow in onset (lag time 3–5 minutes, time to peak 10–15 minutes) and, similarly to the local effect of TEA inside, maximal for the first few EJCs (41). It can be seen that the remote effect of TEA (TEA_o) was not as dependent on the caliber of the electrode, or on temperature, as that of 0 Ca^{2+} ($0Ca_o$).

Local and Remote Effects of Activation of α_2-adrenoceptors

The evoked negative-going EJCs (i.e., ATP release from sites inside the patch) could also be modulated by activation of α_2-adrenoceptors either inside or outside the patch (40). This is shown in Fig. 5D. Addition of tyramine, a sympathomimetic amine that increases the molecular leakage of NA from all regions of the terminals (24), inside the electrode powerfully, within 10–20 seconds, depressed the EJCs during trains at 0.1 Hz. This local effect of diffusely leaking endogenous NA was partially reversed by local application of the competitive antagonist, yohimbine (1 μM). Further addition of the α_2-agonist oxymethazoline (1 μM) outside the electrode (with tyramine and yohimbine still present inside) also strongly depressed the EJCs. In contrast to the local effect of endogenous NA, this remote effect of exogenous agonist began to appear after 5 minutes, was fully developed after 10–15 minutes, and was only partially reversed by further addition of yohimbine to the external medium.

Properties of Remote Release Control

The experiments just described show that the action potential-induced release from sites inside the patch could be manipulated by several specific changes in the external medium. In the case of the effects of 0 Ca^{2+} outside, they show that the remote modulation of release declined exponentially with the square of the internal radius of the electrode tip, (i.e., the area of the patch (Fig. 7B)), suggesting that release from varicosities near the rim of the electrode was preferentially inhibited. The possibility arises that "peripheral" varicosities were more strongly inhibited because they were in fact exposed to bath medium, even though "central" varicosities were not. The following findings argue against that possibility (40): (i) the electrode was internally perfused; anything leaking in would be washed away; (ii) α,β-methylene ATP outside did not block the EJCs, and the presence of Ca^{2+} outside did not sustain the EJCs when the medium inside was Ca^{2+}-free;

α,β-methylene ATP and Ca^{2+} did thus not leak in; (iii) the block of EJCs by α,β-methylene ATP inside, or by 0 Ca^{2+} inside or outside, could be reversed by wash and repeated several times during the same experiment; and (iv) in contrast to all local effects, which were rapid, independent of patch size or temperature, and fully reversible, all remote effects were slow in onset, in many cases limited by patch size and temperature, and often not reversible by wash. The remote effects may thus be biological rather than artifactual and in fact represent the longitudinal interaction between active zones presupposed by the string model. The properties of remote control just described would then provide important information about the mechanisms that control the release probability in sympathetic nerve varicosities, at least in rat tail artery.

Dual Ca^{2+}-dependence

The results suggest that the Ca^{2+}-dependence of nerve impulse-induced transmitter release in rat tail artery may be more complex than imagined, based on analogy with release in nerve terminals in the skeletal neuromuscular endplate (45) or squid giant synapse. (33,46) In rat tail artery as well as in guinea-pig or mouse vas deferens, (40,41) Ca^{2+} apparently has to be present and Cd^{2+}-sensitive, nifedipine-resistant N-type Ca^{2+} channels must be intact both at the release site itself and at regulatory sites some distance (≤400 μm) away in order for the action potential to release a quantum. The "local" Ca^{2+} function triggers exocytosis whenever a releasable vesicle is present; it is fast and independent of patch size or temperature. The remote Ca^{2+} function may be involved in the control of the release probability. It is slow, restricted by patch size, and temperature-sensitive, suggesting that it may operate optimally over short distances, perhaps from one varicosity to the next along the same string, and may require active transport of a mediator (40,41).

Dual K^+-channel Dependence

In many systems, e.g., motor nerve terminals in skeletal neuromuscular junction, (47) sensory terminals in *Aplysia* (48) and presynaptic terminals in squid giant synapse (46,49), K^+ channels and cellular signals or drugs that regulate K^+ channel properties control the duration of the presynaptic action potential and thereby Ca^{2+} influx at the active zones and transmitter release. The K^+ channel blockers such as TEA and aminopyridines, which cause single nerve impulses to exocytotically release the contents of virtually every vesicle docked at active zones, do not change the electron-microscopic appearance of nerve terminals under resting conditions. (50,51) The drugs thus disrupt the mechanisms that normally cause active zones to behave in a binary fashion (i.e., to respond to the nerve impulse by releasing the contents of only one docked vesicle (36). The dramatic increase in the nerve impulse-induced release of transmitter from sympathetic nerves caused by TEA and 4-aminopyridine is also generally ascribed to enhanced local Ca^{2+} influx at active zones, secondary to block of local K^+ channel and broadening of the local nerve terminal action potential (52,53).

The results in Fig. 6C,D show, however, that TEA dramatically increased transmitter release in rat tail artery under conditions when the drug apparently was not in physical contact with the release sites. Similar results have been obtained with both TEA and 4-aminopyridine in guinea-pig and mouse vas deferens (40,41). These findings indicate that the binary release properties of active zones (2) are critically dependent on the operation of K^+ channels at regulatory sites some distance away from the release sites. In contrast to the remote N-type Ca^{2+}-channel-dependent function, functions involving K^+ channel activity are not markedly dependent on temperature or electrode size. The two control functions may thus operate by different mechanisms. Functions involving K^+ channels appear to be effective over larger distances but less dependent on active transport. How the remote K^+ channel-dependent effect induced outside the patch spreads to active zones inside remains to be determined. Possible clues are two reported unusual effects of 4-aminopyridine: (i) it increases the phosphorylation of guanosine 5′-triphosphate-ase (GTPase)-activating protein (GAP) 43, a protein present in nerve (54) and non-myelinating Schwann cells (55), and (ii) it enhances the number of large intramembrane particles in nerve terminals that possibly represent ionic (e.g., Ca^{2+}) channels (56). Whether these effects are involved in the remote control of release from active zones is unclear, but they draw attention to the possibility that K^+ channel blockers may amplify action potential-evoked transmitter release in more than one way.

Prejunctional Autoregulation of Release

That activation of prejunctional α_2-adrenoceptors by endogenous or exogenous agonists inhibits nerve impulse-induced transmitter release from sympathetic terminals is generally accepted (36,57). The fact that exogenous α_2-agonists depress transmitter release from synaptosomes evoked by depolarization (e.g., with high K^+ or veratridine) shows that varicosities possess the receptors and the machinery for agonist-mediated inhibition of release (36). It is commonly assumed, therefore, that NA in released quanta autoinhibits the releasing varicosity via its own autoreceptors. The general validity of this assumption is questionable, however (24,31); it does not fit with the finding in Fig. 3B that the competitive α_2-adrenoceptor antagonist, yohimbine, amplified the EJCs in mouse vas deferens but not in rat tail artery, during 20 Hz trains (31). The reason is not that varicosities in rat tail artery lack the machinery for α_2-adrenoceptor-mediated control of ATP release. At low frequency (0.1–2 Hz) the EJCs in this vessel are dose-dependently depressed by the exogenous α_2-agonists clonidine or xylazine; this effect is prevented or reversed by the competitive α_2-adrenoceptor antagonists, yohimbine or idazoxan (58). When present alone, yohimbine (1 μM) or idazoxan (1 μM) have no effect on EJCs at 0.1 Hz but amplify by 30% to 40% the EJCs during 100 pulses at 2 Hz. The reason why yohimbine did not amplify the EJCs in rat tail artery during 20 Hz trains may thus be that these varicosities are immune to direct autoinhibition by NA in quanta they release during a tetanus. (6) Such local immunity is supported by analysis of "identical" discrete events, the dV/dt of EJPs, (23) or correlogram analysis of EJCs (24) in guinea-pig vas deferens. The hypothesis has therefore been proposed that quantally released NA may not inhibit the releasing varicosity but nearby varicosities (i.e., exert "lateral inhibition") (24,59,60).

This does not imply that active varicosities are immune to inhibition by endogenous NA, however. Both in guinea-pig vas deferens (24) and rat tail artery (Fig. 5D), addition of tyramine inside the electrode inhibits the EJCs by a yohimbine-sensitive mechanism. Tyramine does not inhibit the EJCs when the NA stores are depleted by pretreatment with reserpine (24); the effect is thus secondary to tyramine-induced diffuse molecular leakage of NA from all regions of the terminals in the patch. The reason why the release machinery of active varicosities is inhibited by diffusely leaking endogenous NA but immune to NA in their own released quanta is not known. The experiments just described address the possibility that active varicosities may be sensitive to a spreading effect of activation of α_2-adrenoceptors on nearby varicosities (6). The finding that application of α_2-agonist outside the electrode inhibited release from sites in the patch even though the medium in the recording electrode contained yohimbine (Fig. 5D) suggests that such messenger-mediated remote control of release is feasible.

Conclusions

We hypothesize (Fig. 8) that the concerted actions of several as yet poorly understood local and remote control mechanisms, possibly involving Ca^{2+}-dependent intraaxonal transport (61,62) of messengers (40,41) as well as effects mediated by Schwann cells (63,64), may explain why transmitter release from sympathetic nerve terminals in rat tail artery obeys the string model (6).

HOW DOES THIS RELEASE MACHINERY TRIGGER AND SUSTAIN A NEUROGENIC CONTRACTION?

This is a central question; the mechanical response is what short-term neuromuscular transmission is all about. Briefly, the answer seems to be that release according to the string model yields an effective motor control of the smooth muscle in rat tail artery because of at least four factors: (i) co-transmission; (ii) the kinetics of postreceptor mechanisms; (iii) nerve activity-dependent adaptation of clearance of released transmitter; and (iv) the spatiotemporal pattern of quantal release.

Cotransmission

One feature that characterizes the neurogenic contraction of rat tail artery is (see last section) that it is driven by at least two transmitters, which together address at least three classes of receptors (12). The contraction is triggered by ATP and NA, probably released mainly from small dense-cored vesicles at the preferred release site (active zone) of the varicosities (23). The effects of ATP are mediated via ionotropic P_{2x}-purinoceptors and those of NA via G-protein-coupled, metabotropic α_1- and α_2-adrenoceptors (66). Morphological criteria for the distribution of these receptors in the different layers of the muscle wall are lacking, but functional data suggest that they are all located on innervated smooth-muscle cells at the adventitial surface (7). The P_{2x}-purinoceptors driving this response are probably intrajunctional, judging from the fast kinetics of the EJCs. Direct evidence concerning the intra- and extrajunctional distribution of the NA receptors in blood vessels has been reported only for guinea-pig arterioles. The NA receptors were found to belong to two pharmacologically and regionally distinct categories. Specialized "γ-adrenoceptors" were clustered near the varicosities while normal α-adrenoceptors were evenly distributed over the smooth-muscle surface between the nerve terminals (67). The principle may apply to rat tail artery as well, but here the specialized intrajunctional NA receptors behave pharmacologically as α_2-adrenoceptors and the large majority of those located extrajunctionally as α_1-adrenoceptors (7). Responses mediated by these receptors interact cooperatively in a complex fashion: P_{2x}-purinoceptor- and α-adrenoceptor-effects interact negatively, α_1- and α_2-adrenoceptor-effects positively (20). The relative roles of the three components of the neurogenic contraction mediated by these receptors vary with the stimulus parameters (68,7). The ATP-induced contraction is normally small, fast, and transient; its role is maximal for the response to a single pulse and declines with frequency and train length (68). The NA-induced contractions are slower in onset, more prolonged and mostly much larger. The α_2-component, which has the longest lag time and duration, dominates the response to a single pulse and contributes substantially to responses at low frequency and to short or very long trains at high frequency. The reverse is true for the α_1-component (68,20). A concrete example of the kinetics of the three components of the contractile response is shown in Fig. 9A.

Postreceptor Kinetics

A second feature that characterizes the neurogenic contraction of rat tail artery is that its ATP- and NA-driven components differ in postreceptor kinetics. This point is illustrated in Fig. 9B. Note that the three components of the contractile response to a single pulse peaked much later than the concentration of released ATP and NA at the receptors. The time courses of the contractions are thus mainly determined by the kinetics of the respective postreceptor mechanisms (68). The ATP-driven component is identical in lag time (200 milliseconds) and time course to the non-neurogenic contraction caused by a single pulse or short trains at 20 Hz. Both are blocked by nifedipine and, hence, triggered by activation of voltage-gated L-type Ca^{2+} channels (68). The NA-induced, α_1- and α_2-adrenoceptor-mediated contractions are much more delayed in onset (lag times of 1 or 3 seconds, respectively) and pro-

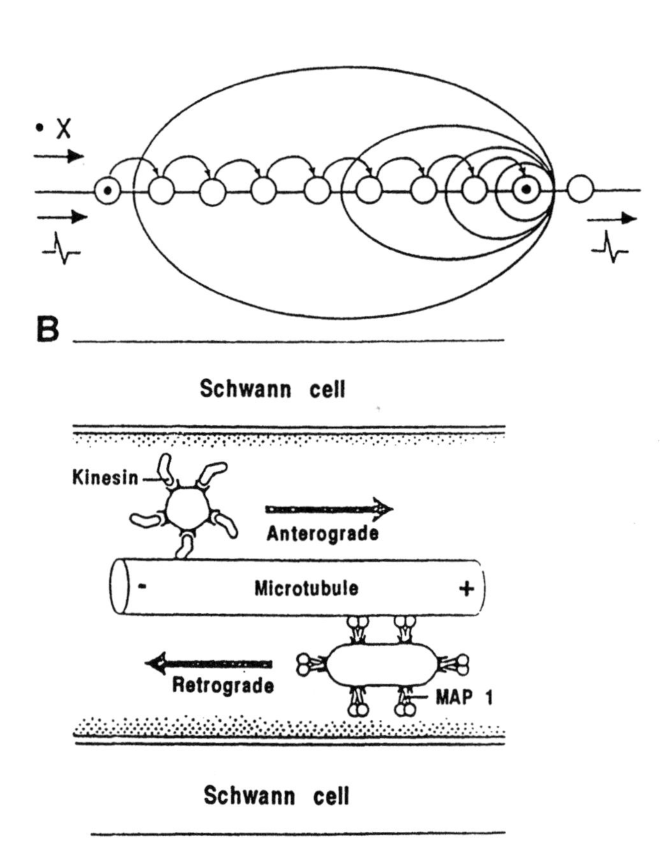

longed, due in part to the slow kinetics of the postreceptor mechanisms (68). A dominant feature of the ATP-driven contraction driven is "speed" (i.e., rapid onset and fast decline). In contrast, the contractions driven by NA are characterized by "inertia" (i.e, their ability to cause the muscle to continue to contract as the ATP response is subsiding and return to baseline tension 20–30 seconds after the exocytotic release (66).

Nerve Activity-dependent Adaptation of Clearance of Released Transmitter

A third feature that characterizes the neurogenic contraction in rat tail artery is that its ATP-component depends mainly on the rate of quantal ATP release but its NA components mainly on the rate of NA removal. The lifetime at the receptors of

NA in single quanta is 1,500–5,000 milliseconds (18,19) (i.e., fifteen- to a hundredfold longer than that of co-released ATP (see Fig. 9B)). Even more important, however, is that ongoing nerve activity appears to actually depress the rate of NA clearance, both by neuronal reuptake (20) and diffusion (7).

Activity-Dependent Plasticity of Neuronal Reuptake

The examples in Fig. 10 show the subtlety of this mechanism. The barely detectable contractile response to a single pulse was mediated mainly by α_2-adrenoceptors. (20) The contractile responses to 20 Hz trains increased steeply in amplitude with the train length and became increasingly mediated by α_1-adrenoceptors (20). The sigmoid shape of the curve suggests positive cooperativity between quanta released by successive pulses, the increasing α_1-component shows that the growth of the contraction was due to recruitment of previously "silent" receptors. Block of NA reuptake by cocaine amplified the single pulse contraction by >20-fold and caused it to possess an α_1-adrenoceptor-mediated component (20). The effect of cocaine declined steeply with the train length; responses to trains of 15–100 pulses were amplified by ≤2-fold (Fig. 10A,B).

The significance of these observations was clarified by amperometric recording of the concomitant increase in the NA concentration at the arterial surface. The recorded signal differed from the contractile response in two respects. In controls,

FIG. 8. A. Hypothesis, cooperative interaction of two factors that travel along sympathetic nerve terminals, the nerve impulse and one or several unknown permissive factors(s) "X" (X_1, X_2, etc.), is required for the action potential in a varicosity to release a quantum. X_1 is a particle-bound permissive factor that travels along the terminals by Ca^{2+}- and α_2-agonist-sensitive, temperature-dependent fast axonal transport (40). Manipulation of X_1 modulates the release probability in nearby varicosities; X_1 may cause each varicosity to be regulatory for the release probability in the next one along the string. X_2 is controlled by K^+ channel activity some distance away from the release site and travels by a temperature-independent mechanism; its functional role is to ensure that release from the active zone is monoquantal (2,41). **B.** Facts to keep in mind when analyzing the regulation of release in sympathetic nerve terminals: 1. Fast and slow transport of material from the soma to the terminals and in the opposite direction is known to play vital roles in the longterm function of nerve terminals. We hypothesize that Ca^{2+}- and temperature-dependent fast axonal transport (61,62) may in addition influence their short-term functions by carrying a regulatory factor (X_1) to the terminals where its presence in the string is a prerequisite for the varicosities to have a finite release probability ($0<p<1$) (i.e., to enable the action potential to trigger monoquantal release), and that the transport of X_1 may be one target of remote control of the release probability in the varicosities. 2. These varicose strings are wrapped in a Schwann sheath except for bare areas at active zones (9). Schwann cells have receptors and reuptake transporters for neurotransmitters, ionic channels, the ability to control the composition (e.g., the Ca^{2+} concentration) of the fluid in the narrow periterminal space, to secrete a variety of regulatory substances and as a result, influence transmitter release as well as clearance in many systems (63,64). We propose that the remote temperature-independent release-enhancing effect of K^+ channel blockers may be propagated as X_2, possibly a Ca^{2+} wave that spreads through the Schwann cell syncytium. (From Okabe and Hirokawa, ref. 65., with permission.)

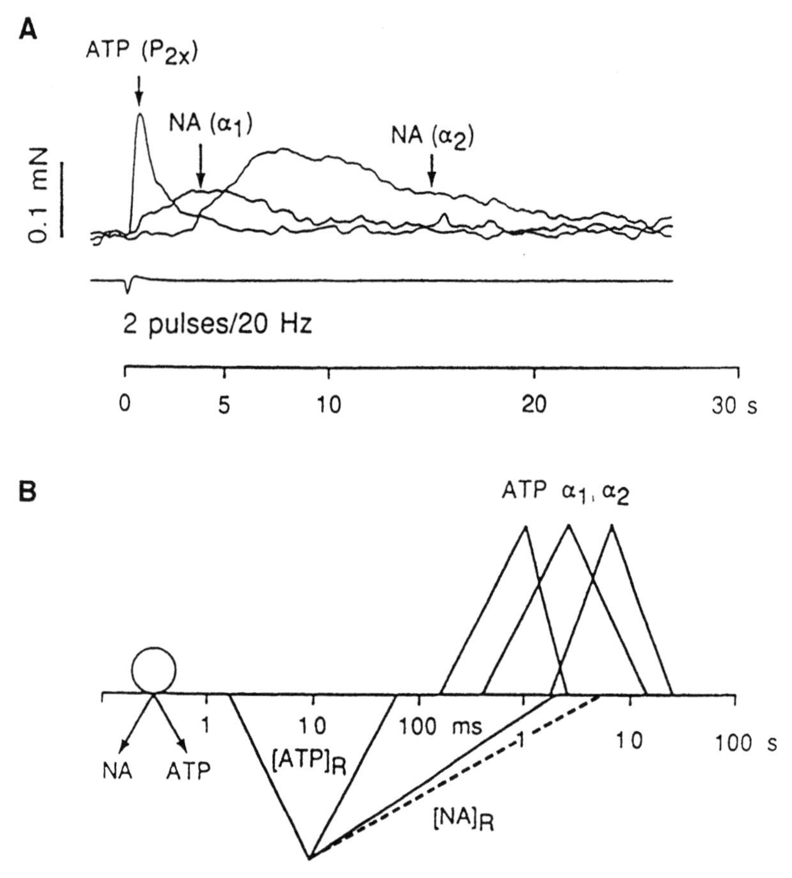

FIG. 9. A. The three pharmacologically isolated components of the contractile response of rat tail artery to 2 pulses at 20 Hz (same kinetics as of the single pulse contraction), mediated by adenosine 5′-triphosphate (ATP) via P_{2x}-purinoceptors and noradrenaline (NA) via α_1- and α_2-adrenoceptors. **B.** Time course (note the log scale) of exocytosis of a quantum, the rise and fall in the concentration of released ATP at P_{2x}-purinoceptors ($[ATP]_R$) and of NA at α_1- and α_2-adrenoceptors ($[NA]_R$); the full line is determined by continuous amperometry, the broken line by differential pulse amperometry (DPA) 40 ms/s, and the three components of the contraction. Note that the contractions occur essentially after the transmitters are eliminated, and how elegantly ATP and NA cooperate to rapidly initiate (within 200 milliseconds) and maintain (for up to half a minute) the contractile response to a single quantum released from 1% of the varicosities. (From Bao et al., ref. 20 and Bao, ref. 69, with permission.)

$\Delta[NA]_{CF}$ increased linearly with the train length, suggesting pure additivity (i.e., that successive pulses added equal amounts of NA to the electrode (18,7)). Furthermore, cocaine amplified $\Delta[NA]_{CF}$ by ≤2-fold, independently of train length (Fig. 10C,D).

The neuronal reuptake transporters that "normally" removed ≥95% of NA in single quanta from the α_2-adrenoceptors mediating the small single-pulse contrac-

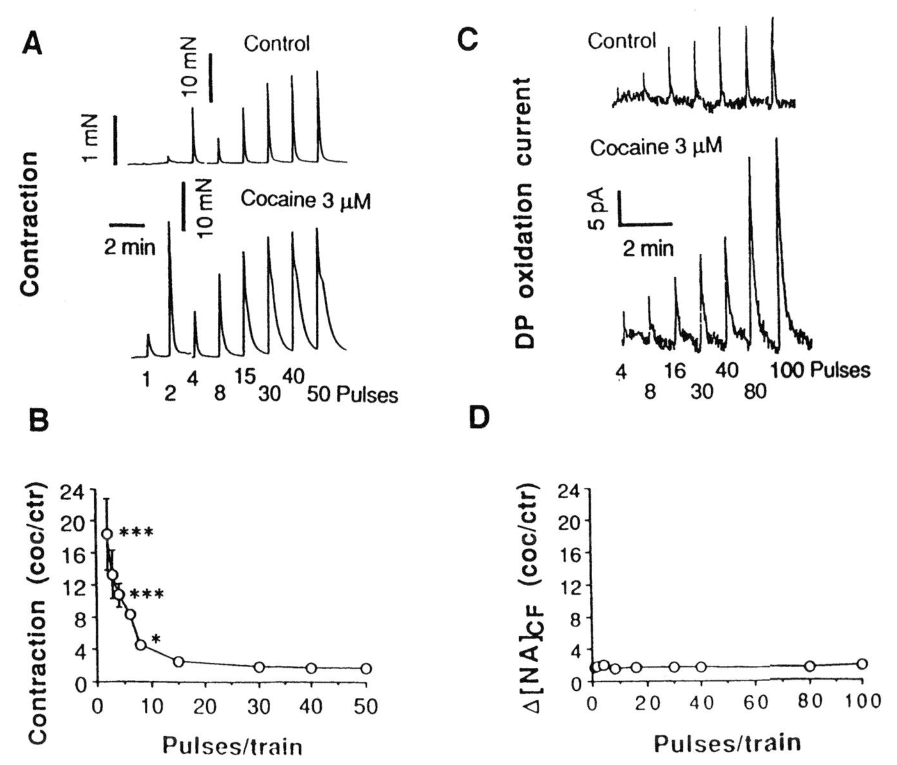

FIG. 10. Effects of cocaine on the (mainly noradrenaline (NA)-mediated) contractions, and on the rise in the concentration of released NA at the carbon-fiber electrode measured by differential pulse amperometry (DPA) ($[NA]_{CF}$), caused by nerve stimulation with 1–50 or 1–100 pulses at 20 Hz. **A,C.** Typical original recordings. Note in **A** the change in calibration. **B,D.** Pooled data from 17 and 7 experiments, respectively. Ordinate: responses with cocaine (coc) divided by those in controls (ctr). The cocaine-induced enhancement of the contractions was inversely related to the number of pulses, and that of $[NA]_{CF}$ independent of train length. In **B,** *$p<0.05$, ***$p<0.001$, when compared to the corresponding values in **D,** unpaired t-test. (From Stjärne et al., ref. 7, with permission.)

tion were thus rapidly saturated by repeated release of quanta during short 20 Hz trains. The reason why their activity was not "seen" by the extrajunctional electrode is probably that it took place intrajunctionally. Conversely, the neuronal reuptake transporters that "normally" removed ≤50% of NA in quanta released during 15–100 pulses at 20 Hz from the α_1-adrenoceptors mediating large contractile responses were not saturated during the 20 Hz trains. The reason why their activity was "seen" by the electrode is probably that it took place extrajunctionally (7). The implications for the "geometry" of neuromuscular transmission in rat tail artery will be discussed later. The point to make at this stage is that it is not facilitation of release (Fig. 3B) but train-length-dependent decline in reuptake of NA, probably into the releasing varicosities, that enables the vessel to virtually ignore single nerve impulses but contract briskly during short high-frequency trains (20,7,6).

Activity-dependent Restriction of the Diffusibility of Released NA

A different mode of "plasticity of clearance" of released NA, particularly apparent during stimulation with a longer tetanus, is illustrated in Fig. 11A,B,C (7).The curve describing the increments in consecutive DPA measurements during the tetanus (Fig. 11B) was closely similar in time course to that of the EJCs under the same conditions (Fig. 3B). This curve was used therefore as a rough approximation of the per pulse neural release of both ATP and NA (31). The ATP- and NA-induced contractile responses differed dramatically in time course (Fig. 11C), in spite of the equally steep decline in the per pulse release of both transmitters. The ATP-induced contraction was small, rapid in onset, but very transient, as expected in view of the fast clearance of released ATP (68). In contrast, the NA-mediated contraction was maintained at a plateau level throughout the tetanus. The reason was not "postreceptor adaptation" but a maintained NA concentration at the receptors, as shown by the fact that $[NA]_{CF}$ did not decline (but in the pocket surrounding the electrode even increased progressively) during the tetanus. Upon cessation of nerve stimulation both $[NA]_{CF}$ and tension returned rapidly to baseline. Cocaine enhanced both $[NA]_{CF}$ and tension during the tetanus and slowed their decline afterwards, showing that neuronal reuptake of NA was fully active. The NA concentration at the adrenoceptors driving the contraction was thus maintained, in spite of the profound decline in per pulse neural NA release, by a mechanism that restricted the availability of released NA for neuronal uptake and washout during, but not after the tetanus (7).

Differential Clearance of Co-released ATP and NA

To explain the findings in Fig. 11A,B,C, we propose in Fig. 11D,E a symmetrical mirror-image model of the intrajunctional fates of co-released ATP and NA (37). The model assumes that the handling of ATP in quanta released from sympathetic varicosities (Fig. 11D) is analogous to that of acetylcholine (ACh) in quanta released from active zones in a well-examined system, cholinergic nerve terminals in skeletal neuromuscular endplate (1). At the site of exocytosis (step 1), the initial local concentration of ACh is the same as in the vesicle (i.e., in the order of 100 mM). To reach the nicotinic receptors, ACh has to pass through a formidable barrier, the basal lamina that is loaded with acetylcholinesterase (AChE), a prototype ACh-binding protein (70) with powerful destructive properties. This enzyme allows ACh time for only a single receptor attachment and channel opening; it is thus capable of destroying, within milliseconds, the ACh molecules in a released quantum. Surprisingly, block of the enzyme increases the initial rate of rise of miniature endplate potentials by less than 50%. One third of the molecules in the small but massive ACh packet is thus enough to saturate the AChE barrier; two thirds bind to the nicotinic receptors and trigger a channel opening (step 2). As a result, the ACh concentration in the cleft falls rapidly; the ACh molecules that asynchronously dis-

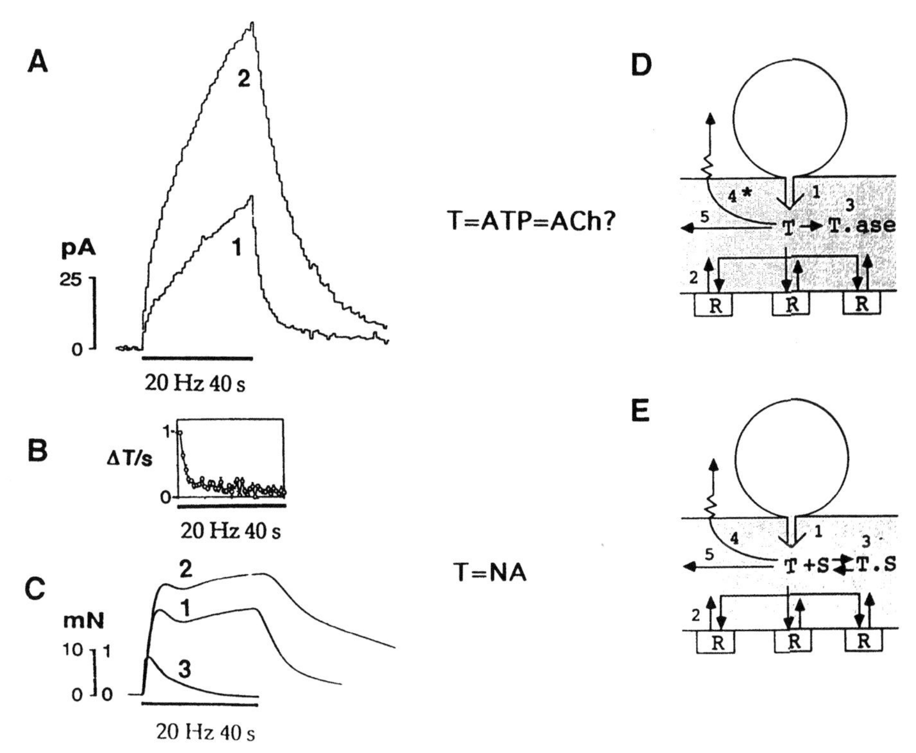

FIG. 11. A-C. Effects of nerve stimulation with 800 pulses at 20 Hz in rat tail artery before and after addition of cocaine (3 μM/l). **A.** Traces (1,2): the noradrenaline (NA) oxidation current measured by differential pulse amperometry (DPA) once per second, in controls and after cocaine, respectively. **B.** Increments in $[NA]_{CF}$ in consecutive DPA measurements, used as a rough measure of the per second neural release of both transmitters (T) and referred to as ΔT/s. **C.** Traces (1,2): the neurogenic (mainly) NA-induced contractions before and after addition of cocaine. Trace (3): the adenosine 5′-triphosphate (ATP)-mediated neurogenic contraction. Amplitude bar: traces (1,2), numbers to the left; trace (3) numbers to the right. **D,E.** A symmetrical, mirror-image model of clearance of released transmitters (T). In **D**, T = ATP; note the proposed analogy with the release and clearance of acetylcholine (ACh) in skeletal neuromuscular junction. In E, T = NA. Step 1: exocytosis. Step 2: reversible binding to postjunctional receptors. Step 3: reversible binding to a low-affinity site in the matrix filling the neuromuscular gap; in **D** it is destructive (the enzyme T-ase); in **E**, it is a protective, buffering site "S." Step 4: neuronal uptake; in **D**, the substrate is a metabolite, the adenosine or choline moieties of ATP or ACh; in **E**, the substrate is intact NA. Step 5: diffusion out of the junction; **D** is designed for rapid elimination of release transmitter; **E** is designed for prolonging its lifetime at the receptors. (From Stjärne et al., ref. 7 and Stjärne, ref. 37, with permission.)

sociate from the receptors are rapidly destroyed by AChE (step 3). The metabolite, choline, is retrieved by a neuronal transporter (step 4); a very small fraction of the released ACh diffuses in intact form out of the junction (step 5). We propose that this sequence may apply as well, although with tenfold slower kinetics, to the intrajunctional fate of ATP in quanta released from sympathetic varicosities. In this case, "T-ase" is an ecto-ATPase (28), "R" an ionotropic P_{2x}-purinoceptor, and the substrate for neuronal uptake is the ATP metabolite, adenosine (68). This strategy of intrajunctional handling clearly promotes rapid elimination of the released transmitter and a fast but brief effector response (68).

In contrast, co-released NA is apparently protected by mechanisms that prolong its presence at the receptors during ongoing nerve activity and help maintain the contractile response even when release has declined to very low levels, and yet upon cessation of nerve impulses promote rapid clearance by reuptake and washout (7). These features are not explainable by unspecific diffusion delay (e.g., due to "tortuosities" in the extracellular space and/or binding of the positively charged NA molecules to polyanionic glycosaminoglycans) (71); they seem to require specific, nerve activity-dependent intrajunctional buffering of released NA (6,7) We propose in Fig. 11E that the extracellular matrix in the sympathetic neuromuscular gap may be enriched both in ecto-ATPase, a low-affinity ATP-binding protein with powerful destructive properties, but also in "S," an NA-binding protein with protective properties (7,6) similar to those of specific transport proteins that store the bulk of several circulating hormones. Thyroxine, for example, exists in the blood stream as a minute "free" pool (0.05% of the total) in which it is diffusible and hence biologically active, in equilibrium with an enormous pool (99.95% of the total) in which it is protein-bound and inactive because it cannot diffuse out of the circulation to reach effector cells (72). An additional qualification, however, is that the NA affinity of "S" has to be nerve activity-dependent.

The Spatiotemporal Pattern of Quantal Release

That this factor is of crucial importance for the NA-induced neurogenic contraction in rat tail artery may be clarified by comparison with the contraction caused by exogenous NA. Both contractions are probably mediated by α-adrenoceptors on innervated "key cells" (21) at the advential border (73,7), but the two responses differ enormously in geometry. The response to bath-applied and therefore homogeneously distributed NA is mediated mainly by α_1-adrenoceptors. During cumulative application this contraction grows to a plateau at each NA concentration; the amplitude is half-maximal at 1 μM and maximal at 10–100 μM NA (12). In contrast, the NA-induced neurogenic contraction is driven by scattered "hotspots" of neurally released NA (Fig. 4B). Depending on the stimulus parameters, it is mediated mainly by intrajunctional α_2- or extrajunctional α_1-adrenoceptors (20). The response consists of an initial rapid and a delayed tonic component that both grow with the frequency and length of stimulus trains. The half-maximum is attained after 15 pulses and the maximum after 50–100 pulses at 20–50 Hz. The amplitude of the maximal NA-induced neurogenic contraction is approximately 50% of that of the maximal response to exogenous NA (4), suggesting that neurally released NA maximally occupies 50% of the smooth-muscle α-adrenoceptors at the advential border. Four features of crucial importance for the neurogenic contraction are summarized schematically in Fig. 12 A-D.

NA in Active Junctions Drives the Neurogenic Contraction

The message in Fig. 12A is that quanta are released into very tiny junctional spaces. (9,15) This implies that free equilibration of the approximately 1,000 NA molecules in the quantum (74) would yield an average NA concentration in this space of 200–300 μM. This is two- to twentyfold higher than the concentration of bath-applied exogenous NA required to cause a maximal contraction (4). The number of NA molecules in a single quantum is thus probably sufficient to saturate the adrenoceptors within the junction (20). Nevertheless, the mainly α_2-adrenoceptor-mediated contractile response to a single pulse is normally extremely small ($\leq 0.1\%$ of the maximal neurogenic contraction). The reasons are that: (i) it is curtailed by neuronal reuptake of $\geq 95\%$ of the released NA in single quanta (20); (ii) all junctional patches (40,000/mm^2) together occupy $\leq 0.5\%$ of the innervated adventitial surface (15); and (iii) each nerve impulse releases a single quantum from one out of a hundred varicosities (i.e., addresses the receptors in only 1% of these patches (10)). The maximal α_2-adrenoceptor-mediated neurogenic contraction, during a high-frequency train that induces repeated release from all varicosities along active strings (Fig. 4B) and increasingly saturates their reuptake transporters, probably amounts to approximately 20% of the neurogenic maximum (20,68). At least 80% of the maximal NA-induced neurogenic contraction is thus mediated by extrajunctional α_1-adrenoceptors (20). This does not lessen the importance of active junctions, however; their most crucial role is to drive outward diffusion of NA to their surround and thereby amplify the neurogenic contraction. The mechanisms controlling the level of free NA in active junctions, the pressure head for diffusion to the surround, are thus of fundamental importance (20,7).

A Functional Two-compartment Model

In Fig. 12B we propose a model for the NA-induced neurogenic contractions in rat tail artery that distinguishes between events in active junctions and their surround (7). Reuptake of $\geq 95\%$ of NA in single quanta into the releasing varicosity ("uptake$_{1a}$") effectively curtails activation of the smooth-muscle α-adrenoceptors inside these junctions. The maximal reuptake per transporter molecule is 2.5 NA molecules per second (75). A release rate that causes the NA load to exceed this level saturates the NA transporter and causes NA to accumulate in and around active junctions (20). An excessive rise in the level of free NA in these junctions is prevented by reversible binding to "S" (7). Re-release from "NA-S"-complexes when release declines during a tetanus maintains the concentration of free NA, sustains a controlled outward diffusion, and prevents or delays the decline of the contraction (7).

The NA concentration at the extrajunctional α_1-adrenoceptors that mediate most of the contractile response to high frequency trains represents the balance between outward diffusion from active junctions and clearance by neuronal uptake in the

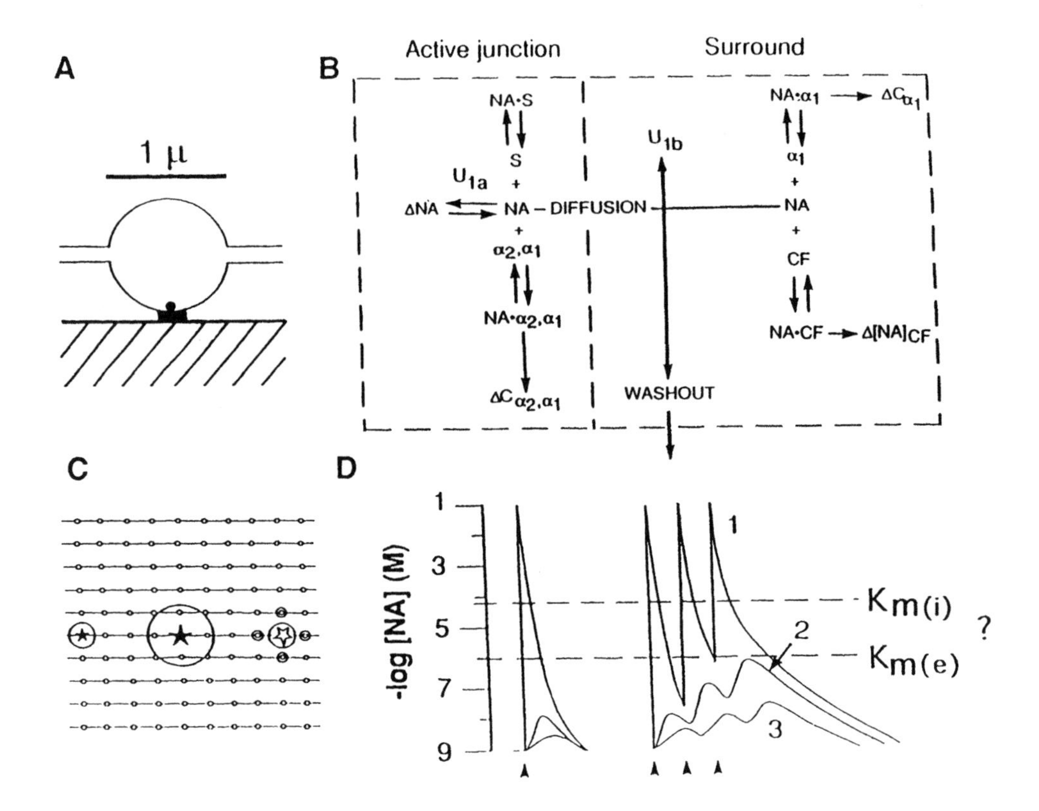

surround ("uptake$_{1b}$") and washout. The NA concentration in this compartment, and the effects of uptake$_{1b}$, are "seen" in real time by the obviously extrajunctional carbon-fiber electrode. Cocaine approximately doubled all $\Delta[NA]_{CF}$ responses to 1–100 pulses at 20 Hz (Fig. 10); the NA level at the extrajunctional α_1-adrenoceptors was thus always reduced by 50% by uptake$_{1b}$. Uptake$_{1b}$ is thus unaffected by nerve activity (i.e., rarely saturated, except possibly near the rim of active junctions during brief high frequency bursts (25)).

The Spatiotemporal Pattern of Quantal Release

The message in Fig. 12C is that the neurogenic contraction depends both on the amount of transmitters released and on how, where, and when they are released. It shows three examples of how NA release according to the string model (Fig. 4B) may control muscle tension in this vessel.

The first concerns the effects of a single pulse (i.e., release of single quanta from 1% of the varicosities). The contractile response is extremely small (0.1% of the neurogenic maximum) and mediated by α_2-adrenoceptors, probably in active junctions. That cocaine amplified the contraction by ≥twentyfold (to 2% of the neurogenic maximum) implies that it is the efficacy of uptake$_{1a}$ that causes this vessel to normally virtually ignore single nerve impulses.

In the second example, 15 pulses at 20 Hz increased the diameter of the NA "hotspot," recruited 25% of all α_1-adrenoceptors at the adventitial surface, and amplified the single-pulse contraction by five hundredfold, from 0.1% to 50% of the neurogenic maximum (4). The per pulse release is only minimally facilitated under these conditions (31); the effect is thus due to suppression of NA clearance (i.e., saturation of uptake$_{1a}$ in varicosities along active strings). During this phase, excessive washout of released NA is prevented by the buffering effect of reversible binding to "S," both in active and surrounding silent junctions (7).

The third example concerns events during longer 20 Hz trains when the per pulse release of NA is fading. Re-release of NA from "NA-S" complexes in previously active and surrounding silent sites now helps maintain the NA concentration at the smooth-muscle α_1- and, especially, α_2-adrenoceptors (68) and thereby the contractile response. In the presence of neuronal uptake blockers, this mechanism prevents,

FIG. 12. A. A junctional contact between a sympathetic varicosity and a smooth-muscle cell in rat tail artery. The basal laminae are fused and the neuromuscular separation 50–100 nm (9). The small dense-cored vesicle, whose internal volume is 2.44×10^{-20} l (if the external diameter is 50 nm and the wall thickness 7 nm), releases its contents into a junctional space (15) with a volume of 6.3×10^{-18} l (if cylindrical with r = 200 nm and h = 50 nm). **B.** A two-compartment model for events in active junctions and their surround. ΔNA is the per pulse exocytotic noradrenaline (NA) release. Released NA may bind to α_2- and α_1-adrenoceptors on the muscle patch inside the junction and via them trigger a small contraction ($\Delta C_{\alpha 2,\alpha 1}$), or to the NA transporter for reuptake into the releasing varicosity (uptake$_{1a}$, "U_{1a}"), or to buffering sites "S" whose NA affinity is increased by nerve activity (see Fig. 11E), or diffuse along its concentration gradient out of the junctional space. In the surround, NA may bind to the more numerous extrajunctional α_1-adrenoceptors and via them drive a much larger component of the neurogenic contraction ($\Delta C_{\alpha 1}$), or to NA transporters to be taken up into nerve (uptake$_{1b}$, "U_{1b}"), or diffuse to the effluent ("washout"). The carbon-fiber (CF) electrode "sees" the NA concentration only in the surround. **C.** Ten strings of varicosities which, according to morphological criteria, junctionally innervate muscle cells at the adventitial surface. The nerve impulse invades all of them, but >90% are at the moment silent ($p = 0$). In <10% of the strings the varicosities have a finite release probability ($0 < p < 1$). Along these active strings, "strong" varicosities (p up to 0.5) alternate with "weak" ones (p down to 0.05). During high-frequency bursts, varicosities in active string release quanta repeatedly into the same junctional gaps. Shown are: **(a)** silent varicosities (unfilled); **(b)** a varicosity that has just released a single quantum (the small filled star denotes the NA concentration in the active junction, the circle that in its surround); **(c)** a "hotspot" (i.e., varicosity releasing quanta repeatedly during a high-frequency burst); and **(d)** a "fading hotspot" in which quantal release is waning (the open star indicates that release fails because the releasable pool is depleted) and the NA concentration is maintained by re-release from "NA-S" complexes both in previously active and nearby silent junctions. **D.** Spatiotemporal fluctuations in the concentration of free NA inside an active junction (trace 1) and in the surround at increasing distances from the releasing varicosity (traces 2,3). Note that the scale is in -log units. Shown are the concentration of free NA after release of a single quantum, and during repeated release of single quanta into the same junction. Arrowheads denote release of single quanta from the same varicosity, the horizontal broken lines K_m for neuronal uptake of NA intrajunctionally ($K_{m(i)}$) and extrajunctionally ($K_{m(e)}$). The extrajunctional carbon-fiber electrode senses only traces 2,3. After a single pulse the NA concentration at the probe peaks at 10 nM (i.e., 7 orders of magnitude below the peak of trace 1). For further comments see the text. (From Stjärne et al., ref. 7, with permission.)

for at least 10 minutes, a decline in the $\Delta[NA]_{CF}$ and contractile responses during nerve stimulation at 20 Hz (7).

Silent strings are by no means redundant, because: (i) all strings are activated in rotation; (37) (ii) "NA-S"-complexes in silent junctions in the neighborhood of active junctions help in buffering the NA concentration at the adrenoceptors; and (iii) the ability of all regions of the terminals to take up NA causes them to act as a huge "NA sink" (4) that, after nerve stimulation, greatly accelerates NA clearance and thereby the relaxation of the vessel (7).

NA Concentration Gradients in and Around Active Junctions

The message in the model in Fig. 12D is that the neurogenic contraction in rat tail artery is driven by steep spatiotemporal gradients of NA inside and outside active junctions. The rise and fall in the concentration of free NA in active junctions (trace 1) is not directly measurable by current methods, but the orders of magnitude may be inferred (73). At the site of exocytosis the initial NA concentration is equal to that in the vesicle (i.e., approximately 68 mM (36)). This is the absolute maximum (i.e., the same after release of a single quantum or a series of quanta into the same junctional space and unaffected, for example, by block of neuronal uptake). The concentration of free NA in the junction is rapidly reduced by many orders of magnitude by; (i) diffusion within and out of the junction; (ii) reuptake into the releasing varicosity; (iii) buffering by binding to "S"; and (iv) binding to adrenoceptors within the junction. Equilibration of NA in a single quantum would yield an average NA concentration of 200–300 μM within the junction (i.e., much above the reported K_m for neuronal uptake, $\leq 1\,\mu M$ (75)). It is surprising, therefore, that the releasing varicosity appears to take up ≥95% of the NA in single quanta (Fig. 10A).

Techniques to study this issue directly (i.e., to monitor in real time the neuronal reuptake of NA into the releasing varicosity) are currently not available. However, a clue may be obtained by analogy with other sytems in which this is possible. In Retzius-P cell synapse in *Hirudo* (76), the kinetics of the electrogenic neuronal reuptake of the transmitter, 5-hydroxytryptamine (5-HT), may be monitored directly by recording the accompanying Na^+-dependent inward current. Application of 5-HT activated this presynaptic current within 1 millisecond; the time to peak was 8 milliseconds. The kinetics of the rising phase were faster than those of the postsynaptic current, but both decayed in parallel. The kinetics of neuronal reuptake thus determined the concentration of 5-HT in released quanta at the postjunctional receptors, and thereby the time course of synaptic transmission. The apparent K_m for this "intrajunctional" 5-HT reuptake was 20–30 μM (i.e., much higher than reported K_m values ($<1\,\mu M$) for the presumably mainly "extrajunctional" neuronal 5-HT uptake in synaptosomes and brain-slice preparations (75,76)). A similar low-NA affinity of intrajunctional neuronal reuptake transporters, in combination with NA buffering by binding to "S," would explain how uptake_{1a} could remove ≥95% of the NA in single quanta in rat tail artery. However, whether or not the intra- and extrajunctional NA reuptake transporters in these nerves actually differ in NA affinity remains to be determined.

The model shows that outward diffusion of NA in a released quantum is driven by a continuously declining concentration gradient. The facts that the NA oxidation current peaked within 200 milliseconds after exocytosis (19), and that cocaine amplified the single pulse concentration by ≥twentyfold, but $\Delta[NA]_{CF}$ by only twofold (Fig. 10) imply that: (i) outward diffusion to the electrode is an early event, independent of $uptake_{1a}$ (i.e., driven mainly at NA concentrations above $K_{m(i)}$), and (ii) that interaction with muscle α_2-adrenoceptors inside active junctions is a much later event, controlled by $uptake_{1a}$, and hence exerted by NA at concentrations well below $K_{m(i)}$.

The model furthermore illustrates the conditions for temporal summation of NA in active junctions. The duration of the NA oxidation current measured by continuous amperometry (19) or by DPA (18) indicates that the total lifetime of free NA inside active junctions after release of a single quantum is between 1,500 milliseconds and 5,000 milliseconds. Repeated release of single quanta into the same junctional space at ≥0.2–0.7 Hz thus causes the NA load on reuptake into the releasing varicosity to exceed its V_{max}. Under such conditions the "tail" NA concentration inside active junctions (trace 1), and at increasing distances in the surround (traces 2,3), inevitably exhibits temporal summation. The "effective" NA concentration driving contractions via α_2-adrenoceptors in active junctions is probably in a low micromolar range because: (i) 1 μM competitive α_2-adrenoceptor antagonist blocks the mainly α_2-adrenoceptor-mediated contractile response to 4 pulses at 20 Hz (20), and (ii) the NA concentration "seen" by the carbon-fiber electrode just outside active junctions rarely exceeds 300 nM, even during high-frequency trains in the presence of cocaine (18).

SUMMARY AND CONCLUSIONS

The recent explosive development in research concerning the fundamental mechanisms of synaptic transmission helps put the present paper in context. It is now evident that not all transmitter vesicles in a nerve terminal, not even all those docked at its active zones, are immediately available for release (36). We watch, fascinated, the unraveling of the amazingly complex cellular mechanisms and molecular machinery that determine whether or not a vesicle is "exocytosis-competent" (77,78,39,79). Studies on quantal release in different systems show that neurons are fundamentally similar in one respect: that transmitter release from individual active zones is monoquantal (2). But they also show that active zones in different neurons differ drastically in the probability of monoquantal release and in the number of quanta immediately available for release (3). This implies that one should not extrapolate directly from transmitter release in one set of presynaptic terminals (e.g., in neuromuscular endplate or squid giant synapse) to that in other nerve terminals, especially if they have a very different morphology. As shown here, one should not even extrapolate from transmitter release in sympathetic nerves in one tissue (e.g., rat tail artery) to that in other tissues or species (e.g., mouse vas deferens). It is noteworthy that most studies of quantal release are based on electrophysiological analysis and therefore deal with release of fast, ionotropic transmitters from small synaptic vesicles at the active zones, especially in neurons in which these events may be examined with high resolution (49,48,46,33,32). Such data are useful as general models of the release of both fast and slow transmitters from small synaptic vesicles at active zones in other systems, provided that these transmitters are released in parallel, as are apparently ATP and NA in sympathetic nerves. They tell us little or nothing, however, about the release of transmitters (e.g., neuropeptides) from the large vesicles, nor about the spatiotemporal pattern of monoquantal release from small synaptic vesicles in the many neurons that have boutons-en-passent terminals. They show that the time course of effector responses to fast, rapidly inactivated transmitters such as ACh or ATP is necessarily release related. But they do not even address the possibility that the effector responses to slow transmitters such as NA, co-released from the same terminals, may obey completely different rules and perhaps rather be clearance related (7).

Regarding neuromuscular transmission in rat tail artery, the evidence described in the present paper enables us to answer, tentatively, the questions asked in the Introduction:

1. That ATP and NA are released in quanta (i.e., as small but massive packets) into narrow junctional spaces is what enables ATP to escape destruction by ecto-ATPase before binding at least once to the intrajunctional P_{2X}-purinoceptors, ensures that NA saturates the intrajunctional α_2-adrenoceptors, and provides the pressure head for outward diffusion that amplifies the contraction by recruiting α_1-adrenoceptors in the surround. That local clearance effectively disposes of ATP and NA in single quanta explains why the vessel nevertheless ignores single nerve impulses.
2. That quanta are released repeatedly into the same junctional spaces is of fundamental importance. It is the saturation of NA reuptake into the releasing varicosities that, by increasing the outward diffusion of NA from the active junctions and recruiting numerous α_1-adrenoceptors in the surround, causes the vessel to contract briskly during brief high frequency bursts.
3. That the spatiotemporal pattern of quantal release apparently obeys the string model, implies that monoquantal release from individual active zones is not entirely probabilistic (i.e., that factors other than the nerve terminal action potential control the availability of exocytosis-competent transmitter vesicles at individual active zones along the terminal branches).
4. The efficacy of the ATP-mediated neurogenic contraction hinges on the per pulse release, but the NA-induced neurogenic contraction depends mainly on the intrinsic slowness and nerve activity-induced plasticity of clearance of released NA.

Our working hypothesis to explain these results is tentative. Its validity for rat tail artery and applicability to other systems, and the molecular and cellular mechanisms involved, remain to be elucidated in future work.

ACKNOWLEDGMENT

The research in this paper was supported by the Swedish Medical Research Council (project B94-14X-03027-25A), Karolinska Institutets Fonder, and the European Science Foundation (twinning grant no. 16).

REFERENCES

1. Katz B. Looking back at the neuromuscular junction. In: Sellin LC, Libelius R, Thesleff S, eds. *Neuromuscular junction*. Amsterdam: Elsevier Science Publishers (Biomedical Division), 1989;3–9.
2. Faber DS, Korn H. Binary mode of release at central synapses. *Trends Neurosci* 1982;5:157–159.
3. Lavidis NA, Bennet MR. Probabilistic secretion of quanta from visualized sympathetic nerve varicosities in mouse vas deferens. *J Physiol (Lond)* 1992;454:9–26.
4. Stjärne L, Bao JX, Gonon F, Msghina M, Stjärne E. On the geometry, kinetics and plasticity of sympathetic neuromuscular transmission. *Jap J Pharmacol* 1992;58 (Suppl II):158–165.
5. Martin AR. Junctional transmission. II. Presynaptic mechanisms. In: Brookhart JM, Mountcastle VB, Section eds. *Handbook of physiology, the nervous system*. vol I. Cellular Biology of Neurons. Part I. *Americ Physiol Soc* 1977;329–355.
6. Stjärne L, Bao J-X, Gonon F, Msghina M, Stjärne E. A nonstochastic string model of sympathetic neuromuscular transmission. *News Physiol Sci* 1993,8:253–260.
7. Stjärne L, Bao J-X, Gonon F, Msghina M. Nerve activity-dependent variations in clearance of released noradrenaline: regulatory roles for sympathetic neuromuscular transmission in rat tail artery. *Neuroscience* 1994;60:1021–1038.
8. Sittiracha T, McLachlan EM, Bell C. The innervation of the caudal artery of the rat. *Neuroscience* 1987;21:647–659.
9. Luff SE, McLachlan EM. Frequency of neuromuscular junction on the arteries of different dimensions in the rabbit, guinea-pig and rat. *Blood Vessels* 1989;26:95–106.
10. Åstrand P, Stjärne L. On the secretory activity of single varicosities in the sympathetic nerves innervating the rat tail artery. *J Physiol (Lond)* 1989;409:207–220.
11. Sneddon P, Burnstock G. ATP as a co-transmitter in rat tail artery. *Eur J Pharmacol* 1984;106:149–152.
12. Bao JX, Eriksson IE, Stjärne L. Neurotransmitters and pre- and postjunctional receptors involved in the vasoconstrictor response to sympathetic nerve stimulation in rat tail artery. *Acta Physiol Scand* 1991;140:467–479.
13. Fried G, Terenius L, Hökfelt T, Goldstein M. Evidence for differential localization of noradrenaline and neuropeptide Y (NPY) in neuronal storage vesicles isolated from rat vas deferens. *J Neurosci* 1985;5:450–458.
14. Hökfelt T. Distribution of noradrenaline storing particles in peripheral adrenergic neurons as revealed by electron microscopy. *Acta Physiol Scand* 1969;76:427–440.
15. Hirst GDS, Bramich NJ, Edwards FR, Klemm M. Transmission at autonomic neuroeffector junctions. *Trends Neurosci* 1992;15:40–46.
16. Zhu PC, Thuresson-Klein J, Klein RL. Exocytosis from large dense cored vesicles outside the active zones of terminal subnucleus caudalis: a possible mechanism for neuropeptide release. *Neuroscience* 1986;19:43–54.
17. Brock JR, Cunnane TC. Electrophysiology of neuroeffector transmission in smooth muscle. In: Burnstock G, Hoyle CHV, eds. *Autonomic Neuroeffector Mechanisms*. Reading, Pennsylvania: Harwood; 1992;121–213.
18. Gonon F, Bao J-X, Msghina M, Suaud-Chagny MF, Stjärne L. Fast and local electrochemical monitoring of noradrenaline release from sympathetic terminals in isolated rat tail artery. *J Neurochem* 1993;60:1251–1257.

19. Gonon F, Msghina M, Stjärne L. Kinetics of noradrenaline released by sympathetic nerves. *Neuroscience* 1993;56:535–538.
20. Bao J-X, Gonon F, Stjärne L. Frequency- and train length-dependent variation in the roles of postjunctional α_1- and α_2-adrenoceptors for the field stimulation-induced neurogenic contraction of rat tail artery. *Naunyn Schmiedebergs Arch Pharmacol* 1993;347:601–616.
21. Burnstock G. Structure of smooth muscle and its innervation. In: Bülbring E, Jones AW, Tomita T, eds. *Smooth Muscle*. London: Edward Arnold Publishers, Ltd, 1970;1–69.
22. Åstrand P, Stjärne L. ATP as a sympathetic co-transmitter in rat vasomotor nerves—further evidence that individual release sites respond to nerve impulses by intermittent release of single quanta. *Acta Physiol Scand* 1989;136:355–365.
23. Cunnane TC, Stjärne L. Transmitter secretion from individual varicosities of guinea-pig and mouse vas deferens: highly intermittent and monoquantal. *Neuroscience* 1984;13:1–20.
24. Brock JA, Cunnane TC. Local application of drugs to sympathetic nerve terminals: an electrophysiological analysis of the role of prejunctional α_2-adrenoceptors in the guinea-pig vas deferens. *Br J Pharmacol* 1991;102:595–600.
25. Msghina M, Gonon F, Stjärne L. Intermittent release of noradrenaline by single pulses and release during short trains at high frequencies from sympathetic nerves of rat tail artery. *Neuroscience* 1993;57:887–890.
26. Wightman RM, Jankowski JA, Kennedy RT, Kawagoe KT, Schroeder TJ, Leszczyszyn DJ, Near JA, Dilberto EJ, Viveros OH. Temporally resolved catecholamine spikes correspond to single vesicle release from individual chromaffin cells. *Proc Natl Acad Sci* 1991;88:10754–10758.
27. Chow RH, von Rüden L, Neher E. Delay in vesicle fusion revealed by electrochemical monitoring of single secretory events in adrenal chromaffin cells. *Nature* 1992;356:60–63.
28. Cunnane TC, Manchanda R. Electrophysiological analysis of the inactivation of sympathetic transmitter in guinea-pig vas deferens. *J Physiol (Lond)* 1988;404:349–364.
29. Kovach PM, Ewing AG, Wilson RL, Wightman RM. In vitro comparison of selectivity of electrodes for in vivo electrochemistry. *J Neurosci Methods* 1984;10:215–227.
30. Bennett MR. An electrophysiological analysis of the uptake of noradrenaline at sympathetic nerve terminals. *J Physiol (Lond)* 1973;229:533–546.
31. Msghina M, Stjärne L. Sympathetic transmitter release in rat tail artery and mouse vas deferens: facilitation and depression during high frequency stimulation. *Neurosci Lett* 1993;155:37–41.
32. Zucker R. Short term synaptic plasticity. *Ann Rev Neurosci* 1989;12:13–31.
33. Swandulla D, Hans M, Zipser K, Augustine GJ. Role of residual calcium in synaptic depression and posttetanic potentiation: fast and slow calcium signalling in nerve terminals. *Neuron* 1991;7:915–926.
34. Elmqvist D, Quastel DMJ. A quantitative study of end-plate potentials in isolated human muscle. *J Physiol (Lond)* 1965;178:505–529.
35. Greengard P, Valtorta F, Czernik AJ, Benfenati F. Synaptic vesicle phosphoproteins and regulation of synaptic function. *Science* 1993;259:780–785.
36. Stjärne L. Basic mechanisms and local modulation of nerve impulse-induced secretion of neurotransmitters from individual sympathetic nerve varicosities. *Rev Physiol Biochem Pharmacol* 1989; 112:1–137.
37. Stjärne L. Possible sites on the neurone for modulatory influences to act. In: Powis D, Bunn S, eds. *Neurotransmitter Release and its Modulation-Biochemical Mechanisms, Physiological Function, Clinical Relevance*. Part II: Presynaptic Modulation of Neurotransmitter Release. Cambridge: Cambridge University Press. (*in press*).
38. Malmfors T. Studies on adrenergic nerves. *Acta Physiol Scand* 1965;64 (Suppl. 248):1–93.
39. Jahn R, Südhoff TC. Synaptic vesicles and exocytosis. *Ann Rev Neurosci* 1993;17:219–246.
40. Stjärne L, Msghina M, Stjärne E. "Upstream" regulation of the release probability in sympathetic nerve varicosities. *Neuroscience* 1990;36:571–587.
41. Stjärne L, Stjärne E, Msghina M, Bao JX. K^+- and Ca^{2+} channel blockers may enhance or depress sympathetic transmitter release via a Ca^{2+}-dependent mechanism "upstream" of the release site. *Neuroscience* 1991;44:673–692.
42. Stjärne L, Stjärne E. Basic features of an extracellular recording method to study secretion of sympathetic co-transmitter, presumably ATP. *Acta Physiol Scand* 1989;135:217–226.
43. Kasakov L, Ellis J, Kirkpatrick K, Burnstock G. Direct evidence for concomitant release of noradrenaline, adenosine 5′-triphosphate and neuropeptide Y from sympathetic nerve supplying the guinea-pig vas deferens. *J Auton Nerv Syst* 1988;22:75–82.

44. Åstrand P, Stjärne L. A calcium-dependent component of the action potential in sympathetic nerve terminals in rat tail artery. *Pflügers Arch* 1991;418:102–108.
45. Katz B, Miledi R. Spontaneous and evoked activity of motor nerve endings in calcium Ringer. *J Physiol (Lond)* 1969;203:689–706.
46. Llinás R, Sugimori M, Simon SM. Transmission by spike-like depolarizations in the squid synapse. *Proc Natl Acad Sci* 1982;79:2415–2419.
47. Katz B, Miledi R. A study of synaptic transmission in the absence of nerve impulses. *J Physiol (Lond)* 1967;192:407–436.
48. Kandel E, Schwartz JH. Molecular biology of learning: modulation of transmitter release. *Science* 1982;229:433–443.
49. Augustine GJ. Regulation of transmitter release at the squid giant synapse by presynaptic delayed rectifier potassium current. *J Physiol (Lond)* 1990;431:343–364.
50. Heuser JE, Reese TS. Structural changes after transmitter release at the frog neuromuscular junction. *J Cell Biol* 1981;88:564–580.
51. Torri-Tarelli F, Grohovaz F, Fesce R, Ceccarelli B. Temporal coincidence between synaptic vesicle fusion and quantal secretion of acetylcholine. *J Cell Biol* 1985;101:1386–1399.
52. Thoenen H, Haefely W, Staehelin H. Potentiation by tetraethylammonium of the response of the cat spleen to postganglionic sympathetic nerve stimulation. *J Pharmacol Exp Ther* 1967;157:532–540.
53. Kirpekar M, Kirpekar SM, Prat JC. Effect of 4-aminopyridine on release of noradrenaline from the perfused cat spleen by nerve stimulation. *J Physiol (Lond)* 1977;272:517–528.
54. Heemskerk FMJ, Schrama LH, Gianotti C, Spierenburg H, Versteeg DHG, De Graan PNE, Gispen WH. 4-Aminopyridine stimulates B-50 (GAP43) phosphorylation and [^{3}H]noradrenaline release in rat hippocampal slices. *J Neurochem* 1990;54:863–869.
55. Curtis R, Stewart HJS, Hall SM, Wilkin GP, Mirsky R, Jessen KR. GAP-43 is expressed by non-myelin forming Schwann cells of the peripheral nervous system. *J Cell Biol* 1992;116:1455–1464.
56. Tokunaga A, Sandri C, Akert K. Increase of large intramembraneous particles in the presynaptic active zone after administration of 4-aminopyridine. *Brain Res* 1979;174:207–219.
57. Starke K. Prejunctional α-autoreceptors. *Rev Physiol Biochem Pharmacol* 1988;107:74–146.
58. Msghina M, Mermet C, Gonon F, Stjärne L. Electrophysiological and electrochemical analysis of the secretion of ATP and noradrenaline from sympathetic nerves in rat tail artery: effects of α_2-adrenoceptor agonists and antagonists and noradrenaline reuptake blockers. *Naunyn Schmiedebergs Arch Pharmacol* 1992;346:173–186.
59. Brock JA, Cunnane TC, Starke K, Wardell CF. α_2-Adrenoceptor-mediated autoinhibition of sympathetic transmitter release in guinea-pig vas deferens studied by intracellular and focal extracellular recording of junction potentials and currents. *Naunyn Schmiedebergs Arch Pharmacol* 1990;343:45–52.
60. Stjärne L. Frequency dependence of presynaptic inhibition of transmitter secretion. In: Vizi ES, ed. *Advances in pharmacological research and practice*, vol II, Modulation of Neurochemical Transmission, Oxford: Pergamon Press; 1980:27–36.
61. Llinás R, Sugimori M, Lin J-W, Leopold PL, Brady ST. ATP-dependent directional movement of rat synaptic vesicles injected into the presynaptic terminals of squid giant synapse. *Proc Natl Acad Sci USA* 1989;86:5656–5660.
62. Breuer AC, Bond M, Atkinson MB. Fast axonal transport is modulated by altering transaxolemmal calcium flux. *Cell Calcium* 1992;13:249–262.
63. Müller CM. A role for glial cells in activity-dependent central nervous plasticity? Review and hypothesis. *Int Rev Neurobiol* 1992;28:215–284.
64. Smith SJ. Do astrocytes process neural information? *Prog Brain Res* 1992;94:119–136.
65. Okabe S. Hirokawa N. Axonal transport. *Curr Opin Cell Biol* 1989;1:91–97.
66. Burnstock G. Noradrenaline and ATP as cotransmitters in sympathetic nerves. *Neurochem Int* 1990;17:357–368.
67. Hirst GDS, Edwards FR. Sympathetic neuroeffector transmission in arteries and arterioles. *Physiol Rev* 1989;69:546–604.
68. Bao JX, Gonon F, Stjärne L. Kinetics of ATP- and noradrenaline-mediated sympathetic neuromuscular transmission in rat tail artery. *Acta Physiol Scand* 1993;149:501–517.
69. Bao J-X. Sympathetic neuromuscular transmission in rat tail artery. *Acta Physiol Scand* 1993;148 (Suppl 610).
70. Sussman JL, Harel M, Frolow F, Oefner C, Goldman A, Toker L, Silman I. Atomic structure of acetylcholinesterase from Torpedo californica: a prototypic acetylcholine binding protein. *Science* 1991;253:872–879.

71. Rice ME, Gerhardt GA, Hierl PM, Nagy G, Adams RN. Diffusion coefficients of neurotransmitters and their metabolites in brain extracellular fluid space. *Neuroscience* 1985;15:891–902.
72. Refetoff S. Thyroid hormone transport. In: De Groot LJ, Cahill JR GF, Odell WD, Martinin L, Potts JR JT, Nelson DH, Steinberger E, Winegrad AI, eds. *Endocrinology*, vol 1, New York: Grune & Stratton; 1979;347–356.
73. Johansson B, Johansson SR, Ljung B, Stage L. A receptor kinetic model of a vascular neuroeffector. *J Pharmacol Exp Ther* 1972;180:636–646.
74. Lagercrantz H, Fried G. Chemical composition of the small noradrenergic vesicles. In: Klein RL, Lagercrantz H, Zimmermann H, eds. *Neurotransmitter Vesicles*. London: Academic Press, 1982; 175–188.
75. Graefe K-H, Bönisch H. The transport of amines across the axonal membrane of noradrenergic and dopaminergic neurons. In: Trendelenburg U, Weiner N, eds. *Catecholamines I. Handbook Exp Pharmacol* vol 90/I. New York: Springer Verlag; 1988;191–245.
76. Bruns D, Engert F, Lux H-D. A fast activating presynaptic reuptake current during serotonergic transmission in identified neurons of *Hirudo*. *Neuron* 1993;10:559–572.
77. Bennett MK, Scheller RH. The molecular machinery for secretion is conserved from yeast to neurons. *Proc Natl Acad Sci USA* 1993;90:2559–2563.
78. Hess SD, Doroshenko PA, Augustine GJ. A functional role for GTP-binding proteins in synaptic vesicle cycline. *Science* 1993;259:1169–1172.
79. Söllner T, Whiteheart SW, Brunner M, Erdjument-Bromage H, Geromanow S, Tempst P, Rothman JE. SNAP receptors implicated in vesicle targeting and fusion. *Nature* 1993;362:318–324.

Molecular and Cellular Mechanisms of Neurotransmitter Release, edited by Lennart Stjärne, Paul Greengard, Sten Grillner, Tomas Hökfelt, and David Ottoson, Raven Press, Ltd., New York © 1994.

28

The Role of Ca^{2+} in Transmitter Release and Long-term Potentiation at Hippocampal Mossy Fiber Synapses

Roger A. Nicoll, Pablo E. Castillo, and Marc G. Weisskopf

Departments of Pharmacology and Physiology, University of California, San Francisco, San Francisco, California, 94143

The synapses made by the Schaffer collateral-commissural fibers onto the dendritic spines of CA1 hippocampal pyramidal neurons have long served as a model for fast excitatory synaptic transmission in the brain. These synapses release the transmitter glutamate, which binds to two subtypes of receptor: N-methyl-D-aspartate (NMDA) and non-NMDA receptors. In addition, repetitive activation of these synapses results in a long-lasting enhancement of synaptic transmission referred to as long-term potentiation (LTP) (1–3). The synapses made by the perforant path fibers on granule cells in the dentate gyrus of the hippocampus, as well as numerous synapses in other areas of the CNS, have similar properties. On the other hand, the synapses made by the mossy fibers onto the proximal dendrites of CA3 hippocampal pyramidal cells are strikingly different. These synapses, unlike those just referred to, are associated with low levels of NMDA receptor binding (4) and exhibit a form of LTP that is entirely independent of NMDA receptors (5). In addition, immunohistochemical studies have shown that these fibers contain not only glutamate, but also contain high levels of neuropeptides, in particular the opioid peptide dynorphin (6).

We have been studying the properties of these synapses for a number of years and have focused on three related aspects of synaptic function: 1) the mechanisms underlying LTP; 2) the possible transmitter role for dynorphin at this synapse; and 3) the types of Ca^{++} channels involved in the release of transmitter at this synapse.

MOSSY-FIBER LTP DEPENDS ON PRESYNAPTIC, BUT NOT POSTSYNAPTIC, Ca^{++} ENTRY

For NMDA receptor-dependent LTP, the entry of Ca^{++} through the NMDA receptor channel serves as the trigger for inducing LTP. It is possible that mossy-fiber LTP utilizes the same basic scheme, but that the route for Ca^{++} entry is different. For instance, the Ca^{++} could enter through voltage-dependent Ca^{++} channels or through a Ca^{++}-permeable non-NMDA subtype of glutamate receptor. To address these possibilities we have performed three types of experiments.

In the first experiment (7), we have loaded the postsynaptic neuron with the Ca^{++} chelator, BAPTA, to prevent any rise in Ca^{++} in the postsynaptic cell during tetanization of the mossy fibers. Since CA3 pyramidal cells, in addition to receiving a mossy-fiber input, also receive an association/commissural (assoc/com) excitatory synaptic input with properties identical to synapses in the CA1 region, it is possible to compare, in the same cell, effects of manipulations on NMDA receptor-dependent and independent forms of LTP. Loading the postsynaptic cell had profoundly different effects on the two forms of LTP (7).While it entirely abolished the LTP evoked in the assoc/com synapses, precisely as predicted from similar studies in the CA1 region (8,9), it had no effect at all on the mossy-fiber LTP. This experiment had two important built-in controls. First, in all experiments we simultaneously recorded the extracellular field potential generated by assoc/com stimulation. This confirmed that the surrounding cells, which had not been filled with BAPTA, generated normal LTP and therefore the absence of LTP in the recorded cell was, in fact, due to the presence of the BAPTA. Second, by comparing the effect of BAPTA on the two forms of LTP in the same cell we were able to demonstrate that levels of BAPTA sufficient to buffer Ca^{++} were, indeed, present, since LTP was blocked at assoc/com synapses, which are considerably farther from the site of injection (i.e., further from the soma than are the mossy fibers).

The second type of experiment involved the use of whole-cell patch-clamp recording so that we could effectively voltage clamp the postsynaptic cell (7). Results from these experiments were in complete accord with the experiments just discussed. Tetanization of the mossy fibers while the cell was clamped at -90 mV did not alter the magnitude of LTP, despite the fact that the membrane was maintained well below the threshold for action potential discharge during the tetanus (Fig. 1). The results just mentioned have been confirmed by Katsuki et al. (10) and Langdon et al. (11) but differ from those of Johnston and his colleagues (12,13) who claim that buffering postsynaptic Ca^{++} or voltage clamping the postsynaptic cell blocks LTP. We have, therefore, done further experiments to resolve this issue.

The third type of experiment (14) involved examining the effects of blocking synaptic transmission during the tetanus (15).The blockade was accomplished either by the removal of extracellular Ca^{++} or by use of high concentrations of the glutamate receptor antagonist, kynurenate (Fig. 2). We first performed these experiments using extracellular recording. After obtaining a baseline the slice was superfused with the nominally 0 Ca^{++} solution. When the synaptic response was completely blocked the mossy fibers were tetanized. No recovery of the response was observed during the tetanus in the "0 Ca" solution. Following washout the responses returned to baseline values. The slice was then superfused with a solution containing 10–20 mM kynurenate, which entirely blocked the synaptic responses. The mossy fibers were again tetanized and no recovery of the responses occurred during

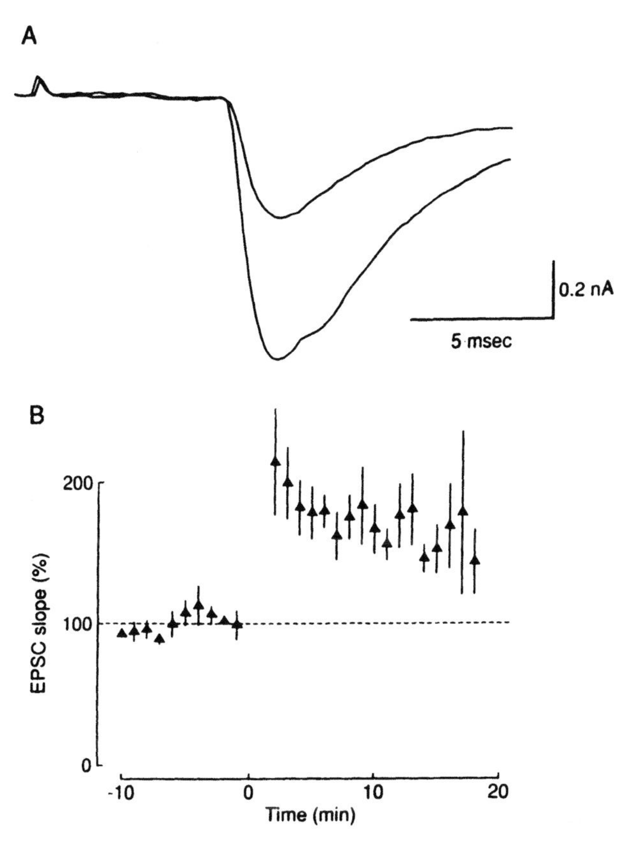

FIG. 1. Postsynaptic BAPTA, (a Ca^{++} chelator) and voltage clamping with a whole-cell pipette during a tetanus fails to block mossy-fiber long-term potentiation (LTP). **A.** An example of excitatory postsynaptic currents (EPSCs) measured with whole-cell voltage clamp recording before and 11 minutes after an LTP-inducing tetanus. The electrode contained 25 mM BAPTA and the cell was held at −90 mV. The mossy fibers were tetanized at 5 Hz for 1 minute. During the tetanus the apparent voltage did not change more than 6 mV and no action potentials were evoked (24). **B.** Summary graph of results from six cells, one of which is illustrated in **A.** A low-frequency LTP-inducing stimulus (5 Hz, 60 seconds) that produced a small amount of mossy-fiber LTP in control slices was used; even with this stimulus the expected magnitude of LTP was obtained, despite the presence of BAPTA and voltage clamping the cell at a holding potential of approximately −100 mV during the tetanus. In all cells the membrane remained well below action potential generation during the tetanus, indicating that the voltage clamp was adequate. Six of six cells showed greater than 20% enhancement at 10 minutes after the tetanus. LTP in the association/commissural (assoc-com) pathyway to normal high-frequency tetani was entirely blocked in these cells. APV was present in four of six cells (7).

the tetanus. Following washout it was clear that the responses had undergone LTP. Indeed, the magnitude of the LTP was indistinguishable from that observed in control slices.

We next repeated the kynurenate experiments using whole-cell patch-clamp techniques so that we could more closely examine the membrane potential during the

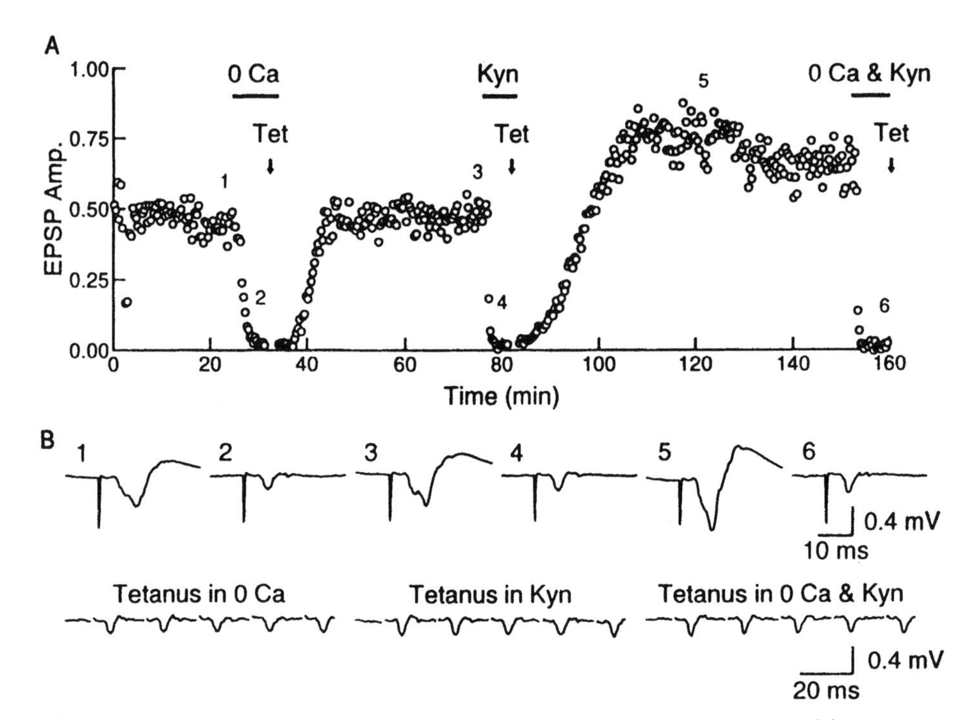

FIG. 2. Mossy fiber long-term potentiation (LTP) is dependent on presynaptic Ca^{++} but not postsynaptic Ca^{++}. **A.** A plot of field potential recordings. Tetanization of the mossy fibers (Tet and downward arrow) in the absence of extracellular Ca^{++} (0 Ca^{++}, 6 mM Mg^{++}, and 100 μM EGTA (a Ca^{++} chelator)) fails to induce LTP. On the other hand, tetanization during blockade of non-N-methyl-D-aspartate (NMDA) receptors with kynurenate (Kyn) (10 mM) evoked normal LTP. **B.** Sample records of the synaptic responses recorded in this experiment. The numbers refer to the time during the experiment (see **A**) when the record was obtained. Note that the responses remaining in "0 Ca" (2), kynurenate (4), and the combined application of "0 Ca" and kynurenate (6) are identical and represent the presynaptic fiber volley. During the tetanus in these three conditions, there was no recovery of the synaptic responses and, as expected for a presynaptic fiber volley, there was no frequency facilitation of the fiber volley response (unpublished observations).

tetanus in kynurenate. We found that even when the membrane potential changed as little as 3 mV during the tetanus, upon washout of the antagonist there was normal LTP. These experiments are in complete accord with our previous experiments, and confirm and extend the work of Ito and Sugiyama (15).They appear to exclude the possibility that the entry of Ca^{++} into the postsynaptic cell or depolarization of the postsynaptic cell play any role in mossy-fiber LTP. They do indicate that entry of Ca^{++} into the presynaptic terminal is critical for LTP induction.

THE EXPRESSION OF MOSSY-FIBER LTP IS PRESYNAPTIC

We have used the phenomenon of paired-pulse facilitation (PPF) to determine if the expression of mossy fiber LTP is presynaptic. PPF is a property of most chemically transmitting synapses in which the size of the second of two closely timed (e.g., 40 milliseconds) excitatory postsynaptic potentials (EPSPs) is enhanced. It is well established that this enhancement is due to an increase in transmitter release and that manipulations that change the release of transmitter change the amount of PPF (16,17). We (7) and others (18) have found that PPF is decreased during mossy-fiber LTP, but not during assoc-com LTP. This is strong evidence that mossy-fiber LTP is expressed presynaptically. Preliminary experiments have been carried out to determine if the sensitivity of synaptic transmission to changing extracellular Ca^{++} is altered by mossy-fiber LTP. Compared to a control, nontetanized pathway or to the responses to elevated Ca^{++} before LTP, the response to elevated extracellular Ca^{++} of a mossy-fiber pathway expressing LTP was very much diminished (Castillo et al., unpublished observations). This is further strong evidence that mossy-fiber LTP is expressed presynaptically. The results summarized thus far in this paper are most consistent with a model in which mossy-fiber LTP is entirely presynaptic, beginning with the induction triggered by the entry of Ca^{++} into the terminal and ending with the maintained enhancement of evoked transmitter release.

THE ROLE OF Ca^{++} CHANNEL SUBTYPES IN MOSSY-FIBER SYNAPTIC TRANSMISSION

Given the evidence that mossy-fiber LTP may be an entirely presynaptic process, our attention next focused on the role of Ca^{++} channels in the control of presynaptic release mechanisms. Ca^{++} channels have been divided into a number of subtypes, based largely on pharmacological agents. Using extracellular recording we have studied the effects of selective antagonists to the different classes of channel on a number of properties of mossy-fiber synapses, including: the release of glutamate that mediates the EPSP; dynorphin, which mediates heterosynaptic inhibition (19); and LTP (14). The L-type channel antagonist, nifedepine, did not affect either glutamate-mediated EPSPs or dynorphin-mediated heterosynaptic inhibition. In addition, it had no effect on LTP. The N-type channel antagonist, ω-conotoxin, reduced mossy-fiber EPSPs by about 75% (Fig. 3A), but had no clear effect on the dynorphin-mediated heterosynaptic inhibition. In addition, LTP could still be evoked after blockade of N-type channels. The P-type channel antagonist, ω-Aga-IVA, entirely blocked the responses remaining in ω-conotoxin. On its own, ω-Aga-IVA blocked the EPSPs a total of 97% (Fig. 3B). Despite the profound depression caused by ω-Aga-IVA, some degree of LTP could still be evoked and ω-conotoxin entirely blocked the responses recorded after inducing LTP. Finally, application of either ω-conotoxin or ω-Aga-IVA after the induction of LTP produced the same degree of antagonism as they did on control, nontetanized pathways. These findings suggest that mossy-fiber terminals contain both N- and P-type Ca^{++} channels, but not L-type. While both channels can contribute to synaptic transmission, P-type channels are primarily responsible for glutamate and probably dynorphin release. The fact that the block by ω-conotoxin and ω-Aga-IVA was considerably greater than 100% is most likely due to the need for a significant amount of Ca^{++} entry to

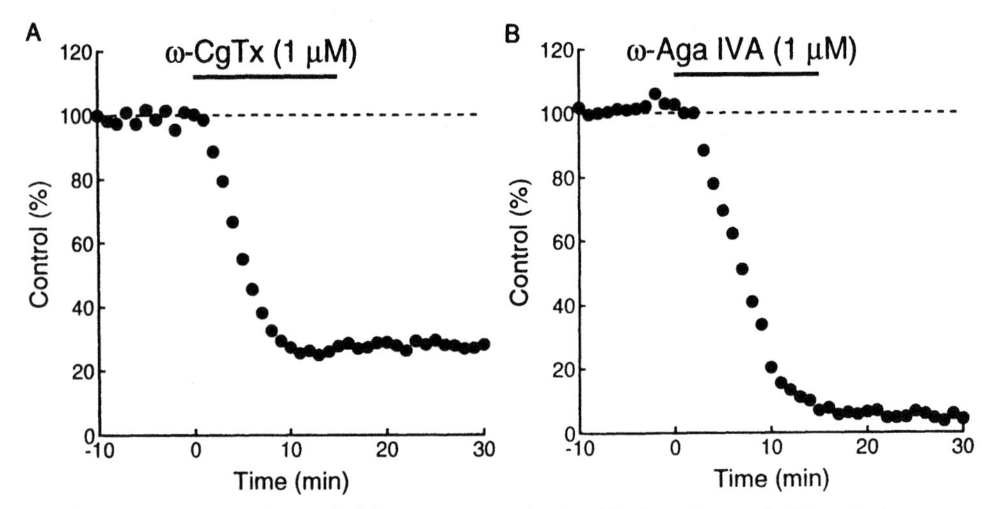

FIG. 3. Both N- and P-type Ca^{++} channels are involved in the release of glutamate from mossy-fiber synapses. **A.** Effect of a saturating concentration of the selective N-type Ca^{++} channel antagonist, ω-conotoxin (ω-CgTx). **B.** Effect of a saturating concentration of the selective P-type Ca^{++} channel antagonist, ω-Aga-IVA (unpublished observations).

reach threshold for transmitter release. No evidence could be obtained that L-, N-, or P-type Ca^{++} channels were selectively modulated during the expression of LTP. This suggests that either the expression of mossy-fiber LTP involves the up regulation of both channel types to the same degree, or more likely, the change occurs downstream from the Ca^{++} entry.

WHAT HAPPENS BETWEEN Ca^{++} ENTRY DURING THE TETANUS AND ENHANCED GLUTAMATE RELEASE

Imaging of Ca^{++} in the mossy-fiber terminals suggests that the maintenance of LTP is not due to a persistent elevation of basal Ca^{++} levels in the terminal (20). We therefore considered a number of possible mechanisms that Ca^{++} might activate. The only positive results that we have obtained thus far are with manipulations that elevate cyclic alenosine monophosphate (AMP) (21). In agreement with previous results (22), we have found that forskolin causes a very large (approximately fourfold) enhancement of mossy-fiber responses. This enhancement is associated with a decrease in PPF, suggesting that, like LTP, the effect is due to an increase in evoked transmitter release. The action of forskolin is mimicked by the analog of cyclic AMP, Sp-cAMPS. Surprisingly, however, it is not mimicked by the analog, 8-bromo-cAMP.

We have carried out three types of experiments to determine if cAMP is involved in mossy-fiber LTP. First, we have examined the time course of the Sp-cAMPS and forskolin. There effects are long-lasting (over 1 hour) but, so far, we have been unable to determine if this is due to the slow washout of the drugs or to a transient rise in cAMP triggering a long-lasting effect. Second, we have examined whether LTP and the forskolin effect share a common mechanism by testing for occlusion. This involved recording the responses from two independent pathways and inducing LTP in one of the pathways. Forskolin was then applied and the effects on the two pathways compared. Finally, the untetanized pathway was tetanized and the magnitude of LTP recorded. We found that LTP was blocked when it was induced during the potentiating action of forskolin. In addition, the action of forskolin was reduced when it was applied after the induction of LTP. These findings indicate that LTP and forskolin share a common site of action. It is interesting to note that a presynaptic form of LTP has been characterized at the crustacean neuromuscular junction in which forskolin has remarkably similar effects (23).

There are two possible ways in which cAMP could be involved in mossy-fiber LTP. Cyclic-AMP could be a necessary step between the entry of Ca^{++} and the enhancement in synaptic transmission, or it could be a parallel pathway that converges onto the same process as LTP. These two alternatives could be distinguished by the use of selective antagonists of the cAMP cascade. Therefore, in the third set of experiments we tested the effects of different classes of inhibitors of the pathway. We have used a number of antagonists, including: 1) calmidazolium to block the action of calmodulin, a mechanism by which Ca^{++} could activate adenylyl cyclase; 2) Rp-cAMPS, to block the action of cAMP; 3) SQ 22,536, to block the adenylyl cyclase; and 4) a number of kinase inhibitors, such as staurosporin and H-7, to block cAMP-dependent protein kinase. Prolonged incubation of slices in these compounds has failed to block mossy fiber LTP. However, we do not have independent evidence that these drugs are effective. Specifically, following these treatments, forskolin is still capable of enhancing mossy fiber responses. In preliminary experiments, the more potent antagonist Rp-8-CPT-cAMPS does appear to block LTP. We are currently testing the effect of this antagonist on the action of forskolin.

CONCLUSIONS

In this chapter we have summarized a number of properties of mossy fiber synapses in the hippocampus that emphasize the marked differences between these synapses and many other type of excitatory synapses. The two most striking differences include the independence of LTP at this synapse on the activation of NMDA receptors and the use of the peptide, dynorphin, as a cotransmitter. Our studies, as well as those of most others, indicate that mossy fiber LTP, unlike NMDA receptor-dependent LTP, is entirely independent of postsynaptic Ca^{++} or postsynaptic membrane potential. Mossy fiber LTP is, however, dependent on the entry of Ca^{++} into the presynaptic terminal. The evidence also indicates that the expression of mossy fiber LTP is presynaptic as well. Therefore, we have examined the role of the L-, N-, and P-types of voltage-dependent Ca^{++} channels in mossy fiber synaptic transmission since they are crucial for transmitter release and are also potential sites of modulation. While both N- and P-type channels are present on the terminals, the P-type is primarily responsible for the release of glutamate and probably dynorphin. In addition, both the induction and expression of mossy-fiber LTP can occur after blockade of L-, N-, or P-type channels. These findings indicate that LTP expression is not due to the selective modulation of presynaptic Ca^{++} channels. Activation of cAMP by forskolin and Sp-cAMPS enhances mossy-fiber synaptic responses, and this effect partially occludes with LTP. This raises the possibility that mossy-fiber LTP involves the activation of cAMP, similar to results obtained at the crustacean neuromuscular junction. Important questions that remain to be addressed concern the second-messenger systems involved and the nature of the cellular changes that account for the lasting enhancement in evoked glutamate release.

REFERENCES

1. Bliss TV, Collingridge GL. A synaptic model of memory: long-term potentiation in the hippocampus. *Nature* 1993;361:31–39.
2. Nicoll RA, Wyllie DJA, Manabe T, Perkel DJ. Current physiological models for long-term potentiation in the CA1 region of the hippocampns. In: Selveston AI, Ascher P, eds. *Cellular and molecular mechanisms underlying higher neural functions*. Chichester, England: John Wiley & Sons Ltd; 1994.
3. Gustafsson B, Wigstrom H. Basic features of long-term potentiation in the hippocampus. *Semin Neurosci* 1990;2:321–333.
4. Monaghan DT, Cotman CW. Distribution of N-methyl-D-aspartate-sensitive L-[^{3}H] glutamate-binding sites in rat brain. *J Neurosci* 1985;5:2909–2919.
5. Harris EW, Cotman CW. Long-term potentiation of guinea-pig mossy fiber responses is not blocked by N-methyl- D-aspartate antagonists. *Neurosci Lett* 1986;70:132–137.
6. McLean S, Rothman RB, Jacobson AE, Rice KC, Herkenham M. Distribution of opiate receptor subtypes and enkephalin and dynorphin immunoreactivity in the hippocampus of squirrel, guinea pig, rat, and hamster. *J Comp Neurol* 1987;255:497–510.
7. Zalutsky RA, Nicoll RA. Comparison of two forms of long-term potentiation in single hippocampal neurons. *Science* 1990;248:1619–1624.
8. Lynch G, Larson J, Kelso S, Barrionuevo G, Schottler F. Intracellular injections of EGTA block induction of hippocampal long-term potentiation. *Nature* 1983;305:719–721.
9. Malenka RC, Kauer JA, Zucker RJ, Nicoll RA. Postsynaptic calcium is sufficient for potentiation of hippocampal synaptic transmission. *Science* 1988;242:81–84.
10. Katsuki H, Kaneko S, Tajima A, Satoh M. Separate mechanisms of long-term potentiation in two input systems to CA3 pyramidal neurons of rat hippocampal slices as revealed by the whole-cell patch-clamp technique. *Neurosci Res* 1991;12:393–402.
11. Langdon RB, Johnson JW, Barrionuevo G. Long-term potentiation of low-amplitude EPSCs elicited by mossy fiber stimulation in rat hippocampus. *Soc Neurosci Abst* 1993;19:433.
12. Williams S, Johnston D. Long-term potentiation of hippocampal mossy fiber synapses is blocked by postsynaptic injection of calcium chelators. *Neuron* 1989;3:583–588.
13. Jaffe D, Johnston D. Induction of long-term potentiation at hippocampal mossy-fiber synapses follows a Hebbian rule. *J Neurophysiol* 1990;64:948–960.
14. Castillo PE, Weisskopf MG, Nicoll RA. The role of Ca^{2+} channels in hippocampal mossy fiber synaptic transmission and long-term potentiation. *Neuron* 1994;12:261–269.
15. Ito I, Sugiyama H. Roles of glutamate receptors in long-term potentiation at hippocampal mossy fiber synapses. *Neuro Report* 1991;2:333–336.
16. Katz B, Miledi R. The role of calcium in neuromuscular facilitation. *J Physiol* 1968;195:481–492.
17. Manabe T, Wyllie DJA, Perkel DJ, Nicoll RA. Modulation of synaptic transmission and long-term potentiation: effects on paired pulse facilitation and EPSC variance in the CA1 region of the hippocampus. *J Neurophysiol* 1993;70:1451–1459.

18. Staubli U, Larson J, Lynch G. Mossy fiber potentiation and long-term potentiation involve different expression mechanisms. *Synapse* 1990;5:333–335.
19. Weisskopf MG, Zalutsky RA, Nicoll RA. The opioid peptide dynorphin mediates heterosynaptic depression of hippocampal mossy fiber synapses and modulated long-term potentiation. *Nature* 1993;362:423–427.
20. Regehr WG, Tank DW. The maintenance of LTP at hippocampal mossy fiber synapses is independent of sustained presynaptic calcium. *Neuron* 1991;7:451–459.
21. Weisskopf MG, Zalutsky RA, Nicoll RA. Cyclic-AMP-mediated enhancement and LTP at mossy fiber synapses in the hippocampus. *Soc Neurosci Abst* 1993;19:1708.
22. Hopkins WF, Johnston D. Noradrenergic enhancement of long-term potentiation at mossy fiber synapses in the hippocampus. *J Neurophysiol* 1988;59:667–687.
23. Dixon D, Atwood L. Adenylate cyclase system is essential for long-term facilitation at the crayfish neuromuscular junction. *J Neurosci* 1989;9:4246–4252.
24. Zalutsky RA, Nicoll RA. Mechanisms of long-term potentiation in CA3 hippocampal pyramidal cells. In: Barnard EA, Costa E, ed. *Transmitter amino acid receptors: structures, transduction, and models for drug development*. New York: Georg Thieme Verlag, 1991;6:415–422.

Molecular and Cellular Mechanisms of Neurotransmitter Release, edited by Lennart Stjärne, Paul Greengard, Sten Grillner, Tomas Hökfelt, and David Ottoson, Raven Press, Ltd., New York © 1994.

29

Communication of Synaptic Potentiation Between Synapses of the Hippocampus

Erin M. Schuman and *Daniel V. Madison

*Division of Biology, California Institute of Technology, Pasadena, California 91125; and *Department of Molecular and Cellular Physiology, Stanford Medical School, Stanford, California 94306*

Many models of basis of learning, memory, and neuronal development include anatomical or functional reorganization that rely on alterations of the strength of synaptic connections between neurons as their basis. In many of these models, the conjoint activity of presynaptic fibers and a postsynaptic target cell can lead to a strengthening of the active synaptic connections (1) that ultimately enhances the function or survival of the synapse. Both theoretical and experimental evidence suggests that groups of synapses of common origin tend to be enhanced together, implying that communication occurs between like synapses such that synapses that are coactive tend to function as a group (2,3).

One phenomenon that has been advanced as a potential physiological mechanism for these forms of plasticity is long-term potentiation (LTP), the long-lasting increase in synaptic transmission induced by intense synaptic activity (4). While LTP is clearly a persistent enhancement of synaptic transmission, it is not certain that it possesses all the properties that would be necessary for mediating the kinds of intercellular communication necessary to underlie the strengthening of coactive afferents. Apparently to the contrary, it has long been known that LTP has the property of "synapse specificity" (5,6). Stated simply, when LTP is induced by a tetanic stimulation in one group of afferent fibers, other nontetanized afferent fibers, even other afferents to the same postsynaptic cell as the tetanized pathway, are not potentiated. This synapse specificity suggests that LTP cannot be communicated between synapses, and thus that LTP is not an appropriate substrate for intersynaptic communication.

Recently, data has appeared that suggests that the induction of LTP may involve some sort of diffusible messenger (7,8). This messenger has been proposed to diffuse from postsynaptic cells where LTP is induced, and to interact with other synaptic elements such as presynaptic terminals. Candidates for such a messenger include arachidonic acid (9), carbon monoxide (10,11), nitric oxide (12,13,14,15), and platelet activating factor (16,17,18). These signals may, in theory, mediate both pre- (13,16,19) and postsynaptic (17,20) processes leading to synaptic enhancement. Such a diffusible messenger could act in a strictly retrograde manner, influencing only the synapses where it is generated, or could also act to enhance the synapses of nearby neurons. Consistent with this view, Bonhoeffer and colleagues (21,22) showed that an LTP induction procedure that involves pairing postsynaptic depolarization of a single neuron with low-frequency stimulation of afferent fibers (23) resulted in a decrease in the action potential latency in both the depolarized cell and a nearby cell. These results suggested that potentiation induced in a single neuron may be communicated to nearby neurons.

METHODS

To test directly the idea that LTP induced in the synapses onto a single neuron could be communicated by a diffusible messenger to synapses on neighboring cells, intracellular recordings were made from two neighboring hippocampal CA1 pyramidal cells simultaneously. These experiments were performed in in vitro hippocampal slices. Hippocampal slices were prepared as described (24). Hippocampal slices were submerged in a stream of solution containing 119 mM NaCl, 2.5 mM KCl, 1.3 mM $MgSO_4$, 2.5 mM $CaCl_2$, 1.0 mM NaH_2PO_4, 26.2 mM $NaHCO_3$, and 11.0 mM glucose. ACSF was maintained at 22°C and was gassed with 95% O_2, 5% CO_2. Intracellular recording electrodes were filled with 2 M cesium acetate. In intracellular injection experiments, L-Me-Arg (100 mM) was dissolved in cesium acetate. Whole-cell recordings were made in the single electrode voltage-clamp mode (Axoclamp 2C) with 75% to 90% series resistance compensation. Whole-cell patch-clamp internal solutions consisted of 100 mM Cesium gluconate, 10 mM BAPTA (a Ca^{2+} chelator), 5 mM MgCl, 2 mM adenosine 5′-triphosphate (ATP), 0.3 mM guanosine 5′-triphosphate (GTP), and 40 mM HEPES. Intracellular excitatory postsynaptic potential (EPSP) or whole-cell excitatory postsynaptic currents (EPSCs), measured in CA1 pyramidal cells, were evoked by stimulation of the Schaffer collateral-commissural afferents (4/min). In dual recording experiments, the presence of synaptic and/or electrical connectivity was tested for by eliciting action potentials in one neuron and observing the response of the second neuron. Each pair was tested for coupling at the beginning of each experiment and also tested at the conclusion of experiments where recordings were maintained for a sufficient period of time. The LTP induction by pairing involved sustained depolarization of the pyramidal neuron by DC current injection in conjunction with low-frequency (1 Hz) stimulation of the test pathway for 30 seconds. The LTP induction by tetanus involved four high-frequency trains of stimulation (100 Hz for 1 second, 30-second interval) delivered at the test intensity. Data were collected and analyzed with software written by us for this purpose under the Axobasic programming environment (Axon Instruments Inc., Foster City, CA). This software measured the amplitude and/or slope of the EPSP or EPSC. Statistical comparisons stated in the text were made with the Student's *t* test, performed on nonnormalized data.

Data are displayed as ensemble average plots representing group means of each EPSP(C), across experiments, aligned with respect to the time of acquisition relative to the time of LTP induction by pairing. Each individual experiment was normalized with respect to the mean value of its EPSP in the 50 responses that preceded the tetanus or pairing. For each group of experiments all data were included that met the following criteria: 1) resting membrane potential of each neuron < -60 mV, and 2) recordings of both cells held for at least 50 minutes following the induction of LTP. The group data shown in the ensemble averages and described in the text reflect all the data that fit these criteria. Experiments where the "paired" neurons failed to exhibit LTP were not excluded. For three experiments in which we had intended to impale neurons that were far apart, the biocytin data revealed that these neurons were in fact close ($<$300 μm). These data were switched to the nearby group.

The basic experimental design was quite simple. After we obtained both intracellular recordings, single shocks were delivered to the Schaffer collaterals at a rate of 1/15 s, until at least 30 minutes of baseline data had been collected. After the baseline period, one of the two postsynaptic cells was depolarized to 0 mV with injected current. During this depolarization, single shocks were delivered through the stimulating electrode at a rate of 1 Hz for approximately 30 pulses. This protocol, known as pairing, has long been known to be a very effective means of activating LTP in the depolarized cell (23). After this stimulation, the paired cell was returned to its resting potential, and stimulation at 1/15 s was resumed to test for changes in synaptic strength (Fig. 1).

RESULTS

Pairing almost always resulted in the production of LTP in the paired cell (17 of 20 trials). The response in the "neighbor" cell that had not undergone pairing was also examined. In 16 of 17 cases where the paired cell showed LTP, the transmission to the neighbor was also persistently enhanced (Fig. 2a). In all cases, the tests for electrical coupling between the paired and neighbor cell were made by inducing action potentials in one and looking for any response in the other. In no case was evidence found for such coupling. Neither was any evidence found that the two cells were synaptically coupled. Since during the pairing the afferents to both cells were activated at 1 Hz, it was also important to test for any effects of 30 seconds of 1 Hz stimulation alone. No persistent effects were observed. In several experiments a transient (~5–10 minutes) depression of synaptic responses was observed.

In this series of 20 experiments, the EPSP amplitude at the paired-cell synapses increased to 183.8 +/− 15.0% of baseline levels (mean percentage of baseline @ 1 hour following LTP induction +/− s.e.m., n = 20). The transmission at the neighboring cell synapses increased to 129.3 +/− 9.1% of baseline levels. The initial slope of the EPSP also increased in the paired cell to a level 183% of baseline, and the EPSPs onto the neighbor cell increased 137%. Seventeen of the paired neurons

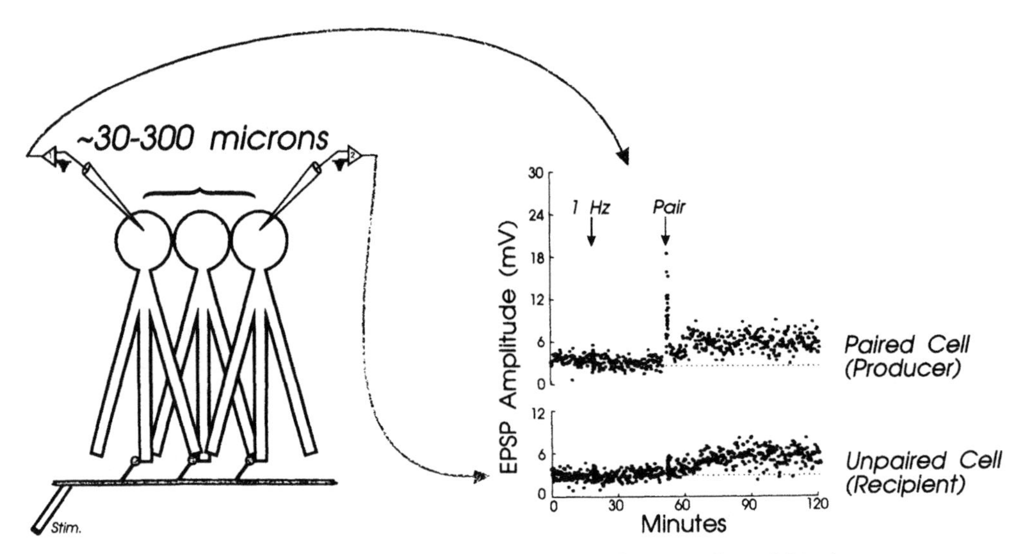

FIG. 1. Experimental schema and individual experiments demonstrating neighboring synapse potentiation. **A.** Schematic diagram showing the position of stimulating and recording electrodes. Two nearby CA1 pyramidal cells were impaled and the excitatory responses to test shocks delivered to the Schaffer collaterals (stim) by a bipolar stimulating electrode were recorded. **B.** Plots of excitatory postsynaptic potential (EPSP) amplitude obtained with intracellular recording from the paired (top) and neighboring (bottom) neuron. Low-frequency stimulation (1 Hz) and low-frequency stimulation paired with depolarization (PAIR) are indicated with arrows. The first two traces in each plot are representative EPSPs before and after longterm potentiation (LTP) induction of the paired neuron, and the third trace is the superimposition of the two. Scale bar is 10 mV and 10 milliseconds. (Figure adapted from Schuman and Madison, ref. 34.)

exhibited potentiation that was at least 130% of baseline levels, with individual values for all 20 cells ranging from no potentiation to 330%. The neighboring neurons exhibited an average enhancement of potentiation that was at least 115% of baseline levels, with individual values for all 20 cells ranging from no potentiation to 253%. These results demonstrate that LTP induction in a single CA1 pyramidal cell can be communicated to synapses onto neighboring neurons. Since 1 Hz stimulation alone produced no change in synaptic efficacy on either the paired or neigh-

FIG. 2. A. Ensemble averages for dual intracellular recordings from nearby pyramidal neurons (n = 20). Top plot shows average data for the paired neurons in which longterm potentiation (LTP) was induced by pairing postsynaptic depolarization with 1 Hz stimulation of afferent fibers (PAIR). Bottom plot shows average data for the enhancement observed in the neighboring neuron in response to LTP induction in the paired neuron. Please note difference in y-axis values in top and bottom plots. **B.** Ensemble averages for dual intracellular recordings from distant (~500 μm) CA1 pyramidal neurons (n = 17). Top plot shows average data for the paired neuron where LTP was induced by pairing (PAIR). Bottom plot shows average data for the synaptic strength at synapses onto the distant neighboring neuron. No enhancement of synaptic transmission was observed. Please note difference in y-axis values in top and bottom plot. (Figure adapted from Schuman and Madison, ref. 34.)

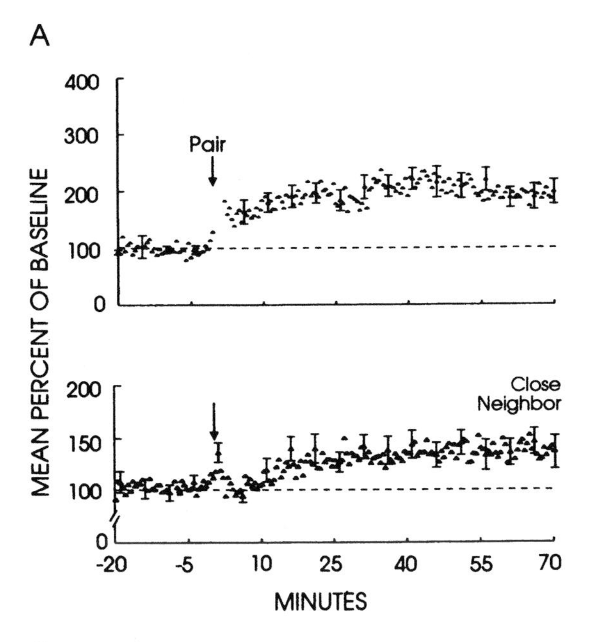
A
Pair
MEAN PERCENT OF BASELINE
400
300
200
100
0
Close
Neighbor
200
150
100
0
-20
-5
10
25
40
55
70
MINUTES

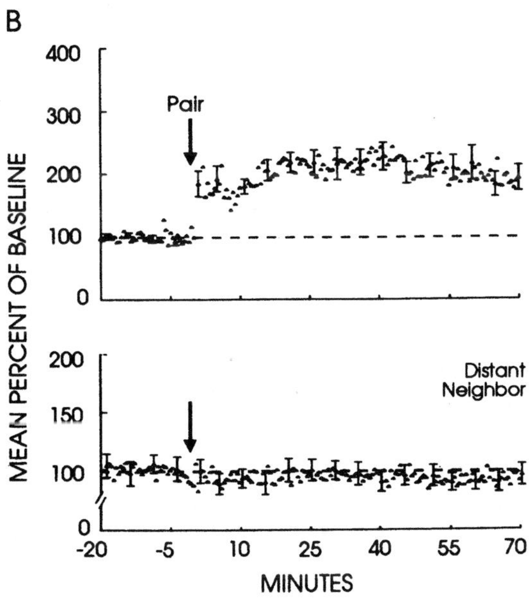
B
Pair
MEAN PERCENT OF BASELINE
400
300
200
100
0
Distant
Neighbor
200
150
100
0
-20
-5
10
25
40
55
70
MINUTES

bor cell, and changes were only observed when one cell was also depolarized, the enhancement of synaptic strength observed at the neighboring cell synapses must be a result of LTP induction in the paired cell.

We have proposed that the communication of potentiation from one cell's synapses to another may result from communication subserved by a diffusible messenger signal. Most of the candidate molecules for a diffusible signal have very limited lifetimes in biological tissues. This leads to the prediction that communication of synaptic potentiation between synapses should occur over a very limited spatial area. To test for a relationship between the amount of synaptic potentiation that is communicated between cells and the distance between those cells, in several experiments biocytin (1%) was included in the recording electrodes, and filled each pyramidal neuron during the course of the experiment (25). Biocytin-filled neurons were later visualized with a light microscope and camera lucida tracings were made. The intersomatic distance between the two recorded cells was measured as an indicator of the distance between them. The mean distance between CA1 cell somata for the initial experiments was 142.1 +/− 36.1 μm (range: 20.0–285.0 μm; n = 9). It was not possible to measure the distance between the dendritic processes due to the complexity of these processes, but in all cases there was extensive overlap of the dendritic processes of the two cells. Thus, the closest approach of the two cells was actually much less than the measured intersomatic distance.

To test directly for limitations in the spatial spread of communicated synaptic potentiation between cells, a second series of experiments was performed in which simultaneous intracellular recordings were made from cells that were located relatively far apart. One of the two cells was then "paired," under experimental conditions that were identical to the first series of experiments except for the distance between the cells. As expected, LTP occurred in the paired cell and enhancement of transmission in the "distant neighbor" cell was never observed (Fig 2b., n = 17). On average the amplitudes of EPSPs onto the paired cell were enhanced 173.6 +/− 10.7% 1 hour after pairing, while the amplitudes of distant EPSPs onto distant neighbor cells were 97.8 +/− 8.6% of control. The average distance between distant neighbor cells was 595 +/− 58.8 μm apart (range: 340–800 μm). In none of these cases was there any detectable dendritic overlap between the two cells.

Taken together, communication of synaptic potentiation observed in close neighbor pairs of cells and the lack of communication onto distant pyramidal cell synapses indicate a spatial restriction in the ability of synapses to communicate potentiation to synapses on other cells. When the relationship between the amount of communicated potentiation and the intersomatic distance between pairs of cells were tested for those pairs where biocytin was present (n = 16, "far" and "near"), there was a significant correlation between the distance and the amount of synaptic enhancement in the neighbor cell (Spearman rank correlation = −.58, $p < .02$). Thus, the degree of potentiation observed at neighboring cell synapses is likely a function of the distance between the neighboring cell synapses and the synapses where LTP induction is occurring (as well as a function of the amount of potentiation induced in the paired cell, data not shown).

We tested for a contribution of one of the putative diffusible messenger candidates, nitric oxide, to the communication of potentiation between cells. This was done by making another series of recordings from pairs of closely spaced CA1 pyramidal cells, and by injecting one of these cells with the nitric oxide synthase inhibitor N-methyl-L-arginine (L-Me-Arg; Fig. 3a). The injection was done by including 100 mM of L-Me-Arg into one of the intracellular recording electrodes (Fig. 3a). The injected cell was then "paired," and observed for the production of LTP. As previously reported (13,14), cells injected with L-Me-Arg into the paired cell prevented long-lasting synaptic potentiation at its own synapses (95.8+/−6.9%). The L-Me-Arg injected into the paired cell also prevented the enhancement of synaptic strength at neighboring cell synapses (102.2+/−10.1%; n=15)

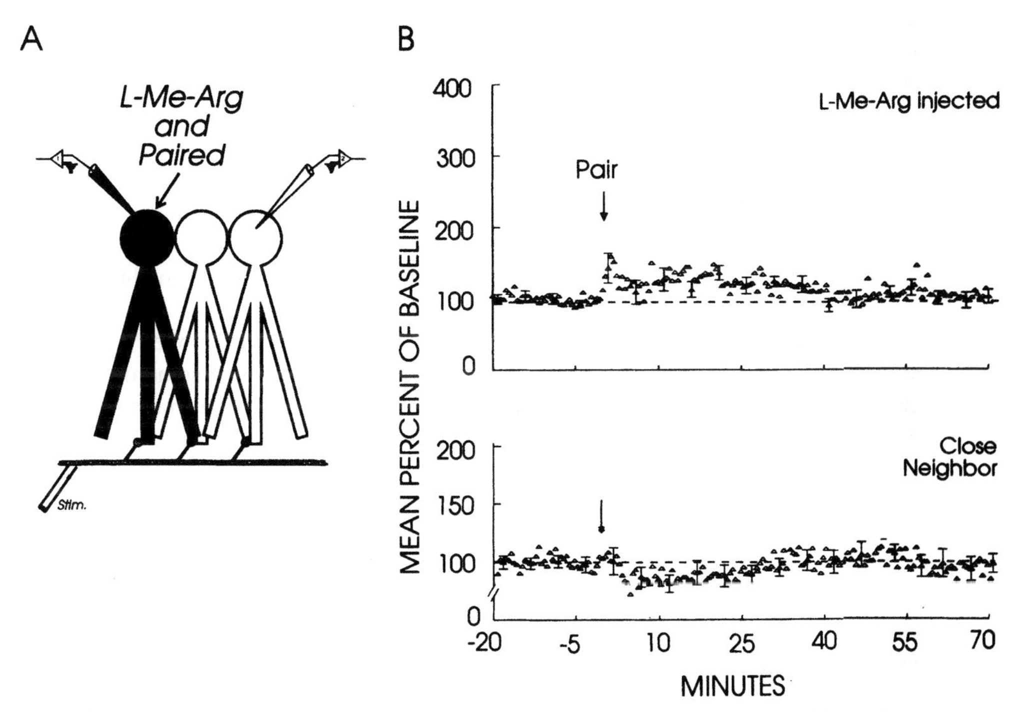

FIG. 3. A. Modified experimental schema, showing that one cell of close neighbor pairs was injected with L-Me-Arginine before pairing. **B.** Ensemble averages for dual intracellular recordings from two nearby pyramidal neurons; the paired cell (top) was filled with the NO synthase inhibitor L-Me-Arg (n=18). Postsynaptic injection of L-Me-Arg blocked not only pairing-induced longterm potentiation (LTP) in the "paired" cell, but also blocked the spread of enhancement in the neighboring cell. Top plot shows average data for the LTP produced by pairing (PAIR). Bottom plot shows average data for the enhancement observed in the neighboring neuron in response to LTP induction in the producer neuron. Please note difference in y-axis values in top and bottom plot. (Figure adapted from Schuman and Madison, ref. 34.)

(Fig. 3b). Camera lucida reconstructions revealed that the mean distance between pairs of pyramidal neurons in these experiments was statistically indistinguishable from the distance between the pairs of neurons that comprise the near group ($x = 143.3 +/- 43.6$ μm; $n = 7$).

In some of the experiments just discussed, the intracellular recordings lasted for a sufficient time to allow us to pair the uninhibited cell (i.e., previously the "neighbor" cell) and to test both for LTP in this cell and a communication of potentiation to the cell injected with L-Me-Arg. In three of five of these cases, pairing depolarization of the neighboring cell with low-frequency stimulation resulted in the enhancement of synaptic strength at synapses onto the cell injected with L-Me-Arg. As expected, the uninhibited cells showed LTP upon pairing, at the usual frequency. This result suggested that potentiation could be reconstituted in cells with inhibited NO synthase by a messenger, perhaps nitric oxide, generated by potentiated near neighbor cells. To test this idea, a series of experiments was performed where one CA1 pyramidal cell was injected with L-Me-Arg, and Schaffer collaterals were tetanized to produce LTP within the entire population of synapses onto other CA1 cells. Such tetanic stimulation resulted in persistent potentiation in synapses onto the injected cell ($159.3 +/- 19.7\%$; $n = 15$). The fact that the pairing protocol could not produce LTP in this cell (13,14) suggested that: 1) when a cell has its NOS inhibited, it alone was not able to support the induction of LTP; and 2) when a cell's neighbors had LTP induced, this potentiation could be communicated to the inhibited cell. It seems unlikely that this difference results from an inability of L-Me-Arg to block tetanus-induced LTP, since under our experimental conditions bath application of the inhibitor is able to prevent it.

The idea that neighboring neurons can reconstitute LTP in NOS inhibited cells, while supported by the data just presented, may be inconsistent with some of the known properties of LTP. Foremost of these is the known ability of several agents to block tetanus-induced LTP when injected into postsynaptic cells. For example, injection of calcium chelators (26,27) into postsynaptic CA1 cells, or hyperpolarization of postsynaptic cells during tetanic stimulation (28,29) are sufficient to prevent tetanic stimulation-induced LTP in the synapses onto those cells. If the nature of the communication of potentiation between cells is essentially retrograde (i.e., the messenger produced by postsynaptic cells and communicated to presynaptic terminals where its action is exerted), then it is difficult to reconcile the ability of a purely postsynaptic manipulation to block tetanus-induced LTP. Put another way, if potentiation in an L-Me-Arg injected cell is produced by the presynaptic action of, say, nitric oxide supplied by neighboring neurons, then in the simplest model the postsynaptic (inhibited) cell should not exert any influence, and any postsynaptic manipulation of that cell should not disrupt the production of tetanus-induced LTP.

Clearly, the simplest model, where a messenger from neighboring cells, acting on the presynaptic terminal, is entirely sufficient to produce potentiation, is an inadequate explanation of all the known data. Rather, the postsynaptic side of the "receiving" synapse must exert some influence. To test this idea, a final series of experiments was conducted. Again, two simultaneous intracellular recordings were made from two, near neighbor, CA1 pyramidal cells activated by the same Schaffer collateral input. One of these cells was treated simultaneously with three manipulations known to disrupt LTP (Fig. 4a). First, it was injected with the calcium chelator BAPTA and was recorded with a whole-cell electrode (rather than a standard intracellular "sharp" electrode) to dialyze the intracellular compartments. The other, untreated cell, was subjected to the "pairing" protocol to induce LTP. During this pairing, the other (BAPTA, dialyzed) cell was voltage-clamped at −95 mV to further disrupt LTP induction. Note that these three manipulations—BAPTA, dialysis, and hyperpolarization—were entirely postsynaptic manipulations. In this series of experiments it was found that the untreated cell produced LTP as expected

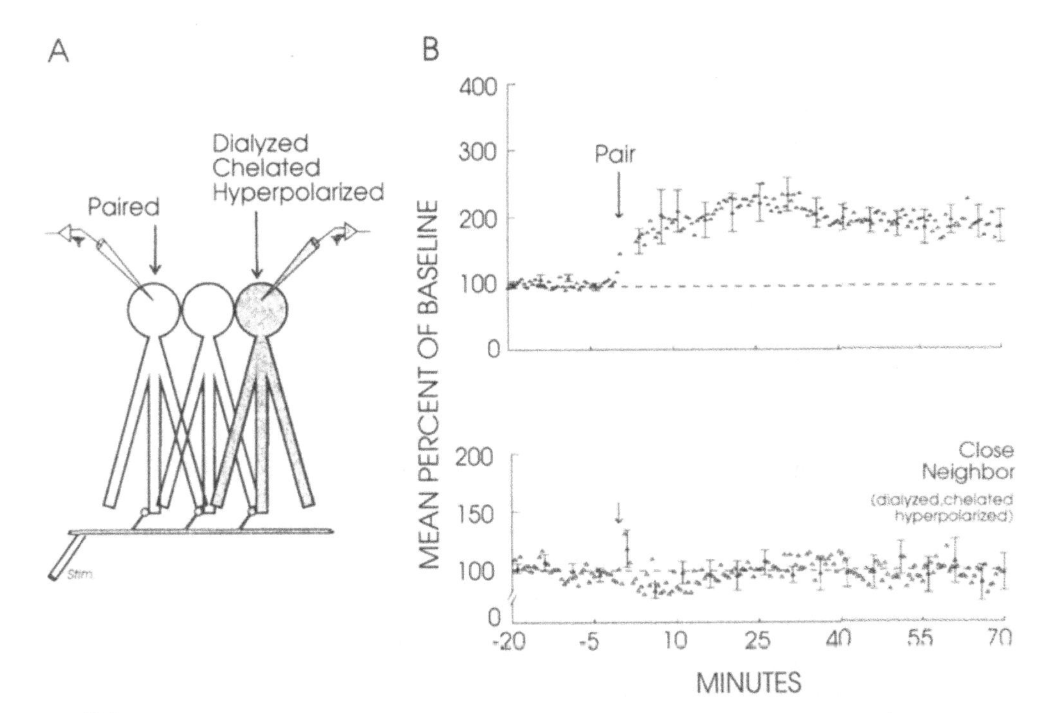

FIG. 4. A. Modified experimental schema, showing that one cell of close neighbor pairs was injected with BAPTA (a Ca^{2+} chelator) and whole-cell dialyzed. The cell was also hyperpolarized during the time its neighbor cell was being paired (depolarized during 1 Hz stimulation). **B.** Ensemble averages for dual recordings from two nearby pyramidal neurons (n = 14); the paired cell was recorded from with a standard intracellular micropipette; the neighboring cell (bottom) was recorded from in whole-cell voltage-clamp mode and filled with BAPTA. During induction of LTP in the paired cell, the neighboring neuron was hyperpolarized to −95 mV. Top plot shows average data for the LTP produced by pairing (PAIR). Bottom plot shows average data for the enhancement observed in the neighboring neuron in response to LTP induction in the producer neuron. The combination of postsynaptic dialysis, Ca^{2+} chelation, and hyperpolarization of the neighboring neuron prevented the synaptic enhancement typically observed. Please note difference in y-axis values in top and bottom plot. (Figure adapted from Schuman and Madison, ref. 34.)

(193.8+/−26.3%; n=14), but that the dialyzed-chelated-hyperpolarized near neighbor cell did not exhibit any potentiation (97.6+/−14.4%). These cells were similarly close together anatomically (x=79.4+/−21.6 μm), showing that inter-cell distance could not explain the lack of communication of potentiation. The inability of the neighboring neuron to exhibit synaptic potentiation under the present conditions suggests that the postsynaptic neuron may actively participate in the enhancement.

DISCUSSION

We have described new evidence for a form of synaptic enhancement that involves spatially limited long-lasting enhancement of synaptic transmission, occurring at synapses near the site of LTP induction. The magnitude of the synaptic potentiation of neighboring cell synapses ranged from no potentiation to a doubling of baseline values; on average, neighboring synapses exhibited potentiation that was roughly one third of the magnitude of the LTP exhibited at "paired"-cell synapses. The variability in the magnitude of the potentiation may, in part, reflect the distance of the neighboring synapses from the site of LTP induction, and thus could be related to the concentration of the diffusible signal that reaches the neighboring synapses. The onset and duration of the neighboring synapse enhancement was similar to that observed in the paired cells, although on average the apparent onset of the potentiation was slightly delayed. We attribute this delay to a transient depression of synaptic transmission that is often caused by the low-frequency stimulation used to induce LTP in the paired neurons.

Previous studies of LTP have suggested that only the synapses of an active afferent pathway become potentiated when a single CA1 neuron is depolarized during low-frequency stimulation; the synapses of a second, independent inactive pathway do not exhibit potentiation (5,6). Under our experimental conditions, both the neighboring and the paired neurons receive synaptic input from a common set of afferent fibers; thus, both sets of synapses were active during LTP induction. It is worthwhile to note that increases in synaptic strength produced by diffusible signals such as arachidonic acid (9), nitric oxide (12,13,14,15), and carbon monoxide (10,11) appear to require activity. Determining whether the neighboring synapse potentiation also occurs in the absence of activity will require a simultaneous examination of both inactive and active synapses near the site of LTP induction. This task promises to be technically challenging, given the difficulty in obtaining independent presynaptic pathways that are in close proximity to one another in the hippocampal slice preparation.

An additional factor that may contribute to the spread of this potentiation appears to be the proximity of the paired and neighbor synapses. Our results indicate that synapses onto distant-neighbor cells are not potentiated, despite receiving the same synaptic input as paired cells. As such, these results are consistent with the synapse specificity that has been reported for LTP in which no spreading of potentiation is observed between two independent afferent pathways that converge on the same postsynaptic neuron. The distance between synapses on neighboring cells is likely to be much less than the intersomatic distances reported here, since there is extensive dendritic overlap in all of these pairs of cells.

The demonstration that LTP induction in a single CA1 pyramidal neuron can potentiate synaptic transmission at neighboring cell synapses in a spatially restricted manner implies the existence of a diffusible signal generated in paired neurons during LTP induction. The injection of L-Me-Arg in the paired cell prevented the neighboring cell enhancement, suggesting that NO production is important for this phenomena, if not as the messenger itself that serves as an essential element upstream from the message production. In addition, it was observed that the blockade of LTP produced by postsynaptic injections of NO synthase inhibitors could be overcome by LTP induction in neighboring cells. Taken together, these two results suggest that during LTP induction NO may diffuse to increase synaptic strength at nearby synapses. The NO may interact with other diffusible signals such as arachidonic acid and carbon monoxide to bring about the potentiation of synaptic strength at neighboring synapses.

Our results do not address whether the enhancement of synaptic transmission at neighboring synapses is mediated by pre- or postsynaptic mechanisms. Since a combination of postsynaptic dialysis, Ca^{2+} chelators, and hyperpolarization of the neighboring cell prevented the enhancement, the postsynaptic neighbor cell is likely to play a crucial role in this communication. Either one of two general possibilities can account for these findings; these are schematized in Fig. 5. First, the communication of potentiation could occur entirely postsynaptically, between the postsynaptic cell emitting the messenger and a neighboring postsynaptic cell receiving it. According to this idea, a single messenger produced in the paired cell would diffuse to the neighboring neuron and interact with a Ca^{2+} and/or voltage-dependent postsynaptic target to bring about the enhancement. Second, the communication of potentiation could occur between the postsynaptic side of the transmitting synapse and the presynaptic side of the receiving synapse. In this latter case, the postsynaptic receiving cell must still play at least a permissive role in this communication. This contribution could be either a constitutive or stimulated process. As one example of a constitutive process, intracellular Ca^{2+} activity that results from the tonic stimulation of postsynaptic N-methyl-D-aspartate (NMDA) channels (30) or window currents through voltage-dependent Ca^{2+} channels, might support the release of a local retrograde messenger that is permissive for the communication of potentiation between synapses. Alternatively, the local messenger may be stimulated by depolarization arising either from the 1 Hz stimulation or from extracellular K^+ produced during depolarization of the paired cell (31,32). According to both of these ideas, both the signal generated in the paired neuron during LTP induction and the Ca^{2+} and/or voltage-dependent contribution of the postsynaptic receiving cell would be required for the communication of potentiation between cells.

Regardless of the actual molecular mechanisms, the clear existence of synaptic potentiation communicated between cells suggests that the formation of synaptic

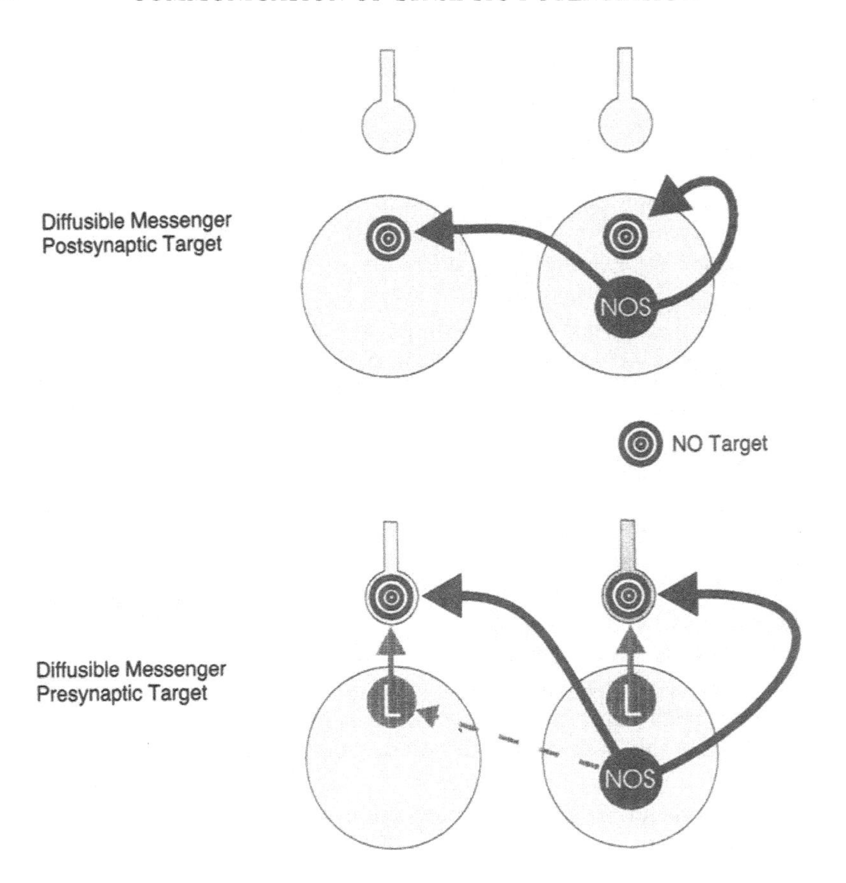

FIG. 5. Schematic representation of two general possibilities of how NO might mediate communication of potentiation between synapses onto different pyramidal cells. In the top panel, NO's action is restricted to postsynaptic cells. This may include both the producing cell and its near neighbors. In the bottom panel, NO travels from its postsynaptic producer cell to its associated presynaptic terminals, and also to the presynaptic terminals of neighboring neurons. To account for the data showing the necessity of postsynaptic involvement in communication of potentiation, it is proposed that the receiving postsynaptic cell also supplies a necessary, perhaps permissive component to the communication. One example of such a contribution might be a local retrograde messenger (L). The NO from the postsynaptic paired cell may also interact with the postsynaptic receiving cell under a variation of this model (dashed gray line) to stimulate production of a postsynaptic permissive factor.

changes previously thought to be entirely synapse specific can also result in synaptic changes at nearby synapses, at least over limited distances. Diffusible signals may act as retrograde synaptic signals or alternatively as effectors of the amplification of synaptic transmission at active synapses in close proximity to the site of messenger generation. These results may be important in the elucidation of plastic processes both in the developing and adult brain (33).

REFERENCES

1. Hebb DO. *The organization of behavior*. New York; John Wiley 1949.
2. Kandel ER, O'Dell TJ. Are adult learning mechanisms also used for development? *Science* 1992; 258:243–245.
3. Goodman CS, Shatz CJ. Developmental mechanisms that generate precise patterns of neuronal connectivity. *Neuron* 1993;10:77–98.
4. Bliss TV, Lomo T. Long-lasting potentiation of synaptic transmission in the dentate area of the anaesthetized rabbit following stimulation of the perforant path. *J Physiol* 1973;232:331–356.
5. Andersen P, Sundberg SH, Sveen O, Wigstrom H. Specific long-lasting potentiation of synaptic transmission in hippocampal slices. *Nature* 1977;266:736–737.
6. Lynch GS, Dunwiddie T, Gribkoff V. Heterosynaptic depression: a postsynaptic correlate of long-term potentiation. *Nature* 1977;266:737–739.
7. Bliss TVP, Douglas RM, Errington ML, Lynch MA. Correlation between long-term potentiation and release of endogenous amino acids from dentate gyrus of anaesthetized rats. *J Physiol* 1986; 377:391–408.
8. Gally JA, Montague PR, Reeke GN, Edelman GM. The NO hypothesis: possible effects of a short-lived, rapidly diffusible signal in the development and function of the nervous system. *Proc Natl Acad Sci USA* 1990;87:3547–3551.
9. Williams JH, Errington ML, Bliss TVP. Arachidonic acid induces a long-term activity-dependent enhancement of synaptic transmission in the hippocampus. *Nature* 1989;341:739–742.
10. Zhuo M, Small SA, Kandel ER, Hawkins RD. Nitric oxide and carbon monoxide produce activity-dependent long-term synaptic enhancement in hippocampus. *Science* 1993;260:1946–1950.
11. Stevens CF, Wang Y. Reversal of long-term potentiation by inhibitors of haem oxygenase. *Nature* 1993;364:147–149.
12. Bohme GA, Bon C, Stutzmann JM, Doble A, Blanchard JC. Possible involvement of nitric oxide in long-term potentiation. *Eur J Pharmacol* 1991;199:379–381.
13. O'Dell TJ, Hawkins RD, Kandel ER, Arancio O. Tests on the roles of two diffusible substances in LTP: evidence for nitric oxide as a possible early retrograde messenger. *Proc Natl Acad Sci USA* 1991;88:11285–11289.
14. Schuman EM, Madison DV. A requirement for the intercellular messenger nitric oxide in long-term potentiation. *Science* 1991;254:1503–1506.
15. Haley JE, Wilcox GL, Chapman PF. The role of nitric oxide in long-term potentiation. *Neuron* 1992;8:211–216.
16. del Cerro S, Arai A, Lynch G. Inhibition of long-term potentiation by an antagonist of platelet-activating factor receptors. *Behav Neur Biol* 1990;54:213–217.
17. Clark GD, Happel LT, Zorumski CF, Bazan NG. Enhancement of hippocampal excitatory synaptic transmission by platelet-activating factor. *Neuron* 1992;9:1211–1216.
18. Wieraszko A, Li G, Kornecki E, Hogan MV, Ehrlich YH. Long-term potentiation in the hippocampus induced by platelet-activating factor. *Neuron* 1993;10:553–557.
19. Herrero I, Miras-Portugal MT, Sanchez-Prieto J. Positive feedback of glutamate exocytosis by metabotropic presynaptic receptor stimulation. *Nature* 1992;360:163–166.
20. Miller B, Sarantis M, Traynelis SF, Attwell D. Potentiation of NMDA receptor currents by arachidonic acid. *Nature* 1992;355:722–725.
21. Bonhoeffer T, Staiger V, Aertsen A. Synaptic plasticity in the rat hippocampal slice cultures: local "Hebbian" conjunction or pre- and postsynaptic stimulation leads to distributed synaptic enhancement. *Proc Natl Acad Sci USA* 1989;86:8113–8117.
22. Kossel A, Bonhoeffer T, Bolz J. Non-Hebbian synapses in rat visual cortex. *Neuro Report* 1990; 1:115–118.
23. Gustafsson B, Wigstrom H, Abraham WC, Huang YY. Long-term potentiation in the hippocampus using depolarizing current pulses as the conditioning stimulus to single volley synaptic potentials. *J Neurosci* 1987;7:774–780.
24. Madison DV, Nicoll RA. Actions of noradrenaline recorded intracellularly in rat hippocampal pyramidal neurones. *J Physiol* 1986;372:221–244.
25. Tseng GF, Parada I, Prince DA. Double-labelling with rhodamine beads and biocytin: a technique for studying corticospinal and other projection neurons in vitro. *J Neurosci Meth* 1991;37:121–131.

26. Lynch G, Larson J, Kelso S, Barrionuevo G, Schottler F. Intracellular injections of EGTA block induction of hippocampal long-term potentiation. *Nature* 1983;305:719–721.
27. Malenka RC, Kauer JA, Zuker RJ, Nicoll RA. Postsynaptic calcium is sufficient for potentiation of hippocampal synaptic transmission. *Science* 1988;242:81–84.
28. Malinow R, Miller JP. Postsynaptic hyperpolarization during conditioning reversibly blocks induction of long-term potentiation. *Nature* 1986;320:529–530.
29. Kelso SR, Ganong AH, Brown TH. Hebbian synapses in hippocampus. *Proc Natl Acad Sci USA* 1986;83:5326–5330.
30. Sah P, Hestrin S, Nicoll RA. Tonic activation of NMDA receptors by ambient glutamate enhances excitability of neurons. *Science* 1989;246:815–818.
31. Ballyk BA, Goh JW. Elevation of extracellular potassium facilitates the induction of hippocampal long-term potentiation. *J Neurosci Res* 1992;33:598–604.
32. Collingridge GL. The Sharpey-Schafer Prize Lecture. The mechanism of induction of NMDA receptor-dependent long-term potentiation in the hippocampus. *Exp Physiol* 1992;77:771–797.
33. Montague PR, Gally JA, Edelman GM. Spatial signaling in the development and function of neural connections. *Cerebral Cortex* 1991;1:99–220.
34. Schuman EM, Madison DV. Diffusible messengers and intercellular signalling: locally distributed synaptic potentiation in the hippocampus. *Science* 1994 (*in press*).

Molecular and Cellular Mechanisms of Neurotransmitter Release, edited by Lennart Stjärne, Paul Greengard, Sten Grillner, Tomas Hökfelt, and David Ottoson, Raven Press, Ltd., New York © 1994.

30

Activity-dependent Modulation of Developing Neuromuscular Synapses

Mu-ming Poo

Department of Biological Sciences, Columbia University, New York, New York 10027

RAPID ONSET OF SYNAPTIC ACTIVITY AT NERVE-MYOCYTE JUNCTION

Studies of neuromuscular synaptogenesis in cell cultures (1–3) and in vivo (4,5) have shown a rapid onset of synaptic activity after nerve-muscle contact. In Xenopus nerve-muscle cultures, spontaneous acetylcholine (ACh) secretion from the nerve terminal was induced by the muscle contact and appears as soon as the nerve-muscle contact is made (6). Functional synaptic transmission between the spinal neuron and embryonic muscle cell is also established within minutes following the contact (7,8). Morphological studies of the identified pre- and postsynaptic cells that exhibited functional synaptic transmission at this early phase of synaptogenesis revealed no sign of synaptic specializations (8). Several ultrastructural studies have clearly shown that presynaptic differentiation of active zone structures and postsynaptic differentiation of receptor clusters and synaptic folds take many days or even weeks to complete (4,9). These results suggest that functional synaptic communication between cells in the developing nervous system is not limited only to regions of cell-cell contacts that exhibit synaptic specializations, and early synaptic communication may be much more widespread than previously realized.

The early onset of synaptic activity may play important roles in the differentiation of postsynaptic cells and the maturation of synaptic connections. In Xenopus cultures, the development of muscle striation is accelerated by the presence of spontaneous synaptic activity at the neuromuscular synapse (10). While clustering of ACh receptors at subsynaptic sites can occur when postsynaptic receptors were blocked (11), it is not clear whether the presence of synaptic activity affects the rate of receptor clustering and other aspects of synapse development. As discussed in the sections that follow, postsynaptic activation of ACh receptors and activity-dependent release of neurotrophic factors from pre- or postsynaptic cells could result in an immediate modulation of the synaptic efficacy. Long-term effect of the activity on synapse maturation remains a possibility.

ELECTRICAL ACTIVITY MODULATES SYNAPTIC EFFICACY

It is well-known that repetitive synaptic activity results in short-term changes in the efficacy of the synaptic transmission at many synapses, including neuromuscular junctions (12). Recent studies of developing synapses in Xenopus nerve-muscle culture showed that persistent changes in both the spontaneous and impulse-evoked synaptic activity can also be induced by a brief episode of synaptic activity. A few action potentials fired at the presynaptic neurons resulted in persistent potentiation of the frequency of spontaneous synaptic currents (13). Repetitive presynaptic stimulation of presynaptic neuron by a train of 100 stimuli at 3 Hz led to significant potentiation of evoked synaptic transmission during the first few hours of synaptic contact. Interestingly, this potentiating effect of the activity disappeared within 1 day in culture, as the synapse matures (Y. Lin and M-m. Poo, unpublished). The existence of a critical period for synaptic potentiation is of particular interest, since critical periods have been found for activity-dependent modulation of synaptic connections in developing visual cortex (14,15).

The robust plasticity of developing neuromuscular synapses may provide a basis for activity-dependent rearrangement of synaptic connectivity during development. (16) Embryonic muscle fibers in vertebrates are initially innervated by multiple motor axons, but become singly innervated as the animal matures (17,18). Previous evidence has shown that the elimination of polyneuronal innervation depends on the presence of synaptic activity. When the activity is blocked presynaptically by tetrodotoxin (19) or postsynaptically by ACh receptor antagonists (20), the synaptic elimination was absent or significantly delayed. How activity affects synapse elimination and whether the active nerve terminal enjoys a selective advantage in the competition for survival remain to be clarified (21–23). To facilitate studies of cellular mechanism underlying synapse competition, several cell culture systems have been developed (24–26). We have used Xenopus nerve-muscle cultures to study activity-dependent synaptic competition between two presynaptic spinal neurons co-innervating the same postsynaptic myocyte (26). The efficacy of the synapses made by the two neurons were measured by recording the evoked synaptic current in the muscle cell when the presynaptic neurons were stimulated to fire action potentials with low-frequency test stimuli. Repetitive stimulation at high frequency to one neuron resulted in immediate suppression of synaptic efficacy of the unstimulated neuron (26). Such heterosynaptic suppression was found to persist in many cells for as long as the recording was made (up to 1.5 hours). An analogous heterosynaptic inhibition has been observed in rat muscle, where stimulation of one motor axon resulted in an immediate and transient inhibition of the synaptic response induced by another co-innervating motor axon (27). Whether persistent synaptic suppression can be induced in vivo by repetitive preferential stimulation and whether such synaptic suppression will eventually lead to morphological changes and withdrawal of the suppressed nerve terminal remain to be examined.

In principle, heterosynaptic suppression could be due to either a direct presynaptic "cross-talk" between active and inactive co-innervating nerve terminals or a suppressive effect of active synapse on the inactive synapse via postsynaptic mechanisms. To test these two possibilities, singly innervated myocytes were used to examine the effect of direct postsynaptic activation on the synaptic efficacy. Repetitive postsynaptic activation was first induced with iontophoretic application of ACh on the surface of the myocyte. Repetitive applications of ACh pulses that mimicked repetitive synaptic inputs produced similar suppression of the existing synapse, as shown by an immediate reduction in the mean amplitude of evoked synaptic currents (28). Synaptic depression was observed for myocytes held in either voltage-clamp or current-clamp conditions during repetitive ACh application, suggesting that depolarization of the myocyte membrane was not required for the observed depression. That the induction of synaptic depression was due to postsynaptic activation of ACh receptors rather than a direct action of ACh on the presynaptic nerve terminal was indicated by the finding that buffering postsynaptic cytosolic Ca^{2+} at a low level with intracellular loading of 1,2-bis(2-aminophenoxy)ethane-N,N,N′,N′-tetraacetic acid (BAPTA) abolished the ACh-induced synaptic depression (28). Similar blockade of heterosynaptic suppression in the doubly innervated myocyte was observed after the BAPTA application in the postsynaptic cell (29). Iontophoretic mapping of postsynaptic ACh sensitivity at extrasynaptic regions before and after induction of synaptic depression by either ACh pulses or heterosynaptic inputs indicated that there was no global change in the myocyte ACh sensitivity. The mean amplitude and the profile of amplitude distribution of spontaneous synaptic currents also showed no significant change following synaptic depression, suggesting that postsynaptic ACh receptor density or sensitivity was not affected. Finally, analysis of the coefficient of variation in the amplitude of evoked synaptic currents also suggested that postsynaptic changes in ACh sensitivity is unlikely to be the sole cause of synaptic depression. Taken together, these results led to the conclusion that synaptic depression induced by both heterosynaptic inputs and ACh pulses was primarily due to a reduction in the presynaptic ACh release. Since Ca^{+} influx into the postsynaptic myocyte is required for the induction of depression, a retrograde signal must be delivered from the postsynaptic cell to presynaptic nerve terminal in order to modulate the ACh secretion mechanisms.

The notion that Ca^{2+} influx through ACh channels provides the inductive signal for synaptic depression is consistent with the observation that the synaptic depression was induced by either repetitive heterosynaptic inputs or ACh pulses when the repetitive stimuli were applied to a voltage-clamped myocyte (at resting potential), since ACh-induced opening of nicotinic ACh channel and the Ca^{2+} influx through the channels can still occur under this condition (30). However, we noted that when repetitive stimuli were applied to a current-clamped myocyte, a slightly higher synaptic depression was induced (29). Depolarization-induced opening of myocyte Ca channels thus may contribute to the induction of synaptic suppression by increasing Ca^{2+} influx.

To further examine the role of postsynaptic membrane depolarization on the synaptic efficacy, repetitive current injections were made in the myocyte to induce repetitive depolarizations of the myocyte (31). Small but significant depression of synaptic efficacy was observed after 30 pulses of repetitive stimulation. The synaptic depression appeared to be induced by Ca^{2+} influx into the myocyte, since it was abolished by buffering myocyte Ca^{2+} at a low level with BAPTA. In comparison with heterosynaptic inputs or ACh applications, repetitive postsynaptic depolarizations alone were much less effective in inducing synaptic depression, suggesting that Ca^{2+} influx through voltage-dependent Ca^{2+} channels provides only a minor contribution to the synaptic depression.

CALCITONIN GENE-RELATED PEPTIDE MODULATES POSTSYNAPTIC RESPONSES

Synaptic activity is likely to involve presynaptic secretion of both neurotransmitter and neuropeptides. Calcitonin gene-related peptide (CGRP) is expressed at developing neuromuscular junctions (32) and is released from the motor-nerve terminal upon nerve excitation (33). It binds to specific muscle surface receptors (34) and elevates adenosine 3′,5′-cyclic phosphate (cAMP) levels (35). Recent in vitro studies have shown that application of CGRP results in phosphorylation of nicotinic ACh channels and alteration in its channel properties, including an increased rate of ACh-induced desensitization of ACh channels in embryonic muscle cells (36–38) and increased burst duration of ACh channels in cultured Xenopus myocytes (39). In the latter study, application of CGRP also resulted in a lengthening of synaptic currents consistent with its effect on ACh channel burst duration. Interestingly, the effect of CGRP on synaptic currents and single ACh channels was observed in 1-day cultures but not 3-day cultures, suggesting that CGRP may exert its potentiating effect only during the initial phase of synaptogenesis. Whether different effects of CGRP reported so far reflect different roles of CGRP at various stages of synapse maturation remains to be determined.

NEUROTROPHINS POTENTIATE PRESYNAPTIC TRANSMITTER SECRETION

Neurotrophins are a group of neurotrophic factors structurally related to nerve growth factor (NGF) that are known to promote the differentiation and survival of specific and overlapping populations of neurons (40,41). Their long-term effects on neurons have been extensively studied in recent years. We have recently found that, in addition to their long-term effects on neurons, some of these neurotrophins produced immediate potentiation of ACh secretion at developing neuromuscular synapses. In 1-day cultures of Xenopus nerve-muscle cells, acute application of neurotrophin-3 (NT-3) and brain-derived growth factor (BDGF) induced within minutes marked elevation of the frequency of miniature excitatory postsynaptic currents (MEPCs) and the amplitude of impulse-evoked synaptic currents (42). The effect was specific to these two neurotrophins, since similar treatment with NGF was without effect, consistent with the specificity of these neurotrophins on the long-term survival of spinal neurons. The potentiation effect of the neurotrophin on synaptic currents appears to be presynaptic in nature, as suggested by the observation that the amplitude distribution and the time course of MEPCs remain unchanged after potentiation.

The finding of the synaptic potentiation by neurotrophins raised the possibility that endogenous neurotrophins secreted from either pre- or postsynaptic cells may play a role in modulating developing synapses. If presynaptic nerve terminals contain neurotrophins, the release could be triggered by electrical activity, similar to that found for other presynaptic neuropeptides. Secretion of neuropeptide from postsynaptic cells could also be activity-dependent. Influx of Ca^{2+} associated with activation of ACh channels may elevate exocytosis of peptide-containing vesicles, as suggested by the recent demonstration of Ca^{2+}-dependent exocytosis from myocytes (43). Studies of neurotrophin expression have shown that the spinal cord and peripheral target tissues of spinal neurons express NT-3 and BDNF in both embryonic and adult rat (44). Thus neurotrophins may participate directly in synaptic modulation, not only in developing systems but also at mature synapses.

CONCLUSION

In many aspects, developing neuromuscular junctions resemble synapses of the adult central nervous system. In both cases, evoked synaptic responses exhibit a high level of fluctuation, the spontaneous synaptic currents are highly variable in amplitudes, and brief periods of electrical activity often induce relatively long-lasting changes in synaptic efficacy. Perhaps only the adult neuromuscular junction can be considered a truly mature synapse, with high fidelity in synaptic transmission. Synapses in the adult central nervous system may have retained the immature status of synaptic functions similar to that of the developing neuromuscular junction. Studies of the activity-dependent modulation of developing neuromuscular synapses may thus reveal cellular mechanisms underlying synaptic plasticity in general.

SUMMARY

Spontaneous and impulse-evoked synaptic currents were observed immediately following nerve-muscle contact in Xenopus cell cultures. The functional significance of this early synaptic activity was examined. Stimulation of pre- and/or postsynaptic cells was found to exert immediate and persistent effects on the efficacy of synaptic transmission. Exogenous application of calcitonin gene-related peptide

(CGRP) and neurotrophins, factors that may be coreleased with ACh in activity-dependent manner at the developing neuromuscular junctions, also modulate either the postsynaptic ACh response or presynaptic ACh release. These results underscore the plasticity of developing neuromuscular synapses and suggest a complex interplay between electrical activity and chemical factors during the formation and maturation of neuronal connections.

REFERENCES

1. Kidokoro Y, Yeh E. Initial synaptic transmission at the growth cone in Xenopus nerve-muscle cultures. *Proc Natl Acad Sci USA* 1982;79:6727–6731.
2. Chow I, Poo M-m. Release of acetylcholine from embryonic neurons upon contact with muscle cell. *J Neurosci* 1985;5:1076–1082.
3. Evers J, Laser M, Sun Y, Xie Z, Poo M-m. Studies of nerve-muscle interactions in Xenopus cell culture: Analysis of early synaptic currents. *J Neurosci* 1989;9:1523–1539.
4. Kullberg R, Lentz T, Cohen MW. Development of myotomal neuromuscular junction in Xenopus laevis : an electrophysiological and fine-structural study. *Dev Biol* 1977;60:101–129.
5. Dennis M. Development of the neuromuscular junction: inductive interactions between cells. *Annu Rev Neurosci* 1981;4:43–68.
6. Xie Z, Poo M-m. Initial events in the formation of neuromuscular synapse. *Proc Natl Acad Sci USA* 1986;83:7069–7073.
7. Sun Y, Poo M-m. Evoked release of acetylcholine from growing embryonic neuron. *Proc Natl Acad Sci USA* 1987;84:2540–2544.
8. Buchanan J, Sun Y, Poo M-m. Studies of nerve-muscle interactions in Xenopus cell culture: fine structure of early functional contacts. *J Neurosci* 1989;9:1540–1554.
9. Takahashi T, Nakajima Y, Hirosawa I, Nakajima S, Onodera K. Structure and physiology of developing neuromuscular synapses in culture. *J Neurosci* 1987;7:473–481.
10. Kidokoro Y, Saito M. Early cross-striation formation in twitching Xenopus myocytes in culture. *Proc Natl Acad Sci USA* 1988;85:1978–1982.
11. Anderson MJ, Cohen MW, Zorychta E. Effects of innervation on the distribution of acetylcholine receptors on cultured muscle cells. *J Physiol* 1977;268:731–756.
12. Zucker RS. Short-term synaptic plasticity. *Annu Rev Neurosci* 1989;12:13–31.
13. Lo Y, Wang T, Poo M-m. Potentiation of spontaneous acetylcholine release by repetitive presynaptic activity. *J Physiol (Paris)* 1991;85:71–78.
14. Hubel DH, Wiesel TN. The period of susceptibility to the physiological effects of unilateral eye closure in kittens. *J Physiol (Lond)* 1970;206:419–436.
15. Goodman, CS, Shatz CJ. Developmental mechanisms that generate precise patterns of neuronal connectivity. *Cell/Neuron* 1993;72/10:77–98.
16. Shatz CL. Impulse activity and the patterning of connections during CNS development. *Neuron* 1990;5:745–756.
17. Redfern PA. Neuromuscular transmission in new-born rats. *J Physiol (Lond)* 1970;209:701–709.
18. Brown MC, Jansen JKS, Van Essen DC. Polyneuronal innervation of skeletal muscle in new-born rats and its elimination during maturation. *J Physiol (Lond)* 1976;329:387–422.
19. Thompson WJ, Kuffler DP, Jansen JKS. The effect of prolonged, reversible block of nerve impulses on the elimination of polyneuronal innervation of new-born rat skeletal muscle fibers. *Neuroscience* 1979;4:271–281.
20. Srihari T, Vrbova G. The role of muscle activity in the differentiation of neuromuscular junctions in slow and fast chick muscles. *J Neurocytol* 1978;7:529–540.
21. Callaway EM, Soha JM, Van Essen DC. Competition favoring inactive over motor neurones during synapse elimination. *Nature* 1987;328:357–363.
22. Ridge RMAP, Betz WJ. The effect of selective, chronic stimulation on motor unit size in developing rat muscle. *J Neurosci* 1984;4:2614–2620.
23. Betz WJ, Ribchester RR, Ridge MAP. Competitive mechanisms underlying synapse elimination in the lumbrical muscle of the rat. *J Neurobiol* 1990;21:1–17.

24. Magchielse T, Meeter E. The effect of neuronal activity on the competitive elimination of neuromuscular junctions in tissue culture. *Dev Brain Res* 1986;25:211–220.
25. Nelson PG, Yu C, Field RD, Neale NA. Synaptic connection in vitro: Modulation of number and efficacy by electrical activity. *Science* 1989;244:585–587.
26. Lo YJ, Poo M-m. Activity-dependent synaptic competition in vitro: heterosynaptic suppression of developing synapses. *Science* 1991;254:1019–1022.
27. Betz, WJ, Chua M, Ridge RMAP. Inhibitory interactions between motoneurone terminals in neonatal rat lumbrical muscle. *J Physiol (Lond)* 1989;418:25–51.
28. Dan Y, Poo M-m. Hebbian depression of isolated neuromuscular synapses in vitro. *Science* 1991; 256:1570–1573.
29. Lo Y, Poo M-m. Heterosynaptic suppression of developing neuromuscular synapses in culture. *J Neurosci (in press).*
30. Decker ER, Dani JA. Calcium permeability of the nicotinic acetylcholine receptor: the single-channel calcium influx is significant. *J Neurosci* 1990;10:3413–3420.
31. Lo Y, Lin Y, Sanes D, Poo M-m. Synaptic depression induced by repetitive postsynaptic depolarizations. *J Neurosci (in press).*
32. Peng HB, Chen Q, De Biasi S, Zhu D. Development of calcitonin gene-related peptide (CGRP) immunoreactivity in relationship to the formation of neuromuscular junction in Xenopus myotomal muscle. *J Comp Neurol* 1989;290:533–543.
33. Uchida S, Yamanoto H, Iio S, Matsumoto N, et al. Release of CGRP-like immunoreactive substance from neuromuscular junction by nerve excitation and its action on striated muscle. *J Neurochem* 1990;54:1000–1003.
34. Roa M, Changeux J. Characterization and developmental evolution of a high-affinity binding site for calcitonin gene-related peptide on chick skeletal muscle membrane. *Neuroscience* 1991;41:563–570.
35. Laufer R, Changeux J. Calcitonin gene-related peptide elevates cyclic AMP levels in chick skeletal muscle: possible neurotrophic role for a coexisting neuronal messenger. *EMBO J* 1987;6:901–906.
36. Huganir RL, Delcour AH, Greengard P, Hess GP. Phosphorylation of the nicotinic acetylcholine receptor regulates its rate of desensitization. *Nature* 1986;321:774–776.
37. Mulle C, Benoit P, Pinset C, Roa M, Changeux JP. Calcitonin gene-related peptide enhances the rate of desensitization of the nicotinic acetylcholine receptor in cultured mouse muscle cells. *Proc Natl Acad Sci USA* 1988;85:5728–5732.
38. Miles K, Greengard P, Huganir RL. Calcitonin gene-related peptide regulates phosphorylation of the nicotinic acetylcholine receptor in rat myotubes. *Neuron* 1989;2:1517–1524.
39. Lu B, Fu W, Greengard P, Poo M-m. Calcitonin gene-related peptide potentiates synaptic responses at developing neuromuscular junction. *Nature* 1993;363:76–79.
40. Thoenen H. The changing scene of neurotrophic factors. *Trends Neurosci* 1991;14:165–170.
41. Lo DC. Signal transduction and regulation of neurotrophins. *Curr Opin Neurobiol* 1992;2:336–340.
42. Lohof AM, Ip NY, Poo M-m. Potentiation of developing neuromuscular synapses by the neurotrophins NT-3 and BDNF. *Nature* 1993;363:350–353.
43. Dan Y, Poo M-m. Quantal transmitter secretion from myocytes loaded with acetylcholine. *Nature* 1992;359:733–736.
44. Maisonpierre PC, Belluscio L, Friedman B, et al. NT-3, BDNF, and NGF in the developing rat nervous system: parallel as well as reciprocal patterns of expression. *Neuron* 1990;5:501–509.

Molecular and Cellular Mechanisms of Neurotransmitter Release, edited by Lennart Stjärne, Paul Greengard, Sten Grillner, Tomas Hökfelt, and David Ottoson, Raven Press, Ltd., New York © 1994.

31

Molecular and Structural Changes Underlying Long-Term Memory Storage in *Aplysia*

*‡Craig H. Bailey, *†Cristina Alberini, *†Mirella Ghirardi, and *†Eric R. Kandel

**Departments of Psychiatry, †Physiology and Cellular Biophysics and Biochemistry and Molecular Biophyics, Center for Neurobiology and Behavior, Columbia University College of Physicians and Surgeons, ‡New York State Psychiatric Institute, New York, New York 10032*

The storage of long-term memory is associated with the growth of synaptic connections (1), new protein synthesis, and altered gene expression (2–4). Despite the association of these neuronal changes with various forms of learning and memory, surprisingly little is known about how each change relates to the other and the relative contribution each may make to the long-term process. In this review we shall address these issues by focusing on recent molecular studies of synaptic growth during long-term sensitization in the marine mollusk *Aplysia californica*. Since molecular and structural changes appear to be signatures of learning-related synaptic plasticity throughout the animal kingdom, principles derived from this approach may be applicable to more complex systems and ultimately to human memory.

The realization that learning and memory are universal features of the nervous system has encouraged the use of several higher invertebrate preparations where the advantages of a tractable central nervous system have facilitated the study of learning and memory using the techniques of modern biology. One such model system, the gill- and siphon-withdrawal reflex of *Aplysia*, has proven particularly useful for correlating changes in cellular and molecular function with simple forms of learning and memory (5–7). One elementary form of nonassociative learning exhibited by this reflex is sensitization. During sensitization, the animal learns to strengthen its defensive reflexes and respond vigorously to a variety of previously neutral or indifferent stimuli after it has been exposed to a potentially threatening or noxious stimulus. In *Aplysia*, sensitization of the gill- and siphon-withdrawal reflex can be induced by a strong stimulus applied to another site, such as the neck or tail. As in the case for other defensive withdrawal reflexes, the behavioral memory for sensitization is graded and retention is proportional to the number of training trials. A single stimulus to the tail gives rise to short-term sensitization lasting minutes to hours. Repeated application of the stimulus results in long-term behavioral sensitization that can last days to weeks (8,9) (Fig. 1).

The memory for both the short- and long-term forms of sensitization is represented on an elementary level by the monosynaptic connections between identified mechanoreceptor sensory neurons and their follower cells. This monosynaptic pathway can be reconstituted in dissociated cell culture (3,10) by serotonin (5-HT), a modulatory neurotransmitter normally released by sensitizing stimuli in the intact animal (11,12). A single application of 5-HT produces short-term changes in synaptic effectiveness, whereas four or five applications of 5-HT over a period of 1.5 hours (or continuous application of 5-HT for 1.5 to 2 hours) produces long-term changes lasting 1 or more days (3). Biophysical studies of this monosynaptic connection suggest that both the similarities and differences in memory reflect, at least in part, intrinsic cellular mechanisms of the nerve cells participating in memory storage. Thus, studies of the sensory-to-motor connection in the intact animal and in culture indicate that a component of the increase in synaptic strength observed during both short- and long-term facilitation (3,9,13) is due to an enhancement in transmitter release by the sensory neuron, accompanied by an increase in excitability attributable to the depression of a specific potassium channel (13–16).

Despite these similarities, the short-term cellular changes differ from the long-term process in two important ways. First, the short-term change involves only covalent modification of preexisting proteins and an alteration of preexisting connections. Both short-term behavioral sensitization in the animal and short-term facilitation in dissociated cell culture do not require ongoing macromolecular synthesis and are not blocked by inhibitors of transcription or translation (3,17). By contrast, these inhibitors selectively block induction of the long-term changes in both the semi-intact animal (18) and in primary cell culture. (3) Most striking is the finding that the induction of long-term facilitation at this single synapse exhibits a critical time window in its requirement for protein and RNA synthesis characteristic of that necessary for learning in both vertebrate and invertebrate animals. (2,3,19, 20,21) Whereas the gene products required for short-term memory are preexisting and must be turned over relatively slowly, some gene products required for long-term memory must be newly synthesized.

Second, the long-term but not the short-term process involves a structural change. Bailey and Chen (22–29) have demonstrated that long-term sensitization training is associated with the growth of new synaptic connections by the sensory neurons onto their follower cells (Fig. 2). This synaptic growth can be induced in the intact ganglion by the intracellular injection of adenosine 3′,5′-cyclic phosphate (cAMP), a second messenger activated by 5-HT(30), and can be reconstituted in sensory-motor cocultures by repeated presentations of 5-HT (31,32).

The finding that long-term memory involves a structural change and requires new macromolecular synthesis and that short-term memory does not raises two fundamental and related questions: 1) what are the mechanisms that convert the short-term form into one of longer duration?; and 2) what are the cellular and molecular

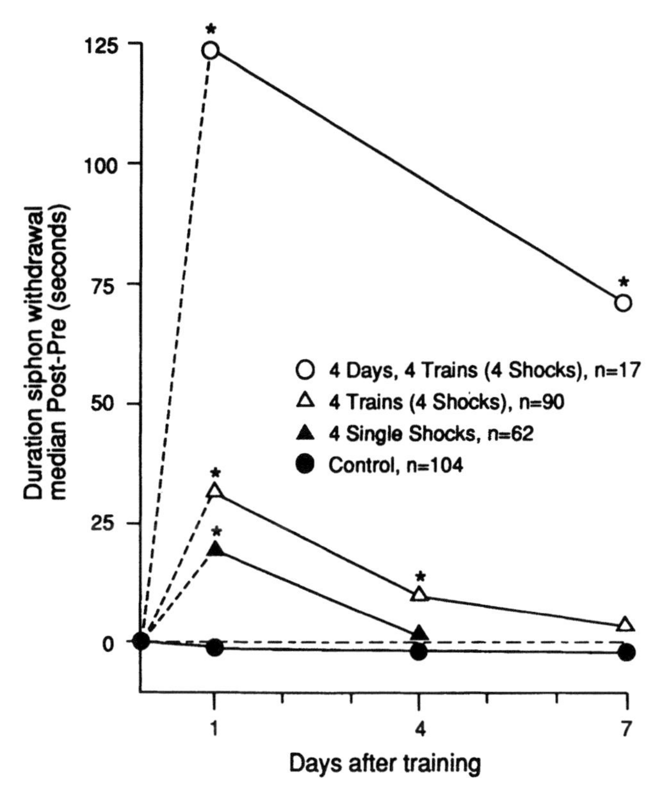

FIG. 1. Behavioral longterm sensitization. A summary of the effects of long-term sensitization training on the duration of siphon withdrawal in *Aplysia californica*. The retention of the memory for sensitization is a graded function proportional to the number of training trials. Experimental animals received either four single shocks for 1 day (filled triangles), four trains of shocks for 1 day (open triangles), or four trains of shocks a day for 4 days (open circles). Control animals were not shocked (filled circles). A pretest determined the mean duration of siphon withdrawal for all animals before training. Posttraining testing was carried out 1, 4, or 7 days after the last day of training. The asterisks indicate a significant difference between the duration of siphon withdrawal for the trained and control animals (Mann-Whitney U tests, $p<.01$). *N* represents the number of animals per group. (From Frost, Castellucci, Hawkins, and Kandel, 1985.)

events that control the induction and maintenance of the long-term form? In this chapter we shall attempt to provide answers to these questions by considering some of the following issues:

1. What are the components of the molecular switch for extending the short-term process, which is independent of protein synthesis, into the long-term process, which requires gene expression?
2. How do the covalent modifications of preexisting proteins and the alteration of preexisting connections during short-term memory become transformed into a structural change?

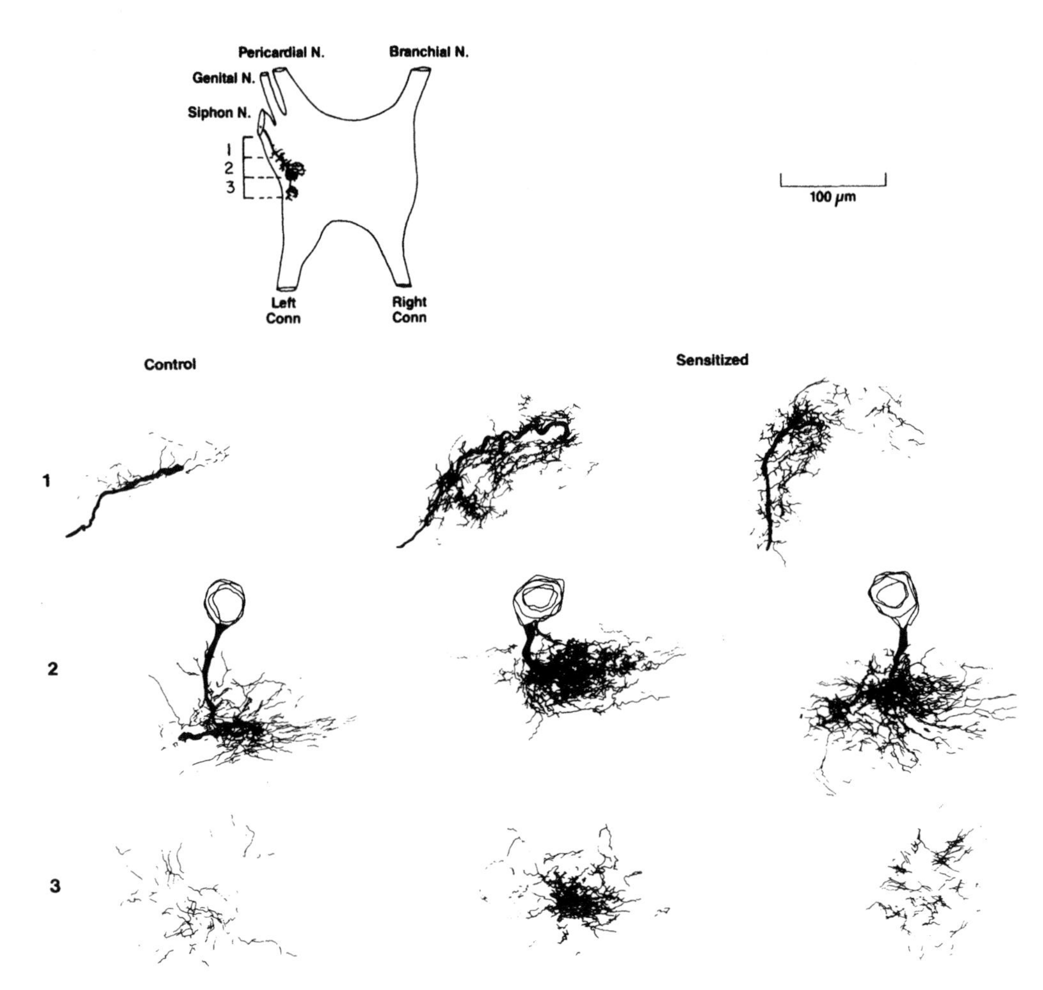

FIG. 2. Serial reconstruction of sensory neurons from long-term sensitized and control animals. Total extent of the neuropil arbors of sensory neurons from one control and two longterm sensitized animals are shown. In each case the rostral (row 3) to caudal (row 1) extent of the arbor is divided rougly into thirds. Each panel was produced by the superimposition of camera lucida tracings of all HRP-labeled processes present in 17 consecutive slab-thick sections and represents a linear segment through the ganglion of roughly 340 μM. For each composite, ventral is up, dorsal is down, lateral is to the left, and medial is to the right. By examining images across each row (rows 1, 2, and 3), the viewer is comparing similar regions of each sensory neuron. In all cases, the arbor of long-term sensitized cells is markedly increased compared with control. From Bailey and Chen, 1988a.

3. Are the structural changes dependent on, and perhaps induced by, the changes in gene and protein expression in the neurons involved?
4. What molecules trigger the gene induction, and can this signaling pathway be traced from the receptor activation at the surface membrane to the genome?

5. How closely do the mechanisms for synaptic plasticity and memory storage in the adult resemble the activity-dependent regulation of synapses that occurs during neuronal differentiation in the developing animal?

TRANSITION FROM SHORT-TERM TO LONG-TERM SENSITIZATION

Both short- and long-term sensitization in the behaving animal and short- and long-term facilitation in more reduced preparations are induced by the same neurotransmitter: serotonin. What is it that produces the transition from a short- to a long-term response? We know that the binding of serotonin to its receptor on the surface membrane of the sensory neuron activates a transmitter-sensitive adenyl cyclase, which in turn increases the concentration of intracellular cAMP and facilitates the dissociation of the catalytic subunit from the regulatory subunit of protein kinase A (PKA). For example, in dissociated cell culture, as in the intact ganglia, the injection of cAMP into presynaptic sensory neurons induces both short- and long-term sensitization (33,34), and inhibitors of cAMP-dependent protein kinase (PKA) block both forms of facilitation (35). A single pulse of 5-HT, which produces short-term facilitation, activates the PKA catalytic subunit that remains in the cytoplasm, largely in the presynaptic terminal (36), where it phosphorylates substrates that contribute to the enhancement of the transmitter release. Among those substrates are K^+ channels (S-channels) whose phosphorylation evokes their closure, producing a broadening of the action potential and a concomitant increase in the amount of Ca^{2+} coming into the cell. In addition, the catalytic subunit of PKA can phosphorylate substrates directly involved in the release of the neurotransmitter (4,13,15,35,37). This increase in the amount of neurotransmitter released is transient and in large part mediates short-term facilitation.

By contrast, long-term facilitation is produced by four or five applications of 5-HT over 1.5 hours (or continuous application of 5-HT for 1.5 to 2 hours) and leads to a cascade of biochemical changes in the cytoplasm, some of which ultimately reach the nucleus where they can induce a long-term alteration in gene expression. This 5-HT-induced activation is accompanied by the translocation into the nucleus of the PKA catalytic subunit (36), where it presumably phosphorylates substrates that regulate gene expression. It is known from other systems that cAMP activation induces a change in gene expression via modification (PKA-dependent phosphorylation) of transcription factors that act on the promoters of the cAMP-inducible genes. The specific sequences on those promoters that bind to these transcription factors are called cAMP-response elements (CREs), and their binding transcription factors are known as CRE-binding proteins (CREBs). In *Aplysia*, Dash et al. (38) have shown that microinjection of oligonucleotides containing CRE into sensory neurons can block long-term facilitation without affecting the short-term process, providing support for the hypothesis that CREBs may play a role in initiating the gene activation required for long-term facilitation. Further evidence has recently been provided by Kaang et al. (39) who microinjected a reporter gene under the control of CRE into

Aplysia sensory neurons and found that its transcription was induced only by repeated pulses of 5-HT—a single pulse did not produce any significant effect. In addition, when a heterologous CREB-GAL4 transcription factor was injected into *Aplysia* sensory neurons, it activated the transcription of a reporter gene via CRE in response to repeated pulses of 5-HT, but again not after a single pulse. This activity is abolished when a single amino-acid residue, that serves as the phosphorylation site for PKA, is mutated. These data serve to reinforce the hypothesis that CREBs transcription factors, activated via PKA-dependent phosphorylation, are required for induction of the long-term process, and suggest that CREBs are involved in regulation of the new gene expression that accompanies long-term sensitization.

What are these target genes? Potential candidates were first suggested by the studies of Sweatt et al. (40) and Barzilai et al. (41) that showed that, during the long-term process, 5-HT induces the synthesis of two classes of proteins in the sensory neurons: early effectors and late effectors. Some of these proteins have now been characterized and their genes have been cloned. One early effector gene, N-terminal ubiquitin hydrolase, seems to be related to the ubiquitin-mediated proteolytic cleavage of the regulatory subunit of PKA, a cleavage that maintains enzymatic activity of the catalytic subunit in the absence of cAMP (42,43) (Inokuchi and Kandel, unpublished data). Other early effector proteins such as clathrin (44) and an NCAM-related cell adhesion molecule (45,46) as well as several late effector proteins, all seem to be related to the structural change (i.e., the growth of new sensory neuron synapses that may carry the facilitation stably into the maintenance phase of long-term memory). We will discuss this in more detail below.

One of the most striking features in the requirement for new macromolecular synthesis during long-term memory is that it is restricted to a very narrow time window, which occurs only during or shortly after the period of training. When inhibitors of RNA or protein synthesis are given 1 hour or later after the completion of training for sensitization or facilitation, the long-term response is not affected. The period that is sensitive to disruption by these inhibitors may be referred to as the *consolidation phase*, and it is during this early phase that transcription and translation appear to be critical for carrying the long-term memory. This observation suggests that there is a rapid and initial expression of genes whose function it might be to activate subsequent mechanisms that ultimately bring the memory to a more stable and self-maintained phase. Are these rapidly expressed early genes a necessary and intermediate step for the later morphological changes? If so, why are constitutively expressed transcription factors activated by post-translational modification not sufficient to regulate the expression of all the effector genes? One attractive hypothesis that can account for a necessary *early phase* is a cascade of gene activation whereby constitutively expressed transcription factors induce early genes that activate late effector genes, which may then contribute to the structural change.

In an attempt to identify primary response genes induced during the consolidation phase, we have looked for inducible transcription factors. Alberini et al. (47) have cloned and studied the *Aplysia* homologue of the mammalian family of C/EBP transcription factors. The *Aplysia* C/EBP (ApC/EBP) has a CRE in its upstream region, and in other contexts these factors also have been shown to be involved in the cAMP regulation of gene transcription in vertebrates (48–50). Alberini et al. (47) have found that the ApC/EBP has the properties of an immediate-early gene (IEG). Its expression is not detected in unstimulated cells, but is rapidly and transiently induced by 5-HT even in the presence of protein synthesis inhibitors such as anisomycin or emetine (i.e., its induction depends on constitutively expressed transcription factors) (Fig. 3). Moreover, blocking the expression of the gene or the activity of the protein blocks long-term facilitation selectively without affecting short-term facilitation (Fig. 4). To demonstrate this, Alberini et al. first blocked expression at the level of DNA by titrating the C/EBP protein in vivo; this prevented the protein from binding to its DNA recognition element (ERP), by means of an oligonucleotide encoding the ERP (Figs. 4A,B). They next blocked the expression of the gene's mRNA with antisense RNA (Figs. 4C,D). Finally, they prevented the action of the protein with a specific antibody that disrupted its binding to the DNA target sequences.

These experiments provide direct evidence for a role for immediate-early genes in synaptic plasticity and suggest that the activation of C/EBP, and perhaps other immediate-early genes, may account for the critical time window of long-term facilitation, as well as for the induction of its maintenance phase. According to this scheme, long-term facilitation requires constitutively expressed transcription factors—perhaps including CRE-binding proteins that activate IEGs, one of which has been shown to be ApC/EBP that then acts on effector genes. Once the early phase is induced, the expression of late target genes may then contribute to the structural changes that appear to be required for the more stable and self-maintained phase of long-term facilitation.

EARLY GENES AND INDUCTION OF THE STRUCTURAL CHANGE

How might this cascade of gene activation contribute to the growth of new sensory neuron synapses that is a signature of long-term facilitation in *Aplysia*? A series of recent studies has begun to explore the roles of early proteins in the induction of the structural change by attempting to characterize and identify each of the 15 proteins that Barzilai et al. (41) observed to be specifically altered in expression during the acquisition of long-term facilitation. Six have now been identified. Surprisingly, two proteins (clathrin and tubulin) that increase and four proteins (NCAM-related cell adhesion molecules) that decrease their level of expression all seem to relate to structural change.

Mayford et al. (45) first focused on the four proteins, D1–D4, that decrease their expression in a transcriptionally-dependent manner following the application of 5-HT or cAMP. Cloning and sequencing cDNAs for these proteins has shown that they encode different isoforms of an immunoglobulin-related cell adhesion mole-

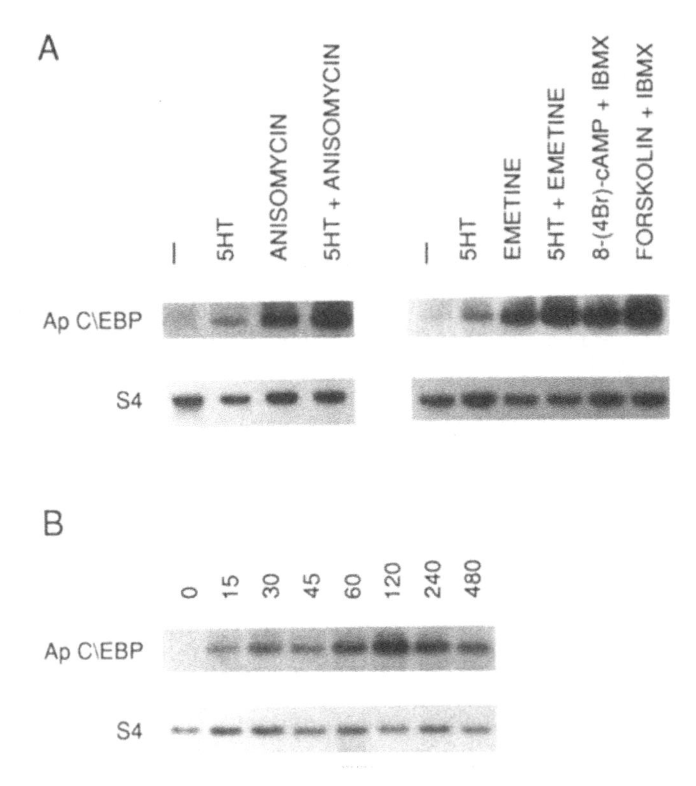

FIG. 3. A. ApC\EBP mRNA expression in CNS of untreated *Aplysia* or *Aplysia* treated with the indicated drugs for 2 hours at 18°C. Ten μg of total RNA extracted from total CNS of in vivo treated *Aplysia* were electrophoresed, blotted, and hybridized with ^{32}P-end-labeled ApC/EBP (top) or S4 (bottom) probes. The latter encodes the *Aplysia* homologue of S4 ribosomal protein, which is constitutively expressed and used as a loading control. **B.** Time course of ApC\EBP mRNA induction following serotonin (5-HT) treatment. Times of treatments are indicated. Five μg of total RNA from total CNS of in vivo treated *Aplysia* were analyzed as described in **A.** From Alberini, Ghirardi, Metz, and Kandel, 1994.

cule, designated apCAM, which shows greatest homology to NCAM in vertebrates and fasciclin II in *Drosophila*. Imaging of fluorescently labeled mAbs to apCAM has demonstrated that not only is there a decrease in the level of expression but that there is also a down-regulation of preexisting protein from the surface membrane of the sensory neurons within 1 hour after the addition of 5-HT. This transient modulation by 5-HT of cell adhesion molecules, therefore, may represent one of the early molecular steps required for initiating learning-related growth of synaptic connections. Blocking the expression of these cell adhesion molecules by mAb results in defasciculation of sensory neuron axons, a step that appears to precede synapse formation in dissociated culture (51,52).

What are the molecular and subcellular mechanisms by which 5-HT modulates apCAM and what significance does this modulation have for the growth of sensory neuron synapses that are induced by 5-HT? To address these questions, Bailey et al. (46) combined thin-section electron microscopy with immunolabeling using a gold-conjugated mAb specific to apCAM. Their results indicate that a 1-hour application of 5-HT can produce a 50% decrease in apCAM at the surface membrane of the sensory neuron. This down-regulation is mediated by a heterologous, protein synthesis-dependent activation of the endocytic pathway, leading to the internalization and presumptive degradation of apCAM. As is the case for the decrease at the level of expression, the 5-HT-induced internalization of apCAM can be simulated by cAMP (53). Indeed, concomitant with the down-regulation of apCAM, Hu et al. (44) have demonstrated that 5-HT and cAMP also induce in the sensory neurons an increase in the number of coated pits and coated vesicles as well as an increase in the expression of the light chain of clathrin (apClathrin), that form of clathrin which in other systems is thought to be involved in regulating the assembly and disassembly of coated pits (54).

Bailey et al. (46) have suggested that the 5-HT-induced internalization of apCAM and consequent membrane remodeling may represent the first steps in the structural program underlying long-term facilitation (Fig. 5). According to this view, learning-related synapse formation is preceded by and perhaps requires endocytic activation, which can then serve a double function. First, endocytic activation could remove cell adhesion molecules from the neuronal surface at sites of apposition, thereby destabilizing adhesive contacts between axonal processes of the sensory neuron and perhaps facilitating defasciculation, a process that may be important in disassembly. Second, the massive endocytic activation might lead to a redistribution of membrane components that favors synapse formation. The assembly of membrane components required for initial synaptic growth may involve insertion, by means of targeted exocytosis, of endocytic membrane retrieved from sites of adhesion and recycled to sites of new synapse formation. Synapse formation may require, in addition, the recruitment of new transport vesicles from the trans-Golgi network. Thus, aspects of the initial steps in learning-related synapse formation may eventually be understood in the context of a novel, heterologous, and targeted form of receptor-mediated endocytosis.

CONCLUSIONS

Aspects of the cellular changes that underlie memory storage in *Aplysia* are represented at the level of individual neurons by changes in the strength and structure of identified synaptic connections. Short-term memory, lasting minutes to hours, involves an alteration in effectiveness of preexisting synaptic connections as a result of the covalent modification of preexisting proteins by second-messenger cascades. By contrast, long-term memory, lasting days or weeks, is associated with the growth of new synaptic connections activated by altered gene expression and new protein synthesis. Recent studies of the synaptic growth that accompanies long-term sensitization in *Aplysia* have begun to characterize the sequence of molecular events responsible for the induction and maintenance of this learning-related structural

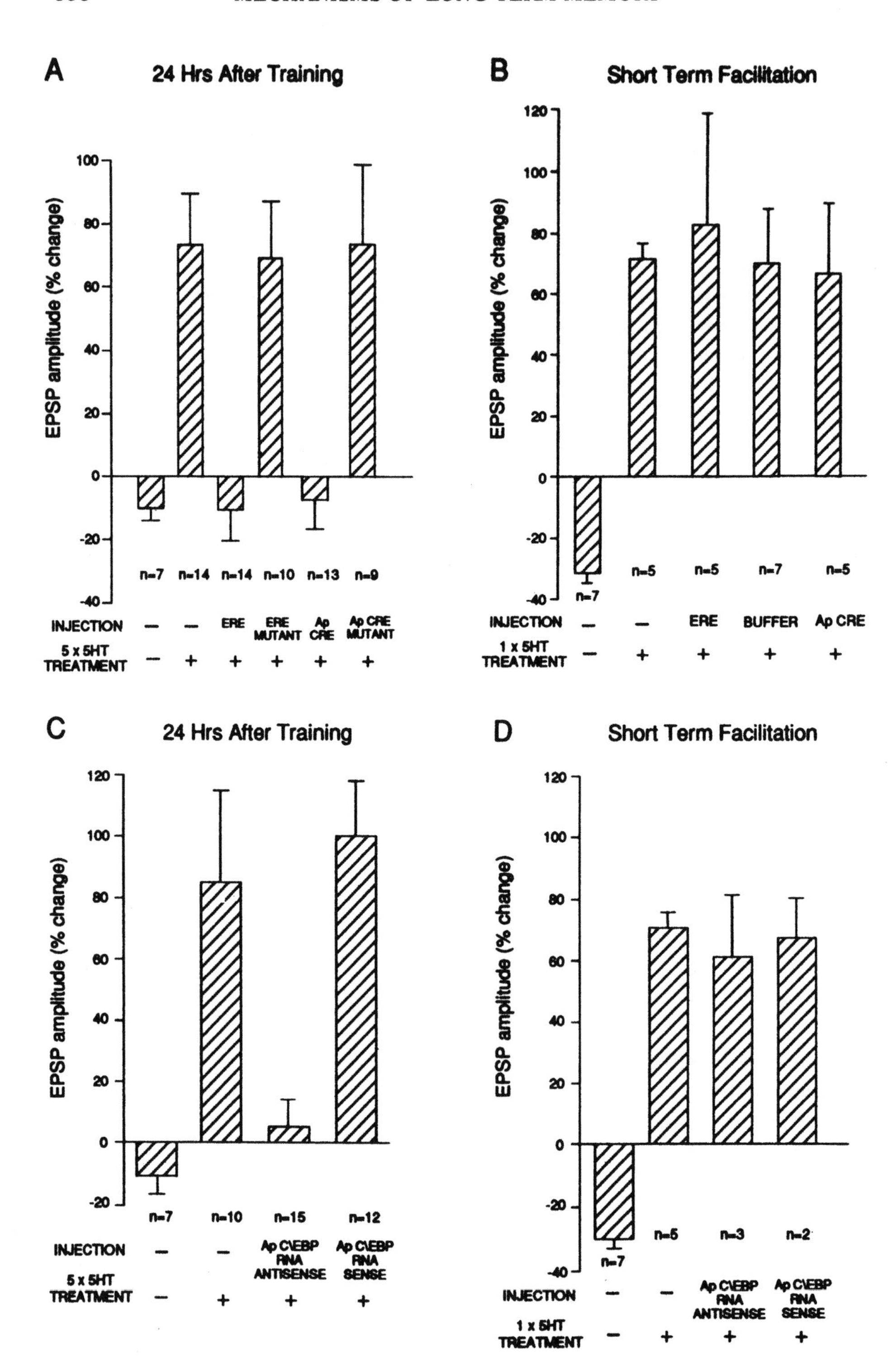

A
24 Hrs After Training
EPSP amplitude (% change)
100
80
60
40
20
0
-20
-40
n=7 n=14 n=14 n=10 n=13 n=9
INJECTION — — ERE ERE MUTANT Ap CRE Ap CRE MUTANT
5 x 5HT TREATMENT — + + + + +
B
Short Term Facilitation
EPSP amplitude (% change)
120
100
80
60
40
20
0
-20
-40
n=7 n=5 n=5 n=7 n=5
INJECTION — — ERE BUFFER Ap CRE
1 x 5HT TREATMENT — + + + +
C
24 Hrs After Training
EPSP amplitude (% change)
120
100
80
60
40
20
0
-20
n=7 n=10 n=15 n=12
INJECTION — — Ap C\EBP RNA ANTISENSE Ap C\EBP RNA SENSE
5 x 5HT TREATMENT — + + +
D
Short Term Facilitation
EPSP amplitude (% change)
120
100
80
60
40
20
0
-20
-40
n=7 n=5 n=3 n=2
INJECTION — — Ap C\EBP RNA ANTISENSE Ap C\EBP RNA SENSE
1 x 5HT TREATMENT — + + +

change. This in turn has indicated that specific molecules and mechanisms important for development of the nervous system can be reutilized in the adult for the purposes of synaptic plasticity and memory storage.

For example, these studies indicate that long-term memory involves the flow of information from receptors on the cell surface to the genome, as seen in other processes of cellular differentiation and growth. Such changes may reflect a recruitment by environmental stimuli of developmental processes that are latent or inhibited in the fully differentiated neuron. Indeed, an increasing body of evidence suggests that the cellular and molecular changes accompanying long-term memory storage share several features in common with the cascade of events that underlie neuronal differentiation and development. In both cases, there is a requirement for new protein and mRNA synthesis. These alterations can be initiated in the long-term process by modulatory transmitters that, in this respect, appear to mimic the effects of growth factors and hormones during the cell cycle and differentiation. Thus, modulatory transmitters important for learning activate not only a cascade of cytoplasmic events required for the short-term process, but also induce a genomic

FIG. 4. Blocking the expression of ApC\EBP at the level of DNA (**A,B**) or mRNA (**C,D**) selectively blocks long-term facilitation. **A.** Injection of ERE oligonucleotide blocks serotonin (5-HT)-induced long-term but not short-term facilitation at the sensory-to-motor-neuron synapse. Bar graph represents the effects of oligonucleotide injections in long-term facilitation. The height of each bar corresponds to the mean percentage change ±SEM in excitatory postsynaptic potential (EPSP) amplitude tested 24 hours after 5-HT treatment. A one-way analysis of variance indicates a difference with treatment ($F=8.21$, df5.6, $p<0.001$). A comparison of the mean (Newman Keul's multiple range test) indicates that 5-HT treatment significantly increases the EPSP amplitude in noninjected cells, as well as in ERE Mutant- or ApCRE Mutant-injected cells, relative to the control (not 5-HT-treated and noninjected cells) ($p<0.01$). On the contrary, the EPSP amplitude change in ERE- or ApCRE-injected cells was not significantly different from that of control cells that were neither injected nor treated. **B.** Bar graph representing the mean percentage change ±SEM of short-term facilitated cells injected with ERE oligonucleotides or with ApCRE, or with buffer. A one-way analysis of variance indicates a difference with treatment ($F=7.04$, df 4.24, $p<0.001$). A comparison of the means shows a significant effect of 5-HT in increasing EPSP amplitude in ERE, in ApCRE, or in buffer-injected cultures compared to the control ($p<0.01$). **C.** Injection of ApC\EBP antisense RNA blocks 5-HT-induced long-term but not short-term facilitation in the sensory motor synapses. Bar graph represents the effects of oligonucleotide injections in long-term facilitation. The height of each bar corresponds to the mean percentage change ±SEM in EPSP amplitude tested 24 hours after 5-HT treatment. A one-way analysis of variance indicates a difference with treatment ($F=8.7$, df3.4, $p<0.001$). A comparison of the mean (Newman Keul's multiple range test) indicates that 5-HT treatment significantly increases the EPSP amplitude in noninjected cells, as well as in ApC/EBP sense RNA-injected cells, relative to the control (not 5-HT-treated and noninjected cells) ($p<0.01$). On the contrary, the EPSP amplitude change in ApC/EBP antisense RNA-injected and 5-HT-treated cells was not significantly different from the control, nontreated, noninjected cultures. **D.** Bar graph representing short-term facilitation of cells injected with ApC\EBP antisense or sense RNA. A one-way analysis of variance indicates a difference with treatment ($F=44.9$, df3.1, $p<0.001$). A comparison of the means shows a significant effect of 5-HT in increasing EPSP amplitude in both ApC\EBP antisense- or sense RNA-injected cells compared to the control cultures that were neither injected nor treated ($p<0.01$). (From Alberini, Ghirardi, Metz, and Kandel, 1994).

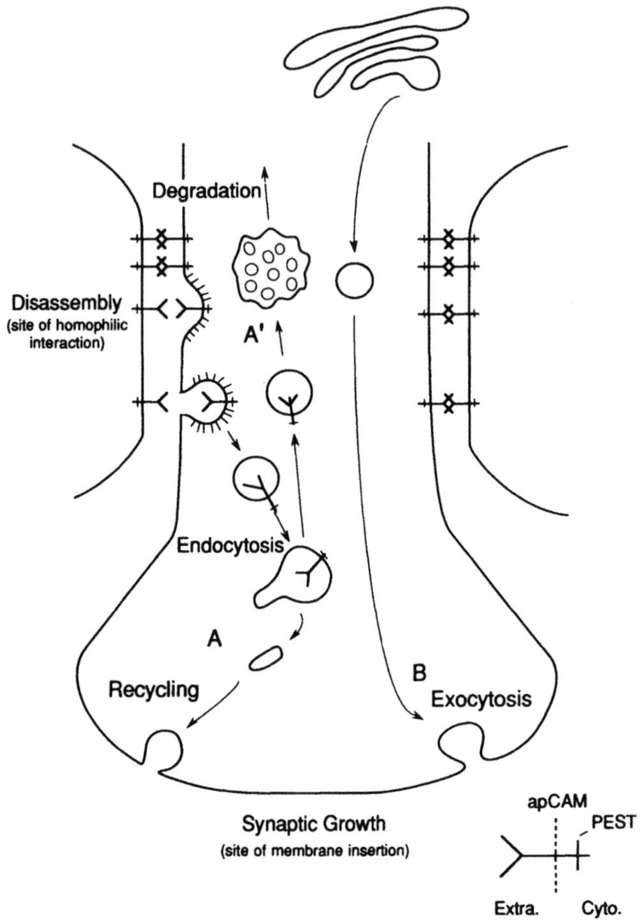

FIG. 5. Internalization, disassembly, and learning-related growth. The onset of synaptic growth is triggered by the facilitating transmitter, serotonin (5-HT). Serotonin leads to a down-regulation of NCAM-related cell adhesion molecules (apCAMs). Part of this down-regulation occurs at the presynaptic membrane, where disassembly is achieved by a transient internalization involving the endosomal pathway. Internalization seems to be particularly active at sites of apposition (depicted here as stable, nongrowing regions where one neurite abuts and adheres to another), and is characterized there by intense endocytic activity. These focal endocytic bursts begin at coated pits and proceed through a series of endosomal precursors, including uncoupling (CURL) vesicles. Here, the internalized plasma membrane can follow one of two pathways. In pathway A, internalized endocytic membrane components are retrieved from sites of apparent adhesion in the tubular extension of the CURL and reinserted at the surface at sites of new synapse formation. In pathway A', the apCAM molecules are targeted for degradation and become segregated within the swollen, vesicular portion of the CURL, which ultimately fuses with or matures into late endosomal compartments such as multivesicular bodies. Additional membrane inserted into the terminal for growth may also come by means of recruitment of new transport vesicles from the trans-Golgi network and the insertion of membrane by exocytosis (pathway B). The cytoplasmic domain of apCAM contains a prominent PEST sequence, which may be proteolytically cleaved and lead to enhanced turnover of newly synthesized protein, as well as contribute to the rapid internalization that accounts for the altered expression of apCAM at the surface membrane. From Bailey, Chen, Keller, and Kandel, 1992.

cascade by which the transmitter can exert a long-term regulation over both the excitability and structure of the neuron through changes in gene expression.

Studies in *Aplysia* have further demonstrated that the earliest stages of long-term memory formation are associated with modulation of an immunoglobulin-related cell adhesion molecule homologous to NCAM. With the emergence of the nervous system, the *Aplysia* NCAM becomes expressed exclusively in neurons and is specifically enriched at synapses. These cell adhesion molecules are maintained into adulthood, at which point they can be down-regulated by serotonin, a transmitter important for sensitization and classical conditioning in *Aplysia* and by cAMP, a second-messenger activated by serotonin. This endocytically mediated down-regulation appears to serve as a preliminary and permissive step for the growth of synaptic connections that accompanies the long-term process. Thus, a molecule used during development for cell adhesion and axon outgrowth is retained into adulthood, at which point it seems to restrain or inhibit growth until it is rapidly and transiently decreased at the cell surface by a modulatory transmitter important for learning. These studies also suggest that processing and storage of information in the nervous system may rely on the same cellular mechanisms utilized by other cells in the body to organize and regulate membrane trafficking.

Finally, insights from the molecular studies of learning and memory in *Aplysia* suggest that the critical time window for macromolecular synthesis that is a ubiquitous feature of long-term memory storage may be explained by a cascade of gene activation whereby one or more immediate-early genes control the transcription of late effector genes. The biological significance of an immediate-early gene-dependent response in long-term plasticity may reside in the necessity of a convergent checkpoint that turns on a genetic program, similar to the cascade of gene activation during cellular differentiation. As is the case for development, in long-term memory a convergent checkpoint and cascade of gene activation may be critical to preserve important functions that ultimately rely on a small number of cells. Critical time windows have been previously described in other contexts, especially as part of developmental processes. For example, establishment of the differentiated state in DNA viruses often requires a sequence of gene activation whereby early regulatory genes lead to the maintained expression of later effector genes. A similar time window is evident in the later stages of *Drosophila* development where the steroid hormone ecdysone induces growth and moulting by altering the expression of early genes whose activity leads to the expression of later genes (55). When ecdysone is given together with a protein synthesis inhibitor, the early genes are induced but the later genes are not.

The similarity between these critical periods and the one found in long-term memory suggests that aspects of the regulatory mechanisms underlying learning-related synaptic plasticity in the adult may eventually be understood in the context of the basic molecular logic used to refine synaptic connections during the later stages of neuronal development. Both processes appear to share a cascade of gene activation with a critical time window during which the differentiated state is still labile and can be modified. That this feature is particularly well-developed in neurons—which characteristically remain plastic throughout most of their life cycle, and can grow and retract their synaptic connections on appropriate target cells in an activity-dependent fashion—may underlie their unique ability to respond to environmental stimuli essential for learning and memory storage.

ACKNOWLEDGMENTS

Work reviewed in this chapter was supported in part by NIH grants MH37134 and GM32099 to C.H.B., and by the Howard Hughes Medical Institute to E.R.K.

REFERENCES

1. Bailey CH, Kandel ER. Structural changes accompanying memory storage. *Annu Rev Physiol* 1993;55:397–426.
2. Davis HP, Squire LR. Protein synthesis and memory: a review. *Psychol Bull* 1984;96:518–559.
3. Montarolo PG, Goelet P, Castellucci VF, Morgan J, Kandel ER, Schacher S. A critical period for macromolecular synthesis in long-term heterosynaptic facilitation in *Aplysia*. *Science* 1986;234: 1249–1254.
4. Goelet, P, Castellucci VF, Schacher S, Kandel ER. The long and the short of long-term memory—a molecular framework. *Nature* 1986;322:419–422.
5. Kandel ER, Schwartz JH. Molecular biology of an elementary form of learning: modulation of transmitter release by cyclic AMP. *Science* 1982;218:433–443.
6. Carew TJ, Sahley CL. Invertebrate learning and memory: from behavior to molecules. *Annu Rev Neurosci* 1986;9:435–487.
7. Byrne JH. Cellular analysis of associative learning. *Physiol Rev* 1987;67:329–439.
8. Pinsker HM, Hening WA, Carew TJ, Kandel ER. Long-term sensitization of a defensive withdrawal reflex in *Aplysia*. *Science* 1973;182:1039–1042.
9. Frost WN, Castellucci VF, Hawkins RD, Kandel ER. Monosynaptic connections from the sensory neurons of the gill- and siphon-withdrawal reflex in *Aplysia* participate in the storage of long-term memory for sensitization. *Proc Natl Acad Sci USA* 1985;82:8266–8269.
10. Rayport SG, Schacher S. Synaptic plasticity in vitro: cell culture of identified *Aplysia* neurons mediating short-term habituation and sensitization. *J Neurosci* 1986;6:759–763.
11. Glanzman DL, Mackey SL, Hawkins RD, Dyke AM, Lloyd PE, Kandel ER. Depletion of serotonin in the nervous system of *Aplysia* reduces the behavioral enhancement of gill withdrawal as well as the heterosynaptic facilitation produced by tail shock. *J Neurosci* 1989;9:4200–4213.
12. Mackey SL, Kandel ER, Hawkins RD. Identified serotonergic neurons LCB1 and RCB1 in the cerebral ganglia of *Aplysia* produce presynaptic facilitation of siphon sensory neurons. *J Neurosci* 1989;9:4227–4235.
13. Dale N, Kandel ER, Schacher S. Serotonin produces long-term changes in the excitability of *Aplysia* sensory neurons in culture that depend on new protein synthesis. *J Neurosci* 1987;7:2232–2238.
14. Klein M, Kandel ER. Mechanism of calcium current modulation underlying presynaptic facilitation and behavioral sensitization in *Aplysia*. *Proc Natl Acad Sci USA* 1980;77:6912–6916.
15. Hochner B, Klein M, Schacher S, Kandel ER. Additional components in the cellular mechanism of presynaptic facilitation contributes to behavioral dishabituation in *Aplysia*. *Proc Natl Acad Sci USA* 1986;83:8794–8798.
16. Scholz KP, Byrne JH. Long-term sensitization in *Aplysia*: biophysical correlates in tail sensory neurons. *Science* 1987;235:685–687.
17. Schwartz JH, Castellucci VF, Kandel ER. Functions of identified neurons and synapses in abdominal ganglion of *Aplysia* in absence of protein synthesis. *J Neurophysiol* 1971;34:939–953.
18. Castellucci VF, Blumenfeld H, Goelet P, Kandel ER. Inhibitor of protein synthesis blocks long-term behavioral sensitization in the isolated gill-withdrawal reflex of *Aplysia*. *J Neurobiol* 1989;20:1–9.

19. Agranoff BW. *The chemistry of mood, motivation and memory*. New York: Plenum Press; 1972.
20. Barondes SH. Protein synthesis dependent and protein synthesis independent memory storage processes. In: Deutsch D, Deutsch JA, eds. *Short-term memory*. New York: Academic Press; 1975; 379–390.
21. Flexner JB, Flexner LB, Stellar E. Memory in mice as affected by intracerebral puromycin. *Science* 1983;141:57–59.
22. Bailey CH, Chen M. Morphological basis of long-term habituation and sensitization in *Aplysia*. *Science* 1983;220:91–93.
23. Bailey CH, Chen M. Long-term memory in *Aplysia* modulates the total number of varicosities of single identified sensory neurons. *Proc Natl Acad Sci USA* 1988a;85:2373–2377.
24. Bailey CH, Chen M. Long-term sensitization in *Aplysia* increases the number of presynaptic contacts onto the identified gill motor neuron L7. *Proc Natl Acad Sci USA* 1988b;85:9356–9359.
25. Bailey CH, Chen M. Time course of structural changes at identified sensory neuron synapses during long-term in *Aplysia*. *J Neurosci* 1989;9:1774–1780.
26. Bailey CH, Chen M. Structural plasticity at identified synapses during long-term memory in *Aplysia*. *J Neurol* 1989;20:356–372.
27. Bailey CH. Morphological basis of short- and long-term memory in *Aplysia*, In: Weingartner H, Lister R, eds. *Perspectives on cognitive neuroscience*. New York: Oxford University Press; 1991; 76–92.
28. Bailey CH, Chen M. Morphological aspects of synaptic plasticity in *Aplysia*: an anatomical substrate for long-term memory. *NY Acad Sci* 1991;627:181–196.
29. Bailey CH, Chen M. The anatomy of long-term sensitization in *Aplysia*: morphological insights into learning and memory, In: Squire L, Lynch G, Weinberger N, eds. *Memory: organization and locus of change*. New York: Oxford University Press; 1992;273–300.
30. Nazif FA, Byrne JH, Cleary LJ. cAMP induces long-term morphological changes in sensory neurons of *Aplysia*. *Brain Res* 1991;539:324–327.
31. Glanzman DL, Kandel ER, Schacher S. Target-dependent structural changes accompanying long-term synaptic facilitation in *Aplysia* neurons. *Science* 1990;249:799–802.
32. Bailey CH, Montarolo PG, Chen M, Kandel ER, Schacher S. Inhibitors of protein and RNA synthesis block the structural changes that accompany long-term heterosynaptic plasticity in the sensory neurons of *Aplysia*. *Neuron* 1992;9:749–758.
33. Schacher S, Castellucci VF, Kandel ER. cAMP evokes long-term facilitation in *Aplysia* sensory neurons that requires new protein synthesis. *Science* 1988;240:1667–1669.
34. Scholz, KP, Byrne, JH. Intracellular injection of cyclic AMP induces a long-term reduction of neuronal potassium currents. *Science* 1988;240:1664–1667.
35. Ghirardi M, Braha O, Hochner B, Montarolo PG, Kandel ER, Dale N. Roles of PKA and PKC in facilitation of evoked and spontaneous transmitter release at depressed and nondepressed synapses in *Aplysia* sensory neurons. *Neuron* 1992;9:479–489.
36. Bacskai BJ, Hochner B, Mahaut-Smith M, Adams SR, Kaang B-K, Kandel ER, Tsien RY. Spatially resolved dynamics of cAMP and protein kinase A subunits in *Aplysia* sensory neurons. *Science* 1993;260:222–226.
37. Dale N, Schacher S, Kandel ER. Long-term facilitation in *Aplysia* involves increases in transmitter release. *Science* 1988;239:282–285.
38. Dash PK, Hochner B, Kandel ER. Injection of the cAMP-responsive element into the nucleus of *Aplysia* sensory neurons blocks long-term facilitation. *Nature* 1990;345:718–721.
39. Kaang BK, Kandel ER, Grant SGN. Activation of cAMP-responsive genes by stimuli that produce long-term facilitation in *Aplysia* sensory neurons. *Neuron* 1993;10:427–435.
40. Sweatt JD, Kandel ER. Persistent and transcriptionally-dependent increase in protein phosphorylation upon long-term facilitation of *Aplysia* sensory neurons. *Nature* 1989;339:51–54.
41. Barzilai A, Kennedy TE, Sweatt JD, Kandel ER. 5-HT modulates protein synthesis and the expression of specific proteins during long-term facilitation in *Aplysia* sensory neurons. *Neuron* 1989;2: 1577–1586.
42. Bergold PJ, Sweatt JD, Kandel ER, Schwartz JH. Protein synthesis during acquisition of long-term facilitation is needed for the persistent loss of regulatory subunits of the *Aplysia* cAMP-dependent protein kinase. *Proc Natl Acad Sci USA* 1990;87:3788–3791.
43. Hegde AN, Goldberg AL, Schwartz JH. Regulatory subunits of cAMP-dependent protein kinases are degraded after conjugation to ubiquitin: A molecular mechanism underlying long-term synaptic plasticity. *Proc Natl Acad Sci USA* 1993;90:7436–7440.

44. Hu Y, Barzilai A, Chen M, Bailey CH, Kandel ER. 5-HT and cAMP induce the formation of coated pits and vesicles and increase the expression of clathrin light chains in sensory neurons of *Aplysia*. *Neuron* 1993;10:921–929.
45. Mayford M, Barzilai A, Keller F, Schacher S, Kandel ER. Modulation of an NCAM-related adhesion molecule with long-term synaptic plasticity in *Aplysia*. *Science* 1992;256:638–644.
46. Bailey CH, Chen M, Keller F, Kandel ER. Serotonin-mediated endocytosis of apCAM: An early step of learning-related synaptic growth in *Aplysia*. *Science* 1992;256:645–649.
47. Alberini C, Ghirardi M, Metz R, Kandel ER. C/EBP is an immediate-early gene required for the consolidation of long-term facilitation in *Aplysia*. *Cell* 1994;76:1099–1114.
48. Christy BA, Nathans D. Functional serum response elements upstream of the growth factor-induced *Zif*/268. *Mol Cell Biol* 1989;9:4889–4895.
49. Herrera RO, Robinson HS, Xanthopulos KG, Spiegelman BM. A direct role for C/EBP and AP-1 binding site in gene expression linked to adipocyte differentiation. *Mol Cell Biol* 1989;9:5331–5339.
50. McKnight SL, Lane MD, Glueckson-Waelsch S. Is CCAA/enhancer binding protein central regulator of energy metabolism? *Genes Dev* 1989;3:2021–2024.
51. Glanzman DL, Kandel ER, Schacher S. Identified target motor neuron regulates neurite outgrowth and synapse formation of *Aplysia* sensory neurons in vitro. *Neuron* 1989;3:441–450.
52. Keller F, Schacher S. Neuron-specific membrane glycoproteins promoting neurite fasciculation in *Aplysia* californica. *J Cell Biol* 1990;111:2637–2650.
53. Bailey CH, Chen M, Kandel ER. Early steps in learning-related synaptic growth: cAMP simulates the heterologous endocytosis of apCAMS induced by 5-HT in sensory neurons of *Aplysia*. *Soc Neurosci Abstr* 1993;18:941.
54. Brodsky FM, Hill BL, Acton SL, Nathke I, Wong DH, Ponnambalam S, Parham P. Clathrin light chains: Arrays of protein motifs that regulate coated-vesicle dynamics. *Trends Biol Chem Sci* 1991;16:208–213.
55. Ashburner M. Puff, genes and hormones revisited. *Cell* 1990;61:1–3.

Subject Index

Page numbers followed by *f* refer to figures; page numbers followed by *t* refer to tables.

D

Q

S

U

V

9780120361298